Machine Learning in VLSI Computer-Aided Design

Ibrahim (Abe) M. Elfadel • Duane S. Boning
Xin Li
Editors

Machine Learning in VLSI Computer-Aided Design

Springer

Editors
Ibrahim (Abe) M. Elfadel
Department of Electrical
and Computer Engineering
and Center for Cyber Physical Systems
Khalifa University, Abu Dhabi, UAE

Duane S. Boning
Massachusetts Institute of Technology
Cambridge, MA, USA

Xin Li
Department of Electrical
and Computer Engineering
Duke University, Durham, NC, USA

ISBN 978-3-030-04665-1 ISBN 978-3-030-04666-8 (eBook)
https://doi.org/10.1007/978-3-030-04666-8

Library of Congress Control Number: 2019930838

This Springer imprint is published by the registered company Springer Nature Switzerland AG.
The registered company address is: Gewerbestrasse 11, 6330 Cham, Switzerland

To Shaza, Adam, Ella, and Lily

Abe

To Peggy, Will, and Tom

Duane

To Karen and Thomas

Xin

Tell me and I forget.
Teach me and I remember.
Involve me and I learn.

Ben Franklin

Foreword

As an active branch of applied computer science, the field of VLSI computer-aided design (VLSI CAD) has always been at the technological forefront in incorporating cutting-edge algorithms in the software tools and methodologies that electronics engineers have used to weave the digital fabric of our world.

This book amply demonstrates that in line with its historical track record, VLSI CAD has also been at the leading edge in making good use of machine-learning technologies to further automate the design, verification, and implementation of the most advanced chips.

Machine learning and VLSI CAD have in common several main characteristics that may have greatly facilitated their interlock. The first is that they are both consumers of Big Data. In fact, Moore's law has essentially guaranteed that chip data grow exponentially big to the point that having tens of billions of transistors in a chip is now so common and almost taken for granted. The second characteristic that they have in common is a structured approach for controlling complexity. In machine learning, this approach is most apparent in the use of layered networks as inference and generalization engines. In VLSI CAD, complexity is controlled through a well-defined abstraction hierarchy, going from the transistor and its technology as raw data to the chip architecture as a model of processing and computation. The third common characteristic of the two fields is their focus on computational efficiency, be it to shorten turn-around time in chip design, as is the case in VLSI CAD, or to promptly detect patterns in time series as is the case in mission-critical cloud analytics. The fourth common characteristic is a focus on automated optimization and synthesis that VLSI CAD has spearheaded, and synthesis is now becoming an important trend for the design of neural networks in machine learning as well.

It is therefore almost natural to think of VLSI CAD engineers as the original data scientists who have been instrumental not only in dealing with big data in the context of chip design but also in enabling the very chips that have ushered the Big Data era and made it a social and business reality.

The various chapters of this timely and comprehensive book should give the reader a thorough understanding of the degree to which machine learning methods have percolated into the various layers of the chip design hierarchy. From lithography and physical design to logic and system design, and from circuit performance estimation to manufacturing yield prediction, VLSI CAD researchers have already brought state-of-the-art algorithms from supervised, unsupervised, and statistical learning to bear on pressing CAD problems such as hotspot detection, design-space exploration, efficient test generation, and post-silicon measurement minimization.

Machine learning in VLSI CAD is expected to play an increasingly important role not only in improving the quality of the models used in *individual CAD tools* but also in enhancing the quality of chip designs that result from the execution of entire *CAD flows and methodologies*.

As the semiconductor industry embraces the rising swell of cognitive systems and edge intelligence, this book could serve as a harbinger and an example of the osmosis that will exist between our cognitive structures and methods, on the one hand, and the hardware architectures and technologies that will support them, on the other.

The value proposition of automation is that it compresses schedules, reduces costs, and eliminates human errors. In the case of VLSI CAD, the automation has achieved not only these objectives but also the infinitely more important outcome of a seamless implementation of a positive feedback loop whereby computers are used to design more powerful computers. This positive feedback loop is the invisible hand of Moore's law.

As we transition from the computing era to the cognitive one, it behooves us to remember the success story of VLSI CAD and to earnestly seek the help of the invisible hand so that our future cognitive systems are used to design more powerful cognitive systems. This book is very much aligned with this ongoing transition from computing to cognition, and it is with deep pleasure that I recommend it to all those who are actively engaged in this exciting transformation.

IBM T. J. Watson Research Center
Yorktown Heights, NY, USA
August 2018

Dr. Ruchir Puri
IBM Fellow, IBM Watson
CTO & Chief Architect

Acknowledgments

We would like to acknowledge and thank the many reviewers who have helped us in getting this book to its present state by closely reading the early versions of its chapters and sharing their valuable comments, through us, with the chapter authors. Their inputs were instrumental in improving the overall quality of the entire book. In alphabetical order, they are:

Bei Yu, Bowen Zhang, Christopher Lang, Haibao Chen, Handi Yu, Hector Gonzalez Diaz, Hongge Chen, Jun Tao, Mark Po-Hung Lin, Nguyen Manh Cuong, Pingqiang Zhou, and Renjian Pan

We also acknowledge the early LaTeX technical support we received from Shahzad Muzaffar as well as the editorial advice and guidance Charles Glaser and the Springer Team provided us with.

The first editor would like to acknowledge the IBM T. J. Watson Research Center, Yorktown Heights, NY, for hosting him on his research leave in Summer 2018, during which the composition of this book was finalized.

Of course, a book of such scope and relevance would not have been possible without the timely contributions of all its authors. To them go our warmest thanks and deepest gratitude.

Abu Dhabi, UAE — Ibrahim (Abe) M. Elfadel
Cambridge, MA, USA — Duane S. Boning
Durham, NC, USA — Xin Li
August 2018

Contents

Contributors

Emrah Acar IBM T. J. Watson Research Center, Yorktown Heights, NY, USA

Ali Ahmadi The University of Texas at Dallas, Richardson, TX, USA

Ronald D. Blanton Department of Electrical and Computer Engineering, Carnegie Mellon University, Pittsburgh, PA, USA

Duane S. Boning Massachusetts Institute of Technology, Cambridge, MA, USA

Paolo Cachecho Department of Electrical and Computer Engineering, American University of Beirut, Beirut, Lebanon

Luca P. Carloni Department of Computer Science, Columbia University, New York, NY, USA

John Carulli GLOBALFOUNDRIES, Malta, NY, USA

Krishnendu Chakrabarty Duke University, Durham, NC, USA

Ali Chehab Maroun Semaan Faculty of Engineering and Architecture, American University of Beirut, Beirut, Lebanon

Hongge Chen Massachusetts Institute of Technology, Cambridge, MA, USA

Suhyeong Choi School of Electrical Engineering, KAIST, Daejeon, South Korea

Ibrahim (Abe) M. Elfadel Department of Electrical and Computer Engineering and Center for Cyber Physical Systems, Khalifa University, Abu Dhabi, UAE

George Gristede IBM T. J. Watson Research Center, Yorktown Heights, NY, USA

Ben Gu Cadence Design Systems, Inc., Austin, TX, USA

Chenjie Gu Strategic CAD Labs, Intel Corporation, Hillsboro, OR, USA

Rehan Hafiz Information Technology University (ITU), Lahore, Pakistan

Muhammad Abdullah Hanif Vienna University of Technology (TU Wien), Vienna, Austria

Jingrui He Arizona State University, Tempe, AZ, USA

Samuel Hertz Electrical and Computer Engineering Department, University of Illinois at Urbana-Champaign, Champaign, IL, USA

Ke Huang San Diego State University, San Diego, CA, USA

Muhammad Usama Javed Information Technology University (ITU), Lahore, Pakistan

Rajiv V. Joshi IBM TJ Watson Labs, Yorktown Heights, NY, USA

Rouwaida Kanj Maroun Semaan Faculty of Engineering and Architecture, American University of Beirut, Beirut, Lebanon

Department of Electrical and Computer Engineering, American University of Beirut, Beirut, Lebanon

Rupesh Raj Karn Department of Electrical and Computer Engineering, Khalifa University of Science and Technology, Abu Dhabi, United Arab Emirates

Asad Khan Texas Instruments, Dallas, TX, USA

Nathan Kupp Yale University, New Haven, CT, USA

Jihye Kwon Department of Computer Science, Columbia University, New York, NY, USA

Mourad Larbi Georgia Institute of Technology, School of Electrical & Computer Engineering, Atlanta, GA, USA

Peng Li Texas Instruments, Dallas, TX, USA

Xin Li Department of Electrical and Computer Engineering, Duke University, Durham, NC, USA

Honghuang Lin Texas Instruments, Dallas, TX, USA

Yibo Lin University of Texas at Austin, Austin, TX, USA

Frank Liu IBM Research Laboratory, Austin, TX, USA

Hongzhou Liu Cadence Design Systems, Inc., Pittsburgh, PA, USA

Hung-Yi Liu Intel Technology and Manufacturing Group, Hillsboro, OR, USA

Lingyi Liu Electrical and Computer Engineering Department, University of Illinois at Urbana-Champaign, Champaign, IL, USA

Kangsheng Luo Cadence Design Systems, Inc., Pittsburgh, PA, USA

Yiorgos Makris The University of Texas at Dallas, Richardson, TX, USA

Maria Malik George Mason University, Fairfax, VA, USA

Amit Nahar Texas Instruments, Dallas, TX, USA

Ricardo Nigaglioni IBM Systems and Technology Group, Austin, TX, USA

Bob Orr Texas Instruments, Dallas, TX, USA

Bruce Owens IBM Systems and Technology Group, Rochester, MN, USA

David Z. Pan University of Texas at Austin, Austin, TX, USA

Michael Pass Texas Instruments, Dallas, TX, USA

Semeen Rehman Vienna University of Technology (TU Wien), Vienna, Austria

Rob A. Rutenbar Department of Computer Science, University of Illinois at Urbana-Champaign, Urbana, IL, USA

Lama Shaer Maroun Semaan Faculty of Engineering and Architecture, American University of Beirut, Beirut, Lebanon

Muhammad Shafique Vienna University of Technology (TU Wien), Vienna, Austria

J. P. Shiely Synopsys, Inc., Mountain View, CA, USA

Seongbo Shim Samsung Electronics, Hwasung, South Korea

Youngsoo Shin School of Electrical Engineering, KAIST, Daejeon, South Korea

Amith Singhee IBM Research, Bangalore, India

Shupeng Sun Department of Electrical and Computer Engineering, Carnegie Mellon University, Pittsburgh, PA, USA

Madhavan Swaminathan Georgia Institute of Technology, School of Electrical & Computer Engineering, Atlanta, GA, USA

Mehdi B. Tahoori Karlsruhe Institute of Technology, Karlsruhe, Germany

Jun Tao State Key Laboratory of ASIC and System, School of Microelectronics, Fudan University, Shanghai, China

Hakki M. Torun Georgia Institute of Technology, School of Electrical & Computer Engineering, Atlanta, GA, USA

Shobha Vasudevan Electrical and Computer Engineering Department, University of Illinois at Urbana-Champaign, Champaign, IL, USA

Arunkumar Vijayan Karlsruhe Institute of Technology, Karlsruhe, Germany

Fa Wang Department of Electrical and Computer Engineering, Carnegie Mellon University, Pittsburgh, PA, USA

Li-C. Wang Department of Electrical and Computer Engineering, University of California, Santa Barbara, Santa Barbara, CA, USA

Constantinos Xanthopoulos The University of Texas at Dallas, Richardson, TX, USA

Jinjun Xiong IBM Thomas J. Watson Research Center, Yorktown Heights, NY, USA

Xuan Zeng State Key Laboratory of ASIC and System, School of Microelectronics, Fudan University, Shanghai, China

Wangyang Zhang Cadence Design Systems, Inc., Pittsburgh, PA, USA

Yada Zhu IBM Thomas J. Watson Research Center, Yorktown Heights, NY, USA

Matthew M. Ziegler IBM T. J. Watson Research Center, Yorktown Heights, NY, USA

About the Editors

Ibrahim (Abe) M. Elfadel is Professor of Electrical and Computer Engineering at Khalifa University, Abu Dhabi, UAE. He is also affiliated with Khalifa University Center for Cyber Phyiscal Systems. Since May 2014, he has been the Program Manager of TwinLab MEMS, a joint collaboration with GLOBALFOUNDRIES and the Singapore Institute of Microelectronics on micro-electromechanical systems. Between May 2013 and May 2018, he was the founding co-director of the Abu Dhabi Center of Excellence on Energy-Efficient Electronic Systems (ACE^{4}S). Between November 2012 and October 2015, he was the founding co-director of Mubadala's TwinLab 3DSC, a joint research center on 3D integrated circuits with the Technical University of Dresden, Germany. He also headed the Masdar Institute Center for Microsystems (iMicro) from November 2013 until March 2016. From 1996 to 2010, he was with the corporate CAD organizations at IBM Research and the IBM Systems and Technology Group, Yorktown Heights, NY, where he was involved in the research, development, and deployment of CAD tools and methodologies for IBM's high-end microprocessors. His current research interests include IoT platform prototyping; IoT communications; energy-efficient edge and cloud computing; power and thermal management of multicore processors; low-power, embedded digital signal processing; 3D integration; and CAD for VLSI, MEMS, and silicon photonics. Dr. Elfadel is the recipient of six Invention Achievement Awards, one Outstanding Technical Achievement Award, and one Research Division Award, all from IBM, for his contributions in the area of VLSI CAD. He is the inventor or co-inventor of 50 issued US patents with several more pending. In 2014, he was the co-recipient of the D. O. Pederson Best Paper Award from the IEEE Transactions on Computer-Aided Design for Integrated Circuits and Systems. Most recently, he received (with Prof. Mohammed Ismail) the SRC Board of Director Special Award for "pioneering semiconductor research in Abu Dhabi." Dr. Elfadel is the co-editor of two Springer books: *3D Stacked Chips: From Emerging Processes to Heterogeneous Systems*, 2016, and *The IoT Physical Layer: Design and Implementation*, 2019. From 2009 to 2013, Dr. Elfadel served as an Associate Editor of the *IEEE Transactions on Computer-Aided Design*. He is currently serving as Associate Editor of the *IEEE Transactions on VLSI Systems* and on the editorial board of the *Microelectronics Journal* (Elsevier). Dr. Elfadel has also served on the Technical Program Committees of several leading conferences, including DAC, ICCAD, ASPDAC, DATE, ICCD, ICECS, and MWSCAS. Most recently, he was the General Co-chair of the IFIP/IEEE 25th International Conference on Very Large Scale Integration (VLSI-SoC 2017), Abu Dhabi, UAE. He received his Ph.D. from MIT in 1993.

Duane S. Boning is the Clarence J. LeBel Professor in Electrical Engineering and Professor of Electrical Engineering and Computer Science in the EECS Department at MIT. He is affiliated with the MIT Microsystems Technology Laboratories and serves as MTL Associate Director for Computation and CAD. From 2004 to 2011, he served as Associate Head of the EECS Department at MIT, from 2011 through 2013 as Director/Faculty Lead of the MIT Skoltech Initiative, and from 2011 to 2018 as Director of the MIT/Masdar Institute Cooperative Program. He is currently the Engineering Faculty Co-Director of the MIT Leaders for Global Operations (LGO) program. Dr. Boning received his S.B. degrees in electrical engineering and in computer science in 1984, and his S.M. and Ph.D. degrees in electrical engineering in 1986 and 1991, respectively, all from the Massachusetts Institute of Technology. He was an NSF Fellow from 1984 to 1989 and an Intel Graduate Fellow in 1990. From 1991 to 1993 he was a Member Technical Staff at the Texas Instruments Semiconductor Process and Design Center in Dallas, Texas, where he worked on semiconductor process representation, process/device simulation tool integration, and statistical modeling and optimization. Dr. Boning is a Fellow of the IEEE and has served as Editor-in-Chief for the *IEEE Transactions on Semiconductor Manufacturing*. He is a member of the IEEE, Electrochemical Society, Eta Kappa Nu, Tau Beta Pi, Materials Research Society, Sigma Xi, and the Association of Computing Machinery.

Xin Li received his Ph.D. degree in Electrical and Computer Engineering from Carnegie Mellon University in 2005. He is currently a Professor in the ECE Department at Duke University and is leading the Institute of Applied Physical Sciences and Engineering and the Data Science Research Center at Duke Kunshan University. His research interests include integrated circuit, signal processing, and data analytics. Dr. Li is the Deputy Editor-in-Chief of *IEEE TCAD*. He was an Associate Editor of *IEEE TCAD*, *IEEE TBME*, *ACM TODAES*, *IEEE D& T*, and *IET CPS*. He was the General Chair of ISVLSI and FAC. He received the NSF CAREER Award in 2012 and six Best Paper Awards from IEEE TCAD, DAC, ICCAD, and ISIC. He is a Fellow of IEEE.

Chapter 1
A Preliminary Taxonomy for Machine Learning in VLSI CAD

Duane S. Boning, Ibrahim (Abe) M. Elfadel, and Xin Li

Taxonomy is described sometimes as a science and sometimes as an art, but really it is a battleground.

Bill Bryson

1.1 Machine Learning Taxonomy

Methods and approaches that fall under the broad description of machine learning are continuing to develop and advance rapidly. A brief taxonomy of machine learning methods is shown in Fig. 1.1. The categories in this tree-oriented taxonomy are approximate; indeed, some methods (such as those involving dimensionality reduction) arise within different portions of the taxonomy. But to help understand the range of methods, this approximate high-level taxonomy enables us to highlight a few key concepts and distinctions among these methods. Our purpose is not to be tutorial in this introduction (rather, we defer to the individual chapters in this book where key concepts are covered in each area of application). For the reader interested in learning the fundamentals of machine learning, we recommend the excellent textbooks such as Bishop [1], Murphy [2], Hastie et al. [3], and Vapnik [4].

D. S. Boning
Massachusetts Institute of Technology, Cambridge, MA, USA
e-mail: boning@mit.edu

I. M. Elfadel (✉)
Department of Electrical and Computer Engineering and Center for Cyber Physical Systems, Khalifa University, Abu Dhabi, UAE
e-mail: ibrahim.elfadel@ku.ac.ae

X. Li
Department of Electrical and Computer Engineering, Duke University, Durham, NC, USA
e-mail: xinli.ece@duke.edu

I. M. Elfadel et al. (eds.), *Machine Learning in VLSI Computer-Aided Design*,
https://doi.org/10.1007/978-3-030-04666-8_1

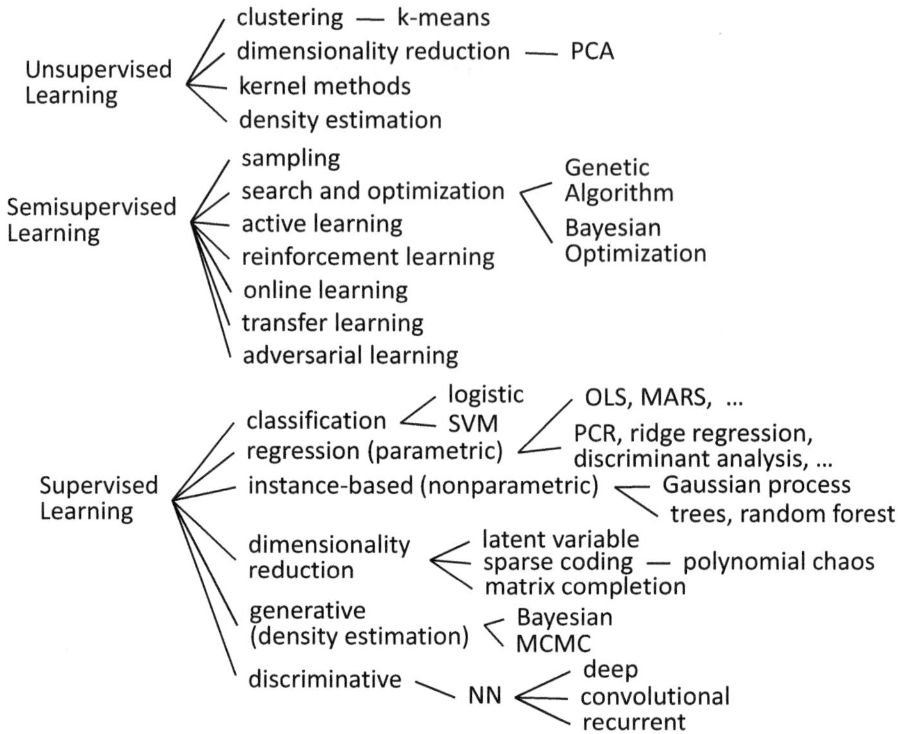

Fig. 1.1 Approximate high-level taxonomy of machine learning methods

1.1.1 Unsupervised, Supervised, and Semisupervised Learning

At the highest level, a distinction between types of machine learning methods is often made with respect to the source of information used to drive learning. *Unsupervised learning* refers to cases where only input data (i.e., independent variables x) are available, rather than input–output pairs (i.e., independent and dependent variables, (x, y)). The goal in unsupervised learning is to discover some structure within the data. One such structure is the grouping of the data in different (typically distinct) clusters. The k-means algorithm is a well-known clustering approach, among many others [5]. A second type of unsupervised learning seeks to discover simpler latent structures in the data, often in order to identify important features or to achieve dimensionality reduction. A classic representative of such methods is principal component analysis (PCA) [6].

In contrast, *supervised learning* methods are driven by input–output data. For example, labels y for input instances x may be available, enabling learning of models to support classification of new or different input instances. A wide range of classification algorithms exist, from basic logistic regression (to identify the probability of membership in some class, together with a decision threshold for deciding or declaring such membership) to other methods such as support vector machine (SVM) approaches [7]. Regression can be viewed as a supervised learning approach, where discrete or continuous output values might be predicted, based on a learned model using continuous or discrete input–output data. Many classical statistical learning methods are regression based, ranging from ordinary least squares (OLS) regression to principal component regression (PCR), ridge regression, and a variety of discriminant-based approaches. In contrast, iterative (e.g., gradient-descent based) rather than regression-based learning methods have gained great interest as efficiencies and computing capability have advanced. In particular, neural network methods have gained enormous recent attention, extending multilayer perceptron (MLP) structures to such approaches as deep neural networks (DNN), convolutional neural networks (CNN), recurrent neural networks (RNN), and other architectures.

An interesting area lies between the two extremes of unsupervised and supervised learning. *Semisupervised learning* methods are those where only some fraction of the available or potential input–output pairs are already available, and a substantial part of the learning effort is to efficiently and effectively decide what new instances of input–output pairs should be gathered. Some classical statistical learning methods can be considered to fall in this area, including design of experiment (DoE), Monte Carlo, and other sampling methods (deciding what set of data points, or next data instance, should be obtained) in order to build input–output models, or to estimate probability density models of inputs and/or outputs. A number of machine learning approaches also fall under the broad semisupervised category. Of these, reinforcement learning (RL) [8] is of increasing interest, in which the system seeks to teach itself based on reward mechanisms and exploration policies, learning both a model to predict results and a model for deciding best moves. Other versions of semisupervised learning have evolved to address related problems. Online learning, for example, seeks to continually update a learned model based on subsequent or continually arriving new data points (selection of which is typically not under the control of the online learning algorithm). Transfer or multi-task learning [9] is another important goal in machine learning, whereby models are generalizable to new scenarios or environments beyond that used to train the model, with limited (or no) additional learning in the new environment. Finally, the broad area of search and optimization can be considered to be a semisupervised learning problem, where the method seeks to find the inputs that achieve one or more good outputs, and typically must decide what next input–output pair to consider when that input–output pair is not already available as a previously learned case. Interesting approaches for such design space exploration include genetic and evolutionary algorithms [10] and Bayesian optimization [11].

1.1.2 Parametric and Nonparametric Methods

Another useful distinction for categorizing and understanding the range of machine learning methods is whether a learned model is *parametric* or *nonparametric*. Parametric methods are those in which the model complexity is fixed, in the sense that the model complexity does not depend on the number of training samples [2]. Note that the complexity of parametric models may still depend on the data, for example, hyperparameters of the learning methods, or regularization approaches that seek to maintain model simplicity based on the data, will often seek to adapt or adjust the number of model parameters or model structure, based on the observed or available training data. In contrast, however, *nonparametric* methods generally depend on the *number* of training samples, in that the model complexity typically grows as more training samples are included, whether or not the underlying relationship is less or more complex. Several instance-based nonparametric learning methods are well-known, such a k-NN (k-nearest neighbors), where decisions about membership may be based on nearness (using some distance measure) to previous or learned examples [12].

1.1.3 Discriminative Versus Generative Methods

A third broad characterization of machine learning methods is to consider whether they are *discriminative* or *generative* [13]. A discriminative model seeks to directly learn the probability density model for the output y given the input x, i.e., to learn $p(y|x)$. In contrast, a generative approach learns the joint pdf $p(x, y)$ and/or the probability of the input conditioned on the observed output, $p(x|y)$, and then uses that model to reason about $p(y|x)$. Many models, particularly classical neural network approaches such as MLPs, are discriminative in nature: the NN adapts weights in order to make direct predictions of outputs y given inputs x. Alternatively, Bayesian approaches embody the indirect or generative approach, where learned models enable one to directly infer what inputs might have been responsible for an observed output (and to then indirectly infer outputs given an input). Bayesian approaches are also a type of density estimation method, which seek to learn the probability density of inputs and/or outputs. Density estimation methods are of interest in unsupervised scenarios as well (e.g., where one would learn the statistical properties of just an input data set, $p(x)$, from which one might be able to estimate, for example, the probability of extremal or low probability events (e.g., failures)).

An interesting emerging machine learning approach that blends generative and discriminative learners together are generative adversarial methods. Among these, generative adversarial networks (GANs) have received substantial recent attention, as a semisupervised learning method [14]. In a GAN, a discriminative model (e.g., a NN predicting an output, such as a classification label) is initially trained on available labeled data. Similarly, a generative model capable of predicting or generating an input based on a desired output is constructed from available training data. The two models are then pitted against each other, where the generative model seeks to generate a new or synthetic input that will "fool" the discriminative model, while the discriminative model likewise continues to train based on its performance in correctly classifying the output corresponding to the new input. Such methods are interesting in being able to generate "realistic" synthetic inputs, such as images corresponding to a set of desired output characteristics or labels.

1.2 VLSI CAD Abstraction Levels

The range of machine learning methods categorized briefly in the previous section provides a rich reservoir of approaches for potential application to the problems and challenges faced in VLSI design. Generally, VLSI CAD involves solving design problems within or between the various design and abstraction levels, from system design to chip fabrication and testing, as pictured in Fig. 1.2. Since the beginning of VLSI CAD, and driven by the increasing complexity both across levels and within each level of design, automatic simulation and generation/synthesis tools have been developed. Typically, simulation tools seek to emulate or produce predictions about performance at a "higher level" based on a "lower level" description of a design. In contrast, synthesis or generators seek to find a lower level design that is optimal or satisfactory with respect to some higher level requirements or goals.

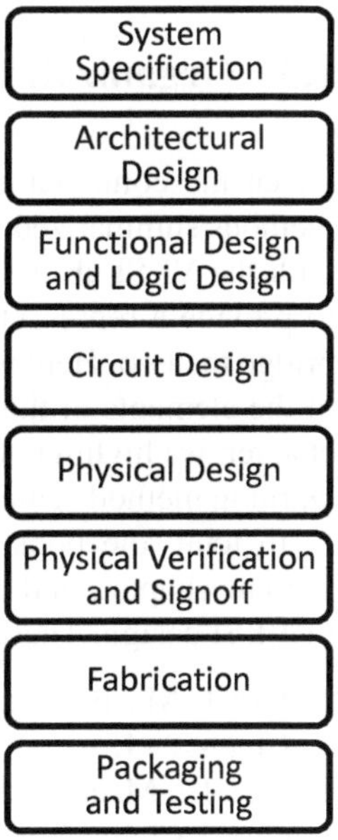

Fig. 1.2 Abstraction and physical levels in VLSI design

Inherent in the VLSI CAD paradigm, then, are numerous models relating inputs to outputs, or utilizing models of inputs. VLSI design also involves large amounts of data and information, stemming in part from the billions of devices or components that can be integrated into a chip. The conceptual connection and attraction to machine learning methods is thus readily apparent. Fundamentally, improved methods for modeling and predicting outputs corresponding to inputs are of great interest at many levels of the VLSI CAD hierarchy. The increased representational capacity of structures such as neural networks over predefined or fixed parametric approximate models is one such attraction. A second area of great interest are density estimation methods, including those that enable representation of inherent variation or uncertainty in VLSI systems, arising from fabrication variations or operating condition and environmental uncertainty (e.g., coverage of a large combinatoric space of possible input signals). Such density models might then serve as inputs to simulation or forward models, to enable generation and estimation of output density distributions. Similarly, dimensionality reduction approaches that identify important and key features and model the inter-relationships between large parameter sets are of great interest.

In addition to the interest in discriminative or simulation-like machine learning methods, it is clear that the generative or synthetic machine learning methods have great potential for VLSI CAD. Similarly, VLSI design space exploration and optimization problems correspond strongly to semisupervised machine learning approaches. Rapid and efficient generation of a satisfactory design, where an enormous number of design alternatives or parameters may be available, is fundamental to many problems in VLSI design.

1.3 Organization of This Book

Given the combination of long-standing and rapidly emerging machine learning methods summarized in Sect. 1.1, and given the many design challenges associated with the VLSI abstraction hierarchy as noted in Sect. 1.2, our goal in this book is to provide a representative (though certainly non-exhaustive) set of examples that bridge the two. Here we overview the parts and chapters in the book, with reference to the elements of the VLSI design and CAD hierarchy discussed above. For each chapter, we highlight the CAD problem or challenge and note the kinds of machine learning methods employed to address that challenge.

In total, the book comprises 21 chapters organized into an introduction, the present chapter, and five parts that are meant to parallel five important domains of VLSI design. These parts are:

- Part I: Machine Learning for Lithography and Physical Design, which covers Chaps. 2–4.
- Part II: Machine Learning for Manufacturing, Yield, and Reliability, which covers Chaps. 5–9.

- Part III: Machine Learning for Failure Modeling, which covers Chaps. 10–13.
- Part IV: Machine Learning for Analog Design, which covers Chaps. 14–17.
- Part V: Machine Learning for System Design and Optimization, which covers Chaps. 18–21.

To help the reader navigate through these parts, we now provide a short description of the content of each chapter.

1.3.1 Machine Learning for Lithography and Physical Design

One of the earliest areas of intersection between machine learning and VLSI CAD has been at the physical design level. We begin in Part I with a focus on this highly vibrant area, with a particular focus on lithography modeling, including optical proximity correction (OPC) and etch proximity correction (EPC) problems. Additional applications to other physical design problems are also discussed.

1.3.1.1 Shiely—Compact Lithographic Process Models

In Chap. 2, Shiely bridges the terminology and concepts in the two fields—lithography compact modeling and machine learning—to highlight common concepts and methods. The chapter then focuses on specific supervised learning approaches leveraging physical simulation and machine learning to generate efficient compact patterning models. In particular, approaches for identification of relevant feature sets, construction of appropriate functional forms, and regularization by way of hyperparameters are described, both for linear models and for nonlinear neural network models for various components in the compact lithographic modeling stack.

1.3.1.2 Shim et al.—Mask Synthesis

In Chap. 3, Shim et al. focus specifically on the mask synthesis problem. The importance of learning parameters, preparation of the compact learning data set, and methods to avoid over-fitting are discussed, with application to both optical proximity correction (OPC) and etch proximity correction (EPC) of mask designs.

1.3.1.3 Lin and Pan—Physical Verification, Mask Synthesis, and Physical Design

Building and expanding on the problem of mask synthesis, Lin and Pan in Chap. 4 explore the application of machine learning to additional problems in physical design and physical verification. In the case of physical verification, machine learning approaches including feature extraction and deep neural network classification methods are applied for lithography hotspot detection. For mask synthesis, supervised learning of improved models enables improvements in sub-resolution assist feature (SRAF) generation and optical proximity correction. Finally, the chapter notes that physical design often involves combinatorial problems. Examples are presented where machine learning prediction and classification models are used to advantage; these include datapath placement, clock optimization, and lithography-friendly routing.

1.3.2 Machine Learning for Manufacturing, Yield, and Reliability

A second major area where machine learning has been applied within VLSI CAD is at the intersection between design and manufacturing, and between design and reliability. In Part II, we present machine learning methods applied to efficient measurement and modeling of variability arising in IC manufacturing, as well as approaches employing machine learning to enhancement manufacturing yield. Similarly, machine learning to support and improve consideration of reliability issues during design are presented.

1.3.2.1 Xanthopoulos et al.—Gaussian Process for Wafer-Level Correlations

The challenges of modeling variation in semiconductor fabrication include multiple granularity levels with variation occurring both spatially and temporally, manifesting in within-die as well as die-to-die variation across the wafer, lots, and facility. In Chap. 5, Xanthopoulos et al. review these modeling challenges, and then develop and discuss the application of Gaussian process machine learning techniques. Applications to test cost reduction, quality improvement, and yield learning are presented.

1.3.2.2 Chen and Boning—Yield Enhancement

The enhancement of yield during fabrication, packaging, and test is of substantial economic importance in IC design and production. The opportunities in using machine learning for packaged memory chip yield enhancement are explored in Chap. 6 by Chen and Boning. In manufacturing, substantial data is often available for supervised learning of classifiers to predict downstream yield, based on available upstream information such as electrical test data for each die prior to packaging. Given a good classifier, die-stack packaged memory yield can be improved substantially. However, machine learning challenges limit this opportunity, including class imbalance due to small numbers of failure training instances, and concept drift in which learned models decay rapidly due to changes in the underlying statistical model. Methods including boosting and online learning are employed and demonstrated to achieve effective classifiers and yield improvement in the face of these challenges.

1.3.2.3 Tao et al.—Virtual Probe

In Chap. 7, Tao et al. present the Virtual Probe as an efficient method for characterizing spatial variation in semiconductor fabrication. The goal is to reduce the number of measurements needed in order to capture or estimate spatially correlated within-die or die-to-die variations. This requires a combination of learning systematic components of that variation, while also recognizing that there are stochastic elements to that variation (in that each wafer or chip may have a different instantiation of systematic variation). The Virtual Probe utilizes a number of machine learning methods to achieve these goals. First, efficient sampling (semisupervised learning) methods are employed, in order to minimize the number of measurements or training examples required. Second, dimensionality reduction approaches are crucial; in Virtual Probe, compressed sensing methods, particularly an efficient function basis using discrete cosine transform (DCT) approximation, are leveraged. Finally, Bayesian learning is employed for the estimation of DCT model coefficients. The Virtual Probe is demonstrated for three applications, including wafer-level silicon characterization (for inter-die variations), chip-level silicon characterization (for intra-die variation), and testing and self-healing of integrated circuits.

1.3.2.4 Xiong et al.—Chip Testing

In Chap. 8, Xiong et al. consider the problem of process variation modeling and VLSI chip testing. They point out that some existing machine learning approaches are not necessarily well positioned to solve the problem of process monitoring and improvement. Extensions to machine learning techniques that leverage temporal, structural, and hierarchical properties are needed. Methods including partial correlation extraction (with family/model selection), hierarchical multi-task learning, and co-clustering of structural and temporal data are described, as steps toward meeting these demands.

1.3.2.5 Vijayan et al.—Aging Analysis

Device and circuit reliability is a key challenge in VLSI design, with well-known problems including bias temperature instability (BTI) effects in which path delays change over the lifetime of the circuit. In Chap. 9, Vijayan et al. consider this problem, and propose a fine-grained workload-induced stress monitoring approach utilizing machine learning techniques. In particular, regression-based support vector machine (SVM) predictors are developed that utilize kernel methods to capture nonlinear dependencies, producing an accurate classifier for this aging problem. These are used in conjunction with space and time sampling of selective flip-flops for runtime monitoring and mitigation of aging.

1.3.3 Machine Learning for Failure Modeling

Modeling of the probabilities of very rare events is important in a number of areas of VLSI design, notably in estimating failure rates for large circuits such as SRAM and other memories. The probabilities associated with the failure of any particular cell are (hopefully, with good design) exceedingly small, but accuracy is important in order to estimate aggregate circuit or system yield or reliability. In Part III, machine learning approaches for modeling of such failures are presented.

1.3.3.1 Singhee—Extreme Statistics in Memories

One approach for modeling of extreme statistics in memories is presented in Chap. 10 by Singhee. Here, extreme value theory methods are employed, including order statistics and the generalized Pareto distribution (GPD) in order to learn the tails of the distribution from the limited data in these tails. The approach is demonstrated for cell and array failure probability estimation in large memories.

1.3.3.2 Kanj et al.—Fast Statistical Analysis Using Logistic Regression

Another approach for fast statistical yield analysis is presented by Kanj et al. in Chap. 11. The approach leverages machine learning modeling and sampling approaches. For modeling, logistic regression-based machine learning techniques are used to model circuit response. Advanced sampling is developed using a mixture importance sampling approach combining a uniform sampling stage and an importance sampling stage. The approach is demonstrated for an industrial 14 nm FinFET SRAM design.

1.3.3.3 Tao et al.—Fast Statistical Analysis of Rare Circuit Failures

In Chap. 12, Tao et al. address the challenge of analysis of rare circuit failures with subset simulation (SUS) and scaled-sigma sampling (SSS) approaches. In particular, the approach leverages machine learning methods related to hierarchical and sequential conditional probability density estimation using Markov chain Monte Carlo (MCMC) methods, with a modified Metropolis sampling approach. The approach is shown to achieve improved accuracy in challenging scenarios, particularly when the dimensionality of the variation space is more than a few hundred.

1.3.3.4 Wang—Learning from Limited Data

Machine learning is sometimes confused with "big data" analytics. In many cases, the availability of enormous data sets does indeed enable machine learning models and algorithms. But in many cases, including cases arising in VLSI CAD, a different machine learning challenge is present—"small" or limited data availability. In Chap. 13, Wang considers this challenge and notes the limitations of using traditional machine learning problem formulations. The chapter proposes an alternative perspective based on an iterative feature search process, a form of semisupervised learning. The approach is demonstrated on two verification and test examples. The first is identification of important signal names for functional verification; the proposed iterative feature search approach is able to text mine design specification documents for this identification. A second example focuses on reduction of trim count based on chip hot tests.

1.3.4 Machine Learning for Analog Design

Analog design is a particularly challenging layer in the VLSI design hierarchy, and thus of particular interest for the development and application of machine learning. In Part IV, approaches are presented for modeling, validation, and optimization of analog designs that leverage advances in machine learning.

1.3.4.1 Tao et al.—Bayesian Model Fusion

An important aspect of circuit design is pre-silicon validation and also post-silicon tuning of circuit models, particularly to account for large-scale process variations. In Chap. 14, Tao et al. highlight the challenges associated with validation and tuning in order to minimize simulation and/or measurement cost. Bayesian model fusion (BMF) is detailed as an approach to reuse simulation and/or measurements from early stages of design or fabrication, in order to minimize the amount of data needed to update or tune models at later stages. The approach is demonstrated for several industrial circuit design examples.

1.3.4.2 Lin et al.—Sparse Relevance Kernel Machine

The design of analog and mixed signal (AMS) circuits is a well-known challenge, particularly demanding in the need for highly accurate models in order to support optimization, verification, and failure diagnosis. In Chap. 15, Lin et al. propose a novel approach based on Bayesian learning, referred to as a sparse relevance kernel machine (SRKM). The approach combines support vector machine (SVM) and relevance methods, in order to build reliable classification models with a limited number of simulation runs. The approach is demonstrated for the design of AMS circuits, including a low-dropout regulator and a built-in self-test charge-pump phase-locked loop circuit.

1.3.4.3 Singhee—Projection Pursuit with SiLVR

Modeling challenges in the functional, logic, and circuit design layers in the VLSI CAD hierarchy have often benefited from simplified or approximate models utilizing response surface modeling (RSM), typically generated from more expensive and intensive simulation at a lower level of abstraction. Traditionally, linear regression approaches with low order polynomial (e.g., linear or quadratic) models have been employed. Enriched by machine learning, the space of model alternatives has expanded dramatically. In particular, alternative basis sets and model building blocks, coupled to effective training algorithms, enable efficient compact models that can substantially reduce model error. Chapter 16 by Singhee notes the limitation of conventional RSM, and reviews advanced methods of latent variable regression (LVR) and projection pursuit regression (PPR). These are combined in SiLVR, with advantages not only in model accuracy, but also in feature identification and interpretability that helps provide insight to the designer. The approach enables modeling of scenarios with large numbers of input variables (as arise in variation analysis problems), as demonstrated in the chapter on a range of digital and analog circuit design problems including master–slave flip-flop design, two-stage RC-compensated opamp design, and bandgap voltage reference design.

1.3.4.4 Torun et al.—Integrated Voltage Regulator Optimization and Uncertainty Quantification

As noted in Sect. 1.2, identification and modeling of complex input–output relationships together with the optimization (or synthesis) of a good design using such models is fundamental to many problems in VLSI CAD. One such example is the design of voltage regulation architectures and circuits, as presented by Torun et al. in Chap. 17. The approach developed in this chapter seeks efficient joint optimization and model construction, by focusing design exploration and prediction only in regions where the optimal design might reside, and to drive the system to that optimum. A number of machine learning techniques are employed toward this end. A key method is two-stage Bayesian optimization (TSBO) to direct the tuning and optimization of the integrated voltage regulator. A second challenge addressed in the chapter is uncertainty quantification (UQ), in order to ensure that the fabricated design will satisfy electrical requirements in the face of manufacturing variations. Here polynomial chaos expansions together with importance sampling are utilized for both probability density estimation and sensitivity analysis of the design.

1.3.5 Machine Learning for System Design and Optimization

At the very highest level, the goal of VLSI design is to find an overall system design solution that meets a complex set of (often competing) performance objectives. In Part V, machine learning approaches for system design and optimization are presented. While this book focuses on machine learning for CAD, it is important for the reader to also be aware of the principles, needs, and opportunities for design of novel machine learning hardware architectures; concepts related to machine learning hardware design are presented in the closing chapter.

1.3.5.1 Ziegler et al.—SynTunSys

System designs are generated by exploration of an enormously complicated design space resulting from both the layering of the design hierarchy and the large number of options, alternatives, and smaller optimization problems within or across each level. Given this large and complex design space, human designers are increasingly challenged to explore and obtain good solutions in a time- and computation-efficient manner. In Chap. 18, Ziegler et al. present SynTunSys, a synthesis parameter auto-tuning system for optimizing high-performance processors. Intelligent search strategies, together with parallel computing, are used to manage the design space exploration process. The system develops and utilizes semisupervised learning methods, including genetic algorithms that address the large number of discrete design alternatives, synthesis and tool parameter selection, and design goal tradeoffs to be explored. The system has been employed successfully for optimizing multiple IBM high-performance server chips. Future opportunities for additional intelligent automation research are also highlighted in this chapter.

1.3.5.2 Karn and Elfadel—Multicore Power and Thermal Proxies

A challenge in the management of multicore processors is effective estimation of power and thermal implications of workload decisions. Various performance counters can be used as proxies; these are typically trained using traditional linear regression techniques. In Chap. 19, Karn and Elfadel consider least angle regression alternatives in order to develop power and thermal proxies, and identify proxy signatures to categorize the workload. The algorithms are demonstrated for real-time multicore power and performance management.

1.3.5.3 Vasudevan et al.—GoldMine for RTL Assertion Generation

Effective and efficient test coverage is a well-known challenge within the VLSI design hierarchy, as pictured in Fig. 1.2. In Chap. 20, Vasudevan et al. present algorithms in the GoldMine system for the automatic generation of assertions for RTL designs. The system leverages machine learning design space exploration methods. A best gain decision forest (BGDF) algorithm that uses static analysis and a novel coverage guided association mining method are presented and compared to existing decision tree and other algorithms.

1.3.5.4 Hanif et al.—Machine Learning Architectures and Hardware Design

Machine learning is not only transforming VLSI CAD tools and techniques, but is also driving the need for novel hardware architectures that can directly and more efficiently execute machine learning tasks. Hardware for machine learning, particularly deep learning, is thus itself an enormous and fast moving area of research and development. Chapter 21 by Hanif et al. provides a brief overview of the most prominent techniques that have been used so far for the design of energy efficient accelerators, and makes the connection between machine learning algorithms and alternative architectures. The chapter also highlights the challenges and research roadmap for future development of adaptable and energy-efficient hardware accelerators for machine learning.

1.3.6 Other Work and Outlook

The area of machine learning for VLSI CAD is fast evolving. While this book seeks to provide a relevant cross-section of such contributions, the reader should understand and be aware that a great deal of other outstanding work has been produced and is underway. In terms of industrial design enablement and practice, most of the major CAD providers have already begun to incorporate machine learning to great advantage in individual tools, or across parts of the design flow. On the research front, many additional machine learning approaches, and many additional design challenges, are being explored. For example, application of matrix completion to VLSI manufacturing and CAD has received recent attention [15]. Other examples include application of Bayesian learning to statistical compact modeling [16, 17] and statistical library characterization [18]. Opportunities for application to statistical timing analysis tool miscorrelation have been noted by Kahng [19], as well as opportunities for prediction and optimization of other physical design problems. The need, and opportunity, for machine learning to help bridge the design cost gap has also been highlighted by Kahng [20]. Large-scale research centers, such as the Center for Advanced Electronics through Machine Learning (CAEML) [21], are also emerging to accelerate advances both in CAD tools and in the exploration and development of relevant machine learning technologies. The future for machine learning and VLSI CAD is exciting, indeed.

References

1. C.M. Bishop, *Pattern Recognition and Machine Learning* (Springer, New York, 2006)
2. K.P. Murphy, *Machine Learning: A Probabilistic Perspective* (MIT Press, Cambridge, 2012)
3. T. Hastie, R. Tibshirani, J. Friedman, *The Elements of Statistical Learning: Data Mining, Inference, and Prediction*, 2nd edn. (Springer, New York, 2009)
4. V.N. Vapnik, *Statistical Learning Theory* (Wiley, New York, 1998)
5. A.K. Jain, Data clustering: 50 years beyond K-means. Pattern Recogn. Lett. **31**(8), 651–666 (2010)
6. I.T. Jolliffe, *Principal Component Analysis*, 2nd edn. (Springer, New York, 2002)
7. V. Vapnik, S. Golowich, A. Smola, Support vector method for function approximation, regression estimation, and signal processing, in *Advances in Neural Information Processing Systems* (1997), pp. 281–287
8. L.P. Kaelbling, M.L. Littman, A.W. Moore, Reinforcement learning: a survey. J. Artif. Intell. Res. **4**, 237–285 (1996)
9. S. Thrun, L. Pratt (eds.), *Learning to Learn* (Springer, New York, 1998)
10. M. Mitchell, *An Introduction to Genetic Algorithms* (MIT Press, Cambridge, 1996)
11. B. Shahriari, K. Swersky, Z. Wang, R.P. Adams, N. de Freitas, Taking the human out of the loop: a review of Bayesian optimization. Proc. IEEE **104**(1), 148–175 (2016)
12. G. Chen, D. Shah, Explaining the success of nearest neighbor methods in prediction. Trends Mach. Learn. **10**(5–6), 337–588 (2018)
13. A.Y. Ng, M.J. Jordan, On discriminative vs. generative classifiers: a comparison of logistic regression and naive Bayes, in *Advances in Neural Information Processing Systems*, vol. 14 (2002), pp. 841–848
14. I.J. Goodfellow, J. Pouget-Abadie, M. Mirza, B. Xu, D. Warde-Farley, S. Ozair, A. Courville, Y. Bengio, Generative adversarial nets, in *Advances in Neural Information Processing Systems* (2014), pp. 2672–2680
15. H. Chen, D. Boning, Z. Zhang, Efficient spatial variation characterization via matrix completion, in *IEEE/ACM Workshop on Variability Modeling and Characterization (VMC)*, Nov 2016
16. L. Yu, S. Saxena, C. Hess, I.M. Elfadel, D.A. Antoniadis, D.S. Boning, Compact model parameter extraction using Bayesian inference, incomplete new measurements, and optimal bias selection. IEEE Trans. Comput. Aided Des. Integr. Circuits Syst. **35**(7), 1138–1150 (2016)

17. L. Yu, S. Saxena, C. Hess, I.M. Elfadel, D.A. Antoniadis, D.S. Boning, Remembrance of transistors past: compact model parameter extraction using Bayesian inference and incomplete new measurements, in *51st ACM/EDAC/IEEE Design Automation Conference (DAC 2014)*, San Francisco, CA, June 2014
18. L. Yu, S. Saxena, C. Hess, I.M. Elfadel, D.A. Antoniadis, D.S. Boning, Statistical library characterization using belief propagation across multiple technology nodes, in *2015 Design, Automation Test in Europe Conference Exhibition (DATE)*, Grenoble, March 2015, pp. 1383–1388
19. A.B. Kahng, Machine learning applications in physical design: recent results and directions, in *International Symposium on Physical Design (ISPD)*, March 2018, pp. 68–73
20. A.B. Kahng, New directions for learning-based IC design tools and methodologies, in *Asia and South Pacific Design Automation Conference (ASP-DAC)*, Jan 2018, pp. 4015-410
21. Center for Advanced Electronics through Machine Learning (CAEML), referenced Aug 2018. https://publish.illinois.edu/advancedelectronics/

Part I
Machine Learning for Lithography and Physical Design

The Crystal Palace, and all its splendid contents, owe their existence to tools *as the physical means—to intellect as the guiding power.*

Charles Babbage

Chapter 2
Machine Learning for Compact Lithographic Process Models

J. P. Shiely

2.1 Introduction

Deep within any modern IC production line flow is an obscure but widespread technology that is an intersection of nano-fabrication and machine learning called computational lithography. This technology uses massive-scale high-performance computing, detailed knowledge of nanometer-scale physics, statistics, signal processing, and electron microscopy to implement a software-driven control loop. By specifying the shapes on photomasks, computational lithography software effectively directs the writing of multilayered patterns, subsequently etched and deposited, on the surfaces of silicon wafers. The GPUs used for today's deep learning training, as well as the CPUs in the cloud or in cell phones, would not be manufacturable without this technology that, at its core, is a machine learning application trained with electron-microscopic images. Despite the intimate connections between machine learning and the field of computational lithography since its earliest incarnations as model-based OPC, most of the practitioners of these arts have developed their work in parallel with little awareness of their interdependencies. As computational lithography experts have developed their own terminology and advanced their own techniques and tools, machine learning techniques have matured and extended their reach across many application domains.

The purpose of this chapter is to bring the communities of nano-fabrication and machine learning together by examining tasks of shared interest to both communities. This may be an awkward introduction for both parties. The risk this chapter runs is that it will include information both too elementary for one camp and too advanced for the other. However, there is a strong motivation for

J. P. Shiely (✉)
Synopsys, Inc., Mountain View, CA, USA
e-mail: shiely@synopsys.com

I. M. Elfadel et al. (eds.), *Machine Learning in VLSI Computer-Aided Design*,
https://doi.org/10.1007/978-3-030-04666-8_2

both communities to attempt to bridge the current divides, which are primarily related to terminology and awareness. Machine learning techniques have now been successfully applied to many domains, bringing to each new community its terminology as well as its techniques. The widespread adoption of words like "inference" or phrases like "supervised learning" in multiple fields from health care to advertising to robotics is evidence that machine learning advances are helping to establish a cross-disciplinary *lingua franca* that describes the construction of predictive models or surrogate functions. Translating computational lithography into the terminology of machine learning will, it is hoped, assist its practitioners to adopt a wider range of machine learning techniques adopted from other fields. For machine learning practitioners, on the other hand, this chapter will primarily make them familiar with an obscure but fascinating field (computational lithography) that stands to benefit enormously from recent advances in supervised learning. For these readers, the chapter hopes to spotlight some of the greatest opportunities for improving computational lithography with new machine learning techniques.

Keeping in mind two potentially very different audiences, and intending to serve as a bridge between two communities, this chapter is organized as follows: In the first section, to explain the domain of computational lithography, with key references, and enable a machine learning researcher to understand the problem being solved and develop intuition for how their skills could be applied; in the second section, to describe the state of the art of computational lithography practice in the current more widely used terminology of "the machine learning community"; and, in the last section, to highlight where these two disciplines can strive forward together to achieve higher-quality outcomes with reduced computational cost.

2.2 The Lithographic Patterning Process

The lithographic patterning process is not a familiar topic for most machine learning researchers, but it has been essential to enable the growth in raw processing power of the chips that have made deep learning cost-effective and practical. A familiar concept to most technologists is Moore's law [33], which predicts a rate of increase in the number of devices on integrated circuits (ICs) over time. In this section, we will explain how the lithographic patterning process facilitated Moore's law and, as a consequence, the current advances in the cost-effectiveness of machine learning and artificial intelligence. We will then provide a description of the lithographic patterning process, sufficient to understand its relationship to machine learning.

2.2.1 Importance of Lithographic Patterning Process to the Economics of Computing

Moore's law is an economic observation that there will be financial and performance benefits for reducing the size of the semiconductor devices that compose ICs. As observed by Dennard et al. [10], the reduction in the size of the semiconductor devices provides multiple reinforcing benefits. First, the smaller devices do not need to exert as much effort to switch their state from on to off, because their state can be determined by a smaller electron charge. The action of a smaller device can be accomplished by less current. The movement of less charge in a device allows it to change state more quickly. Reducing the size of devices therefore improves their primary performance criteria: power consumption and speed.

At the same time, shrinking device size has reduced the cost per device, because more can be manufactured on the same silicon substrate area. Of course, the devices cannot be shrunk to the point where they no longer function. It is equally important that devices not shrink to the point where the probability of successfully manufacturing the integrated circuits (ICs) constructed from them is significantly reduced. When devices are shrunk beyond the point where they can be reliably manufactured, the yield of functional chips on the wafer drops and the economic incentives for shrinking the device disappear.

The critical limiting point of Moore's law, therefore, has for many years been the tradeoff point where further reduction in device size no longer leads to economic advantage for a manufacturer. The consequence and motivation of lithographic patterning process technology research and development has been to push the tradeoff point to smaller, cheaper, faster devices over time. As this progress has proceeded, integrated circuits offering higher memory capacity, higher execution parallelism, higher speed, and lower power have become widely available at low cost, enabling current advances in machine learning and artificial intelligence.

2.2.2 *Representation of the Lithographic Patterning Process*

A complete description of the lithographic patterning process is provided elsewhere and beyond the scope of this work [28, 47]. However, to understand how to apply machine learning technology to improve patterning technology, it is necessary to have a modular understanding of the basic stages that, in sequence, compose a patterning process. This section describes the stages of a patterning step sufficiently to understand the concept of patterning as a communication process that transmits an image from one stage to another [37].

A patterning process is an operation to transform the surface of a silicon wafer. The process can be additive, in which case new material is added to the wafer substrate, or it can be subtractive, in which case material is removed. A process that adds material to the wafer substrate is called a deposition, and a process that removes material is called an etch.

In order to construct ICs it is necessary to control which regions of the wafer substrate are modified by deposition or etching. This control is achieved by a process called photolithography. The stages of the photolithography process are represented in Table 2.1.

Table 2.1 Patterning process stages

Stage	Input	Physical mechanism	Outputs
Mask $f_m : \mathscr{L} \rightarrow \mathscr{M}$	A fracture tool receives a layout $\mathscr{L} \in \{0, 1\}^2$ and converts it into instructions for a mask-writing tool	A mask-writing tool receives instructions to direct an electron beam to deliver energy in selected regions and doses onto a photoresist film deposited on a mask blank. After development, the photoresist selectively protects regions of the mask blank. The mask is etched and materials are removed from the mask blank in regions unprotected by photoresist	The mask writer produces, after etch, a photomask $\mathscr{M}$ which has been partitioned by the etch process into a discrete number of region types
Image $f_i : \mathscr{M} \rightarrow \mathscr{I}$	An exposure tool receives a photomask $\mathscr{M} \in \{0, ..N\}^2$	The exposure tool illuminates the photomask, and the photomask diffracts the light. A projection lens collects a subset of the diffracted orders and forms an image within a thin photoresist layer on the surface of a silicon wafer	The exposure tool projects light onto the wafer surface, depositing energy in a three-dimensional distribution $\mathscr{I}$ within the light-sensitive photoresist
Resist $f_r : \mathscr{I} \rightarrow \mathscr{R}$	A series of photoresist processing tools receives the wafer, which has been exposed to light energy in a three-dimensional distribution $\mathscr{I} \in \mathbb{R}^3$	The photoresist processing tools submit the photoresist-coated wafer to various thermal and chemical treatments	The photoresist will dissolve during the development process at a rate that depends on the amount of energy received. The process partitions the wafer substrate into $\mathscr{R}$, defining a region protected by the photoresist and a region that is unprotected
Etch $f_i : \mathscr{R} \rightarrow \mathscr{E}$	An etch reactor receives a wafer that is partially masked by a film of photoresist, in a pattern defined by $\mathscr{R} \in \{0, 1\}^2$ or $\mathscr{R} \in \{0, 1\}^3$	The etch reactor bombards the wafer with etching particles that react with the unmasked regions on the surface of the wafer substrate	The etching particles remove material from the unmasked regions, transferring a pattern $\mathscr{I}$ into the wafer substrate

Generically, each stage of a patterning process can be considered a mapping from one representation of the transmitted message to another. For example,

$$\mathscr{B} = f_B(\mathscr{A}) \tag{2.1}$$

represents a stage as a transfer function $f_B : \mathscr{A} \rightarrow \mathscr{B}$, where $\mathscr{A}$ is the input distribution and $\mathscr{B}$ is the transmitted distribution. Distributions can be either two-dimensional or three-dimensional, and discrete-valued or real-valued.

For simplification, we will describe only subtractive patterning processes, which begin with a designed layout and end with an etched wafer substrate. Using the transfer function concept, we can represent the patterning process as sequence of pattern transforms:

- The **mask** stage, f_m, which communicates the polygon shapes $\mathscr{L}$ in a design layout to a distribution of shapes on the photomask $\mathscr{M}$.
- The **image** stage, f_i, which communicates the shapes on the photomask $\mathscr{M}$ into an image $\mathscr{I}$ distributed within the thin resist film on the wafer surface.
- The **resist** stage, f_r, which transforms the image distribution $\mathscr{I}$ into a distribution of shapes representing regions protected by the developed photoresist.
- The **etch** stage, f_e, which transforms the shapes of regions protected by photoresist $\mathscr{R}$ into the final etched shapes transmitted into the wafer substrate.

We will now discuss each pattern transform in more detail.

2.2.2.1 Mask Transfer Function

The mask transfer function f_m is the first stage in the patterning process. Photomasks are manufactured with their own lithographic process, beginning with a design layout and a mask blank, and ending with an etched photomask. A mask blank is an unpatterned photomask. Photoresist is applied uniformly to the surface of the mask blank. The pattern is transmitted into the photoresist using a mask writer. The mask writer reads a file representing the layout $\mathscr{L}$ and precisely controls the deflection of electron beams to deliver a patterned distribution of energy into the photoresist on the mask blank surface. The resist is processed and developed, which renders some regions of the mask blank exposed and other regions protected by the photoresist film. Finally, a pattern is etched into the mask through the regions unprotected by the resist.

The input to this stage is the layout $\mathscr{L} \in \{0, 1\}^2$, a two-dimensional discrete distribution that indicates which regions on the mask should be etched and which should not. The mask manufacturing stage is represented as the transfer function f_m. The output of this stage is a two-dimensional discrete distribution $\mathscr{M} \in \{0, 1\}$ representing a partition of the mask into regions that have been etched and those that have not. In an ideal mask manufacturing process, the communication of the layout to the mask would be perfect and $\mathscr{L}$ would be perfectly reproduced in $\mathscr{M}$. However, the electron beam, although very high resolution, still has a resolution limit that serves as a low-pass filter and prevents corners from being accurately transmitted to the mask [7]. The photoresist and etch processes are also not ideal [21]. The mask transfer function f_m captures the nonidealities introduced by this process, which result in a photomask that is not a perfect reproduction of the design.

2.2.2.2 Imaging Transfer Function

The image transfer function f_i represents the mask diffraction and projection processes. This stage achieves the transmission of the photomask pattern $\mathscr{M}$ in the form of an image $\mathscr{I}$ distributed through the volume of the photoresist. Astonishing progress in this stage has been, since the 1970s, one of the key enablers of IC miniaturization.

The input to the image transfer function f_i is the photomask $\mathscr{M}$. Although a single mask transfer function can only partition the photomask into two regions, it is possible that a photomask has been constructed from multiple mask transfer functions. For example, some photomasks use three or more regions to engineer interference effects that improve the image contrast [25]. For simplicity, we will only discuss binary masks in this chapter.

The imaging transfer function represents the most complicated and costly portion of the IC manufacturing process. It depends on three elements: an exposure tool, a silicon wafer substrate, and a photoresist film sitting atop the wafer substrate.

The exposure tool transmits light from its illuminator through the photomask, which diffracts the light. A portion of the diffracted light is collected by the exposure tool's projection lenses or mirrors, which reduce the size of the pattern and transmit it to the thin photoresist layer that lies atop the silicon wafer substrate.

The output of the image transfer function is a three-dimensional distribution of energy delivered, in varying doses, into volume of the photoresist. We represent this energy distribution as $\mathscr{I} \in \mathbb{R}^3$.

When the patterns on the photomask have dimensions that are relatively large compared to the wavelength of the illumination, the image transmitted through the projection system to the wafer looks like a projected shadow of the photomask pattern. However, when the photomask feature sizes approach the wavelength of the light provided by the illuminator, then the projected image will less faithfully resemble the pattern on the photomask. Such a system is called "diffraction-limited," belonging to a class that includes microscopes, telescopes, and cameras. The physics of diffraction-limited systems are well-understood [1, 19] and can even account for reflections and standing waves within the thin film of the resist [52]. Commercial software tools are available to simulate the transmission of photomask images and can achieve the task with very high accuracy.

It is possible to observe the input to the image transfer function, the photomask $\mathscr{M}$, using a scanning electron microscope (SEM). The SEM produces a two-dimensional image from which, with some image processing, a representation of $\mathscr{M}$ can be directly extracted [41]. It is also possible to observe the output image distribution $\mathscr{I}$ of the image transfer function using an Aerial Image Measurement

System (AIMS) [16]. It is therefore technically possible to collect both the input data and output data from this transfer function and to train a system, using a supervised learning process.

2.2.2.3 Resist Transfer Function

The resist transfer function f_r represents a series of chemical and thermal process steps that convert the three-dimensional image distribution $\mathscr{I}$ within the photoresist into a binary partition $\mathscr{R}$. For many applications, the partition is considered two dimensional, $\mathscr{R} \in \{0, 1\}^2$, indicating the regions of the wafer plane that are protected, or not protected, by the photoresist. However, it is more physically complete to consider the partition in three dimensions, $\mathscr{R} \in \{0, 1\}^3$, enclosing the volume of the photoresist, since the height of the remaining photoresist is not constant over the wafer surface. This can sometimes be important for patterning since thin photoresist may not be sufficient to protect the substrate during etch.

Unlike the image transfer function, which can be derived from electromagnetic first principles, the resist transfer function represents several physical processes that are still not fully understood. The most common formulations of the resist transfer function [27, 30] rely on several parameters that must be inferred indirectly. The transfer function wraps together several processes, including: the initial response of the photoresist when it is exposed to energy from the projection transfer function, which induces chemical and mechanical changes in the film; thermal processes including diffusion and reaction of species in the resist; and selective removal of the resist during development, rendering some regions of the substrate accessible to the etch process and other regions masked. Each of these processes has several variants, and the manufacturers of photoresists do not operate from a first-principles understanding that connects the chemical composition of their materials to certain outcomes. Modifications to the resist chemistry can break models without warning [38] and take many years to fix [13].

The input to the resist transfer function is the distribution of the deposited image $\mathscr{I}$ which is observable using an AIMS tool [16]. The output of the transfer function is observable from multiple methods including scanning electron microscopes (SEM), atomic force microscopes (AFM), and focused ion beams (FIBs). AFM and FIB measurements are more accurate for capturing resist profile and three-dimensional information, but they are also expensive compared to SEMs. On the other hand, when using an SEM to measure photoresist, several artifacts and calibration issues must be considered [4, 40] to correctly interpret the results. The most common SEM measurements capture the image parallel to the wafer surface, or "top-down," and it is not possible to see some of the three-dimensional structure of the resist. SEM measurements of photoresist structures also typically include a significant amount of noise, and it is difficult to reduce the noise without introducing other measurement problems.

The SEM measurements considered to be most precise are extracted from simple one-dimensional test structures. This is partly because a single two-dimensional image of a one-dimensional test structure can be averaged along the length of the line to reduce noise. Another factor in favor of one-dimensional structure measurements is the simplified alignment of symmetric one-dimensional SEM images with the original design layout data.

Determining a resist transfer function is quite different from determining the image transfer function: the physical theory is not comprehensive, and complete accurate measurements are difficult to acquire. The earliest incarnations of resist transfer function models were hand-crafted tables that documented the combined effects of the image and resist transfer functions in systematically distorting the pattern. Eventually, the construction of the rules from these tables was automated [36].

As devices continued to shrink, the rules tables became more complicated to design and to populate. Model-based OPC proposed a continuous-valued representation of the image and resist transfer functions [45], learning a "behavioral model" from measured data using linear regression on empirically tuned features. This practice became standard for the OPC and computational lithography industry [8], remains the dominant technology, and will be discussed in detail in Sect. 2.3.

While the resist transfer function is currently implemented using a linear regression model and learning from SEM image data, there are two significant opportunities for newer machine learning methods to advance beyond the state of the art. First, linear regression models are highly dependent on well-engineered input features. Designing new features, or even merely selecting which features to use, is an art rather than a science. The potential of deep neural networks to learn input feature representations offers the possibility of automating this expensive task, improving productivity. Second, deep learning techniques offer greater representational capacity and are likely more capable of effectively utilizing two-dimensional SEM contour data than conventional linear models. We will examine these possibilities in Sect. 2.4.3.

2.2.2.4 Etch Transfer Function

Following the selective masking of regions of the substrate surface by the resist transfer function, the wafer is etched. The etch transfer function f_e represents this stage, which is achieved inside of a tool called an etch reactor. The input to the etch process is the silicon wafer substrate masked by a photoresist pattern represented by $\mathscr{R}$. The etch attacks both the photoresist and the unmasked substrate, but the photoresist ideally protects regions of the substrate such that the etch transfers the desired pattern into the unprotected regions. The output of the etch process is the final result of the patterning process, $\mathscr{E} \in \{0, 1\}^2$, which partitions the wafer surface into regions that have been etched or not-etched.

The etch process is more complicated and less well understood than the photoresist process. Modern plasma etches use highly engineered chemistries to etch with high selectivity (meaning that the etch rate is faster for some materials than others) as well as anisotropy (meaning that the etch occurs faster in some direction than in others.) Well-controlled etch processes are critical for enabling high-aspect ratio memories, interconnects, and finFET transistor technologies [42]. An etch process is also usually a sequence of processes with very different chemistries.

TCAD topography simulation tools provide the capability of modeling etch processes but they are computationally expensive. Also, some etch effects have very long influence ranges on the range of many microns and are not easily integrated into device-level topography simulation.

It is relatively easier to get accurate measurements of the post-etch contour $\mathscr{E}$ using an SEM, compared to measurement of photoresist, because the substrate material is usually more stable and interacts less with the measurement process. It is usually sufficient, for large-area computational lithography, for an etch model to predict the two-dimensional contour of the etched regions of the substrate. In some situations, it is desirable to predict the sidewall angles of the etched apertures, which may influence the reliability of interconnections between the layers in a device.

Compact etch models have long been used in production for etch proximity correction [5]. The models are linear regressions trained on line and space widths extracted from SEM images of simple test structures [46]. Sometimes, the etch and resist processes are combined together, for the purpose of modeling, into one function, called a "lumped model."

2.2.3 *Summary*

A patterning process communicates two-dimensional information onto the surface of a silicon wafer through a series of stages, including mask-making, lithographic projection, photoresist processing, and etching. The stages represent substantially different physical processes. Some of the stages, like optical projection, operate with well-understood physical mechanisms, whereas others, like resist and etch, must be learned from empirical data. A schematic of the patterning process and its stages is shown in Fig. 2.1.

The key point of this section is that, although the patterning process is a composite of several different mechanisms that are physically diverse, it can be represented as a sequence of transfer functions. Abstracting the patterning process in this way prepares us to discuss how models of patterning can be constructed and trained, which will be the subject of the next section.

2.3 Machine Learning of Compact Process Models

In the previous section, we described the lithographic patterning process as a series of transfer functions, mapping from a layout to a photomask, then to a projected image, next into a patterned mask of photoresist, and finally into an etched pattern that has altered the structure of the silicon wafer surface. We also described the inputs, outputs, and physical mechanisms of each of these transfer functions.

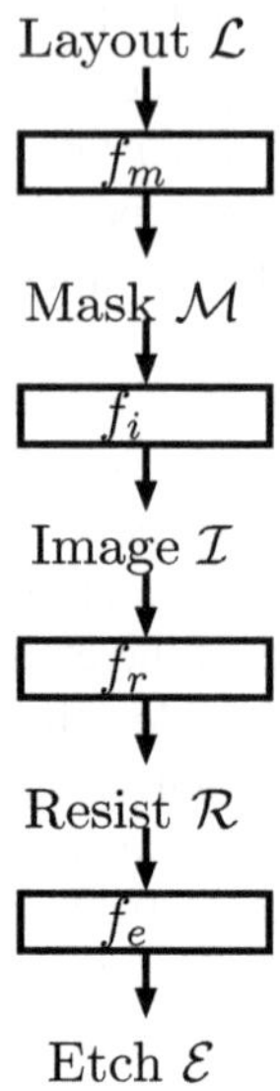

Fig. 2.1 The lithography patterning process is a communication process, transmitting the shapes of a layer in an IC design, represented by the layout $\mathscr{L}$, to etched patterns on the silicon wafer substrate, represented by $\mathscr{E}$. The communication process may be further decomposed into a series of transfer functions: f_m representing the communication of the two-dimensional layout onto a planar photomask $\mathscr{M}$, f_i the transmission of energy into a three-dimensional volume on the surface of a silicon wafer in a distribution represented by $\mathscr{I}$, f_r the distribution $\mathscr{R}$ of photoresist on the wafer surface after development, and $\mathscr{E}$ the distribution of the pattern etched into the silicon wafer surface resulting from etch transfer function f_e

Although the physical mechanisms for each stage are diverse, abstracting each stage into similar image-to-image transforms enables us to describe a common procedure for training models to represent these stages.

In this section, we will describe a procedure for training a compact process model (CPM), a particular kind of model of the lithographic patterning process. CPMs are used to estimate the outcomes of patterning processes for IC designs. Applications of CPMs include design-for-manufacturing, mask synthesis, and lithography verification.

Lithographic patterning processes are also modeled by a category of tools called rigorous or TCAD simulators. These tools apply the best available physical theories to the task of making models that are interpretable in terms of causes. They can provide insight and also can predict outcomes prior to the existence of real implementations of patterning processes, for example during the early phases of technology development. CPMs, however, have very different requirements and purpose from rigorous simulators: they must be capable of simulating the entire chip area with a practical computational budget, they are built using real measured data from manufacturing processes, and they must capture behaviors for which physical understanding may be lacking.

This section describes a typical procedure for training CPMs. The intention is to reinterpret this procedure, which has developed as its own specialty since the late 1990s, using the more widely understood terminology of machine learning. This translation should help assist cross-disciplinary learning between these two fields.

The section is broken into two parts: first, we formally describe the CPM machine learning problem statement; then, we examine the components of the supervised learning procedure which trains most industrial CPMs currently in service.

2.3.1 *The Compact Process Model Machine Learning Problem Statement*

Our starting point for terminology will be to apply to CPMs the definition of machine learning provided by Mitchell [31]:

> **Definition:** A computer program is said to **learn** from experience E with respect to some class of tasks T and performance measure P, if its performance at tasks in T, as measured by P, improves with experience E.

In this section, we will define the tasks T that CPMs accomplish, the performance metrics P by which they are graded, and the nature of the available training data that provides the experience E to the learner.

2.3.1.1 The Compact Process Model Task

The most common task for a CPM is

- **Task T:** Infer from mask layout data $\mathscr{L}$ the pattern transferred to the substrate by a patterning process.

Figure 2.2 illustrates the role of a CPM in context and clarifies the task it performs. It shows the relationship between a design target $\mathscr{D}$, a mask layout $\mathscr{L}$, and the pattern transferred to the wafer, either the resist or etch contours, $\mathscr{R}$ or $\mathscr{E}$. We have already described the patterning process in Sect. 2.2 as the communication to the silicon substrate of a two-dimensional layout, representing a single layer in the design of an integrated circuit.

When training a CPM, the task T is usually to begin with an input layout $\mathscr{L}$ that has been generated by a mask synthesis process such as optical proximity correction (OPC) and reticle enhancement. Mask synthesis usually also uses a CPM, so this procedure is potentially recursive, with each iteration using a more refined CPM to generate the layout used for training the next iteration. There are strong economic incentives to complete the learning and converge to a good CPM in as few iterations as possible, however.

In practice, the layout $\mathscr{L}$ may be generated using an approximated CPM, which could be mostly optical and use a resist model that has been calibrated on a previous

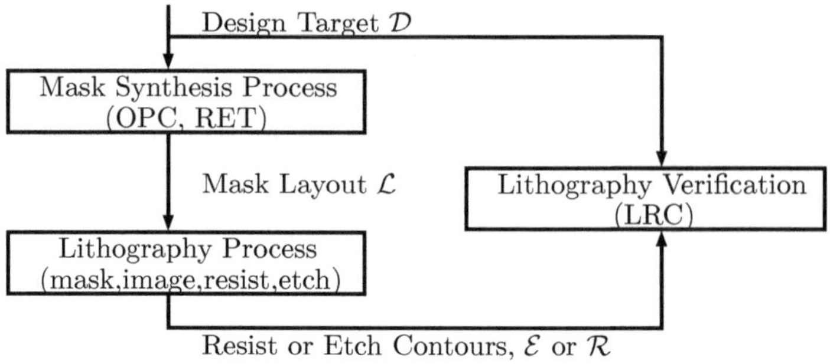

Fig. 2.2 The task of the CPM is shaped by the lithographic process, mask synthesis applications and lithography verification. The intention of the lithographer is to communicate the information in the design target $\mathscr{D}$ onto the wafer. The task of the CPM is to play the role of a virtual lithography process, and infer the resist or etch contours $\mathscr{R}$ or $\mathscr{E}$ from the layout $\mathscr{L}$. The layout will be generated by a mask synthesis process such as OPC or RET so as to counteract the systematic distortions of the lithography process and improve the correspondence between the design target and the wafer result, which will be evaluated during lithography rule check (LRC)

Table 2.2 Compact process model tasks

Task description	Mapping
Resist contour model	$\mathscr{L} \in \{0, 1\}^2 \rightarrow \mathscr{R} \in \{0, 1\}^2$
3D resist model	$\mathscr{L} \in \{0, 1\}^2 \rightarrow \mathscr{R} \in \{0, 1\}^3$
Etch contour model	$\mathscr{L} \in \{0, 1\}^2 \rightarrow \mathscr{E} \in \{0, 1\}^2$
Etch bias model	$\mathscr{R} \in \{0, 1\}^2 \rightarrow \mathscr{E} \in \{0, 1\}^2$

similar process. In some cases, the test layout may be generated from the design target using a simple rules-based correction. For a small enough test set of well-chosen test structures, it is also possible to use a rigorous TCAD lithography simulator to perform the mask synthesis that generates the input layout $\mathscr{L}$.

Once the layout data $\mathscr{L}$ has been determined, the task is to infer the pattern transferred to the silicon wafer substrate. It is most common to map from layout $\mathscr{L}$ to photoresist contour $\mathscr{R}$. Historically, the lithography teams that build CPMs have prioritized modeling the photoresist pattern. Although the most important outcome of the lithography process is the distribution of the etch contour $\mathscr{E}$, the mapping from $\mathscr{R}$ to $\mathscr{E}$ has been handled separately using a rules-based correction, etch process adjustments, or sometimes an additional staged correction with a model that maps directly from the photoresist contour to the etch contour, $\mathscr{R} \rightarrow \mathscr{E}$.

In summary, the task that a CPM may perform depends on the needs of the application that will utilize it. The most common types of CPM tasks are shown in Table 2.2: a resist contour model maps from a layout to a two-dimensional boundary between regions masked and not-masked by the resist; a 3D resist model from layout to a three-dimensional boundary between the inside and the outside of the photoresist film; an etch contour model from the layout to the two-dimensional boundary between regions etched and not-etched; and an etch bias model from the resist contour to the etch contour, without reference to the original layout.

Of the tasks in Table 2.2, the most common is the resist contour model, which will be the focus of the remainder of this section. However, the concepts apply to the other tasks, with proper adjustments of the input and output data in the training set.

It is typical in industrial applications to measure the photoresist contour $\mathscr{R}$ using a top-down SEM image which is then analyzed to find a two-dimensional boundary of the transferred pattern. In practice, the top-down SEM does not capture all of the information about the pattern transfer. Some work on CPMs has demonstrated benefits in modeling the three-dimensional characteristics of the resist [49]. Motivations include the potential for resist undercut that cannot be observed by top-down SEM, and the impact of shallowly sloped sidewalls or excessive resist thickness on the etch pattern transfer. But, the most common task required for a CPM is prediction of the two-dimensional contour boundary of the transferred pattern.

2.3.1.2 The CPM Training Experience

We define the CPM experience as:

- **Experience *E*:** Test pattern {input, output} training data pairs including mask layout polygon inputs and SEM image outputs.

The specific nature of the input or the output in the training data is determined by the CPM task. In this section, we will describe the data experienced during the training of a CPM that models a resist contour.

The input to a resist contour CPM is always a mask layout $\mathscr{L}$. For the common case of a binary photomask, we can consider the input as partitioning a two-dimensional plane into two different mask regions, each with different optical characteristics. In a real manufacturing environment, a photomask will be imaged several times over the surface of the wafer. We will consider only a single mask exposure on a single region of the wafer. There is a one-to-one mapping between each point on the photomask and each point on the wafer.

Following fabrication of the test structures, a printed feature is observed in the vicinity of a calibration gauge, as illustrated in Fig. 2.3. The distance between edges of the mask layout intersected by the gauge is the mask critical distance, or mask CD. The distance between observed edges of the printed feature is the printed critical dimension, or printed CD. The displacement of an observed edge from its corresponding mask edge is the edge placement error, or EPE. The difference between the mask CD and the printed CD is the CD error. These CD error measurements, extracted from scanning electron microscope measurements, are the raw input to the training process. The task of the model during training is usually to predict the CD error or EPE for every gauge in the training set.

As is typical of most machine learning problems, the greatest challenge lies in acquiring high-quality examples of the desired output values, also called "labeled data." Because it is costly to gather measurements, a smaller dataset that contains

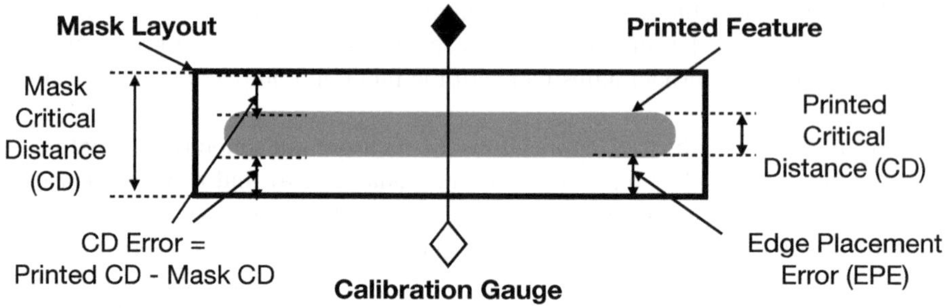

Fig. 2.3 The data experienced by the CPM during the training process is illustrated. The input is the mask layout, a set of polygons representing the features on the photomask. Following fabrication of the test structures, a printed feature is observed in the vicinity of a calibration gauge. The distance between edges of the mask layout intersected by the gauge is the mask critical distance, or mask CD. The distance between observed edges of the printed feature is the printed critical dimension, or printed CD. The displacement of an observed edge from its corresponding mask edge is the edge placement error, or EPE. The difference between the mask CD and the printed CD is the CD error

sufficient information is preferred. Historically, datasets begin with sample sizes in the range of hundreds, and trended into the low thousands over the years.

In addition to cost, there are significant technical challenges in acquiring accurate data from the SEM measurement process, which impact the lower bound of unavoidable error for the CPM. To understand these accuracy limitations, we will briefly describe the operation of the SEM. The direct output of the SEM tool is a two-dimensional matrix of measurements of electron scattering off of the structures on the surface of the silicon wafer, resembling an image. The raw data are processed to infer the printed features.

There are many subtleties that must be considered to accurately interpret this data [4], and we shall only discuss a few of the more prominent. One source of uncertainty in interpreting the SEM images is the inherently stochastic nature of the electron scattering that forms the image, which introduces noise. This uncertainty can be partly mitigated through repeated measurements. For one-dimensional test structures, a single SEM image provides several directly comparable measurements, and the noise can be averaged out.

Accurately mapping from the design layout graphics to the SEM measurement can also be difficult. To infer the position of the measured contour $\mathscr{R}$ in relation to the coordinate space of layout $\mathscr{L}$, it is necessary to possess an accurate coordinate mapping relationship, including the position, scale, rotation, and several other transform variables. Most of the test structures are designed so that these variables can be determined in a straightforward fashion. For example, if the test pattern is symmetric, and the lithography process preserves symmetry, then this reduces the likelihood of placement error. Also, if the test pattern has a clearly identifiable pitch, which is common for one-dimensional structures, then an accurate overlay of the SEM image to the design coordinates is more easily achieved. However, these considerations can be overly restrictive for ensuring that the test patterns used in the training dataset are sufficiently representative of the test conditions, and can lead to overfitting.

Another complication for interpreting SEM images of photoresist is the interaction between the energetic electrons in the electron beam, and the chemicals and polymers in the photoresist. The electrons deliver energy into the photoresist, inducing chemical reactions that cause the resist to shrink and lose volume. Also, the electrons may accumulate on the substrate and induce a charge that interferes with the measurement process.

With these measurement limitations in mind, it has become common to populate the training dataset with samples that measure CDs of simple symmetric one-dimensional features, which introduce the lowest unavoidable error. The training dataset is then usually supplemented with samples of CDs measured from symmetric two-dimensional test structures. The population size of the two-dimensional dataset samples will be related to the type of geometries that are being communicated by the given lithographic process. For example, training datasets for lithography processes that print patterns of holes will include more two-dimensional samples, compared to training datasets for lithography processes that print lines.

Because the samples in the dataset are collected from test structures that are designed to reduce measurement error, they do not completely represent the distribution of shapes that will be encountered in a real layout. In order to determine the degree of overfitting that may have been introduced during CPM training, it is common to hold out a subset of samples in a separate "dev" dataset.

The dev dataset will include additional samples from one-dimensional and two-dimensional symmetric test structures to detect overfitting and will also introduce new test structures that are representative of real designs. They may be derived from critically important cells, or from patterns that are known to be very sensitive, such as those that have previously caused hotspots. These "design-based" test structures can be compared manually or automatically on a contour-to-contour basis and used to ensure that the model has not overfit on the training dataset.

It is also possible to inspect the contour prediction of the CPM on design-based test structures and identify specific regions where the model has underfit, due to a lack of sufficiently general samples during the training process. This may inform a better selection of test structures for a future CPM training exercise.

Examples of symmetric one-dimensional, symmetric two-dimensional, and design-based test structures are shown in Fig. 2.4.

2.3.1.3 CPM Performance Metrics

We describe the CPM performance metric as:

- **Performance *P*:** Correspondence between the boundary of the patterned region as measured and as modeled.

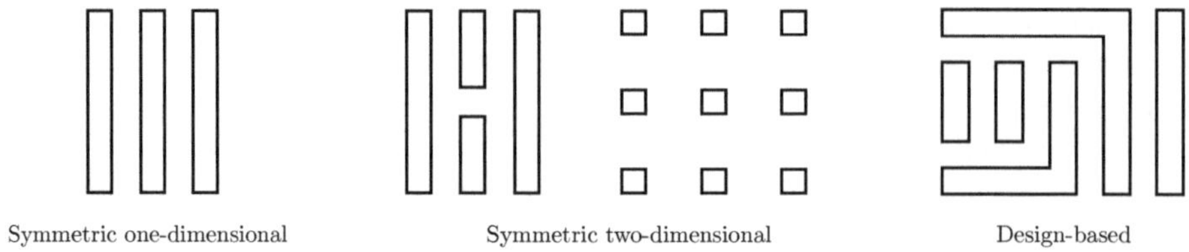

Fig. 2.4 Three different categories of test structures that are used to generate sample data, which supplies the experience necessary for the CPM to learn the behavior of the lithography process. The symmetric one-dimensional test structures produce sample data with the smallest unavoidable measurement uncertainty, and typically have made up the large bulk of training datasets. Symmetric two-dimensional test structures are also used to generate samples for the training dataset, in increasing proportions if the lithography process is intended to communicate hole-shaped patterns. These structures have larger uncertainty because it is more difficult to reduce noise through averaging. Design-based test structures can introduce more unavoidable error in the training set due to uncertainties in the placement or scale. Design-based test structures are more often used in the dev set rather than the training set, and are very valuable for detecting overfitting, or for helping to identify gaps in the training dataset

Table 2.3 CPM performance metrics

Metric	Description	Symbol
1D CD RMSE	Root-mean-square error of difference between predicted and measured CDs for samples from one-dimensional test structures	$P_{1D,\mathrm{RMSE}}$
2D CD RMSE	Root-mean-square error of difference between predicted and measured CDs for samples from two-dimensional test structures	$P_{2D,\mathrm{RMSE}}$
Absolute anchor gauge CD error	Absolute value of the largest prediction error for one or more gauges designated as anchors, which are preferred by the modeler to have the smallest error	P_{anchor}
Absolute CD error range	Sum of the largest positive-valued prediction error and the absolute value of the largest negative prediction error	P_{range}

Several different implementations for achieving this performance metric are considered during the training process, as shown in Table 2.3.

The simplest and most common method of comparison is to evaluate the root-mean-square error (RMSE) of the CDs predicted by the model y' and the CDs from the corresponding structures extracted from the SEM measurement y, as provided in the equation:

$$\mathrm{RMSE} = \sqrt{\frac{\Sigma_{i=1}^{n}(y_i' - y_i)^2}{n}}. \tag{2.2}$$

Because of the difference in unavoidable error associated with the measurement of two-dimensional test structures compared with one-dimensional, the two populations of gauges are often considered separately.

For integrated circuit manufacturing, there are often particular configurations of shapes on the input layout $\mathscr{L}$ that are of paramount performance due to their frequency of appearance and their functional role in the device. A particular spacing distance between neighboring lines, for example, may serve as a key characteristic of a particular lithography process. For these structures, it is very important that the bias in the model be minimized. To assist in reducing bias on selected features, it is common to specify a metric of maximum absolute prediction error on a small number of the so-called anchor gauges.

Finally, it is common to make the absolute range of the model prediction error a performance metric. Evaluating performance by the absolute range can have two effects: first, the role of the hardest-to-fit samples in the dataset is magnified, inviting further attention. It may be that the point is an outlier or mislabeled, for example, and this can be assessed on a sample-by-sample basis after an initial CPM has been generated. A model that keeps the worst-case CD prediction error below 2 nm may be preferred over a model with a low RMSE but a maximum CD prediction error of 5 nm. Using the range as a metric during training allows the error to redistribute over the dataset so as to reduce the worst-case error, and can be interpreted as a kind of tolerance band of allowable error. There are many variants on this kind of tolerance band metric, and one interesting example is an asymmetric error range band. By this technique, the training can encourage the CPM to bias its prediction error systematically in a particular direction where it might be more easily mitigated in manufacturing. For example, a metric could be evaluated on line-ends to encourage the model to systematically predict lines longer than they are actually printed, rather than risk a mis-prediction that will shorten the line and lead to electrical connectivity failure.

2.3.1.4 Summary of CPM Problem Statement

In summary, we have provided the following machine learning problem statement for CPMs:

- **Task T:** Infer the measurement of the pattern transferred to the substrate by a patterning process.
- **Performance P:** Correspondence between the boundary of the patterned region as measured and as modeled.
- **Experience E:** Test pattern data pairs including mask layout polygon inputs and SEM image outputs representing the same structures.

In the next section, we will discuss how this problem statement can be resolved using the supervised learning methodology.

2.3.2 *Supervised Learning of a CPM*

Previously, we described the CPM machine learning problem statement. We now describe in detail a widely used implementation which resolves the CPM problem statement using the supervised learning procedure. Supervised learning may be defined as:

> **Supervised Learning**: the machine learning task of inferring a function from a set of training examples, where each example is a pair consisting of an input object and a desired output value.

The supervised learning procedure provides a recipe for resolving the CPM machine learning problem, with the following ingredients [17]:

- A Dataset
- A Cost Function
- An Optimization Procedure
- A Model

The recipe, and the communication between the ingredients, is represented in the schematic Fig. 2.5. The **dataset** is a collection of paired data: a collection of input examples x and, for each, an associated output example y. The **model** infers an output y' when it receives input x. A **cost function** is used to evaluate the fitness of the model in predicting y', relative to the provided example outputs y. An **optimization procedure** finds the values of the model parameters θ and feature-generation parameters ϕ that minimize the cost function. Feature-generation

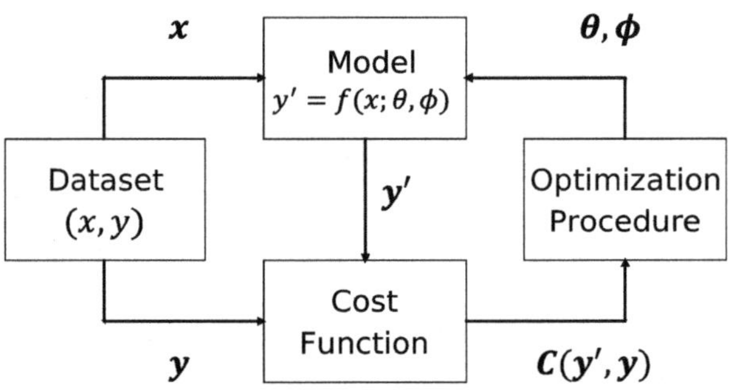

Fig. 2.5 The supervised learning process may be abstracted into a common recipe with the following ingredients: a dataset, a cost function, an optimization procedure, and a model. The dataset provides training examples, which are pairs of inputs x and desired outcomes y. The model receives an input x and predicts y', and is parameterized by learned variables θ and hyperparameters ϕ. The cost function C is some metric of the difference between the model result y' and the target value y from the dataset. The optimization procedure seeks to find the values of θ and ϕ that minimize the cost function C. The specific implementations for each of the ingredients may be considered separately and replaced individually, although an optimized solution often must consider the combined effects

parameters are not usually included in descriptions of supervised learning but are important for CPMs and will be discussed in Sect. 2.3.2.2.

In this section, we will describe each of these ingredients, in relation to training CPMs, in detail. We will initially limit our description to CPMs that use the linear regression model form, because the models are currently the most widespread in industry and the easiest to describe. In Sect. 2.4, we will replace the linear regression ingredient with a variety of other model forms, but much of the other supervised learning framework discussed in this section will remain relevant even when newer model forms are adopted.

2.3.2.1 CPM Model Form

The linear regression model form is the most common currently in use for constructing CPMs. The model form is given by the equation:

$$\mathbf{y}' = \boldsymbol{\theta}^T \mathbf{x}. \tag{2.3}$$

Each model input is represented as a vector of numbers $\mathbf{x}$, and each model output is represented by a vector of values $\mathbf{y}'$. The values of elements composing the vector $\boldsymbol{\theta}$ are determined during the model training process. In Eq. (2.3), we have chosen the convention that the first element x_0 in $\mathbf{x}$ is equal to 1, which means that the first element θ_0 in $\boldsymbol{\theta}$ is equal to the model bias, denoted as b in some alternative formulations of the linear regression model. The other elements of θ serve as weights that determine the relative magnitude of contribution from each of the features in $\mathbf{x}$, sometimes denoted by a weight vector $\mathbf{w}$ in alternative formulations.

It is surprising that such a simple model form can represent a process as complicated as lithographic patterning. Much of the complexity of CPMs that enables them to mimic nonlinear behaviors with a simple linear model lies in the feature-generating functions that produce the vector x from the original raw pattern input, as well as the interpretation of the output vector y'. We will discuss input representation and output interpretation of the dataset in Sect. 2.3.2.2.

2.3.2.2 CPM Supervised Learning Dataset

We have previously discussed the training data experienced by the learning algorithm, in the form of an input/output pair, and the task as inferring from the mask layout the resist or etch boundary contour ($\mathscr{L} \rightarrow \$\mathscr{R}$ or $\mathscr{L} \rightarrow \$\mathscr{E}$), or inferring the etch from the resist boundary contour ($\mathscr{R} \rightarrow \mathscr{E}$). We have also indicated that the model form is a linear regression.

The linear regression model form takes, as input, a vector of scalar values. For a CPM, we need to convert the input representation from a layout or contour, usually represented as collections of polygons, into a vector of values that is consumable by the linear regression. The output of a linear regression is a single scalar value.

We also must, therefore, explain how the output can be interpreted to reconstruct the desired outcome of the task, the resist or etch contours. We will now explain how the polygon inputs are represented to the model, and how the model outputs are interpreted to reconstruct polygon contours.

Input Feature Generation

We consider the task of inferring the resist field $\mathscr{R}$ from the input layout $\mathscr{L}$. Specifically, we want to know whether a specific observation point R on the resist field is within, or without, the boundary of the photoresist. We can consider the possible outcomes of the inference at R to be either **inside** or **outside**.

We further assume that data influencing the outcome at R will be the region of the layout $\mathscr{L}$ within the vicinity of an observation point L that corresponds, in the layout coordinate space, to the location of the observation point R in the resist coordinate space. Further, we define the "vicinity of L" as a region enclosed by a circle of radius A_L, which we will call the layout ambit.

To determine whether R is inside or outside the photoresist contour boundary, we provide as input the layout graphics within ambit of the corresponding point L on the layout, as shown in Fig. 2.6.

The input to the linear regression model must be a vector of real-valued elements, $\mathbf{x}$, not a circular region of polygons clipped near the vicinity of L. We will call each of these elements a feature of L. Every i-th element x_i in $\mathbf{x}$ is created by a feature-generating function $g_i(L)$ that operates on the region of $\mathscr{L}$ within ambit A_L of observation point L, so that:

$$x_i = g_i(L). \tag{2.4}$$

For simplicity, we consider an idealized case, where the mask transfer function t_m and the image transfer function t_i have been determined, and it is possible to observe the intermediate result of the projected image $\mathscr{I}$:

$$\mathscr{I} = f_i\left(f_m\left(\mathscr{L}\right)\right), \tag{2.5}$$

and $\mathscr{I} \in \mathbb{R}^2$. Every observation point L on the mask layout $\mathscr{L}$ as well as on the resist image $\mathscr{R}$ now maps to a corresponding point I in the image field $\mathscr{I}$. Rather than extracting features from the raw polygons of the layout, we will train the CPM with features extracted from the projection image. We may then represent the feature vector $\mathbf{x}$ as a function of feature generator operating on I rather than L:

$$x_i = g_i(I). \tag{2.6}$$

An example of a feature that could be extracted from $\mathscr{I}$ at observation point I could be the value of $\mathscr{I}$ itself at that location, or $x_i(L) = g_i(I) = \mathscr{I}(I)$.

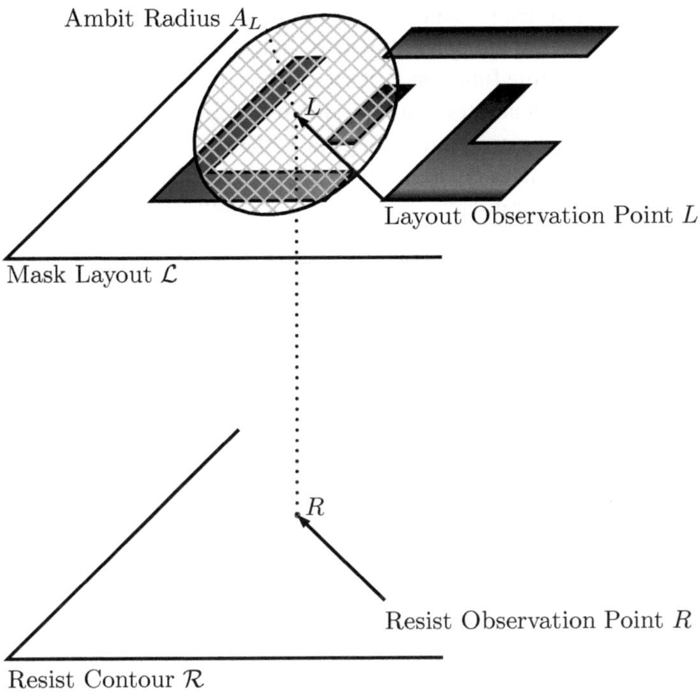

Fig. 2.6 The task of the $\mathscr{L} \rightarrow \mathscr{R}$ CPM is to infer whether point R is inside or outside of the resist contour boundary, using as input the cross-hatched region of the layout within the ambit radius A_L of observation point L on the layout $\mathscr{L}$

This feature would be the aerial image intensity at the point on the image field I corresponding to point L on the layout.

Another example of a feature that could be extracted from $\mathscr{I}$ could be a physically motivated quantity, like the acid concentration of the photoresist after the processing step called post-exposure bake, or PEB [27]. Various physically motivated terms are included in PCMs if they can be calculated quickly and if they are able to help improve the model fitness.

Features do not need to be necessarily physically motivated, however. Sometimes, the feature generators operate more like image processing operators, detecting the local density of $\mathscr{I}$ within the vicinity of I, or the value of edge-detection or corner-detection filters. Frequently, a convolution of a kernel K evaluated at point I in the image will be a feature generator. A radially symmetric kernel can be used to sample local projection image density. An antisymmetric kernel can be used to detect change in the projection image intensity, indicating potential edges or corners.

Much of the literature for CPMs explores the effectiveness of adding new features, generated from new feature generators. The virtues of the different feature generators used in CPMs are beyond the scope of the chapter. The key point is that feature generation, using hand-tuned feature-generating functions, is the main mechanism for improving the performance of linear regression CPMs.

Each feature-generating function g_i is parameterized by a vector of values $\boldsymbol{\phi}_i$. For example, if the feature generator is a two-dimensional radially symmetric Gaussian kernel convolved with the projection image at I, then it may be parameterized by its standard deviation σ. The intensity of the image $\mathscr{I}$ at I can be considered a feature with a large number of parameters, including every value used to specify the f_m and f_i transfer functions.

For simpler notation, we will combine all of the individual feature-generating functions g_i into a feature-generating vector evaluated at observation point I of projection image $\mathscr{I}$. We denote it as **g**, so that:

$$\mathbf{x} = \mathbf{g}(I; \boldsymbol{\phi}) \tag{2.7}$$

where $\boldsymbol{\phi}$ is a vector composed of all of the parameters necessary to initialize all of the feature-generating functions.

Considering the many feature-generator parameters necessary to define the optical projection system, as well as the parameters dedicated to photoresist, etch, or measurement behaviors, the overall number of parameters that require co-optimization can grow into the range of one hundred, although it is preferred to keep the number much smaller.

CPMs based on the linear model have proven surprisingly resilient in capturing lithographic processes, through the efforts of engineers to craft new feature-generating functions. However, each new feature adds more parameters. The complexity of determining the appropriate features to use for a given model as well as to find the best values for all of the necessary parameters has complicated the task of building CPMs. We will see in Sect. 2.4 that one of the opportunities for neural networks in the CPM domain is to reduce the reliance on hand-crafted feature engineering and optimization of a large number of parameters for feature generators.

Output Representation

The output of a linear regression model is a scalar value. The outcome we intend to interpret from the model is whether an observation point R in the resist image is inside or outside of the resist contour boundary. There are three different methods by which this information is encoded: variable bias, constant threshold, and variable threshold. The three output encodings are illustrated in Fig. 2.7.

Part a) of Fig. 2.7 illustrates the information we are trying to encode in the model output. The input to the model is a layout $\mathscr{L}$. We measure the width of a feature on the layout and find the layout CD, $CD_{\mathscr{L}}$. The output of the model indicates that points inside the resist contour $\mathscr{R}$ are confined within a narrower line, $CD_{\mathscr{R}}$.

If we are using the resist bias encoding, in Fig. 2.7b, we evaluate the model on the edge of the line in the layout $\mathscr{L}$. The model returns a value B that tells us the distance from the layout edge to the edge of the resist contour.

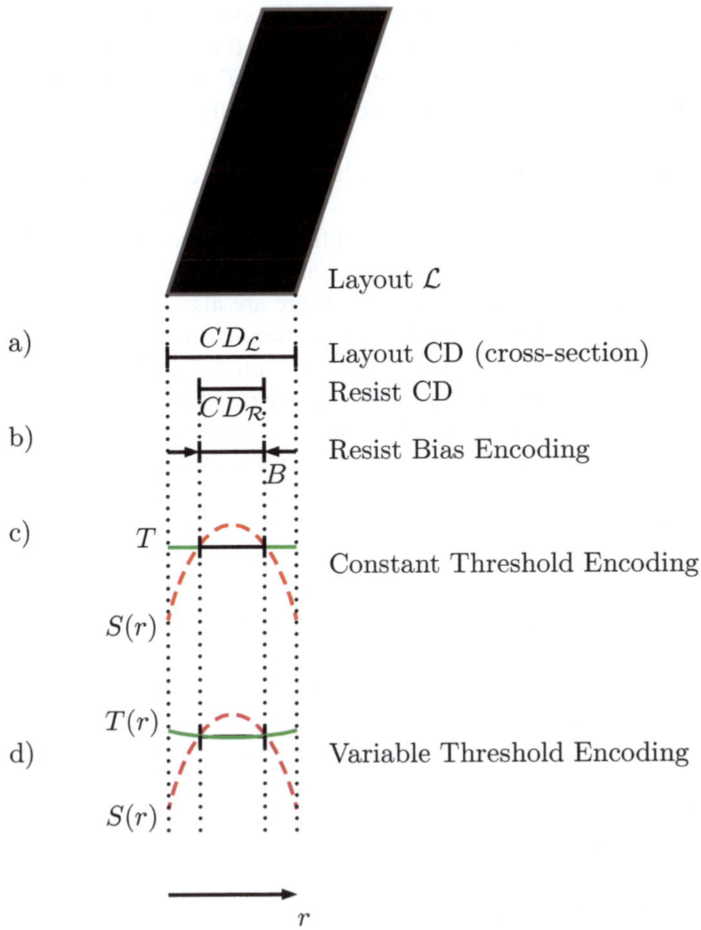

Fig. 2.7 The output of the linear regression model is a scalar value evaluated at each position r on the resist field. The output must be interpreted to indicate the position of the resist contour boundary, and whether a given point r is inside or outside of the photoresist. In (**a**), we see how the width of a figure on a layout $\mathscr{L}$ may be measured to find the layout CD, $CD_{\mathscr{L}}$. We have trained the model so that it predicts a narrower resist CD, shown by $CD_{\mathscr{R}}$. In (**b**), we see that the location of the resist contour can be encoded by a bias value B returned by the model, which can be evaluated on the layout edge and is a vector pointing to the resist contour edge. In (**c**), the constant threshold encoding is shown, which trains a signal function $S(r)$ to intersect the threshold function at the desired contour locations. In (**d**), we see the variable threshold encoding, which often uses the image intensity as the signal $S(r)$ and trains a threshold function T to intercept the signal at the locations of contours

If we are using the constant threshold encoding, in Fig. 2.7c, the model provides a scalar value T, and a function $S(r)$, a function of position r on the resist image. The intersection points between S and T indicate the resist contour boundaries. When $T > S$, the observation point r is inside the resist contour. When $T < S$, the observation point is outside of the resist contour.

The variable threshold encoding in Fig. 2.7d is similar to the constant threshold encoding, except that the value of T is now a function of the observation point r. The interpretation remains the same: the roots of $S - T$ are contour points, the region for which $T < S$ is outside of the resist contour, and the region $T > S$ is inside the resist contour.

The bias encoding is easy to interpret but, in practice, can introduce difficulties for mask synthesis tools when the biases are large compared to the minimum allowable spaces on the layout. Also, it is difficult to encode the resist CD using the bias encoding if there is not a one-to-one relationship between edges on the layout features and edges on the resist contours. There are also difficulties in translating discontinuous points on the layout, such as near corners and jogs, into a bias representation. However, when kept within its optimal application parameters, it can provide an efficient encoding for inferring the resist contour from the layout.

A more robust representation is the constant threshold output encoding, which can smoothly represent mapping between arbitrary two-dimensional layouts and contours. It can be evaluated across the entire domain of r rather than just at the edges of the layout.

The variable threshold representation produces two output channels from the linear regression model. This should provide identical contour results to the constant threshold model. For convenience, the signal S can be the unmodified projection image intensity $I(r)$, and the threshold function T can be the polynomial, trained by linear regression, that crosses the signal at the desired contour points. This provides computational lithography tools not only access to the resist contour but also simultaneous access to the original image intensity. With this encoding, a single model can provide information on the purely optical behavior of the process, as well as the final resist contour position.

Summary of CPM Dataset Representation

The task of the model is to map from layout polygons to a resist contour, but these data cannot be directly input to, or output from, a linear regression model. It is necessary to extract features from the layout in order to feed the input to the linear regression. It is also necessary to define how the output of the linear regression is to be interpreted in order to construct the desired final output result, the resist contour $\mathscr{R}$.

2.3.2.3 CPM Supervised Learning Cost Function

We have already discussed the optimization metrics of CPMs, as well as the satisficing metrics that are used to select between alternative models that have satisfied the optimization target, in Sect. 2.3.1.3. These metrics usually evaluate the difference between measured and predicted CDs. However, the output representation of the linear regression model provides the signal value y'. The signal must be analyzed in order to determine actual contour positions and CDs, as described in Sect. 2.3.2.2. To avoid the additional computational cost of CD analysis, it is common to construct the supervised learning per-element loss cost function in terms of the signal differences between the expected and predicted signal, $(y' - y)$, rather than the difference between the expected and predicted contour position or CD.

We assume that the dataset has been divided into a training set and a test set. The training set will be used serve as examples available during the training process and will influence the values of $\boldsymbol{\theta}$. The test set will be withheld from the training process and used to assess how well the model generalizes and predicts the outputs for inputs that it has not previously experienced.

Let the set of input feature vectors for every example in the training set be denoted as $\mathbf{x}^{(\mathrm{train})}$, and the input feature vector for the i-th example in the training set as $\mathbf{x}^{(i)}$. Similarly for the output $y'^{(\mathrm{train})}$ and $y^{(i)}$. The loss function, for example, i is given as:

$$L(y'^{(i)}, y^{(i)}) = \frac{1}{2}(y'^{(i)} - y^{(i)})^2. \tag{2.8}$$

From Eq. (2.3), we know that $\mathbf{y}'$ is a function of $\boldsymbol{\theta}$, so that:

$$L\left(y'\left(\boldsymbol{\theta}\right)^{(i)}, y^{(i)}\right) = \frac{1}{2}(\boldsymbol{\theta}^T\mathbf{x}^{(i)} - y^{(i)})^2. \tag{2.9}$$

The cost function J is merely the average of the loss function for all m examples in the training set:

$$\begin{aligned} J(\boldsymbol{\theta}) &= \frac{1}{m}\sum_{i=1}^{m} L\left(y'^{(i)}, y\left(\boldsymbol{\theta}\right)^{(i)}\right) \\ &= \frac{1}{m}\sum_{i=1}^{m}\frac{1}{2}(\boldsymbol{\theta}^T\mathbf{x}^{(i)} - y^{(i)})^2, \end{aligned} \tag{2.10}$$

which is also called the **mean squared error** on the training set, or $\mathrm{MSE}_{\mathrm{train}}$.

Alternative cost functions are used to **regularize** the model, which reduces risk of overfitting by penalizing model complexity [18]. This can be achieved by incorporating the values of the weights $\mathbf{w}$ in the parameter vector $\boldsymbol{\theta}$ into the cost function. To avoid penalizing the bias term, θ_0, we will modify our representation of the parameter vector into a bias value b and a weight vector $\mathbf{w}$:

$$\boldsymbol{\theta} = \begin{bmatrix} b \\ \mathbf{w} \end{bmatrix}. \tag{2.11}$$

The cost function may then penalize the model complexity by incorporating the L_1 and L_2 norms of the weight vector **w**, such that:

$$J(\mathbf{w}, b) = \frac{1}{m}\sum_{i=1}^{m}\frac{1}{2}(\mathbf{w}^T\mathbf{x}^{(i)} + b - y^{(i)})^2 + \lambda_2||\mathbf{w}||_2 + \lambda_1||\mathbf{w}||_1, \tag{2.12}$$

where λ_1 and λ_2 are model hyperparameters that are determined during optimization:

$$||\mathbf{w}||_1 = \sum_{j=1}^{n}|w_j|, \tag{2.13}$$

and

$$||\mathbf{w}||_2 = \sum_{j=1}^{n}w_j^2. \tag{2.14}$$

The cost function J is optimized for fixed values of λ_1 and λ_2, which must be determined during the hyperparameter tuning process.

As shown in Eq. (2.7), the input to the model is represented by the feature vector **x**, which is generated from the image intensity input $\mathscr{I}$ by a vector of feature-generation functions, $\mathbf{g}(\mathscr{I}; \boldsymbol{\phi})$. In order to optimize the parameter values $\boldsymbol{\phi}$ of the feature-generating functions, we evaluate a performance function $P(\boldsymbol{\phi})$. This function is usually some linearly weighted combination of the performance metrics described in Sect. 2.3.1.3. A typical example would be

$$\begin{aligned} P(\boldsymbol{\phi}) =& \gamma_{1D}P_{1D,\mathrm{RMSE}}(\boldsymbol{\phi}) + \gamma_{2D}P_{2D,\mathrm{RMSE}}(\boldsymbol{\phi}) \\ &+ \gamma_{\mathrm{anchor}}P_{\mathrm{anchor}}(\boldsymbol{\phi}) + \gamma_{\mathrm{range}}P_{\mathrm{range}}(\boldsymbol{\phi}), \end{aligned} \tag{2.15}$$

where γ_{1D}, γ_{2D}, γ_{anchor}, and γ_{range} are chosen by the modeler and become, effectively, the steering mechanism to direct the overall optimization. We will discuss how the feature-generator parameter vector $\boldsymbol{\phi}$ is found to minimize P in the following section.

A summary of the various functions, including the loss, cost, and performance functions, is shown in Table 2.4.

2.3.2.4 CPM Supervised Learning Optimization Algorithm

The final ingredient in the supervised learning recipe is the optimization algorithm, which is used to find the parameter values that provide the best performance metrics. Compact process models require three different types of parameter values to be determined during optimization: the linear regression parameters $\boldsymbol{\theta}$, the

Table 2.4 Functions used to evaluate the model

Name	Symbol	Purpose
Loss	$L(y' - y)$	The deviation, for a single example, between the predicted and expected outcome
Cost	$J(\boldsymbol{\theta})$	The average of the loss function for all examples in the training set, parameterized by the linear regression parameters θ. May also include additional regularization cost penalties to reduce model complexity
Performance	$P(\boldsymbol{\theta})$	The effectiveness of the model in meeting the performance metrics, parameterized by the feature-generation parameters $\boldsymbol{\phi}$. Typically evaluated on the training set and the test set

Table 2.5 Parameter types, and their optimization goals and methods

Parameter type	Optimization goal	Method
ϕ	Find the best feature-generator parameter values ϕ to minimize P	Simulated annealing, genetic algorithm, Nelder–Mead
λ	Find the best regularization hyperparameter values λ to reduce model generalization error	K-fold cross-validation to estimate Var(J), heuristics to tradeoff vs Bias(J)
θ	Find the best linear regression bias and weights θ to minimize the cost J	Gradient descent, Moore–Penrose pseudo-inverse, coordinate descent

regularization hyperparameters λ_1 and λ_2, and the feature-generation parameters $\boldsymbol{\phi}$. Each of these different parameter types are determined by different methods, as shown in Table 2.5. The procedure for determining all three sets of parameters is nested. In the outer loop, the feature parameter vectors ϕ are chosen. In the middle loop, the values of the cost function hyperparameters λ are determined. In the inner loop, the linear regression parameters θ are determined. Each level of this nested loop will now be discussed in detail.

Determining Feature-Generator Parameters ϕ

Before the model cost function can be evaluated, the input representations $\mathbf{x}$ for each training sample must be generated. Recall from Sect. 2.3.2.2 that the raw input for each training sample is $\mathscr{L}$, usually a set of polygons, or a two-dimensional array of numbers composing an image intensity $\mathscr{I}$. The CPM does not directly use the raw input representations but rather extracts a set of features that will represent the inputs to the model. The features are generated by a collection of feature generators, which are parameterized. If $\mathbf{g}(X, \boldsymbol{\phi})$ is a vector of feature-generating functions, parameterized by $\boldsymbol{\phi}$, then:

$$\mathbf{x} = \mathbf{g}(X, \boldsymbol{\phi}). \tag{2.16}$$

Generally, we seek $\boldsymbol{\phi}$ that minimizes the cost function J:

$$\underset{\phi}{\arg\min}\, J\left(\mathbf{x}^{(\text{train})}\left(\phi\right);\theta,\lambda\right). \tag{2.17}$$

It is necessary to make an initial guess for the values of $\boldsymbol{\phi}$ before we can generate the feature vector **x**. In some cases, the parameters in $\boldsymbol{\phi}$ correspond to physical quantities about which the user has some prior knowledge. For example, the feature generator corresponding to the aerial image intensity is parameterized by many values that relate to the projection lithography equipment, that are either measurable or provided by the tool manufacturer. These parameters can be initialized to physically meaningful values with some degree of confidence. Other parameters do not relate to physical quantities and must be guessed. Values that were found from prior models may serve as good starting points.

Following the generation of the first model, its performance P will be evaluated by Eq. (2.15). The outer-loop $\boldsymbol{\phi}$-optimizer will then choose a new location in $\boldsymbol{\phi}$-space. The optimization is complicated by the highly nonlinear response surface of the performance function P in ϕ-space. Typical optimizers include Nelder–Mead, simulated annealing, and genetic algorithms.

The outer loop will terminate based on criteria specific to the chosen optimizer, typically after a fixed number of iterations, or when P meets its performance requirements, or when P has not improved noticeably for several iterations.

Determining Regularization Hyperparameters λ

The purpose of regularization is to improve the generalization performance of the model and reduce the likelihood of overfitting. It is not possible to determine the regularization hyperparemeters without making models and testing them to assess their variance. Since data can be scarce for CPMs, a pragmatic approach to measuring model variance as a function of the hyperparameter value is to use k-fold cross-validation [18]. The training set is split into k equal parts called folds. Then, k models are fit, and for each model a different fold is held out for validation. Each of the models' performance can be evaluated by comparing the prediction error on the held-out validation data to the model bias on the training data.

The hyperparameters are set by sweeping through a range of values, making a large number of models following the k-fold cross-validation methodology, and calculating the model prediction variance for each value of λ. It is typical for k to equal 10. Values for λ are selected that indicate the model as simple as it can be without increasing prediction bias.

Determining Linear Regression Parameters θ

We will usually have more training samples than we have trainable θ parameters for linear regression CPMs. For very large datasets, a gradient-descent method, including batch gradient descent and stochastic gradient descent, may be used to minimize the cost function. This approach works very generally and efficiently to find $\boldsymbol{\theta}$.

For CPMs with merely thousands of training samples, which is a common case, it can be even more efficient to calculate the θ parameters analytically. Finding model parameters to minimize the RMSE is equivalent to solving the problem of ordinary least squares, for which there are many resources [6] including singular value decomposition (SVD). Some complications arise in efficiently minimizing the cost functions if the λ_1 hyperparameter is nonzero, since this can no longer be posed an ordinary least squares problem, but this problem has been solved through the method of least angle regression [12].

The efficiency and robustness of algorithms for finding the parameters of a linear regression model are a strong motivation for using this simple form. In the practice of training CPMs, the selection of parameters θ can be almost fully automated and out of the view of the modeling process.

Summary of Supervised Learning Optimization Algorithm

To recap, three different functions are used to drive optimization: the loss function, which is defined per-element; the cost function, which is chosen to be RMSE so that a rapid analytical solution can be used as the linear regression training algorithm; and the objective function, which is used to optimize the feature-generation parameters ϕ. The historical motivation for this approach is to keep the inner machine learning optimization loop fast and robust, and postpone the calculation of more costly performance metrics to the outer loop, where they will be used to drive the feature generator parameters rather than directly influencing the linear regression parameter w and t. In cases where the least squares algorithm cannot find a sufficient minimum, this could lead to nonideal performance of the model since the performance metrics P are indirectly related to the optimized cost function J. However, as the optimization converges, the cost function is usually a sufficient proxy for the performance metrics.

2.4 Neural Network Compact Patterning Models

Compact patterning models utilize machine learning techniques to optimize the manufacturing process for integrated circuits, putting them at the intersection of two technology revolutions. Section 2.2 described the patterning process used to manufacture each layer of an integrated circuit. Section 2.3 explained how linear regression models of the patterning process are trained. Linear regression models have been an appropriate model form for many years, considering the relatively small number of measurements collected for training compact patterning models. However, advances in computational architecture and training algorithms for neural networks open up many new possibilities for overcoming some of the limitations of conventional patterning process models.

In this chapter, we will survey prior work adapting neural networks to implement transfer functions in compact process models. Applications include the mask, image, resist, and etch transfer functions.

2.4.1 Neural Network Mask Transfer Function

The mask transfer function is the first stage in the patterning process, representing the transformation of the layout from a digital representation to a physical, manufactured photomask. There are several papers that demonstrate effective linear regression models of electron beam mask writing [20, 55]. The earliest paper relating lithography to neural networks was Frye, Cummings, and Rietman [9, 14]. The paper appears to be the first to apply machine learning and backpropagation [39] to the field of computational lithography. The input features were from an array of binary-valued pixels representing the intended image to be written on the mask. The first layer in the network calculated the average intensity integrated over nine radial distances from the evaluation point as well as the center intensity. Engineering the features in this way is a kind of radial weight sharing that reduces the number of parameters and also ensures that the output will be radially symmetrical. This provided ten features to a linear regression model.

Training data was generated by using an analytic expression for the electron scattering and a nonlinear resist model to capture the thresholding behavior of resist. A conventional iterative correction was performed on a test pattern, and each pixel of the corrected mask constituted the output y of the training sample, with a 19×19 square of pixel values representing the mask target pixel values as inputs.

The output of the network was a single linear activation neuron that indicated a dose adjustment that should be applied to the electron beam writing at each point on the mask. The output of the linear regression model represented the degree to which the observation point was underdosed. In effect, this early simple network was designed to learn an inverse of the mask transfer function $\mathscr{L} = f_m^{-1}(\mathscr{M})$, modifying the representation of the layout $\mathscr{L}$ to improve the mask $\mathscr{M}$. The benefit of the neural network implementation was runtime, which the paper claimed had reduced the time to optimize the mask from 6 years/mm^2 to 18 h/mm^2. The speedup was achieved by eliminating the need for solving the inverse problem iteratively, and instead using learning an inverse filter which will be convolved with the mask target pattern to return the ideal dose adjustment.

More recent work [35] follows the example of Frye, learning a radially symmetric convolution kernel to model ebeam proximity effects, but also offers an innovation to correct for resist surface charging. The electron beam's trajectory is deflected by the influence of electrons that accumulate, during mask writing, on the photoresist, leading to systematic errors in the measured edge placements of the transmitted pattern [34]. Figure 2.8 illustrates the three-layer resist surface charging model form,

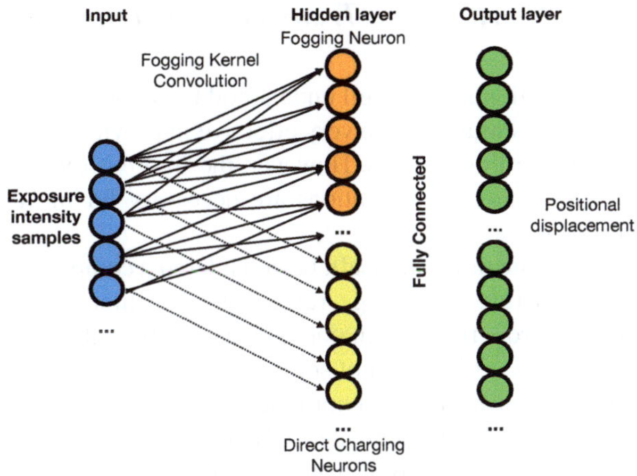

Fig. 2.8 A novel multilayer perceptron to simulate the mask transfer function f_m. The model receives the intensity of the electron beam exposure at the inputs and models two different effects: fogging, which is affected by the nearest neighbors, uses a fogging kernel to pass weighted sums of nearby neurons to the middle layer; and direct charging, which uses a direct nonlinear mapping from every input neuron to a neuron in the middle layer [35]

which has several novel characteristics. Charging can be modeled as a composite of two processes: direct charging and fogging. To model both effects and their interactions, the input values, representing the exposure intensity of the mask to the ebeam, are passed into two different middle layers. One middle layer merely applies a nonlinear activation on each input pixel, representing the direct charging as a kind of thresholding operation. The other middle layer is constructed by learning an asymmetric point-spread function to capture the fogging charging. Fogging charging can lead to asymmetric displacement on the mask and also may require an elliptical shape to the point-spread function. The output layer is fully connected to both middle layers. The output nodes represent a displacement map from the input exposure intensity map, from which can be interpreted the magnitude and direction of positional error for the printed edges. The models are trained using supervised learning from CD SEM measurements of manufactured photomasks.

In summary, linear regression models and neural networks have been applied to model electron beam proximity and charging effects. The proximity effect models are relatively simple in structure, learning symmetric point-spread functions that are used as convolution kernels, resulting in essentially linear models. Charging effect models learn how to represent asymmetric distortion effects, mapping from a scalar exposure intensity map to a vector-valued positional error displacement map.

There remains room for further work in this area. Specifically, there has not yet been a demonstration of a mask transfer function that succeeds in capturing the etch component of the mask manufacturing process, which can cause significant distortions that cannot be compensated by point-spread functions. More details on attempts to capture etch effects with neural networks can be found in Sect. 2.4.4.

2.4.2 Neural Network Image Transfer Function

The image transfer function is very well understood by theoretical models from first principles, and rigorous simulators are very effective in providing predictive models of this stage of the patterning process. However, the rigorous tools are too slow for large area application, and approximations are necessary for most computational lithography applications.

The image transfer function can, itself, be considered a composite of two transfer functions: mask diffraction and image refraction. The first stage represents the diffraction of the light from the illuminator as it passes through the photomask. The second stage represents the refraction of some of the diffracted orders from the first stage by the projection lenses, focused onto the wafer substrate to form an image.

The second stage can be implemented efficiently and with high precision in CPMs using an approximation derived from rigorous principles [15]. Consequently, there is no motivation for machine learning for this part of the image transfer function. On the other hand, the first stage, mask diffraction, though it can be modeled rigorously [50], is very difficult to model accurately with a CPM [32]. At the same time, the computational expense of the rigorous mask diffraction model is many orders of magnitude too slow for use in large-area computational lithography. Many efforts have been made to generate compact representations of the photomask diffraction, usually relying on hand-generated features and a segmentation of regions of the mask into spatial domains [3, 26, 54]. These efforts have been critical to enabling production-level OPC and lithography verification. The runtime of these compact model forms is generally considered acceptable for full-chip applications. The models are essentially perturbation models, with each new term representing a higher-order effect of smaller magnitude. Higher-order terms involve scattering interactions from more than one location on the mask. When multiple regions of the mask are interacting electromagnetically, the hand-crafted features become more complicated. As a result, it appears beyond the capability of hand-engineered feature generation to craft a compact mask transmission model that can capture the high-order interactions of some photomask geometries. Examples of difficult mask geometries for CPMs include densely packed chromeless masks, and EUV masks with assist features.

2.4.2.1 Neural Network Mask Diffraction Model

The task for the image transfer function, therefore, is to accurately represent high-order interactions of light scattering off the densely packed edges on photomasks with very fine geometries. An early approach attempted to train a neural network to approximate the mask transmission function from input mask layouts [2]. The training data supplied to the model was rigorously simulated results from the waveguide method, which is effectively a frequency-domain representation of the information transmitted through the photomask. The performance metric was the error in the predicted value for each sampled point in the simulated spectrum. The RMSE of all of the signal prediction errors is used as the cost function.

The input representation is the scalar diffraction spectrum, which can be quickly calculated. Each point in the scalar diffraction spectrum represents the magnitude and phase of a wave with a particular spatial frequency through the mask. This can be represented as a complex number in the frequency domain. As the light interacts with the edges of the features on the photomask, it will become polarized. Polarization can be represented as the propagation of two waves, the transverse magnetic (or TM) and the transverse electric (or TE) waves. The output representation is, therefore, effectively two complete spectrum maps, complex valued at each point, in the frequency domain.

The model form utilizes a pair of neural networks, each with two to four hidden layers. Each of the independent networks is responsible for predicting the spectrum of one of either the TE or TM polarizations, which are accessible on the output nodes of the network. The hidden nodes use sigmoid activation. The model is trained by gradient descent, and the weights are updated by backpropagation. The number of hidden layers, and the number of neurons per layer are tunable hyperparameters. A genetic algorithm is used to identify the hyperparameter values that balances the minimization of the prediction error and the variance. The best candidate models are then validated on highly sensitive test data to detect overfitting and prevent selection of a model with high variance.

Unfortunately, the network did not demonstrate strong prediction results based on frequency-domain data. When space-domain information was added to the input representation, such as the size of a space or feature, the prediction of the network improved. But, extracting space-domain feature information like the sizes of spaces or lines eliminates many of the intended advantages of neural networks over conventional domain decomposition methods, which can also construct accurate tables for specific lines and spaces.

More recently, a team demonstrated the application of a deep neural network (DNN) to model the diffraction effects resulting from mask topography [23]. Whereas conventional domain decomposition methods are restricted to specific simple mask geometries that directly relate to the pre-computed tables in their mask library, the proposed DNN approach can be trained with input mask layout data of any shape. This is increasingly important in leading-edge lithography applications, which can benefit from more flexible mask shapes. Training data is generated by feeding mask layout samples into a rigorous electromagnetic solver for mask diffraction. This generates for each sample a corresponding set of images that represent a complex-valued mask transfer function. Each point on the illuminator approaches the mask from a different angle, requiring a different rigorous simulation. For efficiency, an approximation is made by selecting a few points in the source for which to generate simulated data.

The first stage of the implementation rasterizes the mask layout $\mathscr{L}$ into a matrix of pixel values, which serve as the input to the DNN. The output is a pair of images representing the complex-valued mask transmission function. These two output images are then input to the lens refraction model, which will generate the final image by convolving a set of kernels that characterize the interaction of the illuminator and projection lenses.

The architecture of the neural network is not disclosed in the cited work. Results demonstrate that the neural network is capable of learning the mask transmission function with sufficient accuracy to keep the total CD error for the entire image transfer function f_i in the range of $+5/-3$ nm, as compared to a rigorous simulation. Further accuracy improvement may be required for industrial application, but the work illustrates the potential of neural networks to approximate complex imaging functions. The runtime is not disclosed in the work either, but experience with rigorous mask simulation leads to a reasonable estimate that the relative speedup is greater than three orders of magnitude, which brings it into the range of potential large-scale application in computational lithography.

2.4.2.2 Summary of Neural Network Image Transforms

The image transform can be subdivided into the mask diffraction and projection components, the first of which is an attractive target for machine learning due to the computational cost of rigorous simulation, whereas the second is less attractive due to an existing efficient physical theory. Attempts to use neural networks to represent mask diffraction processes have shown promise when space-domain information about the mask layout has been supplied to the input.

2.4.3 Neural Network Resist Transfer Function

We have already described the generation of a linear regression resist model in Sect. 2.3.2, which represents the dominant form for CPMs in current production. To recap, the conventional resist CPM is a transfer function f_r mapping $\mathscr{I} \rightarrow \mathscr{R}$. It is trained to learn a function called a "threshold" which intersects the intensity signal $\mathscr{I}$ at the location of resist contour edges. An SEM image provides the data that determines the target resist contour edges for the model. The performance metric for the model is typically either the RMSE of the predicted edge positions or the max edge position prediction error across the entire training set. The model form is usually a linear regression, but the model is capable of representing some nonlinear behaviors through extensive use of hand-engineered features. The features are generated from the input image intensity distributions, resulting in a set of feature maps. For a given point, the model samples every feature map and learns the relative weights of each through a pseudo-inverse of the linear model, solving for the weights $\mathbf{w}$.

2.4.3.1 Motivation for Neural Network Resist Transfer Functions

Despite the prevalence of the linear threshold resist model form in industry, it suffers from several defects, primarily related to its representational capacity and the extensive feature engineering that must be performed to capture the observed behaviors. Selection of the appropriate features remains an art, and the optimization of all of the parameters for the feature-generating functions is computationally intensive. The opportunity for neural networks is to expand the behavioral range of the resist transform and reduce the need for human intervention and computation to select and optimize features.

2.4.3.2 Earliest Neural Network Resist Model

There have been many attempts to apply neural networks to the problem of photoresist modeling. Mardiris provided the earliest demonstration of resist modeling with a neural network [29]. In this application, the task of the neural network is to approximate the well-known ABC model from Dill [11]. The model provides a mapping from the image intensity $\mathscr{I}$ at the resist surface to a quantity called "dissolution inhibitor concentration" distributed throughout the three-dimensional volume of the resist. The inhibitor concentration at each point in the resist can then be transformed into a development rate using a simple calibrated model. From the development rate, it is possible to iteratively evolve a "development front" which represents the boundary between resist and non-resist regions.

The development rate at a particular location in the resist z can be simplified as:

$$R(I_0, z) = \exp(E_1 + E_2 M(I_0, z) + E_3 M(I_0, z)^2) \tag{2.18}$$

where I_0 is the image intensity at the surface of the resist, M normalized concentration of the dissolution inhibitor, and E_1, E_2, and E_3 are fitted to experimental data. The function M must be solved numerically, and the motivation for the neural network is to provide a faster alternative.

The training data is generated from simulations using Eq. (2.18). The performance metric is the difference between the predicted and simulated development rate, and the cost function is the RMSE for the training data set. The input is represented as a pair (I_0, z) of real-valued numbers, and the output is a scalar value representing R. The model form is a neural network using a single hidden layer with sigmoidal activation on the hidden and output layers as shown in Fig. 2.9. The model is trained by gradient descent with backpropagation.

The purpose of Mardiris' model is to learn a function approximation of a system that otherwise requires numerical solution of a system of partial differential equations. Although the paper does not provide runtime comparisons, it claims that the neural network implementation is "much faster" than the conventional method, which is plausible. Unfortunately, the output of the model is a development rate, which is not directly consumable in an OPC tool or lithography verification tool.

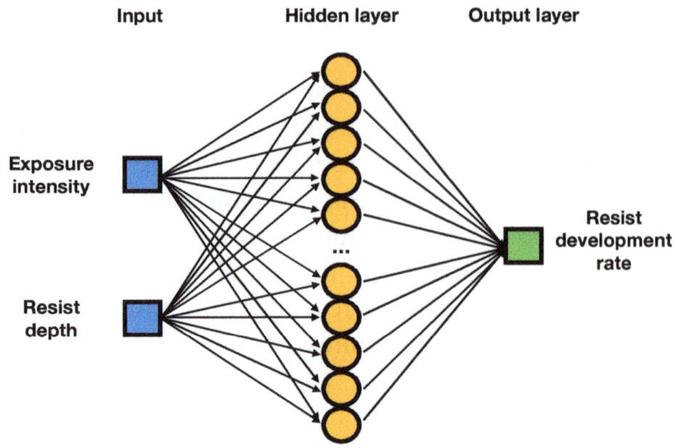

Fig. 2.9 A single hidden-layer neural network to approximate Dill's model, which represents the rate of resist dissolution (the development rate) as a function of the exposure intensity and the depth below the resist surface [29]

Instead, the output of the Mardiris neural network is fed into a cellular automaton model to evolve, over several time steps, the final resist contour boundary $\mathscr{R}$.

2.4.3.3 Neural Network Resist Model That Maps to a Contour

Since Mardiris' work, several neural network models have been designed to infer the resist contour boundary after development, which simplified their integration into large-area computational lithography systems. Zach [53] made an important contribution with a neural network that provides a variable threshold function as its output. By representing the output of the resist model as a variable threshold, the resist contour can be found following the same procedure described in Fig. 2.7 in Sect. 2.3.2.2. The task of Zach's model is to predict the resist contour locations, using training data extracted from 2D SEMs. The performance metric is the RMSE of the predicted contours at selected locations, compared to the measure edge positions.

Zach's work was motivated by the observation that the image intensity function $\mathscr{I}$ includes all of the information necessary to determine $\mathscr{R}$, but the engineered features available to him in his commercial OPC tool were too limited. That the feature-space was insufficient could be shown by observing that they could not discriminate fundamentally different patterns.

The input representation for Zach's neural network model was a vector of image intensity values $\mathbf{x}$ sampled along a cutline extending a micron from the point of interest $\mathbf{p}$. The output representation of the model is a shorter vector $\mathbf{y}$ that constructs a threshold function near the point of interest $\mathbf{p}$. Interpreting $\mathbf{y}$ as a threshold function t, and interpolating t near $\mathbf{p}$ to find intersections with the image intensity $\mathscr{I}$, the contour point of $\mathscr{R}$ along the cutline is found.

It is difficult to evaluate the performance of the model based on the published data, and it is worthy of note that the author found the neural network prediction to be roughly equivalent in quality to a linear regression model trained using the same procedure. Although the results were inconclusive regarding the benefits of neural networks, Zach did provide an effective supervised learning framework for finding a transfer function $f_r : \mathscr{I} \rightarrow \mathscr{R}$ which could have been easily integrated into commercial OPC tools. However, the limitation of the input representation to one-dimensional cutlines may have prevented the neural network from providing differentiation from linear models.

2.4.3.4 Neural Networks of Two- and Three-Dimensional Resist Models

More recently, work has extended neural network resist models beyond one-dimensional cutlines. Shim et al. [44] demonstrated a three-dimensional and a two-dimensional resist transfer function f_r implemented with a multilayer perceptron. The task of the two-dimensional resist contour model is to predict the resist contour locations, and the task of the three-dimensional resist model is to predict the resist heights. The training data for the network was generated from rigorous lithographic simulation.

For the three-dimensional resist neural network, the performance metric was the difference between the predicted and simulated resist height as a function of position on the wafer substrate plane. The cost function was the RMSE of this difference. A satisficing performance metric was the maximum observed error.

The two-dimensional resist contour neural network was implemented as a classifier, and its performance metric was its accuracy in predicting whether a given location on the wafer contour plane was "inside" or "outside" of the photoresist contour boundary. The cost function appears to have been the negative log likelihood of the output nodes.

Both the three-dimensional and two-dimensional resist neural networks make an interesting choice for their input representation. These models do not use the intensity $\mathscr{I}$ as inputs. Instead, they use features g_L extracted from the layout $\mathscr{L}$, so that $f_r : g_L(\mathscr{L}) \rightarrow \mathscr{R}$. This effectively lumps the mask transform f_m and image transform f_i into the resist model. The network will then, effectively, learn the representations of the combined effects of the mask, image, and resist transforms at the same time. This kind of approach is known as "lumped modeling."

The feature extraction method used for both the three-dimensional resist height model and the contour classifier is shown in Fig. 2.10. The input features were 33 measurements of the local pattern density, distributed in a polar grid sampled at discrete angles on multiple concentric rings centered on the observation point. Also provided as input features were the convolution results ϕ_{nm} of several sampling functions over the layout $\mathscr{L}$. The sampling functions are a family of kernel function Ψ_{nm} indexed by n and m and given by:

$$\Psi_{nm}(r, \theta) = J_n(r) \cos(m\theta), \tag{2.19}$$

where J_n is the nth Bessel function and θ the angle relative to the center of the kernel. This family of kernel functions can form a complete two-dimensional

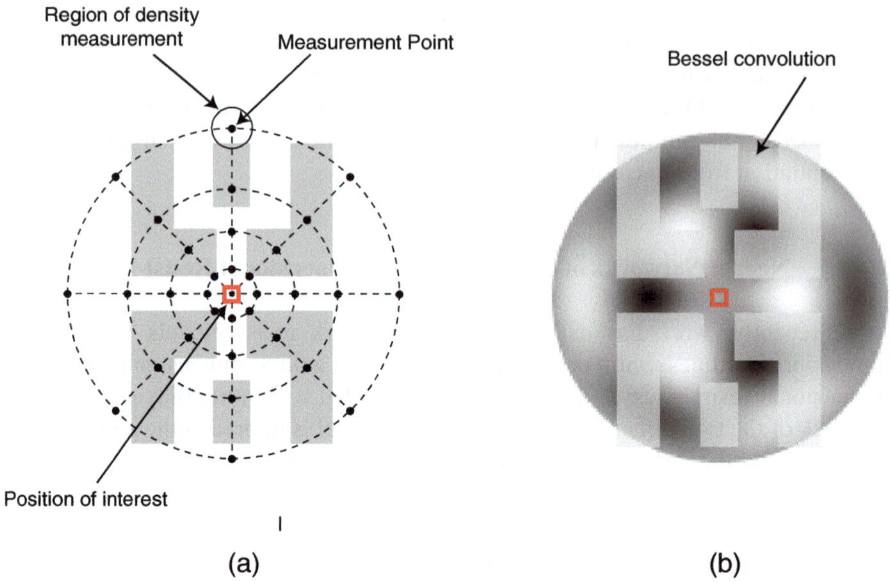

Fig. 2.10 Feature extraction $g_L(\mathscr{L})$ for the three-dimensional resist height neural network, and the two-dimensional resist contour classifier, both from Shim et al. [44]. In (**a**), a set of 49 or 65 measurements of pattern density positioned radially around the point of interest **p** are extracted from the layout $\mathscr{L}$. In (**b**), an example of a convolution of a radial Bessel function with the layout $\mathscr{L}$. A series of convolutions with 20 or 50 members of this family of functions is used to help represent the optical transfer function, without directly calculating the mask $\mathscr{M}$ or the image $\mathscr{I}$. The network learns a mapping to the resist boundary from features extracted from the layout $\mathscr{L}$, effectively lumping the mask, image, and resist transformations into a single learned transform. Figure used with permission from [44] and modified for notational consistency

band-limited representation of the input pattern $\mathscr{L}$, and therefore serves as a convenient feature generator. Since we know that the optical projection system filters out high-order, and hence high-frequency, terms from the mask diffraction spectrum, we can actually completely represent the layout using samples of convolution maps generated from this basis. It therefore is feasible that the neural network could learn the image transfer function during its training process.

For both the three-dimensional and two-dimensional resist models, a multilayer perceptron is used for the network architecture, with sigmoid activations on all of the hidden layers, as illustrated in Fig. 2.11.

The output node of the three-dimensional model represents the height of the resist at the observation point, and is implemented by a single linear neuron. The two-dimensional classification model represents the output as two logistic regression neurons, one for each potential class label $y, \in \{\text{inside resist, outside resist}\}$.

The training algorithm for both models is stochastic gradient descent, to reduce memory requirements of optimizing all training samples simultaneously and to reduce the risk of getting trapped in a saddle point of the cost function. Back

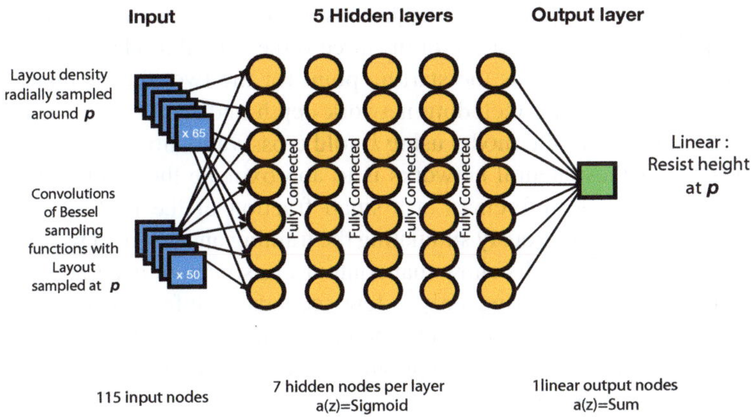

(a)

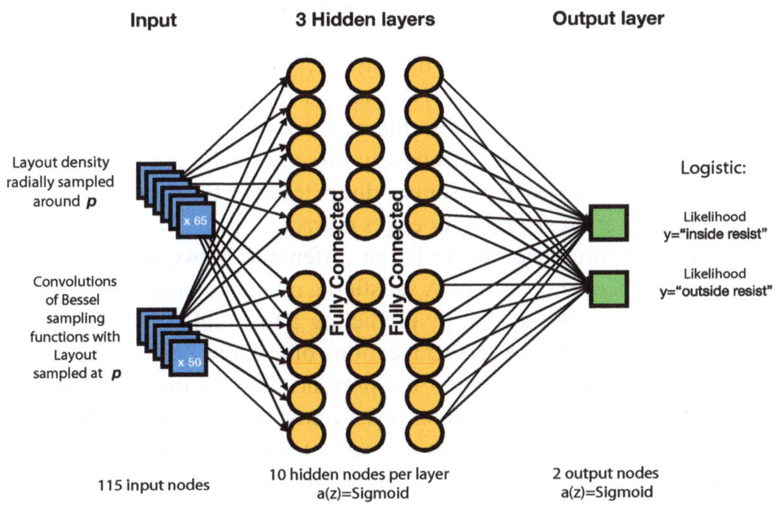

(b)

Fig. 2.11 Network architecture for: (**a**) the resist height regression model and (**b**) the resist contour classification model, both by Shim et al. [44]. The input features and hidden-layer architectures are the same for both models. The input features are up to 65 density samples of layout local to observation point **p** and convolutions of up to 50 members of the Bessel sampling function family with layout $\mathscr{L}$ evaluated at **p**, as illustrated in Fig. 2.10. The network is a multilayer perceptron with sigmoid activations on all hidden-layer neurons. For the three-dimensional resist model in (**a**), the output is a linear neuron representing the resist height at **p**. For the contour classification model in (**b**), the output layer is two logistic neurons. One neuron represents the likelihood that observation point **p** is labeled "inside resist," while the other neuron the likelihood of **p** labeled "outside resist" propagation is used to update the weights, using the regression cost function as the starting point for differentiation for the three-dimensional model and the inverse log-likelihood cost function as the starting point for the two-dimensional contour classifier. Hyperparameters, including network depth and width, were selected by analyzing the variance of the model using k-fold cross-validation.

The purpose of these neural networks is to approximate the results of rigorous simulations, but provide inferences at a small fraction of the runtime. The total computational time for rigorously simulating the training and test data would require approximately 2 h. The results show that contour classification accuracy can reach 94.9% on the test set, requiring only a few seconds to inference all of the test scenarios, implying a speedup of several orders of magnitude. The performance of the three-dimensional resist model to predict resist height appears to be within 5% of the maximum resist height.

2.4.3.5 Convolutional Neural Network Resist Model

The final resist transform model we will consider, from Watanabe et al. [48] is based on a convolutional neural network (CNN) similar to the AlexNet architecture [22]. The aim of this architecture is to train the network to learn the necessary features for mapping from the input image intensity directly to the output image. It is the architecture that most closely represents a direct transfer function $f_r : \mathscr{I} \to \mathscr{R}$.

The training data are resist contour edges extracted from 2D SEMs of developed photoresist. The performance metric is the difference between the predicted and measured edge at selected positions. A satisficing metric is the maximum prediction error. The model achieves the task by predicting a threshold function that is equal to the image intensity at the observed locations of contour edges (and not equal to the image intensity when the edge is not present). This is the same methodology used for conventional linear regression resist transfer functions, such as described in Sect. 2.3. The cost function is also the same for both methods, specifically the RMSE of the differences between the image intensities and the modeled thresholds at all of the locations where the resist contour edge has been measured.

The exciting aspect of CNNs is that the input representation is simply a 2D pixel map of the image intensity. There is no concern with extracting features from the raw image data, because it is the purpose of the CNN to learn the feature generators in the earlier convolutional layers of the network. A high-level representation of the network is shown in Fig. 2.12. A CNN directly accepts a two-dimensional map of the image intensity $\mathscr{I}$ without any further feature engineering and learns the necessary nonlinear feature extractions to complete the mapping to $\mathscr{R}$ in the lower layers of the network. The example from Watanabe et al. [48] uses three convolutional layers. Each convolutional layer is composed of a **conv** layer followed by a **pool** layer. The first **conv** layer is a collection of images generated by convolving the input image map with a collection of kernels, the shapes of which are learned during the training process. The result is a collection of feature maps, much like the feature maps generated in a conventional variable threshold model such as shown in Sect. 2.3.2.2.

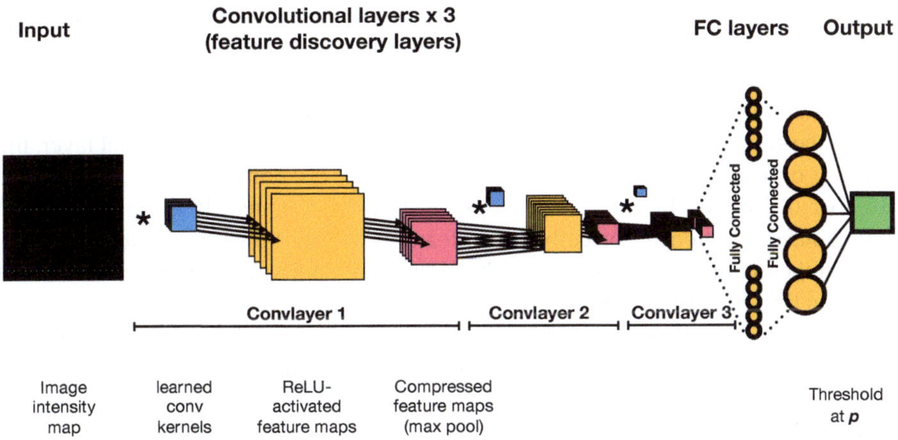

Fig. 2.12 A convolutional neural network (CNN) directly accepts a two-dimensional map of the image intensity $\mathscr{I}$ without any further feature engineering and learns the necessary nonlinear feature extractions to complete the mapping to $\mathscr{R}$ in the lower layers of the network. The example from Watanabe et al. [48] uses three convolutional layers. Each convolutional layer is composed of a **conv** layer followed by a **pool** layer. The first **conv** layer is a collection of images generated by convolving the input image map with a collection of kernels, the shapes of which are learned during the training process

Each of the maps in the first layer can be considered functionally identical to the feature maps generated in a linear regression CPM; however, the feature generation is learned automatically during training.

The CNN departs from the conventional linear regression CPM when each pixel on the feature maps is activated, element-wise. This is usually done using a rectified linear unit, or ReLU, which maps from a pixel z in a feature map to a pixel a in an activation map by:

$$a = \max(0, z). \tag{2.20}$$

Through experience, many practitioners of model building with deeper networks such as CNNs have found that the optimization proceeds more efficiently with ReLU activation.

Following the method first demonstrated by LeCun et al. [24], the image maps are then subsampled by a pooling operation. The pooling stage of a convolutional layer compresses the activation map to reduce the volume of data passed to the next layer. In the case of Watanabe's model, a "max-pooling" subsampling method is used, where the largest-valued activation pixel for a tile will represent the whole tile. This reduces the input volume to the next layer by a factor of four.

The second convolutional layer of a CNN operates the same as the first one. However, image intensity that served as input to the first layer was only a single map. Subsequent layers receive the entire collection of pooled activation maps generated from the preceding layer. Because the input volume of the second, and later, stages is now more than two dimensions, the convolutions become volumetric. Instead of merely weighting neighboring pixels on the same image, the second and third convolution layers must integrate over the feature map index as well.

As the size of the feature maps is compressed in each new convolutional layer, the number of maps per layer typically increases. At the last convolutional layer, a large number of low-resolution feature maps have been generated. If the convolutional layers have functioned correctly, these features should be sufficient to replace or exceed the quality of the laboriously hand-engineered filters used in a conventional variable threshold resist CPM.

The collection of final feature maps are then fed into a series of two fully connected layers, which successively reduce the number of neurons until finally being collected at a single neuron. This neuron is a linear output representing the threshold.

The network is trained using backpropagation, and optimized by Adam to help manage the learning rate. Mini-batch SGD is used to keep the memory requirements manageable and also reduce the likelihood of getting trapped in a saddle point of the cost function. Dropout method is used for regularization to help prevent overfitting.

The resulting model was demonstrated to reduce the RMS error of the resist model to predict CDs measured by an SEM by 70% compared to a conventional compact resist model with hand-engineered features, such as the model described in Sect. 2.3.2. More specifically, the conventional model achieved an RMS error of approximately 5 nm, while the CNN resist model achieved 1.5 nm on the training data set, and 1.64 mn on the test data set, indicating good generalization.

The demonstration of CNNs for resist modeling demonstrates two key values machine learning can offer to the construction of resist transfer functions: greater representational capacity, resulting in a smaller model error; also greater productivity, by learning the necessary features automatically during the training process, using only aerial image data as input and no hand-crafted features.

2.4.3.6 Summary of Neural Network Resist Transfer Functions

We have discussed the evolution of neural network resist models. Because the output is observable by an SEM image, or simulated by rigorous simulators, and because the internal physical mechanisms of photoresist processing are not fully understood, the resist stage of the patterning process has long been a subject of new model forms. We have discussed model forms ranging from linear regressions to binary classifiers as well as MLPs and deep CNNs. Much that we have learned from this section is directly transferrable to the next stage of the patterning process, which is etch.

2.4.4 Neural Network Etch Transfer Functions

As described in Sect. 2.2.2.4, etch is the removal of substrate material caused by the interaction of an etching particle with the substrate. The photoresist selectively masks some regions of the substrate from interaction with the etching particles.

Etch CPMs can be used for verification, to help detect etch hotspots that are likely to lead to pattern transfer problems and impact the yield or reliability of a device. Etch CPMs can also be used to inform etch proximity correction (EPC), which constructs a resist target. The resist target generated by EPC will then determine the success criteria for OPC and RET. CPMs for the etch stage have typically been linear regression models, trained to predict the displacement of the post-etch pattern from the resist edge position. Production-grade model-based etch correction has been deployed since at least 2004 [5].

Rigorous treatments of the etching process exist but are not applicable to large-area simulation. Etch CPMs are phenomenological, motivated by some simple physical insights. The discussion that follows will overlook many important physical aspects of the etch process, including depositions that occur incidentally to the etch process. To first order, the etch CPM determines a probability that an etch particle will interact with the substrate. A region is more likely to interact with the substrate if the substrate is not masked by the photoresist, and if the population of etch particles that can access the region is greater. Etch CPMs therefore usually include some short-range density measurement as a proxy for the likelihood of substrate interaction with an etching particle, and a longer-range density measurement that may be a proxy for the local population density of etching particles.

The probability that the substrate may interact with an etching particle is also increased if the photoresist protection layer is thin, since the resist may be eroded away during the etching process. When the resist sidewall angle is shallow near the contour boundary, resist erosion and unwanted etching will be higher. For this reason, etch CPMs have also included the photoresist sidewall angle as a feature [51].

Compact etch models may also add a nonlinear feature called visibility to the model [5]. Visibility emulates the three-dimensional character of the photoresist mask that protects the substrate from etch, as illustrated in Fig. 2.13. In this case, a nonlinear function of the pattern is required since only edges within visual communication with each other can effectively screen each other. A visibility kernel is used as a quick approximation of the proportion of incoming angles, from the point of view of a substrate region, that are screened by the local environment of the photoresist. It is easier to grasp the meaning of a visibility kernel by imaging yourself on the sidewalk in a city with tall buildings in a rainstorm. Being away from buildings makes you more likely to get rained on. Being near the outside corner of a building reduces the probability of getting wet. When you are alongside a building's edge, the probability goes down further. If you are in an alley between two buildings,

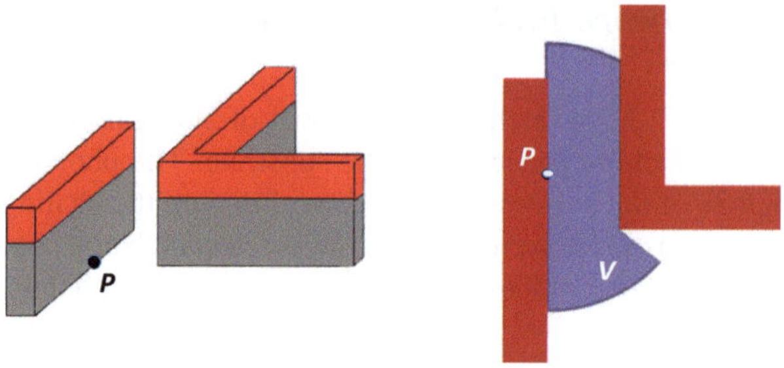

Fig. 2.13 An example of a visible area extraction from a layout. On the left is a three-dimensional representation of an observation point P at the base of a protective photoresist wall. It is screened from etch by the neighborhood of resist walls that are directly "visible" to it. Represented from the top-down in two dimensions, the solid region V represents the region of points from which point P would be "visible" to etching particles. The extracted visible area V is then further integrated with a weighting function to produce a scalar value with respect to location P. An efficient learned representation of the visible area, as well as density and resist sidewall and potentially other interactions, is approximated by the neural network in Shim et al. [43]

the probability goes down further still, because two walls are screening you from raindrops.

The visible kernel is computationally costly. Densities are linear and can be computed by high-speed convolutions, but the visible kernel is inherently nonlinear. The presence of a neighboring edge completely screens out all of the edges behind it. Implementations of the visible kernel have either had to approximate the effect with linear terms or rely on geometric algorithms to trace out the boundary of the visible area region and then perform integration. In the first case, the accuracy suffers, since the required nonlinearity cannot be supported. In the latter case, the runtime suffers and the scaling can be poor as the complexity of the visible area region increases.

In summary, the etch CPM conventionally is a linear regression composed of zero or more of each of the following feature types:

- Density measurements
- Visible kernel features
- Resist sidewall measurements

Recently, the training of an etch transfer function using a neural network has been demonstrated by Shim et al. [43] The method uses a multilayer perceptron network, of the same architecture shown in Fig. 2.11a. The only difference is that the output linear neuron, rather than representing the resist height at the observation point **p**, is a single linear neuron representing the magnitude of the etch bias. The features fed into the MLP are similar to those illustrated in Fig. 2.10: a polar mesh of layout density samples, and a vector of convolution values between the layout and a family of sampling function kernels provided by Eq. (2.19).

The choice of features is motivated by the need to capture linear and nonlinear effects: the polar map of density samples can be used to learn a linear convolution kernel that can replicate the traditional density measurements; the same map can be used to construct a nonlinear response to the input pattern that approximates the visible kernel signal; finally, the information from convolutions with polar sampling functions can be used to approximate the aerial image contrast and, therefore, the resist sidewall measurements.

While the input features to the MLP are likely to be sufficient to replicate the capabilities of the conventional etch CPM, the MLP has much greater capacity to combine all of these features into more complete representations of the etch process. The intention is to reduce the error of the model and, ideally, speed up the model by replacing the costly visible calculation.

Shim trains the model using measured etch biases, which are the displacements of edges between the resist and etch stages. The model is an etch transfer function f_e, mapping $\mathscr{R} \rightarrow \mathscr{E}$. The supervised learning procedure minimizes the RMSE between predicted and measured etch biases. It is trained by stochastic gradient descent, so as to manage memory consumption and reduce the likelihood that the optimizer will get trapped in the saddle point of cost functions. The model hyperparameters, including the layer width and the number of hidden layers, are selected to prevent overfitting, which is observed using k-fold cross-validation.

The results of the model demonstrate an improvement in the representational capacity compared to a conventional CPM, as evidenced by a reduction in the RMSE of the etch bias prediction of more than 20%, compared to a conventional etch CPM. The implementation, however, did not demonstrate a runtime improvement, although it is likely that exploitation of greater potential parallelism in an implementation such as a CNN should provide significant speedup.

2.4.5 Summary of Neural Network Compact Patterning Models

We have surveyed several applications of machine learning to the problem of constructing CPMs, each of which is represented in Table 2.6. Every stage of the patterning process is represented, including the mask, image, resist, and etch transfer functions. The task in all cases is to train a model that approximates a function. Sometimes, the model is trained by experiencing measured training data, and sometimes from simulated training data. In almost every case, the performance of the model is determined by how well the model predicts a real-valued target function, trained to minimize RMSE. The exception is a model that is trained to maximize the likelihood of accurate binary classification. Input representations vary, but the most common case is a 2D pixel map representing the spatial distribution of a quantity. The output representations are usually very compact, between one and four scalar values, or two different classification labels. Model forms cover a very wide range, from linear regressions to CNNs. Training algorithms are usually a variant of gradient descent, although in at least one case it is possible to solve for the learned parameters directly using the Moore–Penrose pseudo-inverse. The benefit provided by the neural network is usually related to the nature of the training data: if the input data is simulated, the benefit is to provide a faster calculation time; if the

Table 2.6 Machine learning compact patterning models

Stage	Task	Experience (training data)	Performance metrics	Cost function	Input representation	Output representation	Model form	Training algorithm	Benefit
Mask [14]	Approximate the dose adjustment to improve $\mathscr{M}$	Pre-corrected dose maps	Error in predicted pixel value	RMSE	Binary 2D pixel map	Scalar real (dose adjustment)	Linear regression	GD w/ backprop	Speed
Mask [35]	Approximate $\mathscr{L} \to \mathscr{M}$	2D SEM images of $\mathscr{M}$	Difference between the predicted and measured edge positions	RMSE	Real-valued 2D pixel map	Vector-valued position error $(\Delta p_x, \Delta p_y)$	Parallel single hidden-layer neural network with linear outputs	GD w/ backprop	Accuracy
Image [2]	Approximate the diffracted spectrum of $\mathscr{M}$	Rigorously simulated diffraction spectra	Difference between predicted and simulated spectra	RMSE	Two real-valued 2D pixel maps	Four real-valued 2D pixel maps	Two multilayer perceptrons. Sigmoid activations. Linear outputs	GD w/ backprop	Speed
Resist (Sect. 2.3)	Approximate $\mathscr{I} \to \mathscr{R}$	2D SEM images of $\mathscr{R}$	Difference between predicted and measured edge positions	RMSE of $s - t$ at measured edge locations	Multiple hand-generated feature maps with tunable parameters	One scalar (variable threshold)	Linear regression. L1 regularization and k-fold cross-validation	Moore–Penrose pseudo-inverse	Speed and accuracy
Resist [29]	Approximate the transform from $\mathscr{I}$ to resist development rate	Simulated development vs exposure intensity curves	Difference between predicted and simulated development rates	RMSE	Two scalars (exposure intensity and resist depth)	One scalar (etch rate)	Single hidden layer. Sigmoidal or tanh activation. Linear output	GD w/ backprop	Speed
Resist [53]	Approximate $\mathscr{I} \to \mathscr{R}$	2D SEM images of $\mathscr{R}$	Difference between predicted and measured edge positions	RMSE	1D vector (aerial image intensity cutline)	1D vector (resist 1/dose cutline)	Single hidden layer. Nonlinear activation. Linear output	GD w/ backprop	Accuracy

Resist [48]	Approximate $\mathscr{I} \rightarrow \mathscr{R}$	2D SEM images of $\mathscr{R}$	Difference between predicted and measured edge positions	RMSE of $s - t$ at measured edge locations	Real-valued 2D pixel map (aerial image)	One scalar (variable threshold)	CNN. Three conv+ max_pool layers. Two FC layers feeding to linear output	Mini-batch GD w/ backprop. Adam optimizer. Dropou	Accuracy
Resist [44]	Approximate $\mathscr{L} \rightarrow \mathscr{R}$	3D resist simulations	Difference between predicted and simulated resist height	RMSE	Radial pattern density, Bessel function convolutions	One scalar (resist height)	Multilayer perceptron. Sigmoid activation	SGD w/ backprop. k-fold cross-validation	Speed
Resist [44]	Approximate $\mathscr{L} \rightarrow \mathscr{R}$	Simulated resist contours	Classification accuracy	Likelihood	Radial pattern density, Bessel function convolutions	One scalar (resist height)	Multilayer perceptron. Sigmoid activation	SGD w/ backprop. k-fold cross-validation	Speed
Etch [43]	Approximate $\mathscr{R} \rightarrow \mathscr{E}$	Measured etch bias ($\mathscr{R} \rightarrow \mathscr{E}$)	Difference between predicted and simulated etch bias	RMSE	Radial pattern density, Bessel function convolutions	one scalar (etch bias)	Multilayer perceptron. Sigmoid activation	SGD w/ backprop. k-fold cross-validation	Accuracy

input data is measured, the benefit is to more accurately reflect observed behavior in the absence of a completely physical understanding. The publication dates for the surveyed works range from 1990 to 2017.

2.5 Conclusions

Machine learning and computational lithography share a long history. The aim of computational lithography is to improve the fidelity and yield of the integrated circuit manufacturing process. Machine learning is useful for finding efficient representations of the patterning processes that construct integrated circuits. We call these efficient representations compact patterning models, or CPMs, and they are built using a supervised learning process. The forms of these models range from linear regressions to convolutional neural networks.

In the first section, we described the patterning process and the stages that compose it, including mask-making, image projection, photoresist processing, and etching. It is hoped that the presentation motivates more machine learning researchers to explore the field of computational lithography, and in particular the construction and training of CPMs, more closely.

In the second section, we introduced we described in detail the supervised training of a linear regression lumped process model, the most common model form and training procedure for a CPM. We attempted to apply the current terminology of machine learning to the problem. It is hoped that the presentation motivates experts in the state of the art of CPMs to delve more deeply into the machine learning literature.

In the final section, we surveyed the long shared history of neural networks and CPMs. After a lot of early interest in the 1990s, interest in neural networks for CPMs waned. Recently, however, the maturing training methodologies and improved parallel computational infrastructure have reinvigorated interest in neural networks for computational lithography. Early indications are that the reacquaintance of computational lithography and neural networks will lead to faster, more accurate CPMs that are built with less user intervention and fewer hand-crafted features.

Acknowledgements Special thanks to Mike Rieger and John Stirniman for introducing me to this fascinating field.

References

1. E. Abbe, Beitrage zur Theorie des Mikroskops und der mikroskopischen Wahrnehmung. Arch. Mikrosk. Anat. **9**, 413–418 (1873)
2. Agudelo et al., Application of artificial neural networks to compact mask models in optical lithography simulation. J. Micro/Nanolith, MEMS MOEMS **13**(1), 0110022-1–16, (2014)

3. J.T. Azpiroz, Analysis and modeling of photomask near-fields in sub-wavelength deep ultraviolet lithography with optical proximity correction, Dissertation, University of California, Los Angeles, 2004
4. S. Babin et al., Modeling of charge and discharge in scanning electron microscopy. Proc. SPIE **7378** (2009). https://doi.org/10.1117/12.828575
5. D. Beale et al., Etch modeling for accurate full-chip process proximity correction. Proc. SPIE **5754** (2004). https://doi.org/10.1117/12.600815
6. Å. Björk, *Numerical Methods for Least Squares Problems* (Society for Industrial and Applied Mathematics, Philadelphia, 1996). https://doi.org/10.1137/1.9781611971484
7. S.-Y. Chou et al., Study of mask corner rounding effects on lithographic patterning for 90-nm technology and beyond. Proc. SPIE **5446** (2004). https://doi.org/10.1117/12.557745
8. N. Cobb, Fast optical and process proximity correction algorithms for integrated circuit manufacturing, Dissertation, University of California, Berkeley, 1998
9. K. Cummings et al., Using a neural network to proximity correct patterns written with a Cambridge electron beam microfabricator 10.5 lithography system. Appl. Phys. Lett. **57**, 1431 (1990)
10. R. Dennard et al., Design of ion-implanted MOSFET's with very small physical dimensions. IEEE J. Solid State Circuits (1974). https://doi.org/10.1109/JSSC.1974.1050511
11. F.H. Dill, Modeling projection printing of positive photoresists. IEEE Trans Electron Devices **22**, 456–464 (1975)
12. B. Efron et al., Least angle regression. Ann. Stat. **32**, 407–499 (2004)
13. C. Fang et al., A physics-based model for negative tone development materials. J. Photopolym. Sci. Technol. **27**, 53–59 (2014)
14. R. Frye et al., Proximity effect corrections in electron beam lithography using a neural network, in *IEEE International Conference on Systems, Man, and Cybernetics Conference Proceedings* (1990). https://doi.org/10.1109/ICSMC.1990.142210
15. H. Gamo, Matrix treatment of partial coherence, in *Progress in Optics*, ed. by E. Wolf (1964). https://doi.org/10.1016/S0079-6638(08)70571-7
16. A. Garetto et al., Aerial imaging technology for photomask qualification: from a microscope to a metrology tool. Adv. Opt. Technol. (2012). https://doi.org/10.1515/aot-2012-0124
17. I. Goodfellow, Y. Bengio, A. Courville, *Deep Learning* (The MIT Press, Cambridge, 2016)
18. T. Hastie, R. Tibshirani, J. Friedman, *The Elements of Statistical Learning: Data Mining, Inference, and Prediction*, 2nd edn. (Springer, New York, 2009)
19. H. Hopkins, On the diffraction theory of optical images. Proc. Roy. Soc. A **217**, 408 (1953)
20. A. Isoyan, L. Melvin, Full-chip high resolution electron-beam lithography proximity effect correction modeling. Proc. SPIE **7637** (2010). https://doi.org/10.1117/12.846681
21. J. Kotani et al., Mask CD uniformity improvement by dry etching loading effect correction. Proc. SPIE **5256**. https://doi.org/10.1117/12.524430
22. A. Krizhevsky, ImageNet classification with deep convolutional neural networks. Adv. Neural Inf. Proces. Syst. **25**, 1097–1105 (2012)
23. S. Lan et al., Deep learning assisted fast mask optimization. Proc. SPIE **10587** (2018). https://doi.org/10.1117/12.2297514
24. Y. LeCun et al., Gradient-based learning applied to document recognition. Proc. IEEE **86**(11), 2278–2324 (1998)
25. M. Levenson et al., Improving resolution in photolithography with a phase shifting mask. IEEE Trans. Electron Devices **29**, 1828–1836 (1982)
26. P. Liu et al., A full-chip 3D computational framework. Proc. SPIE **83260A** (2012). https://doi.org/10.1117/12.916076
27. C. Mack, New kinetic model for resist dissolution. J. Electrochem. Soc. **139**, L34–L39 (1992)
28. C. Mack, *Fundamental Principles of Optical Lithography* (Wiley, Hoboken, 2007)
29. V. Mardiris, Neural networks for the simulation of photoresist exposure process in integrated circuit fabrication. Model. Simul. Mater. Sci. Eng. **5**, 439–450 (1997)
30. D. Matiut et al., New models for the simulation of post-exposure bake of chemically amplified resists. Proc. SPIE **5039** (2003). https://doi.org/10.1117/12/485080

31. T. Mitchell, *Machine Learning* (McGraw-Hill, New York, 1997)
32. M.G. Moharam et al., Rigorous coupled-wave analysis of planar-grating diffraction. J. Opt. Soc. Am. A **71** (1981). https://doi.org/10.1364/JOSA.71.000811
33. G. Moore, Cramming more components onto integrated circuits. Electronics **38**, 114–117 (1965)
34. N. Nakayamada et al., Modeling of resist surface charging effect on EBM-8000 and its comparison with EBM-6000. Proc. SPIE **8701** (2013). https://doi.org/10.1117/12.2030095
35. N. Nakayamada et al., Electron beam lithography modeling assisted by artificial intelligence technology. Proc. SPIE **10454** (2017). https://doi.org/10.1117/12/2282841
36. O. Otto et al., Automated optical proximity correction: a rules-based approach. Proc. SPIE **2197** (1994). https://doi.org/10.1117/12.175422
37. M. Rieger, Communication theory in optical lithography. J. Micro/Nanolithogr. MEMS MOEMS **11**(1) (2012). https://doi.org/10.1117/1.JMM.11.1.013003
38. S. Robertson, Negative tone development: gaining insight through physical simulation. Proc. SPIE **7972** (2011). https://doi.org/10.1117/12.879506
39. D. Rumelhart, G. Hinton, R. Williams, Learning representations by backpropagating errors. Nature **323**, 533 (1986)
40. V. Rutigiliani et al., Setting up a proper power spectral density (PSD) and autocorrelation analysis for material and process characterization. Proc. SPIE **10585** (2018). https://doi.org/10.1117/12.2297264
41. I. Santo et al., Accurate contour extraction from mask SEM image. Proc. SPIE **9050** (2014). https://doi.org/10.1117/12.2046530
42. D. Shamiryan et al., Dry etching process for bulk finFET manufacturing. Microelectron. Eng. **86**(1), 96–98 (2009)
43. S. Shim et al., Etch proximity correction through machine-learning driven etch bias model. Proc. SPIE **9782** (2016). https://doi.org/10.1117/12.2219057
44. S. Shim et al., Machine learning-based resist 3D model. Proc. SPIE **10147** (2017). https://doi.org/10.1117/12.2257904
45. L. Stirniman, M. RIeger, Fast proximity correction with zone sampling. Proc. SPIE (1994). https://doi.org/10.1117/12.175423
46. I. Stobert et al., Etch correction and OPC: a look at the current and future of etch correction. 1493 Proc. SPIEE **8685** (2013). https://doi.org/10.1117/12.2015000
47. L.F. Thompson et al. (eds.), *Introduction to Microlithography*, 2nd edn. (American Chemical Society, Washington, 1994)
48. Y. Watanabe et al., Accurate lithography simulation model based on convolutional neural networks. Proc. SPIE **10147** (2017). https://doi.org/10.1117/12.2257871
49. F. Weisbuch, A.S. Naranaya, Assessing SEM contour based OPC models quality using rigorous simulation. Proc. SPIE **9051** (2014). https://doi.org/10.1117/12.2047826
50. A.K. Wong et al., Massively parallel electromagnetic simulation for photolithographic applications. IEEE Trans. Comput. Aided Des. Integr. Circuits Syst. **14** (1995). https://doi.org/10.1109/43.466339
51. C. Wu et al., Photoresist 3D profile related etch process simulation and its application to full chip etch compact modeling. Proc. SPIE **9426** (2015). https://doi.org/10.1117/12.2086048
52. M. Young, Modeling high numerical aperture optical lithography. Proc. SPIE **922** (1988). https://doi.org/10.1117/12.968409
53. F. Zach, Neural network based approach to resist modeling and OPC. Proc. SPIE **5377** (2004). https://doi.org/10.1117/12.535931
54. H. Zhang et al., An accurate ILT-enabling full-chip mask 3d model for all-angle patterns (2013). Proc. SPIE **8880**. https://doi.org/10.1117/12.2026468
55. R. Zimmerman et al., Predictive modeling for EBPC in EBDW. Proc. SPIE **7488** (2009). https://doi.org/10.1117/12.833482

Chapter 3
Machine Learning for Mask Synthesis

Seongbo Shim, Suhyeong Choi, and Youngsoo Shin

3.1 Introduction

With semiconductor technology scaling, mask synthesis tasks become more and more challenging. Popular model-based OPC (MB-OPC) relies on the iteration of lithography simulation and mask image correction [1]; it requires more iterations and each simulation takes more time, and resulting OPC time often becomes unmanageable. Conventional EPC methods become more inaccurate due to the simple rules and empirical models they employ [2–5].

Machine learning (ML) has recently been applied to mask synthesis tasks for their efficient implementation. ML technique consists of two phases: training and actual testing as shown in Fig. 3.1. In the training phase, ML model receives a vector of parameters ($\mathbf{x}$), such as local densities, extracted from a layout pattern and outputs a predicted value ($f(\mathbf{x})$), such as a mask bias for OPC application. ML model is optimized so that the difference between predicted value and desired value (y) can be minimized for all test patterns for the training. In testing phase, parameters extracted from unknown pattern of interest are provided to the trained ML model, which then outputs predicted value.

This chapter addresses two applications of machine learning in mask synthesis process: ML-OPC and ML-EPC. In Sect. 3.2, some prior works on ML-OPC are reviewed, which are followed by our own proposal. The idea is implemented and demonstrated using 28-nm industrial technology. Section 3.3 addresses EPC problem. Some conventional approaches to EPC and their limitations are introduced.

S. Shim (✉)
Samsung Electronics, Hwasung, South Korea
e-mail: sb47.shim@samsung.com

S. Choi · Y. Shin
School of Electrical Engineering, KAIST, Daejeon, South Korea
e-mail: suhyeung93@kaist.ac.kr; youngsoo@ee.kaist.ac.kr

I. M. Elfadel et al. (eds.), *Machine Learning in VLSI Computer-Aided Design*,
https://doi.org/10.1007/978-3-030-04666-8_3

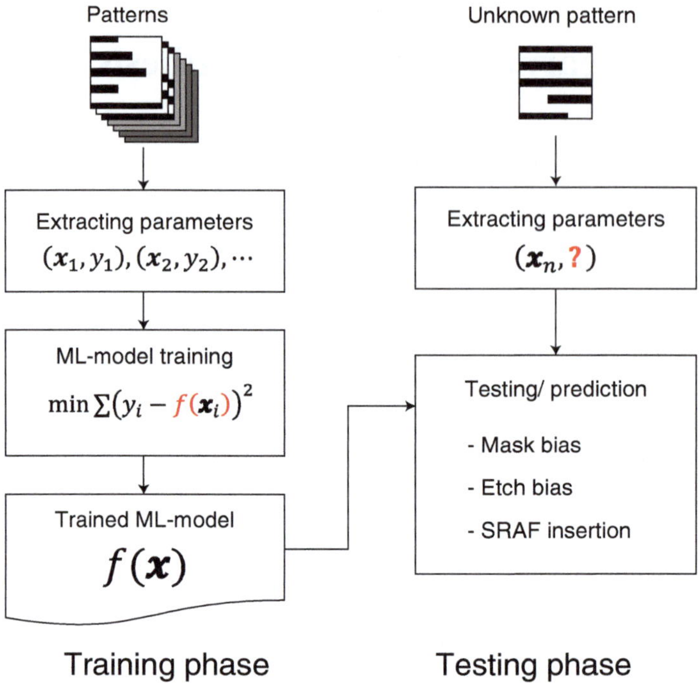

Fig. 3.1 Typical procedure of machine learning

How ML technique is applied to EPC is presented and experimentally demonstrated using 20-nm node DRAM gate layer. In both applications of machine learning, we discuss some key components: choice of model parameters, systematic reduction of training data, and the method to avoid over-fitting. Section 3.4 summarizes this chapter.

3.2 Machine Learning-Guided OPC

Conventional MB-OPC is based on iterative lithography simulation and mask correction by inspecting the simulation result. As feature sizes scale down, a lithography simulation takes longer time because of denser polygons, longer optical influence range, and more kernels. In addition, smaller feature size requires more OPC iterations to satisfy smaller edge placement error (EPE) tolerance. Thus, MB-OPC in 20-nm technology takes about 180 times longer than it does in 40-nm technology (Samsung Electronics Corp., OPC Principal Engineer, Personal communication, May 2016).

Recently, ML technique has been applied to OPC to overcome the limitation of MB-OPC. It consists of training and testing phases. In the training phase (Fig. 3.2a), an interesting segment of the layout (and its surroundings) is parameterized by a set of parameters ($\mathbf{x}$), e.g., local pattern densities that are measured around the interesting segment. The parameters are inputted to a mask bias model (F) that outputs a predicted value of mask bias ($F(\mathbf{x})$) for the input segment, which is then used to synthesize a mask image. The model F is trained so that $F(\mathbf{x})$ can approach a desired (target) value of mask bias, which is obtained by conventional MB-OPC in advance. In the subsequent testing phase (Fig. 3.2b), a layout segment to be corrected is parameterized using the same set of parameters that were used in the training phase, and the trained model F predicts the mask bias for the segment, which is then used to synthesize mask image.

The accuracy and runtime of ML-OPC are affected by the method of model training and the choice of parameters. A few existing studies adopt hierarchical Bayesian model (HBM) [6], multilayer perceptron (MLP) [7], and support vector regression (SVR) [8] for model training and local pattern densities or pixel values of rasterized layout as parameters. The parameters may be extracted and optimized by an information-theoretic approach with a dynamic programming model [9]. The

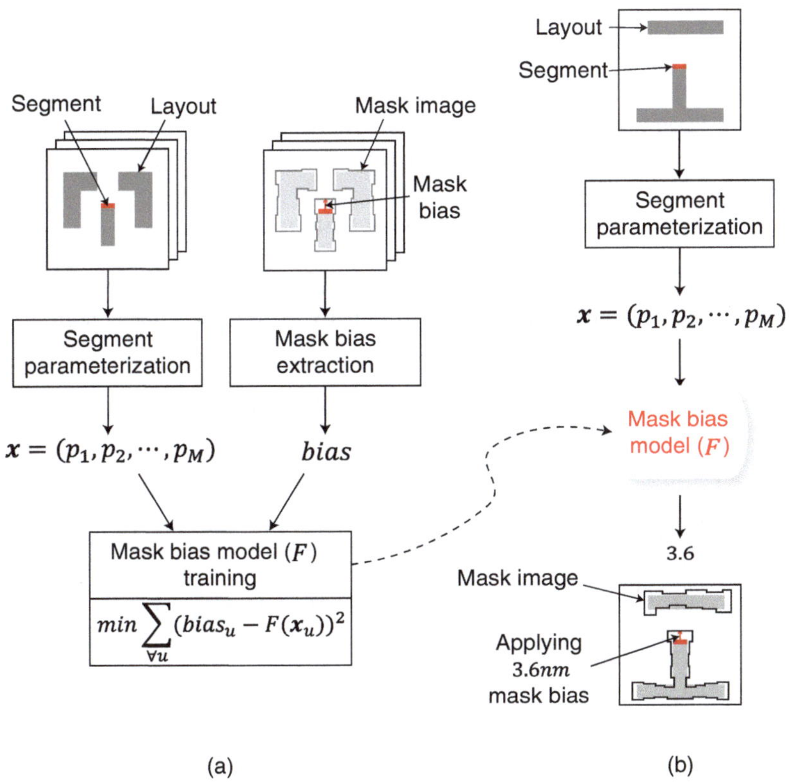

Fig. 3.2 ML-OPC: (**a**) training and (**b**) actual testing

number of parameters is however typically very large (e.g., some hundreds), which negatively affects both runtime and accuracy.

This section introduces the state-of-the-art ML-OPC [10, 11], in which polar Fourier transform (PFT) signals are used as input parameters. A PFT signal is obtained by multiplying a PFT basis function with layout image near a segment of interest. Due to orthogonality of PFT basis functions, the PFT signals have little redundancy which allows us to reduce the number of parameters. Reduced number of parameters also helps in accuracy provided that unnecessarily large number of parameters in the conventional approach often causes over-fitting that negatively affects accuracy.

3.2.1 Prior Works

One popular parameter type in ML-OPC is a set of pixel values from a rasterized layout (Fig. 3.3a) [7, 8, 12, 13]. A local layout surrounding the segment of interest is extracted and rasterized; if more than half of pixel area is overlapped with layout, 1 is assigned to the pixel and, otherwise, 0 is assigned. The number of parameters drastically increases for larger layout and smaller pixel, which in turn increase OPC runtime and often cause over-fitting due to many unnecessary parameters. Small layout with large pixels, on the other hand, yields small number of parameters, but it may lose necessary geometry information, thereby resulting in inaccurate OPC result.

Another popular parameter type is local pattern density (Fig. 3.3b) [6, 14, 15]. A few concentric circles with their auxiliary lines crossing centers of the circles are drawn around an interesting segment. At each point where the circles and lines intersect (measurement point), a pattern density within each local region (region

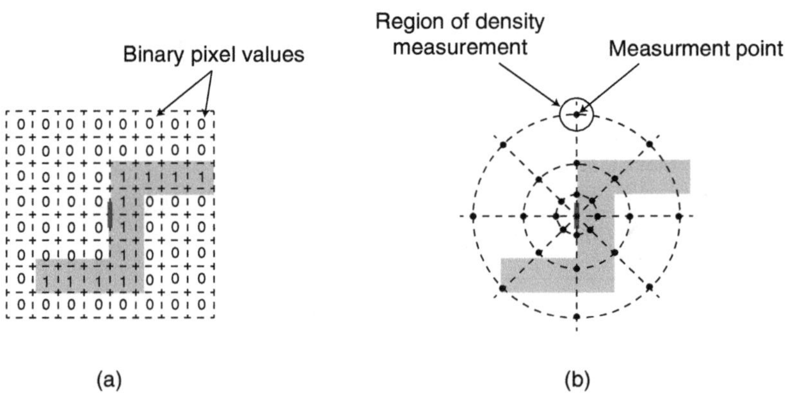

Fig. 3.3 Parameters of conventional ML-OPC: (**a**) binary pixel value and (**b**) local pattern density

of density measurement) is measured and used as a parameter. The number of parameters is determined by the numbers of the concentric circles and lines that are used. This method yields smaller number of parameters than the pixel value parameters because measurement positions are rather sparsely distributed.

In both methods, the region where parameters are extracted is usually smaller than optical influence range, a circular region of about 1 μm or longer radius. If the region is expended for the benefit of accuracy in a way as to extract parameters within the optical influence range, the number of parameters becomes unrealistic, e.g., 32 times in pixel value parameters and twice in local density parameters.

3.2.2 *ML-OPC with PFT Signals as Model Parameters*

3.2.2.1 PFT Signal

The polar Fourier transform (PFT) is widely used in modeling optical diffraction and interference [16, 17]. The real part of the PFT basis function is given by:

$$\Psi_{nm}(r, \varphi) = J_n(r)\cos(m\varphi), \tag{3.1}$$

where the n-th Bessel function (J_n) and $\cos(m\varphi)$ are, respectively, its radial and angular components,[1] n is the number of critical points in the radial direction as shown in Fig. 3.4, and m is the number of periods in the direction of angular φ. PFT basis function becomes more complicated as n and m increase as shown in Fig. 3.5. PFT basis functions are orthogonal to one another due to the orthogonality of the Bessel and cosine functions.

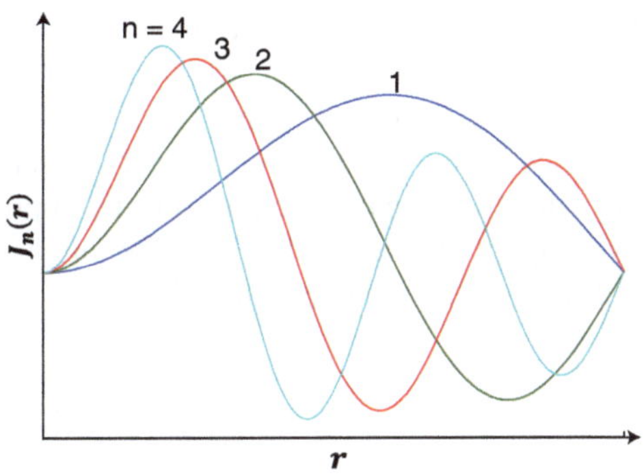

Fig. 3.4 Some Bessel functions with zero boundary condition

[1]The complete form of the Bessel function is $\Psi_{nm}(r, \varphi) = J_n(r)e^{im\varphi}$, but the real part is only considered in this chapter.

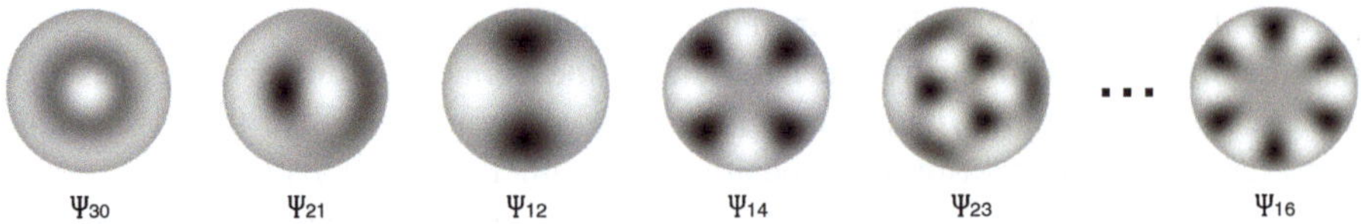

Fig. 3.5 Some basis functions of PFT

Fig. 3.6 Obtaining PFT signal

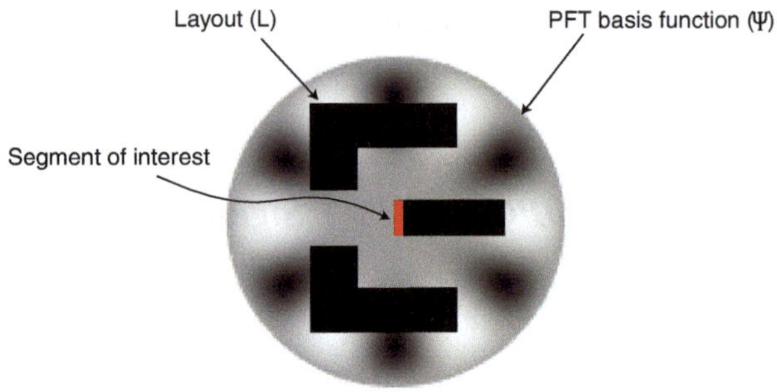

Using PFT basis functions, the PFT signal (ϕ_k) of a layout segment can be expressed as follows:

$$\phi_k = \sum_{\forall x,y} \Psi_k(x, y)L(x, y), \tag{3.2}$$

where Ψ_k is the k-th PFT basis function in Cartesian coordinates (nm is replaced by k for the sake of simplicity), which has its origin at the center of the layout segment as shown in Fig. 3.6. The segment (and its surroundings) is represented by an array of pixels, which is called the layout image (L). A pixel in this image is 1 if it is within a polygon of the layout, or 0 otherwise. Note that an ambit of PFT basis function is close to optical diameter, so that influence of surrounding patterns on a segment of interest is all taken into account; optical diameter of 45-nm technology, for instance, is around 2 μm [18].

In commercial OPC tools, PFT signals are commonly used to determine the printability of a pattern on a wafer by approximating the variation in light intensity over layout segments [19, 20]. The intensity of light (I) at the center of a layout segment can be expressed as follows:

$$I = \sum_{\forall k} c_k \phi_k^2, \tag{3.3}$$

where c_k is the contribution of the k-th PFT basis function. PFT basis functions and corresponding c_ks are predetermined once lithographic condition is given. Critical dimensions (CDs) of some sample patterns on a wafer are measured from SEM image after actual lithography process, and c_ks are calibrated to yield simulated CDs as close as possible to measured ones [21, 22]. At this moment, some candidates of PFT basis functions with both n and m increasing from 0 are taken into account. Final PFT basis functions are then picked in descending order of corresponding c_ks until target accuracies of CDs are satisfied for the sample patterns [23]; at last, around 20 PFT signals are determined.

There are two benefits if PFT signals are used as representative parameters of a layout segment. First, they are readily available from commercial OPC tools, which therefore can serve as a platform to build ML-OPC. Second, a layout segment can be modeled by fewer parameters by using PFTs instead of pattern densities or pixel values, which allows ML-OPC to be trained more quickly [10].

3.2.2.2 MLP Construction

PFT signals obtained from each segment are arranged as a vector, which is inputted to ML-OPC model. Multilayer perceptron (MLP) is employed to build new ML-OPC model as shown in Fig. 3.7. The input layer of MLP has n nodes, each of which receives a parameter from an interesting segment. Each input node propagates the signal that it receives to every node in the first hidden layer. Each connection is made by an edge, which has a weight, by which the signal is multiplied. The weighted signals received by each hidden node are summed: if the total is larger than some threshold, the hidden node outputs 1, but otherwise 0. Thresholding is realized by a sigmoid function $h(x)$, with the following form:

$$h(x) = \frac{1}{1 + e^{-(x - x_o)}}, \tag{3.4}$$

where x is the sum of the weighted signals, and x_o is the threshold. Each node in the first hidden layer then propagates its output value to every node in the next layer

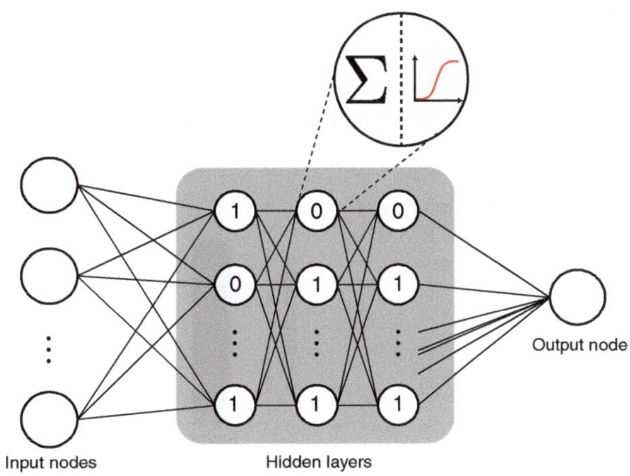

Fig. 3.7 An MLP network as an ML-OPC model

in the same manner as the nodes in the input layer, and this process is repeated until the output layer is reached. The output layer consists of one node, which receives the weighted signals and outputs the summed value as a predicted mask bias.

Edge weights and threshold in each node are appropriately adjusted in advance so that mask bias can be more accurately predicted. Sample segments are extracted from some test layouts for training, and each segment is associated with a vector of PFT signals together with a mask bias result from MB-OPC. MLP predicts mask bias once the vector is applied to ML-OPC, and the difference of mask bias from MB-OPC and ML-OPC is defined as an error. Cost function is a sum of squared errors for all segments used in training process. Gradient descent method is applied to minimize the cost function by adjusting edge weights and threshold in each node; because the function is differentiable if thresholding is done by using sigmoid:

$$\frac{dh(x)}{dx} = h(x)(1 - h(x)), \tag{3.5}$$

the minimization is fast. The numbers of layers and nodes in each layer are determined in empirical fashion. The resulting MLP is new mask bias model.

3.2.3 Experiments

New ML-OPC is implemented using Python for MLP construction and commercial OPC package (Progen and Proteus [19, 20]) for PFT signal computation. ArF immersion lithography (1.2NA) with annular illumination is assumed; corresponding optical influence range is about 1 μm. MLP consists of three hidden layers, each of which contains ten nodes. The MLP is trained for 20,000 training segments that are extracted from metal 1 test layouts in 20-nm logic technology. Other test layouts containing about two million segments are also prepared for testing the trained MLP.

Assessment of Parameters

A standard ML-OPC using local pattern densities is also implemented as a reference of comparison. It uses 80 local pattern densities as parameters, which are compared with only ten parameters in the new ML-OPC as illustrated in Fig. 3.8a. Runtime of

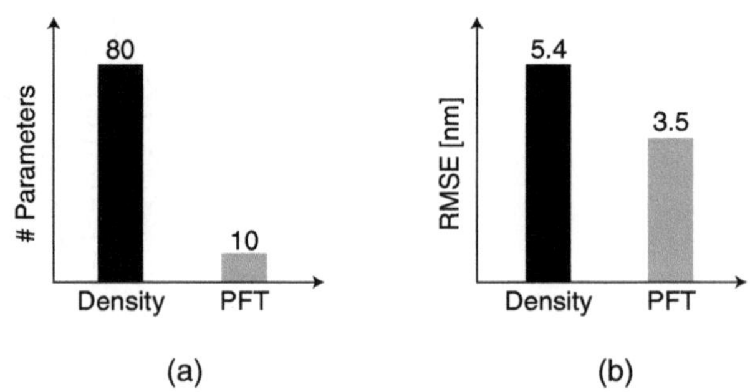

Fig. 3.8 (**a**) The number of parameters and (**b**) RMSE of conventional and new ML-OPC

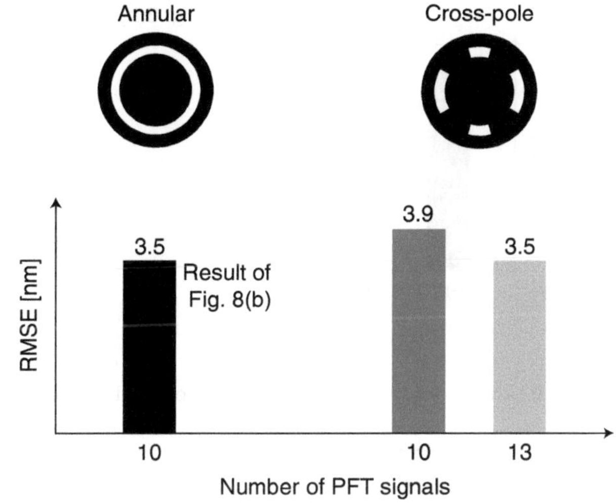

Fig. 3.9 RMSE of new ML-OPC with different illumination type

MLP construction is reduced by 26% in the new ML-OPC due to smaller number of parameters. There is even larger reduction of runtime in actual OPC, by 80%.

The accuracy of ML-OPC is assessed by using root mean square error (RMSE) of predicted mask bias, with mask bias from MB-OPC as a reference. Figure 3.8b shows that new ML-OPC achieves 3.5-nm RMSE, which is 1.9 nm smaller than that of standard one, even though smaller number of parameters are used; this is understandable consequence of less over-fitting that occurs in new ML-OPC.

Complexity of Illumination
Illumination types affect optical proximity effect. To investigate the impact of illumination complexity on new ML-OPC, the same experiment is re-performed assuming cross-pole illumination. As shown in Fig. 3.9, RMSE increases to 3.9 nm due to high complexity of cross-pole (compared to annular), which may cause more complex optical proximity effects. To reduce RMSE to 3.5 nm, the same quantity of annular is assumed, three more PFT signals are required (see bright gray bar).

Layer and Device
To investigate the impact of layer type, new ML-OPC is applied to metal 1 and contact layouts. As shown in Fig. 3.10a, metal 1 causes larger RMSE because its layout is typically more complex than contact layout. For the same reason, ML-OPC shows larger RMSE in logic devices than in memory devices, as shown in Fig. 3.10b.

Hybrid ML- and MB-OPC
The new ML-OPC is much faster than conventional MB-OPC, but a little RMS error is left. In industry, therefore, it is proposed to perform ML-OPC first to reduce EPE dramatically and the rest is eliminated by MB-OPC, namely hybrid ML- and MB-OPC. As illustrated in Fig. 3.11, for the layout in metal 1 layer, while MB-OPC

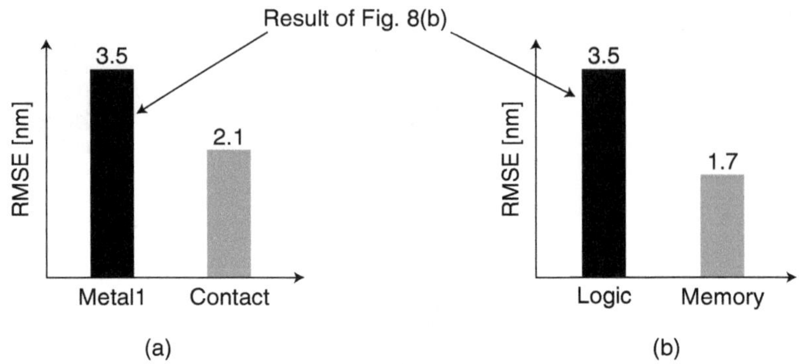

Fig. 3.10 RMSE of new ML-OPC with different design type: layer and device

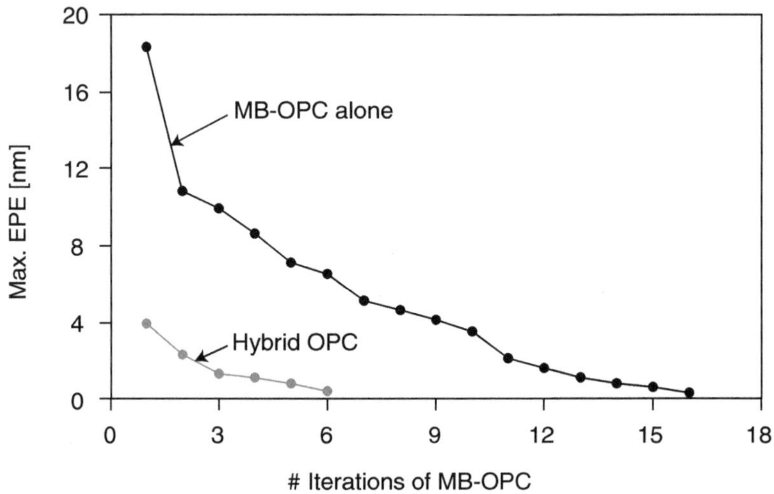

Fig. 3.11 Maximum EPE with the number of MB-OPC iterations for stand-alone MB-OPC and hybrid ML- and MB-OPC

(black line) takes 16 iterations to meet EPE criteria, MB-OPC starting from the result of ML-OPC (gray line) takes only six iterations; hybrid ML- and MB-OPC is almost three times faster than casual MB-OPC remaining negligible EPE.

3.3 Machine Learning-Guided EPC

Rule- and model-based methods of etch proximity correction (EPC) are widely used, but they are insufficiently accurate for technologies below 20 nm. Simple rules are no longer adequate for the complicated patterns in layouts, and models

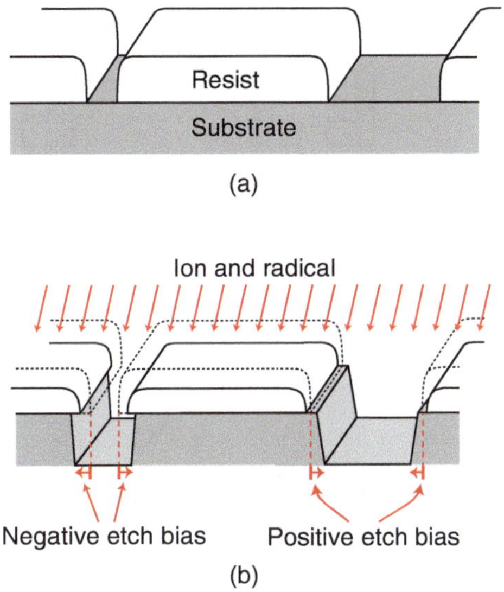

Fig. 3.12 (**a**) Resist image with small and large spaces and (**b**) corresponding etch result with negative and positive etch biases

based on a few empirically determined parameters cannot reflect etching phenomena physically. This section introduces how machine learning technique is applied to EPC (ML-EPC). Each segment of interest, together with its surroundings, is characterized by geometric and optical parameters, which are then inputted to MLP that predicts the etch bias. The ML-EPC is implemented using a commercial OPC tool, and applied it to a DRAM gate layer in 20-nm technology, achieving predictions that are 34% more accurate than model-based EPC (MB-EPC).

3.3.1 Preliminaries

3.3.1.1 Etch Bias and EPC

The etching process is a key step in semiconductor fabrication: Resist pattern is created on a wafer by optical lithography (Fig. 3.12a); and, during the subsequent etching process, etching particles (e.g., ions and radicals) collide with the surface of the resist and the substrate (Fig. 3.12b). The problems during etching process include etch proximity effects, which can produce an over-etched substrate (negative etch bias) due to lateral erosion of the photo resist, or an under-etched substrate (positive etch bias) due to excessive deposition (Fig. 3.12b). The extent of the over- or under-etching is called the etch bias.

To obtain a pattern of correct size on a wafer after both lithography and etching process, a design layout has to be modified through a process called etch proximity correction (EPC). EPC compensates for etch bias before the design layout is inputted to optical proximity correction (OPC) as shown in Fig. 3.13.

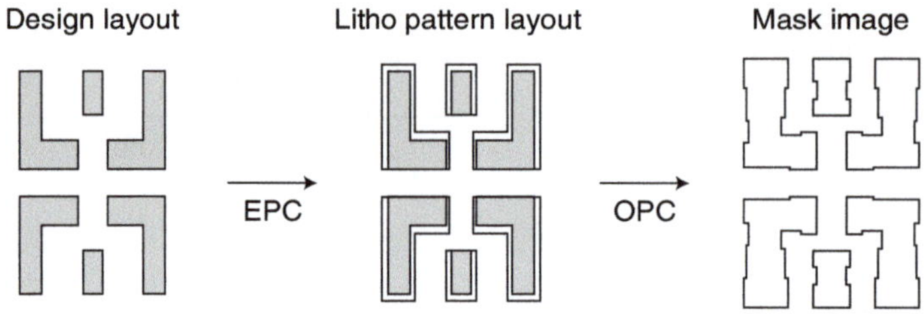

Fig. 3.13 EPC and OPC

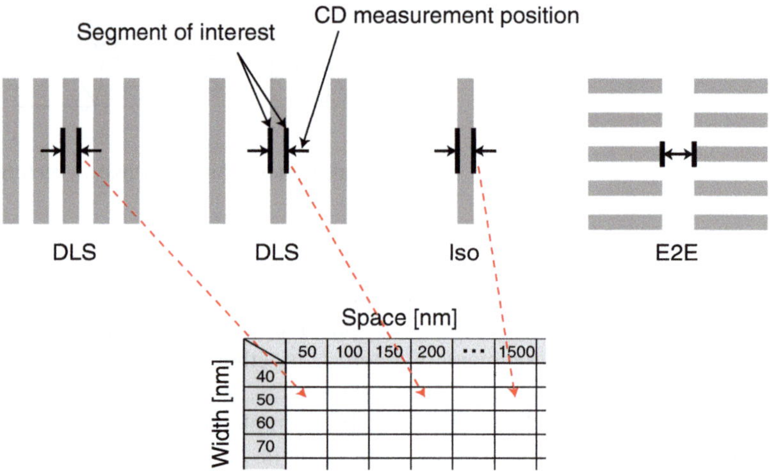

Fig. 3.14 Synthetic patterns and etch bias rule table used in RB-EPC

3.3.1.2 Rule- and Model-Based EPC

Etch bias is affected by the amount of etching particles and their angle and direction of their collision with the surface of the substrate [24]. It is not feasible to apply a rigorous simulation of the etching process over an entire chip because it may take a few days, which is not practical. So, EPC has to be based on heuristics, which may be empirical rules (rule-based EPC) or simplified models (model-based EPC).

In rule-based EPC (RB-EPC), a set of typical test patterns, such as the dense line and space (DLS), isolated line (Iso), and line-end to line-end (E2E) patterns shown in Fig. 3.14, are patterned on a wafer; the etch bias is then measured at the center of each pattern; and the results are summarized in a table against the parameters of the patterns, e.g., their width and space. RB-EPC has been used for circuits with some technology nodes (Samsung Electronics Corp., OPC Principal Engineer, Personal communication, May 2016) [2], but the rules are too simplistic to cover the range of pattern shapes that may occur in a real layout. It is estimated [2] that on-chip variation (OCV) in 10-nm technology remains 28% out of tolerance after RB-EPC has been applied.

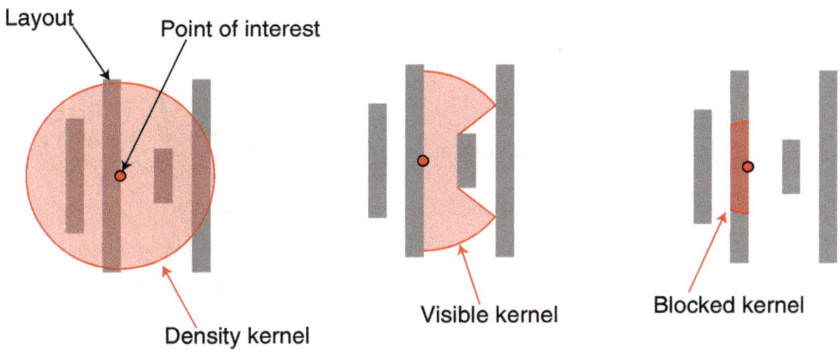

Fig. 3.15 Kernels for MB-EPC

In model-based EPC (MB-EPC), etch bias is modeled as a simple function of a few empirical parameters:

$$\text{Etch bias} = C_0 + C_1\mathbf{Den} + C_2\mathbf{Vis} + C_3\mathbf{Blo} + C_4\mathbf{Den}^2 + \cdots, \tag{3.6}$$

where **Den** is the density of layout within a region of density kernel; **Vis** is the area of the open space not hidden by the edges that are neighbors of a point of interest (i.e., space beyond the nearest edge is ignored); and **Blo** is the area of the nearest polygon that overlaps with the blocked kernel, as shown in Fig. 3.15 [3]. The coefficients C_i and the radii of the kernels are adjusted through regression. The number of terms in the model function is determined in empirical fashion. MB-EPC can deal with a greater range of patterns than RB-EPC, but it still fails to achieve a satisfactory OCV. It is estimated [3] that OCV in 20-nm DRAM device is still 15% of gate size after MB-EPC has been applied.

3.3.2 ML-EPC

Machine learning technique is applied to etch bias model for more accurate prediction of etch bias. Figure 3.16 illustrates new etch bias model. A litho pattern layout is divided into edge segments. A number of parameters are extracted from a segment of interest by examining its surroundings (Fig. 3.16a). The parameters are then submitted to MLP, which outputs the predicted etch bias of the segment (Fig. 3.16b). This MLP is trained on a number of training segments that are prepared in advance so that their etch biases can be predicted as correctly as possible. The trained MLP is effectively the new etch bias model.

There are three key factors affecting accuracy and efficiency of ML-EPC: the set of training segments used for MLP construction, the input parameters, and the complexity of the MLP. Following sections describe ML-EPC steps with how such key factors are determined.

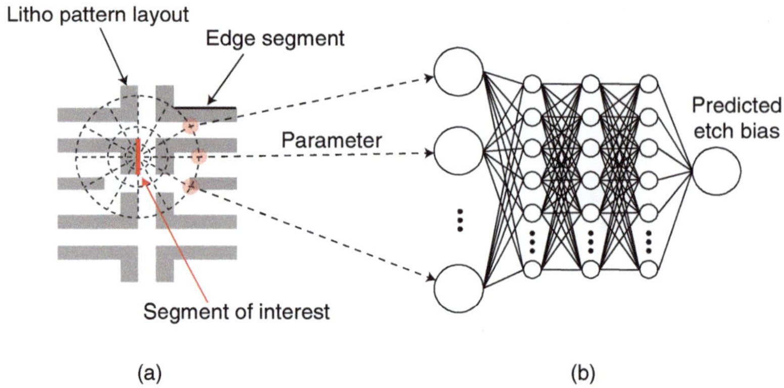

Fig. 3.16 ML-guided etch bias model: (**a**) each segment in a layout is parameterized, and (**b**) MLP is constructed to predict etch bias from input parameters

3.3.2.1 Preparation of Training Segments

Two types of pattern are used to prepare training segments: typical synthetic patterns and complicated patterns extracted from layouts of real circuits. Typical synthetic patterns (see Fig. 3.14) can be analyzed to see how etch bias changes in response to changes in geometry, e.g., a space of DLS patterns. For example, if etch bias appears to change abruptly in response to a gradual change in geometry, we can infer that there are errors in the measured data, and that we need to reexamine the measurement process.

The etch bias of a synthetic pattern is determined as follows: A lithography process is performed using a photomask that includes mask images of the training patterns. Critical dimension (CD), which is the width of a pattern (or space) between two segments placed face-to-face, is measured as shown in Fig. 3.14, after lithography (CD_{litho}). This measurement process is performed again after the subsequent etching (CD_{etch}) process. Since all the synthetic patterns are symmetric, a pair of segments can be assumed to have the same etch bias, and thus the etch bias of one segment can be expressed as $(CD_{etch} - CD_{litho})/2$.

More complicated patterns are obtained from actual circuit layouts as follows: Test layouts are prepared by shrinking layouts of earlier technologies, or synthesizing designs from a preliminary version of the new technology library. Pairs of segment for CD measurement (see candidate position of CD measurement) are then identified on parallel edges, as shown in Fig. 3.17a. Each candidate position is then examined to determine whether the two associated segments and their surroundings are symmetric; if they are symmetric (e.g., CD2 and CD3 in Fig. 3.17a), then they should acquire the same etch bias, which is expressed as $(CD_{etch} - CD_{litho})/2$; otherwise (e.g., CD1 in Fig. 3.17a), they are likely to have different etch biases, and neither bias can be readily identified: such points are rejected. Identical, or very similar, points are filtered out through a classification process. The segments

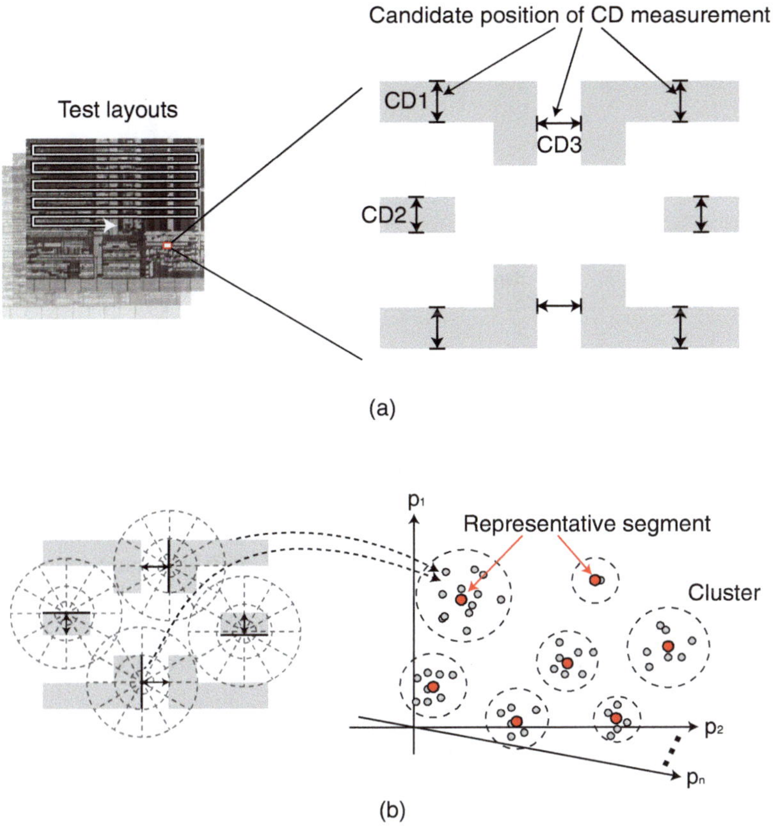

Fig. 3.17 Preparations of complicated training segments: (**a**) segments extracted from test layout and (**b**) representative segments after classifying the segments

corresponding to the remaining points are parameterized and mapped to points in an n-dimensional parameter space, as shown in Fig. 3.17b. Some close points are then grouped through modified K-mean clustering [25, 26], and representative points are selected from each cluster: the corresponding segments are used to train new etch bias model.

3.3.2.2 Extracting Parameters

Two types of parameter are used to characterize a segment and its surroundings. The first parameter is a local pattern density, which is measured at points sampled around the segment of interest. Figure 3.18 shows how to draw lines passing through the center of the segment, and place a series of touching circles along each line. The radii r of these circles, which are the regions in which pattern density will be measured, increase by a user-defined ratio as circles are placed further from the

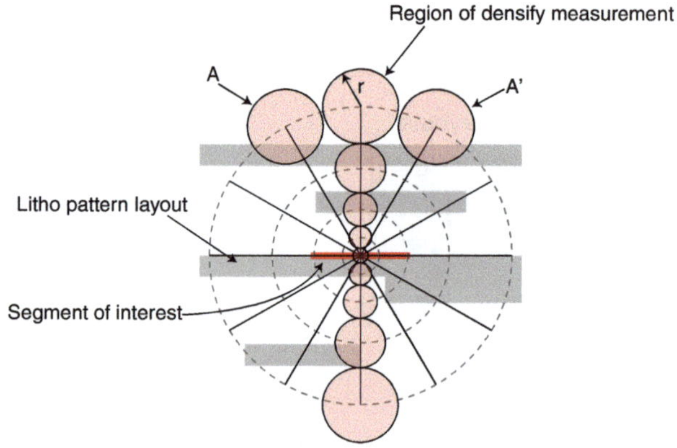

Fig. 3.18 Parameterization of a segment using local pattern densities

segment. The total number of circles determines the number of parameters. A local pattern density is associated with the quantity of etching particles, and the position at which local pattern density is measured is associated with incident angle and direction of the particles. Because etching particles with the same incident angle and the symmetric directions equally affect etch bias, density values at symmetric locations (e.g., A and A′ in Fig. 3.18) can be summed and regarded as a single parameter, which reduces the number of parameters, making it faster to train the MLP.

To take account of sidewall angle of resist, an optical kernel signal ϕ_{nm} is chosen as the second parameter type. The PFT signals, which are used as parameters of new ML-EPC presented in Sect. 3.2.2, can be used as the optical kernel signals. Optical kernel signals model constructive or destructive interference of light at the center of the segment, and thus are associated with the vertical shape of the resist, which affects the etch bias [27]. Since each function makes a separate contribution to the resulting light intensity, those functions that make the largest contribution can be chosen.

3.3.2.3 Construction of Etch Bias Model

MLP is employed to build new etch bias model as shown in Fig. 3.19. The input layer of MLP has n nodes, each of which receives a parameter from an interesting segment. Each input node propagates the signal that it receives to every node in the first hidden layer. Each connection is made by an edge, which has a weight, by which the signal is multiplied. The weighted signals received by each hidden node are summed: if the total is larger than some threshold, the hidden node outputs 1, but otherwise 0; the thresholding is realized by a sigmoid function. Each node in the first hidden layer then propagates its output value to every node in the next layer in the same manner as the nodes in the input layer, and this process is repeated until the output layer is reached.

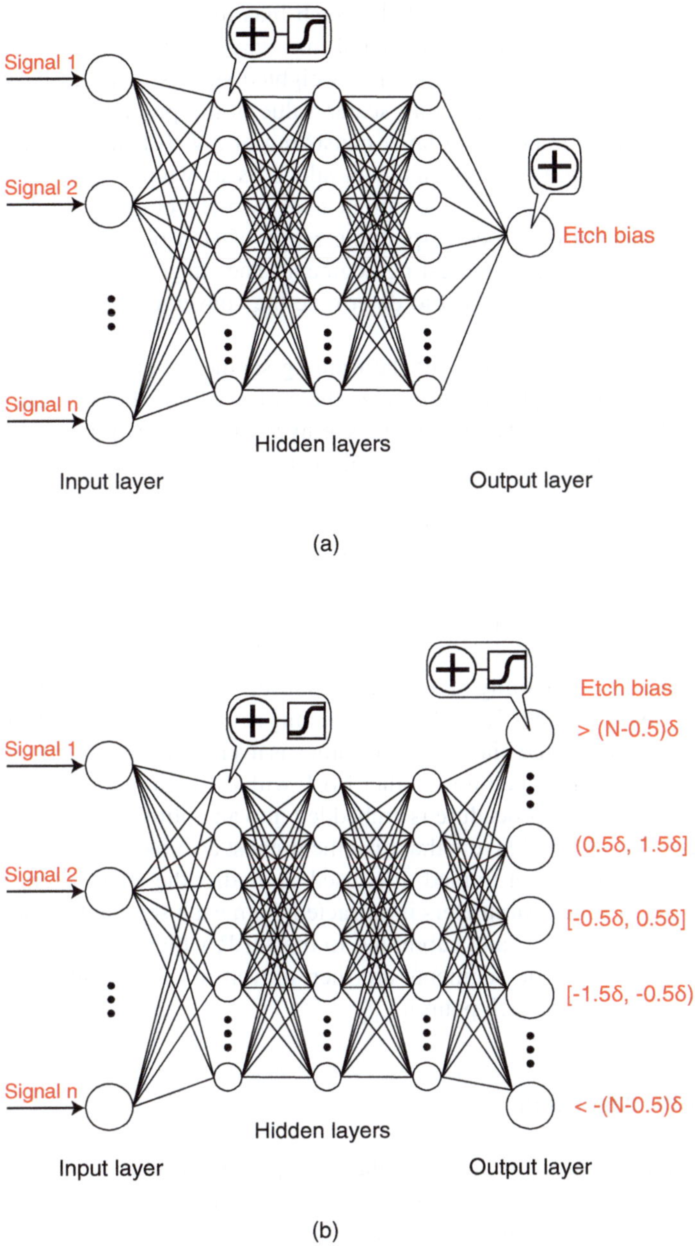

Fig. 3.19 MLP for (**a**) regression and (**b**) classification

Two types of MLP are constructed as etch bias model: regression and classification. MLP for regression is constructed in a way that the output layer consists of one node, as shown in Fig. 3.19a. The weighted signals received by this output node are summed, and the resulting scalar value is the predicted etch bias. MLP for classification, with an output layer consisting of $2N + 1$ nodes, each of which corresponds to a certain range of etch bias values, as shown in Fig. 3.19b. This MLP maps an input segment to a single output node, which returns 1 while the others return 0; the median of the corresponding range of etch bias values becomes the predicted etch bias of the interesting segment. The range covered by each node has an extent of δ, which is used to determine the convergence criteria in the EPC algorithm in Sect. 3.3.2.4.

Each MLP is evaluated over all the training segments, and optimizes the edge weights and threshold values to minimize the cost, which is the difference between the predicted and the actual etch bias. This minimization is performed by the method of steepest descent in the same manner of Sect. 3.2.2.2. The resulting MLP is new etch bias model.

The accuracy of prediction is related to the complexity of the MLP. An MLP with insufficient nodes and layers cannot be trained successfully. Conversely, an overcomplex MLP may be too specific to the training segments, making it overfitted, and produces poor results for segments with different configurations. The numbers of nodes and layers are adjusted through k-fold cross-validation. The set of training segments is randomly divided into k subsets of equal size; $k - 1$ of these sets are used to construct an MLP, while the remaining set is used to assess the proportion of correctly predicted etch biases. The accuracy is averaged over k iterations, and the whole iterations are repeated for another MLP with different numbers of nodes and layers. This adjustment procedure is embedded into an optimization that determines the numbers of nodes and layers that give the highest accuracy.

Once an MLP has been trained, it is used to predict the etch pattern for a litho pattern layout. A set of parameters is extracted from each segment in the layout, as described in Sect. 3.3.2.2, and submitted to the MLP, which then predicts the etch bias of that segment. The segments are then moved in or out following the bias, which yields a predicted etch pattern.

3.3.2.4 EPC Algorithm

EPC is to find a litho pattern layout that is expected to produce the desired etch pattern, as shown in Fig. 3.21. Since no analytical function exist to model etch proximity effect, EPC cannot be performed in analytical fashion, so trial-and-error is the only way for this.

Figure 3.20 shows new EPC algorithm, which receives a design layout $\mathscr{D}_{\text{in}}$, which becomes the initial litho pattern layout $\mathscr{L}$ (L1), and outputs a corrected litho pattern layout $\mathscr{L}_{\text{out}}$ that is expected to produce the desired etch pattern. $\mathscr{L}$ is modified iteratively (L2–L7), with the aim of making the expected etch pattern $\mathscr{D}$ as similar as possible to the desired etch pattern $\mathscr{D}_{\text{in}}$. Each segment of the litho

Fig. 3.20 The EPC algorithm

Input: A design layout $\mathscr{D}_{\text{in}}$
Output: A litho pattern layout $\mathscr{L}_{\text{out}}$

L1: $\mathscr{L} \leftarrow \mathscr{D}_{\text{in}}$
L2: **repeat** for max_iterations
L3: A set of biases $\leftarrow$ *MLP*(a set of segments from $\mathscr{L}$)
L4: $\mathscr{D} \leftarrow ETCH(\mathscr{L}$, a set of biases)
L5: A set of EPEs $\leftarrow Measure\_EPE(\mathscr{D}_{\text{in}}, \mathscr{D})$
L6: **if** $EPE_{\text{max}} \leq \varepsilon$ **then** Exit loop
L7: $\mathscr{L} \leftarrow CORRECT(\mathscr{L}, -\alpha \times$ a set of EPEs)
L8: **return** $\mathscr{L}_{\text{out}} \leftarrow \mathscr{L}$

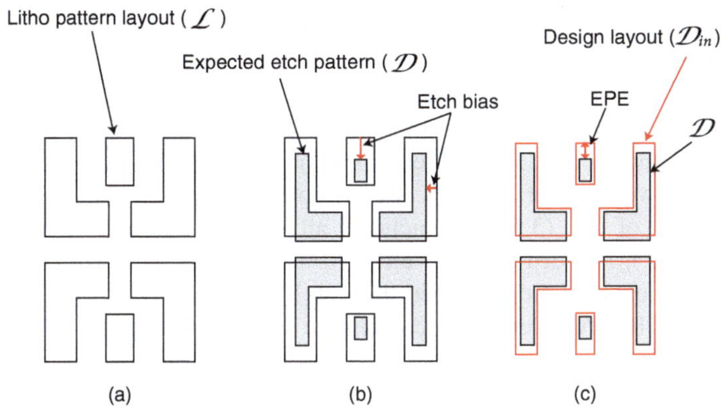

Fig. 3.21 (**a**) Current litho pattern layout ($\mathscr{L}$), (**b**) expected etch pattern ($\mathscr{D}$), and (**c**) EPE measured between the expected etch pattern ($\mathscr{D}$) and a design layout ($\mathscr{D}_{\text{in}}$)

pattern layout $\mathscr{L}$ is parameterized and inputted to the MLP, which returns (L3) a set of expected etch biases for all the segments. Each edge of $\mathscr{L}$ is moved following the expected etch bias, which yields expected etch pattern $\mathscr{D}$ (L4), as shown in Fig. 3.21b. The expected etch pattern is then compared with the design layout; the difference between them is modeled by edge placement errors (EPEs), which are signed distances between the edges of $\mathscr{D}_{\text{in}}$ and $\mathscr{D}$, as shown in Fig. 3.21c. The EPE is measured at the center of every segment of $\mathscr{D}_{\text{in}}$ (L5). A new $\mathscr{L}$ is constructed (L7) by moving each segment of the current $\mathscr{L}$ in the opposite direction to the EPE through a distance of $\alpha \times EPE$, where α is a user-defined coefficient that improves convergence. If the magnitude of the largest EPE is less than a threshold ϵ, then iteration stops; otherwise, it continues until a user-defined iteration count is reached. Note that ϵ needs to be larger than δ, which is the precision of the predicted etch bias obtained from the MLP, as discussed in Sect. 3.3.2.4.

3.3.3 Experiments

ML-EPC was implemented using Proteus [20] for parameter extraction and Python to construct the MLP. For MLP training, 1.2-k training segments are extracted from both synthetic and actual layouts; a further 1.5-k test segments are extracted from the layouts and used to assess accuracy. A segment is represented by 35 parameters: 29 local densities and 6 optical kernel signals, so MLP has to have 35 input nodes. To avoid over-fitting, tenfold cross-validation is performed for two alternative MLPs (see Fig. 3.19a and b). The MLP for regression had 4 hidden layers, each of 10 hidden nodes, and the MLP for classification had 3 hidden layers, each of 14 hidden nodes. In the MLP for classification, N and δ, respectively, are set to 75 nm and 0.2 nm, yielding 151 output nodes to cover etch biases from -15 nm to 15 nm, with 0.2-nm precision.

A rule-based (RB) and a model-based (MB) EPC are also implemented for comparisons. Each rule covers 28 increments in pattern space and 20 increments in pattern width; each of the 560 combinations that result is associated with an etch bias. Two versions of MB-EPC were implemented: MB-2nd was constructed with a version of Eq. (3.6) that includes all six second-order terms ($\mathbf{Vis}^2$ and $\mathbf{Vis} \times \mathbf{Den}$, etc.); MB-3rd includes all third-order terms. The size of each kernel and the coefficients were determined through linear regression using Progen [19]. The RB- and MB-EPCs were trained using the same 1200 segments.

3.3.3.1 Assessment of Accuracy

Accuracy is evaluated in terms of root mean square (RMS) errors for the 1200 training segments (see white bars in Fig. 3.22). RB-EPC has very large RMS errors, indicating that a few simple rules cannot represent complicated pattern shapes adequately. With the MB-EPC, the RMS errors decrease as higher-order terms are introduced. MLPs for regression and classification, respectively, produced results with average RMS errors of 2.1 nm and 1.6 nm, which are placed between MB-2nd (2.3 nm) and MB-3rd (1.4 nm).

The RMS errors for the 1500 test segments that were not used in model training are also compared (see black bars). RB now produces even larger errors, probably due to its poor coverage of complicated patterns. The errors produced by MB-2nd increase only slightly, but the errors from MB-3rd become almost three times larger. This may be attributed to the inability of empirical kernels to model etching phenomena accurately, which may cause over-fitting when too many terms are involved in model training.

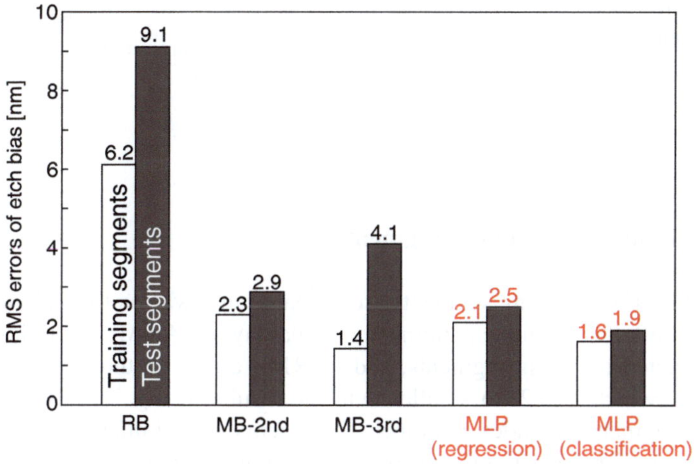

Fig. 3.22 RMS error of etch bias for training (white) and test (black) patterns

Table 3.1 Errors with MLPs for regression and classification

	RMS errors (nm)			
	Large etch bias (Fig. 3.22)		Small etch bias	
Segments	Regression	Classification	Regression	Classification
Training	2.1	1.6	0.8	1.1
Test	2.5	1.9	1.2	1.6

3.3.3.2 Comparison of Two MLP Types

In general, using MLP for classification results in a slightly smaller RMS errors than MLP for regression, as shown in Fig. 3.22. However, it is observed that the MLP for regression achieves smaller errors than the MLP for classification when the etch bias is small.

A second set of training and test segments is prepared in a way as to include 1200 training and 1500 test segments with an etch bias ranging from −10 nm to 5 nm, which is little more than half of its range in the experiments which generated the results shown in Fig. 3.22. As before, two types of MLP were constructed for these training segments. This time, the MLP for regression produced smaller errors than classification (see columns 4 and 5 in Table 3.1). This suggests that MLP for regression may be more suitable for a week etch process, which produces small etch biases; whereas classification is likely to be more suitable for a strong etch process, which produces large etch biases.

Table 3.2 Errors with different parameter sets

Parameter sets	# Parameters	RMS errors (nm)	
		Training	Test
Den	49	3.3	3.5
Den+Opt	59	1.5	1.9
Den+Opt (reduced)	35	1.6	1.9

3.3.3.3 Sampling of Training Segments

A third set of 1200 training segments are prepared in a way as to randomly extract them from both the synthetic patterns and actual layout. An MLP for classification was constructed for these segments, and its RMS errors on the training segments was 1.4 nm, which is 0.2 nm smaller than the results in Fig. 3.22 because the set of training segments includes a number of segments with similar surroundings. But the RMS errors for the test segments become 2.5 nm, which is 0.6 nm larger than the results in Fig. 3.22, because the randomly sampled segments do not include a range of patterns comprehensively.

3.3.3.4 Changing the Parameter Sets

Three different combinations of parameter types are tried with MLPs for classification. Table 3.2 shows the RMS errors for the training and test segments classified using the three parameter sets. The significant contribution of optical kernel signal to accuracy is largely independent of that of the local pattern densities. If no optical kernel signals are considered (see the second row), the RMS error is very large for both the training and test segments. Accuracy is much improved when optical kernel signals are introduced (see the third row).

The number of parameters is further reduced by summing pairs of densities in symmetric locations, and using the result as a single parameter, and the six most significant optical kernel signals are only used, as discussed in Sect. 3.3.2.2. The number of parameters is in turn reduced by 40% (last row), and the runtime of the MLP is reduced by 49%, with an RMS accuracy of 1.6 nm and 1.9 nm for the training and test segments, respectively.

3.3.3.5 Effect of Cross-validation

To investigate the effect of cross-validation, two more MLPs for classification are constructed. One is a simple MLP, consisting of a single hidden layer with ten nodes, which produces very high RMS errors for both training (6.7 nm) and test segments (7.8 nm); this is an understandable consequence of its simplicity. The second MLP is more complicated, with 5 hidden layers, each consisting of 15 nodes. It produces very small RMS errors for training segments (1.1 nm), which is expected with this

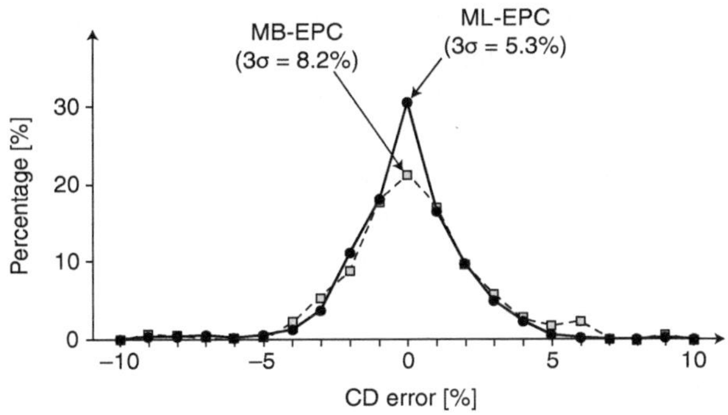

Fig. 3.23 Histogram of CD errors on a wafer after MB-EPC and ML-EPC

number of degrees of freedom; but, its error with test segments is 2.8 nm, which is 0.9 nm larger than the errors produced by MLP constructed using cross-validation. It can be attributed that the cross-validation declines to over-fitting.

3.3.3.6 Assessment of ML-EPC

ML-EPC and MB-2nd are compared in terms of the critical dimension (CD); MB-2nd is used for etch bias prediction in L3 of Fig. 3.20. MB-EPC was applied to a 20-nm DRAM gate layer, and the gate CD was measured in the core and peripheral regions. The distribution of CD errors is shown (dashed curve) in Fig. 3.23, in which the 3-sigma value is 8.2% of the target CD. The same experiment was performed using ML-EPC with the MLP for classification, and the resulting CD error distribution is shown as a continuous curve, and the 3-sigma value is reduced to 5.3%.

ML-EPC requires about eight times more runtime than MB-2nd because many local densities are extracted around every interesting segment. It is expected that runtime can significantly be improved if commercial tools are optimized to measure local densities.

3.4 Conclusions

Machine learning technique has recently been applied to mask synthesis problems for higher speed and better accuracy. MLP is used to build a mask bias model for ML-OPC. The choice of parameters is very important in its accuracy as well as runtime. New ML-OPC uses PFT signals as parameters. Since they are orthogonal while they model light interference very well, the number of parameters can significantly be reduced without loss of accuracy, which was demonstrated using 2X-nm node logic metal layer. Another application of ML is EPC, in which a model is built to predict etch bias. The key components of new etch bias model are the methods of preparing training segments and parameterizing them. Training segments are systematically prepared to keep their number small, while ensuring sufficient coverage to handle the shapes of patterns in real layouts. Local density distributions and optical kernel signals have been proposed as parameters. MLP as the etch bias model is optimized through cross-validation to avoid over-fitting. ML-EPC was demonstrated using 20-nm DRAM technology and showed more accurate predictions of etch bias compared to conventional RB- and MB-EPC.

References

1. N. Cobb, Fast optical and process proximity correction algorithm for integrated circuit manufacturing, Ph.D. Dissertation, University of California at Berkeley, 1998
2. M. Salama, A. Hamouda, Efficient etch bias compensation techniques for accurate on-wafer patterning, in *SPIE Advanced Lithography*, March 2015, pp. 1–7
3. J. Park, S. Kim, S. Shim, S. Suh, H. Oh, The effective etch process proximity correction methodology for improving on chip CD variation on 20 nm node DRAM gate, in *SPIE Advanced Lithography*, March 2011, pp. 1–10
4. S. Shang, Y. Granik, M. Niehoff, Etch proximity correction by integrated model-based retargeting and OPC flow, in *SPIE Advanced Lithography*, March 2007, pp. 225–232
5. Q. Liu, R. Cheng, L. Zhang, Study of model based etch bias retarget for OPC, in *SPIE Advanced Lithography*, March 2010, pp. 1–6
6. T. Matsunawa, B. Yu, D.Z. Pan, Optical proximity correction with hierarchical Bayes model, in *SPIE Advanced Lithography*, March 2015, pp. 1–10
7. R. Luo, Optical proximity correction using a multilayer perceptron neural network. J. Opt. **15**(7), 075708–075713 (2013)
8. K.-S. Luo, Z. Shi, X.-L. Yan, Z. Geng, SVM based layout retargeting for fast and regularized inverse lithography. J. Zhejiang Univ. Sci. C **15**(5), 390–400 (2014)
9. H. Zhang, B. Yu, E. Young, Enabling online learning in lithography hotspot detection with information-theoretic feature optimization, in *IEEE/ACM International Conference On Computer-Aided Design (ICCAD)*, November 2016, pp. 1–8
10. S. Choi, S. Shim, Y. Shin, Machine learning (ML)-guided OPC using basis functions of polar Fourier transform, in *SPIE Advanced Lithography*, March 2016, pp. 1–8
11. M. Jeong, J.W. Hahn, Prediction of biases for optical proximity correction through partial coherent identification. J. Micro/Nanolithogr. MEMS MOEMS **15**(1), 1–12 (2016)
12. A. Gu, A. Zakhor, Optical proximity correction with linear regression. IEEE Trans. Semicond. Manuf. **21**(2), 263–271 (2008)
13. T. Shah, O. Dabeer, Fast inverse lithography using machine learning, in *Indian Workshop on Machine Learning*, July 2013, pp. 21–22
14. S. Shim, Y. Shin, Etch proximity correction through machine-learning-driven etch bias model, in *SPIE Advanced Lithography*, March 2016, pp. 1–10
15. S. Shim, Y. Shin, Machine learning-guided etch proximity correction. IEEE Trans. Semicond. Manuf. **30**(1), 1–7 (2017)
16. Q. Wang, O. Ronneberger, H. Burkhardt, Fourier analysis in polar and spherical coordinates, Albert-Ludwigs-Universität Freiburg, Institut für Informatik, Technical Report, 2008
17. M. Born, E. Wolf, *Principles to Optics*, 4th edn. (Pergamon Press, London, 1970)
18. N. Cobb, D. Dudau, Dense OPC and verification for 45 nm, in *Proceedings of SPIE Advanced Lithography*, March 2006, pp. 1–7
19. Synopsys, Progen, User Guide (2016)
20. Synopsys, Proteus, User Guide (2016)

21. M. Feldman, *Nanolithography: The Art of Fabricating Nanoelectronic and Nanophotonic Devices and Systems* (Woodhead, Oxford, 2014)
22. J. Li, X. Li, R. Lugg, L.S. Melvin III, Kernel count reduction in model based optical proximity correction process models. Jpn. J. Appl. Phys. **48**(6S), 1–5 (06FA05) (2009)
23. P. Gong, S. Liu, W. Lv, X. Zhou, Fast aerial image simulations for partially coherent systems by transmission cross coefficient decomposition with analytical kernels. J. Vac. Sci. Technol. B: Nanotechnol. Microelectron. Mater. Process. Meas. Phenom. **30**(6), 1–7 (06FG03) (2012)
24. Y. Granik, Correction for etch proximity: new models and applications, in *SPIE Advanced Lithography*, March 2001, pp. 98–112
25. J.B. MacQueen, Some methods for classification and analysis of multivariate observations, in *Fifth Berkeley Symposium on Mathematical Statistics and Probability*, vol. 1 (1967), pp. 281–297
26. R. Ng, J. Han, CLARANS: a method for clustering objects for spatial data mining. IEEE Trans. Knowl. Data Eng. **14**(5), 1003–1016 (2002)
27. C. Wu, W. Yang, L. Luan, H. Song, Photoresist 3D profile related etch process simulation and its application to full chip etch compact modeling, in *SPIE Advanced Lithography*, March 2015, pp. 1–8

Chapter 4
Machine Learning in Physical Verification, Mask Synthesis, and Physical Design

Yibo Lin and David Z. Pan

4.1 Introduction

The extreme scaling of VLSI circuits has reached to the manufacturing limitation. Various challenges have been raised from the printability issues due to lithography resolution, process variation, etc. Thus a design needs to be highly optimized and extensively verified for manufacturability. In addition, design quality and manufacturability in late stages of the design flow become increasingly sensitive to the changes in early stages, increasing turn-around time and slowing down the design closure. Thus, early stage prediction of valid designs is becoming more and more critical.

In this chapter, we focus on the backend design challenges, including physical design, mask synthesis, and physical verification. Physical design implements a gate-level netlist to a layout with physical locations and geometries. It typically includes placement, clock tree synthesis, routing, etc. Mask synthesis is a follow-up step to improve the printability utilizing resolution enhancement techniques (RETs), such as optical proximity correction (OPC) and sub-resolution assist features (SRAFs). Besides optimization stages like physical design and mask synthesis, physical verification, e.g., hotspot detection [12, 20], photoresist modeling [25, 46], is another crucial step to validate the manufacturability of a design.

The aforementioned stages have been developing for decades and are encountering various challenges in performance and efficiency in advanced technology nodes. The advances in machine learning stimulate new opportunities to boost the design closure of the backend flow. General machine learning can be categorized as supervised learning and unsupervised learning. Supervised learning is essentially a

Y. Lin (✉) · D. Z. Pan
University of Texas at Austin, Austin, TX, USA
e-mail: yibolin@utexas.edu; dpan@ece.utexas.edu

I. M. Elfadel et al. (eds.), *Machine Learning in VLSI Computer-Aided Design*,
https://doi.org/10.1007/978-3-030-04666-8_4

modeling technique in which underlying correlation between features and labels is extracted to build a model. The model is then able to predict labels when given new features. Typical algorithms for supervised learning include boosting, logistic regression, support vector machines, deep neural networks, and so on [14]. Unsupervised learning does not require labels from data samples, such as a clustering problem. It aims at learning the hidden structures in the data.

This chapter will survey recent practices and researches on the machine learning applications to the backend design automation, from late stages to early, i.e., physical verification, mask synthesis, and physical design. We will introduce the motivations of each problem and briefly explain the learning techniques to solve the problem. Section 4.2 focuses on the challenges in physical verification; Sect. 4.3 illustrates the problems in mask synthesis; Sect. 4.4 talks about the applications in physical design; Sect. 4.5 concludes the chapter.

4.2 Machine Learning in Physical Verification

Physical verification refers to various validation steps in the manufacturing process, such as lithography hotspot detection. Machine learning mainly helps to decrease turn-around time and manufacturing costs.

Lithography hotspots denote specific patterns that tend to fail in printing even after RETs, which are becoming more and more common due to the complexity of lithography system and process variation. The early detection of lithography hotspot remains to be a critical challenge to enhance manufacturability and reduce costs. Generally, expensive lithography simulation is required for accurate hotspot detection, often leading to long turn-around time. Therefore, developing efficient hotspot detection is desired and machine learning is a suitable technique due to its efficiency in prediction.

Hotspot detection is a classification task in which layout clips labeled as either hotspots or non-hotspots are provided. The objective is to construct a model that can accurately classify clips. Figure 4.1 shows two layouts with hotspots. In practice, the performance of hotspot detection is mainly evaluated with two metrics, i.e., detection accuracy and false alarm. *Detection accuracy* is defined as the ratio between the number of correctly detected hotspots and the number of real hotspots. *False alarm* is defined as the number of non-hotspots that are recognized as hotspots. The modeling objective is to first improve detection accuracy and then minimize false alarms. In other words, it is tolerable to mis-classify a few non-hotspots into hotspots, while the opposite is not encouraged.

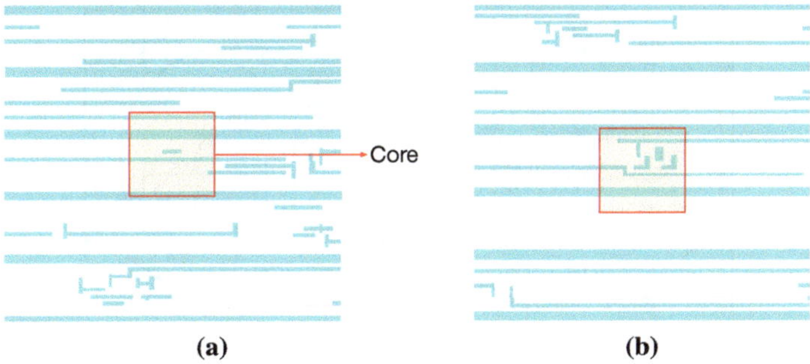

Fig. 4.1 Example of hotspot patterns marked in red [12]. (**a**) A relatively simple pattern with hotspot; (**b**) a relatively complex pattern with hotspot

4.2.1 Layout Feature Extraction and Encoding

The performance of hotspot detection is highly dependent to layout feature extraction and model selection. Layout feature represents the layout attributes to determine hotspots and non-hotspots, which is fundamental to the detection. Typical feature representations include density based feature [28, 47], fragmentation based feature, and concentric circle area sampling (CCAS) [29, 59, 60].

Figure 4.2a illustrates the fragmentation based feature extraction. Each fragment F within a circle with an effective radius r is considered. Geometric characteristics of fragments covered by the circle are extracted as the representation of F, e.g., pattern shapes, distances between layout and corner information. Figure 4.2b explains the concentric circles with area sampling to capture the layout information that matches the diffraction of lights. In this feature representation, a layout is sampled with different number of positions along concentric circles. Figure 4.2c shows the density-based feature extraction. The feature is a vector of pattern densities computed from the layout density within each grid. As features are generally extracted to a feature vector, the spatial information between elements of the vector is lost [24, 54]. For example, geometrically close sampling points may not correspond to elements close to each other in indices. Hence, a feature tensor representation is proposed to keep such spatial information, as shown in Fig. 4.2d. The original clip is converted to a hyper-image after feature tensor extraction. That is, the clip is divided into 12×12 blocks with each block holding a part of the clip, shown as the "Division" process in the figure. Then each block is converted to an image with 100×100 pixels. Discrete cosine transformation (DCT) is then applied to each block and a transformed 100×100 image with each pixel representing the coefficient of a specific frequency is obtained, shown as the "DCT" process. Due to the nature of DCT, high frequency coefficients are nearly zero. Thus, only the coefficients of lowest 32 frequencies are kept for each block. Combining on the coefficients of all blocks forms the feature tensor representation with the dimensionality of $12 \times 12 \times 32$, shown as the "Encoding" process.

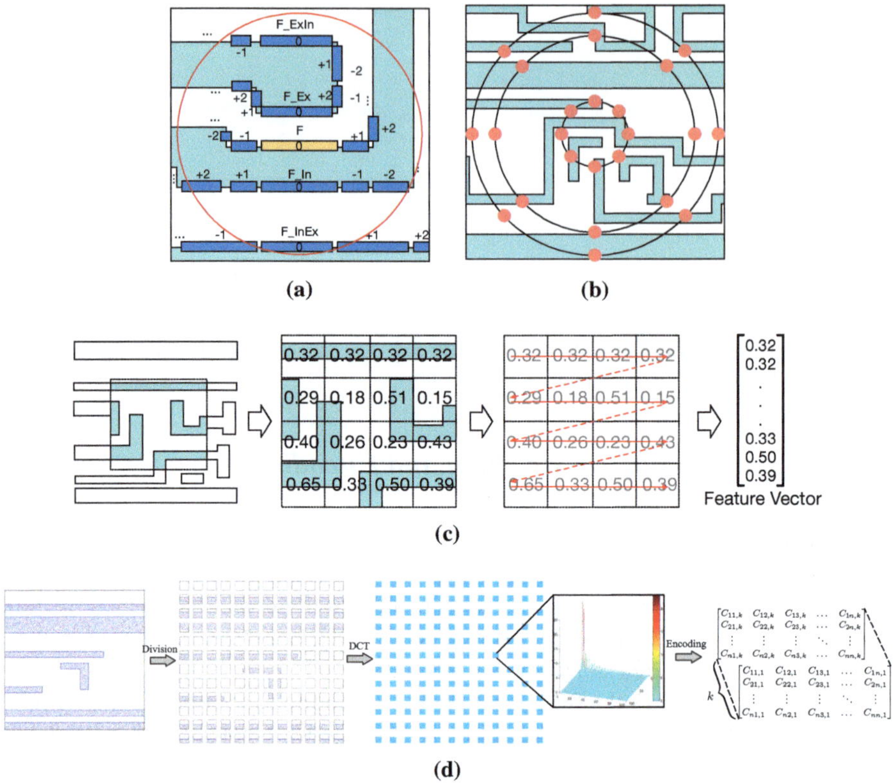

Fig. 4.2 (**a**) Fragmentation based hotspot signature extraction [8]. (**b**) CCAS feature extraction [23]. (**c**) Density-based pattern representation [60]. (**d**) Feature tensor generation [54]

4.2.2 Machine Learning Models for Hotspot Detection

Conventionally pattern matching is widely used in hotspot detection, as shown in Fig. 4.3a, while it cannot handle the situation when a pattern is not found in the pre-built library [20, 56]. Then fuzzy pattern matching is developed to dynamically tune the regions around the known hotspots, as shown in Fig. 4.3b [23]. Machine learning based approaches recently demonstrate even better generality, as shown in Fig. 4.3c.

Besides feature extraction, various machine learning models have been used as hotspot detection kernels to achieve high accuracy and low false alarms, including support vector machine (SVM) [9, 57], artificial neural network (ANN) [9], and boosting methods [28, 59]. Zhang et al. [59] also propose an online learning scheme to verify newly detected hotspots and incrementally update the model.

To mitigate the impacts from feature representation to the detection accuracy, deep neural network (DNN) has been proposed for hotspot detection [30, 40]. DNN

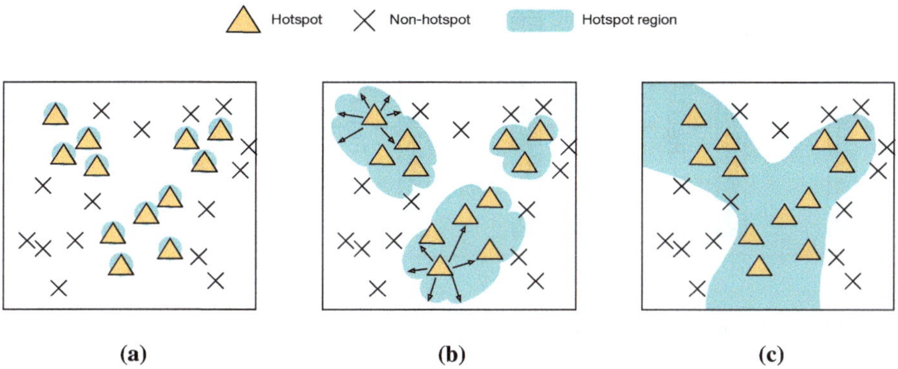

Fig. 4.3 A 2D-space example of hotspot region decision. (**a**) Pattern matching. (**b**) Fuzzy pattern matching. (**c**) Machine learning [23]

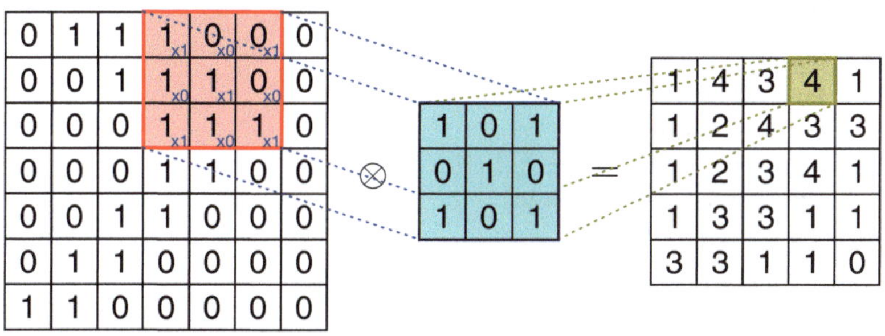

Fig. 4.4 Example of convolution with one kernel filter

can avoid the manual efforts for the selection of feature extraction approaches, because it takes high-dimensional layout as input and perform automatic feature extraction during training. Promising empirical results have been observed with DNN in several papers [30, 40, 52, 53]. A typical configuration of DNN structure is shown in Fig. 4.5. It consists of four convolution layers and two fully connected layers. Each convolution layer uses a set of kernels to perform convolution on an input tensor. Figure 4.4 shows the operation of convolution for a 7×7 tensor with a 3×3 kernel filter. The kernel filter sweeps through the input tensor and fills each element in the 5×5 output tensor. Each fully connected layer performs linear transformation to an input vector as follows,

$$x \mapsto \mathbf{W}x, \tag{4.1}$$

where $x \in \mathscr{R}^{m\times 1}$ is the input vector and $\mathbf{W} \in \mathscr{R}^{n\times m}$ represents the neuron weights. The activation function ReLU layer is defined as follows,

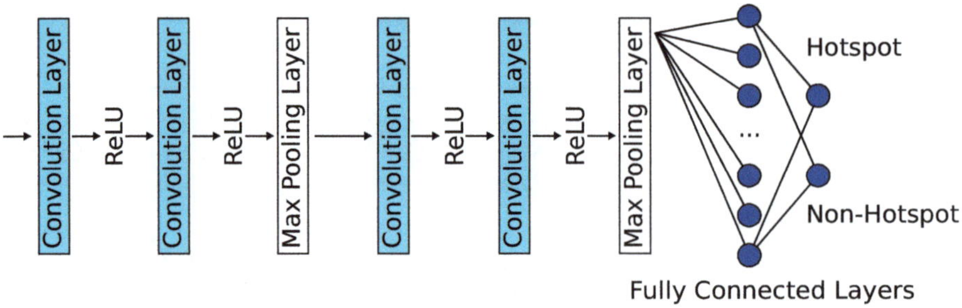

Fig. 4.5 An example of a neural network for hotspot detection [54]

Algorithm 1 Mini-batch gradient descent (MGD)

1: **function** MGD($(\mathbf{W}, \lambda, \alpha, k, \mathbf{y}_h^*, \mathbf{y}_n^*)$)
2:     Initialize parameters $j \leftarrow 0, \mathbf{W} > 0$;
3:     **while not** stop condition **do**
4:         $j \leftarrow j + 1$;
5:         Sample m training instances $\{\mathbf{F}_1, \mathbf{F}_2, \ldots, \mathbf{F}_m\}$;
6:         **for** $i = 1, 2, \ldots, m$ **do**
7:             $\mathscr{G}_i \leftarrow backprop(\mathbf{F}_i)$;
8:         Calculate gradient $\bar{\mathscr{G}} \leftarrow \frac{1}{m}\sum_{i=1}^{m} \mathscr{G}_i$;
9:         Update weight $\mathbf{W} \leftarrow \mathbf{W} - \lambda \bar{\mathscr{G}_i}$;
10:        **if** j mod $k = 0$ **then**
11:            $\lambda \leftarrow \alpha\lambda, j \leftarrow 0$;
12:    **return** Trained model f;

$$x \mapsto \begin{cases} x, & \text{if } x > 0, \\ 0, & \text{otherwise,} \end{cases} \tag{4.2}$$

where $x \in \mathscr{R}^{m \times 1}$ is the input vector.

The DNN in Fig. 4.5 is trained with mini-batch gradient descent (MGD) algorithm, as shown in Algorithm 1 [54]. Variable $\mathbf{W}$ is the neuron weights, λ is the learning rate, $\alpha \in (0, 1)$ denotes the decay factor of the learning rate, k denotes the decay step, $\mathbf{y}_h^*$ is the hotspot ground truth, and $\mathbf{y}_n^*$ is the non-hotspot ground truth. In each iteration, m training instances $\{\mathbf{F}_1, \mathbf{F}_2, \ldots, \mathbf{F}_m\}$ are randomly sampled. The gradients are computed from line 6–8. The neuron weights $\mathbf{W}$ is updated with the gradients and the learning rate in line 14. In every k iterations, the learning rate λ is decayed to $\alpha\lambda$, as shown in line 10–11. MGD function returns the best model for the training set.

Despite the convenience in automatic feature extraction, the best configuration of DNN still requires manual trial and error process, such as searching for the number and types of layers. Later, Matsunawa et al. [30] and Yang et al. [54] further propose two different DNN structures that can improve the accuracy and reduce false alarms.

Table 4.1 Comparison between the state-of-the-art hotspot detectors [52, 54]

	Train		Test		AdaBoost [28]		Online [59]		Deep [54]		Deep [52]	
Bench	HS#	NHS#	HS#	NHS#	FA#	Accu (%)	FA#	Accu (%)	FA#	Accu (%)	FA#	Accu (%)
ICCAD	1204	17,096	2524	13,503	2919	84.2	4497	97.7	3413	98.2	1776	97.36
Industry1	34,281	15,635	17,157	7801	557	93.2	1136	89.9	680	98.9	307	98.41
Industry2	15,197	48,758	7520	24,457	1320	44.8	7402	88.4	2165	93.6	793	90.56
Industry3	24,776	49,315	12,228	24,817	3144	44.0	8609	82.3	4196	91.3	1723	83.63
Avg.	-	-	-	-	2397	66.6	5411	89.6	2613	95.5	1150	92.49
Ratio	-	-	-	-	0.92	0.70	2.07	0.94	1.0	1.0	0.44	0.97

Table 4.1 shows the comparison between various state-of-the-art hotspot detectors on both ICCAD 2012 contest benchmarks and industrial designs [24, 52, 54]. The number of clips with hotspots is represented as "HS#" and the number of clips without hotspots is represented as "NHS#". Column "Accu" denotes the accuracy and column "FA" denotes the false alarm. While different hotspot detectors may have different objectives in their problem formulations, the table reports the accuracy and false alarm for reference. Generally, deep learning achieves high accuracy with relatively low false alarm [52, 54]. Detectors like the online boosting algorithm [59] mainly try to reduce the overall detection and simulation time (ODST) using online learning with reasonable accuracy.

4.3 Machine Learning in Mask Synthesis

As the technology nodes scale to the limit of light wavelength, various resolution enhancement techniques (RETs), such as optical proximity correction (OPC), source mask co-optimization, and sub-resolution assist features (SRAFs), become a necessity. Machine learning can be applied to various RETs to improve the turn-around time of mask synthesis.

4.3.1 Mask Synthesis Flow

Figure 4.6a gives a standard mask synthesis flow in which target patterns (layout) are taken as input and mask patterns are generated after iterative optimization procedures including SRAF generation, OPC, mask rule check (MRC), and lithography compliance check (LCC) [24]. In SRAF generation, sub-resolution assist features are inserted to benefit the printing of target patterns. In OPC, the edge segments of target patterns are optimized for robust lithography printing. To ensure mask manufacturing friendliness, mask manufacturing rules should be checked after these

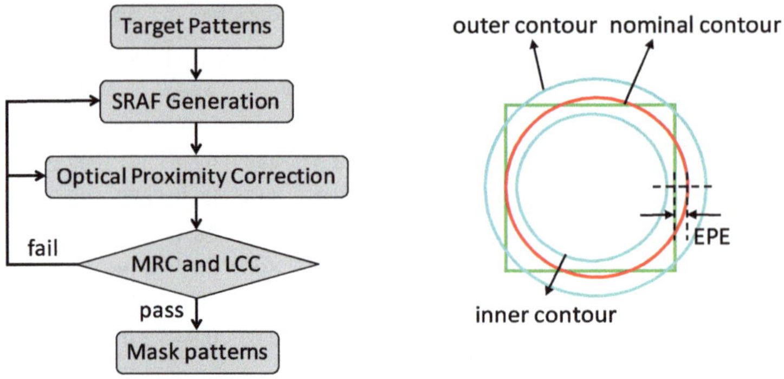

(a) **(b)**

Fig. 4.6 Mask synthesis: (**a**) a standard mask synthesis flow, (**b**) printing contours under different {focus, dose} conditions [50]

optimization procedures in MRC. Then, LCC performs the lithography simulation under a set of process windows to check printability. Here process windows denote different {focus, dose} conditions to generate printing contours, such as nominal, inner, and outer contour, as shown in Fig. 4.6b. To quantify the process windows, two metrics are introduced: edge placement error (EPE) evaluates the distance between the target pattern contour and the nominal contour; process variation (PV) band evaluates the area between the inner and outer contour. A typical objective of RETs is to minimize EPE and PV band.

4.3.2 Machine Learning for Sub-resolution Assist Features

SRAFs are small rectangles within the sub-resolution domain to assist the printing of target patterns. In other words, they will not be actually printed, even though they are on masks. The effectiveness of SRAFs for an isolated contact is illustrated in Fig. 4.7. It can be seen that Fig. 4.7c (with SRAFs) achieves much smaller PV band than Fig. 4.7b (without SRAFs). This is because SRAFs deliver light to the positions of target patterns in a proper phase, improving the robustness of printing. In advanced technology nodes, developing fast yet high-quality SRAF generations is increasingly critical to the yield [49, 50].

There are two types of conventional SRAF generation approaches, model-based and rule-based. Model-based SRAF generation ensures high-quality and robustness but is computationally expensive [21, 35, 38, 39, 42, 55]. Thus, it is not scalable to large layout designs. On the other hand, rule-based SRAF generation enables superfast turn-around time by complicated look-up-tables [19, 22, 37], while its performance highly depends on the quality of look-up-tables which require adjustment with significant engineering efforts [50].

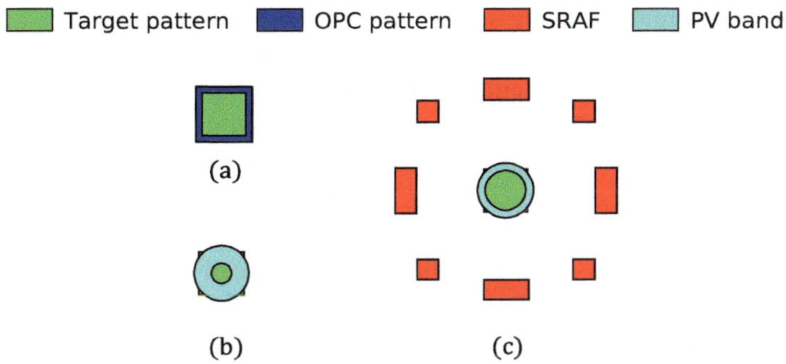

Fig. 4.7 (**a**) An isolated contact, (**b**) printing with OPC only, (**c**) printing with SRAF generation and OPC [50]

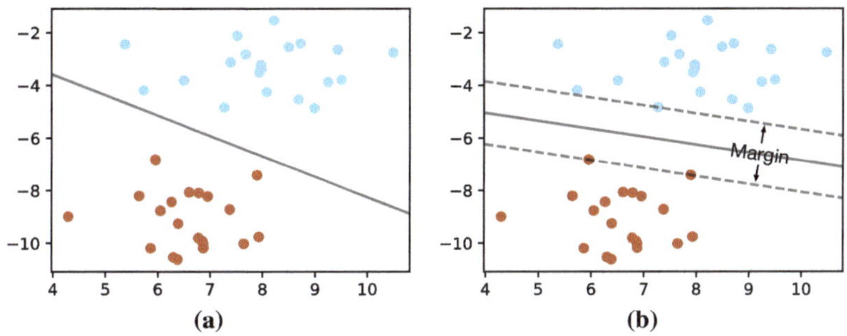

Fig. 4.8 Example of separation hyperplane in (**a**) logistic regression and (**b**) SVM

Supervised learning is promising to efficiently approximate model-based SRAF generation to improve turn-around time and meanwhile maintain high quality [50]. The training data comes from model-based SRAF generation. The model is trained to predict whether a pixel should be covered by SRAFs. The actual SRAFs are generated with the guidance of the model, subjecting to SRAF rules. Learning-based SRAF generation is formulated into a classification problem in which feature vectors are extracted with CCAS and the kernel models adopt both logistic regression and SVM.

Intuitively, logistic regression incorporates the logistic function to map the output of a linear regression into a class label where the fitted model predicts the class probability and the objective is to maximize the probability of correct classification. Figure 4.8a gives an example of separation hyperplane for logistic regression. The mathematical formulation of logistic regression is as follows [10]:

$$\min_{\boldsymbol{w}} \ \frac{1}{2}\boldsymbol{w}^T\boldsymbol{w} + C\sum_i \log(1 + e^{-y_i \boldsymbol{w}^T \boldsymbol{x}_i}), \tag{4.3}$$

where $\boldsymbol{w}$ is the weight parameters determined during training, $\boldsymbol{x_i}$ and y_i are features and label (-1 or 1 for two-class classification) for ith data sample, respectively. The first term $\frac{1}{2}\boldsymbol{w}^T\boldsymbol{w}$ is the $L2$ regularization to avoid overfitting utilizing maximum likelihood method [16]. The second term is the overall error cost. Parameter C sets the importance of the regularization term. Thus the objective for training is to minimize the overall error cost with $L2$ regularization.

On the other hand, SVM defines a hyperplane to maximize the margin between the decision boundaries. Figure 4.8b shows an example of the separation hyperplane in SVM and the margin to maximize in the objective. The mathematical formulation of support vector machine with linear kernel is as follows [3]:

$$\min_{\boldsymbol{w},b,\xi} \ \frac{1}{2}\boldsymbol{w}^T\boldsymbol{w} + C\sum_i \xi_i, \tag{4.4a}$$

$$\text{s.t.} \ \ y_i(\boldsymbol{w}^T\boldsymbol{x_i} + b) \geq 1 - \xi_i, \tag{4.4b}$$

$$\xi_i \geq 0, \ \ \forall i, \tag{4.4c}$$

where $\boldsymbol{w}, b, \xi$ are variables to be determined during training. Variable b is the bias for the hyperplane and ξ denotes the error for the ith data sample. The objective function contains two terms: one term for error minimization; the other for $L2$ regularization like that in logistic regression.

The comparison of various SRAF generation approaches in EPE, PV band, and runtime is shown in Fig. 4.9. The model-based SRAF generation uses Mentor Calibre with industrial-strength setup. As shown in Fig. 4.9a, SVM based classification leads to better approximation to the model-based approach than does logistic regression in terms of PV band. The differences in EPE are marginal, as shown in Fig. 4.9b. The major benefit of the learning-based approaches comes from the runtime, as shown in Fig. 4.9c. Over 3$\times$ speedup for a layout clip with 10 μm $\times$ 10 μm size can be achieved by both logistic regression and SVM due to efficient prediction.

4.3.3 Machine Learning for Optical Proximity Correction

OPC is another important RET to improve the performance of advanced lithography. Figure 4.10 demonstrates the effectiveness of OPC, where the edges of target patterns are fragmented and each segment is shifted in a way that the target patterns can be robustly printed, i.e., the EPE values are minimized. Conventional model-based OPC approaches are notorious for their runtime overhead [15, 33]. To overcome the runtime issue, regression models are proposed to enable fast full-chip OPC with an acceptable performance loss, such as linear regression [15, 18] and nonlinear regression [26, 27].

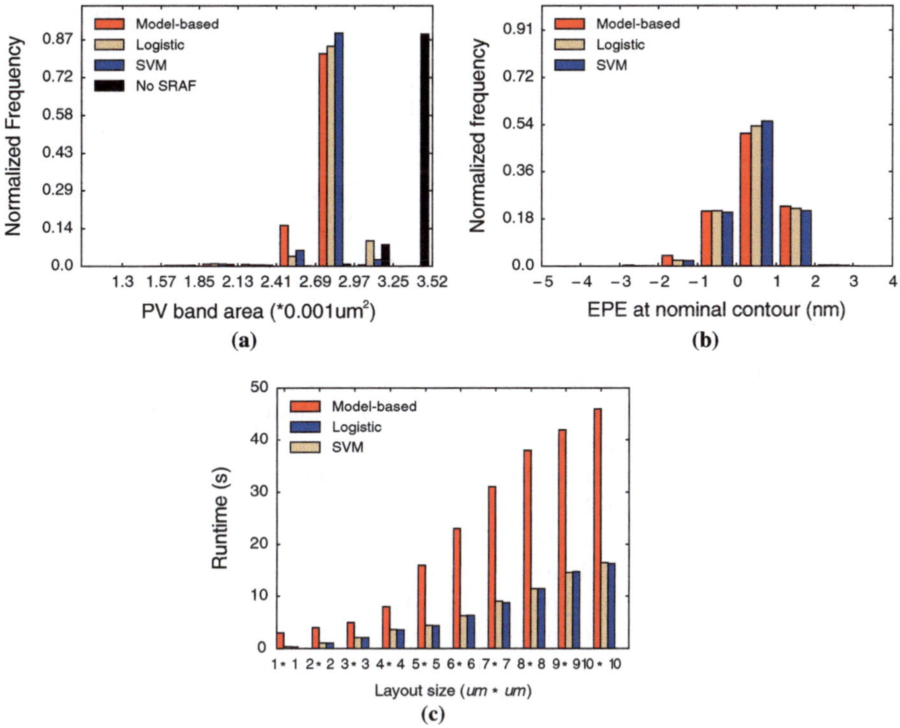

Fig. 4.9 Comparison among different schemes in terms of (**a**) PV band distribution, (**b**) EPE distribution at nominal conditions, and (**c**) runtime [50]

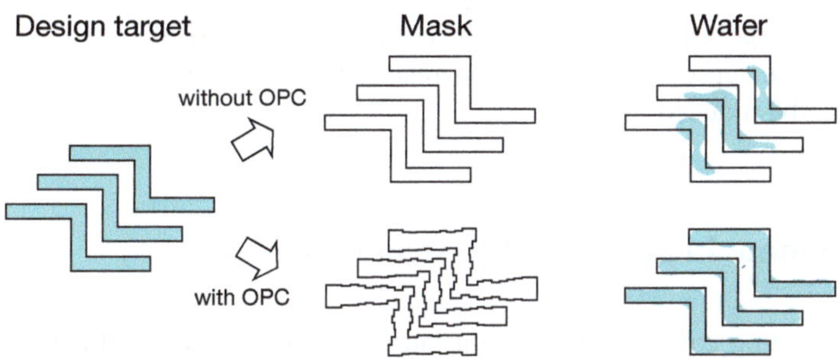

Fig. 4.10 Wafer patterns w./w.o. OPC [31]

Figure 4.11 shows the flow of regression-based OPC, which consists of training and testing phases. Besides standard steps for both model-based OPC and machine learning based approaches like edge fragmentation, training phase requires both model-based OPC and feature extraction for model calibration, while testing phase only needs feature extraction for model validation. Current regression-based techniques suffer from overfitting issues, degrading the accuracy of OPC results in

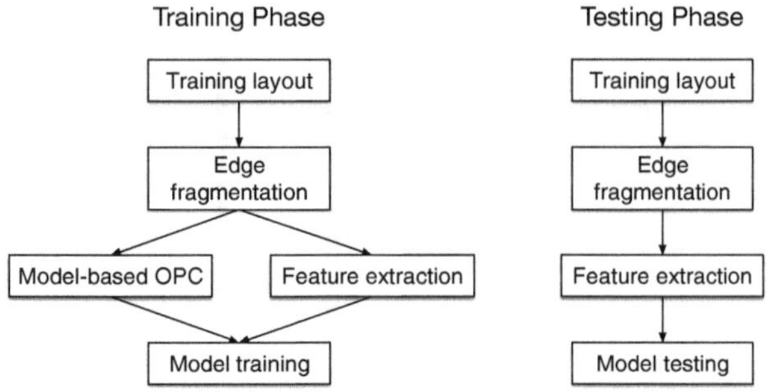

Fig. 4.11 Machine learning based OPC flow [31]

the testing phase. In addition, increasingly complicated designs result in complex optical proximity effects toward the sub-resolution domain, causing the difficulty in achieving accurate regression models.

To overcome the aforementioned challenges, a hierarchical Bayes model (HBM) is proposed for the OPC problem with CCAS feature extraction [31]. In the HBM, a generalized linear mixed model (GLMM) is trained with explicit consideration of different edge types, including normal, convex, concave, line-end edge, etc. GLMM handles these edge types by regarding them as a random effect with a random variance. For unknown variables, the HBM assumes a non-informative prior distribution, thus avoiding the lack of prior information. Therefore, better OPC results can be generated by HBM compared with previous regression approaches.

The comparison between HBM-based approach and model-based (MB) approach is demonstrated in Fig. 4.12. MB_ik denotes OPC results from the *k*th iteration of the MB-approach. The HBM-based approach can achieve EPE results comparable to that of the MB-approach at the 10th iteration, while the former is much faster. Hence, it is suggested to initialize OPC conditions with HBM-based approach and use MB-based approach to finish the rest OPC iterations, such that the overall runtime can be reduced [31].

4.4 Machine Learning in Physical Design

Machine learning can not only benefit late stages in the backend design flow, but also early stages, such as physical design. Physical design contains many difficult combinatorial problems that are hard to solve optimally. These problems are getting even more complicated due to increasing constraints from design configurations and manufacturing in advanced technology nodes.

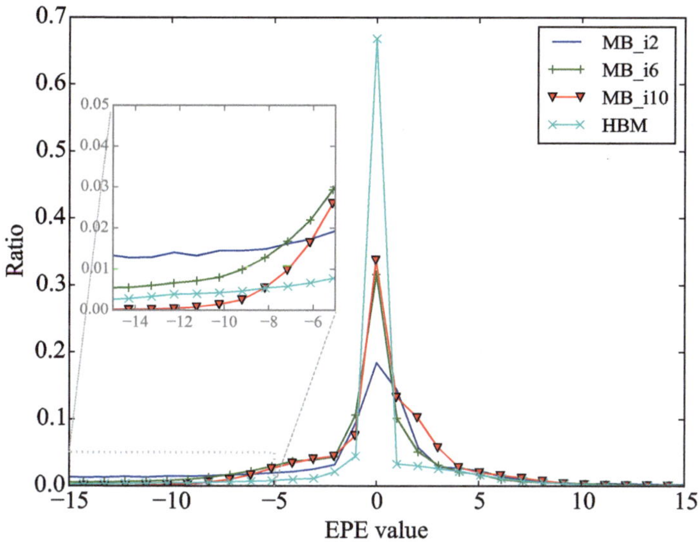

Fig. 4.12 Compare HBM-based and model-based OPC in terms of EPE distributions [31]

4.4.1 Machine Learning for Datapath Placement

Wirelength, such as Steiner tree wirelength (StWL), is a widely used metric in VLSI placement for random logic designs. It is the first-order approximation to interconnect delay and capacitance. As minimizing HPWL is generally NP-hard [6], analytical placement with iterative optimizations is developed. Logic designs often contain datapaths in which cells are characterized with a high degree of bit-wise parallelism. Conventional analytical placement usually handles them sub-optimally [43]. For designs with many embedded datapaths, it is critical to extract and place them appropriately for high quality placement [5, 44, 48].

An example in which modern placers fail to handle datapaths effectively is shown in Fig. 4.13 [44, 58]. Figure 4.13a sketches a datapath circuit in which there are three bit-stacks of cells, {2, 3, 4, 5}, {6, 7, 8, 9}, and {10, 11, 12, 13}, with fixed I/O pins. Two placement solutions, one from a datapath-aware placer, PADE [44], and the other from a conventional placer, Fast-Place3 [41], are shown in Fig. 4.13b, c, respectively. PADE is able to achieve smaller StWL because it packs and aligns each bit-stack more tightly than does Fast-Place3.

The automatic extraction of datapath is critical to eventual placement quality, in which machine learning techniques can bring benefits. In the datapath extraction proposed by Ward et al. [44], datapaths are evaluated with SVM and ANN techniques and then extracted according to their importance. A combined SVM and ANN learning approach is developed to classify datapath and non-datapath patterns in the initial netlist for efficient modeling. SVM is able to achieve global optimal in maximizing the separation margin, while it is susceptible to data noise. ANN, on

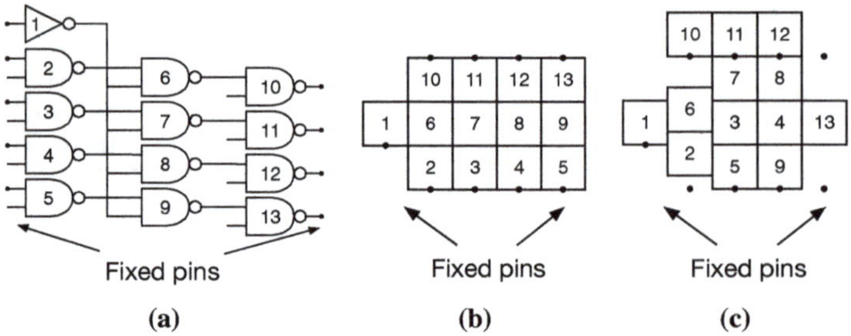

Fig. 4.13 An example of datapath-aware placement (PADE) achieves a 14% improvement in StWL compared to conventional placement (Fast-Place3) [44]. (**a**) Logic circuit; (**b**) PADE results: total StWL= 524; (**c**) FastPlace3 results: total StWL= 612

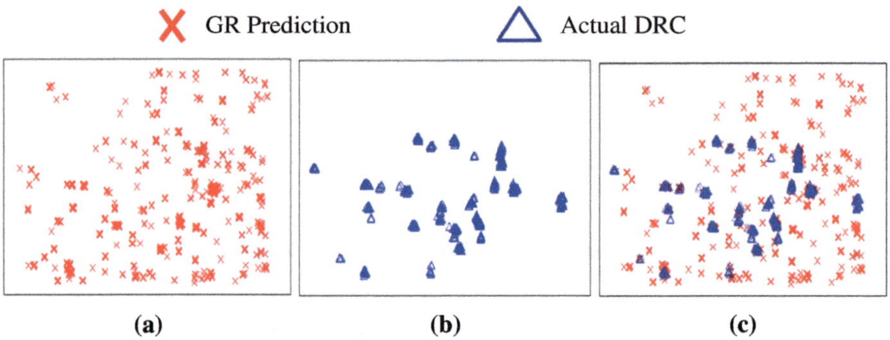

Fig. 4.14 (**a**) Routing hotspots predicted by global routing and (**b**) actual routing hotspots and (**c**) an overlay of predicted and actual routing hotspot [2]

the other hand, is more robust to noise, but more difficult to achieve optimal training accuracy. A pattern is considered to be datapath if and only if both the SVM model and the ANN model output positive predictions, utilizing the advantages of SVM and ANN.

4.4.2 Machine Learning for Routability-Driven Placement

Machine learning can also be applied to routability prediction in placement, which is very critical to the solution quality of modern placement. This is motivated by the gap between conventional routing congestion metrics, such as global routing congestion and the actual detailed routing violations [1, 2]. Figure 4.14 shows an example in which the routing hotspots reported by global routing are quite different from those in actual detailed routing. Figure 4.15 shows a similar comparison in which an SVM model with radius basis function (RBF) is adopted to build the correlation between layout features and routing hotspots. The routing hotspots

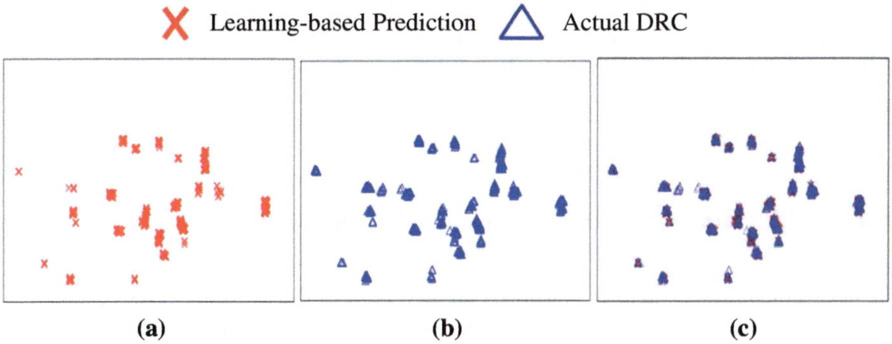

Fig. 4.15 (**a**) Routing hotspots predicted by machine learning and (**b**) actual routing hotspots and (**c**) an overlay of predicted and actual routing hotspot [2]

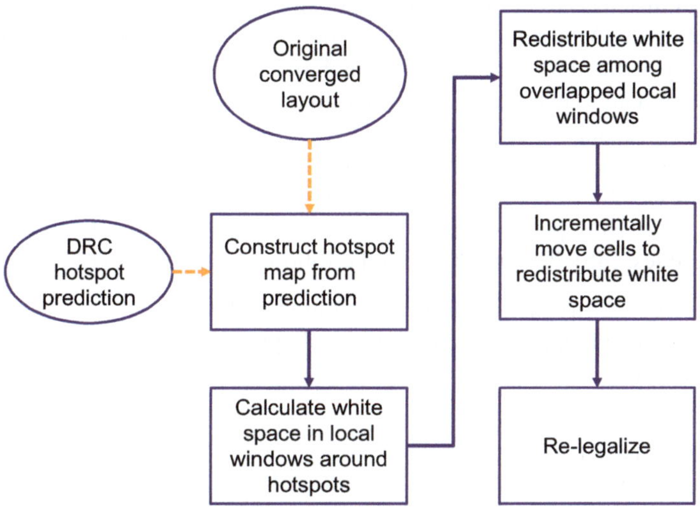

Fig. 4.16 Routability optimization guided by routing hotspot prediction [2]

reported by machine learning models are much more consistent with the actual routing hotspots. Various features in placement are extracted to train an accurate routing congestion predictor, including density parameters (e.g., local pin and cell density), global routing parameters (e.g., local overflow, demand, and capacity), pin proximity, cells tending to result in congestion (e.g., multi-height, sequential cells), connectivity parameters (e.g., number of nets in local windows), and structural parameters (e.g., number and depth of fanin and fanout logic stages). Hence, congestion prediction is very effective in placement.

Based on the routing hotspot predictor, a routability optimization algorithm is developed to redistribute white space, as shown in Fig. 4.16. White space around hotspots is extracted and redistributed by incrementally moving cells to improve routability. Incremental legalization is needed to remove overlaps between cells. Experimental results demonstrate an average of 20.6% and a maximum of 76.8% reduction in the number of DRC violations with negligible degradation in wirelength and timing [2].

4.4.3 Machine Learning for Clock Optimization

Besides placement, machine learning can also benefit the design of clock networks. Latch optimization including clustering and placement is of significant importance in modern VLSI designs to optimize skew and power consumption in clock networks. Latches in one cluster share a common local clock buffer (LCB) and are generally placed physically together [4, 17, 36]. Figure 4.17 shows that latches are tightly clustered around LCBs, dramatically reducing the overall wirelength of local clock trees. A learning based latch optimization methodology is proposed in [45] with a genetic algorithm for initial latch placement and decision tree induction for latch template matching. It is reported that 20–30% average reduction in local clock tree capacitance is achieved in industrial designs.

Classification with decision tree predicts a target class F based on an input vector E in which $(E, F) = (e_1, e_2, e_3, \ldots, e_k, F)$ with $(1 \leq k \leq |E|)$. A decision tree learns by recursively partitioning the source data into subsequent subsets based on the attribute test [45]. This method has the following advantages [34]:

- no prior probability distributions are required to data;
- greedy induction approaches provide good approximation to finding an optimal decision tree;

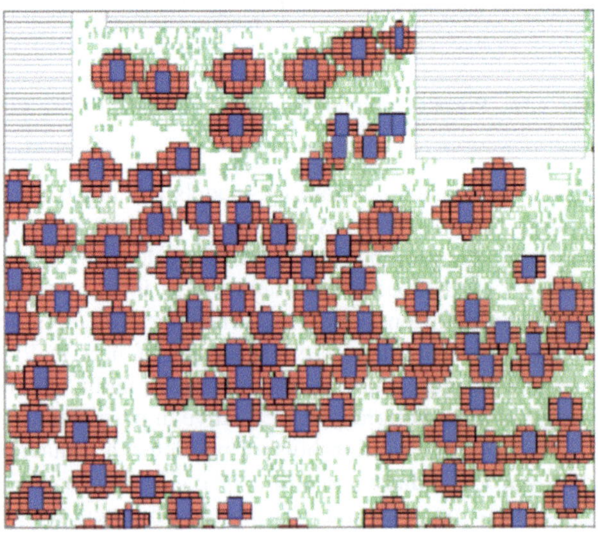

Fig. 4.17 Multi-GHz design showing clustered latches, where red cells are latches and purple cells are local clock buffers (LCB) [45]

Algorithm 2 Decision tree induction

```
1: function TREEBUILD(E, F)
2:     if stopping_condition(E, F) = TRUE then
3:         Create a new node from the leaf;
4:         Classify the leaf;
5:         Return leaf;
6:     else
7:         Create a node root;
8:         Find the best split and set it equal to the root test condition;
9:         Let V = the set of possible outcome test conditions of the root node;
10:        for each v ∈ V do
11:            E_v = training records given the root test condition;
12:            child = TREEBUILD(E_v, F);
13:            Add child as descendent of root and label;
14:    Return root;
```

- fast prediction with worse case $\mathcal{O}(\omega)$ complexity is possible, where ω is the depth of the tree.

The decision tree induction algorithm is mainly used for structured template selection. This problem requires a quick decision on the best template given even an unknown set of input requirements. In addition, the algorithm can handle categorical variables with multiple classes and is easy to implement and maintain. The template selection problem is a suitable application of this algorithm.

The decision tree induction algorithm is presented in Algorithm 2 [45]. When the stopping condition is met, a new node is created with either a test condition or a class label, and then classify and return the final decision, in line 3–5. Let V be the set of possible class labels in node $leaf$. When the stopping condition is not yet met, a new node called $root$ is created and the best split based on specific metrics is searched in line 7–8. For each possible label in $root$, line 11–13 recursively call for evaluating a new node.

4.4.4 Machine Learning for Lithography Friendly Routing

Despite other applications, the last application of machine learning we introduce in this chapter is to improve lithography friendliness in early stage. Lithography hotspot mitigation in the post-routing stage lacks the flexibility and thus requires early consideration [32]. Rule-based approaches have been developed for hotspot correction [11, 51]. However, with learning-based hotspot prediction, hotspots can be identified with high accuracy to guide routing effectively.

The major challenge of hotspots detection in routing lies in the requirement of early prediction before a real routing path is obtained. A layout region with metal blockages and unrouted pins, Pin1 to Pin4, is shown in Fig. 4.18a. Due to the existence of unrouted nets, general hotspot detection approaches fail to work, while potential hotspots may be caused by the routing segments from Pin1 to

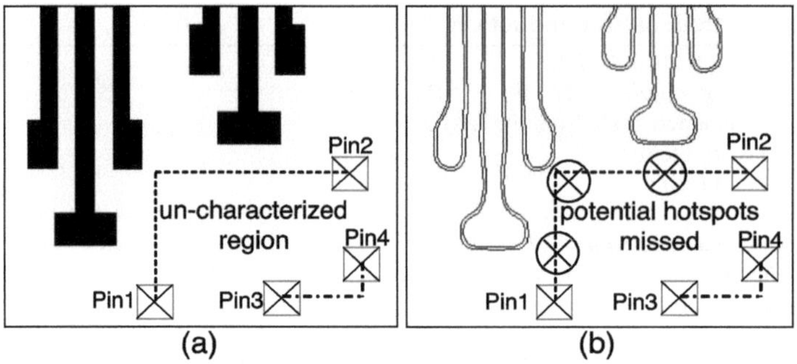

Fig. 4.18 The challenge of hotspot detection in detailed routing [7]. (**a**) A layout region with metal blockages and unrouted pins; (**b**) potential hotspots caused by the routing segments from Pin1 to Pin2

Pin2, as shown in Fig. 4.18b. To tackle this problem, a pre-built hotspot prediction model and a routing path prediction model are developed for hotspot detection in routing [7]. The routing path prediction model explores possible routing solutions with given available routing resources and identify preferable routes according to routing congestion and hotspots predicted by the hotspot prediction model. Due to complexity of data, ANN is adopted as the classifier. The techniques report an average of 50% reduction on lithography hotspots with 18–30% runtime overhead compared with existing lithography friendly routing works.

4.5 Conclusions

Machine learning has demonstrated promising benefits to various key steps in the VLSI backend design flow. This chapter surveys several critical issues in physical verification, mask synthesis, and physical design, such as hotspot detection, SRAF generation, OPC, placement, and routing optimization. There is also a tremendous ongoing research in this area to develop effective and efficient techniques based on machine learning. This will enable fast closure of the backend design flow as well as increasing the quality of backend synthesis solutions such that timing, power, area, and yield can be eventually improved.

In addition, backend VLSI design is still at its early stage in applying machine learning techniques. For example, in the SRAF generation, pixel-by-pixel prediction is required and only linear models are used, limiting the application of more complicated models due to high computational expense. Similarly, OPC is only affordable to adopt linear models as well. Such kind of optimization problems essentially need to generate a new mask image with a given layout image. It is worth exploring whether generative learning techniques can be applied [13]. For placement and routing problems, manual selection of important features is still required, while it is not clear whether general representation of layout information exists and whether automatic feature selection can be developed. Furthermore, unlike fields with extensive research on machine learning like image recognition in which large amount of data is available, it is generally difficult and expensive to obtain enough data in VLSI design for training robust and accurate models. Therefore, it is critical to develop techniques to improve modeling accuracy with relaxed requirement of big data so that machine learning can be widely adopted. All these challenges remain to be explored in the future.

References

1. W.T.J. Chan, Y. Du, A.B. Kahng, S. Nath, K. Samadi, Beol stack-aware routability prediction from placement using data mining techniques, in *IEEE International Conference on Computer Design (ICCD)* (2016), pp. 41–48
2. W.T.J. Chan, P.H. Ho, A.B. Kahng, P. Saxena, Routability optimization for industrial designs at sub-14nm process nodes using machine learning, in *ACM International Symposium on Physical Design (ISPD)* (2017), pp. 15–21
3. C.C. Chang, C.J. Lin, LIBSVM: a library for support vector machines. ACM Trans. Intell. Syst. Technol. **2**, 27:1–27:27 (2011). Software available at http://www.csie.ntu.edu.tw/~cjlin/libsvm
4. M. Cho, H. Xiang, H. Ren, M.M. Ziegler, R. Puri, LatchPlanner: latch placement algorithm for datapath-oriented high-performance VLSI designs, in *IEEE/ACM International Conference on Computer-Aided Design (ICCAD)* (2013), pp. 342–348
5. S. Chou, M.K. Hsu, Y.W. Chang, Structure-aware placement for datapath-intensive circuit designs, in *ACM/IEEE Design Automation Conference (DAC)* (2012), pp. 762–767
6. S. Chowdhury, Analytical approaches to the combinatorial optimization in linear placement problems. IEEE Trans. Comput. Aided Des. Integr. Circuits Syst. **8**(6), 630–639 (1989)
7. D. Ding, J.R. Gao, K. Yuan, D.Z. Pan, AENEID: a generic lithography-friendly detailed router based on post-RET data learning and hotspot detection, in *ACM/IEEE Design Automation Conference (DAC)* (2011), pp. 795–800
8. D. Ding, J.A. Torres, D.Z. Pan, High performance lithography hotspot detection with successively refined pattern identifications and machine learning. IEEE Trans. Comput. Aided Des. Integr. Circuits Syst. **30**(11), 1621–1634 (2011)
9. D. Ding, B. Yu, J. Ghosh, D.Z. Pan, Epic: efficient prediction of IC manufacturing hotspots with a unified meta-classification formulation, in *IEEE/ACM Asia and South Pacific Design Automation Conference (ASPDAC)* (2012), pp. 263–270
10. R.E. Fan, K.W. Chang, C.J. Hsieh, X.R. Wang, C.J. Lin, LIBLINEAR: a library for large linear classification. J. Mach. Learn. Res. **9**(Aug), 1871–1874 (2008)
11. J.R. Gao, H. Jawandha, P. Atkar, A. Walimbe, B. Baidya, O. Rizzo, D.Z. Pan, Self-aligned double patterning compliant routing with in-design physical verification flow, in *Proceedings of SPIE*, vol. 8684 (2013), p. 868408
12. J.R. Gao, B. Yu, D.Z. Pan, Accurate lithography hotspot detection based on PCA-SVM classifier with hierarchical data clustering, in *Proceedings of SPIE*, vol. 9053 (2014), p. 90530E
13. I. Goodfellow, J. Pouget-Abadie, M. Mirza, B. Xu, D. Warde-Farley, S. Ozair, A. Courville, Y. Bengio, Generative adversarial nets, in *Advances in Neural Information Processing Systems* (2014), pp. 2672–2680
14. I. Goodfellow, Y. Bengio, A. Courville, *Deep Learning* (MIT press, Cambridge, 2016)
15. A. Gu, A. Zakhor, Optical proximity correction with linear regression. IEEE Trans. Semicond. Manuf. **21**(2), 263–271 (2008)
16. T. Hastie, R. Tibshirani, J. Friedman, J. Franklin, The elements of statistical learning: data mining, inference and prediction. Math. Intell. **27**(2), 83–85 (2005)
17. S. Held, U. Schorr, Post-routing latch optimization for timing closure, in *ACM/IEEE Design Automation Conference (DAC)* (2014), pp. 7:1–7:6

18. N. Jia, E.Y. Lam, Machine learning for inverse lithography: using stochastic gradient descent for robust photomask synthesis. J. Opt. **12**(4), 045601:1–045601:9 (2010)
19. J.H. Jun, M. Park, C. Park, H. Yang, D. Yim, M. Do, D. Lee, T. Kim, J. Choi, G. Luk-Pat, et al.: Layout optimization with assist features placement by model based rule tables for 2x node random contact, in *Proceedings of SPIE* (2015), p. 94270D
20. A.B. Kahng, C.H. Park, X. Xu, Fast dual graph based hotspot detection, in *Proceedings of SPIE*, vol. 6349 (2016), p. 63490H
21. B.S. Kim, Y.H. Kim, S.H. Lee, S.I. Kim, S.R. Ha, J. Kim, A. Tritchkov, Pixel-based SRAF implementation for 32nm lithography process, in *Proceedings of SPIE* (2008), p. 71220T
22. C. Kodama, T. Kotani, S. Nojima, S. Mimotogi, Sub-resolution assist feature arranging method and computer program product and manufacturing method of semiconductor device (2014). US Patent 8809072
23. S.Y. Lin, J.Y. Chen, J.C. Li, W.y. Wen, S.C. Chang, A novel fuzzy matching model for lithography hotspot detection, in *ACM/IEEE Design Automation Conference (DAC)* (2013), pp. 1–6
24. Y. Lin, X. Xu, J. Ou, D.Z. Pan, Machine learning for mask/wafer hotspot detection and mask synthesis, in *Photomask Technology*, vol. 10451 (International Society for Optics and Photonics, Bellingham, 2017), p. 104510A
25. Y. Lin, Y. Watanabe, T. Kimura, T. Matsunawa, S. Nojima, M. Li, D.Z. Pan, Data efficient lithography modeling with residual neural networks and transfer learning, in *ACM International Symposium on Physical Design (ISPD)* (2018), pp. 82–89
26. R. Luo, Optical proximity correction using a multilayer perceptron neural network. J. Opt. **15**(7), 075708 (2013)
27. K.S. Luo, Z. Shi, X.L. Yan, Z. Geng, SVM based layout retargeting for fast and regularized inverse lithography. J. Zhejiang Univ. Sci. C **15**(5), 390–400 (2014)
28. T. Matsunawa, J.R. Gao, B. Yu, D.Z. Pan, A new lithography hotspot detection framework based on AdaBoost classifier and simplified feature extraction, in *Proceedings of SPIE*, vol. 9427 (2015)
29. T. Matsunawa, B. Yu, D.Z. Pan, Optical proximity correction with hierarchical Bayes model, in *Proceedings of SPIE*, vol. 9426 (2015)
30. T. Matsunawa, S. Nojima, T. Kotani, Automatic layout feature extraction for lithography hotspot detection based on deep neural network, in *Proceedings of SPIE* (2016)
31. T. Matsunawa, B. Yu, D.Z. Pan, Optical proximity correction with hierarchical Bayes model. J. Micro/Nanolithogr. MEMS MOEMS **15**(2), 021009 (2016)
32. J. Mitra, P. Yu, D.Z. Pan, RADAR: RET-aware detailed routing using fast lithography simulations, in *ACM/IEEE Design Automation Conference (DAC)*, pp. 369–372 (2005)
33. S. Miyama, K. Yamamoto, K. Koyama, Large-area optical proximity correction with a combination of rule-based and simulation-based methods. Jpn. J. Appl. Phys. **35**(12S), 6370 (1996)
34. M.J. Moshkov, Time complexity of decision trees, in *Transactions on Rough Sets III* (Springer, Berlin, 2005), pp. 244–459
35. L. Pang, Y. Liu, D. Abrams, Inverse lithography technology (ILT): a natural solution for model-based SRAF at 45nm and 32nm, in *Proceedings of SPIE*, p. 660739 (2007)
36. D. Papa, C. Alpert, C. Sze, Z. Li, N. Viswanathan, G.J. Nam, I.L. Markov, Physical synthesis with clock-network optimization for large systems on chips. IEEE Micro **31**(4), 51–62 (2011)
37. Y. Ping, S. McGowan, Y. Gong, Y.M. Foong, J. Liu, J. Qiu, V. Shu, B. Yan, J. Ye, P. Li, et al.: Process window enhancement using advanced ret techniques for 20nm contact layer, in *Proceedings of SPIE*, p. 90521N (2014)
38. K. Sakajiri, A. Tritchkov, Y. Granik, Model-based SRAF insertion through pixel-based mask optimization at 32nm and beyond, in *Proceedings of SPIE*, p. 702811 (2008)
39. S.D. Shang, L.Swallow, Y. Granik, Model-based SRAF insertion (2011). US Patent 8037429
40. M. Shin, J.H. Lee, Accurate lithography hotspot detection using deep convolutional neural networks. J. Micro/Nanolithogr. MEMS MOEMS **15**(4), 043507 (2016)

41. N. Viswanathan, M. Pan, C. Chu, FastPlace 3.0: a fast multilevel quadratic placement algorithm with placement congestion control, in *IEEE/ACM Asia and South Pacific Design Automation Conference (ASPDAC)* (2007), pp. 135–140
42. R. Viswanathan, J.T. Azpiroz, P. Selvam, Process optimization through model based SRAF printing prediction, in *Proceedings of SPIE* (2012), p. 83261A
43. S.I. Ward, D.A. Papa, Z. Li, C.N. Sze, C.J. Alpert, E. Swartzlander, Quantifying academic placer performance on custom designs, in *ACM International Symposium on Physical Design (ISPD)* (2011), pp. 91–98
44. S. Ward, D. Ding, D.Z. Pan, PADE: a high-performance placer with automatic datapath extraction and evaluation through high dimensional data learning, in *ACM/IEEE Design Automation Conference (DAC)* (2012), pp. 756–761
45. S.I. Ward, N. Viswanathan, N.Y. Zhou, C.C. Sze, Z. Li, C.J. Alpert, D.Z. Pan, Clock power minimization using structured latch templates and decision tree induction, in *IEEE/ACM International Conference on Computer-Aided Design (ICCAD)* (2013), pp. 599–606
46. Y. Watanabe, T. Kimura, T. Matsunawa, S. Nojima, Accurate lithography simulation model based on convolutional neural networks, in *Proceedings of SPIE*, vol. 10147 (2017), p. 101470K
47. W.Y. Wen, J.C. Li, S.Y. Lin, J.Y. Chen, S.C. Chang, A fuzzy-matching model with grid reduction for lithography hotspot detection. IEEE Trans. Comput. Aided Des. Integr. Circuits Syst. **33**(11), 1671–1680 (2014)
48. H. Xiang, M. Cho, H. Ren, M. Ziegler, R. Puri, Network flow based datapath bit slicing, in *ACM International Symposium on Physical Design (ISPD)* (2013), pp. 139–146
49. X. Xu, T. Matsunawa, S. Nojima, C. Kodama, T. Kotani, D.Z. Pan, A machine learning based framework for sub-resolution assist feature generation, in *ACM International Symposium on Physical Design (ISPD)* (2016), pp. 161–168
50. X. Xu, Y. Lin, M. Li, T. Matsunawa, S. Nojima, C. Kodama, T. Kotani, D.Z. Pan, Sub-resolution assist feature generation with supervised data learning. IEEE Trans. Comput. Aided Des. Integr. Circuits Syst. **37**(6), 1225–1236 (2018)
51. J. Yang, N. Rodriguez, O. Omedes, F. Gennari, Y.C. Lai, V. Mankad, DRCPlus in a router: automatic elimination of lithography hotspots using 2D pattern detection and correction. Proc. SPIE **7641**, 76410Q (2010)
52. H. Yang, Y. Lin, B. Yu, F.E. Young, Lithography hotspot detection: from shallow to deep learning, in *IEEE International System-on-Chip Conference (SOCC)* (2017), pp. 233–238
53. H. Yang, L. Luo, J. Su, C. Lin, B. Yu, Imbalance aware lithography hotspot detection: a deep learning approach. Proc. SPIE **16**(3), 033504 (2017)
54. H. Yang, J. Su, Y. Zou, B. Yu, F.E. Young, Layout hotspot detection with feature tensor generation and deep biased learning, in *ACM/IEEE Design Automation Conference (DAC)* (2017), p. 62
55. J. Ye, Y. Cao, H. Feng, System and method for model-based sub-resolution assist feature generation (2011). US Patent 7882480
56. Y.T. Yu, Y.C. Chan, S. Sinha, I.H.R. Jiang, C. Chiang, Accurate process-hotspot detection using critical design rule extraction, in *ACM/IEEE Design Automation Conference (DAC)* (2012), pp. 1167–1172
57. Y.T. Yu, G.H. Lin, I.H.R. Jiang, C. Chiang, Machine learning based hotspot detection using topological classification and critical feature extraction, in *ACM/IEEE Design Automation Conference (DAC)* (2013), p. 67
58. B. Yu, D.Z. Pan, T. Matsunawa, X. Zeng, Machine learning and pattern matching in physical design, in *IEEE/ACM Asia and South Pacific Design Automation Conference (ASPDAC)* (2015), pp. 286–293
59. H. Zhang, B. Yu, Y.F. Evangeline, Enabling online learning in lithography hotspot detection with information-theoretic feature optimization, in *IEEE/ACM International Conference on Computer-Aided Design (ICCAD)* (2016), p. 47
60. H. Zhang, F. Zhu, H. Li, F.E. Young, B. Yu, Bilinear lithography hotspot detection, in *ACM International Symposium on Physical Design (ISPD)* (2017), pp. 7–14

Part II
Machine Learning for Manufacturing, Yield, and Reliability

The Committee expressed their confident opinion of the adequacy of the machinery to work under all the friction and strain to which it can be exposed; of its durability, strength, solidity, and equilibrium; of the prevention of, or compensation for, wear by friction; of the accuracy of the various adjustments; and of the judgment and discretion displayed by the inventor, in his determination to admit into the mechanism nothing but the very best and most finished workmanship; as a contrary course would have been false economy, and might have led to the loss of the whole capital expended on it.

Charles Babbage

Chapter 5
Gaussian Process-Based Wafer-Level Correlation Modeling and Its Applications

Constantinos Xanthopoulos, Ke Huang, Ali Ahmadi, Nathan Kupp, John Carulli, Amit Nahar, Bob Orr, Michael Pass, and Yiorgos Makris

5.1 Introduction

As the complexity of modern integrated circuits (ICs) increases and the feature sizes continue to shrink, undesirable process variations are becoming more and more challenging for semiconductor manufacturers. These variations often result in marginalities or even failures, impacting both quality and yield. Therefore, understanding, monitoring, and minimizing these variations is critical for ensuring the production of modern semiconductor devices. In order to understand how process variations affect the performance and functionality of each manufactured device, several types of tests are performed after fabrication. In this chapter, we focus on the tests that are performed at the wafer level, which are usually categorized into two major types.

C. Xanthopoulos · A. Ahmadi · Y. Makris (✉)
The University of Texas at Dallas, Richardson, TX, USA
e-mail: constantinos.xanthopoulos@utdallas.edu; ali.ahmadi@utdallas.edu; yiorgos.makris@utdallas.edu

K. Huang
San Diego State University, San Diego, CA, USA
e-mail: khuang@mail.sdsu.edu

N. Kupp
Yale University, New Haven, CT, USA

J. Carulli
GLOBALFOUNDRIES, Malta, NY, USA
e-mail: john.carulli@globalfoundries.com

A. Nahar · B. Orr · M. Pass
Texas Instruments, Dallas, TX, USA
e-mail: a-nahar2@ti.com; borr@ti.com; m-pas@ti.com

I. M. Elfadel et al. (eds.), *Machine Learning in VLSI Computer-Aided Design*,
https://doi.org/10.1007/978-3-030-04666-8_5

5.1.1 Types of Wafer-Level Test Measurements

The first type of tests utilizes e-test measurements, also known as inline or kerf measurements. E-test measurements provide a signature that indicates the wafer's position in the process space and can be used as a measure of health for each wafer under inspection. These e-test measurements are performed on simple circuit components located across the surface of each wafer. To optimize wafer area efficiency, these structures are typically constructed in the wafer scribe lines which are destroyed during wafer dicing. Collecting e-test measurements from the scribe lines adjacent to every die would be prohibitively time-consuming. As a result, there are generally only a handful of e-test sites measured on a wafer which can have hundreds or thousands of die. In rare cases, when a high spatial sampling frequency is necessary, e-test structures are placed inside each die, allowing for a more detailed representation of the wafer's health at the expense of longer test times.

The second type of tests performed at the wafer level is commonly referred to as specification tests. During these tests, the performances of every fabricated circuit are checked against its design specification, in order to capture devices that do not function properly due to random manufacturing defects or due to the effects of extreme process variations. Specification testing consumes the majority of time spent across all testing processes, yet at the same time, it identifies the bulk of failing devices. After this stage, the wafers are sliced, the passing die get packaged and functionally tested one last time, in a stage called final test, to make sure that the process of packaging did not severely affect them.

Current test flows and modern automated test equipment (ATE) allow the mapping of a sliced and packaged die back to its corresponding wafer location, through tracking of the unique die and wafer IDs. Despite this capability, the full wafer signature of final tests cannot be constructed due to the missing die locations that did not pass the previous stage. Moreover, since packaging can affect each sliced die in a non-systematic way, any wafer-level analysis based on final test measurements could be severely skewed and thus ineffective. For these reasons, in this chapter, we only focus on the kerf and specification test measurements, which provide abundant information for analyzing the wafer-level variation.

5.1.2 Wafer-Level Statistical Correlations

By univariately analyzing the variation of the measurements for each test performed at the wafer level, we can adequately identify the majority of failing devices. This is currently the de facto testing strategy, in which every measurement is compared against known specification limits. This comparison determines the pass or fail decision for every device under test (DUT). While these strategies have proven successful, they can be improved by employing wafer-level correlation models, resulting in increased testing accuracy as well as the reduction of test time. There are multiple such correlations that we can exploit to achieve this. These correlations may be spatial, between different die on the same wafer, temporal, between the same die locations on different wafers, or both. Additionally, these correlations may include e-tests variables, specification tests variables, or both. In recent years, several statistical methodologies have been proposed that utilize such correlations. These studies sought to reduce the number of specification tests required (i.e., test compaction) [7, 8], optimize the test flow at wafer-level optimization [2, 3], introduce new alternative tests [5, 31, 32], or build machine learning models to learn classification boundaries separating passing and failing populations of devices in a multidimensional space of low-cost measurements [26, 28].

All these studies take advantage of the statistical correlations between tests at wafer-level, but not the spatiotemporal correlations that are common for many e-test and probe measurements. Spatial correlation refers to the relationship a test measurement exhibits according to the location, on the surface of the wafer, of the site that is being tested, and its proximity to other sites. These correlations exist due to the fact that proximate locations on a wafer have very similar physical properties, as they have been similarly influenced by the same series of manufacturing steps. Likewise, temporal correlation refers to the relationship between consecutively manufactured wafers, as they have been affected by very similar process characteristics. A manufacturing lot is a set of wafers that have been sourced from subsequent slices of the same silicon ingot and have gone through all fabrication steps as a unit. Due to this, temporal correlations between wafers within a lot are usually powerful, though such correlation is less evident across lots.

Using these wafer-level spatial correlations, we can predict measurements at unobserved die locations, allowing us to collect wafer measurements from only a sparse subset of die. The sampling rate conversely affects the error of the predictions and the test time (cost). Figure 5.1 depicts an overview of the spatial interpolation methodology that we will be discussing in this chapter. As shown in Fig. 5.1, the statistical model is trained using sampled measurements across the surface of the wafer and it predicts the measurements for the remaining locations of the wafer. As

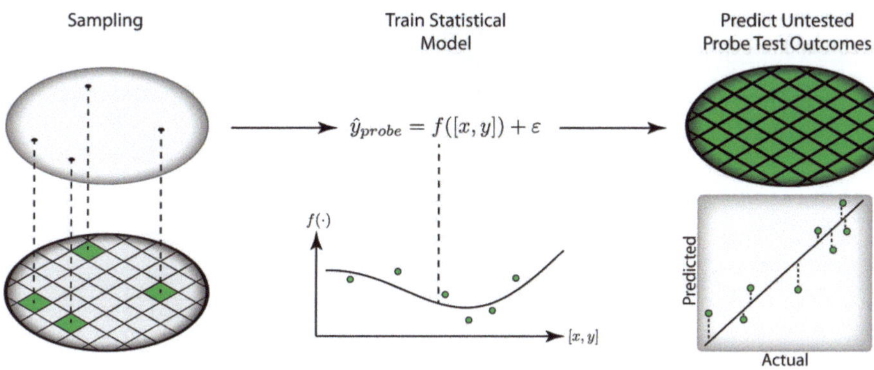

Fig. 5.1 Overview of wafer measurement spatial interpolation

it is expected for any kind of prediction, there is an error associated with it and this is represented by the deviation from the 45-degree diagonal in the actual-vs-predicted graph.

Figure 5.2 shows the trade-off curve between the two key metrics, where 100% test coverage reflects the commonly applied approach where all wafer sites are measured. At the other extreme, 0% test coverage reflects a static model that has no knowledge of the wafer under test. Test cost in the first case would be maximal, while in the second case the prediction error is maximized. By tuning the sampling rate, we can dramatically reduce test cost while incurring minimal test error. After determining the optimal sampling rate, depending on the application, one can utilize the wafer-level spatiotemporal correlation to achieve various goals.

The benefit of spatial and spatiotemporal correlation models is threefold: First, sparse wafer-level e-test measurements can be accurately extrapolated to every die location, which enables us to efficiently monitor process variations based only on a handful of explicitly measured e-test sites. Second, the use of spatial correlation modeling can significantly reduce the test cost of analog/RF circuits, by measuring performances on a sparse subset of die on each wafer and, subsequently, using spatial correlation models to predict the performance values at unobserved die locations. Third, the HVM yield of a device can be predicted from measurements obtained on a sample of die from a small set of wafers available in early production. This description resembles the typical problem faced in post-silicon validation, where the performance distribution of HVM devices needs to be extrapolated from a few sample die obtained from a few engineering wafers produced over the period of a few months.

Several techniques that exploit wafer-level spatial correlations have been proposed. In [23], the expectation-maximization (EM) algorithm is used to estimate spatial wafer measurements, assuming that data comes from a multivariate normal distribution. The Box-Cox transformation is used in case data is not normally distributed. The "virtual probe" (VP) approach [10, 34–36] models spatial variation via a discrete cosine transform (DCT) that projects spatial statistics into the frequency

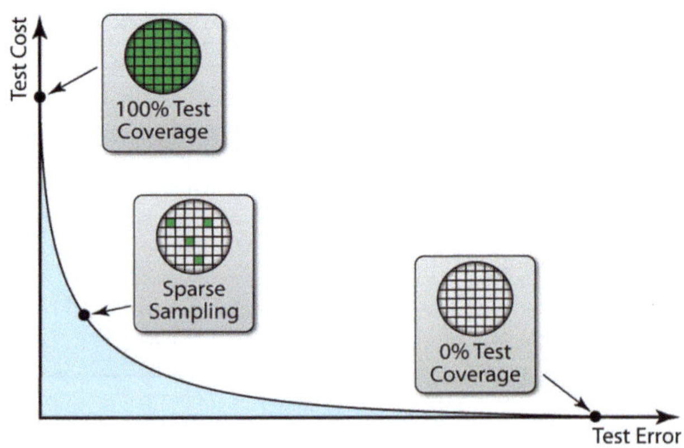

Fig. 5.2 Trade-off between test cost and test error

domain. In this chapter, we focus on the Gaussian process (GP) regression-based modeling, which is a method of interpolation often used in geo-statistics and spatial analysis. The author of [18] lays the groundwork for applying Gaussian process (GP) models to spatial interpolation of semiconductor data based on generalized least square fitting and a structured correlation function. As recently shown in [16, 17], using such GP models can dramatically improve both prediction accuracy and computational time, as compared to the previous state-of-the-art modeling techniques.

5.2 Gaussian Process-Based Regression Models

In this section, we articulate the theoretical underpinnings of Gaussian process models as well as the various enhancements for modeling spatial variation of wafer-level measurements. GP modeling [22] is a regression methodology that enables the learning of functions over Gaussian random fields based on sampled data. Such a regression approach is well suited for spatial modeling of semiconductor test data, as it is incredibly flexible in modeling data and imposes none of the traditional a priori assumptions about the underlying form of the generative function that tend to bias ordinary linear regression models.

Gaussian process models originate from the union of Bayesian statistics and the kernel theory of support vector machines [24, 29, 30]. The fundamental concept of Gaussian processes is to model function outputs as drawn from a prior distribution with a fixed mean and a kernel-based covariance function. In this section, we outline the theoretical basis for GP models and subsequently explain the Gaussian process model-based methodology for learning from semiconductor test data, and, more specifically, extrapolating wafer-level measurements that are sparsely sampled across a wafer.

We begin this discussion of GP model theory by considering the monolithic linear regression formulation $t = f(\mathbf{x}) + \varepsilon$, where $\mathbf{x}$ is the d-dimensional input vector, t is the output target value, $f(\mathbf{x}) = \mathbf{x}^\top \mathbf{w}$, and ε represents independent and identically distributed (i.i.d.) additive noise. In the traditional formulation, $\mathbf{w}$ is a vector of weights associated with each dimension of $\mathbf{x}$. This means that, in the case of normalized data, large values of $\mathbf{w}$ correspond to more "important" features in the model, when the features being used have being previously standardized. Prescribing a model form that is linear in the parameters will reduce model variance at the expense of introducing higher bias. As a result, linear models perform well when the true generative function happens to be linear, but this is often not the case. In cases where the high bias of linear regression is inappropriate, we seek an alternative bias-variance trade-off, one that is proffered by GP models.

With Gaussian processes, we do not presume the generative function $f(\mathbf{x})$ is of linear form in the original feature space, as shown by the one-dimensional input space curve on the left side of Fig. 5.3. Instead, we define a Gaussian process as a collection of random variables $f(\mathbf{x})$ indexed by coordinates $\mathbf{x}$, such that every

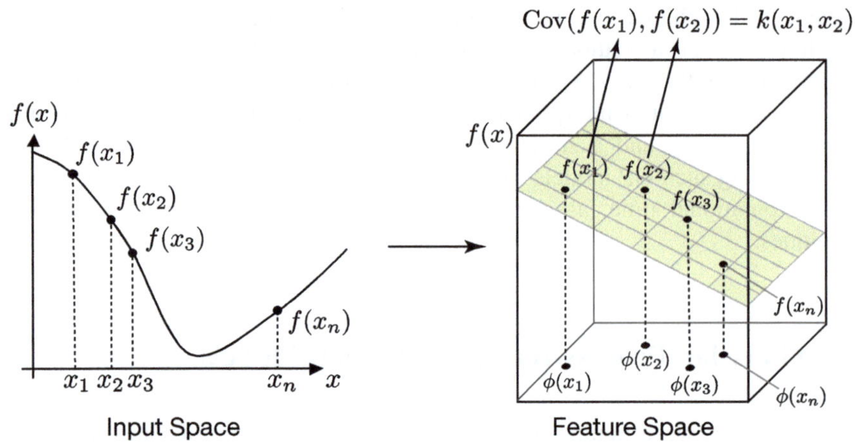

Fig. 5.3 Overview of Gaussian process modeling

finite set of n function evaluations $\{f(\mathbf{x}_1), f(\mathbf{x}_2), \ldots, f(\mathbf{x}_n)\}$ over the coordinates is jointly Gaussian distributed. To derive a GP model for regression, we first consider a noise-free linear model, shown by the right side of Fig. 5.3, which has the form:

$$t = f(\mathbf{x}) = \phi(\mathbf{x})^\top \mathbf{w}_f \tag{5.1}$$

where $\phi(\mathbf{x})$ is a function of the inputs mapping the input columns into some high-dimensional feature space, shown by the bottom plane on the right side of Fig. 5.3, and $\mathbf{w}_f$ is the weight vector associated with $\phi(\mathbf{x})$ in the feature space. For example, a scalar input $\mathbf{x}$ could be projected into the feature space: $\phi(\mathbf{x}) = (1, \mathbf{x}, \mathbf{x}^2)^\top$. We assign a Bayesian prior on the weights such that $\mathbf{w}_f \sim \mathcal{N}(0, \Sigma_p)$. As the realizations of the Gaussian process at points $\{f(\mathbf{x}_1), f(\mathbf{x}_2), \ldots, f(\mathbf{x}_n)\}$ are jointly Gaussian, we can fully specify the Gaussian process with mean and covariance functions:

$$\mathbb{E}[f(\mathbf{x})] = \phi(\mathbf{x})^\top \mathbb{E}[\mathbf{w}_f] = 0, \tag{5.2}$$

$$\begin{aligned} \operatorname{cov}\big(f(\mathbf{x}), f(\mathbf{x}')\big) &= \mathbb{E}[f(\mathbf{x}) f(\mathbf{x}')] \\ &= \phi(\mathbf{x})^\top \mathbb{E}[\mathbf{w}_f \mathbf{w}_f^\top] \phi(\mathbf{x}') \\ &= \phi(\mathbf{x})^\top \Sigma_p \phi(\mathbf{x}') \end{aligned} \tag{5.3}$$

5.2.1 Modeling Covariance with Kernel Functions

Recall that our ultimate goal of building a Gaussian process-based regression model is to somehow capture spatial variation in t as a function of the coordinates $\mathbf{x}$. The following discussion demonstrates how we can accomplish this task by modeling our data as drawn from a process with a covariance function that depends on spatial location. By taking this approach, proximal data points are modeled as being highly covariant, and distant points are modeled with low covariance. This property codifies our intuition and a priori knowledge of the domain; we expect the variation of test measurements to strongly correlate to the wafer coordinates of the tested sites.

Consider the covariance function specified in (5.3). Now, since covariance matrices are by definition positive semi-definite (see proof in the section "Appendix 1: Proof of Positive Semi-Definite for Covariance Matrix"), we can redefine Σ_p as $(\Sigma_p^{1/2})^2$ and rewrite (5.3) as:

$$\begin{aligned}\operatorname{cov}\big(f(\mathbf{x}), f(\mathbf{x}')\big) &= \phi(\mathbf{x})^\top \Sigma_p \phi(\mathbf{x}') \\ &= \phi(\mathbf{x})^\top (\Sigma_p^{1/2})^\top \Sigma_p^{1/2} \phi(\mathbf{x}')\end{aligned} \tag{5.4}$$

We now introduce the parameter $\psi(\mathbf{x})$ by defining $\psi(\mathbf{x}) = \Sigma_p^{1/2}\phi(\mathbf{x})$ and subsequently rewrite the covariance of (5.3) as:

$$\begin{aligned}\operatorname{cov}\big(f(\mathbf{x}), f(\mathbf{x}')\big) &= \phi(\mathbf{x})^\top (\Sigma_p^{1/2})^\top \Sigma_p^{1/2} \phi(\mathbf{x}') \\ &= \langle \psi(\mathbf{x}), \psi(\mathbf{x}') \rangle\end{aligned} \tag{5.5}$$

Crucially, this covariance function is formed as an inner product, permitting us to express (5.5) as a kernel function [4]: $\operatorname{cov}\big(f(\mathbf{x}), f(\mathbf{x}')\big) = k(\mathbf{x}, \mathbf{x}')$. In other words, the covariance between any outputs can be written as a function of the inputs using the kernel function without needing to explicitly compute $\phi(\mathbf{x})$. Thus, we use the terms *covariance function* and *kernel function* interchangeably in this paper. Many kernel functions exist, and any function $k(\cdot, \cdot)$ that satisfies Mercer's condition [29] is a valid kernel function. However, only a handful of kernels are commonly used. For example, one can use the squared exponential, also known as the radial basis function kernel:

$$k(\mathbf{x}, \mathbf{x}') = \exp\left(-\frac{1}{2l^2}|\mathbf{x} - \mathbf{x}'|^2\right) \tag{5.6}$$

where l is some characteristic length scale of the squared exponential kernel. Employing this kernel is equivalent to training a linear regression model with an infinite-dimensional feature space. Substituting our squared exponential covariance function into the definition of the Gaussian process, we arrive at a Gaussian process formulation as:

$$t = f(\mathbf{x}) \sim \mathcal{GP}(0, k(\mathbf{x}, \mathbf{x}')) \tag{5.7}$$

The following section describes how to employ this process to derive predictive distributions as well as how to manage the inclusion of additive noise in the model.

5.2.2 *Training and Prediction*

Suppose that we are provided a training set of n data points $X = \{\mathbf{x}_1, \mathbf{x}_2, \ldots, \mathbf{x}_n\}^\top$ observed in a d-dimensional space, e.g., each vector in X is $\mathbf{x}_i = \{x_1, x_2, \ldots, x_d\}$, therefore X is an $n \times d$ matrix of inputs. With these input points, we are also given a set of predictive targets, $\mathbf{t} = \{t_1, t_2, \ldots, t_n\}^\top$. Now, we wish to model the observed data as a noise-free Gaussian process and define, as before, $y = f(\mathbf{x}) \sim \mathcal{GP}(0, k(\mathbf{x}, \mathbf{x}'))$.

To derive the predictive distribution of this Gaussian process, we first write the joint distribution of the training set targets and a new test function value as:

$$\begin{bmatrix} \mathbf{t} \\ f_* \end{bmatrix} \sim \mathcal{N}\left(\mathbf{0}, \begin{bmatrix} K & \mathbf{k}_* \\ \mathbf{k}_*^\top & k(\mathbf{x}_*, \mathbf{x}_*) \end{bmatrix}\right) \tag{5.8}$$

where $\mathbf{x}_*$ is a location we wish to extrapolate to, and where we have defined $K = K(X, X')$ as the matrix of the kernel function $k(\mathbf{x}, \mathbf{x}')$ evaluated at all pairs of training locations. We have also defined $\mathbf{k}_* = K(X, \mathbf{x}_*)$ as the column vector of kernel evaluations between the test point and the entire set of training points, and lastly, $k(\mathbf{x}_*, \mathbf{x}_*)$ as the variance of the test function value at the observation point $\mathbf{x}_*$. With this distribution, we can condition the test function value on the observed data to obtain the predictive distribution (see derivation in the section "Appendix 2: Marginal and Conditional Distribution of Multivariate Normal Distribution"):

$$f_*|X, \mathbf{t}, \mathbf{x}_* \sim \mathcal{N}(\mathbf{k}_*^\top K^{-1}\mathbf{t}, k(\mathbf{x}_*, \mathbf{x}_*) - \mathbf{k}_*^\top K^{-1}\mathbf{k}_*) \tag{5.9}$$

In this work, we primarily concern ourselves with point predictions, and so we use simply the distribution mean $\bar{f}_* = \mathbf{k}_*^\top K^{-1}\mathbf{t}$ to generate a point prediction from the predictive distribution. This corresponds to decision-theoretic risk minimization [13, 29, 30] using a squared-loss function.

5.2.3 *Regularization*

To avoid overfitting, a technique known as regularization [13] is often employed in decision-theoretic empirical risk minimization. In a traditional linear regression model, regularization typically involves penalizing the L_1 or L_2 norm of the model coefficient estimates $\hat{\beta}$ to ensure that the "slope" of the model is not too large. This ensures that extrapolative predictions are not too extreme.

GP models handle regularization somewhat differently. Instead of penalizing coefficients, we consider our predictive targets $\mathbf{t} = \{t'_1, t'_2, \ldots, t'_n\}$ as affected by additive noise such that $t'_i = t_i + \varepsilon$, where we make the usual i.i.d. assumptions about the additive noise, $\varepsilon \sim \mathcal{N}(0, \sigma_n^2)$. To incorporate this into our GP model, we update (5.7) to model additive noise in the observations:

$$y = f(\mathbf{x}) + \varepsilon \sim \mathcal{GP}(0, k(\mathbf{x}, \mathbf{x}') + \sigma_n^2 \delta_{\mathbf{x},\mathbf{x}'}) \tag{5.10}$$

where $\delta_{\mathbf{x},\mathbf{x}'}$ is the Kronecker delta function. This, in turn, affects the joint distribution of (5.8):

$$\begin{bmatrix} \mathbf{t} \\ f_* \end{bmatrix} \sim \mathcal{N}\left(\mathbf{0}, \begin{bmatrix} K + \sigma_n^2 I & \mathbf{k}_* \\ \mathbf{k}_*^\top & k(\mathbf{x}_*, \mathbf{x}_*) \end{bmatrix}\right) \tag{5.11}$$

as well as the predictive distribution:

$$f_* | X, \mathbf{t}, \mathbf{x}_* \sim \mathcal{N}(\mathbf{k}_*^\top (K + \sigma_n^2 I)^{-1} \mathbf{t}, k(\mathbf{x}_*, \mathbf{x}_*) - \mathbf{k}_*^\top (K + \sigma_n^2 I)^{-1} \mathbf{k}_*) \tag{5.12}$$

resulting in a point prediction for new observations of $\bar{f}_* = \mathbf{k}_*^\top (K + \sigma_n^2 I)^{-1} \mathbf{t}$.

This constrains the fitted model to avoid extreme predictions. For example, consider the univariate fit of Fig. 5.4, shown with four monotonically increasing noise parameters $\sigma_n^2 = \{0, 0.0001, 0.01, 0.5\}$. The blue line is the fitted model, the red dots are the training data, and the dotted red line is the true generative function. As this noise parameter increases, the model gradually flattens, and for very large σ_n^2, approaches a constant fit. Applying a model with a $\sigma_n^2 = 0$ is equivalent to the hypothesis that our observations are noise free. Therefore, employing a nonzero σ_n^2 captures our intuition that real-world data measurements are affected by noise, and that we should not expect to fit a model exactly through each observed data point. As will be shown later, this parameter can be elegantly learned during the model selection and parameter adaptation phase of GP training.

In Fig. 5.5, we show the effects of incorporating additive noise on example wafer data, with $\sigma_n^2 = \{0, 0.00001, 0.01, 0.1\}$. As can be seen from the figure, modeling observations as noise free leads to extreme variation in the model as it fits the response surface exactly through each point observation; relaxing this constraint leads to smoother response surfaces.

5.2.4 Modeling Radial Variation

In semiconductor manufacturing, many operations include the rotation of the wafers which leads to a radially shaped variation on several wafer measurements. Such radial variation can be modeled efficiently by incorporating the radius feature to the input vector $\mathbf{x}$ during training of GP model, which is achieved by updating the

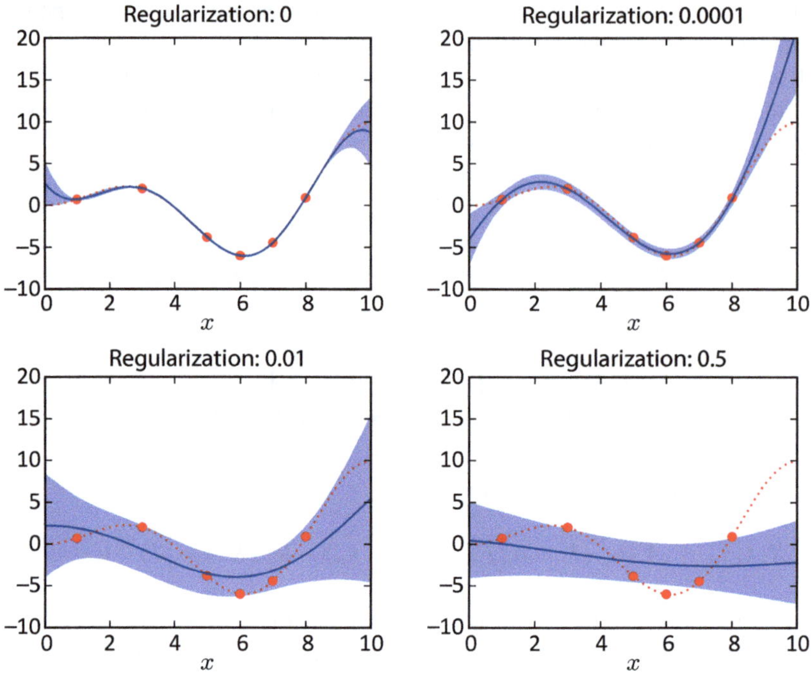

Fig. 5.4 Effects of the regularization parameter when learning from training data

Cartesian coordinates from $\mathbf{x} = [x, y]$ to include a radius $r = \sqrt{x^2 + y^2}$: $\mathbf{x} = [x, y, r]$.

Now, applying the Gaussian process regression model over this space will result in a model that takes radial variation patterns into account. In Fig. 5.6, we show the impact on the prediction outcomes.

5.2.5 *Model Selection and Adaptation of Hyperparameters*

In Sect. 5.2.1, we have shown the training of GP models with a fixed covariance (kernel) function, which is a squared exponential kernel specified in (8) with fixed characteristic length scale value l. However, in many practical applications the default kernel function, as well as the default hyperparameters, is not appropriate for the particular case study. While some properties such as stationarity of kernel function may be easy to determine from the context, we typically have only rather vague information about other properties, such as the value of length scale parameter l and regularization parameter σ_n^2. Thus, it is very important to automatically adapt GP models to fit each particular application. In this section, we introduce a

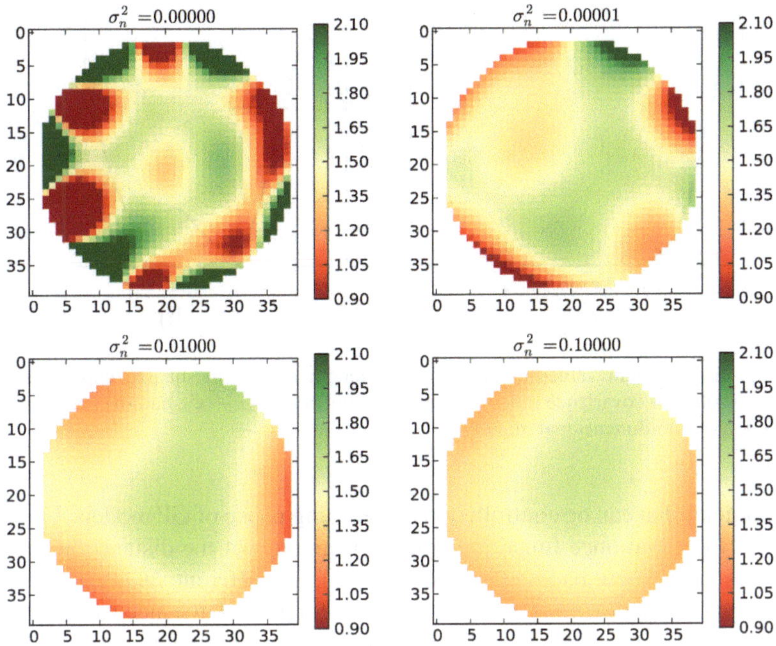

Fig. 5.5 Effects of the regularization parameter when modeling wafer variation

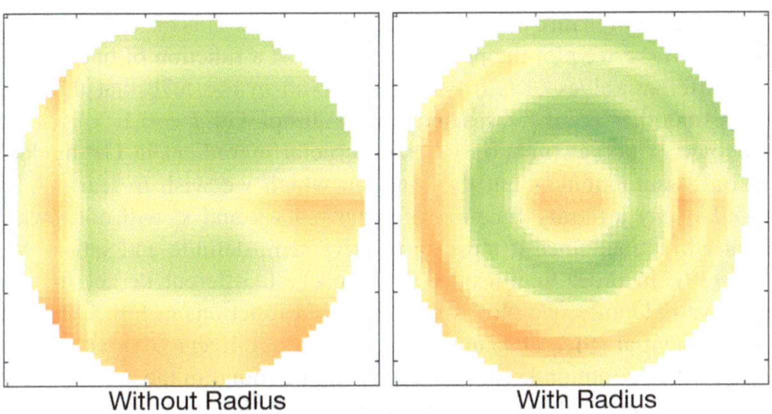

Fig. 5.6 Radial modeling

method that addresses the model selection and adaptation of hyperparameters in GP modeling.

In Sect. 5.2.3, we discussed the impact of the regularization parameter on the spatial model. In order to illustrate the impact that other hyperparameters have on the predictive function, we first use the squared exponential kernel function, as specified in (5.6). In this function, the characteristic length scale parameter l is a

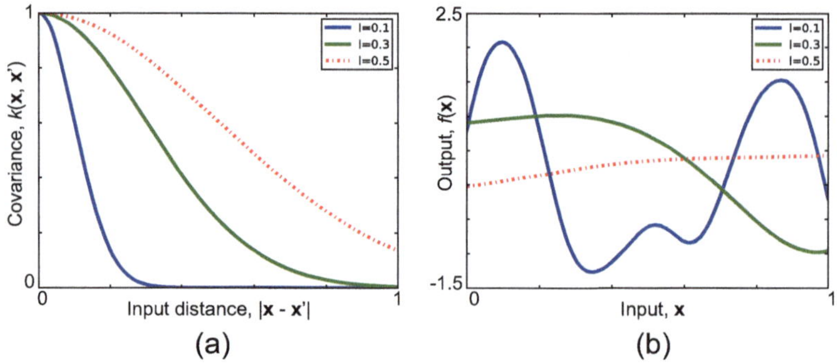

Fig. 5.7 Using different values of characteristic length scale l in squared exponential kernel function, we plot (**a**) covariance function value as a function of $|\mathbf{x} - \mathbf{x}'|$, and (**b**) generative output as a function of one-dimensional input $\mathbf{x}$

free parameter that can be controlled in the learning phase of GP models. Figure 5.7a illustrates the covariance function value as a function of the distance between any two points in the input parameter space $|\mathbf{x} - \mathbf{x}'|$, for different values of l. Note that the covariance function value remains the same for a fixed distance value, regardless of the individual input values $\mathbf{x}$ and $\mathbf{x}'$. This is named *kernel stationarity property* and allows us to draw the kernel function as a function of the distance. As can be observed in Fig. 5.7a, as the value of l increases, the covariance function tends to be smoother, which will in turn affect the predictive output. This observation is justified by Fig. 5.7b, where we plot the predictive output as a function of one-dimensional input for different values of l. It can be observed in Fig. 5.7b that the predictive function is smoother as l increases and is very abrupt with $l = 0.1$.

The choice of kernel function is also a crucial ingredient in GP models, as it encodes our assumptions about the function which we wish to learn. As shown in Sect. 5.2.1, any arbitrary function of input pairs $\mathbf{x}$ and $\mathbf{x}'$ will not always be a valid kernel function, since it must be positive semi-definite and satisfy Mercer's condition [29]. In order to illustrate the impact of different kernel functions on the predictive outcomes, we plot the set of kernel functions in Fig. 5.8a–c, namely squared exponential (SE), absolute exponential (AE), linear (LI), and polynomial (POLY). A list of common kernel functions can be found in the section "Appendix 3: Summary of Commonly Used Kernel Functions." Note that LI and POLY kernel functions are plotted in 3-D space as a function of $\mathbf{x}$ and $\mathbf{x}'$, since these are *nonstationary* kernel functions. As can be observed in Fig. 5.8, the AE kernel function has a smaller value than SE kernel when input distance is small; as a consequence, proximal data are less correlated when building GP models using AE kernel function. The generative output in Fig. 5.8d can further justify this observation, where the output generated using AE is very abrupt, while SE kernel function has a very smooth output. We also observe that the smoothness of the output using LI and POLY kernel functions is in between that obtained using SE and AE.

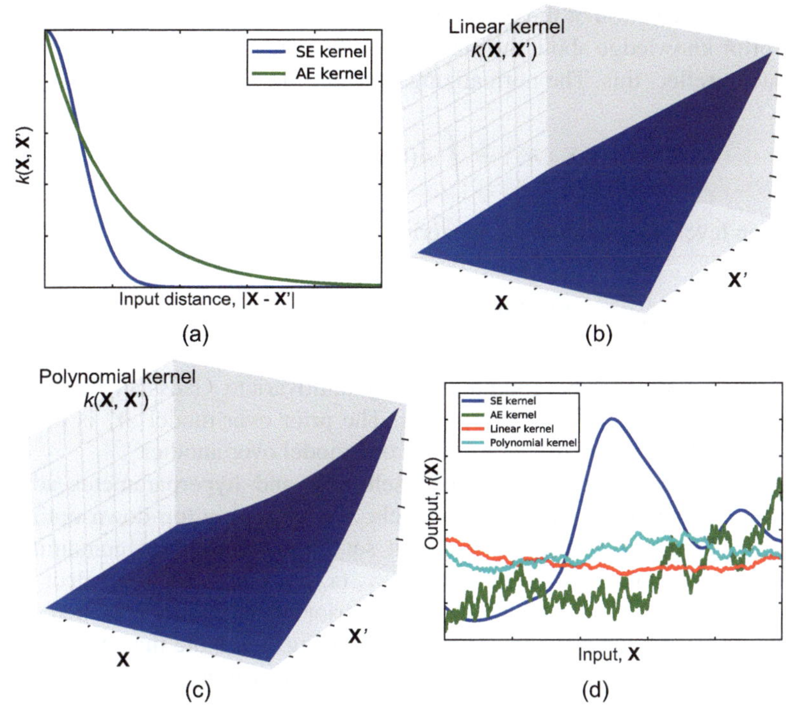

Fig. 5.8 Impact of choice of kernel function on the generative outcomes, (**a**) squared exponential (SE) and absolute exponential (AE) kernel function, (**b**) linear (LI) kernel function, (**c**) polynomial (POLY) kernel function, and (**d**) generative output using different kernel function

As discussed before, the choice of kernel function and hyperparameters has a significant impact on the predictive output using GP models; thus, learning an optimal model is crucial in building accurate spatial correlation models. In order to effectively learn the GP models, we first consider a hierarchical model that includes all the parameters to be optimized, as suggested by Rasmussen and Williams [22]. At the lower level, we have the hyperparameter vector $\boldsymbol{\theta}$, which can include the length scale parameter l, regularization parameter σ_n^2, and other model parameters [33]: $\boldsymbol{\theta} = (l, \ldots, \sigma_n^2)$. At the higher level, we have a (discrete) set of possible model structures, i.e., kernel functions, $\mathscr{H}_i, i = 1, \ldots, n_k$ under consideration, where n_k is the number of considered kernels functions. Then, we can express the posterior over the hyperparameter vector $\boldsymbol{\theta}$ using Bayesian inference:

$$p(\boldsymbol{\theta}|\mathbf{t}, X, \mathscr{H}_i) = \frac{p(\mathbf{t}|X, \boldsymbol{\theta}, \mathscr{H}_i)\, p(\boldsymbol{\theta}|\mathscr{H}_i)}{p(\mathbf{t}|X, \mathscr{H}_i)} \tag{5.13}$$

where $p(\mathbf{t}|X, \boldsymbol{\theta}, \mathscr{H}_i)$ is the *likelihood* which can be computed using multivariate Gaussian distribution, $p(\boldsymbol{\theta}|\mathscr{H}_i)$ is the *hyper prior* (the prior for the hyperparameters), and $p(\mathbf{t}|X, \mathscr{H}_i)$ is the normalizing constant. The prior $p(\boldsymbol{\theta}|\mathscr{H}_i)$ encodes as a probability distribution of the parameters independently of the data. If we have only vague prior knowledge about the parameters, then the prior distribution is chosen to be broad to reflect this. The normalizing constant $p(\mathbf{t}|X, \mathscr{H}_i)$ is given by:

$$p(\mathbf{t}|X, \mathscr{H}_i) = \int p(\mathbf{t}|X, \boldsymbol{\theta}, \mathscr{H}_i)p(\boldsymbol{\theta}|\mathscr{H}_i)d\boldsymbol{\theta} \tag{5.14}$$

At the top level, we compute the posterior for the model:

$$p(\mathscr{H}_i|\mathbf{t}, X) = \frac{p(\mathbf{t}|X, \mathscr{H}_i)p(\mathscr{H}_i)}{p(\mathbf{t}|X)} \tag{5.15}$$

where $p(\mathbf{t}|X, \mathscr{H}_i)$ can be computed using multivariate Gaussian distribution as before, $p(\mathbf{t}|X) = \sum_i p(\mathbf{t}|X, \mathscr{H}_i)p(\mathscr{H}_i)$. The prior over model $\mathscr{H}_i$ is often taken to be flat, so that a priori we do not favor one model over another.

Figure 5.9 summarizes the model selection and hyperparameter adaptation procedure for learning optimal GP models. We propose a top-down optimization procedure. First, we consider a hold-out set S_1 of n_1 samples containing input and target data points: $S_1 = \{(\mathbf{x}_1, t_1), \ldots, (\mathbf{x}_{n1}, t_{n1})\}$. Then, we select the j-th model as the optimal one such that the posterior of the model specified in (5.15) is maximized, as shown in the second block of Fig. 5.9. Note that since we have a rather limited set of possible kernel functions in the search space, we only have to assess explicitly the posterior of the model for each considered kernel function by setting the corresponding hyperparameters to their nominal values.

Once the optimal kernel function is chosen, we can further tune the model by optimizing the hyperparameter vector $\boldsymbol{\theta}$, as shown in the third block of Fig. 5.9. The optimization consists of searching in the hyperparameter space such that the posterior of the hyperparameter $p(\boldsymbol{\theta}|\mathbf{t}, X, \mathscr{H}_i)$ is maximized. As shown in (5.14), this step requires the evaluation of the integral over $\boldsymbol{\theta}$, which may not be analytically tractable in practice. Several approximation methods can be used to compute the integral in (5.14) such as Markov chain Monte Carlo [11]. As an approximation,

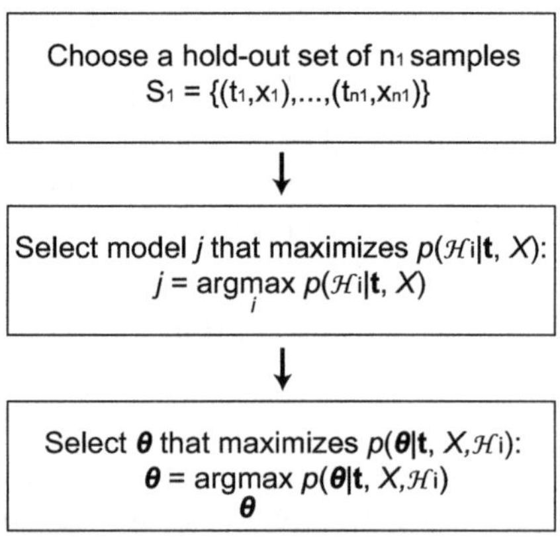

Fig. 5.9 Model selection and hyperparameter adaptation procedure for learning optimal GP models

one can also shy away from using the hyperparameter posterior in (5.13), and instead maximize the likelihood $p(\mathbf{t}|X, \boldsymbol{\theta}, \mathcal{H}_i)$ w.r.t. the hyperparameters, $\boldsymbol{\theta}$. This approximation is known as type II maximum likelihood (ML-II) [22]. Let $\mathbf{t}_l$ denote the set of target values in the hold-out set S_l, $\mathbf{t}_l = \{t_1, \ldots, t_{nl}\}$. Then, the likelihood function can be written as:

$$\mathcal{L}(\boldsymbol{\theta}|\mathbf{t}_l, X, \mathcal{H}_i) = p(\mathbf{t}_l|X, \boldsymbol{\theta}, \mathcal{H}_i) \tag{5.16}$$

Finally, we use the maximum-likelihood estimation (MLE) to choose an optimal value of $\boldsymbol{\theta}$:

$$\hat{\boldsymbol{\theta}} = \arg\max_{\boldsymbol{\theta}} \big(\mathcal{L}(\boldsymbol{\theta}|\mathbf{t}_l, X, \mathcal{H}_i)\big) \tag{5.17}$$

Several methods exist in order to estimate the optimal value of $\hat{\boldsymbol{\theta}}$. Authors in [22] proposed a marginal likelihood gradient method by computing the partial derivatives of the marginal likelihood w.r.t. the hyperparameter. This method is rather computationally intensive. In this work, we propose to use simulated annealing (SA) algorithm, which is very efficient for locating a good approximation to the global optimum, while maintaining reasonable computational time [6, 15]. The procedure of SA algorithm is depicted in Algorithm 1.

Algorithm 1 Simulated Annealing (SA) algorithm for choosing optimal $\boldsymbol{\theta}$

1: Randomly choose $\boldsymbol{\theta}$ between $[\boldsymbol{\theta}_l, \boldsymbol{\theta}_u]$ as initial value, where $\boldsymbol{\theta}_l/\boldsymbol{\theta}_u$ denotes the lower/upper bound vector for $\boldsymbol{\theta}$. The initial value is denoted by $\boldsymbol{\theta}_0$. Define the cooling schedule $T(t) = 1/(e^t)$, set $t = 1$
2: Compute the likelihood $\mathcal{L}(\boldsymbol{\theta}_0|\mathbf{t}_l, X, \mathcal{H}_i)$
3: Randomly choose $\boldsymbol{\theta}_t$ between $[\boldsymbol{\theta}_{t-1} - \mathbf{v}, \boldsymbol{\theta}_{t-1} + \mathbf{v}]$, where $\mathbf{v}$ is a vector of small values in $[\boldsymbol{\theta}_l, \boldsymbol{\theta}_u]$. Compute $\mathcal{L}(\boldsymbol{\theta}_t|\mathbf{t}_l, X, \mathcal{H}_i)$
4: **if** $\mathcal{L}(\boldsymbol{\theta}_t|\mathbf{t}_l, X, \mathcal{H}_i) \geq \mathcal{L}(\boldsymbol{\theta}_{t-1}|\mathbf{t}_l, X, \mathcal{H}_i)$ **then**
5: Accept $\boldsymbol{\theta}_t$ as the new optimal value
6: **else**
7: The probability of accepting $\boldsymbol{\theta}_t$ is
8: $\mathrm{P}_t = \exp\Big(-\big(\mathcal{L}(\boldsymbol{\theta}_{t-1}|\mathbf{t}_l, X, \mathcal{H}_i) - \mathcal{L}(\boldsymbol{\theta}_t|\mathbf{t}_l, X, \mathcal{H}_i)\big)/T(t)\Big)$
9: Update and assign $t = t + 1$
10: Repeat steps 3–9, until the assignments do not change

5.2.6 Handling Discontinuous Effect in GP Modeling

While an optimal choice of kernel function and hyperparameters can greatly improve the ability of the GP models to adapt different types of spatial patterns (e.g., nonlinear variation, noise-affected variation, and smooth or abrupt variation), some essential assumptions still must hold when applying the method. Indeed, the success

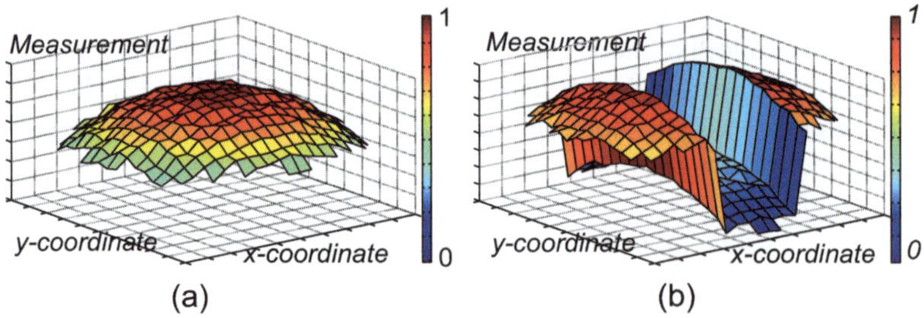

Fig. 5.10 Spatial wafer measurement data that (**a**) can be modeled efficiently by GP models and (**b**) cannot be modeled efficiently by GP models

of the GP models relies on the assumption that the generative function f underlying the spatial variation should be *continuous*, albeit affected by noise. This assumption can be seen in the structure of GP models: proximal data points are modeled as being highly covariant, and distant points are modeled with low covariance. An example of wafer measurement data that can be modeled efficiently by GP model is shown in Fig. 5.10a. As can be observed, despite the nonlinear behavior of the measurements, the GP models are able to capture the spatial correlation. Similar assumptions are made when applying the VP and EM approaches. For example, a small number of dominant DCT coefficients at low frequencies are assumed to represent most spatial patterns of process variations in VP models.

However, our experience with production data shows that these assumptions may not hold for certain measurements. For example, spatial features of certain test data on a wafer can involve localized effects such as discontinuous trends. These localized discontinuous effects may be caused by a variety of sources with the most prominent ones being the reticle shot and the multisite testing strategy. The first is part of the lithography processes which produces rectangular regions with discrete characteristics. The second refers to the parallelization of the probe testing that aims to reduce the overall time. This multisite testing strategy leads to systematic variations between the die tested simultaneously, due to minor variations in the measurement mechanisms of each site. Our experience shows that, for certain measurements and for certain periods of time, these localized discontinuous effects (a) dominate spatial variations on the wafer and (b) are stationary, i.e., most wafers have very similar spatial discontinuous patterns. Figure 5.10b illustrates an example of the simulated spatial discontinuous effects on a wafer measurement.

In order to efficiently handle discontinuous effect in GP modeling, the k-means clustering algorithm is used. A preliminary work that uses the k-means algorithm in modeling discontinuous effects can be found in [14]. Figure 5.11 shows an overview of the method, which consists of two main stages, namely pretraining and training. The objective of the first stage is to partition the wafer into k clusters, which reflect the k "levels" of wafer measurements induced by discontinuous effects. Partitioning of the wafer is accomplished by a k-means clustering algorithm applied on a wafer, for which all measurements for all die locations are explicitly collected.

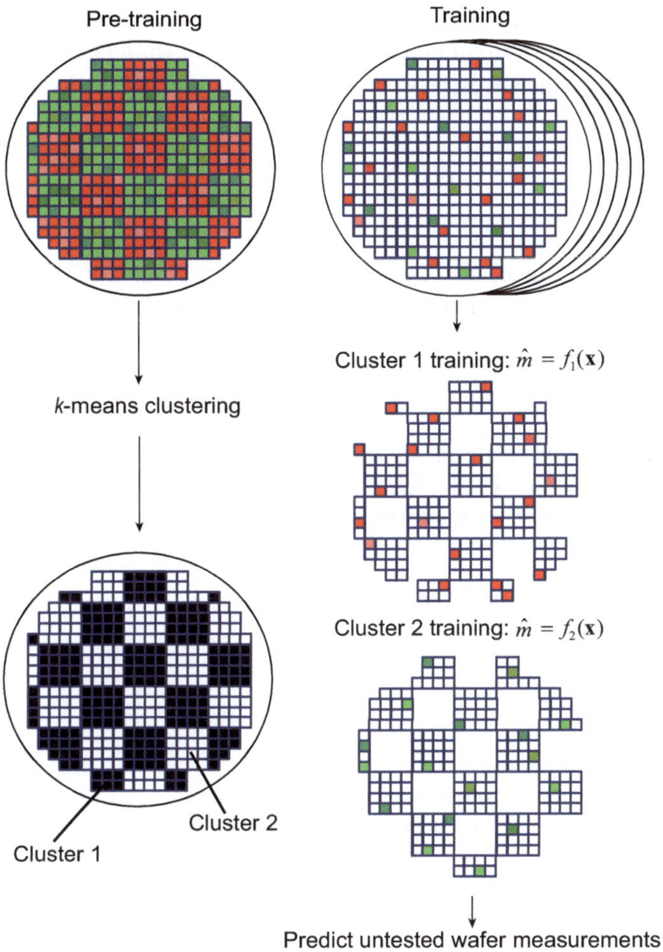

Fig. 5.11 Overview of proposed methodology

The left-hand side of Fig. 5.11 shows an example of k-means clustering applied to a simulated wafer map which has 2 clusters. We note that the clusters might be different for each measurement, yet we assume stationarity across wafers, so identifying the clusters is a one-time effort. Furthermore, the number of wafers needed to compute k-means clustering is typically very small.

The details of the partitioning procedure are explained as follows. Let the set:

$$M^i = \{m_1, m_2, \ldots, m_n\} \tag{5.18}$$

include the values of the i-th measurement on all die of a wafer, with m_j denoting the measurement on the j-th die and n denoting the total number of die which are to be clustered. The k-means clustering algorithm aims to partition M^i into k sets ($k \leq n$): $\{S_1, S_2, \ldots, S_k\}$ so as to minimize the expected *distortion* D, which is defined as the sum of squared distances between each observation and its dominating cluster mean:

$$D = \sum_j \|\bar{m}_{k(j)} - m_j\|^2 \tag{5.19}$$

where $\bar{m}_{k(j)}$ denotes the nearest cluster mean value for observation m_j. In this work, we use the most common iterative refinement technique to refine the choices of cluster means in order to reduce the distortion D. The technique is presented in detail in Algorithm 2 [19]:

Algorithm 2 k-Means clustering algorithm for discontinuous effects

1: Set k cluster means $\{\bar{m}_1, \bar{m}_2, \ldots, \bar{m}_k\}$ to random values.
2: Each measurement in M^i is assigned to the cluster with the nearest cluster mean. The assigned p-th cluster is denoted by S_p:

$$S_p = \{m_j : \|m_j - \bar{m}_p\|^2 \leq \|m_j - \bar{m}_q\|^2, \forall 1 \leq q \leq k\} \tag{5.20}$$

3: Compute the new cluster means:

$$\bar{m}_p = \frac{1}{n_p} \sum_{m_j \in S_p} m_j \tag{5.21}$$

where n_p is the number of observations in the p-th cluster.
4: Repeat steps 2 and 3 until the assignments do not change.

The question that naturally arises next concerns the choice of k, which is the central hyperparameter in the k-means clustering algorithm. Underestimating k would result in clusters that still contain discontinuous patterns, while overestimating k would reduce the amount of available data in each cluster during training of spatial correlation models. The authors of [21] conducted a very comprehensive comparative study of 30 methods for determining the number of clusters in data. Among the variety of examined methods, the approach suggested in [9] generally outperformed the others. This approach consists of choosing an optimal value for k by maximizing the between-cluster dispersion and minimizing the within-cluster dispersion. Formally, the optimal value for k is defined as [9]:

$$k = \underset{g}{\operatorname{argmax}}\ CH(g) \tag{5.22}$$

where $CH(g)$ is the Calinski and Harabasz index when g clusters are considered and is defined as:

$$CH(g) = \frac{B(g)(g-1)}{W(g)(n-g)} \tag{5.23}$$

where n is the total number of die on the wafer, $B(g)$ and $W(g)$ are the between- and within-cluster sums of squared errors computed as:

$$B(g) = \sum_{p=1}^{g} n_p(\bar{m}_p - \bar{m})(\bar{m}_p - \bar{m})^T \tag{5.24}$$

$$W(g) = \sum_{p=1}^{g}(\sum_{m_j \in S_p} (m_j - \bar{m}_p)(m_j - \bar{m}_p)^T) \tag{5.25}$$

where n_p denotes the number of samples in the p-th cluster, $\bar{m}_p$ denotes the cluster mean of the p-th cluster, and $\bar{m}$ denotes the mean of all measurement samples in M^i. Equation (5.22) allows us to automatically choose an optimal value for k for a particular measurement without making any assumptions about its discontinuity trends.

Once the k clusters are identified on the wafer for each particular measurement, we proceed to the second stage, where we capture spatial correlation within each cluster. In other words, we train GP models within each individual cluster as shown on the right-hand side of Fig. 5.11. Individually training and predicting measurements in each cluster allows us to avoid modeling spatial correlation on an entire wafer, which can lead to erroneous spatial models due to the discontinuous effects.

5.2.7 Progressive Sampling (GP-PS)

Selecting a sample of die locations that accurately reflect the spatial variation across a wafer is crucial in any wafer-level spatial correlation modeling method. To date, most methods rely on a set of randomly selected die locations. Attempts to guarantee sufficient coverage by employing a Latin hypercube sampling approach to choose random sample points over the entire wafer evenly have also been made [36]. Usually, the selection of the sample location on a wafer is performed once without taking into consideration any a priori knowledge of spatial variation patterns on the wafer. In contrast, an iterative progressive sampling approach can select the training samples that better represent the spatial variation pattern across a wafer. To achieve this, we leverage the ability of the GP model to provide a confidence level for all predicted samples in each iteration. Algorithm 3 outlines the progressive sampling approach [1].

In particular, the sampling procedure starts with n' samples randomly chosen from the wafer. Note that n' is set to be significantly smaller than the total number of samples, n, allowed in our budget for building the model. Using these samples, we build a GP model, and we predict the values along with the prediction confidence level at each unobserved die location. This confidence level is then used to guide our sampling at the next iteration, towards reducing the uncertainty and improving the accuracy of the GP model. Specifically, we identify a set of k locations for which the predictions have the highest uncertainty and lowest confidence level, we sample them to obtain the true values, and we then use them to augment our training sample. Note that the selection of these k new locations also uses their Euclidean distance as a metric, in order to distribute the new samples across many areas of low confidence. This progressive sampling process is repeated until a stopping criterion is reached. In Fig. 5.12, we illustrate an example of progressive sampling in 2 iterations. As can be observed, carefully selecting a new sample in the training set can significantly improve the accuracy of the spatial correlation model.

Algorithm 3 Progressive sampling of information-rich training locations

1: Randomly select n' samples on the wafer as initial training set: $S = \{\mathbf{x}_1|t_1, \cdots, \mathbf{x}_{n'}|t_{n'}\}$
2: Build spatial GP model using set S and predict values and confidence at unobserved die locations (set U)
3: **while** until k locations are added to the training set **do**
4: For each $\mathbf{x}_i$ in U, calculate $\mathbf{d}_i = min\{|\mathbf{x}_i - \mathbf{x}_j|^2, \forall \mathbf{x}_j \in S\}$
5: Select location $\mathbf{x}_i$ which has the highest variance and maximum Euclidean distance from current training set
6: Add $\mathbf{x}_i$ to the set S and remove it from set U and obtain corresponding true value $\mathbf{t}_{x_i}$
7: **end**
8: Augment the training set $S = \{S, \mathbf{x}_{h_1}|t_{h_1}, \cdots, \mathbf{x}_{h_k}|t_{h_k}\}$
9: Repeat steps 2–8, until stopping criterion is reached

The stopping criterion of Algorithm 3 depends on the application and prediction problem. There are two standard methods: (1) when the highest uncertainty of prediction drops below a given threshold, or (2) when a given budget of samples, n, is reached. The number of samples added in each iteration, k, is also problem-dependent.

5.2.8 Spatiotemporal Feature Inclusion (GP-ST)

During the manufacturing process, wafers processed together, such as those within the same lot, will exhibit similar intra-wafer spatial variation and will also exhibit time-dependent inter-wafer variation which would be beneficial to include in our correlation model. Thus, we extend the GP modeling over spatial coordinates to a joint spatiotemporal space, capturing our intuition that wafer-level spatial variance is also time-dependent. Essentially, we conjecture that *a single spatiotemporal model learned from samples across all wafers in a lot could be more accurate*

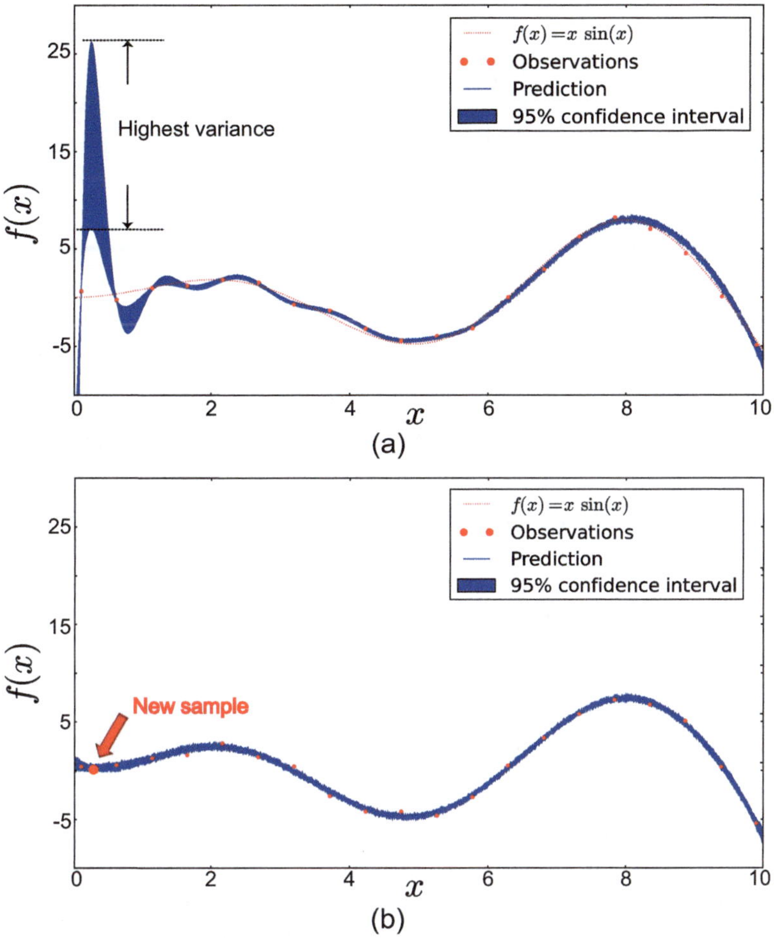

Fig. 5.12 Example of progressive sampling with prediction at (**a**) iteration i and (**b**) iteration $i + 1$

than individual models learned from the same samples for each wafer [1]. To accommodate this in our GP model, we can directly update our coordinates from $\mathbf{x} = [x, y]$ to include a time feature t:

$$\mathbf{x} = [x, y, t]$$

Applying Gaussian process regression over this space will result in a model that takes the time-dependent variation into account. In Fig. 5.13, we illustrate the concept of time-dependent wafer-level spatial variation. Evidently, the spatial variation exhibited by each of these three consecutively produced wafers is not only similar but also exhibits a temporal trend (i.e., a linear gradient that intensifies over time).

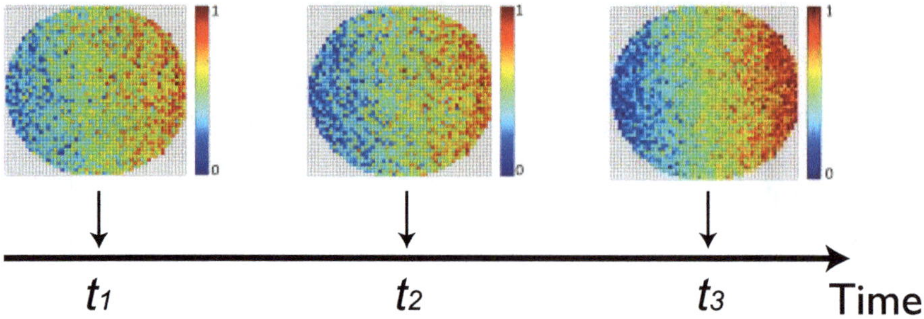

Fig. 5.13 Time-dependent spatial variation

5.2.9 Spatiotemporal GP w. Progressive Sampling (GP-ST-PS)

Combining progressive sampling with spatiotemporal correlation modeling is also a plausible direction. Specifically, it is possible that the accuracy of a spatiotemporal model can be improved by employing progressive sampling. Indeed, these two enhancements are orthogonal, in the sense that progressive sampling seeks to provide a training set which better reflects the underlying statistics of the problem, while spatiotemporal modeling explores the correlation which exists both within a wafer (spatial variation) and across wafers (time-dependent variation), to establish a more comprehensive model. In order to assist GP in learning time-dependent variation, we modify the sampling strategy such that the initial portion of the sampling budget covers common locations over all the wafers in a lot. Then, the progressive sampling method is applied to the entire lot as a whole for the remaining portion of the sampling budget. Again, our conjecture is that wafers in a lot have both spatial and temporal correlations which can be efficiently learned by a spatiotemporal GP model; hence, we train one GP model for all the wafers in a lot [1].

5.2.10 Considerations for Production Test Deployment

Spatial correlation models have primarily been seeking to achieve test-cost/test-time reduction at probe, by landing on a limited set of die locations and then using the correlation models to predict the performances of the unobserved die. The progressive sampling method outlined in Sect. 5.2.7 complicates this flow, since the wafer will need to remain on the probe equipment while the models are constructed and applied, and samples will need to be obtained iteratively through multiple passes. This iterative procedure may become overly time consuming and may eventually counterbalance the time reduction achieved by the spatial correlation model. Clearly, this trade-off between test time and prediction accuracy needs to be considered in order to decide whether progressive sampling makes sense in an HVM test.

Spatiotemporal modeling, on the other hand, as described in Sect. 5.2.8, brings about a different challenge. Specifically, after measurements on a sparse sample of die are obtained at probe, the wafer will leave the probe equipment without the ability to make decisions on the unobserved locations. Indeed, one will have to wait until all wafers have passed through the probe equipment so that the obtained samples can be used to build the spatiotemporal model, which will then be used to predict the unobserved measurements for all wafers in the lot. While the technical capabilities for coding this in modern ATE are available, if the overall confidence of the model is low and the rest of the locations on each wafer need to be explicitly measured, then the wafers will have to be subjected to a second test insertion.

5.2.11 High-Volume Manufacturing Yield Estimation

In the process of developing a new product and prior to high-volume manufacturing (HVM), die samples from early silicon wafers are obtained and characterized. The objective of such characterization is mainly the post-silicon design validation as well as HVM performance and yield estimation. Indeed, such tasks are crucial as they rely on a few die in order to make both manufacturing and marketing decisions that affect the production lifetime of a device. In reality, only few engineering wafers are available for such analysis, and only a small number of die are measured from each such wafer. Typically, only a few iterations of this early silicon are produced in a relatively short period prior to HVM. However, analysis of such die is extensive and is done in dedicated characterization labs without the time and throughput constraints and requirements of production testing. This setting lends itself naturally to the spatiotemporal Gaussian process with progressive sampling (GP-ST-PS) method. With this, it is possible to progressively sample die across available wafers and repeat the process every time new engineering wafers are received, in order to obtain a coherent picture of the performance distribution and expected yield of the product [1].

Essentially, given a small set of observations over a few wafers, the objective is to infer the probability density function (PDF) of the entire population from the limited observations available. In general, density estimation approaches are categorized into parametric and nonparametric. In parametric methods, assumptions about the form of density are made, while in the nonparametric no such assumptions are made. Herein, the latter will be used, since usually there is no information regarding the form of the density function. Two options are explored, namely a simple histogram-based estimation method and a more advanced kernel density estimation method.

5.2.11.1 Histogram with Random Sampling

A simple and straightforward method for density estimation from a small set of samples is a histogram. Creating a histogram requires the range of n observations to be divided into bins and the number of samples which fall in each bin to be counted. The probability density is, then, estimated as:

$$f'(\mathbf{x}) = \frac{n_b}{n * h} \; for \; l_b \leq x \leq u_b \tag{5.26}$$

where n_b is the number of observations in bin b, u_b and l_b are the upper and lower limits of bin b, and $h = u_b - l_b$.

Evidently, the accuracy of this estimate depends on the choice of the random sample and, more specifically, by how accurately it reflects the statistics underlying the observed phenomenon. In the context of our problem, the randomly chosen die may not adequately represent the statistical distribution across the available wafers, let alone across the entire HVM production, especially since no notion of spatial or temporal correlation is captured in a histogram.

5.2.11.2 Histogram with GP-ST-PS

To address this limitation, we can improve the quality of the sample that is used to generate the histogram by employing the spatiotemporal Gaussian process method with progressive sampling (GP-ST-PS). Instead of estimating the histogram using a random set of n samples from few available wafers, we employ the GP-ST-PS method to intelligently select these n samples in order to build an accurate spatiotemporal GP model which can subsequently be used to accurately predict the values of all unobserved die locations across the available wafers. In this sense, we increase the number and utility of available points (i.e., both observed and predicted) for building the histogram and estimating the probability density function, without increasing the cost of sampling as expressed by the total number of sampled locations. Assuming that the GP-ST-PS method provides accurate estimates by leveraging the spatiotemporal correlation across the available wafers, the new histogram should yield a more accurate density function than the one obtained from only the n observed die samples.

5.2.11.3 Kernel Density Estimation

While the histogram method is easy to implement and straightforward to interpret, it has several disadvantages. First, the size and the number of bins dominantly affects the estimation. Furthermore, the tails of the distribution, which have a very low likelihood of being represented in the sample, will remain insufficiently modeled. To address these issues, the concept of kernel density estimation (KDE) has been proposed for density estimation and enhanced synthetic population generation. In order to generate a large volume of synthetic data which accurately reflects the distribution of measurements in high-volume production from a small number of observations, we use a nonparametric KDE method [27]. This method relies on the estimation of the densities $\hat{f}(\mathbf{t})$, using the available observations $\mathbf{t}_i$, $i = 1, \cdots, M$, where M is the number of available samples used to build the density. We do not make any assumption regarding its parametric form (e.g., normal). Instead, we use nonparametric KDE, which allows the observations to speak for themselves. The kernel density estimate is defined as [25]:

$$\hat{f}(\mathbf{t}) = \frac{1}{M \times h^d} \sum_{i=1}^{M} K_e\left(\frac{1}{h}(\mathbf{t} - \mathbf{t}_i)\right) \tag{5.27}$$

where h is a parameter called bandwidth, $d = n_m$ is the dimension of $\mathbf{t}$, and $K_e(m)$ is the Epanechnikov kernel:

$$K_e(m) = \begin{cases} \frac{1}{2}c_d^{-1}(d+2)(1 - m^T m) & \text{if} \quad m^T m < 1 \\ 0 & \text{otherwise} \end{cases} \tag{5.28}$$

and $c_d = 2\pi^{d/2}/(d \cdot \Gamma(d/2))$ is the volume of the unit d-dimensional sphere. The kernel density estimate can be interpreted as the normalized sum of a set of identical kernels centered on the available observations for the 1-dimensional case. The bandwidth h corresponds to the distance between the center of the kernel and the kernel's edge, where the kernel's density becomes zero.

In this work, we use an adaptive version of (5.27). In particular, we allow the bandwidth h to vary from one observation $\mathbf{t}_i$ to another, allowing larger bandwidths for the observations that lie at the tails of the distribution. The adaptive kernel density estimate is defined as [25]:

$$\hat{f_\alpha}(\mathbf{t}) = \frac{1}{M} \sum_{i=1}^{M} \frac{1}{(h \cdot \lambda_i)^d} K_e\left(\frac{1}{h \cdot \lambda_i}(\mathbf{t} - \mathbf{t}_i)\right) \tag{5.29}$$

where the local bandwidth factors λ_i are defined as:

$$\lambda_i = \{\hat{f}(\mathbf{t}_i)/g\}^{-\alpha}, \tag{5.30}$$

$\hat{f}(\mathbf{t}_i)$ is the pilot density estimate given in (5.27), g is the geometric mean:

$$\log g = M^{-1} \sum_{i=1}^{M} \log \hat{f}(\mathbf{t}_i) \tag{5.31}$$

and α is a parameter which controls the local bandwidths. The larger the α, the larger the diagnostic measurement space where the density $\hat{f}(\mathbf{t})$ is nonzero.

Once the probability density $\hat{f}(\mathbf{t})$ is estimated, we can sample $\hat{f}(\mathbf{t})$ to generate a large synthetic dataset $s_n = \{\mathbf{t}_1, \ldots, \mathbf{t}_{M'}\}$, $M' \gg M$, which will better reflect the distribution tails.

Of course, in the context of our problem, KDE can be used either with the initial n die samples as the starting point (i.e., KDE with random sampling) or with the enhanced sample as predicted by the GP-ST-PS method (KDE with GP-ST-PS sampling). The expectation here is that GP-ST-PS method will provide a better starting point for KDE, which will ultimately lead to a more accurate estimation of the probability density.

5.2.12 Prediction Evaluation

To evaluate effectiveness of predictive models, we compute the mean absolute percent error across all predictions:

$$\epsilon_{jh} = \frac{1}{n_{test}} \sum_{i=1}^{n_{test}} \left| (\hat{m}_i^{(h)} - m_i^{(h)}) / m_i^{(h)} \right| \tag{5.32}$$

where ϵ_{jh} represents the mean percent error of predicting the h-th measurement for all unmeasured die locations on a particular wafer j, and n_{test} denotes the number of predicted die locations on the j-th wafer. Then, we can summarize the mean prediction error over all considered wafers as:

$$\epsilon_h = \frac{1}{N_{wafers}} \sum_{i=1}^{N_{wafers}} \epsilon_{ih} \tag{5.33}$$

where N_{wafers} denotes the number of considered wafers.

To gain insight into the prediction outcome, it is also worthwhile to compute the Test Escape (TE) and Yield Loss (YL) incurred by applying the spatial correlation models. For a particular measurement, let the indicator functions $I_1^{(i)}$ and $I_2^{(i)}$ be equal to "1" if the predicted value of the i-th die location passes/fails its specification, while the actual value fails/passes the specification, and let $I_1^{(i)}$ and $I_2^{(i)}$ be equal to "0" otherwise. Then, the overall TE and YL are defined as:

$$\hat{TE} = \frac{1}{N} \sum_{i=1}^{N} I_1^{(i)} \tag{5.34}$$

$$\hat{YL} = \frac{1}{N}\sum_{i=1}^{N} I_2^{(i)} \tag{5.35}$$

where N is the number of predicted die locations on all wafers.

5.3 Applications

Having talked about the theoretical underpinnings of the Gaussian process-based modeling, we will now focus on a number of actual applications that seek to reduce test cost, improve test quality, and predict yield. Evaluation of all these applications has been performed using industrial datasets, containing thousands of devices each.

5.3.1 *Spatial Correlation Modeling of E-Test Measurement*

The first application involves spatial correlation modeling of e-test data collected from industry high-volume manufacturing. E-test measurements are a set of process characterization parameters collected from scribe line test structures [20]. These test structures are drawn in the areas of the wafer that are destroyed during wafer dicing, as shown in Fig. 5.14, and therefore can only be measured at wafer level, before dicing. A subset of e-test measurements may be typically collected from these structures after completing a layer or two of metallization, and the remainder are collected later, during wafer acceptance testing. The e-test structure is often associated with several die within a reticle, as it is typically drawn to the full height of the reticle. For example, the illustration of Fig. 5.14 shows a 2×2 die reticle and, consequently, the e-test structure spans two die in the y-dimension. The e-test measurements collected from these scribe-line structures are designed to capture a very broad set of process statistics and, generally, include parameters such as:

- V_{th}, T_{ox}, G_m, I_{off}, and measurements for all types of NMOS/PMOS transistors (e.g., high V_{th} and low V_{th})
- N-well/poly resistor measurements
- Diode V_f and parasitics
- Capacitor unit cell measurements and parasitics
- Inductor L and Q and parasitics
- General parasitics
- Physical process dimensions

These types of measurements capture a great deal of information regarding the health of the process and particular wafers. However, most semiconductor manufacturers are collecting only a few samples of these measurements from around the wafer. As shown in Fig. 5.15, the e-test sites can be sparsely measured with

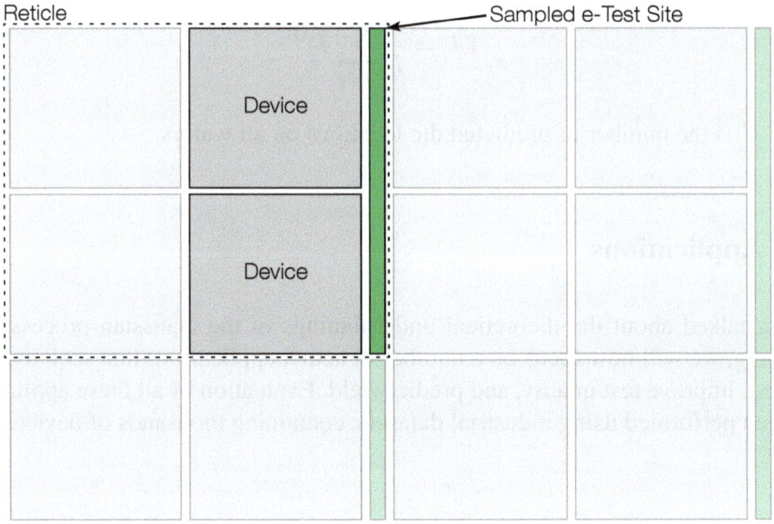

Fig. 5.14 e-Test measurement site diagram

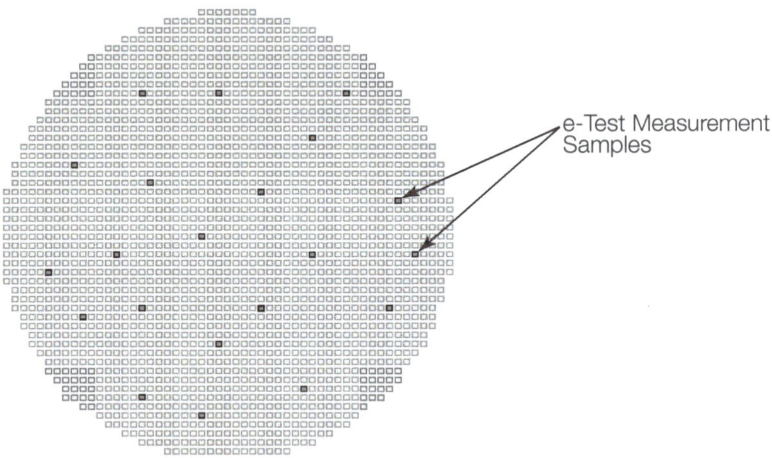

Fig. 5.15 e-Test measurements

sampling rates of few percentage points being the standard practice. This has effectively limited the scope of statistical models that can be applied using this data to perform wafer-level correlation or zonal analysis of the wafer. By extrapolating the sparsely sampled e-test measurements to every die location on a wafer, we enable test engineers to examine e-test measurement outcomes at the die level and make no assumptions about wafer-to-wafer similarity or stationarity of process statistics over time.

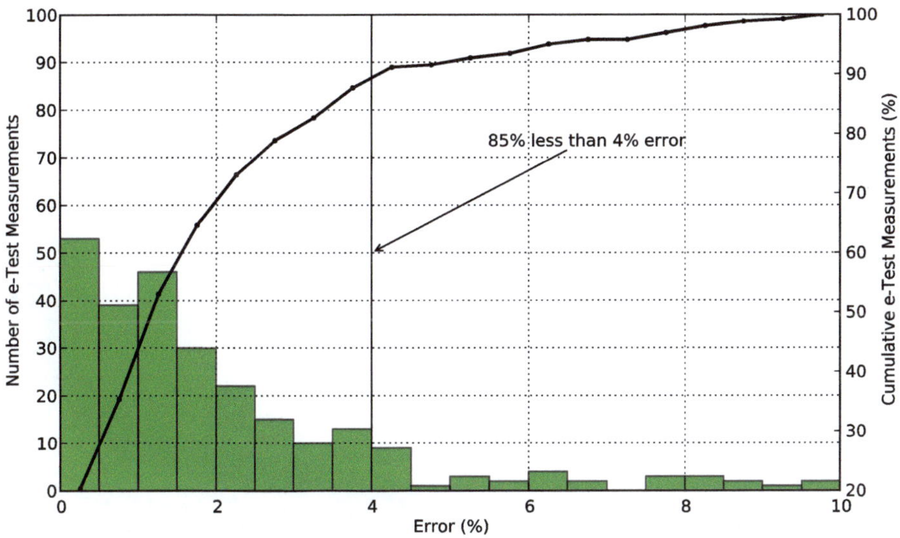

Fig. 5.16 Virtual probe prediction error across all wafers

The dataset that was used for the experimental results presented here has in total 8440 wafers. Each wafer has 269 e-test measurements collected from 21 sites sampled across the wafer. Increasing the number of sites per wafer would provide more information about the health of the process and particular wafers, at the expense of longer testing time. The solution presented here bridges this gap by enhancing the granularity of information without affecting the time spent testing each wafer.

Leave-one-out cross-validation was used to characterize the prediction error at each of the 21 sites. Specifically, we train the spatial model using 20 available sites, predict on the 21st site, and compute the mean prediction error across all wafers using Eq. (5.33). Then, we repeat the process for every e-test site, leaving out a single observation as a test set and training on the rest to produce an estimate for the test observation, and the mean cross-validation error across all 21 sites was collected for each e-test measurement.

5.3.1.1 Virtual Probe

As a baseline reference, let's first provide experimental results for virtual probe (VP) [10]. In Fig. 5.16, we present a histogram of mean prediction errors across all wafers. The black line overlaid on the histogram represents the cumulative percent of e-test measurements included in each successive bin. Virtual probe performs quite well, with more than 85% of the e-test measurements realizing less than 4% mean prediction error.

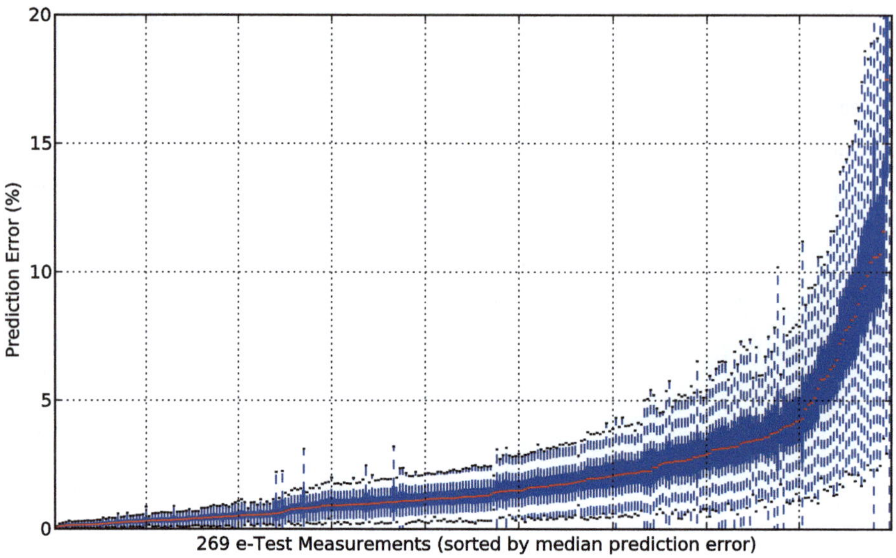

Fig. 5.17 Virtual probe prediction error for each e-test measurement

In Fig. 5.17, we present an overview of the prediction errors with 10–90% error bars shown for all 269 e-test measurements, sorted by median VP prediction error. Each index on the x-axis corresponds to a single e-test measurement, and the y-axis shows the prediction error in percent incurred by VP.

5.3.1.2 Proposed Method: Gaussian Process Models

In Fig. 5.18, we display a histogram of the mean e-test measurement prediction errors across all wafers. As can be seen from the figure, the Gaussian process model prediction errors are even lower than VP, with more than 90% of e-test measurement predictions below 4% error.

In Fig. 5.19, we present the GP model prediction errors with 10–90% error bars. Again, a qualitative comparison to Fig. 5.17 shows that the errors are generally lower and the error bars tighter than for VP. The error ranges are quite small across the majority of e-test measurements, demonstrating that the variance of the errors is low over the entire set of 8440 wafers. Importantly, this shows that the models are insensitive to process shifts over time, a result that is attributable to the fact that we train and deploy our models on a per-wafer basis.

Figure 5.20 presents a zoomed-in version of Fig. 5.19 with the best-predicted 30 measurements, e.g., the left-hand side of Fig. 5.19. The x-axis shows the 30 e-test parameters; these include a mixture of diode, capacitor, and NMOS/PMOS transistor parameters. The y-axis is, again, percent error for these 30 measurements,

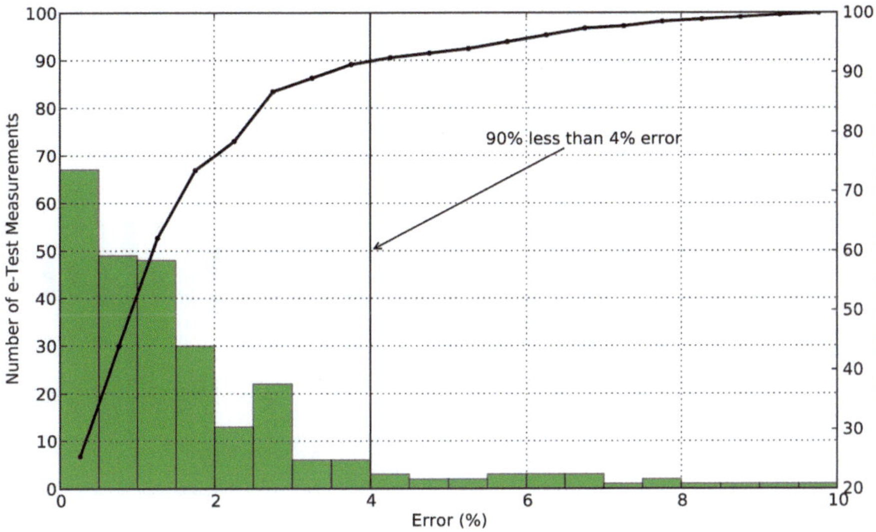

Fig. 5.18 Gaussian process model prediction error across all wafers

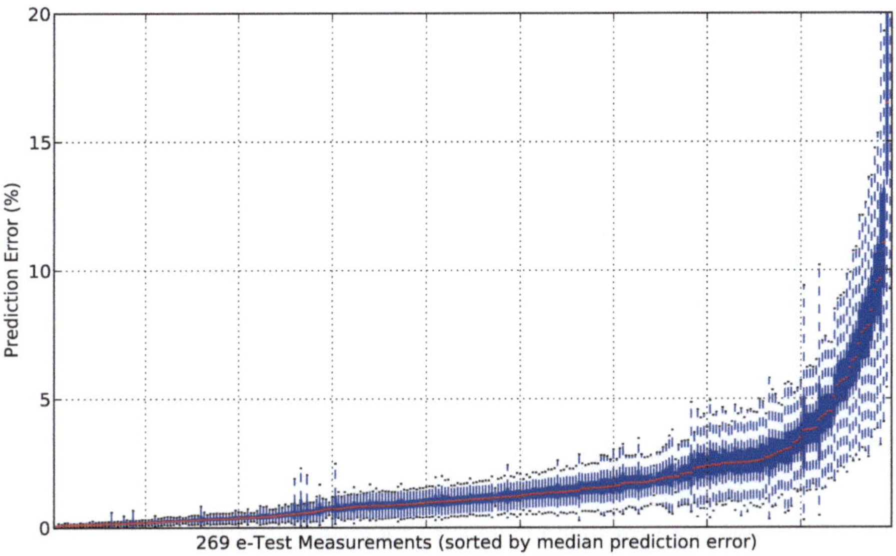

Fig. 5.19 Gaussian process prediction error for each e-test measurement

and the mean prediction error is less than 0.2%, indicating that we can very successfully interpolate these parameters across the surface of the wafer.

In Fig. 5.21, we present another zoomed-in version of Fig. 5.19, in this case with the worst-predicted 30 e-test measurements, or the right-hand side of Fig. 5.19. The parameters shown here are the most challenging for us to predict with our GP

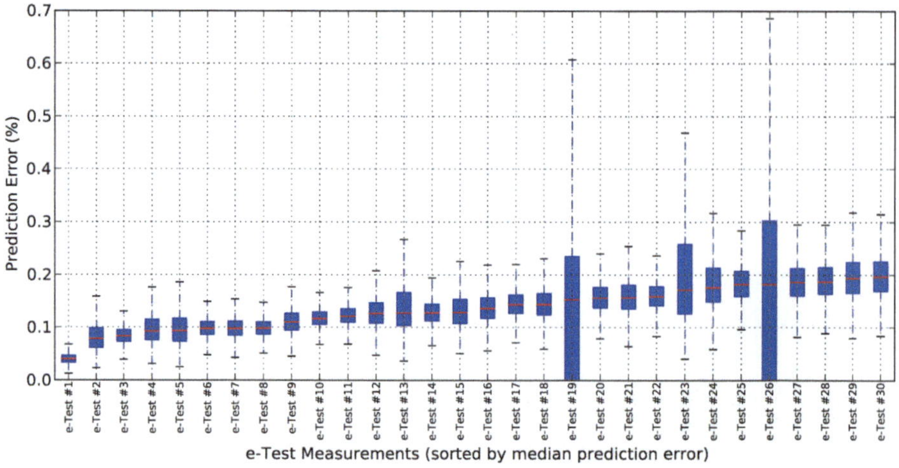

Fig. 5.20 Prediction error for best 30 e-test measurements

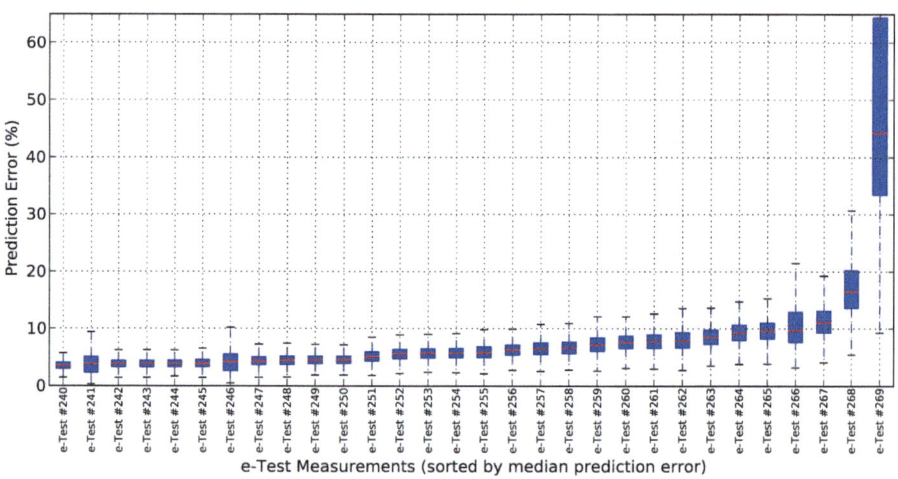

Fig. 5.21 Prediction error for worst 30 e-test measurements

models, in some cases reaching more than 40% prediction error. These parameters are largely comprised of: (a) G_{ds} drain–source conductance measurements of various transistors, and (b) various resistance measurements. Both tend to have high cross-wafer or even die-to-die variation. Consequently, the high prediction error for these parameters is relatively unsurprising. However, for most of these worst-case prediction errors, the error is under 10%, which is still within an acceptable range for most use cases. A possibility to improve the prediction errors for these parameters is to increase the number of samples during the training.

5.3.1.3 Comparison to Virtual Probe

In Fig. 5.22, we present a comparison of the two methodologies, with VP set as the baseline at 0%. The proposed methodology consistently outperforms VP by an average of 0.5%, and in some cases, by almost 5%. The overall mean prediction error of VP across all e-test measurements and all wafers is 2.68%, and the overall mean prediction error for Gaussian process-based spatial models is 2.09%.

A tabular comparison of the proposed GP model approach versus VP is provided in Table 5.1, with overall mean error reported alongside mean training and prediction time per wafer across all e-test measurements. The timing measurements were collected on a 2010 Core i5 2.4 GHz, and represent the mean total time required to construct and predict with all 269 × 21 = 5649 models for a given wafer. Note that the proposed methodology consistently has a lower error than VP, while incurring an order of magnitude less runtime to construct and evaluate the predictive models.

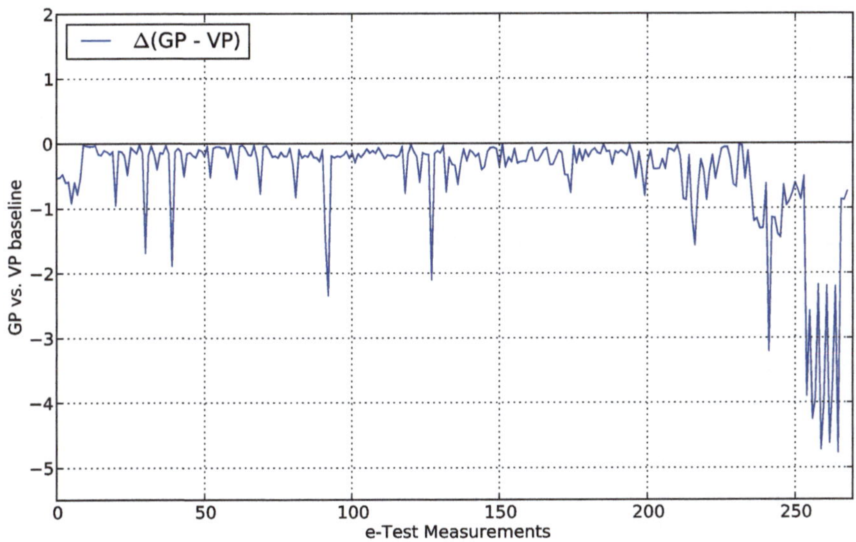

Fig. 5.22 Comparison between Gaussian process regression and virtual probe

Table 5.1 Virtual probe and Gaussian process models comparison

Method	Overall mean percent error	Avg. running time (per wafer)
Virtual probe	2.68%	116.2 s
Gaussian process model	2.09%	16.43 s

5.3.2 Spatial Correlation Modeling of Probe-Test Specification Measurements

The second application presented here involves spatial correlation modeling of probe-test specification measurements of an RF transceiver with multiple radios, built in a 65-nm technology. The dataset contains a total of 3406 wafers, each of which has approximately 2000 devices, with 71 probe-test specification measurements collected on each device.

Discontinuous spatial patterns are present for specific measurements in this case study. Thus, we apply the k-means clustering algorithm shown in Algorithm 2, to partition the wafer into k distinct regions when a measurement exhibits discontinuous spatial effects. The number of clusters, k, and the shape of the clusters are computed as explained in Sect. 5.2.6 during the pretraining stage, *using only the first wafer*. The minimum and maximum values of k obtained for the 71 measurements are 2 and 5, respectively, with an average of 3.9. Once the k clusters are known, we use a randomly chosen training sample of 20 devices from each cluster in order to train the spatial correlation models. Thus, a total number of $20 \times k$ devices are used for training in each wafer. The trained models are then used to predict the untested probe-test outcomes at the remaining die coordinates, and the mean prediction error across all wafers is computed through Eq. (5.33).

5.3.2.1 Stationarity Verification of Discontinuous Spatial Patterns

The procedure outlined above relies on the assumption that the discontinuous spatial patterns are stationary over time, i.e., the clusters for a given measurement remain the same across wafers. To illustrate that this is, indeed, the case, Fig. 5.23 shows the wafer maps of a measurement exhibiting discontinuous patterns, on randomly chosen wafers from each of the four quarters of the total 3406 wafers. As can be observed, the discontinuous patterns remain very similar across wafers. To further investigate this issue, we performed the following experiment: for each of the four wafers shown in Fig. 5.23, we ran the k-means clustering algorithm and used the obtained clusters, instead of the ones learned from the first wafer, to build the correlation models. What we found is that the number of clusters remained the same (i.e., 4) and that for the vast majority of die, the cluster that they ended up belonging to was the same. Most importantly, as shown in Table 5.2, the differences in error rate, TE, and YL were insignificant, corroborating stationarity.

5.3.2.2 Prediction Errors Using the Proposed Approach

Figure 5.24 summarizes the prediction error, with 10–90% error bars shown, for each of the 71 measurements. As can be observed in Fig. 5.24, the prediction error

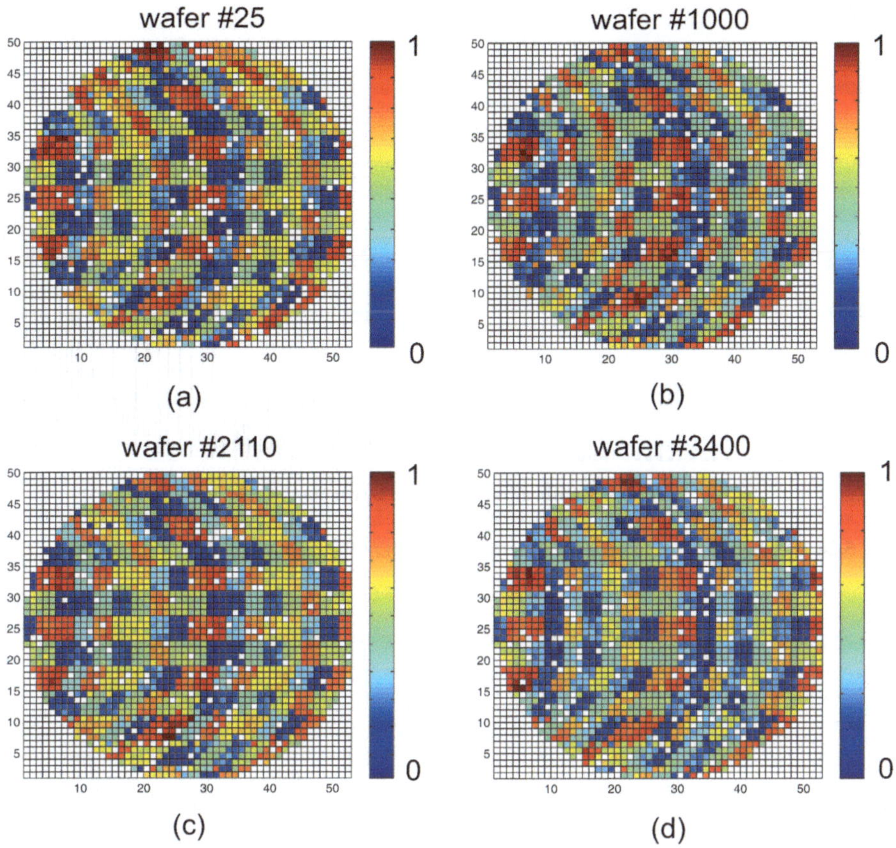

Fig. 5.23 Wafer maps of measurement 69 sampled randomly in the: (**a**) 1st, (**b**) 2nd, (**c**) 3rd, and (**d**) 4th quarter of the 3406 wafers

Table 5.2 Stationarity verification

	Wafer # 25		Wafer # 1000		Wafer # 2110		Wafer # 3400	
Clustering on wafer	#1	#25	#1	#1000	#1	#2110	#1	#3400
ϵ (%)	4.9	4	4.4	3.3	3.5	3.3	4.4	4.2
TE	0.03	0.03	0.02	0.02	0.03	0.03	0.04	0.03
YL	0	0	0.001	0.001	0	0	0.001	0

for most measurements is below 5%, verifying the excellent capability of the method in capturing spatial wafer correlation.

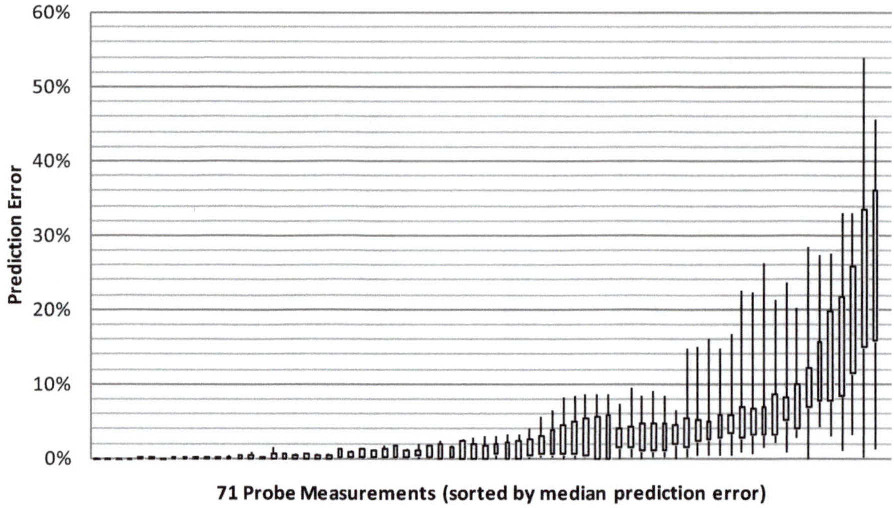

Fig. 5.24 Prediction error of proposed method for each probe measurement

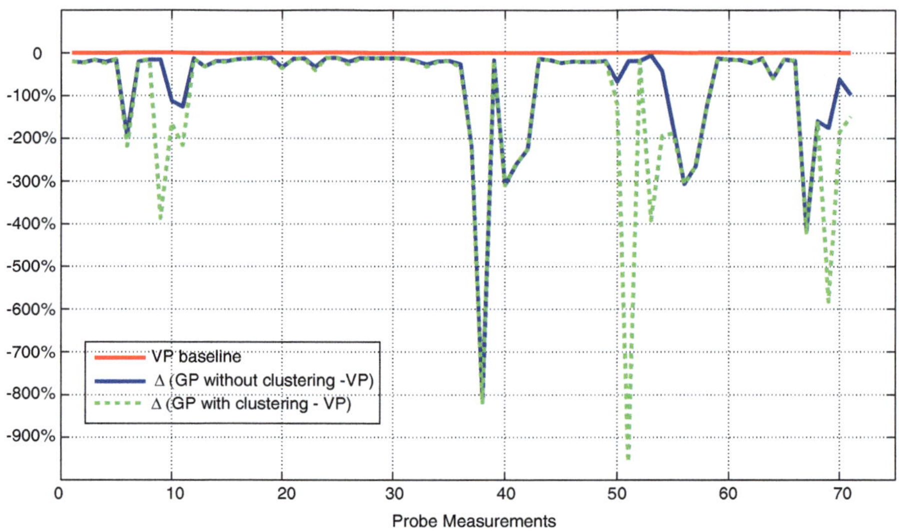

Fig. 5.25 Prediction error comparison of VP, GP without clustering and GP with clustering

5.3.2.3 Comparison to Existing Approaches

In Fig. 5.25, we compare the prediction error, of the proposed approach to the error of the VP [10, 34–36] and the GP models [16] without clustering, using the same number of training samples per wafer and setting the VP error as the baseline at 0%. As can be observed, the GP approach without clustering outperforms the VP approach, and GP with clustering achieves a significant further improvement in prediction error for several measurements. The measurements that enjoy a substantial improvement in prediction error are those which exhibit spatial discontinuous effects, yet we stress that even for the measurements which show no discontinuous patterns, GP with clustering performs at least as well as GP without clustering.

Figure 5.26 shows the prediction wafer map of measurement 11 which exhibits discontinuous effects. Figure 5.26a shows the normalized actual wafer map, and Fig. 5.26b/c/d show the prediction wafer map using the GP model with clustering/VP model/GP model without clustering, respectively. As can be observed, the discontinuous effect is correctly captured by the GP model with clustering, while the VP and GP models without clustering are not effective when such discontinuous patterns exist on the wafer. This observation is further supported by the large improvement in percentile prediction error.

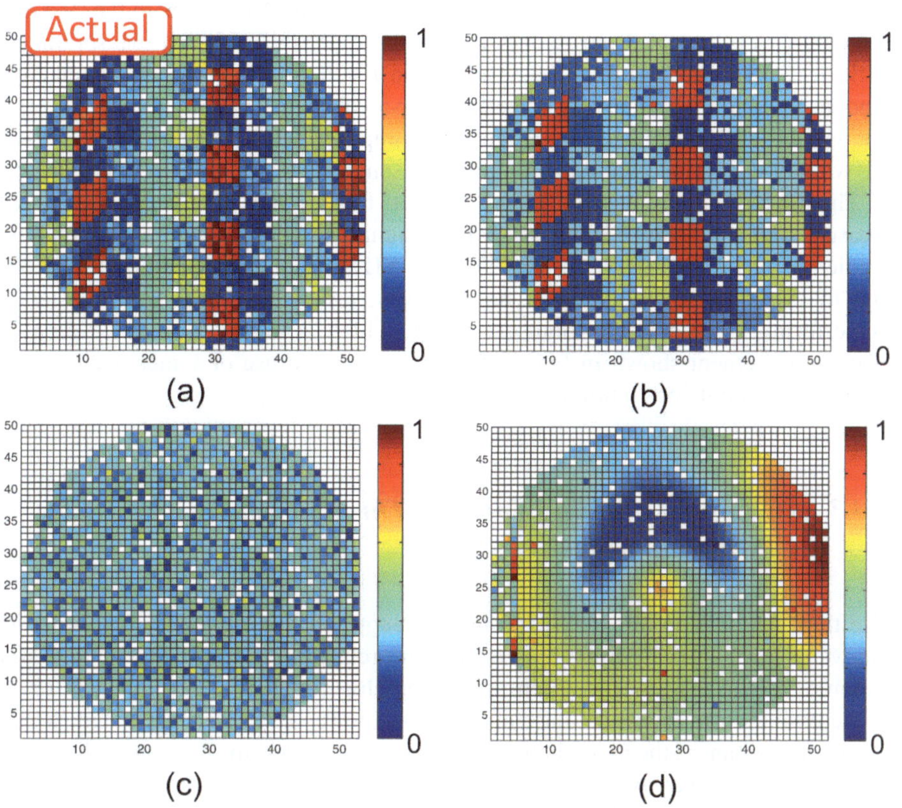

Fig. 5.26 Prediction of measurement 11: (**a**) Actual wafer map, (**b**) predicted wafer map using the GP with clustering, (**c**) predicted wafer map using VP, and (**d**) predicted wafer map using GP without clustering

Table 5.3 Test Escape (TE) and Yield Loss (YL) in parts per million (ppm) comparison

	All measurements		Measurement 53		Measurement 69	
	TE	YL	TE	YL	TE	YL
VP	28,000	85,000	4.4	48	6.6	81
GP–	28,000	**4400**	3.9	7	**6.3**	32
GP+	**14,000**	**4400**	**0.19**	**2.8**	**6.3**	**0.19**

Bold formatting denotes the best value per column

5.3.2.4 Test Escape and Yield Loss Improvement

To further elucidate the impact of the improved prediction error, in the 2nd and 3rd columns of Table 5.3 we compare the Test Escape (TE) and Yield Loss (YL) computed using (5.34) and (5.35) for the VP, the GP without clustering (denoted by GP-), and the GP with clustering (denoted by GP+), for all 71 measurements and for all die of the 3406 wafers. Note that the high TE and YL values in the 2nd and 3rd column of Table 5.3 are mainly due to some poorly predicted measurements such as those in the right-hand side shown in Fig. 5.24. For these measurements, random variation dominates systematic wafer-level spatial variation. Nevertheless, GP with clustering achieves a significant TE improvement as compared to VP and GP without clustering, while maintaining the same YL as GP without clustering.

The 4th to 7th columns of Table 5.3 report the TE and YL for two individual measurements, namely 53 and 69, where significant improvement in percentile prediction error is observed. The TE and YL are computed using Eqs. (5.34) and (5.35), respectively. As can be observed, GP with clustering consistently outperforms VP and GP without clustering, which is in-line with the percentile error improvement shown in Fig. 5.25, and justifies the use of k-means algorithm in building spatial correlation models.

5.3.3 *IC Laser Trimming Speedup Through Spatial Correlation Modeling*

The third application aims to speed up the lengthy trimming process by utilizing the wafer-level spatial correlations of the trim-related probe measurements. Laser trimming is the process by which several on-chip resistors are trimmed in order to increase their resistance. Consequently, specific IC parameters are tuned, which essentially enhances the manufacturing yield of a given design. To achieve this, laser trimming burns away portions of these resistors using a laser trimming machine, in order to raise their resistance values until a target is reached. This is a lengthy procedure which is typically performed in a closed loop, where the laser is iteratively fired and the IC parameter in question is monitored until a target condition is satisfied.

In this special case of probe-test spatial correlation modeling, the goal is to predict the trimming lengths of the laser-trimmable resistors from a small sample of die locations on each wafer, thereby eliminating the time-consuming, closed-loop, iterative measure-trim steps for the majority of the devices. Our dataset consists of one wafer with 1924 devices, each of which contains two laser-trimmable resistors that have been already trimmed. As is commonly practiced, each resistor is divided into two parts that correspond to different trim stages, the coarse trim where the trimming is performed faster but with limited accuracy and the fine trim which is slower but with the best possible resolution allowing for the final target to be reached without overshooting. The trimmable performances in these devices are two current measurements which have to be finely tuned to a specific value in order to meet the device specifications. For each die, the pre- and post-trim current values are collected as well as the trim lengths and times for both coarse and fine trimming. Two spatial correlation approaches have been proposed for predicting the trimming lengths of the coarse trim, one in which the spatial model is directly trained by using the trim lengths of the sampled locations and one where the trim rates (i.e., the current difference over the trim length) are modeled and used to indirectly predict the trim lengths. The latter approach allows for adaptive calibration of the coarse target, resulting in further time savings especially when the hard-coded target has been conservatively selected.

Figure 5.27 depicts the current closed-loop-based method as well as the two spatial modeling-based methods described above. As explained earlier, the trimming process is split into two stages, the coarse and the fine. During the coarse trimming, for each laser pulse more material is removed in order to rapidly approach the final

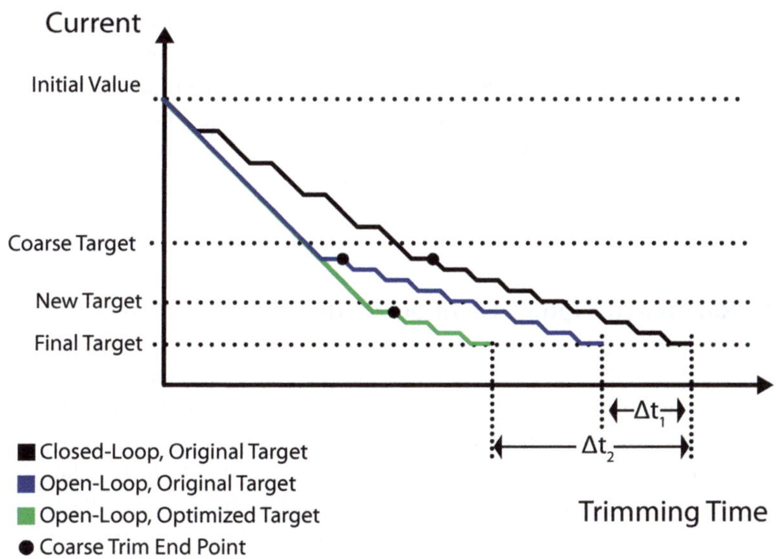

Fig. 5.27 Current value vs trimmed time in a two-stage trimming process

target with the expense of lower accuracy. On the other hand, during fine trim each pulse has the minimum possible length, thus achieving the highest possible resolution. The horizontal sections of the progress lines represent the settling time that is required after each laser pulse in order to accurately measure the resulting current value. For all die locations that are sampled in order to train the wafer-level spatial model, the normal closed-loop process is followed, and all the required information is collected. Once we have trained the spatial models, we then have the ability to predict the coarse trim length for each die, therefore eliminating the settling times. This is shown in both the blue and green progress lines in Fig. 5.27, as we reach the coarse target faster. The difference between the original and optimized target open-loop methods is that in the latter the coarse target has been optimized to be closer to the final, minimizing the trim time spent in the slow fine trimming stage.

In order to illustrate the effectiveness of the proposed method, we split our dataset into training and validation sets and simulated the testing process. The training set includes 10% of devices which were uniformly sampled over the wafer. Figures 5.28b and 5.29b show the sampled die locations for each resistor.

5.3.3.1 Length-Based, Original Target Prediction

As shown in Figs. 5.28a and 5.29a, the measurements of interest have variation that is smoothly spatially correlated, with a prominent radial component. The existence of spatial variation allows us to sample the length measurements sparsely and build the GP model. Then, by using the trained model, we predict measurements for the non-sampled die. Figure 5.28c shows the predicted length wafer map to be visually compared with Fig. 5.28a. For further visualization, the prediction error, which is the difference between the actual wafer map (Fig. 5.28a) and the predicted wafer map (Fig. 5.28c), is shown in Fig. 5.28e. The first observation from the error map is that the variance of error is balanced over the wafer. Moreover, positive errors (those towards red color) can be interpreted as overpredicted lengths. Similar results for resistor B are shown in Fig. 5.29. The maximum prediction error for resistors A and B (which corresponds to the yellow colored die on the error maps) is less than 3%.

5.3.3.2 Rate-Based, Optimized Target Prediction

In order to evaluate the possible speedup between the standard closed-loop laser trim, the length-based original-target prediction method, and the rate-based optimized-target one, we performed a preliminary experiment only to measure the various trimming times. Due to test time limitations, we collected data for only 200 die locations on a new untrimmed wafer. Since the goal of this experiment was to assess the amount of speedup and not the prediction accuracy or potential yield loss, the coarse trim target was selected by using an intelligent guess of the trim rate, from the data collected for the previous experiments. Table 5.4 shows the measured average trim times per die in each stage. As expected, there is a significant

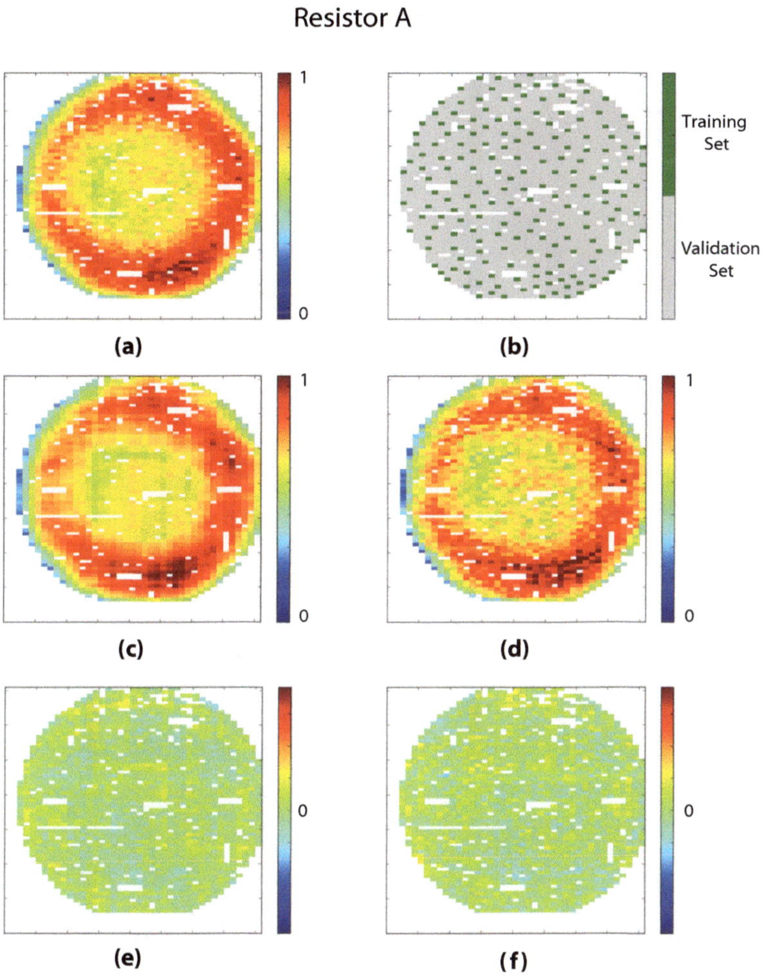

Fig. 5.28 Actual and predicted lengths using both methods for resistor A. (**a**) Actual trim length. (**b**) Sampled locations. (**c**) Length-based, original target predicted length. (**d**) Rate-based, original target predicted length. (**e**) Length-based, original target prediction error. (**f**) Rate-based, original target prediction error

improvement of the coarse trim time, as a result of eliminating the time-consuming measurement loop. The overall speedup is shown in Table 5.5, which is 1.25 for Resistor A and 1.32 for B.

In addition to the results above, Tables 5.4 and 5.5 show the extrapolated trimming times and speedup for rate-based prediction of length for a new optimized target. This target was selected based on the previously reported maximum prediction errors. The estimated speedup by using the rate-based, optimized target prediction is 1.98 and 1.89 for each resistor, respectively.

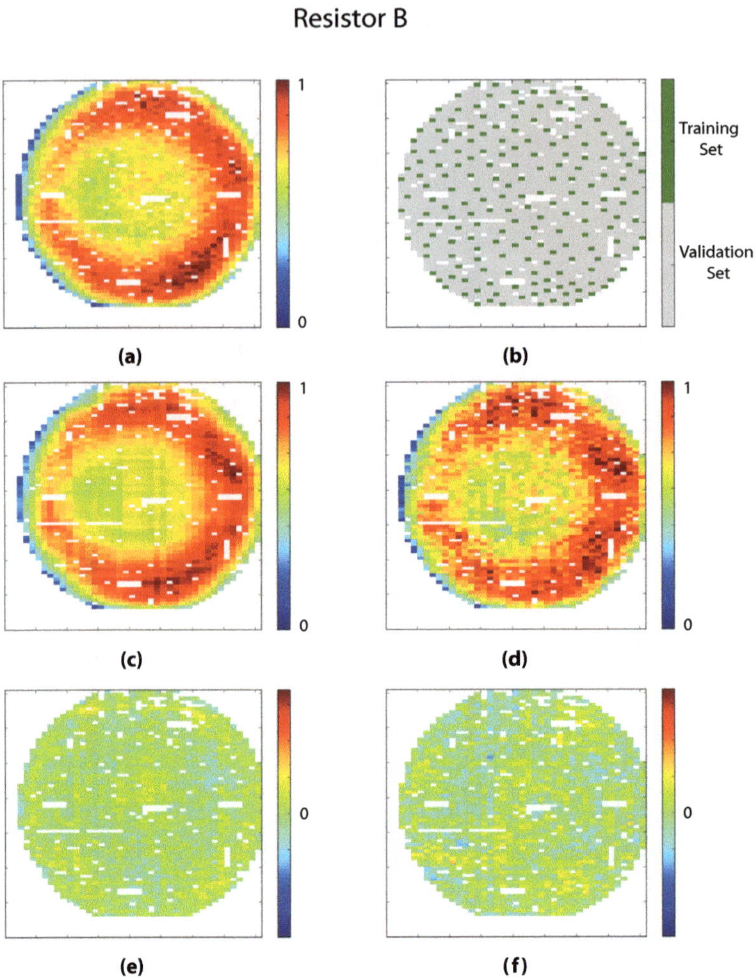

Fig. 5.29 Actual and predicted lengths using both methods for resistor B. (**a**) Actual trim length. (**b**) Sampled locations. (**c**) Length-based, original target predicted length. (**d**) Rate-based, original target predicted length. (**e**) Length-based, original target prediction error. (**f**) Rate-based, original target prediction error

Table 5.4 Average trimming times per die

	Closed-loop		Length-based original target		Rate-based optimized target	
Resistor	Coarse	Fine	Coarse	Fine	Coarse	Fine
A	312 ms	281 ms	190 ms	281 ms	233 ms	66 ms
B	338 ms	231 ms	200 ms	231 ms	244 ms	57 ms

Table 5.5 Total average trimming times and speedup per die

		Open-loop			
	Closed-loop	Original target		Optimized target	
Resistor	Total time	Total time	Speedup	Total time	Speedup
A	593 ms	471 ms	1.25	299 ms	1.98
B	569 ms	431 ms	1.32	301 ms	1.89

5.3.4 HVM Yield Estimation

In this section, we demonstrate the utilization of correlation modeling for HVM yield estimation. As we mentioned in Sect. 5.2.11, the spatiotemporal GP model with progressive sampling is used to enhance yield estimation. Thus, we first evaluate the effectiveness of progressive sampling and the inclusion of temporal feature in enhancing the prediction accuracy of correlation models. Then, the spatiotemporal GP model is used to estimate high-volume yield for a 65-nm analog/RF device. Our dataset comprises a total of 300 time-stamped wafers, grouped in lots of 8–12 wafers, and produced over a period of 6 months. Each wafer has approximately 5500 devices, on which 39 parametric probe-test measurements are collected. Our experiments seek to: (1) quantify the accuracy improvement achieved by progressive sampling and spatiotemporal modeling in predicting probe tests using wafer-level spatial correlation, and (2) assess the ability of the spatiotemporal model in predicting HVM yield based on measurements from a small number of early production wafers.

5.3.4.1 Accuracy Improvement of Enhanced Model

In all of the experiments described below, we sample 10% of the die locations on each wafer for training the correlation models, which are subsequently used to predict the values on the remaining 90% of non-sampled die locations. The predicted values are then compared to the actual values that we have in our dataset, in order to calculate and report the prediction error of each of the compared models. In the case of the baseline spatial correlation model, which is used as a reference point, the 10% sample is randomly selected across the wafer. In the case of progressive sampling, this 10% sampling budget is reached in multiple steps, where the selection of samples added in each step is guided by the models constructed using the samples of all previous steps. In the case of the spatiotemporal models, each lot of wafers is treated holistically. Specifically, we use the 10% sample of (randomly or progressively selected) die from each wafer in the lot to construct one spatiotemporal correlation model for the entire lot, which is subsequently used to predict the values on the remaining 90% of die locations on each of the wafers in the lot. In summary, the four prediction models that we compare in our experiment are the following:

- **Gaussian Process (GP):** Given a wafer, randomly select 10% of the die on this wafer, measure the parameter of interest, and train a GP model to predict this parameter as a function of die coordinates on the wafer. Then, use this model to predict this parameter for the remaining 90% of die on the wafer.
- **GP with Progressive Sampling (GP-PS):** Given a wafer, select die locations using the progressive sampling algorithm (Algorithm 3) in increments of 2.5%. In each iteration, use all available samples to identify the die locations where model confidence is low, in order to select the next die sample increment. Once the overall sampling budget of 10% is reached (i.e., after 4 iterations), use the final model to predict this parameter for the remaining 90% of die on the wafer.
- **GP with Spatiotemporal Features (GP-ST):** Given a lot of wafers, randomly select 10% of die from each wafer and train one spatiotemporal GP model (i.e., a model that expresses a parameter as a function of die coordinates as well as the temporal order of a wafer within its lot) for the entire lot. Then, use this spatiotemporal model to predict this parameter for the remaining 90% of die on all wafers in the lot.
- **GP with Spatiotemporal Features and Progressive Sampling (GP-ST-PS):** Given a lot of wafers, select die locations on each wafer using the progressive sampling algorithm in increments of 2.5% per lot. In the first increment, the same random locations are chosen on each wafer, in order to help the model capture time-dependent correlation. Subsequent sample increments are chosen across all wafers in the lot based on the prediction confidence level of the spatiotemporal correlation model constructed using all previous samples from all the wafers. Once the overall sampling budget of 10% is reached (i.e., after 4 iterations), use the final spatiotemporal correlation model to predict this parameter for the remaining 90% of die on all wafers in the lot.

In order to assess the accuracy of each model, we compare the values for the predicted die locations (90%) to the actual probe-test outcomes and we capture the discrepancy using the absolute percentile error metric:

$$\epsilon = \frac{|\mathbf{t}' - \mathbf{t}|}{\text{Specification Range}} \tag{5.36}$$

where $\mathbf{t}$ is the probe-test outcome for a die, $\mathbf{t}'$ is the corresponding predicted value for that die, and *Specification Range* is the range of that measurement for which the device is operational.

Figure 5.30 presents the impact of the introduced correlation model enhancement methods on a randomly chosen parameter (i.e., measurement 16) among the 39 probe tests in our dataset, on one randomly chosen wafer. Figure 5.30a shows the actual wafer map, while Fig. 5.30b–e shows the predicted wafer maps using each of the four methods, i.e., GP, GP-PS, GP-ST, and GP-ST-PS, respectively. Even though the differences are subtle, they are still visible through the wafer maps, where it may be observed that the predictions by GP-PS, GP-ST, and GP-ST-PS are better than the prediction of the original GP and become progressively more accurate with respect

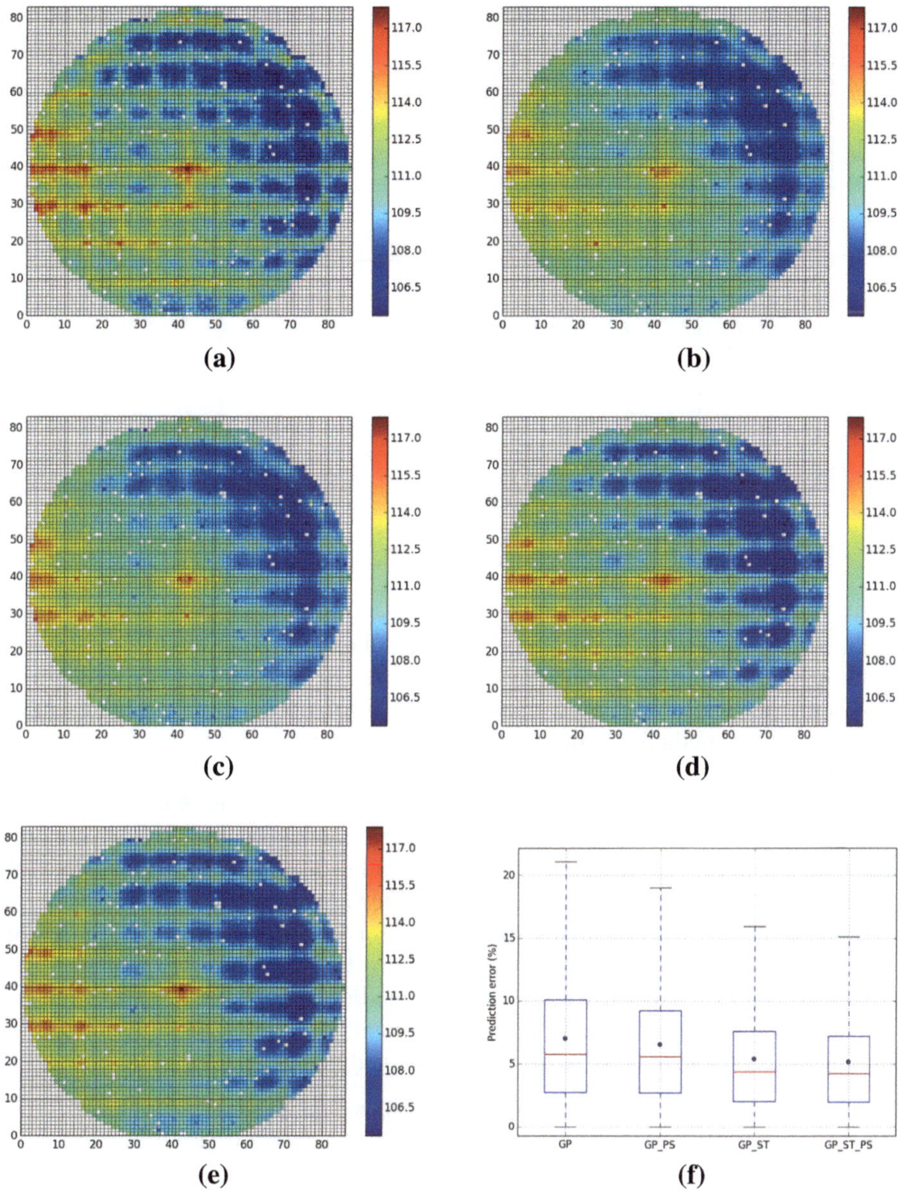

Fig. 5.30 Actual and predicted wafer maps and prediction error for measurement 16. (**a**) Actual wafer map. (**b**) Predicted wafer map (GP). (**c**) Predicted wafer map (GP-PS). (**d**) Predicted wafer map (GP-ST). (**e**) Predicted wafer map (GP-ST-PS). (**f**) Prediction error

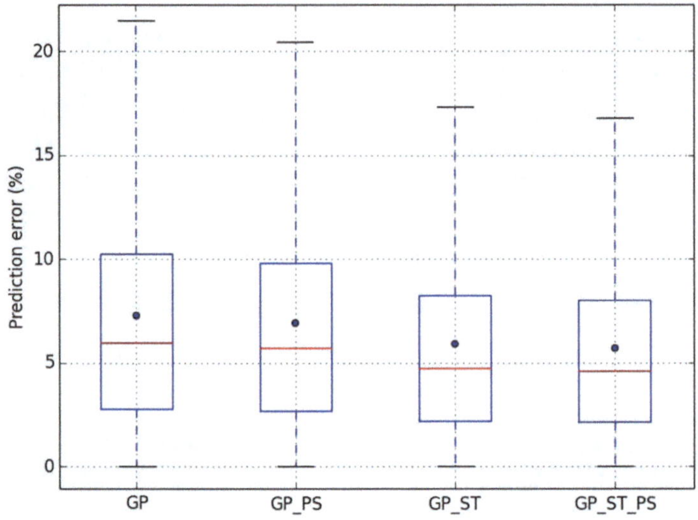

Fig. 5.31 Measurement 16 prediction error over all wafers

to the actual wafer map. The box plot in Fig. 5.30f makes all these subtle differences evident by reporting the prediction error for each of the four models. The ***y*-axis** shows the percentile prediction error, with the black dot on each bar representing the mean of the prediction error across all die on the wafer.

Figure 5.31 shows the prediction error of the four methods for measurement 16 across all 300 wafers. The mean error is **7.3**%, **6.9**%, **5.9**%, and **5.6**% for GP, GP-PS, GP-ST, and GP-ST-PS, respectively. We note that the proposed models not only generate a lower prediction error in the test data but also result in tighter error bars than the baseline model, which indicates that the variance of error is also smaller.

Finally, Fig. 5.32 summarizes the mean error of the four methods for all 39 measurements across all 300 wafers. As may be observed, the enhancements presented herein consistently result in lower prediction error for all measurements. In order to demonstrate the average improvement over the original GP model, we use the difference in mean relative error as defined below:

$$\delta\text{-MRE} = |*| \frac{\text{GP Error} - \text{Proposed Method Error}}{\text{GP Error}} \times 100$$

Accordingly, we calculate the δ-MRE between GP and GP-PS, GP-ST, and GP-ST-PS as **2.6**%, **11**%, and **16**%, respectively. Based on the above results, we observe the following:

- Enhancing the original Gaussian process-based wafer-level spatial correlation method with temporal inter-wafer information and progressive sampling results in a notable improvement in the accuracy of the prediction model.

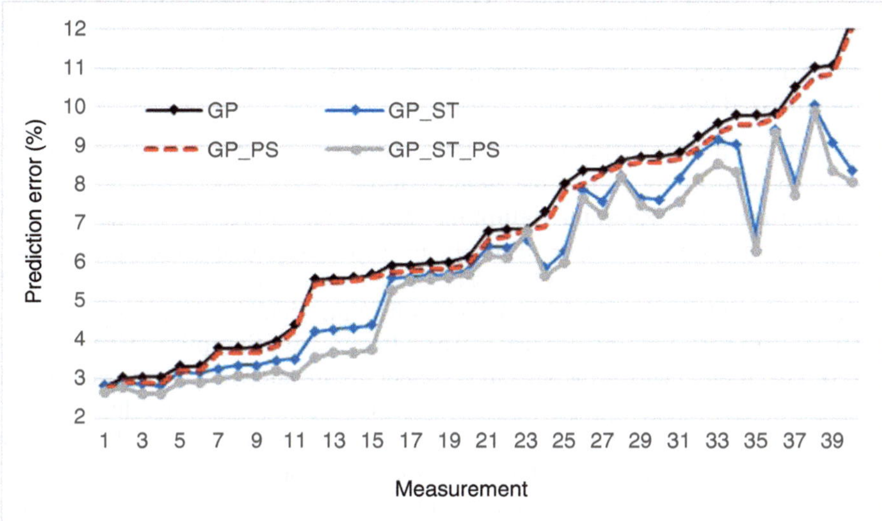

Fig. 5.32 Prediction error for 39 measurements over all wafers

- The improvement obtained by spatiotemporal modeling is significantly higher than the improvement obtained by progressive sampling. This is attributed to the fact that the locations sampled on each wafer in a lot are different; therefore, since wafers in the same lot are expected to be affected by the same systematic variation sources, sampled locations in a wafer carry valuable information regarding the same locations on other wafers, which were not included in the sample. Thereby, the collective statistics for the entire wafer are significantly improved.

5.3.4.2 Yield Estimation Results

In the experiments described below, which seek to demonstrate the effectiveness of the proposed method in estimating HVM yield from a few early silicon wafers, we use the first wafer from each of the first **5** lots—chronologically—in our dataset, as our early silicon samples. Our sampling budget remains **10%** of the die locations as in the previous experiments. Using these samples, we estimate the probability density function (PDF), the cumulative distribution function (CDF), and the yield for each of the 39 probe tests in our dataset using the methods described previously. For the entire dataset of the remaining **295** wafers, we also calculate the actual density and the actual yield for each of the 39 probe tests. For the purpose of this study, since the specification limits for each probe test were not disclosed to us, the actual yield is computed by setting the lower and upper specification limits at the $[\mathbf{L}, \mathbf{U}] = \pm 3\sigma$ range of each measurement across the 295 HVM wafers. The same limits are also used to predict the yield from the estimated PDF/CDF and compare to the actual yield so that the accuracy of the proposed methods can be assessed.

The actual data and the four estimation methods considered in our experiment are summarized below:

- **Actual:** All die of all HVM wafers (295 wafers) are used to compute the actual yield and density for each of the 39 probe tests.
- **Histogram with Random Sampling (Hist-RS):** Given the 5 early silicon wafers, randomly select 10% of the die from each wafer and create a histogram with 20 uniformly distributed bins. The percentage of sampled die across these 5 wafers that falls in the [**L**, **U**] range will reflect the yield for each probe test.
- **Histogram with Spatiotemporal GP and Progressive Sampling (Hist-GP-ST-PS):** Given the 5 early silicon wafers, select die locations on every wafer using the progressive sampling algorithm in increments of 2.5%. Ensure that the same die locations are picked on each wafer in the first iteration and guide the selection of subsequent iterations using the prediction confidence level of a spatiotemporal GP model built using all previously selected samples. Once the 10% sample budget is reached (i.e., after 4 iterations), use the final GP-ST-PS model to predict the parameter of interest for the unobserved 90% die across the 5 early wafers. Finally, use both the sampled and the predicted die to create a histogram with 20 uniformly distributed bins. The percentage of all die across these 5 wafers that falls in the [**L**, **U**] range will reflect the yield for each probe test.
- **KDE with Random Sampling (KDE-RS):** Given the 5 early silicon wafers, randomly select 10% of the die from each wafer and estimate the density of these samples using KDE. Then, sample the estimated distribution in order to generate one million synthetic die instances. The percentage of synthetic die instances that falls in the [**L**, **U**] range will reflect the yield for each probe test.
- **KDE with Spatiotemporal GP and Progressive Sampling (KDE-GP-ST-PS):** Given the 5 early silicon wafers, select die locations on every wafer using the progressive sampling algorithm in increments of 2.5% until the 10% sampling budget is reached, following the same procedure described above for Hist-GP-ST-PS. Then, use the final GP-ST-PS model to predict the parameter of interest for the unobserved 90% die across the 5 early wafers. Subsequently, use all sampled and predicted values to estimate the density of the distribution using KDE. Then, sample the estimated distribution in order to generate one million synthetic die instances. The percentage of synthetic die instances that falls in the [**L**, **U**] range will reflect the yield for each probe test.

To evaluate the effectiveness of these methods, we first compare the estimated distributions. Figure 5.33a shows the actual PDF and the estimated PDFs using the Hist-RS and the Hist-GP-ST-PS methods for a randomly selected measurement among the 39 probe tests (i.e., measurement 24). Similarly, Fig. 5.33b shows the actual PDF and the estimated PDFs using the KDE-RS and the KDE-GP-ST-PS methods for the same measurement. Although the differences are subtle, one can still observe that the GP-ST-PS method provides a more accurate sample and results in a better estimation both with the simple histogram method and with the advanced

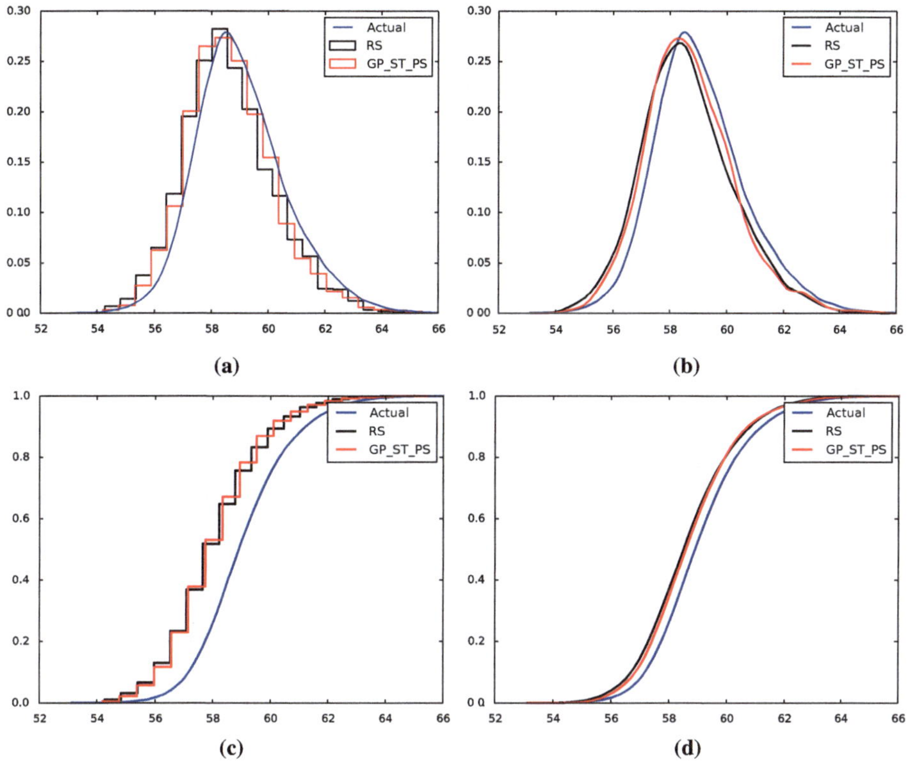

Fig. 5.33 PDF and CDF estimated by histogram-based and KDE methods for measurement 24. (**a**) Estimated PDF using histogram-based approach. (**b**) Estimated PDF of KDE-based approach. (**c**) CDF of histogram-based estimation approach. (**d**) CDF of KDE-based estimation approach

KDE method. As expected, one can also observe that KDE is a compelling method for estimating the actual distribution. Nevertheless, it still benefits from having a better starting point, as provided by the GP-ST-PS method.

In order to quantitatively compare the quality of estimation, we use the Kolmogorov–Smirnov (KS) test [12] as a goodness-of-fit metric. KS test is a nonparametric test for one-dimensional probability distributions that can be used to compare a sample to a reference. In KS, the comparison metric is the maximum distance between the CDF of the estimated density and the actual CDF. A smaller distance (i.e., closer to 0) indicates a better fit between the real distribution and the estimated one. Figure 5.33c shows the actual CDF and the estimated CDFs using the Hist-RS and the Hist-GP-ST-PS methods for measurement 24, while Fig. 5.33d shows the actual CDF and the estimated CDFs using the KDE-RS and the KDE-GP-ST-PS methods for the same measurement. Based on these CDFs, in Table 5.6 we compute the KS metric for each of the four predicted CDFs in contrast to the actual. The results corroborate our claim that the information added by GP-ST-PS

Table 5.6 KS metric of estimated CDFs for measurement 24

	Histogram		KDE	
	RS	GP-ST-PS	RS	GP-ST-PS
Distance	0.25	0.23	0.11	0.087

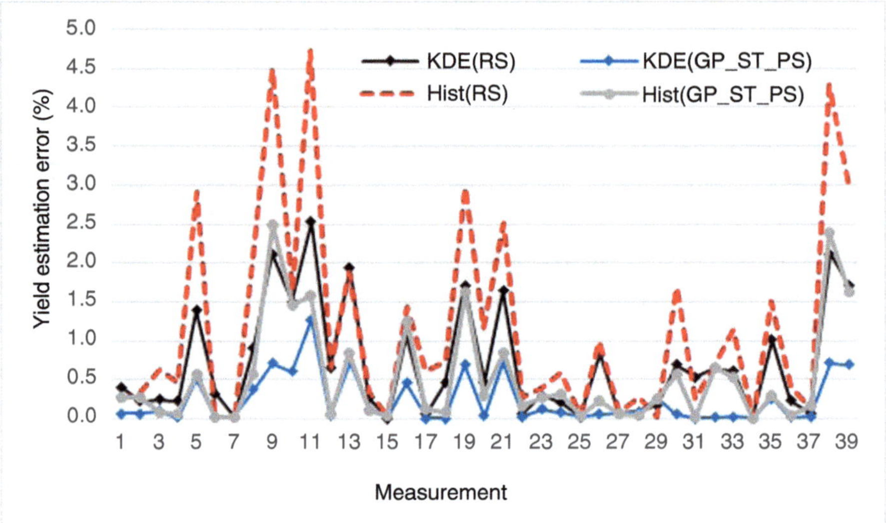

Fig. 5.34 Yield estimation error for 39 measurements

Table 5.7 Average error of HVM yield estimation

	Histogram		KDE	
	RS	GP-ST-PS	RS	GP-ST-PS
Yield error	1.16%	0.63%	0.61%	0.21%

helps in better estimating the actual distribution, both for the histogram-based and for the KDE-based method.

Finally, we compare the yield estimated by each of the four methods to the actual HVM yield and we compute the corresponding yield estimation error as the absolute difference between the two. Figure 5.34 shows the yield estimation error for each of the four aforementioned methods for each of the 39 probe tests, as a percentage on the ***y*-axis**. Additionally, Table 5.7 shows the average yield error for all 39 measurements. Once again, the results confirm the effectiveness of enhancing the initial sample using the GP-ST-PS method towards improving the accuracy of HVM yield estimation.

5.4 Conclusions

In the beginning of this chapter, we presented the benefits of spatial correlation modeling that enable the extrapolation of wafer-level measurements via sparse sampling of test data. The Gaussian process models were presented in detail, with emphasis to the various parameters as well as to the domain-specific improvements that affect the modeling accuracy. A k-means clustering algorithm was presented to handle discontinuous effects in modeling spatial correlation, which partitions the wafer into separate regions to avoid training discontinuous data using a unique continuous model. Progressive sampling and inclusion of spatiotemporal feature constitute powerful enhancements to spatial correlation modeling. Finally, we discussed application of GP in estimating HVM yield. In Sect. 5.3, several case studies of high-volume manufacturing data demonstrated the effectiveness of the proposed approach in modeling various wafer-level measurements. These applications included extrapolation of sparsely sampled e-test measurements, test cost reduction for probe-test specification measurements, speedup of the IC laser trimming process, and HVM yield estimation. The effectiveness of each of these applications has been evaluated using sizable industrial datasets. Employing spatial and spatiotemporal correlation models has been shown to provide significant insight into the variation of the semiconductor manufacturing process, thus enabling previously inaccessible, optimal regions of the quality-cost trade-off.

Appendix 1: Proof of Positive Semi-Definite for Covariance Matrix

Proof To show that covariance matrices are always positive semi-definite, let Σ be the covariance matrix for some vector $\mathbf{x}$, and μ the mean of $\mathbf{x}$. Then, by definition, $\Sigma = \mathbb{E}[(\mathbf{x} - \mu)^\top(\mathbf{x} - \mu)]$. Now, proof that Σ is positive semi-definite requires us to show that $\mathbf{a}^\top\Sigma\mathbf{a} \geq 0$ for all vectors $\mathbf{a}$. So, we have

$$
\begin{aligned}
\mathbf{a}^\top\Sigma\mathbf{a} &= \mathbf{a}^\top\mathbb{E}[(\mathbf{x} - \mu)(\mathbf{x} - \mu)^\top]\mathbf{a} \\
&= \mathbb{E}[\mathbf{a}^\top(\mathbf{x} - \mu)(\mathbf{x} - \mu)^\top\mathbf{a}] \\
&= E[\mathbf{b}^2] \geq 0,
\end{aligned}
$$

where we have defined $\mathbf{b} = \mathbf{a}^\top(\mathbf{x} - \mu)$.

Appendix 2: Marginal and Conditional Distribution of Multivariate Normal Distribution

The multivariate Gaussian (or Normal) distribution has a joint probability density given by:

$$p(\mathbf{x}|\boldsymbol{\mu}, \Sigma) = (2\pi)^{d/2} \exp\left(-\frac{1}{2}(\mathbf{x} - \boldsymbol{\mu})^T \Sigma^{-1} (\mathbf{x} - \boldsymbol{\mu})\right), \tag{5.37}$$

where $\boldsymbol{\mu}$ is the *mean* vector (of length d) and Σ is the (symmetric, positive definite) *covariance* matrix (of size $d \times d$). As a shorthand, we write $\mathbf{x} \sim \mathcal{N}(\mathbf{m}, \Sigma)$.

Let $\mathbf{x}$ and $\mathbf{y}$ be jointly Gaussian random vectors:

$$\begin{bmatrix} \mathbf{x} \\ \mathbf{y} \end{bmatrix} \sim \mathcal{N}\left(\mathbf{0}, \begin{bmatrix} A & C \\ C^\top & B \end{bmatrix}\right) \tag{5.38}$$

where A and B are the covariances for $\mathbf{x}$ and $\mathbf{y}$, respectively, and C is the covariance between $\mathbf{x}$ and $\mathbf{y}$.

Now, we would like to be able to write out the form for the inverse covariance matrix in (5.38). We can make use of the Schur complement and write this as:

$$\begin{bmatrix} A & C \\ C^\top & B \end{bmatrix}^{-1} = \begin{bmatrix} I & 0 \\ -B^{-1}C^\top & I \end{bmatrix} \begin{bmatrix} (A - CB^{-1}C^\top)^{-1} & 0 \\ 0 & B^{-1} \end{bmatrix} \begin{bmatrix} I & 0 \\ -B^{-1}C^\top & I \end{bmatrix} \tag{5.39}$$

Now, the joint distribution $p(\mathbf{x}, \mathbf{y})$ can be written as:

$$p(\mathbf{x}, \mathbf{y}) \propto \exp\left(-\frac{1}{2} \begin{bmatrix} \mathbf{x} \\ \mathbf{y} \end{bmatrix}^\top \begin{bmatrix} A & C \\ C^\top & B \end{bmatrix}^{-1} \begin{bmatrix} \mathbf{x} \\ \mathbf{y} \end{bmatrix}\right) \tag{5.40}$$

Then, we can substitute in the above expression of the inverse of the block covariance matrix defined in (5.39), and we obtain

$$p(\mathbf{x}, \mathbf{y}) \propto \exp\left(-\frac{1}{2} \begin{bmatrix} \mathbf{x} - CB^{-1}\mathbf{y} \\ \mathbf{y} \end{bmatrix}^\top\right.$$

$$\begin{bmatrix} (A - CB^{-1}C^{\top})^{-1} & 0 \\ 0 & B^{-1} \end{bmatrix}$$

$$\begin{bmatrix} \mathbf{x} - CB^{-1}\mathbf{y} \\ \mathbf{y} \end{bmatrix}\Bigg) \tag{5.41}$$

Using the fact that the center matrix is block diagonal, we have

$$p(\mathbf{x}, \mathbf{y}) \propto \exp\Big(-\frac{1}{2}(\mathbf{x} - CB^{-1}\mathbf{y})^{\top}(A - CB^{-1}C^{\top})^{-1} (\mathbf{x} - CB^{-1}\mathbf{y})\Big) \exp\Big(-\frac{1}{2}\mathbf{y}^{\top}B^{-1}\mathbf{y}\Big) \tag{5.42}$$

If we condition equation (5.42) on $\mathbf{y}$, then the second exponential term drops out as a constant, and we have

$$p(\mathbf{x}|\mathbf{y}) \sim \mathcal{N}\Big(CB^{-1}\mathbf{y}, A - CB^{-1}C^{\top}\Big) \tag{5.43}$$

Appendix 3: Summary of Commonly Used Kernel Functions

A kernel function $k(\mathbf{x}, \mathbf{x}')$ can be either expressed as a function of the distance between any input pair: $r = |\mathbf{x} - \mathbf{x}'|$, if it is *stationary*, or as a function individual $\mathbf{x}$ and $\mathbf{x}'$, where $\mathbf{x}$ is a D-dimensional vector in the input space: $\mathbf{x} = (x_1, x_2, \ldots, x_D)$. Some commonly used include:

- Linear (LI) kernel function:

$$k(\mathbf{x}, \mathbf{x}') = \sum_{d=1}^{D} \sigma_d^2 x_d x_d' \tag{5.44}$$

  where σ_d is the hyperparameter of the function.
- Polynomial (POLY) kernel function:

$$k(\mathbf{x}, \mathbf{x}') = (\mathbf{x} \cdot \mathbf{x}' + \sigma_0^2)^p \tag{5.45}$$

  where σ_0 and p are the hyperparameters of the function.
- Squared exponential (SE) kernel function:

$$k(\mathbf{x}, \mathbf{x}') = \exp\left(-\frac{1}{2l^2}|\mathbf{x} - \mathbf{x}'|^2\right) \tag{5.46}$$

  where l is the hyperparameter of the function.

– Absolute exponential (AE) kernel function:

$$k(\mathbf{x}, \mathbf{x}') = \exp\left(-\frac{1}{2l^2}|\mathbf{x} - \mathbf{x}'|^p\right) \tag{5.47}$$

where l and p are the hyperparameters of the function.

References

1. A. Ahmadi, K. Huang, S. Natarajan, J. Carulli, Y. Makris, Spatio-temporal wafer-level correlation modeling with progressive sampling: a pathway to HVM yield estimation, in *IEEE International Test Conference* (2014), pp. 1–10
2. A. Ahmadi, A. Nahar, B. Orr, M. Pas, Y. Makris, Wafer-level process variation-driven probe-test flow selection for test cost reduction in Analog/RF ICs, in *IEEE VLSI Test Symposium* (2016), pp. 1–6
3. A. Ahmadi, C. Xanthopoulos, A. Nahar, B. Orr, M. Pas, Y. Makris, Harnessing process variations for optimizing wafer-level probe-test flow, in *IEEE International Test Conference* (2016)
4. M.A. Aizerman, E.A. Braverman, L. Rozonoer, Theoretical foundations of the potential function method in pattern recognition learning. Autom. Remote Control **25**, 821–837 (1964)
5. S.S. Akbay, A. Chatterjee, Fault-based alternate test of RF components, in *IEEE International Conference on Computer Design* (2007), pp. 518–525
6. D. Bertsimas, J. Tsitsiklis, Simulated annealing. Stat. Sci. **8**(1), 10–15 (1993)
7. S. Biswas, R.D. Blanton, Test compaction for mixed-signal circuits using pass-fail test data, in *IEEE VLSI Test Symposium* (2008), pp. 299–308
8. J.B. Brockman, S.W. Director, Predictive subset testing: optimizing IC parametric performance testing for quality, cost, and yield. IEEE Trans. Semicond. Manuf. **2**(3), 104–113 (1989)
9. T. Calinski, J. Harabasz, A dendrite method for cluster analysis. Commun. Stat. **3**(1), 1–27 (1974)
10. H.M. Chang, K.T. Cheng, W. Zhang, X. Li, K.M. Butler, Test cost reduction through performance prediction using virtual probe, in *IEEE International Test Conference* (2011), pp. 1–9
11. S. Duane, A.D. Kennedy, B.J. Pendleton, D. Roweth, Hybrid monte carlo. Phys. Lett. B **195**(2), 216–222 (1987)
12. J. Durbin, *Distribution Theory for Tests Based on Sample Distribution Function*, vol. 9 (SIAM, Philadelphia, 1973)
13. T. Hastie, R. Tibshirani, J. Friedman, *The Elements of Statistical Learning: Data Mining, Inference, and Prediction* (Springer, New York, 2001)
14. K. Huang, N. Kupp, J. Carulli, Y. Makris, Handling discontinuous effects in modeling spatial correlation of wafer-level analog/RF tests, in *IEEE Design, Automation and Test in Europe Conference* (2013), pp. 553–558
15. S. Kirkpatrick, C.D. Gelatt, M.P. Vecchi, Optimization by simulated annealing. Science **220**(4598), 671–680 (1983)
16. N. Kupp, K. Huang, J. Carulli, Y. Makris, Spatial correlation modeling for probe test cost reduction, in *IEEE/ACM International Conference on Computer-Aided Design* (2012), pp. 23–29
17. N. Kupp, K. Huang, J. Carulli, Y. Makris, Spatial estimation of wafer measurement parameters using Gaussian process models, in *IEEE International Test Conference* (2012), pp. 1–8
18. F. Liu, A general framework for spatial correlation modeling in VLSI design, in *IEEE/ACM Design Automation Conference* (2007), pp. 817–822

19. D. MacKay, *Information Theory, Inference and Learning Algorithms* (Cambridge University Press, Cambridge, 2003)
20. W. Maly, A.J. Strojwas, S.W. Director, VLSI yield prediction and estimation: a unified framework. IEEE Trans. Comput. Aided Des. Integr. Circuits Syst. **5**(1), 114–130 (1986)
21. G.W. Milligan, M.C. Cooper, An examination of procedures for determining the number of clusters in a data set. Psychometrica **50**(2), 159–179 (1985)
22. C.E. Rasmussen, C.K.I. Williams, *Gaussian Processes for Machine Learning* (MIT Press, Cambridge, 2006)
23. S. Reda, S.R. Nassif, Accurate spatial estimation and decomposition techniques for variability characterization. IEEE Trans. Semicond. Manuf. **23**(3), 345–357 (2010)
24. B. Schölkopf, A.J. Smola, *Learning with Kernels: Support Vector Machines, Regularization, Optimization, and Beyond* (MIT Press, Cambridge, 2001)
25. B.W. Silverman, *Density Estimation for Statistics and Data Analysis* (Chapman and Hall/CRC, Boca Raton, 1986)
26. H.G. Stratigopoulos, Y. Makris, Error moderation in low-cost machine-learning-based Analog/RF testing. IEEE Trans. Comput. Aided Des. Integr. Circuits Syst. **27**(2), 339–351 (2008)
27. H.G. Stratigopoulos, S. Mir, A. Bounceur, Evaluation of analog/RF test measurements at the design stage. IEEE Trans. Comput. Aided Des. Integr. Circuits Syst. **28**(4), 582–590 (2009)
28. H.G. Stratigopoulos, P. Drineas, M. Slamani, Y. Makris, RF specification test compaction using learning machines. IEEE Trans. Very Large Scale Integration Syst. **18**(6), 998–1002 (2010)
29. V. Vapnik, *The Nature of Statistical Learning Theory* (Springer, New York, 1995)
30. V. Vapnik, *Statistical Learning Theory* (Wiley, New York, 1998)
31. P.N. Variyam, S. Cherubal, A. Chatterjee, Prediction of analog performance parameters using fast transient testing. IEEE Trans. Comput. Aided Des. Integr. Circuits Syst. **21**(3), 349–361 (2002)
32. R. Voorakaranam, S.S. Akbay, S. Bhattacharya, S. Cherubal, A. Chatterjee, Signature testing of analog and RF circuits: algorithms and methodology. IEEE Trans. Circuits Syst. I **54**(5), 1018–1031 (2007)
33. C.K.I. Williams, C.E. Rasmussen, Gaussian processes for regression, in *Advances in Neural Information Processing Systems 8* (MIT Press, Cambridge, 1996), pp. 514–520
34. W. Zhang, X. Li, E. Acar, F. Liu, R. Rutenbar, Multi-wafer virtual probe: minimum-cost variation characterization by exploring wafer-to-wafer correlation, in *IEEE/ACM International Conference on Computer-Aided Design* (2010), pp. 47–54
35. W. Zhang, K. Balakrishnan, X. Li, D. Boning, R. Rutenbar, Toward efficient spatial variation decomposition via sparse regression, in *IEEE/ACM International Conference on Computer-Aided Design* (2011), pp. 162–169
36. W. Zhang, X. Li, F. Liu, E. Acar, R. Rutenbar, R.D. Blanton, Virtual probe: a statistical framework for low-cost silicon characterization of nanoscale integrated circuits. IEEE Trans. Comput. Aided Des. Integr. Circuits Syst. **30**(12), 1814–1827 (2011)

Chapter 6
Machine Learning Approaches for IC Manufacturing Yield Enhancement

Hongge Chen and Duane S. Boning

6.1 Introduction

Semiconductor manufacturing processes are highly automated with large amounts of available data. Due to the uncertainty in nanoscale fabrication and the growing complexity of the process, various machine learning and data mining techniques have been proposed to improve different steps of manufacturing [1–3]. In this chapter, we consider machine learning to predict and enhance the yield of integrated circuit (IC) manufacturing. More specifically, we want to predict the results of final device tests from early stage data. The results of the final test are assumed to be binary, for example, pass/fail or good/bad. Two main challenges, class imbalance and concept drift, in building machine learning models for integrated circuit yield enhancement are discussed and addressed in this chapter, based on and extending the work presented in [4] and [5].

6.1.1 Challenge One: Imbalanced Classification

In high-volume semiconductor manufacturing, high manufacturing yield is often achieved. As a result, the number of failure devices may only be a very small fraction of the sample set. The issue of extreme differences in prior class probabilities is called class imbalance [6]; building classifiers on highly imbalanced datasets is challenging. In scenarios we consider here, class imbalance arises because we have much smaller numbers of dies belonging to the "failure" class compared to

H. Chen · D. S. Boning (✉)
Massachusetts Institute of Technology, Cambridge, MA, USA
e-mail: chenhg@mit.edu; boning@mtl.mit.edu

I. M. Elfadel et al. (eds.), *Machine Learning in VLSI Computer-Aided Design*,
https://doi.org/10.1007/978-3-030-04666-8_6

those in the "good die" class. In this chapter, we discuss batch and online learning ensemble methods to learn on highly imbalanced datasets from IC manufacturing. Our approach will be presented in Sects. 6.4.1 and 6.4.2.

6.1.2 Challenge Two: Concept Drift

The data from a manufacturing process is often available in a data stream manner. Due to subtle changes in the process, the underling probability distributions of the data are expected to change over time. This issue is called concept drift, with the unfortunate effect that learning models trained on old data may be inconsistent with new data [7]. To overcome this challenge, we present an incremental learning approach to learn in nonstationary environments. Our approach for incremental learning and continuous classifier updating is discussed in Sect. 6.4.3.

6.1.3 Application

As an application, our learning methods for class imbalance and concept drift are tested on a flash memory packaging process. There are two main reasons why a packaging process, the last step of semiconductor device fabrication, is a good and important object of study. First, packaging is almost the last chance for a semiconductor manufacturing company to improve its product quality before the final test. Second, at packaging, we have more input features, including unpackaged die electrical test results, for the learning model than at most other previous manufacturing stages. Opportunities for machine learning-based yield improvement in packaging processes have not attracted much attention with few previous works reported. In this chapter, we build on our previous discussion in [4] and [5] to show that with a good machine learning model classifying the flash memory dies before packaging, we can improve the final part yield significantly. Our key ideas in this chapter are as follows:

1. Batch and online ensemble machine learning techniques are explained to address the class imbalance problem.
2. An incremental learning framework is designed to overcome the problem of concept drift.
3. We build a mathematical model and demonstrate the possibility of yield improvement using a classifier to detect bad dies before packaging.
4. The experimental results show the potential for significant improvement of yield using industrial data.

The rest of this chapter is organized as follows. Section 6.2 is an introduction to the flash memory manufacturing process. Section 6.3 provides the preliminaries of the problem, including evaluation metrics and mathematical formulation. In Sect. 6.4, the batch, online, and incremental learning frameworks are discussed. Then, the experimental results are presented in Sect. 6.5. Section 6.6 concludes this chapter.

6.2 Background of the Manufacturing Process

As illustrated in Fig. 6.1, there are six high-level steps in the IC memory manufacturing process [8], Wafer Fabrication, Cherry Pick, KGD, Assembly (Packaging), Die Preparation, and Memory Test. In flash memory chip manufacturing, wafers are fabricated in wafer fab facilities, after which dies on each wafer go through a standard electrical test process. Based on the results of these tests, wafers and dies are marked with different parameters and bins. The two main tests in the fabrication facility are SME1 and SME2. These two tests are various electrical tests at two different temperatures. Note that in order to reduce cost, SME1 and SME2 are performed in a sequential manner, meaning that dies that fail SME1 will not be tested in SME2. Dies that fail either SME1 or SME2 are marked on the wafer by the fab. Then, the wafers, together with their data, are shipped to an assembly and test facility. This facility can then bin or "cherry pick," by sorting the wafers into different quality tiers according to the test data from the fab. A further testing step called the known good die (KGD) test is carried out. KGD is product-oriented: dies from wafers in different tiers undergo different KGD procedures. Those dies on the "prime" wafers identified in the cherry pick go through a strict KGD test and will be packaged into high-end products. Then, the packaging process begins. After necessary preparations and processing steps, the wafers are diced into singular dies. Then, wire bonding attaches the dies to the substrate, with other passive components mounted on. Importantly, multiple dies are typically stacked together. Then, the device is encapsulated and separated [8]. After packaging, there is a final memory test which verifies the functionality of the product in the extreme environment of the real world. Though the dies are packaged into products, memory test also still assesses the performances of each die within the package. For each die, the output of the memory test is pass (good) or fail (bad). Note that all dies selected for

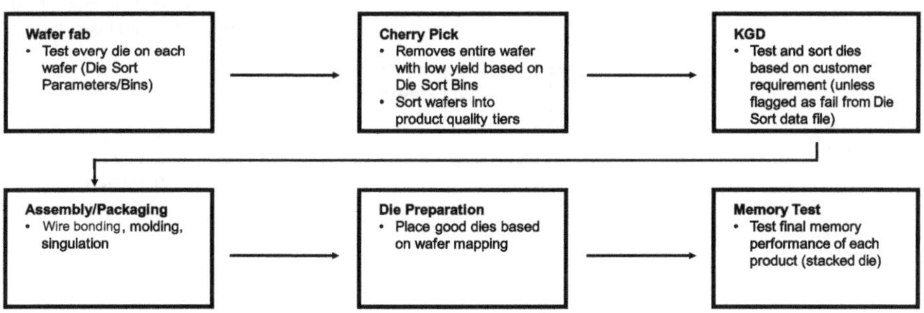

Fig. 6.1 Flash memory chip manufacturing process as discussed in [8]

packaging have already passed earlier strict die electrical tests, but some of them still fail the high-performance final memory test after packaging. In other words, there exists "actually" bad or low-performance dies despite being judged by the earlier tests as good or high-performance, and given the limitations of the existing test decision process they can only be detected later by the final memory test. Because the dies are sequentially connected in the package, if one die in the package fails, the whole package is considered as fail. The company does not necessarily discard such "failed" packages, but rather downgrades them to low-end products with less profit. Since the total number of packages is fixed, to achieve high profit, the company desires more high-end products. Because a single fail die will lead to the failure (downgrading) of the whole package, good (high-performance) dies stacked in that package may be wasted. If we know that the failure probability of a single die is $p_{\text{die fail}}$, which is usually very small, the failure rate of the final packages will be

$$p_{\text{package fail}} = 1 - (1 - p_{\text{die fail}})^s \approx s p_{\text{die fail}}, \tag{6.1}$$

where s (= 4, 8, or even 16) is the number of dies in a package. Thus, the real failure rate of the product can be much higher than the failure rate of dies.

The conceptual goal of a machine learning-based classifier is shown in Fig. 6.2. Each die from wafer fab (shown at left in Fig. 6.2) has a large number of features based on electrical test measurements. Based on these measurements, some dies are identified by the test program to be bad (e.g., red dies in Fig. 6.2), while the rest are assumed to be good dies and are sent to packaging, having passed earlier SME1 and SME2 electrical tests. So, the packaging process proceeds on the assumption that all of these dies are good and will meet high-performance specifications. But, actually some of these dies are not good (pictured as orange dies in Fig. 6.2), and presently these failures can only be detected by the final memory test. If we can build a classifier to predict the result of the final memory test based on the available die electrical test data from the fab, more good dies (shown as blue dies in Fig. 6.2) will be stacked together for more high-end products.

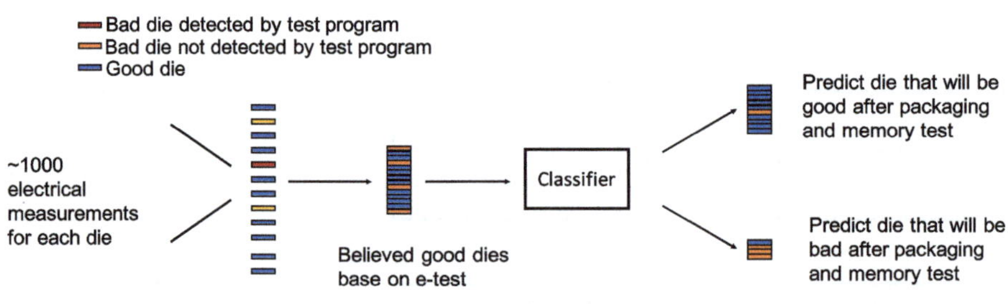

Fig. 6.2 Manufacturing process with a classifier

6.3 Preliminaries

As mentioned in Sect. 6.1, a good classifier for memory test results before packaging will help to increase yield and profit for the following reasons.

- One can stack predicted good dies together for high-end products and package predicted lower-performance ("bad") dies together for low-end products. If the classifier is good enough, the number of good packages will increase.
- Dies predicted as "bad" can go through a lower-cost low-end packaging and test process directly and the cost of downgrading is saved.

For each die, all of its early test results and fabrication parameters can be used as the input of the classifier. In our data, approximately 1000 measured features (including results from SME1, SME2, and KGD) are available for each die. The output of the classifier is a prediction of the die's memory test result. Figures 6.3 and 6.4 illustrate the packaging and testing process without and with a classifier, respectively. In these two simple examples, 4 out of 16 incoming dies (all of which passed the fab die tests) are actually bad. If we have a fairly good classifier (even if imperfect as illustrated in Fig. 6.4) to predict the memory test result and package dies predicted as good together as high-end products and dies predicted as bad together as low-end products, our high-end yield can be improved from 25% to 50% in this example. A similar issue also exists in the novel 3D integrated circuit packaging technology: if one chip in a stack fails, the whole package fails. In 3D IC packaging cases, where different types of chips or devices are stacked in a package, we would need classifiers for each kind of chip.

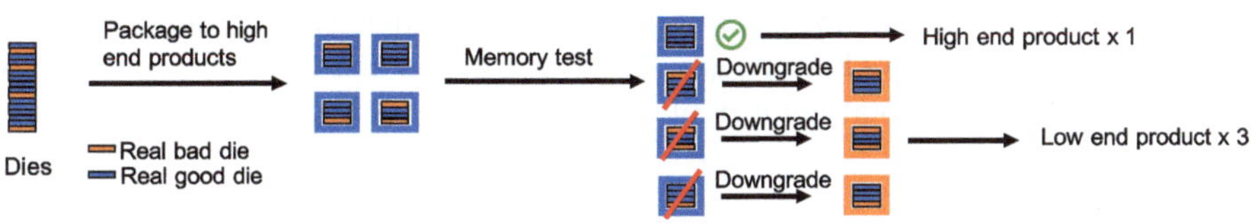

Fig. 6.3 The packaging and memory testing process without a classifier

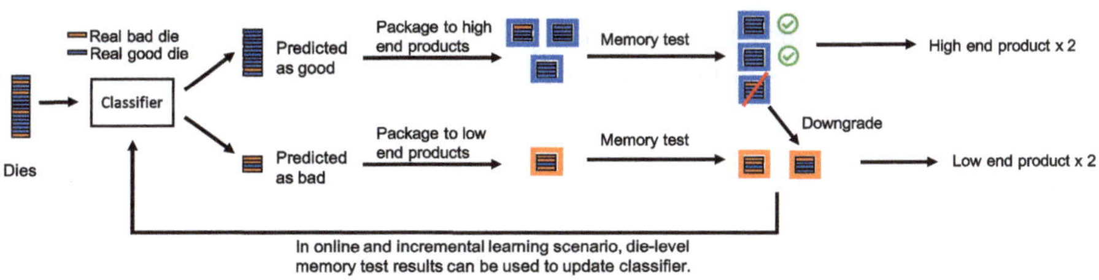

Fig. 6.4 The packaging and memory testing process with a classifier

In Sects. 6.3.1 and 6.3.2, we next introduce terms and evaluation metrics used in this work. Then, the mathematical insight of the yield improvement and profit gain will be provided in Sect. 6.3.3.

6.3.1 Evaluation Metric: Confusion Matrix

Confusion matrices are perhaps the easiest and most widely used performance metric in classification problems. Suppose we have a classification problem with two possible classes; the confusion matrix is a 2×2 matrix visualizing the performance of a supervised learning method. Each column of a confusion matrix represents a possible predicted condition (output of the classifier), while each row of the matrix represents a possible true condition (ground truth). The confusion matrix and corresponding terminology with respect to a binary classification problem are shown in Table 6.1. In our problem, fail dies are labeled as 1 (positive or signaling a failure) in the final memory test and pass dies are labeled as 0 (negative). The confusion matrix is constructed based on the results of a classifier and known final memory test on a test set of N dies. Each die in the test set must be placed into one cell in the matrix. Fail dies that are correctly predicted as fail are placed into the true positive (TP) cell, and those that are incorrectly predicted as pass are placed into the false negative (FN) cell. Good dies that are correctly predicted as pass are placed into the true negative (TN) cell, while those incorrectly predicted as fail are placed into the false positive (FP) cell. Then, we sum up the number of dies in each cell and obtain the confusion matrix of the classifier. FN is also called Type II error and FP is also called Type I error.

Based on the confusion matrix, some terminology and derivations are developed. First, it is obvious that we have $N_{\text{fail}} = \text{TP} + \text{FN}$ true fail dies and $N_{\text{pass}} = \text{FP} + \text{TN}$ true pass dies, where total number of dies $N = N_{\text{fail}} + N_{\text{pass}}$. The classification accuracy is defined as:

$$Acc = \frac{\text{TP} + \text{TN}}{\text{TP} + \text{TN} + \text{FP} + \text{FN}} = \frac{\text{TP} + \text{TN}}{N}.$$

In our high manufacturing yield and highly imbalanced classification scenario, classification accuracy is in fact not a good metric since we can achieve fairly high

Table 6.1 The confusion matrix and corresponding terminology with respect to a binary classification problem

		Predicted condition	
		Predicted positive (fail)	Predicted negative (pass)
True condition	Condition positive (fail)	True positive (TP)	False negative (FN)
	Condition negative (pass)	False positive (FP)	True negative (TN)

accuracy by simply labeling all dies as pass. For example, if in our data $N_{\text{fail}} = 10$ and $N_{\text{pass}} = 990$ (which means that the yield is 99%), then a trivial classifier predicting all dies as good can achieve 99% classification accuracy since TN = 990, FN = 10, and TP = FP = 0. However, this trivial classifier cannot give any useful information.

To overcome this, we instead need to consider both true and false positive rates. The true positive rate (TPR) is defined as:

$$\text{TPR} = \frac{\text{TP}}{\text{TP} + \text{FN}},$$

which gives the fraction of positive instances that are correctly detected as positive. The false positive rate (FPR) is defined as:

$$\text{FPR} = \frac{\text{FP}}{\text{FP} + \text{TN}},$$

which gives the fraction of negative instances that are sacrificed or lost to false alarms. In the aforementioned example, the trivial classifier has TPR = FPR = 0. If our classifier detects 8 out of 10 bad dies while mistakenly labels 99 good dies as bad, we have TPR = 0.8 and FPR = 0.1. In general, a good classifier is expected to have both low FPR and high TPR.

6.3.2 *Evaluation Metric: ROC Curves*

The receiver operating characteristic (ROC) curve is a graphic illustration of the performance of a binary classifier based on TPR and FPR. We denote "positive" or fail by 1, and "negative" or pass by 0.

For a given die, the vector of input features of our classifiers is denoted as $x \in X$, where X is the space of input features. The prediction result y belongs to Y, the set of all possible prediction results. In our binary classification, $Y = \{0, 1\}$. A classifier $h(x, y)$ is assumed to output a nonnegative real value, the score of an instance with features x to be classified in a class $y \in Y$. Moreover, we have

$$\sum_{y} h(x, y) = 1 \text{ and } h(x, y) \in [0, 1]. \tag{6.2}$$

Then, $h(x, y)$ can be interpreted as the probability of x to be classified into class y. To make a prediction, a threshold η is set to map the scores to $\{1, 0\}$. For example, when $\eta = 0.6$, a die with feature vector x is predicted to be a member of class 1 if $h(x, 1) > 0.6$ and to be a member of class 0 otherwise. Under each threshold, we thus have different TPR and FPR. The receiver operating characteristic (ROC)

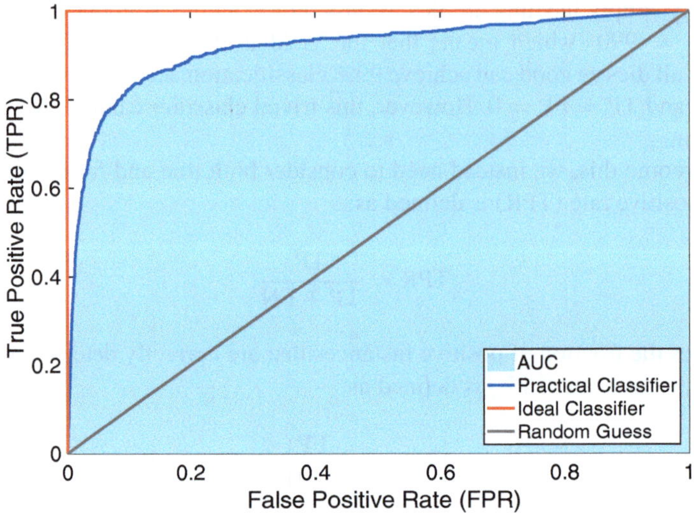

Fig. 6.5 Receiver operating characteristic (ROC) curves and the area under ROC curve, or AUC

curve is obtained by plotting TPR against FPR under different thresholds. Note that when $\eta = 1$, all instances will be classified as 0, thus TPR $=$ FPR $= 0$; and when $\eta = 0$, we have TPR $=$ FPR $= 1$. So, an ROC curve must go through $(0, 0)$ and $(1, 1)$. Figure 6.5 shows different example ROC curves. An ideal ROC curve goes through TPR $= 1$, FPR $= 0$, which means that under some threshold, we can achieve 100% accuracy in classification with an ideal classifier. A classifier based purely on a random guess is a straight line, while a practical classifier is somewhere in between these two cases.

The area under an ROC curve (AUC) is a useful overall metric for classifier performances with all possible thresholds, as presented in Fig. 6.5. For an ideal classifier, AUC $= 1$; for random guess, AUC $= 0.5$; and for a practical classifier, AUC $\in (0.5, 1)$.

6.3.3 *Mathematical Formulation*

With a large number of testing dies, we can estimate the underlying positive and negative probabilities by the relative frequency. Let H be the true condition and y be the predicted condition. For given input features x and threshold η, $y = 1$ if $h(x, y) > \eta$ and 0 otherwise. Then, the joint distribution of true condition H and predicted condition y can be estimated by:

$$p(H=1, y=1) = \frac{\text{TP}}{N}, \; p(H=1, y=0) = \frac{\text{FN}}{N}$$
$$p(H=0, y=1) = \frac{\text{FP}}{N}, \; p(H=0, y=0) = \frac{\text{TN}}{N}. \quad (6.3)$$

Here, $p(E)$ indicates the probability of event E. Without any classifiers, if we randomly package the dies into packages of s chips in each stack the failure rate of the packages is

$$p_{\text{package fail}} = 1 - p(H=0)^s. \quad (6.4)$$

Then, if n dies are produced, we can produce n/s packages in total. The expected number of good packages is

$$\mathbb{E}[m_1(s)] = \frac{n}{s} p(H=0)^s, \quad (6.5)$$

where round-off error is neglected (number of leftover dies $< s$ not packaged). With a classifier to predict the memory test result before packaging, we can stack the dies predicted as fail together and the dies predicted as pass together. Recall that our "fail" classification still indicates that the die will work in low-performance products, while "pass" dies are required for high-performance products. Using the law of total expectation, the expected number of good (high-performance) packages in this case is given by:

$$\mathbb{E}[m_2(s)] = \mathbb{E}_k[\mathbb{E}[m_2(s)|k]] = \mathbb{E}[kp(H=0|y=0)^s]$$
$$= \frac{n}{s} p(H=0|y=0)^s p(y=0), \quad (6.6)$$

where k is the number of stacks packaged as high-end products and is subject to a binomial distribution $B(\frac{n}{s}, \; p(y=0))$. Again, round-off error is neglected. If the classifier is ideal, all good dies are packaged into high-end products and the maximum number of high-end products is achieved:

$$\mathbb{E}[m_2^*(s)] = \frac{n}{s} p(y=0) = \frac{n}{s} p(H=0). \quad (6.7)$$

It is easy to prove that $\mathbb{E}[m_2^*] \geq \mathbb{E}[m_1]$ and $\mathbb{E}[m_2^*] \geq \mathbb{E}[m_2]$. The expected yield improvement is $\mathbb{E}[m_2(s)] - \mathbb{E}[m_1(s)]$ which can be either positive or negative. Note that TP, TN, FP, and FN are uniquely determined by N_{pass}, N_{fail}, FPR, and TPR. Subtracting (6.5) from (6.6), plugging in (6.3), and with a little manipulation, we have the expected yield improvement:

$$
\begin{aligned}
&\frac{\mathbb{E}[m_2(s) - m_1(s)]}{n/s} \\
&= \frac{p(H=0, y=0)^s}{p(y=0)^{s-1}} - p(H=0)^s \\
&= \frac{(p(H=0)p(y=0|H=0))^s}{[(p(H=0)p(y=0|H=0) + (p(H=1)p(y=0|H=1)]^{s-1}} \\
&\quad - p(H=0)^s \\
&= \frac{[N_{\text{pass}}(1-\text{FPR})]^s}{N[N_{\text{pass}}(1-\text{FPR}) + N_{\text{fail}}(1-\text{TPR})]^{s-1}} - \frac{N_{\text{pass}}^s}{N^s},
\end{aligned} \tag{6.8}
$$

which is a function of TPR and FPR. In this work's imbalanced classification scenario, we have $N_{\text{fail}} \ll N_{\text{pass}}$. As shown in Fig. 6.5, TPR $\geq$ FPR should hold, so we also have $N_{\text{fail}}(1 - \text{TPR}) \ll N_{\text{pass}}(1 - \text{FPR})$. Then, (6.8) can be linearized as:

$$
\begin{aligned}
\frac{\mathbb{E}[m_2(s) - m_1(s)]}{n/s} &= \frac{[N_{\text{pass}}(1-\text{FPR})]^s}{N[N_{\text{pass}}(1-\text{FPR}) + N_{\text{fail}}(1-\text{TPR})]^{s-1}} - \frac{N_{\text{pass}}^s}{N^s} \\
&\approx \frac{N_{\text{pass}}}{N}(1-\text{FPR})\left[1 - (s-1)\frac{N_{\text{fail}}(1-\text{TPR})}{N_{\text{pass}}(1-\text{FPR})}\right] \\
&\quad - \left(1 - \frac{sN_{\text{fail}}}{N}\right) \\
&= (s-1)\frac{N_{\text{fail}}}{N}\text{TPR} - \frac{N_{\text{pass}}}{N}\text{FPR} \\
&= \frac{1}{N}[(s-1)\text{TP} - \text{FP}],
\end{aligned} \tag{6.9}
$$

where we use the fact that $(1 + x)^n \approx 1 + nx$ when $|x| \ll 1$. This gives us an easy way to estimate the yield improvement. For example, when we have four dies in a package, the yield improvement will be positive if the number of bad dies correctly detected is at least $\frac{1}{3}$ of the number of good dies mistakenly predicted as bad.

Here, we implicitly assume that the underlying joint probability distribution $p(H, y)$ does not change over time. If it does, as in Sects. 6.1.2 and 6.4.3, Eq. (6.8) from old data may not be valid for future prediction.

6.4 Learning Models

Here, we describe the learning models used to overcome imbalanced classification and concept drift. The goal is to increase TPR and reduce FPR so as to maximize the expected yield improvement as in Eq. (6.8).

6.4.1 Imbalanced Classification and Batch RUSBoost Learning

The imbalanced classification problem is pervasive in many real-world problems, especially when connected with anomaly detection [6, 9], such as in Email fraud detection, medical diagnosis, or computer intrusion detections. Since most conventional learning techniques assume fairly balanced distributions, their performances may be compromised on a highly imbalanced dataset. Class imbalance is a classical problem with some state-of-the-art solutions [10]. In our flash memory data, the fail dies only consist of around 2% of total dies and the problem is highly imbalanced.

A fundamental approach is to not make any changes in the learning algorithm, and instead to modify the dataset itself to provide a balanced training set for the classifier. The easiest and most intuitive ways to alleviate class imbalance are oversampling the minority class or undersampling the majority class. Though these seem to be equivalent in function, both oversampling and undersampling have problems that may be harmful to the performance of learning. For undersampling, reducing the number of the training instances means that some important patterns or information may be lost and the data becomes noisy. Oversampling, on the other hand, may lead to an increase in computation time and overfitting due to a large number of identical minority training instances. To tackle the problems of simple oversampling and undersampling, more sophisticated techniques are available.

Sampling techniques can be integrated with ensemble methods to interact with the models. Ensemble methods combine the results from multiple base classifiers (simple classifiers such as decision trees and naive Bayes classifiers) to achieve better performance. One of the most well-known ensemble methods to combine classifiers is Adaptive Boosting (AdaBoost) [11]. In AdaBoost, the first base classifier is trained on the original dataset. Then, when training the subsequent base classifiers, instances misclassified by previous classifiers will be assigned larger weights (i.e., boosted). After training all the base classifiers, the ensemble classifier's final result is a weighted average of the base classifiers' output. The individual base classifiers may be very weak, but as long as the classification performances of each of them are slightly better than random guessing, the final ensemble classifier can be proven to converge to a strong classifier.

In our problem, RUSBoost is used to tackle the problem of imbalanced dataset in classification. Here, RUS stands for "random undersampling" [12], which means removing instances in the majority class at random. RUSBoost is a widely used hybrid approach embedding under-sampling in the AdaBoost framework. Before training each new base classifier, the instances in the majority class are randomly undersampled without replacement as in step 4 of Algorithm 1. For example, if the desired class ratio for training is 50 : 50, then instances in the majority class are randomly removed until the number of majority instances equals to the number of minority instances. The class ratio here is a hyper-parameter to be determined by cross validation. The main advantage of random undersampling is its simplicity. With multiple classifiers in an ensemble, the problem of noisy base classifiers can be alleviated. Random undersampling the majority class generates a new training set S'_t with weight distribution D'_t, which is used to train base classifier h_t. Then, the pseudo-loss on the original dataset S with weights D_t is calculated as in step 7. In step 9, we update the weights. Misclassified instances at t will have large weights at $t+1$, so the training of the base classifier at $t+1$ will concentrate more on these instances.

Algorithm 1 Batch RUSBoost [12]

1: **input**: S instances: $(x_1, y_1), (x_2, y_2), \ldots, (x_m, y_m)$, $y_i \in \{1, 0\}$ with majority class $y_i = 0$ and minority class $y_i = 1$; base classifiers $h_1, \ldots, h_T$
2: **initialize** $D_1(i) = \frac{1}{m}$ for any i
3: **for** $t = 1, 2, \ldots, T$ **do**
4: Create new training set S'_t by undersampling the majority class, the weight distribution of the remaining instances is D'_t
5: Train base classifier h_t with S'_t and D'_t
6: Test h_t on S, $h_t(x_i, y_i) \in [0, 1]$
7: Calculate pseudo-loss on S and D_t: $\epsilon_t = \sum_{i:y_i \neq y} D_t(i)(1 - h_t(x_i, y_i) + h_t(x_t, y))$.
8: $\alpha_t = \frac{\epsilon_t}{1-\epsilon_t}$
9: Update $D_{t+1}(i) = D_t(i)\alpha_t^{\frac{1}{2}(1+h_t(x_i,y_i)-h_t(x_t,y))}$
10: Normalize $D_{t+1}(i) : D_{t+1}(i) = \frac{D_{t+1}(i)}{\sum_j D_{t+1}(j)}$
11: **output**: $H(x, y) = \sum_{t=1}^{T} h_t(x, y)\log\frac{1}{\alpha_t}$ with normalization

6.4.2 Online RUSBoost Learning

Conventional batch machine learning models are trained on a fixed training set. In our problem, however, all dies, no matter predicted as pass or fail, go through the memory test as either high-end or low-end products and thus we eventually have access to the true labels for all test instances. This enables us to further improve our classifier with new test data in an online approach, as shown in Fig. 6.4. Due to the large number of dies produced each day, it is expensive to store all the old training data and retrain a new classifier with both old and new data when more data is available. Therefore, we use an online learning model to update our existing model when the information about new dies arrives. A similar model updating technique has also been studied in the CAD community, such as for hotspot detection in [13].

Online machine learning is a vibrant subarea in machine learning research. Online learning methods have been proven to be very powerful in applications where training data becomes available with a sequential order. In contrast with traditional batch learning which needs to see the entire training dataset before giving an output, online learning requires instant output after seeing each datum. The online classifier is built to output prediction p_t for input x_t at each time step t, $t = 1, 2, \ldots, T$. At time t, classifier receives x_t from "nature." In our work, "nature" refers to the early test data from SME1, SME2, or KGD. The classifier is asked for a prediction p_t for this x_t. Then at some later point, the classifier receives the true label y_t. Comparing p_t and y_t, the classifier then updates its internal model.

The online learning model used in our work is an online version of RUSBoost proposed in [14], as shown in Algorithm 2. This Online RUSBoost is based on the online boosting framework proposed in [15]. The online versions of bagging and boosting approaches are discussed in [15]. The key idea of these approaches is to approximate the corresponding batch bagging and boosting version with the training dataset observed as a data stream. Mathematically, this approximation depends on the well-known limiting case of a binomial distribution (when the number of samples is very large and the probability of success is very small) approaching a Poisson distribution. Indeed, for a given data point, if we draw N independent samples with probability of success $\frac{\lambda}{N}$, the probability of having n copies is

$$p_n = \binom{N}{n}\left(\frac{\lambda}{N}\right)^n\left(1-\frac{\lambda}{N}\right)^{N-n} \to \frac{e^{-\lambda}\lambda^n}{n!} \tag{6.10}$$

when $N \to \infty$. The reason for this approximation is that in the online learning, we may not know the length of the whole data stream before observing all the data. But with this approximation, we do not need the number of data points N when it is very large. Then instead of changing the probability distribution on the samples as in batch RUSBoost, we simply change the λ values as in Algorithm 2. Experimental results have shown that the online learning can achieve comparable performance to the corresponding batch learning method [15].

6.4.3 *Incremental Learning for Concept Drift and Class Imbalance*

The classical setup of a batch machine learning problem is based on learning from data drawn from an unknown but fixed distribution. However, problems in the real world are usually more complicated, especially with parameters in the learning scenario involving time. The data we receive may not be generated at the same time, and a critical issue is that the underling probability distribution of the training data

Algorithm 2 Online RUSBoost [14]

input: Data stream: $(x_1, y_1), (x_2, y_2), \ldots, (x_m, y_m)$, $y_i \in \{1, 0\}$ with majority class $y_i = 0$ and minority class $y_i = 1$; sampling rate C
Base classifiers $h_1, \ldots, h_K, \lambda_k^{SC} = \lambda_k^{SW} = 0$ for all k.
Initialize $\lambda_k^{SC} = \lambda_k^{SW} = \lambda_k^{POS} = \lambda_k^{NEG} = 0$ for all k, $n^+ = n^- = 0$
when (x_i, y_i) **arrives**:
Reset $\lambda = 1$
for $k = 1, 2, \ldots, K$ **do**
    if $y_i = 1$ **then**
        $\lambda_k^{POS} \leftarrow \lambda_k^{POS} + \lambda, n^+ = n^+ + 1$
        $\lambda_k^{RUS} \leftarrow \lambda \frac{n^+}{n^+ + n^-} / \frac{\lambda_k^{POS}}{\lambda_k^{POS} + \lambda_k^{NEG}}$
    else
        $\lambda_k^{NEG} \leftarrow \lambda_k^{NEG} + \lambda, n^- = n^- + 1$
        $\lambda_k^{RUS} \leftarrow \lambda \frac{Cn^+}{n^+ + n^-} / \frac{\lambda_k^{NEG}}{\lambda_k^{POS} + \lambda_k^{NEG}}$
    Generate random variable a_i according to $Poisson(\lambda^{RUS})$
    Do a_i times: train base classifier h_k with (x_i, y_i)
    if $h_k(x_i) = y_i$ **then**
        $\lambda_k^{SC} \leftarrow \lambda_k^{SC} + \lambda$
        $\epsilon_k \leftarrow \frac{\lambda_k^{SW}}{\lambda_k^{SW} + \lambda_k^{SC}}$
        $\lambda \leftarrow \frac{\lambda}{2(1-\epsilon_k)}$
    else
        $\lambda_k^{SW} \leftarrow \lambda_k^{SW} + \lambda$
        $\epsilon_k \leftarrow \frac{\lambda_k^{SW}}{\lambda_k^{SW} + \lambda_k^{SC}}$
        $\lambda \leftarrow \frac{\lambda}{2\epsilon_k}$
output: $H(x) = \arg\max_y \sum_{t=1}^{T} \mathbf{1}_{\{h_k(x)=y\}} \log \frac{1-\epsilon_k}{\epsilon_t}$

may change with time [7]. The issue of concept drift in the statistics community refers to the unexpected or unknown changes in the statistical properties of the dataset used to train and test the model. The most obvious problem caused by concept drift is that the model built on previous training data may lose its accuracy over time, as illustrated in Fig. 6.6. In this figure, the blue crosses and circles are data points at t_1, while the red crosses and circles are data points at t_2. The blue dashed line is the decision boundary at t_1, which successfully classifies most blue data points at t_1. However, as the distribution of the data changes in t_2 (concept drift), if we still use the blue dashed line as the decision boundary, red data points in the right upper corner will be misclassified. In fact, at t_2, our decision boundary should be changed to the red dashed line. In the data stream from a semiconductor manufacturing process, concept drift also exists. One obvious reason is inherent change in the manufacturing process, due to continued improvement, equipment aging, or environmental changes in the fabrication, in assembly, or in the testing process.

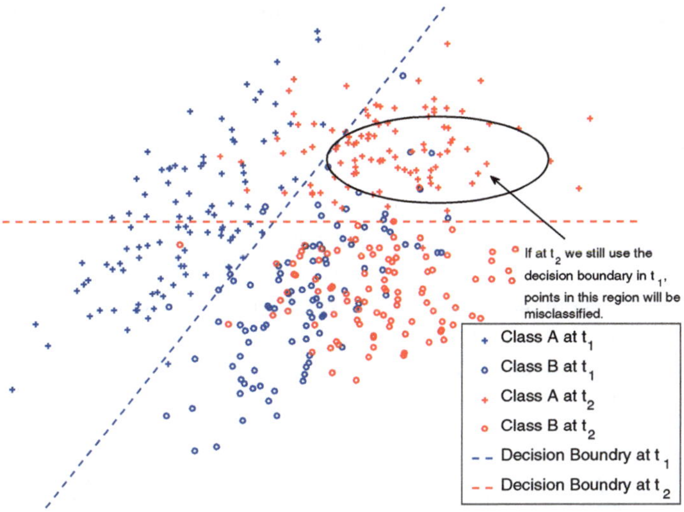

Fig. 6.6 Example of concept drift at two time steps t_1 and t_2. If at t_2 we still use the decision boundary at t_1, points in the ellipse will be misclassified

It is believed that dealing with concept drift and class imbalance simultaneously is an inherently difficult problem [16]. Due to the scarcity of fail dies (instances from minority class), the time interval between two consecutive fail dies can be large. Substantial changes in the manufacturing process may happen in this interval, and the underlying distribution of these two fail dies' parameters may not be identical. With this problem formulation, Online RUSBoost in Algorithm 2 is not the best choice to handle a nonstationary environment since it is designed to approximate the batch learning algorithm.

Intuitively, in an evolving environment, learning models should only be trained on the more "recent" data and should only be used to predict the near future. However, as the concept drift can be very fast, the qualified recent data may be inadequate to train a good classifier due to overfitting. This problem is amplified in the class imbalance scenario where the qualified recent minority instances are even fewer. To learn in a nonstationary environment, the idea of incremental learning has been proposed. In contrast to the online learning scenario, where the information from old instances is discarded after training, incremental learning emphasizes extending the model's knowledge by leveraging new data while preserving information from old data. To implement this idea, we store only the classifiers trained on old data (since storing all old data is expensive) and invoke those classifiers together with the new classifiers to help us to make predictions in the future. While a new classifier trained only on recent data may be incapable of making a good prediction, old classifiers trained on a similar pattern may participate in decision-making and improve performance.

In this area, the Learn++.NIE [17] algorithm is a state-of-the-art ensemble method. Learn++.NIE assumes that there is a data stream of chunks (mini batches). For example, in our case, each data chunk can be dies produced on one day. At each time step t, a new chunk $\mathscr{D}_t = \{(x_1, y_1), (x_2, y_2), \ldots, (x_{m^{(t)}}, y_{m^{(t)}})\}$ arrives. A new base classifier h_t is trained on $\mathscr{D}_t$ in a batch manner, as shown in Fig. 6.7. The final ensemble combines the base classifiers by calculating a weighted average of each base classifier's output. The weight of the base classifier trained at time step $k \leq t$ is determined by that base classifier's overall performance on $\mathscr{D}_k, \mathscr{D}_{k+1}, \ldots, \mathscr{D}_t$. Then, a sigmoid error weighting is used to emphasize the performance on recent data chunks as shown in Fig. 6.8. The sigmoid values a and b are hyper-parameters to be tuned by cross validation, where a controls the slope and b controls the halfway crossing point. A detailed introduction of the Learn++.NIE is available in [17].

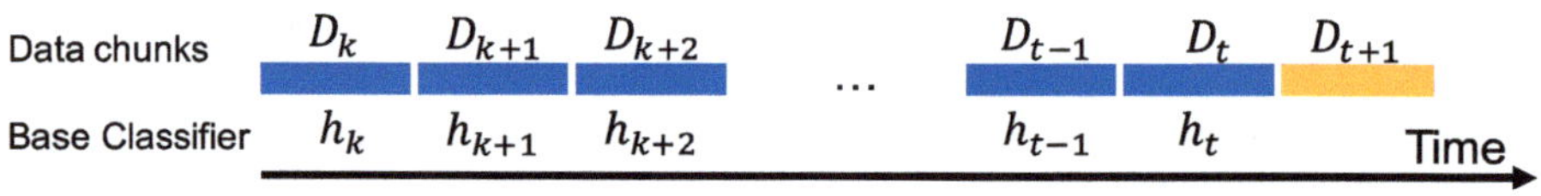

Fig. 6.7 The idea of incremental learning is that we divide our data stream into data chunks. Then, we train base classifiers on each data chunk and combine them to predict dies in the new data chunk

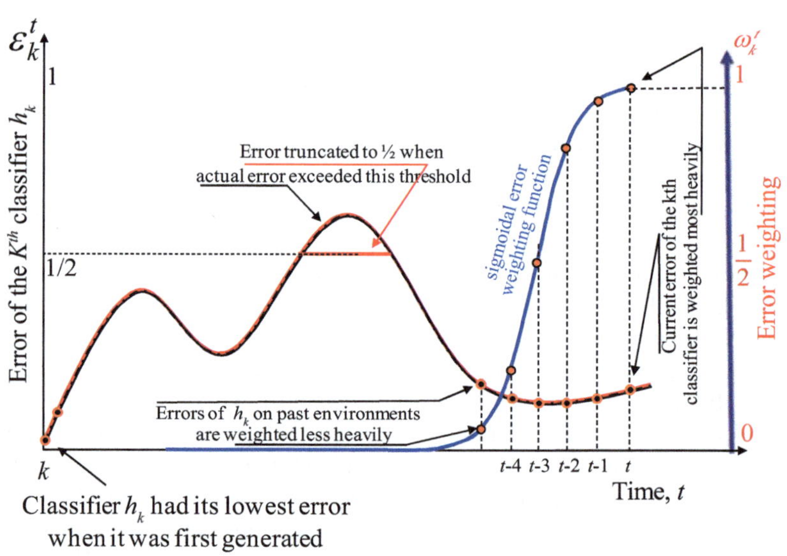

Fig. 6.8 An illustration of the sigmoidal error weighting in [18]. Learn++.NIE in [17] uses a similar error weighting

Algorithm 3 Modified Learn++.NIE

1: **input**: Data chunks: $\mathscr{D}_t = \{(x_i, y_i)\}$, $i = 1, 2, \ldots, m^{(t)}$ with majority class $y_i = 0$ and minority class $y_i = 1$; positive sigmoid parameters a and b.
2: **for** $t = 1, 2, \ldots$ **do**
3:     **when** $\mathscr{D}_t$ **arrives**:
4:     Call RUSBoost on $\mathscr{D}_t$ to generate a new base classifier $h_t(\cdot, \cdot) : X \times Y \rightarrow [0, 1]$
5:     For $k \in \{1, 2, \ldots, t\}$, evaluate $\text{auc}_k^{(t)}$, area under ROC curve obtained by h_k on $\mathscr{D}_t$.
6:     For $k \in \{1, 2, \ldots, t\}$, calculate pseudo-loss $\epsilon_k^{(t)} = 1 - \text{auc}_k^{(t)}$.
7:     **if** $\epsilon_t^{(t)} > 0.5$ **then**
8:         Discard h_t and train a new one as in line 4 and redo lines 5 and 6.
9:     **if** $\epsilon_k^{(t)} > 0.5$ **and** $k < t$ **then**
10:         Set $\epsilon_k^{(t)} = 0.5$
11:     $\beta_k^{(t)} = \epsilon_k^{(t)}/(1 - \epsilon_k^{(t)})$
12:     For $k \in \{1, 2, \ldots, t\}$, calculate weights:
    Define: $\omega_k^{(t)} = [1 + \exp(-a(t - k - b))]^{-1}$
    Normalize: $\hat{\omega}_k^{(t)} = \omega_k^{(t)} / \sum_{j=0}^{t-k} \omega_k^{(t-j)}$
    $\hat{\beta}_k^{(t)} = \sum_{j=0}^{t-k} \hat{\omega}_k^{(t-j)} \beta_k^{(t-j)}$
13:     $W_k^{(t)} = \log \frac{1}{\hat{\beta}_k^{(t)}}$
14: **output**: $H^{(t)}(x, y) = \sum_{k=1}^{t} W_k^{(t)} h_k(x, y)$ with normalization

In order to make it more suitable to our problem, we discuss a modified version of Learn++.NIE in this chapter, as expressed in Algorithm 3. Key ideas in this approach include the following:

- Instead of undersampling bagging as in [17], we use RUSBoost as our base classifier. Note that each RUSBoost base classifier is again an ensemble model with its own base classifiers.
- Since the output of the base classifier is a probability, we use $\text{auc}_k^{(t)}$, the area under ROC curve, to evaluate the performance of the base classifier. The pseudo-loss is obtained by $\epsilon_k^{(t)} = 1 - \text{auc}_k^{(t)} \in [0, 1]$.
- The final output of the ensemble is a weighted average of the probability from each base classifier.

6.5 Experimental Results

The proposed batch, online, and incremental learning methods are implemented and applied to memory die test data from our industrial partner. We choose a decision tree [19] as the base classifier in this work due to its simplicity, minimal assumptions required, and potential robustness against class imbalance with sampling [20].

One RUSBoost ensemble model consists of 40 decision tree base classifiers. In Sects. 6.5.1 and 6.5.2, as the batch and online learning methods in this chapter are not designed for concept drift, we use the data of around 100 k dies (2.5% actual failure rate) produced in a short period of time, 8 days, to implement the algorithms. We assume that no major change in the manufacturing process takes place in this period and thus the concept drift in the data stream over this interval is negligible. In Sect. 6.5.3, the data of dies produced in a larger time span, 1 month, is used (around 370 k dies in total), and the modified Learn++.NIE is used to handle concept drift and class imbalance simultaneously.

6.5.1 RUSBoost on Imbalanced Dataset

This section examines the performance of RUSBoost on the imbalanced memory test dataset. Our first experiment is to show the effectiveness of RUSBoost on an extremely imbalanced memory test dataset. We use 40 decision trees in an ensemble, and the learning rate in fitting is set as 0.15. Figure 6.9 shows typical ROC curves of RUSBoost and AdaBoost on an imbalanced memory test dataset; AdaBoost largely fails on the dataset, while RUSBoost maintains high performance. Since AdaBoost tends to classify nearly all dies as pass, its TPR is very low. For example, typical confusion matrices by setting threshold to 0.5 are shown in Tables 6.2 and 6.3. As expected, AdaBoost tends to label nearly all dies as pass to achieve both simplicity

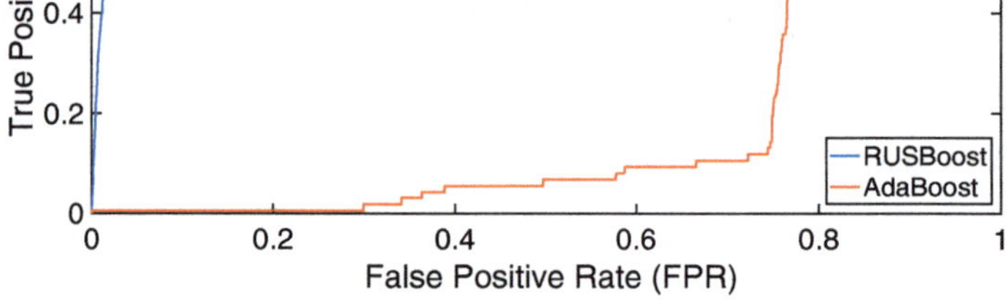

Fig. 6.9 The ROC curves of RUSBoost and AdaBoost on an imbalanced memory test dataset

Table 6.2 A typical confusion matrix using AdaBoost, threshold 0.5

	Predicted as fail	Predicted as pass	
True fail	1	633	TPR = 0.2%
True pass	6	29,360	FPR = 0.0%

Table 6.3 A typical confusion matrix using RUSBoost, threshold 0.5

	Predicted as fail	Predicted as pass	
True fail	523	111	TPR = 82.5%
True pass	2518	26,848	FPR = 8.8%

and high accuracy. Then, only 0.2% (1 out of 634) of the real bad dies are detected by the classifier, and the economic benefit generated is negligible, or even negative. In contrast, RUSBoost identifies 523 of 634 bad dies, with significant potential economic saving as a result.

Recalling Eq. (6.8), the expected packaged memory yield improvement is a function of TPR and FPR. However, FPR and TPR are constrained by the ROC curve of the classifier. Then, the optimal point (FPR*, TPR*) is where the contour plot of expected yield improvement is tangent to the ROC curve, and the corresponding optimal threshold η^* is determined for future prediction. Figure 6.10 gives the ROC curve and contour plots of expected yield improvement $\frac{\mathbb{E}[m_2(s)-m_1(s)]}{n/s}$ with different s (number of dies in a package) for the data in our application. As expected, with larger s, we can achieve a larger yield improvement with more dies in a package. When $s = 16$, the maximum yield improvement is as large as 20.48% for this case of no concept drift, over an 8-day period. With smaller s, the yield without a classifier itself is very high, so even 1% of improvement is difficult, though still economically valuable. Also, the optimal threshold η^* decreases as s increases, which means that more dies will be classified as fail. Intuitively, as s increases, the cost of mistakenly labeling a real bad die as a good die increases, since more good dies in its package will be wasted. Then, the optimal classifier tends to decrease the threshold to detect more real bad dies. Additionally, note that the contour plots of the expected yield improvement using (6.8) are approximately straight lines, which proves that the linearization in (6.9) is valid in our case.

6.5.2 *The Effectiveness of Online Learning*

With the same dataset, once we know the memory test result of a die in the test set, it becomes our new training instance for updating the online learning classifier in Algorithm 2. As online boosting methods are designed to approximate the batch learning asymptotically, each die in the test set is predicted by a classifier trained on the training set and all the test instances before it. As shown in Fig. 6.11, all base classifiers of both batch and online models are trained on the training set. In the batch learning scenario, these base classifiers are combined to test the dies in the test set without any further training. In the online learning scenario, however, the dies in the test set are received sequentially. The test results are then used to update the online classifiers to achieve higher prediction accuracy. The online classifier is tested on the same dataset we use in Sect. 6.5.1. As Fig. 6.12 indicates, online learning

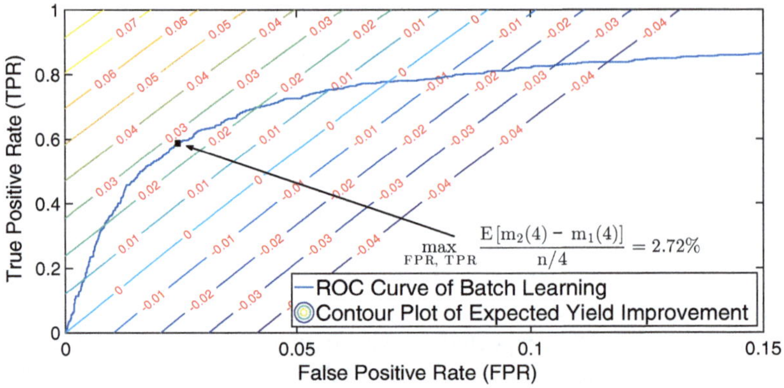

$s = 4$, yield without a classifier is 88.06%, yield with a classifier is 90.78%, $\eta^* = 0.75$.

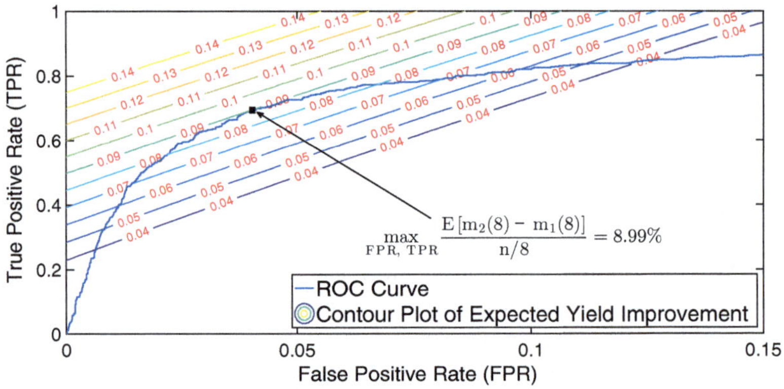

$s = 8$, yield without a classifier is 77.54%, yield with a classifier is 86.53%, $\eta^* = 0.68$.

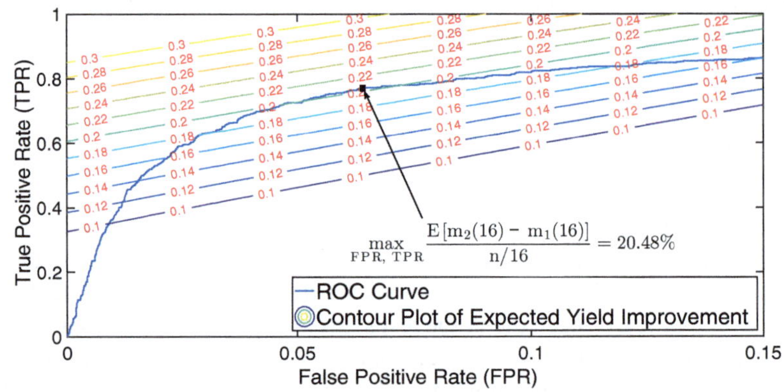

$s = 16$, yield without a classifier is 60.12%, yield with a classifier is 80.60%, $\eta^* = 0.60$.

Fig. 6.10 Contour plots of yield improvement $\frac{\mathbb{E}[m_2(s)-m_1(s)]}{n/s}$ with the number of dies in a package $s = 4$, $s = 8$, and $s = 16$. The yield improvement is optimized under the constraint of ROC curve that is determined by the classifier. With more dies in a package, we can potentially achieve larger yield improvement

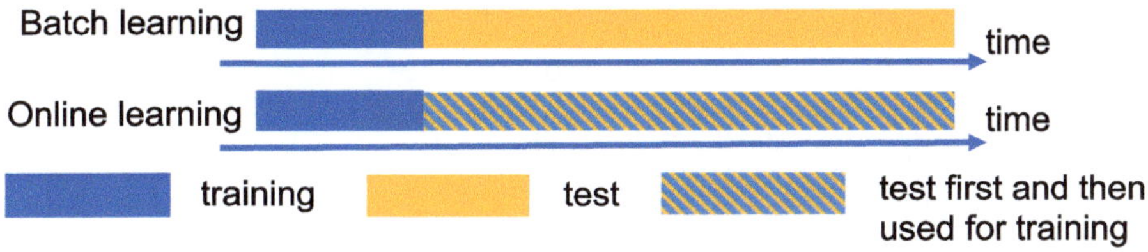

Fig. 6.11 The batch and online learning scenarios. All base classifiers of both batch and online models are trained on the training set. In the batch learning scenario, these base classifiers are combined to test the dies in the test set without any further training. In the online learning scenario, however, the test results are then used to update the model

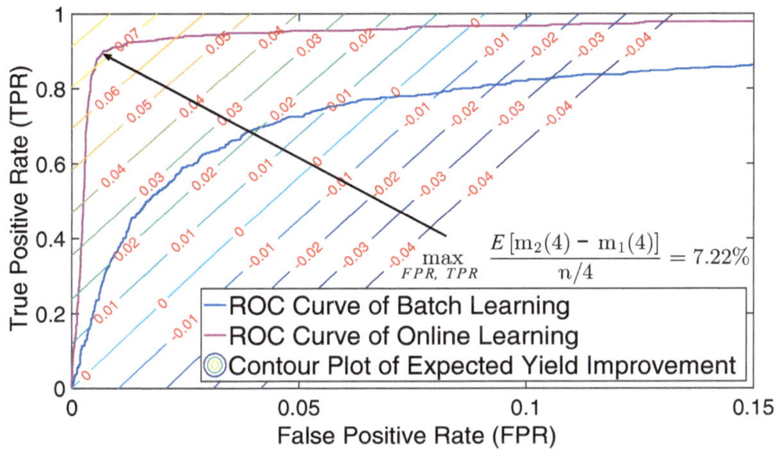

Fig. 6.12 Contour plots of yield improvement with the number of dies in a package $s = 4$. Yield with a batch classifier is 90.78%, while yield with an online classifier is 95.28%, compared to yield with no classifier of 88.06%

could enable as much as an extra 4.5% of expected yield improvement compared to the batch classifier when $s = 4$, primarily by a much improved classifier and corresponding ROC curve, when there is no concept drift present.

6.5.3 Incremental Learning with Concept Drift

Finally, we consider concept drift. We first show that concept drift is indeed present in our data within relevant time intervals. We then demonstrate the effectiveness of our incremental learning approach with this concept drift.

We consider the following experiment to prove the existence of concept drift. In this experiment, our larger data stream (about 370 k dies) is divided into five sequential chunks, each covering approximately one week. Half of the third chunk

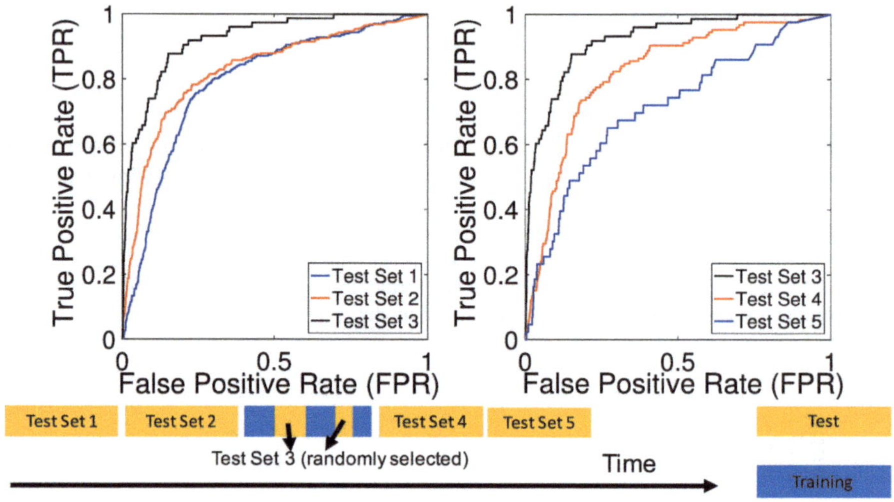

Fig. 6.13 Existence of concept drift. The data stream is divided into five chunks. The training set is randomly selected from chunk 3. The remaining dies in chunk 3 are used as test set 3. The classifier is tested on all five test sets. The performance of the classifier decays over time (both forward and backward) as the ROC curves show, confirming substantial concept drift

is randomly selected as the training set. The rest of the dies in chunk 3 are used as test set 3. Then, the classifier is tested on all five test sets. Figure 6.13 shows the ROC curves obtained on the five test sets. Clearly, the classifier's performance on test sets 1, 2, 4, and 5 are worse than on test set 3, especially in the low false alarm region that we care about most. This phenomenon indicates the existence of concept drift, where the earlier time periods (1 and 2) and the later time periods (4 and 5) have different underlying statistical structures than period 3 used to build the classifier; indeed, the degradation in the ROC is quite large in our data. Figure 6.14 shows the degradation of AUC value over time. A RUSBoost classifier is trained with data from week 0 and tested on data from week 1, 2, ..., 9. If we choose 0.85 as the threshold of major concept drift, only data within two weeks may be used.

As mentioned in Sect. 6.3.3, in a drifting environment, expected yield improvement $\frac{\mathbb{E}(m_2 - m_1)}{n/s}$ in the future can no longer be estimated by old data. The choice of the optimal threshold in classification then requires expert advice or evaluation of cost tradeoff [8]. Here, we assess the overall performance of the classifiers by their AUC values over time. We compare the AUC value of our modified Learn++.NIE ensemble (incremental learning) against a single RUSBoost classifier trained only on the most recent data chunk. Note that instead of using decision trees, the incremental learning takes RUSBoost classifiers trained on each data chunk as base classifiers, since data chunks are imbalanced. We divide our 370 k dataset from one month into 14 time steps with two to three days in a time step. As the number of dies from each day may differ, we make sure that in each data chunk we have approximately the same number of dies. Then, we test the incremental learning

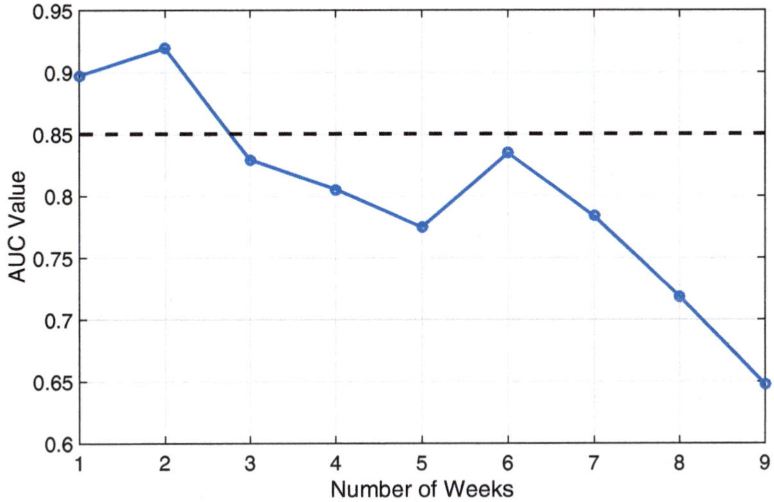

Fig. 6.14 The AUC values obtained on test sets from each week. The model is trained on data from week 0. The threshold of major concept drift is set as 0.85 (dashed line)

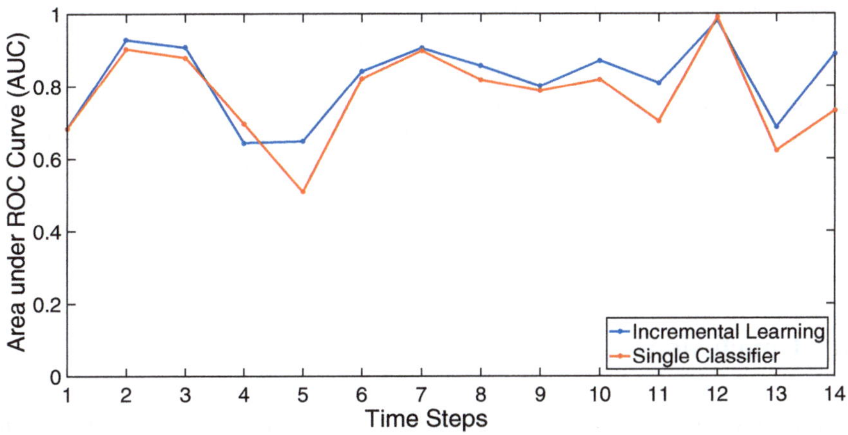

Fig. 6.15 Average area under ROC curves obtained on data chunks by incremental learning, and by a single classifier trained only on the most recent data chunk

method and single classifier on this data stream. This procedure is repeated 30 times, and the average AUC results are shown in Fig. 6.15.

With old base classifiers stored, incremental learning outperforms a single RUSBoost classifier on the whole. The average AUC value with a 95% confidence interval of the single classifier is 0.79 ± 0.002, while that of the incremental learning is 0.82 ± 0.001 (a 4% improvement). If we do have expert advice for the optimal threshold, average expected yield improvement with a 95% confidence interval by our incremental learning and a single classifier is given in Table 6.4, where our

Table 6.4 Average expected yield improvement with a 95% confidence interval by incremental learning and by a single classifier when optimal thresholds are given

s	Incremental learning	Single classifier	Improvement factor
4	$0.36\% \pm 0.03\%$	$0.08\% \pm 0.02\%$	$4.2\times$
8	$1.4\% \pm 0.03\%$	$0.6\% \pm 0.1\%$	$2.2\times$
16	$3.4\% \pm 0.05\%$	$2.4\% \pm 0.1\%$	$1.4\times$

Table 6.5 Average expected yield improvement with a 95% confidence interval by incremental learning and by a single classifier when FPR = 5% thresholds are given

s	Incremental learning	Single classifier	Improvement factor
4	$-3.6\% \pm 0.02\%$	$-4.2\% \pm 0.10\%$	–
8	$-2.9\% \pm 0.03\%$	$-3.4\% \pm 0.08\%$	–
16	$0.0\% \pm 0.03\%$	$-1.9\% \pm 0.08\%$	–

incremental learning increases the average expected yield improvement by $1.4\times$ to $4.2\times$ compared to the single classifier.

It is important to note that, in a nonstationary environment with class imbalance, a sophisticated method to choose the threshold for classification is necessary. For example, an intuitive way is to fix the false positive rate (FPR). Though this method also uses assumed or historical information about the joint distribution of H (true condition) and y (predicted condition), we may not achieve expected yield improvement with this intuitive naive approach. The result of expected yield improvement using thresholds with fixed FPR at 5% for our dataset is shown in Table 6.5, which indicates negative yield improvement for both incremental learning and single classifier. In contrast, optimal selection of thresholds can generate substantial yield improvements, as shown in Table 6.4.

6.6 Conclusions

In this chapter, we identify two key challenges in building practical learning models for semiconductor manufacturing process yield improvement, class imbalance, and concept drift. Batch, online, and incremental learning frameworks are presented to resolve these challenges. As an application, we study the packaging process of flash memory. We first show the possibility of memory package yield improvement in expectation by adding a classifier before packaging. The effectiveness of the algorithms are demonstrated on real data from industry. In the future, we will investigate methods to determine the optimal classification threshold in nonstationary environments. We will also devise more sophisticated models to predict defect categories. We hope that our work will attract more researchers in the CAD and machine learning community to the semiconductor manufacturing area.

Acknowledgements We thank SanDisk Semiconductor (Shanghai) Co. Ltd. (SDSS) for assistance and support. We also thank Professor Roy Welsch from MIT Sloan School of Management, and Honghong Chen, Fangfang You, Wenting Zhang, and Yew Wee Cheong from SDSS for helpful discussions.

References

1. A. Kusiak, Rough set theory: a data mining tool for semiconductor manufacturing. IEEE Trans. Electron. Packag. Manuf. **24**(1), 44–50 (2001)
2. C.K. Shin, S.C. Park, A machine learning approach to yield management in semiconductor manufacturing. Int. J. Prod. Res. **38**(17), 4261–4271 (2000)
3. S.M. Weiss, A. Dhurandhar, R.J. Baseman, Improving quality control by early prediction of manufacturing outcomes, in *Proceedings of the 19th ACM SIGKDD International Conference on Knowledge Discovery and Data Mining* (2013), pp. 1258–1266
4. H. Chen, D.S. Boning, Online and incremental machine learning approaches for IC yield improvement, in *IEEE/ACM 36th International Conference on Computer-Aided Design* (2017)
5. H. Chen, Novel machine learning approaches for modeling variations in semiconductor manufacturing, Master's thesis, Massachusetts Institute of Technology, 2017
6. N. Japkowicz, S. Stephen, The class imbalance problem: a systematic study. Intell. Data Anal. **6**(5), 429–449 (2002)
7. A. Tsymbal, The problem of concept drift: definitions and related work, Technical Report, vol. 106, no. 2, Computer Science Department, Trinity College Dublin, 2004
8. N.A. Arnold, Wafer defect prediction with statistical machine learning, Master's thesis, Massachusetts Institute of Technology, 2016
9. Y. Sun, A.K. Wong, M.S. Kamel, Classification of imbalanced data: a review. Int. J. Pattern Recognit. Artif. Intell. **23**(04), 687–719 (2009)
10. H. He, E.A. Garcia, Learning from imbalanced data. IEEE Trans. Knowl. Data Eng. **21**(9), 1263–1284 (2009)
11. Y. Freund, R.E. Schapire, Experiments with a new boosting algorithm, in *ICML*, vol. 96 (1996), pp. 148–156
12. C. Seiffert, T.M. Khoshgoftaar, J. Van Hulse, A. Napolitano, RUSBoost: a hybrid approach to alleviating class imbalance. IEEE Trans. Syst. Man Cybern. Part A Syst. Humans **40**(1), 185–197 (2010)
13. H. Zhang, B. Yu, E.F. Young, Enabling online learning in lithography hotspot detection with information-theoretic feature optimization, in *IEEE/ACM 35th International Conference on Computer-Aided Design* (2016)
14. B. Wang, J. Pineau, Online bagging and boosting for imbalanced data streams. IEEE Trans. Knowl. Data Eng. **28**(12), 3353–3366 (2016)
15. N.C. Oza, Online bagging and boosting, in *IEEE International Conference on Systems, Man and Cybernetics, 2005*, vol. 3 (2005), pp. 2340–2345
16. T.R. Hoens, R. Polikar, N.V. Chawla, Learning from streaming data with concept drift and imbalance: an overview. Prog. Artif. Intell. **1**(1), 89–101 (2012)
17. G. Ditzler, R. Polikar, Incremental learning of concept drift from streaming imbalanced data. IEEE Trans. Knowl. Data Eng. **25**(10), 2283–2301 (2013)
18. R. Elwell, R. Polikar, Incremental learning of concept drift in nonstationary environments. IEEE Trans. Neural Netw. **22**(10), 1517–1531 (2011)
19. S.R. Safavian, D. Landgrebe, A survey of decision tree classifier methodology. IEEE Trans. Syst. Man Cybern. **21**(3), 660–674 (1991)
20. J. Van Hulse, T.M. Khoshgoftaar, A. Napolitano, Experimental perspectives on learning from imbalanced data, in *Proceedings of the 24th ICML* (2007), pp. 935–942

Chapter 7
Efficient Process Variation Characterization by Virtual Probe

Jun Tao, Wangyang Zhang, Xin Li, Frank Liu, Emrah Acar, Rob A. Rutenbar, Ronald D. Blanton, and Xuan Zeng

7.1 Introduction

As integrated circuits (ICs) scale to finer feature size, it becomes increasingly difficult to control process variations for nanoscale technologies [1, 2]. The increasing fluctuations in manufacturing process introduce unavoidable and significant uncertainties in circuit performance. Hence, modeling and analyzing these variations to ensure manufacturability and improve parametric yield has been identified as a top priority for today's IC design.

J. Tao (✉) · X. Zeng (✉)
State Key Laboratory of ASIC and System, School of Microelectronics, Fudan University, Shanghai, China
e-mail: taojun@fudan.edu.cn; xzeng@fudan.edu.cn

W. Zhang
Cadence Design Systems, Inc., Pittsburgh, PA, USA

X. Li (✉)
Department of Electrical and Computer Engineering, Duke University, Durham, NC, USA
e-mail: xinli.ece@duke.edu

F. Liu
IBM Research Laboratory, Austin, TX, USA
e-mail: frankliu@us.ibm.com

E. Acar
IBM T. J. Watson Research Center, Yorktown Heights, NY, USA
e-mail: emrah@us.ibm.com

I. M. Elfadel et al. (eds.), *Machine Learning in VLSI Computer-Aided Design*,
https://doi.org/10.1007/978-3-030-04666-8_7

Towards this goal, various techniques have been proposed for statistical IC analysis and optimization, e.g., design centering, statistical timing analysis [3–6], and post-silicon tuning [7–9]. These techniques aim to predict and, consequently, minimize circuit-level performance variations in order to create a robust design with high parametric yield. The efficiency of these methods relies heavily on the accuracy of the variation model (e.g., distribution and correlation) that provides the important information about manufacturing uncertainties.

Accurately extracting the variation model, however, is not trivial. Silicon wafers/chips must be carefully tested and characterized using multiple test structures (e.g., I-V structures and ring oscillators) deployed in wafer scribe lines and/or within product chips [10–14]. The traditional silicon characterization suffers from three major issues:

- **Large area overhead:** Today's advanced microprocessor chips typically contain hundreds of on-chip ring oscillators to characterize and monitor parametric variations, resulting in significant overhead in silicon area [11].
- **Long testing time:** Physically measuring all test structures through a limited number of I/O ports consumes a large amount of testing time [12]. In nanoscale technologies, IC testing has contributed to a significant portion of the total manufacturing cost [15].
- **Low testing reliability:** IC testing may even damage the wafer/chip being tested. For instance, wafer probe test may permanently damage the wafer due to mechanical stress [12].

The combination of these critical issues results in continuously growing silicon characterization cost, as more and more test structures must be added to capture the complicated spatial variations of small devices. Even though silicon characterization has been extensively studied in the past, there is an immediate need to revisit this area and develop a more efficient methodology to reduce cost.

To this end, we ask the following fundamental question: *How many test structures are minimally required to fully capture the spatial variation information*? A quick answer to this question can be made based on the well-known Nyquist–Shannon sampling theorem [16]. Namely, if the variations contain no spatial frequency higher than f_{MAX}, the sampling frequency must be at least $2 \cdot f_{\mathrm{MAX}}$, i.e., test structures must be spaced at most $1/(2 \cdot f_{\mathrm{MAX}})$ apart.

The Nyquist sampling theorem generally assumes that all frequency components below the maximum frequency f_{MAX} may exist. However, this is not true

R. A. Rutenbar
Department of Computer Science, University of Illinois at Urbana-Champaign, Urbana, IL, USA
e-mail: rutenbar@illinois.edu

R. D. Blanton
Department of Electrical and Computer Engineering, Carnegie Mellon University, Pittsburgh, PA, USA
e-mail: blanton@ece.cmu.edu

for most silicon characterization applications. As will be demonstrated by the industrial measurement data in Sect. 7.5, spatial variations typically have a sparse representation in frequency domain (i.e., a large number of Fourier coefficients are almost zero). In this case, simply sampling at Nyquist rate generates a large number of redundant data. Such redundancy has been observed in many other application domains. For example, the key idea of image compression is to remove the redundancy and represent the information in a compact form [17]. However, our silicon characterization problem is substantially different from image compression, as we do not want to fully sample spatial variations at Nyquist rate and then "compress" them. Instead, we want to avoid redundant sampling in the first place to reduce characterization cost. The challenging issue here is how to efficiently sample few test structures on a wafer/chip and then accurately recover the essential spatial variation information.

In this chapter, we exploit recent advances in statistics (known as *compressed sensing* [18–21]) to develop a novel framework of *virtual probe* (VP) for low-cost silicon testing and characterization. Our goal is to accurately predict the spatial variations of a wafer/chip by measuring very few test structures at a set of selected locations. The proposed VP algorithm is derived from *maximum* a posteriori (MAP) estimation [22]. It is mathematically formulated as a linear programming problem that can be solved both robustly and efficiently. Most importantly, several theoretical studies from the statistics community have proved that by exploring the sparse pattern in spatial frequency domain, VP can fully reconstruct the spatial variations with probability nearly equal to 1, even if the spatial sampling frequency is much lower than the Nyquist rate [18–21]. As will be demonstrated by the industrial examples in Sect. 7.5, VP shows superior accuracy over several traditional methods, e.g., two-dimensional interpolation [23], Kriging prediction [24], and k-LSE estimation [25].

The remainder of this chapter is organized as follows. In Sect. 7.2, we develop the mathematical formulation of VP and, then, discuss the implementation details in Sect. 7.3. Next, several possible applications of VP are briefly discussed in Sect. 7.4. The efficacy of VP is demonstrated by a number of examples with industrial measurement data in Sect. 7.5. Finally, we conclude in Sect. 7.6.

7.2 Virtual Probe

The key idea of virtual probe (VP) is to deploy and measure very few test structures at a set of selected locations of a wafer/chip. The parametric variations at other locations are not directly measured by silicon testing. Instead, virtual probes are conceptually added at these locations to predict the variation information through the use of a statistical algorithm, as shown in Fig. 7.1. In other words, unlike the traditional approach that uses a large number of test structures, we propose to physically monitor the variability at very few locations and then apply a "smart" algorithm to accurately predict the complete spatial variation. In this section, we

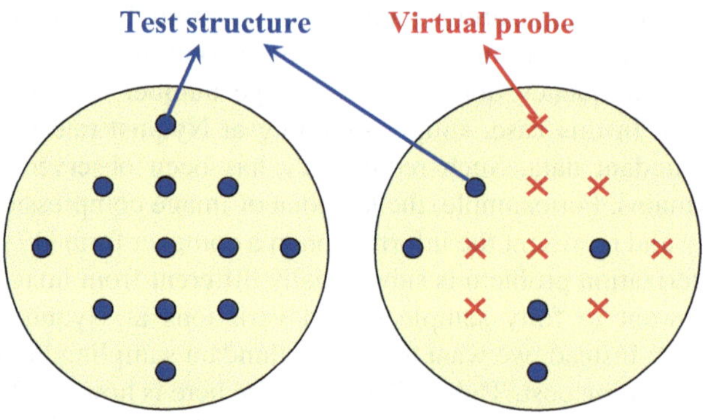

Fig. 7.1 An example of the proposed virtual probes. (Left) Traditionally, a large number of test structures are deployed and measured to fully characterize process variations. (Right) We propose to deploy and measure very few test structures, and virtual probes are conceptually added to fully recover spatial variations through the use of a statistical algorithm

first derive the mathematical formulation of VP based on spatial frequency-domain analysis. Next, we derive a maximum a posteriori (MAP) algorithm to solve the VP problem by exploring the unique sparse pattern in frequency domain. Finally, the accuracy of the MAP estimation is justified by studying several important theorems recently developed in the statistics community [18–21].

7.2.1 Mathematical Formulation

Mathematically, the spatial variations of a performance of interest (e.g., the frequency of a ring oscillator) can be expressed as a two-dimensional function $g(x, y)$, where x and y represent the coordinates of a spatial location on a wafer or chip. If $g(x, y)$ contains no spatial frequency higher than f_{MAX}, the Nyquist–Shannon sampling theorem [16] tells us to sample $g(x, y)$ with a frequency of $2 \cdot f_{MAX}$ in order to perfectly recover the continuous function $g(x, y)$.

Mathematically, the function $g(x, y)$ can be mapped to the frequency domain by a number of two-dimensional linear transforms such as Fourier transform [16], discrete cosine transform (DCT) [17], and wavelet transform [17]. In this chapter, we use DCT to illustrate the basic idea of VP. It should be noted, however, that the proposed VP framework can also be implemented with other linear transformations.

We discretize the two-dimensional function $g(x, y)$ at a spatial frequency higher than the Nyquist rate. Without loss of generality, we denote the coordinates x and y as integers $x \in \{1, 2, \ldots, P\}$ and $y \in \{1, 2, \ldots, Q\}$ after discretization. The DCT transform can be represented as [17]:

$$G(u,v)=\sum_{x=1}^{P}\sum_{y=1}^{Q}\alpha_u\cdot\beta_v\cdot g(x,y)\cdot\cos\frac{\pi\cdot(2x-1)\cdot(u-1)}{2\cdot P}\cdot\cos\frac{\pi\cdot(2y-1)\cdot(v-1)}{2\cdot Q},\tag{7.1}$$

where:

$$\alpha_u=\begin{cases}\sqrt{1/P} & (u=1)\\ \sqrt{2/P} & (2\le u\le P)\end{cases},\tag{7.2}$$

$$\beta_v=\begin{cases}\sqrt{1/Q} & (v=1)\\ \sqrt{2/Q} & (2\le v\le Q)\end{cases}.\tag{7.3}$$

In (7.1), $\{G(u,v); u=1,2,\ldots,P; v=1,2,\ldots,Q\}$ represents a set of DCT coefficients. Equivalently, the sampling values $\{g(x,y); x=1,2,\ldots,P; y=1,2,\ldots,Q\}$ can be expressed as the linear combination of $\{G(u,v); u=1,2,\ldots,P; v=1,2,\ldots,Q\}$ by the inverse discrete cosine transform (IDCT) [17]:

$$g(x,y)=\sum_{u=1}^{P}\sum_{v=1}^{Q}\alpha_u\cdot\beta_v\cdot G(u,v)\cdot\cos\frac{\pi\cdot(2x-1)\cdot(u-1)}{2\cdot P}\cdot\cos\frac{\pi\cdot(2y-1)\cdot(v-1)}{2\cdot Q}.\tag{7.4}$$

From (7.1)–(7.4), it is easy to verify that once the sampling values $\{g(x,y); x=1,2,\ldots,P; y=1,2,\ldots,Q\}$ are known, the DCT coefficients $\{G(u,v); u=1,2,\ldots,P; v=1,2,\ldots,Q\}$ are uniquely determined, and vice versa.

The proposed VP framework, however, will go one step further. Our objective is to accurately recover $\{g(x,y); x=1,2,\ldots,P; y=1,2,\ldots,Q\}$ from a *very small* number of (say, M) samples at the locations $\{(x_m,y_m); m=1,2,\ldots,M\}$ where $M\ll PQ$. In other words, the recovery can be formulated as the following linear equation:

$$\mathbf{A}\cdot\boldsymbol{\eta}=\mathbf{b},\tag{7.5}$$

where:

$$\mathbf{A}=\begin{bmatrix}A_{1,1,1} & A_{1,1,2} & \cdots & A_{1,P,Q}\\ A_{2,1,1} & A_{2,1,2} & \cdots & A_{2,P,Q}\\ \vdots & \vdots & \ddots & \vdots\\ A_{M,1,1} & A_{M,1,2} & \cdots & A_{M,P,Q}\end{bmatrix},\tag{7.6}$$

$$A_{m,u,v} = \alpha_u \cdot \beta_v \cdot \cos\frac{\pi \cdot (2x_m - 1) \cdot (u-1)}{2 \cdot P} \cdot \cos\frac{\pi \cdot (2y_m - 1) \cdot (v-1)}{2 \cdot Q}, \tag{7.7}$$

$$\boldsymbol{\eta} = \begin{bmatrix} G(1,1) & G(1,2) & \cdots & G(P,Q) \end{bmatrix}^T, \tag{7.8}$$

$$\mathbf{b} = \begin{bmatrix} g(x_1, y_1) & g(x_2, y_2) & \cdots & g(x_M, y_M) \end{bmatrix}^T. \tag{7.9}$$

In (7.5)–(7.9), the DCT coefficients $\{G(u, v); u = 1, 2, \ldots, P; v = 1, 2, \ldots, Q\}$ are the problem unknowns. In other words, we need to determine $\{G(u, v); u = 1, 2, \ldots, P; v = 1, 2, \ldots, Q\}$ based on the measurement data $\{g(x_m, y_m); m = 1, 2, \ldots, M\}$. Once the DCT coefficients $\{G(u, v); u = 1, 2, \ldots, P; v = 1, 2, \ldots, Q\}$ are known, the function $\{g(x, y); x = 1, 2, \ldots, P; y = 1, 2, \ldots, Q\}$ can be easily calculated by the IDCT in (7.4).

Solving the linear equation $\mathbf{A} \cdot \boldsymbol{\eta} = \mathbf{b}$ in (7.5), however, is not trivial, since M (i.e., the number of equations) is vastly less than PQ (i.e., the number of unknowns). Namely, the linear equation in (7.5) is profoundly underdetermined. While Eq. (7.5) cannot be uniquely solved by a simple matrix inverse, its solution can be statistically determined by considering additional prior information via Bayesian inference, as will be discussed in the next subsection.

7.2.2 Maximum A Posteriori (MAP) Estimation

In this subsection, we describe an efficient algorithm using maximum a posteriori (MAP) estimation to statistically solve the linear equation in (7.5). Although the result of this subsection can be derived by applying a number of elegant statistics theorems [18–22, 26], we attempt to describe the MAP algorithm at a level that is intuitive to the CAD community. More mathematical details of MAP can be found in [18–21, 26] and [22].

To solve (7.5), we first define a so-called *prior distribution* for $\boldsymbol{\eta}$ [22]. Intuitively, the prior distribution represents our prior knowledge about $\boldsymbol{\eta}$ without seeing any measurement data. Such prior information helps us to further constrain the underdetermined linear equation $\mathbf{A} \cdot \boldsymbol{\eta} = \mathbf{b}$ in (7.5) so that a meaningful solution can be uniquely found. At first glance, this seems impossible, since we would expect that the spatial variations and, hence, the DCT coefficients in $\boldsymbol{\eta}$ are substantially different from wafer to wafer and from chip to chip. However, we will show in this chapter that $\boldsymbol{\eta}$ has a unique property that we can exploit to define the prior distribution.

Before moving forward, let us first examine the following example of an industrial IC design. We measure the flush delay of this circuit from 17 wafers, each containing 282 chips. Flush delay is the time for a transition to propagate across a scan chain. We calculate the DCT coefficients and plot the histogram of them in Fig. 7.2. We notice that the distribution has a sharp peak at zero. This implies that

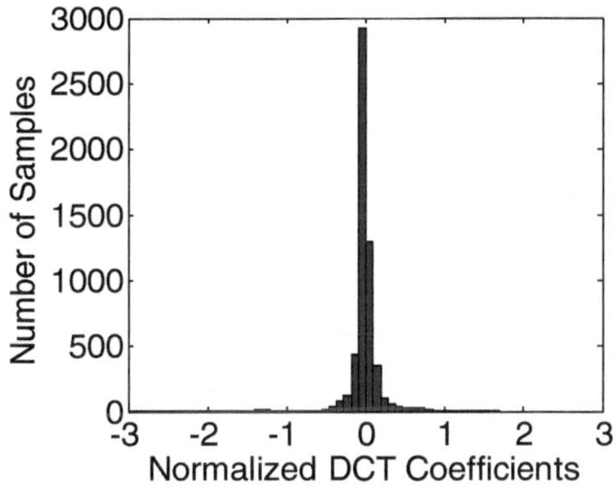

Fig. 7.2 Histogram of the normalized DCT coefficients calculated from 17 wafers for an industrial IC design example

most DCT coefficients are close to zero. In general, if the performance variations $\{g(x, y); x = 1, 2, \ldots, P; y = 1, 2, \ldots, Q\}$ present a spatial pattern, i.e., the variations are spatially correlated, the vector $\boldsymbol{\eta}$ that contains the corresponding DCT coefficients $\{G(u, v); u = 1, 2, \ldots, P; v = 1, 2, \ldots, Q\}$ is sparse. This unique property of sparseness has been observed in many image processing tasks [17], and has motivated the compressed sensing research for image recovery using a minimum number of samples [18–21, 26]. The previous research in compressed sensing shows that if most of these DCT coefficients are expected to be zero, we can reconstruct the image from a surprisingly small (i.e., compressed) set of samples. As will be demonstrated by several industrial examples in Sect. 7.5, this assumption of sparseness is also valid for our silicon characterization application.

While we assume that a large number of DCT coefficients are close to zero, we do not know the exact locations of these zeros. Otherwise, solving $\boldsymbol{\eta}$ from the linear equation $\mathbf{A} \cdot \boldsymbol{\eta} = \mathbf{b}$ in (7.5) becomes trivial. To find the unique solution $\boldsymbol{\eta}$ of the underdetermined linear equation $\mathbf{A} \cdot \boldsymbol{\eta} = \mathbf{b}$, we need to statistically model the histogram in Fig. 7.2 by using a zero-mean Laplace distribution to approximate the probability density function (PDF) of each DCT coefficient $\{\eta_i; i = 1, 2, \ldots, PQ\}$ [22]:

$$pdf(\eta_i) = \frac{1}{2\lambda} \cdot \exp\left(-\frac{|\eta_i|}{\lambda}\right) \qquad (i = 1, 2, \ldots, PQ), \tag{7.10}$$

where $pdf(\eta_i)$ stands for the PDF of η_i, and $\lambda > 0$ is a parameter that controls the variance of the distribution. The parameter λ in (7.10) can be optimally found by maximum likelihood estimation (MLE) [22]. Figure 7.3 shows the optimally fitted Laplace distribution for the data set in Fig. 7.2 . In practice, however, it is not necessary to know the value of λ. As will be shown in (7.17), the solution $\boldsymbol{\eta}$ is independent of the actual value of λ.

To completely define the prior distribution, we further assume that all DCT coefficients in the vector $\boldsymbol{\eta} \in \mathbb{R}^{PQ}$ are mutually independent. Hence, the joint PDF of $\boldsymbol{\eta}$ is represented as:

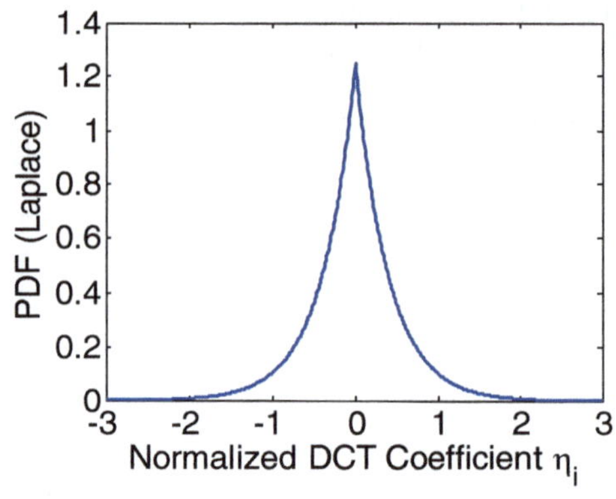

Fig. 7.3 Optimally fitted Laplace distribution for the normalized DCT coefficients calculated from 17 wafers for an industrial IC design example

$$\begin{aligned} pdf(\boldsymbol{\eta}) &= \left(\frac{1}{2\lambda}\right)^{PQ} \cdot \prod_{i=1}^{PQ} \exp\left(-\frac{|\eta_i|}{\lambda}\right) \\ &= \left(\frac{1}{2\lambda}\right)^{PQ} \cdot \exp\left(-\frac{\|\boldsymbol{\eta}\|_1}{\lambda}\right) \end{aligned}, \tag{7.11}$$

where $\| \bullet \|_1$ denotes the L_1-norm, i.e., the summation of the absolute value of all elements in the vector. The prior PDF in (7.11) has a threefold meaning:

1. The DCT coefficients $\{\eta_i; i = 1, 2, \ldots, PQ\}$ have a high probability to equal zero. This, in turn, implies the sparseness of $\boldsymbol{\eta}$.
2. The prior PDF in (7.11) treats each η_i equally. In other words, the prior PDF does not tell us which η_i is zero or nonzero. We need a "smart" algorithm to automatically find the nonzero coefficients based on a limited number of sampling points $\{g(x_m, y_m); m = 1, 2, \ldots, M\}$.
3. The independence assumption in (7.11) simply means that we do not know the correlation of $\boldsymbol{\eta}$ in advance. The correlation information will be taken into account by the posterior distribution (see (7.14)), once the measurement data are available.

Next, to derive the MAP algorithm, we need to introduce another important terminology—*likelihood function*—that is mathematically defined as the conditional probability $pdf(\mathbf{b}\,|\boldsymbol{\eta})$. It models the relationship between the measurement data in $\mathbf{b}$ and the unknown DCT coefficients in $\boldsymbol{\eta}$. Given the linear equation $\mathbf{A}\cdot\boldsymbol{\eta} = \mathbf{b}$ in (7.5), the measurement data in $\mathbf{b}$ and the DCT coefficients in $\boldsymbol{\eta}$ must satisfy the linear equation $\mathbf{A} \cdot \boldsymbol{\eta} = \mathbf{b}$ in (7.5). In other words, it is impossible to observe a set of variables $\boldsymbol{\eta}$ and $\mathbf{b}$ for which the equation $\mathbf{A} \cdot \boldsymbol{\eta} = \mathbf{b}$ does not hold. Hence, the likelihood function is a Dirac delta function where the conditional probability $pdf(\mathbf{b}\,|\boldsymbol{\eta})$ is nonzero if and only if $\mathbf{A} \cdot \boldsymbol{\eta}$ equals $\mathbf{b}$:

$$pdf\left(\mathbf{b}\left|\boldsymbol{\eta}\right.\right)=\begin{cases}\infty & (\mathbf{A}\cdot\boldsymbol{\eta}=\mathbf{b})\\ 0 & (\mathbf{A}\cdot\boldsymbol{\eta}\neq\mathbf{b})\end{cases}, \tag{7.12}$$

where

$$\int_{\mathbf{A}\cdot\boldsymbol{\eta}=\mathbf{b}} pdf(\mathbf{b}\left|\boldsymbol{\eta}\right.)\cdot d\mathbf{b}=1. \tag{7.13}$$

After defining the prior distribution in (7.11) and the likelihood function in (7.12), we are now ready to describe the MAP algorithm to uniquely determine the unknown DCT coefficients in $\boldsymbol{\eta}$. The key idea of MAP is to find the optimal solution $\boldsymbol{\eta}$ that maximizes the posterior distribution, i.e., the conditional PDF $pdf(\boldsymbol{\eta}\left|\mathbf{b}\right.)$. Namely, it aims to find the solution $\boldsymbol{\eta}$ that is most likely to occur. Based on Bayes' theorem [22], the posterior distribution $pdf(\boldsymbol{\eta}\left|\mathbf{b}\right.)$ is proportional to the prior distribution $pdf(\boldsymbol{\eta})$ and the likelihood function $pdf(\mathbf{b}\left|\boldsymbol{\eta}\right.)$:

$$pdf\left(\boldsymbol{\eta}\left|\mathbf{b}\right.\right)\propto pdf(\boldsymbol{\eta})\cdot pdf\left(\mathbf{b}\left|\boldsymbol{\eta}\right.\right). \tag{7.14}$$

Hence, the MAP algorithm attempts to solve the following optimization problem:

$$\max_{\boldsymbol{\eta}} \quad pdf(\boldsymbol{\eta})\cdot pdf\left(\mathbf{b}\left|\boldsymbol{\eta}\right.\right). \tag{7.15}$$

In our case, the likelihood function is a Dirac delta function, as shown in (7.12). Therefore, maximizing the posterior probability in (7.14) is equivalent to maximizing the prior probability in (7.11) subject to the linear constraint $\mathbf{A}\cdot\boldsymbol{\eta}=\mathbf{b}$:

$$\begin{array}{ll}\max\limits_{\boldsymbol{\eta}} & \dfrac{1}{(2\lambda)^{PQ}}\cdot\exp\left(-\dfrac{\|\boldsymbol{\eta}\|_1}{\lambda}\right)\\ s.t. & \mathbf{A}\cdot\boldsymbol{\eta}=\mathbf{b}\end{array}. \tag{7.16}$$

Since the exponential function $\exp(-\|\boldsymbol{\eta}\|_1/\lambda)$ where $\lambda > 0$ monotonically decreases in $\|\boldsymbol{\eta}\|_1$, the optimization in (7.16) can be rewritten as:

$$\begin{array}{ll}\max\limits_{\boldsymbol{\eta}} & \|\boldsymbol{\eta}\|_1\\ s.t. & \mathbf{A}\cdot\boldsymbol{\eta}=\mathbf{b}\end{array}. \tag{7.17}$$

Note that the optimization in (7.17) is independent of the parameter λ in (7.11).

Equation (7.17) is referred to as L_1-norm regularization in the literature [18–21, 26]. To illustrate the connection between L_1-norm regularization and sparse solution, we consider a simple two-dimensional example (i.e., $\boldsymbol{\eta}=[\eta_1\ \eta_2]^T$), as shown in Fig. 7.4. In this example, the equality constraint only consists of one linear equation and, hence, the feasible space of the constrained optimization problem can

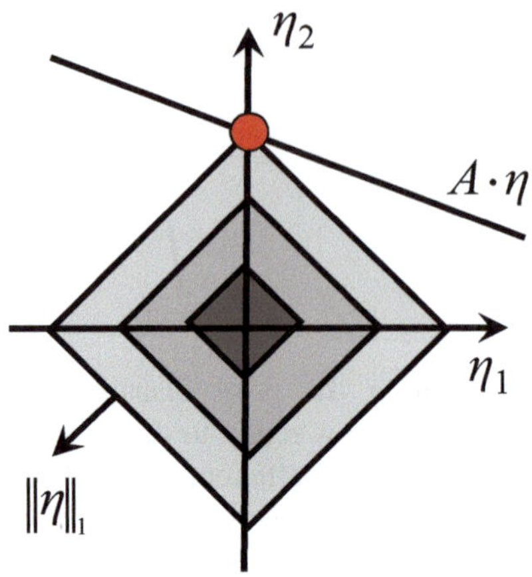

Fig. 7.4 The proposed L_1-norm regularization results in a sparse solution $\boldsymbol{\eta}$, as illustrated by the simple two-dimensional example

be represented by a line $\mathbf{A} \cdot \boldsymbol{\eta} = \mathbf{b}$ in the two-dimensional space. On the other hand, the contour lines of the cost function $\|\boldsymbol{\eta}\|_1$ correspond to a number of rotated squares. It can be seen from Fig. 7.4 that the optimal solution solved by L_1-norm regularization is located at one of the vertices of the contour lines. This observation implies that one of the coefficients (i.e., η_1 in this example) is exactly zero and, therefore, a sparse solution $\boldsymbol{\eta}$ is achieved.

Equation (7.17) can be converted to an equivalent linear programming problem and solved both robustly (i.e., with guaranteed global optimum) and efficiently (i.e., with low computational cost). The detailed algorithm of solving (7.17), as well as several other implementation issues, will be discussed in Sect. 7.3.

7.2.3 Accuracy of MAP Estimation

Given the prior distribution in (7.11), the MAP estimation, i.e., the L_1-norm regularization in (7.17), is statistically optimal, since it finds the solution $\boldsymbol{\eta}$ that maximizes the posterior probability, as shown in (7.15). However, it remains an open question if the accuracy of the MAP estimation can be quantitatively measured. In other words, we need to answer the following two questions:

1. Can the MAP estimation find the exact solution $\boldsymbol{\eta}$ for the underdetermined linear equation $\mathbf{A} \cdot \boldsymbol{\eta} = \mathbf{b}$?
2. If the answer is yes, what are the sufficient conditions to guarantee the finding of the exact solution $\boldsymbol{\eta}$?

In this subsection, we will answer these open questions by studying several important statistics theorems.

It has been proven in [18–21] that given the linear equation $\mathbf{A} \cdot \boldsymbol{\eta} = \mathbf{b}$ in (7.5), the accuracy of the MAP estimation depends on the orthonormality of the column vectors of the matrix $\mathbf{A} \in \mathbb{R}^{M \times PQ}$. To intuitively illustrate this concept, we first consider a trivial case where the number of equations (i.e., M) equals the number of unknowns (i.e., PQ) and, hence, $\mathbf{A}$ is a square matrix. Furthermore, we assume that all column vectors of $\mathbf{A}$ are orthonormal, i.e., $\mathbf{A}$ is an orthogonal matrix with $\mathbf{A}^T \cdot \mathbf{A} = \mathbf{I}$ where $\mathbf{I}$ is an identity matrix. In this trivial case, the exact solution η of $\mathbf{A} \cdot \boldsymbol{\eta} = \mathbf{b}$ can be accurately determined as:

$$\boldsymbol{\eta} = \mathbf{A}^T \cdot \mathbf{b}. \tag{7.18}$$

In practice, since VP aims to predict the spatial variations from very few samples, the linear equation $\mathbf{A} \cdot \boldsymbol{\eta} = \mathbf{b}$ in (7.5) is underdetermined and the matrix $\mathbf{A} \in \mathbb{R}^{M \times PQ}$ has more columns than rows (i.e., $M < PQ$). It is impossible for all columns of $\mathbf{A}$ to be orthonormal. In this case, it turns out that the solution $\boldsymbol{\eta}$ can be accurately found if the columns of $\mathbf{A}$ are approximately orthonormal. Based on the theorems of compressed sensing [18–21], the "orthonormality" of a matrix $\mathbf{A}$ can be quantitatively measured by its restricted isometry property (RIP).

Definition 7.1 A matrix $\mathbf{A}$ satisfies the restricted isometry property (RIP) of order K with constant $\delta_K < 1$, if the inequality:

$$(1 - \delta_K) \cdot \|\boldsymbol{\eta}\|_2^2 \leq \|\mathbf{A} \cdot \boldsymbol{\eta}\|_2 \leq (1 + \delta_K) \cdot \|\boldsymbol{\eta}\|_2^2 \tag{7.19}$$

holds for every vector $\boldsymbol{\eta}$ that contains only K nonzero elements. In (7.19), $\| \bullet \|_2$ denotes the L_2-norm, i.e., the square root of the summation of the squares of all elements in the vector.

If all columns of the matrix $\mathbf{A} \in \mathbb{R}^{M \times PQ}$ are almost orthonormal, RIP should be satisfied with a large K and a small δ_K. In the extreme case where $\mathbf{A}$ is exactly an orthogonal matrix, $\|\mathbf{A} \cdot \boldsymbol{\eta}\|_2$ is equal to $\|\boldsymbol{\eta}\|_2$ for every vector $\boldsymbol{\eta} \in \mathbb{R}^{PQ}$, since the linear transformation by an orthogonal matrix does not change the L_2-norm of the vector $\boldsymbol{\eta}$ [27]. Hence, RIP is satisfied with $K = PQ$ and $\delta_K = 0$.

The concept of RIP has been successfully applied to assess the inherent difficulty of finding the exact solution $\boldsymbol{\eta}$ from the underdetermined linear equation $\mathbf{A} \cdot \boldsymbol{\eta} = \mathbf{b}$ in (7.5). From example, the following theorem has been shown in [19].

Theorem 7.1 *The L_1-norm regularization in (7.17) guarantees to find the exact solution $\boldsymbol{\eta}$ of the underdetermined linear equation $\mathbf{A} \cdot \boldsymbol{\eta} = \mathbf{b}$ in (5), if the following three conditions are all satisfied:*

1. *The solution vector $\boldsymbol{\eta}$ contains at most S nonzeros.*
2. *The matrix $\mathbf{A}$ satisfies the RIP of order $2S$ with constant $\delta_{2S} < 1$ and the RIP of order $3S$ with constant $\delta_{3S} < 1$.*
3. *The two RIP constants δ_{2S} and δ_{3S} further satisfy the inequality $\delta_{2S} + \delta_{3S} < 1$.*

Note that the conditions in Theorem 7.1 are sufficient but not necessary. A number of other sufficient conditions have also been derived in the literature. More details can be found in [19].

While RIP offers a solid theoretical foundation to assess the accuracy of the MAP estimation, computing the RIP constant δ_K for a given matrix $\mathbf{A}$ is an NP-hard problem [18–21]. For this reason, an alternative metric, coherence, has been proposed to measure the orthonormality of a matrix $\mathbf{A}$ [20].

Definition 7.2 Given a matrix $\mathbf{A}$ for which every column vector has unit length (i.e., unit L_2-norm), its coherence is defined as:

$$\mu = \max_{i \neq j} \left| \langle \mathbf{a}_i , \mathbf{a}_j \rangle \right| , \tag{7.20}$$

where $\mathbf{a}_i$ and $\mathbf{a}_j$ denote the ith and jth columns of $\mathbf{A}$ respectively, and $\langle \bullet, \bullet \rangle$ stands for the inner product of two vectors.

Similar to RIP, the coherence value μ in (7.20) offers a quantitative criterion to judge if the columns of the matrix $\mathbf{A}$ are approximately orthonormal. For instance, if all columns of $\mathbf{A}$ are orthonormal, the coherence value μ reaches the minimum (i.e., zero); otherwise, the coherence value μ is always greater than zero.

While the RIP constant δ_K in (7.19) is difficult to compute, the coherence value μ in (7.20) can be easily calculated by the inner product of column vectors. Once μ is known, the RIP constant δ_K is bounded by [20]:

$$\delta_K \leq \mu \cdot (K - 1) , \tag{7.21}$$

where K denotes the order of RIP. In other words, while the exact value of the RIP constant δ_K is unknown, its upper bound can be efficiently estimated by coherence. This, in turn, offers a computationally tractable way to verify the sufficient conditions in Theorem 7.1. More details on coherence and its applications can be found in [20].

The aforementioned discussions summarize the theoretical framework to justify the accuracy of the MAP estimation. It demonstrates a number of sufficient conditions which guarantee to find the exact sparse solution $\boldsymbol{\eta}$ from the underdetermined linear equation $\mathbf{A} \cdot \boldsymbol{\eta} = \mathbf{b}$. In our application of spatial variation characterization, the number of nonzeros in the vector $\boldsymbol{\eta}$ is not known in advance. Hence, it can be difficult to verify the conditions in Theorem 7.1 and then determine if the exact solution $\boldsymbol{\eta}$ is accurately solved. However, the theoretical results summarized in this subsection demonstrate the importance of column orthonormality for the matrix $\mathbf{A}$ in (7.5). This, in turn, motivates us to develop efficient techniques to improve the column orthonormality and, hence, enhance the accuracy of VP. The details of these implementation issues will be discussed in Sect. 7.3.

7.3 Implementation Details

Our proposed VP technique is made of practical utility by carefully addressing several important implementation issues, including:

1. A column normalization scheme to improve the orthonormality of the matrix **A** for the underdetermined linear equation $\mathbf{A} \cdot \boldsymbol{\eta} = \mathbf{b}$ in (7.5);
2. A linear programming formulation to efficiently solve the L_1-norm regularization problem in (7.17);
3. A modified Latin hypercube sampling (M-LHS) scheme to randomly select the spatial sampling locations with small coherence; and
4. A DCT coefficient preselection scheme to further improve the prediction accuracy by carefully removing noncritical high-frequency DCT coefficients.

In this section, we describe these implementation details and highlight their novelties.

7.3.1 Normalization

As discussed in Sect. 7.2.3, all columns of the matrix **A** of the linear equation $\mathbf{A} \cdot \boldsymbol{\eta} = \mathbf{b}$ should be approximately orthonormal so that the MAP estimation is capable of accurately finding the solution $\boldsymbol{\eta}$. The requirement on orthonormality has a twofold meaning. First, the columns of the matrix **A** should be approximately orthogonal. Second, all these columns should have unit length (i.e., unit L_2-norm).

It is important to note that the column orthogonality of the matrix **A** cannot be enhanced by applying a simple orthogonalization algorithm, e.g., the Gram–Schmidt orthogonalization [27]. Such an orthogonalization process will change the solution vector $\boldsymbol{\eta}$ and compromise its unique sparse pattern.

The requirement on unit length, however, can be easily satisfied, if we normalize each column of the matrix **A** by its L_2-norm. It has been demonstrated by the statistics community that the aforementioned normalization can efficiently improve the accuracy of the MAP estimation [20]. Hence, it is adopted in this chapter and applied to (7.17), before the L_1-norm regularization problem is solved by a numerical solver.

7.3.2 Linear Programming

Once all columns of the matrix **A** are normalized to unit length, a numerical solver should be applied to solve the L_1-norm regularization problem in (7.17) and find the optimal solution $\boldsymbol{\eta}$. From (7.17), we would notice that the cost function $\|\boldsymbol{\eta}\|_1$ is not smooth and, hence, it cannot be easily minimized by a simple gradient-based algorithm [28]. To address this issue, we introduce a set of slack variables $\{\theta_i; i = 1, 2, \ldots, PQ\}$ and rewrite (7.17) as:

$$\begin{aligned} \min_{\boldsymbol{\eta},\boldsymbol{\theta}} \quad & \theta_1 + \theta_2 + \cdots + \theta_{PQ} \\ s.t. \quad & \mathbf{A} \cdot \boldsymbol{\eta} = \mathbf{b} \\ & -\theta_i \leq \eta_i \leq \theta_i \qquad (i = 1, 2, \ldots, PQ) \end{aligned} \tag{7.22}$$

Intuitively, by minimizing the cost function in (7.22), all constraints $\{\theta_i \leq \eta_i \leq \theta_i; i = 1, 2, \ldots, PQ\}$ will become active, i.e., $\{|\eta_i| \leq \theta_i; i = 1, 2, \ldots, PQ\}$. For this reason, the optimizations in (7.17) and (7.22) are equivalent, i.e., they share the same optimal solution $\boldsymbol{\eta}$. This conclusion can be formally proven based on the Karush–Kuhn–Tucker condition from optimization theory [28].

Note that both the cost function and the constraints in (7.22) are linear. Therefore, it is a linear programming problem and can be solved both robustly (i.e., with guaranteed global optimum) and efficiently (i.e., with low computational cost), e.g., by using the interior-point method with polynomial complexity [28]. For large-scale problems, there exist a number of fast algorithms (e.g., TNIPM [29] with the complexity $O((PQ)^{1.2})$) that can solve (7.22) with millions of variables in a few minutes.

7.3.3 *Latin Hypercube Sampling*

The study in Sect. 7.2.3 shows that the accuracy of the MAP estimation depends on the orthonormality of the matrix **A** in (7.5). According to the definition of the matrix **A**, it can be easily seen that the value of **A** is determined by the sampling locations $\{(x_m, y_m); m = 1, 2, \ldots, M\}$. In other words, different choices of sampling locations will provide different values of the matrix **A** and, hence, different results of the MAP estimation. It, in turn, motivates us to develop an efficient algorithm to find a set of "good" sampling locations. As such, the orthonormality of the matrix **A** is well approximated, thereby resulting in high prediction accuracy for the MAP estimation.

While directly optimizing the orthonormality of the matrix **A**, e.g., minimizing the coherence value μ in (7.20), is not trivial, it has been proven that random sampling is able to result in a good matrix **A** [18–21]. In particular, the theoretical results in [18–21] demonstrate that if the vector $\boldsymbol{\eta} \in \mathbb{R}^{PQ}$ contains at most S ($S \ll PQ$) nonzeros and M sampling locations are randomly selected where M is in the order of $O\left(S \cdot \log\left(PQ\right)\right)$, the sufficient conditions in Theorem 7.1 are almost guaranteed to hold (i.e., with probability nearly equal to 1), implying that the exact value of $\boldsymbol{\eta}$ can be accurately determined with extremely high probability. To the best of our knowledge, there is no other sampling scheme that clearly outperforms random sampling.

Based on these observations, we adopt the random sampling strategy in this chapter. Our objective is to evenly distribute M random sampling points over the entire wafer/chip. To achieve this goal, we borrow the idea of Latin hypercube sampling (LHS) from the statistics community [30] and develop a modified

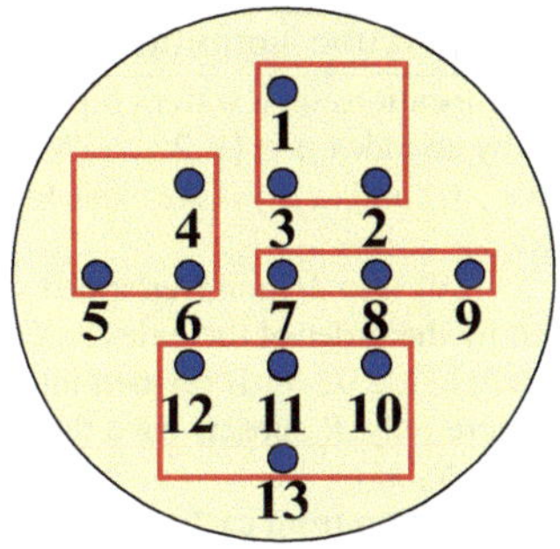

Fig. 7.5 A simple partition example for the modified Latin hypercube sampling (M-LHS) algorithm where 13 possible sampling locations are labeled as $\{(x_n, y_n); n = 1, 2, \ldots, 13\}$ and divided into 4 subsets $\{(x_n, y_n); n = 1, 2, 3\}$, $\{(x_n, y_n); n = 4, 5, 6\}$, $\{(x_n, y_n); n = 7, 8, 9\}$, and $\{(x_n, y_n); n = 10, 11, 12, 13\}$

Latin hypercube sampling (M-LHS) algorithm to generate well-controlled random samples.

Starting from N possible sampling locations on a wafer/chip, we first label each sampling location by an index $n \in \{1, 2, \ldots, N\}$. For illustration purpose, Fig. 7.5 shows a simple example where 13 sampling locations are sequentially labeled. This simple example reveals two important properties of our labeling scheme. First, if two different sampling locations (x_i, y_i) and (x_j, y_j) are in the same neighborhood, their indexes i and j should be close to each other. Second, the number of possible sampling locations (i.e., N) can be less than the total number of DCT coefficients (i.e., PQ), since the shape of a wafer is close to a circle, instead of a rectangle.

Next, to evenly distribute the M sampling locations among the N possible choices, we partition the index set $\{1, 2, \ldots, N\}$ into M nonoverlapped subsets. There are two possible scenarios when we construct these subsets:

1. M is a factor of N (i.e., M divides N without leaving a remainder). In this case, the partition of the set $\{1, 2, \ldots, N\}$ is simply determined by the order of the indexes. Namely, the M subsets are: $\{1, 2, \ldots, N/M\}$, $\{N/M + 1, N/M + 2, \ldots, 2 \cdot N/M\}$, etc.
2. M is not a factor of N and M divides N leaving a nonzero remainder R. In this case, we again divide the set $\{1, 2, \ldots, N\}$ into M nonoverlapping subsets based on the order of the indexes. However, the sizes of all subsets are different: either $\lfloor N/M \rfloor$ (the largest integer that is less than or equal to N/M) or $\lfloor N/M \rfloor + 1$. The choices between $\lfloor N/M \rfloor$ and $\lfloor N/M \rfloor + 1$ are randomly selected for each subset with the constraint that only R subsets have the size of $\lfloor N/M \rfloor + 1$ and the other $M - R$ subsets have the size of $\lfloor N/M \rfloor$. Hence, each of the N possible sampling locations belongs to one of the M subsets.

Figure 7.5 shows a simple partition example with 13 possible sampling locations and 4 subsets. In this case, since $M = 4$ is not a factor of $N = 13$, the sizes of these 4 subsets are not identical. Studying Fig. 7.5, we would notice that each of the

Algorithm 4 Modified Latin hypercube sampling (M-LHS)

1: Start from N possible sampling locations on a wafer/chip.
2: Label each sampling location by an index $n \in \{1, 2, \ldots, N\}$ so that the indexes i and j of two different sampling locations (x_i, y_i) and (x_j, y_j) are close to each other if (x_i, y_i) and (x_j, y_j) are in the same neighborhood.
3: Partition the index set $\{1, 2, \ldots, N\}$ into M nonoverlapped subsets. If M is a factor of N, the partition is simply determined by the order of the indexes. Otherwise, if M divides N leaving a nonzero remainder R, the set $\{1, 2, \ldots, N\}$ is divided into M nonoverlapped subsets based on the order of the indexes where only R subsets have the size of $\lfloor N/M \rfloor + 1$ and the other $M - R$ subsets have the size of $\lfloor N/M \rfloor$.
4: Randomly select one sampling location from each of the M subsets, resulting in M sampling locations in total.

M subsets generated by our partition scheme contains the sampling locations in the same neighborhood. In other words, the M subsets conceptually represent M local clusters spatially distributed over the wafer/chip.

As the final step of the proposed M-LHS algorithm, we randomly select one sampling location from each of the M subsets, thereby resulting in M sampling locations in total. Algorithm 4 summarizes the major steps of our M-LHS algorithm. Unlike the traditional LHS algorithm that aims to sample continuous random variables [30], Algorithm 4 has been particularly tuned to randomly select M choices out of N possible candidates. Hence, it is referred to as modified Latin hypercube sampling (M-LHS) in this chapter.

Compared to brute-force random sampling, M-LHS guarantees to distribute the M sampling locations over different local regions on the wafer/chip. Namely, it eliminates the possibility that many sampling locations are selected from the same local space. This, in turn, leads to superior prediction accuracy over brute-force random sampling, as will be demonstrated by the numerical examples in Sect. 7.5.

7.3.4 DCT Coefficient Preselection

It is shown in the previous subsection that the number of samples required for VP depends on the logarithm of the total number of DCT coefficients, i.e., $\log(PQ)$. If we know that a subset of DCT coefficients must be zero, we no longer need to consider these DCT coefficients as problem unknowns. Given a limited number of sampling points, such a preselection strategy for DCT coefficients can further improve the accuracy of the proposed VP algorithm.

As is demonstrated in the literature [14], spatial patterns of process variations are often smooth. It, in turn, implies that the spatial variation patterns may be accurately represented by a few dominant DCT coefficients at low frequencies. Namely, the low-frequency DCT coefficients are more important than the high-frequency ones when predicting spatial variations.

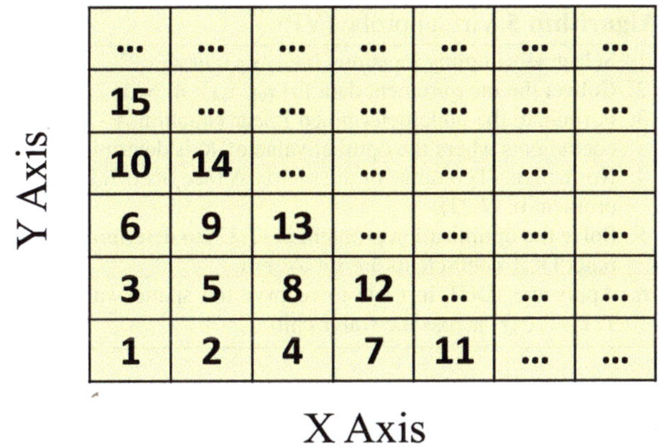

Fig. 7.6 The importance ranking of DCT coefficients defined in [25] where the rank "1" corresponds to the most important DCT coefficient

Motivated by this observation, we follow the idea proposed in [25] to rank the importance of all DCT coefficients, as shown in Fig. 7.6. When solving the linear equation $\mathbf{A} \cdot \boldsymbol{\eta} = \mathbf{b}$ by (7.17), we only consider the first K low-frequency DCT coefficients as problem unknowns. All other high-frequency DCT coefficients are simply set to zero. In other words, the optimization in (7.17) only needs to solve the sparse solution for K (instead of PQ) problem unknowns.

In our implementation, cross-validation [22] is further used to estimate the optimal value of K. An F-fold cross-validation partitions the sampling points into F groups. Prediction error is estimated from F independent runs. In each run, one of the F groups is used to estimate the prediction error, and all other groups are used to solve the L_1-norm regularization problem in (7.17) to determine the unknown DCT coefficients. Note that the training data for coefficient estimation and the testing data for error estimation are not overlapped. Hence, over-fitting can be easily detected. In addition, different groups are selected for error estimation in different runs. As such, each run results in an error value ε_f (where $f \in \{1, 2, \cdots, F\}$) that is measured from a unique set of testing data. The final prediction error is computed as the average of $\{\varepsilon_f; f = 1, 2, \ldots, F\}$, i.e., $\varepsilon = (\varepsilon_1 + \varepsilon_2 + \cdots + \varepsilon_F)/F$. The optimal value of K is determined to minimize the cross-validation error ε.

7.3.5 Summary

Algorithm 5 summarizes the major steps of the proposed VP method. It starts from very few (i.e., M) sampling locations $\{g(x_m, y_m); m = 1, 2, \ldots, M\}$ determined by M-LHS. Next, it formulates an underdetermined linear equation based on these measurement data and solves all DCT coefficients. Finally, the spatial variations $\{g(x, y); x = 1, 2, \ldots, P; y = 1, 2, \ldots, Q\}$ are recovered by the IDCT in (7.4).

In summary, the proposed VP method offers a number of important advantages over other traditional techniques:

Algorithm 5 Virtual probe (VP)

1: Select M sampling locations $\{(x_m, y_m); m = 1, 2, \ldots, M\}$ by M-LHS (i.e., Algorithm 4).
2: Collect the measurement data $\{g(x_m, y_m); m = 1, 2, \ldots, M\}$ at these locations.
3: Formulate the underdetermined linear equation $\mathbf{A} \cdot \boldsymbol{\eta} = \mathbf{b}$ in (7.5)–(7.9) for the first K DCT coefficients where the optimal value of K is determined by cross-validation (see Sect. 7.3.4).
4: Normalize all columns of the matrix $\mathbf{A}$ (see Sect. 7.3.1) and formulate the linear programming problem in (7.22).
5: Solve the optimization problem in (7.22) to determine $\boldsymbol{\eta}$, i.e., the first K DCT coefficients. All other DCT coefficients are set to zero.
6: Apply the IDCT in (7.4) to recover the spatial variations $\{g(x, y); x = 1, 2, \ldots, P; y = 1, 2, \ldots, Q\}$ across the wafer/chip.

- **Low cost**: VP is developed to minimize the number of test structures required to fully extract the spatial variation information. It, in turn, reduces the testing and measurement cost, e.g., area overhead, testing and characterization time, and yield loss during testing. In addition, the VP formulation in (7.22) is a linear programming problem and it can be solved both robustly (i.e., with guaranteed global optimum) and efficiently (i.e., with low computational cost).
- **High accuracy**: The accuracy of VP is guaranteed by the theoretical studies from the statistics community [18–21, 26], as discussed in Sect. 7.2.3. Namely, with several general assumptions, VP can fully reconstruct the spatial variations with probability nearly equal to 1. In addition, the accuracy of VP can be verified in real time by the cross-validation method mentioned in Sect. 7.3.4. This error estimation scheme is extremely important, since it provides a practical, quantitative criterion to determine whether the result of VP is sufficiently accurate or not. Additional sampling points can be further collected to improve accuracy, until the prediction error is sufficiently small.
- **General purpose**: VP can be used to predict the spatial pattern of both inter-die and spatially correlated intra-die variations. The prediction by VP is based on the measurement data collected from the current wafer/chip only. It does not require any historical data for training and, hence, can efficiently handle the nonstationary effects, e.g., process drifting caused by equipment aging. The only assumption posed by VP is that the spatial variations must have a sparse representation in frequency domain. This assumption is valid for spatially correlated process variations. In other words, the variations of interest are not dominated by independent random mismatches (e.g., random dopant fluctuations). As will be demonstrated by the experimental examples in Sect. 7.5, such a sparseness assumption holds for a number of performance variations (e.g., ring oscillator delay and full-chip leakage). The impact of independent random mismatches on these performance metrics is averaged out and, hence, becomes nondominant. In practice, the sparseness assumption can be verified by the error estimation scheme we previously mentioned. Namely, if the frequency-domain representation is not sparse, we will observe a large prediction error reported by VP.

7.4 Applications of Virtual Probe

The proposed VP method can be applied to a broad range of applications related to integrated circuits. In this section, we will briefly discuss these possible applications, including: (1) wafer-level silicon characterization (for inter-die variations), (2) chip-level silicon characterization (for intra-die variations), and (3) testing and self-healing of integrated circuits. Note that the main objective of this section is to motivate several possible future research directions based upon our VP technique. The details of these new research problems are beyond the scope of this chapter and, hence, are not discussed here.

7.4.1 Wafer-Level Silicon Characterization

To characterize parametric variations at wafer level (i.e., inter-die variations), test structures are deployed in wafer scribe lines [10–12], as shown in Fig. 7.7. These test structures do not have area overhead, as they are not within a product chip. However, it does not simply mean that the characterization is free. Instead, wafer-level characterization can still be expensive due to the following two reasons.

First, test structures in scribe lines must be measured by wafer probe test, as these devices will be completely destroyed during wafer dicing before packaging. Within this testing process, a probe card will contact the I/O pads of the test structures to measure currents, voltages, or frequencies. Such a wafer probe testing, however, is not perfectly safe. It may break the wafer being tested due to mechanical stress, create additional yield loss, and eventually increase manufacturing cost. Second, wafer probe test (e.g., aligning the probe card with the I/O pads and collecting all measurement data) is time consuming. This, in turn, further increases manufacturing cost, as the overall manufacturing time is increased.

For these two reasons, it is crucial to reduce the number of measured test structures so that the overall testing and characterization cost is minimized. Our proposed VP method perfectly fits this need. Namely, we propose to deploy and

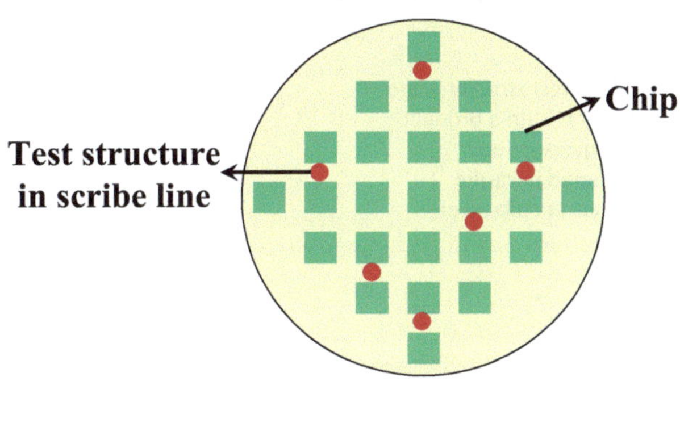

Fig. 7.7 Test structures are deployed in wafer scribe lines to measure and characterize inter-die variations at wafer level

measure very few test structures randomly distributed over the scribe lines of a wafer. Once the measurement data are collected, Algorithm 5 is applied to reconstruct the spatial variations across the wafer. Note that since the test structures are constrained within scribe lines, the aforementioned wafer-level characterization may not provide sufficient resolution to predict intra-die variations. It, therefore, implies that additional test structures are required for chip-level silicon characterization, as will be discussed in the next subsection.

7.4.2 Chip-Level Silicon Characterization

On-chip test structures are typically used to characterize intra-die variations at chip level [10–12], as shown in Fig. 7.8. The cost of chip-level characterization consists of two major portions: (1) area overhead and (2) testing time.

First, on-chip test structures are deployed within a product chip at a number of preselected locations. If too many test structures are used, they lead to significant area overhead and, hence, become financially intractable. Second, all on-chip test structures must be measured through a limited number of I/O pads. This testing process is time consuming and directly increases manufacturing cost.

Motivated by these observations, we propose to deploy and measure very few on-chip test structures and then apply VP to reconstruct the complete spatial variation pattern at chip level. As such, the characterization cost is substantially reduced.

7.4.3 Beyond Silicon Characterization

The silicon characterization results extracted by VP can be efficiently applied to a number of practical applications. In this subsection, we briefly discuss two important application examples: (1) speed binning and (2) post-silicon tuning.

In traditional speed binning, all manufactured chips are tested individually to determine the maximum operation frequency [15]. This is expensive, since each chip must be repeatedly tested with different speed setups. Given the proposed VP

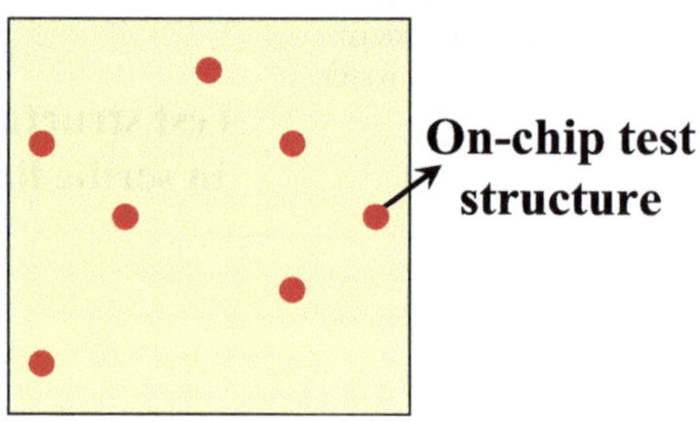

Fig. 7.8 Test structures are deployed within a product chip to measure and characterize intra-die variations at chip level

framework, we can potentially test a small number of chips to find their speed bins and then use VP to predict the speed of other chips on the same wafer. Note that even if the prediction by VP is not exact, it can still be used to optimize the testing scheme to reduce cost. For instance, if the speed of an untested chip is estimated by VP, the speed test should start from the nearest bin since this chip is most likely to fall in that speed bin. Such a strategy helps us to find the appropriate speed bin quickly and, hence, reduce testing cost.

On the other hand, post-silicon tuning is a recently developed technique to improve parametric yield in the presence of large-scale process variations [7–9]. It adaptively configures a number of tunable parameters (e.g., supply voltage and body bias) so that a given circuit can work properly under different process conditions. An important component of post-silicon tuning is to accurately measure the process condition of a given chip so that the tunable parameters can be appropriately configured to adjust the circuit behavior. Such measurement, however, is not trivial, since it often requires a large number of on-chip sensors. We believe that the proposed VP framework can be used to predict the process condition from a significantly reduced number of on-chip sensors. By minimizing the number of required sensors, both the design complexity and the manufacturing cost can be reduced.

7.5 Numerical Experiments

In this section, we demonstrate the efficacy of VP using several examples based on industrial measurement data. All numerical experiments are performed on a 2.8-GHz Linux server.

7.5.1 Flush Delay Measurement Data

We consider the flush delay values measured from 282 industrial chips on the same wafer, as shown in Fig. 7.9. In this example, the measured delay significantly varies from chip to chip due to process variations. Our goal is to capture these wafer-level delay variations. We use a two-dimensional function $g(x, y)$ to model the delay, where $x \in \{1, 2, \ldots, 18\}$ and $y \in \{1, 2, \ldots, 19\}$. Each coordinate point (x, y) corresponds to a chip. Next, we apply a two-dimensional DCT to $g(x, y)$, yielding the frequency-domain components $G(u, v)$ shown in Fig. 7.10.

Two important observations can be made from the result in Fig. 7.10. First, $G(u, v)$ contains substantial high-frequency components, implying that the spatial sampling rate cannot be drastically reduced according to the well-known Nyquist–Shannon sampling theorem. Second, $G(u, v)$ is sparse, as its magnitude is almost zero at a large number of frequencies. This sparse pattern is the essential necessary condition that makes the proposed VP framework applicable to this example.

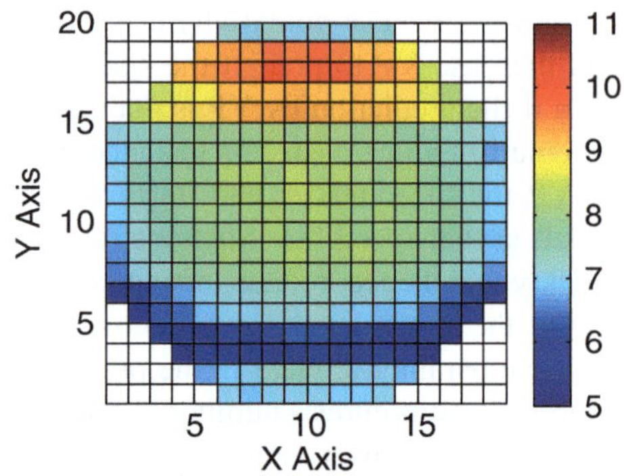

Fig. 7.9 Measured flush delay values (normalized by a randomly selected constant) of 282 industrial chips from the same wafer show significant spatial variations

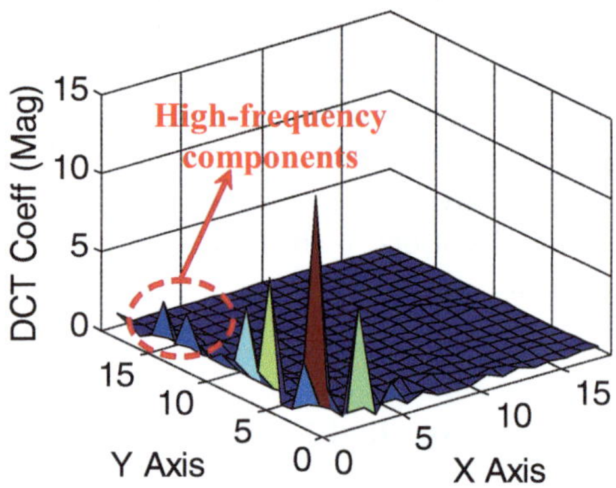

Fig. 7.10 Discrete cosine transform (DCT) coefficients (magnitude) of the normalized flush delay measurement show a unique sparse pattern

In what follows, we first use the data set in Fig. 7.9 to compare the modified Latin hypercube sampling algorithm (i.e., M-LHS summarized in Algorithm 4) with two brute-force sampling methods (i.e., grid sampling and random sampling), thereby demonstrating the superior accuracy achieved by M-LHS. Next, we further apply the proposed VP technique (i.e., Algorithm 5) and several traditional methods to predict the spatial delay variations and compare the accuracy of these different approaches.

7.5.1.1 Spatial Sample Generation

For testing and comparison purposes, we implement three different sampling schemes to select the spatial locations for silicon testing: (1) grid sampling, (2) brute-force random sampling, and (3) modified Latin hypercube sampling (M-LHS). Grid sampling deterministically picks up a set of spatial locations from a uniform two-dimensional grid. Brute-force random sampling simply selects random spatial locations by using a pseudo-random number generator. Finally, M-LHS follows the partition and selection steps summarized in Algorithm 4 to determine random sampling locations.

We apply the aforementioned three sampling schemes to select 128 chips out of 282 possible candidates on the same wafer. Next, we collect the flush delay data for these 128 selected chips and apply the MAP estimation (see Algorithm 5) to predict the spatial variations of the entire wafer. In Step 5 of Algorithm 5, the linear optimization is efficiently solved by the $l1$-MAGIC package developed by California Institute of Technology, taking about 8 s to finish in this example. Such an experiment is repeated for 1000 times in order to accurately estimate the statistics of prediction error.

Figure 7.11 shows the coherence (see Definition 7.2) and the average error for these 1000 repeated runs. As discussed in Sect. 7.2.3, the value of coherence provides a quantitative measure to assess the orthonormality of the matrix **A** in (7.5). On the other hand, the average error of the MAP estimation is calculated by:

$$\text{Error}_{\text{AVG}} = \sqrt{\frac{\sum_{x}\sum_{y}\left[g(x,y) - \tilde{g}(x,y)\right]^2}{\sum_{x}\sum_{y}\left[g(x,y)\right]^2}}, \tag{7.23}$$

where $g(x, y)$ and $\tilde{g}(x, y)$ denote the exact value and the estimated value of the flush delay at the location (x, y), respectively. Note that since grid sampling is deterministic, all 1000 runs yield the same result. Hence, only one data point is plotted in Fig. 7.11. Table 7.1 further shows the statistics of both coherence and error calculated from these 1000 repeated runs.

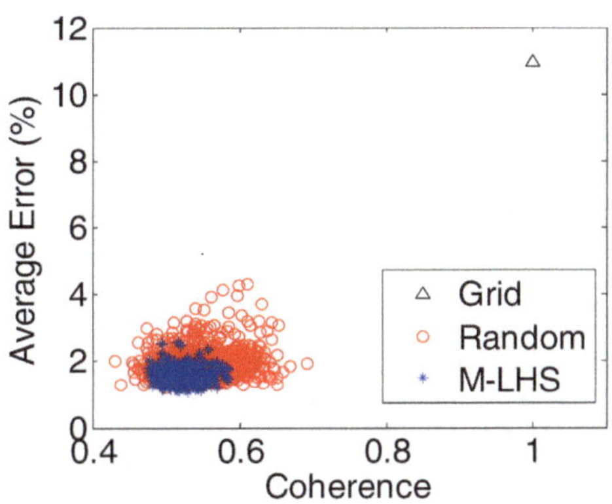

Fig. 7.11 Coherence and average error calculated from 128 chips with 1000 repeated runs for three different sampling techniques: (1) grid sampling (Grid), (2) brute-force random sampling (Random), and (3) modified Latin hypercube sampling (M-LHS)

Table 7.1 Statistics of coherence and average error calculated from 128 chips with 1000 repeated runs

	Coherence		Average error (%)	
Method	Mean	Std	Mean	Std
Grid	1.000	N/A	10.96	N/A
Random	0.545	0.040	1.90	0.41
M-LHS	0.521	0.020	1.54	0.20

Studying Fig. 7.11 and Table 7.1, we would have two important observations. First, grid sampling results in a large coherence value. It, in turn, implies that the columns of the matrix **A** in (7.5) are not approximately orthonormal. According to the discussions in Sect. 7.2.3, the MAP estimation cannot accurately predict the spatial variations in this case. This conclusion is consistent with the results in Fig. 7.11 where the average error associated with grid sampling is extremely large.

Second, both brute-force random sampling and M-LHS yield small coherence values and, consequently, small prediction errors. These results demonstrate the fact that compared to grid sampling, randomized sampling algorithms can efficiently generate "good" spatial samples and need less samples to achieve similar prediction accuracy. These results demonstrate the fact that a randomized algorithm can efficiently generate good spatial samples. In addition, comparing brute-force random sampling with M-LHS, we would notice that both methods result in similar mean values for coherence and error. However, M-LHS is able to reduce the standard deviation of both coherence and error by about 2× in this example. Such a reduction in standard deviation occurs, because M-LHS well controls the random samples by the partition and selection steps summarized in Algorithm 4. This is an important benefit offered by M-LHS, since it reduces the probability that a set of "bad" samples are randomly selected and, hence, result in large prediction error.

7.5.1.2 Spatial Variation Prediction

To quantitatively evaluate the accuracy of the proposed VP technique, we repeatedly apply Algorithm 5 with M-LHS to predict the wafer-level spatial variations with different numbers of spatial samples. For testing and comparison purposes, we implemented a number of traditional methods: (1) the two-dimensional interpolation method with uniform grid sampling [23], (2) the Kriging method with exponential correlation function [24], (3) the k-LSE method based on DCT analysis [25], and (4) the simple VP implementation without DCT coefficient preselection [31].

Figure 7.12 shows the average error calculated by (7.23) as a function of the number of samples (i.e., M) for different algorithms. M-LHS is applied to select

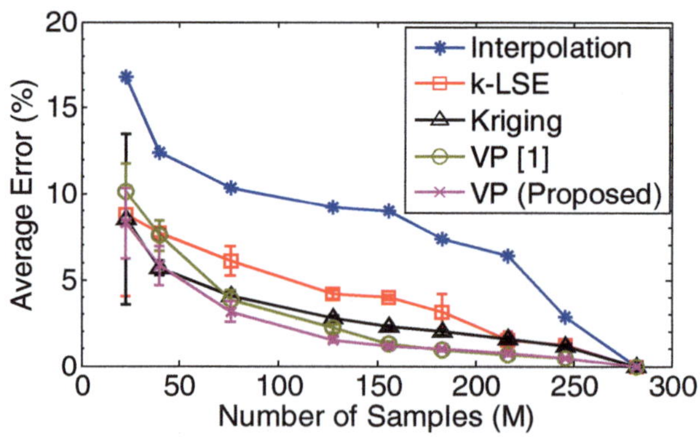

Fig. 7.12 Average prediction error (both mean and standard deviation) of different algorithms estimated by 100 repeated runs

the sampling locations for all methods except two-dimensional interpolation. To account for the inherent randomness of M-LHS sampling, we repeatedly run each algorithm for 100 times and plot the mean and the standard deviation of the average error in Fig. 7.12. Note that Algorithm 5 achieves the highest accuracy in this example. Compared to the simple VP implementation developed in [31], our new implementation (i.e., Algorithm 5) achieves superior accuracy by carefully preselecting the important DCT coefficients at low frequencies. Such a preselection scheme is particularly important, if the number of available samples (i.e., M) is small and, hence, it is difficult to accurately find the nonzeros from all DCT coefficients by L_1-norm regularization. On the other hand, the proposed VP algorithm outperforms other traditional techniques (i.e., two-dimensional interpolation, Kriging prediction, and k-LSE estimation), because all these traditional methods assume a smooth spatial variation pattern and, therefore, cannot accurately capture the high-frequency components of our measurement data shown in Fig. 7.10.

Figure 7.13 shows the flush delay values predicted from 60 tested chips (i.e., $M = 60$) by the proposed VP algorithm. In this example, 120 DCT coefficients are preselected by the cross-validation algorithm in Sect. 7.3.4, before the final l_1-norm regularization step is applied. Figure 7.14 further plots the DCT coefficients associated with the spatial variation pattern in Fig. 7.13. Comparing Figs. 7.10

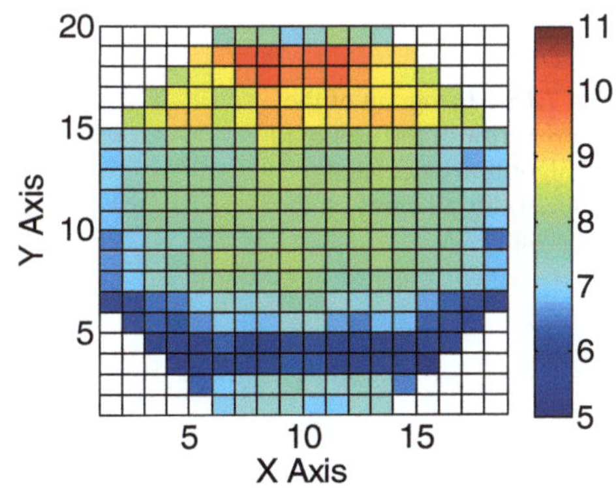

Fig. 7.13 Flush delay values predicted from 60 tested chips by the proposed VP algorithm

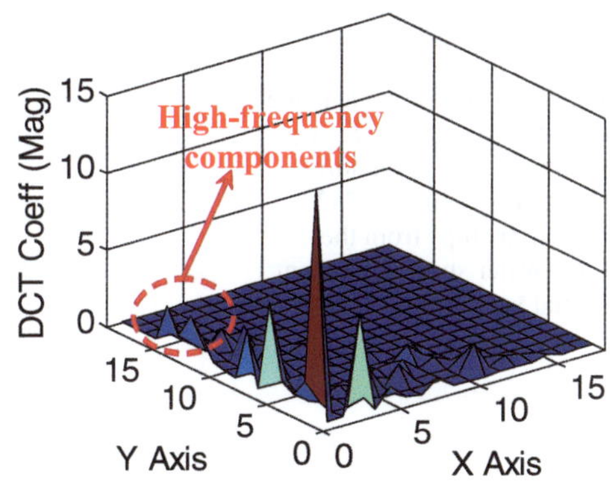

Fig. 7.14 The proposed VP algorithm accurately captures the low-frequency and high-frequency DCT coefficients by using 60 tested chips

and 7.14, we would notice that both the low-frequency and the high-frequency DCT coefficients are accurately captured in this example.

To quantitatively assess the prediction accuracy of each chip, we calculate the following relative error:

$$\text{Error}_{\text{REL}}\left(x, y\right) = \left| \frac{g(x, y) - \tilde{g}(x, y)}{g(x, y)} \right|, \tag{7.24}$$

where $g(x, y)$ and $\tilde{g}(x, y)$ are similarly defined as those in (7.23). The error metric in (7.24) measures the difference between the measurement data (i.e., Fig. 7.9) and the prediction results (i.e., Fig. 7.13) for every chip. Figure 7.15 shows the histogram of the relative error calculated for all chips on the same wafer. Note that the relative error is less than 10% for most chips in this example.

7.5.2 Leakage Current Measurement Data

We consider the leakage current measurement data collected by IDDQ test for the same industrial circuit design. Figure 7.16 shows the normalized leakage current values $\log_{10}(I_{\text{LEAK}})$ (after logarithmic transform) as a function of the location (x, y). Figure 7.17 further shows the frequency-domain components after DCT.

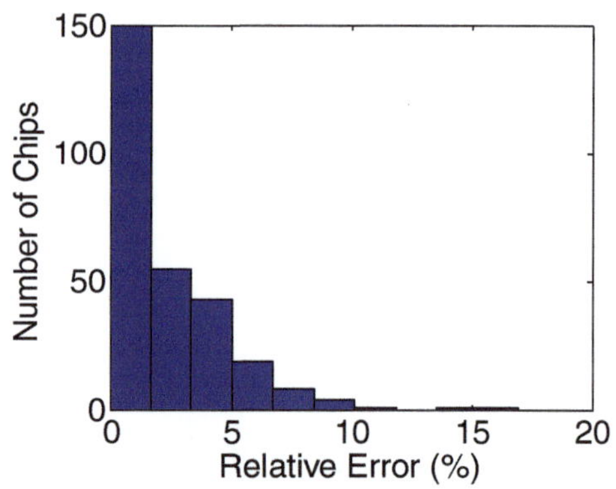

Fig. 7.15 Histogram of the relative error of the proposed VP algorithm calculated by (7.24) for all chips on the same wafer

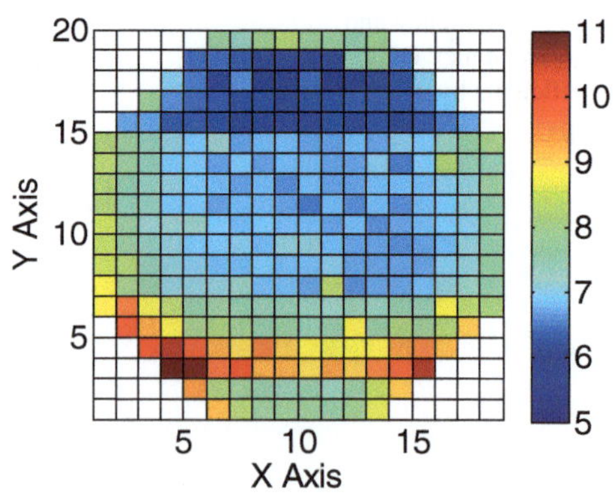

Fig. 7.16 Measured leakage current values $\log_{10}(I_{\text{LEAK}})$ (normalized by a randomly selected constant) of 282 industrial chips from the same wafer show significant spatial variations

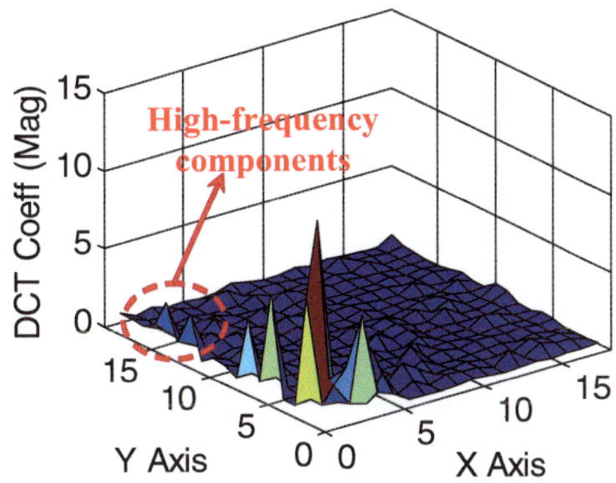

Fig. 7.17 Discrete cosine transform (DCT) coefficients (magnitude) of the normalized leakage current measurement $\log_{10}(I_{LEAK})$ show a unique pattern that is approximately sparse

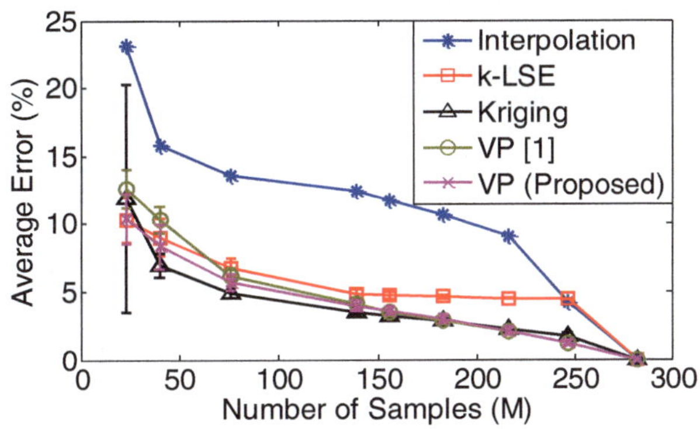

Fig. 7.18 Average prediction error (both mean and standard deviation) of different algorithms estimated by 100 repeated runs

Similar to the flush delay example, the DCT coefficients contain important high-frequency components. In addition, a large number of small DCT coefficients are observed and, hence, the frequency-domain representation is approximately (but not exactly) sparse. This observation is consistent with the fact that the full-chip leakage current partially depends on V_{TH} mismatches that are not spatially correlated.

We apply different algorithms to predict the spatial variations based on a few (i.e., M) sampling points. Figure 7.18 shows the average error calculated by (7.23). Similar to the previous example, both the mean and the standard deviation of the average error are calculated from 100 repeated runs and they are plotted in Fig. 7.18. Note that the proposed VP method (i.e., Algorithm 5) achieves better accuracy than three traditional techniques: (1) two-dimensional interpolation, (2) k-LSE estimation, and (3) the simple VP implementation in [31]. However, the accuracy of VP is slightly worse than Kriging prediction, because the frequency-domain representation is not exactly sparse in this example.

Figure 7.19 shows the leakage current values $\log_{10}(I_{LEAK})$ (after logarithmic transform) predicted from 100 tested chips by the proposed VP algorithm. In this example, 240 DCT coefficients are preselected by the cross-validation algorithm in Sect. 7.3.4, before the final l_1-norm regularization step is applied. Figure 7.20 further shows the histogram of the relative error calculated for all chips using (7.24). Note that the relative error is less than 10% for most chips in this example.

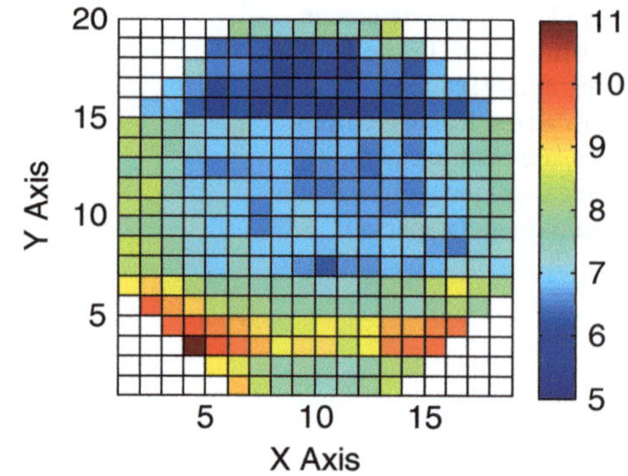

Fig. 7.19 Leakage current values $\log_{10}(I_{\text{LEAK}})$ predicted from 100 tested chips by the proposed VP algorithm

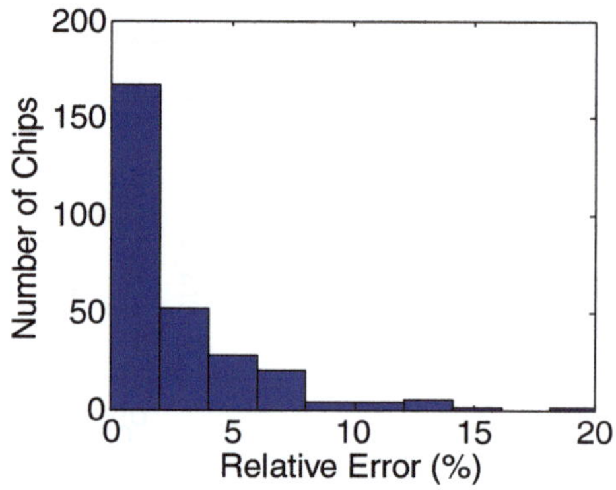

Fig. 7.20 Histogram of the relative error of the proposed VP algorithm calculated by (7.24) for all chips on the same wafer

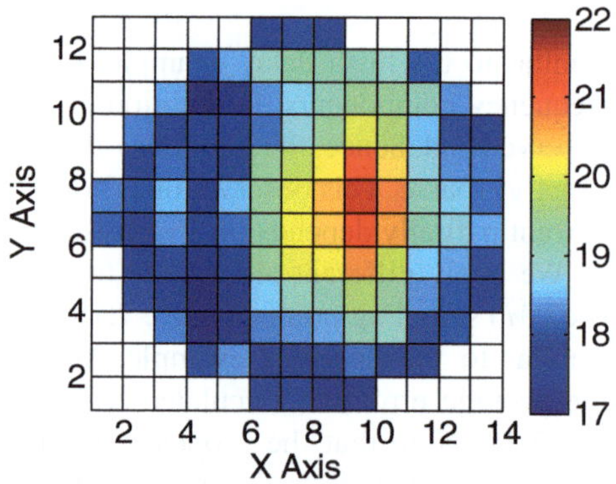

Fig. 7.21 Measured ring oscillator (RO) period values (normalized by a randomly selected constant) of 117 ROs from the same wafer show significant spatial variations

7.5.3 Ring Oscillator Period Measurement Data

We consider the ring oscillator (RO) period measurement data collected from a wafer at an advanced technology node. These RO measurement data are strongly correlated with the final chip performance and, hence, are often used for process monitoring and control [10, 11]. Our wafer contains 117 ROs distributed over different spatial locations. Figure 7.21 shows the normalized RO period values as

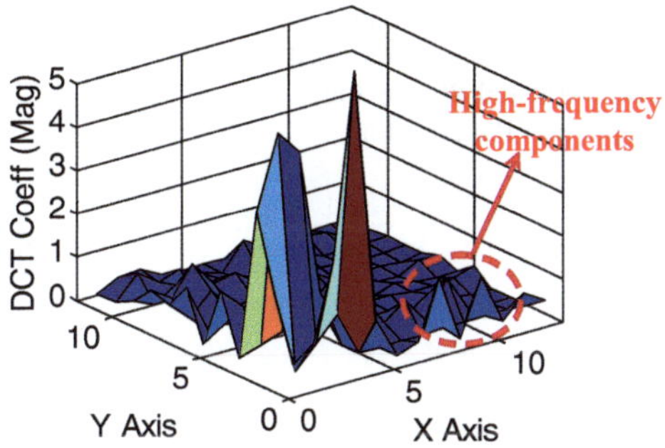

Fig. 7.22 Discrete cosine transform (DCT) coefficients (magnitude) of the normalized RO period show a unique pattern that is approximately sparse

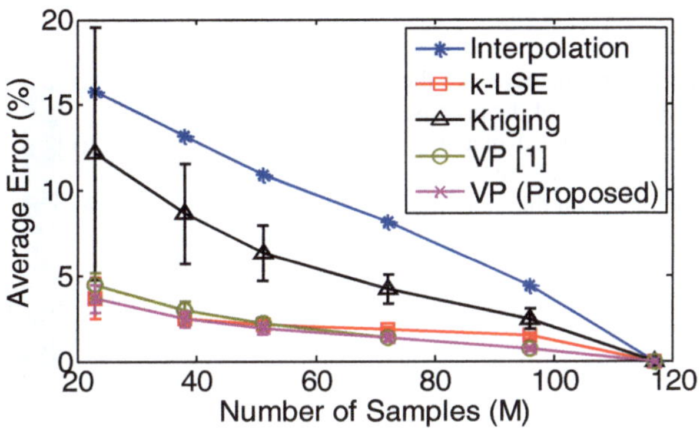

Fig. 7.23 Average prediction error (both mean and standard deviation) of different algorithms estimated by 100 repeated runs

a function of the location (x, y). Figure 7.22 further shows the frequency-domain components after DCT. Similar to the leakage current example, the DCT coefficients are approximately sparse and contain significant high-frequency components.

We apply different algorithms to recover the spatial variations based on a few (i.e., M) sampling points. Figure 7.23 compares the average error calculated by (7.23) for different methods. Both the mean and the standard deviation of the average error are calculated from 100 repeated runs and they are plotted in Fig. 7.23. Note that the proposed VP technique (i.e., Algorithm 5) achieves the best accuracy in this example. The Kriging method shows large error, because it assumes an exponential correlation model while the actual spatial correlation does not match the model template. In general, the Kriging method needs to know the correlation template in advance. If the prior knowledge of the correlation function is not correct, the Kriging method may fail to predict the spatial variations accurately.

Figure 7.24 further shows the RO period values predicted from 40 tested ROs by the proposed VP algorithm. The cross-validation algorithm in Sect. 7.3.4 preselects all DCT coefficients in this example. Figure 7.25 shows the histogram of the relative error calculated for all ROs using (7.24). Note that the relative error is less than 5% for most chips in this example.

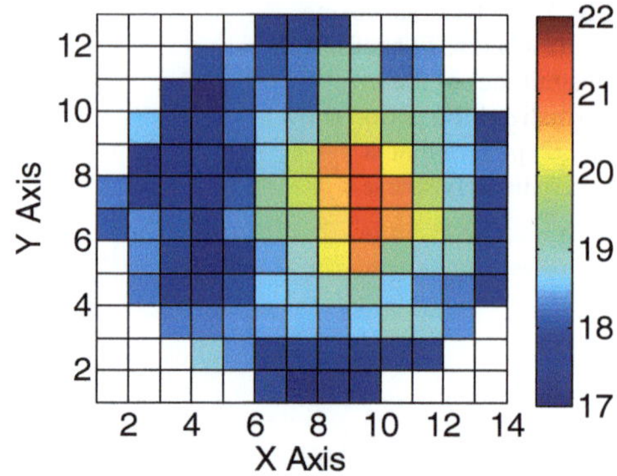

Fig. 7.24 Ring oscillator (RO) period values predicted from 40 tested ROs by the proposed VP algorithm

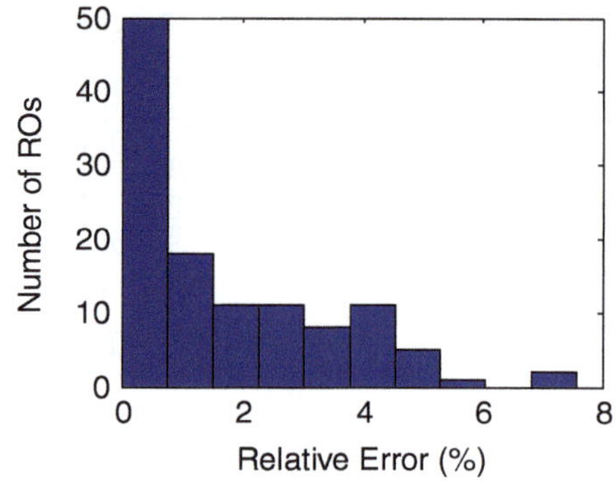

Fig. 7.25 Histogram of the relative error of the proposed VP algorithm calculated by (7.24) for all ring oscillators on the same wafer

7.6 Conclusions

In this chapter, we propose a novel *virtual probe* (VP) framework to efficiently and accurately recover full-wafer/chip spatial variations from an extremely small set of measurement data, thereby reducing the cost of silicon characterization and testing. VP exploits recent breakthroughs in compressed sensing [18–21]. It is formulated as a maximum a posteriori (MAP) problem and can be efficiently solved via linear programming. Our numerical examples based on industrial measurement data demonstrate that VP provides superior accuracy over other traditional methods, including two-dimensional interpolation, Kriging prediction, and k-LSE estimation.

References

1. S. Nassif, Delay variability: sources, impacts and trends, in *IEEE International Solid-State Circuits Conference* (2000), pp. 368–369
2. Semiconductor Industry Associate, *2007 International Technology Roadmap for Semiconductors (ITRS)* (Semiconductor Industry Association, Washington, 2007)
3. H. Chang, S. Sapatnekar, Statistical timing analysis under spatial correlations. IEEE Trans. Comput. Aided Des. Integr. Circuits Syst. **24**(9), 1467–1482 (2005)

4. C. Visweswariah, K. Ravindran, K. Kalafala, S. Walker, S. Narayan, First-order incremental block-based statistical timing analysis, in *Design Automation Conference* (2004), pp. 331–336
5. Y. Zhan, A. Strojwas, X. Li, L. Pileggi, D. Newmark, M. Sharma, Correlation aware statistical timing analysis with non-Gaussian delay distributions, in *Design Automation Conference* (2005), pp. 77–82
6. K. Heloue, F. Najm, Statistical timing analysis with two-sided constraints, in *International Conference on Computer-Aided Design* (2005), pp. 829–836
7. M. Mani, A. Singh, M. Orshansky, Joint design-time and post-silicon minimization of parametric yield loss using adjustable robust optimization, in *International Conference on Computer-Aided Design* (2006), pp. 19–26
8. S. Kulkarni, D. Sylvester, D. Blaauw, A statistical framework for post-silicon tuning through body bias clustering, in *International Conference on Computer-Aided Design* (2006), pp. 39–46
9. Q. Liu, S. Sapatnekar, Synthesizing a representative critical path for post-silicon delay prediction, in *IEEE International Symposium on Physical Design* (2009), pp. 183–190
10. M. Ketchen, M. Bhushan, D. Pearson, High speed test structures for in-line process monitoring and model calibration, in *IEEE International Conference on Microelectronic Test Structures* (2005), pp. 33–38
11. M. Bhushan, A. Gattiker, M. Ketchen, K. Das, Ring oscillators for CMOS process tuning and variability control. IEEE Trans. Semicond. Manuf. **19**(1), 10–18 (2006)
12. W. Mann, F. Taber, P. Seitzer, J. Broz, The leading edge of production wafer probe test technology, in *IEEE International Test Conference* (2004), pp. 1168–1195
13. F. Koushanfar, P. Boufounos, D. Shamsi, Post-silicon timing characterization by compressed sensing, in *International Conference on Computer-Aided Design* (2008), pp. 185–189
14. S. Reda, S. Nassif, Analyzing the impact of process variations on parametric measurements: novel models and applications, in *Design, Automation & Test in Europe* (2009), pp. 375–380
15. M. Bushnell, V. Agrawal, *Essentials of Electronic Testing for Digital, Memory and Mixed-Signal VLSI Circuits* (Kluwer Academic Publishers, Norwell, 2000)
16. A. Oppenheim, *Signals and Systems* (Prentice Hall, Upper Saddle River, 1996)
17. R. Gonzalez, R. Woods, *Digital Image Processing* (Prentice Hall, Upper Saddle River, 2007)
18. D. Donoho, Compressed sensing. IEEE Trans. Inf. Theory **52**(4), 1289–1306 (2006)
19. E. Candes, Compressive sampling, in *International Congress of Mathematicians* (2006)
20. J. Tropp, S. Wright, Computational methods for sparse solution of linear inverse problems. Proc. IEEE **98**(6), 948–958 (2010)
21. D. Donoho, J. Tanner, Precise undersampling theorems. Proc. IEEE **98**(6), 913–924 (2010)
22. C. Bishop, *Pattern Recognition and Machine Learning* (Prentice Hall, Upper Saddle River, 2007)
23. W. Press, S. Teukolsky, W. Vetterling, B. Flannery, *Numerical Recipes: The Art of Scientific Computing* (Cambridge University Press, Cambridge, 2007)
24. F. Liu, A general framework for spatial correlation modeling in VLSI design, in *Design Automation Conference* (2007), pp. 817–822
25. A. Nowroz, R. Cochran, S. Reda, Thermal monitoring of real processors: techniques for sensor allocation and full characterization, in *Design Automation Conference* (2010), pp. 56–61
26. R. Tibshirani, Regression shrinkage and selection via the Lasso. J. R. Stat. Soc. **58**(1), 267–288 (1996)
27. G. Golub, C. Loan, *Matrix Computations* (Johns Hopkins University Press, Baltimore, 1996)
28. S. Boyd, L. Vandenberghe, *Convex Optimization* (Cambridge University Press, Cambridge, 2004)
29. S. Kim, K. Koh, M. Lustig, S. Boyd, D. Gorinevsky, An interior-point method for large-scale l_1-regularized least squares. IEEE J. Sel. Top. Sign. Proces. **1**(4), 606–617 (2007)
30. M. McKay, R. Beckman, W. Conover, A comparison of three methods for selecting values of input variables in the analysis of output from computer code. Technometrics **42**(1), 55–61 (1979)
31. X. Li, R. Rutenbar, R. Blanton, Virtual probe: a statistically optimal framework for minimum-cost silicon characterization of nanoscale integrated circuits, in *International Conference on Computer-Aided Design* (2009), pp. 433–440

Chapter 8
Machine Learning for VLSI Chip Testing and Semiconductor Manufacturing Process Monitoring and Improvement

Jinjun Xiong, Yada Zhu, and Jingrui He

8.1 Introduction

The recent disruptive advancement in mobile and information technologies has fundamentally transformed business, industries, and society at large and has spurred an exponential growth in data volume [1]. Machine learning and big data analytics have thus become some of the hottest buzz phrases these days, and their growth in popularity continues to expand from the mobile and IT industries into many other industries, including finance, health care, medicine, science, and entertainment. They were considered one of the fastest growing technology trends in 2014 [2], and numerous anecdotes have led us to believe that the glamour of machine learning and big data analytics has attracted many best and brightest away from their traditional technological profession, including the $336 billion industry of semiconductor.

In this chapter, we argue that the semiconductor industry, however, also offers many challenging and interesting machine learning and big data analytic problems, and solutions to these problems would result in much higher business impact than many of those trendy analytics business. The argument is almost self-evident. For example, semiconductor manufacturing represents one of the most complex manufacturing processes in the world [3], and large amount of data associated with process variables monitored over time (e.g., temperature, impedance, gas flow, electric bias) are collected and stored in a huge database. If those data are analyzed for a variety of purposes, such as anomaly detection, quality control, and fault diagnosis, it may lead to significant reduction in the manufacturing cost [4, 5].

J. Xiong (✉) · Y. Zhu
IBM Thomas J. Watson Research Center, Yorktown Heights, NY, USA
e-mail: jinjun@us.ibm.com; yzhu@us.ibm.com

J. He
Arizona State University, Tempe, AZ, USA
e-mail: jingrui.he@asu.edu

I. M. Elfadel et al. (eds.), *Machine Learning in VLSI Computer-Aided Design*,
https://doi.org/10.1007/978-3-030-04666-8_8

Similar arguments can be made to other semiconductor-related businesses, including electronic design automation (EDA) and fabless design houses where large amounts of design data and reports are generated and processed on a regular basis. Deep insights can be gained about designs and design process itself if those data can be subject to machine learning and big data analytics.

This paper tries to illustrate this point through a number of practical but challenging problems arising from the latest semiconductor manufacturing process.

We first show how machine learning techniques, especially those regression-related problems, often under the "disguise" of optimization problems, have been used frequently (often with nontrivial modeling skills and mathematical sophistications) to solve the semiconductor problems. For some other types of semiconductor problems, such as manufacturing process monitoring and improvement, we show that some existing machine learning algorithms are not necessarily well positioned to solve them, and novel machine learning techniques involving temporal, structural and hierarchical properties need to be further developed. In either scenario, we convey the message that machine learning and existing semiconductor industry research are closely related, and researchers often contribute to and benefit from each other.

The remainder of this chapter is organized as follows. We first give some background related to the topics of this chapter in Sect. 8.2. We then show how the well-known nonlinear regression techniques from machine learning are used to solve process variation modeling and chip testing problems in Sect. 8.3. Two examples are then discussed in Sects. 8.4 and 8.5, respectively, to show how existing machine learning techniques such as multitask learning and clustering need to be extended to solve the manufacturing wafer quality prediction and process monitoring problems, thus contributing back to the machine learning communities. We conclude this chapter in Sect. 8.6.

8.2 Background

We briefly review a number of data mining and machine learning techniques that are commonly used in solving many big data analytic problems. And, we will show later that these techniques can also be used, sometimes with nontrivial extensions, to solve problems resulting from semiconductor manufacturing process.

At a high level, machine learning can be thought of as minimizing some kind of measure of errors for a model based on data such that the model can help to make predictions.

One of the most frequently used machine learning techniques is regression, where the model is in the form of some mathematical form, be it a linear function or nonlinear function, to establish the relationship between input and output numerical variables. As such, the regression problem is often formulated as a nonlinear optimization problem. In semiconductor manufacturing, predicting wafer quality based on the process monitoring data is typically formulated as a regression problem.

Another problem arising frequently in the big data world is to learn data representations that are common across multiple related learning tasks, such as collaborative E-mail spam filtering in open membership systems [6], web search ranking data sets from several countries [7], detecting multiple objects from images [8], prediction of student test results for a collection of schools based on school demographics [9], and the prediction of survival of patients in different clinics [10]. These problems are typically formulated as a multitask learning (MTL) problem, where the goal is to build good models for all target tasks. In doing so, the key is to leverage task relatedness that affects the information shared among various tasks. To predict wafer quality in semiconductor manufacturing, process monitoring data generated from multiple machines (or chambers) that follow the same process receipts can also be formulated as an MTL problem in order to improve prediction performance on future unseen examples.

Another well-known machine learning technique is clustering, an unsupervised learning technique to group objects such that objects in the same cluster are more similar to each other than to those in other clusters. (It can be applied to detect process anomalies based on the size of clusters or the change of cluster memberships.) It has found applications, such as text, web-log, and market-basket data analysis. Typically, those data can be arranged into a two-dimensional array with some kind of co-occurrence relationship, for example, word-document or webpage-user browsing co-occurrence arrays. Most existing clustering algorithms focus on one-way clustering, i.e., cluster data along one dimension of the array based on similarity measures of data along the other dimension.

Time series data is another popular subject in machine learning because of its pervasive applicability to many modeling scenarios. As such, it is hardly surprising to see that time series data mining has attracted significant attention and research effort. Time series analysis aims at extracting meaningful information from sequences of data points, measured at successive time stamps. For time series data, researchers have studied a variety of problems, such as classification [11, 12], clustering [13–15], search and indexing [16, 17], forecasting [18], and outlier detection [19]. (In semiconductor manufacturing, process monitoring data is high-dimensional time series that has been widely studied to improve process monitoring, yield, and wafer quality control.)

In contrast to the recent booming phenomena of machine learning and big data in many industries, the semiconductor industry, manufacturing process in particular, has long (though less publicized than the former) embraced the challenges of machine learning and big data and has developed many interesting machine learning algorithms to address those challenges. One such example is called virtual metrology [20], which develops methods to predict properties of a wafer based on a large set of measured machine parameters and sensor data of the manufacturing process, without performing the (costly) physical measurement of wafer properties. Another example is statistical analysis of chip design quality, such as process variability modeling and statistical chip test for yield improvement.

In the following sections, we will present a number of techniques developed resulting from research in semiconductor chip designs and manufacturing. We will show that some of those techniques are mainly published in semiconductor-related conferences, while others are mainly published in machine learning-related conferences. But in essence, their approaches and methodologies are very similar, i.e., both bring in deep domain knowledge to model the issue in some mathematical form, formulate the problem as an optimization problem, and then utilize the data to solve the optimization problem. The major difference, however, is that techniques published in the semiconductor area typically focus on applying existing machine learning techniques, including nontrivial optimization techniques, to solve the problem, while the machine learning papers focus more on the special structure of the problem where extending existing machine learning techniques become necessary. But, such distinction is not a major hurdle for many researchers working in traditional semiconductor areas, as they easily understand the lingo from both sides. For them, whether or not to catch the trend to shift their research focus to machine learning is not that important. What is more important is to let research problems themselves drive the right course of actions.

8.3 Machine Learning for Chip Testing and Yield Optimization

In this section, we present a number of techniques developed in a subfield of the semiconductor industry called electronic design automation (EDA), and, in particular, an even smaller area within EDA called statistical chip testing and yield optimization considering process variations. Though those techniques are traditionally not being considered as machine learning techniques, we show, in fact, that many techniques developed in that domain are closely related to machine learning under the "disguise" of various optimization problems.

8.3.1 Robust Spatial Correlation Extraction

One of the main concerns in the semiconductor industry recently is the impact of technology scaling. Aggressive scaling down of transistors and interconnects has resulted in miraculous achievements in chip performance and functionality. This deep scaling of semiconductor technology, however, has introduced the problem of uncontrollable process variations. That is, we are unable to make transistors and interconnects with accurately predictable characteristics, let alone to make transistors the same on different copies of the same chip and even at different locations of the same chip. Thus, the only way to cope with variability is to model the variability based on data, a task closely related to machine learning. We discuss a technique to robustly model spatial correlation based on measurement data in this section.

8.3.1.1 Problem Formulations

There are two orthogonal ways to classify process variations. The first one is to classify the variations according to the scope of their occurrence as: (1) die-to-die variation, which describes the variation that affects parameters in different dies differently, but affects parameters within a die equally, and (2) within-die variation (also called intra-die variation, across-chip variation, on-chip variation, *spatial variation*, or *spatial correlation*), which describes the variation that affects process parameters at different locations of the same die differently. The second one is to classify the variations according to the scale of their causes, that is: (1) systematic variation, which describes the deterministic portion of the variation, and (2) random variation, which describes the variation that is independent of any other conditions.

We denote F as the measurable process parameter of interest, which can be either a physical parameter, like channel length, channel width, silicon oxide thickness, and wire thickness, or a parametric quantity, such as gate delay and threshold voltage. Because of manufacturing process variations, these process parameters are no longer fixed values. We model the parameter as a random variable, which is a complicated function of die-to-die (D2D) systematic and random variations, and within-die (WID) systematic and random variations. Conceptually, we can represent it as:

$$F = h(Z_{\mathrm{D2D}}, \mathrm{sys}, Z_{\mathrm{D2D}}, \mathrm{rnd}, Z_{\mathrm{WID}}, \mathrm{sys}, Z_{\mathrm{WID}}, \mathrm{rnd}), \tag{8.1}$$

where $Z_{\mathrm{D2D}}, \mathrm{sys}$ models the die-to-die systematic variation; $Z_{\mathrm{D2D}}, \mathrm{rnd}$ models the die-to-die random variation; $Z_{\mathrm{WID}}, \mathrm{sys}$ models the within-die systematic variation; and $Z_{\mathrm{WID}}, \mathrm{rnd}$ models the within-die random variation. All variation components are complicated functions of the manufacturing process, feature's relative location in the wafer, feature's relative location in the die, and feature's local geometry patterns, to name just a few. Assuming that the impact of each variation component is linear, we write (8.1) as follows:

$$F = h_0 + h_1(Z_{\mathrm{D2D}}, \mathrm{sys}) + h_2(Z_{\mathrm{D2D}}, \mathrm{rnd}) + h_3(Z_{\mathrm{WID}}, \mathrm{sys}) + h_4(Z_{\mathrm{WID}}, \mathrm{rnd}) + X_r,$$

where h_0 is a function that models the nominal value of F under nominal manufacturing conditions without any variation; h_1, h_2, h_3, and h_4 are functions that model the impact of respective variation component on F; and X_r is a residual part that models the purely independent random variation that is not explainable by other variation components. We further define

$$F_s = h_1(Z_{\mathrm{D2D}}, \mathrm{sys}) + h_3(Z_{\mathrm{WID}}, \mathrm{sys}),$$

$$F_r = h_2(Z_{\text{D2D}}, \text{rnd}) + h_4(Z_{\text{WID}}, \text{rnd}) + X_r,$$

where F_s models the systematic variation of F, while F_r is a zero-mean random variable that models the random variation of F. Hence, we have

$$F = h_0 + F_s + F_r. \tag{8.2}$$

The variance of F, σ_F^2, is also called the *overall chip variance*.

Because the systematic variation is more like a deterministic variation, we lump it with the nominal value h_0, that is:

$$f_0 = h_0 + F_s, \tag{8.3}$$

where f_0 is the mean value of F with the systematic variation considered. The extraction of mean value f_0 is relatively easy and is essentially done through averaging. In the following, we mainly concern ourselves in extracting the random variation parts F_r. Towards this end, we rewrite the random variation F_r as follows:

$$F_r = X_g + X_s + X_r,$$

where $X_g = h_2(Z_{\text{D2D}}, \text{rnd})$ and $X_s = h_4(Z_{\text{WID}}, \text{rnd})$. Therefore, we have

$$F = f_0 + F_r = f_0 + X_g + X_s + X_r. \tag{8.4}$$

The three types of random variations, X_g, X_s, and X_r, are *independent* by definition. Hence, the variance of F_r is given by:

$$\sigma_{F_r}^2 = \sigma_G^2 + \sigma_S^2 + \sigma_R^2, \tag{8.5}$$

where σ_G^2, σ_S^2, and σ_R^2 are the variances of X_g, X_s, and X_r, respectively. When the systematic variation is excluded, the *overall chip variance* is equivalent to the random variance, i.e., $\sigma_F^2 = \sigma_{F_r}^2$.

It has been observed that devices that are physically close to each other are more likely to have similar characteristics than devices that are far apart. This phenomenon is captured by the modeling of *spatial correlation*.

We model the random part of process variation F_r as a *homogeneous and isotropic* random filed, which is a real random function $F(x, y)$ of position (x, y) in the 2-dimensional space $\mathscr{R}^2$ with its mean and variance being constants, and its correlation function $\rho(x_i, x_j, y_i, y_j)$ between any two points depending only on the distance v between them, that is:

$$\rho(x_i, x_j, y_i, y_j) = \rho(v_{i,j}), \tag{8.6}$$

where $v_{i,j} = \sqrt{(x_i - x_j)^2 + (y_i - y_j)^2}$. If the spatial variation follows a homogeneous and isotropic random field, the same distance $v_{i,j}$ always corresponds to the same $\rho(v_{i,j})$, regardless of their locations. Therefore, for simplicity, we denote $\rho(v_{i,j})$ as $\rho(v)$ in the following whenever there is no ambiguity.

Formally, a *valid spatial correlation function* $\rho(v)$ is a function such that the correlation matrix generated from $\rho(v)$ for arbitrary number of points on the two-dimensional space is always *positive semidefinite*.

In its simplest way, a valid spatial correlation function should be a monotonically decreasing function. But, not all monotonically decreasing functions qualify for being a valid spatial correlation function. In fact, per [21], a necessary and sufficient condition for the function $\rho(v)$ to be a valid spatial correlation function of a homogeneous and isotropic random field is that it can be represented in the form of:

$$\rho(v) = \int_0^{\infty} J_0(\omega v) d(\Phi(\omega)), \tag{8.7}$$

where $J_0(t)$ is the Bessel function of order zero and $\Phi(\omega)$ is a real nondecreasing function on $[0, \infty)$ such that for some nonnegative p:

$$\int_0^{\infty} \frac{d\Phi(\omega)}{(1 + \omega^2)^p} < \infty. \tag{8.8}$$

Based on this, we propose the following problem formulation:

Extraction of Spatial Correlation Function Given noisy measurement data for the parameter of interest with possible inconsistency, extract the inter-chip global variation component σ_G^2, the intra-chip spatial variation component σ_S^2, the random variation component σ_R^2, and the spatial correlation function $\rho(v)$, so that the extracted variation components accurately capture the underlying variation model, and the spatial correlation function is always a valid correlation function satisfying condition (8.7).

It's easy to see that we can formulate the extraction of the spatial correlation function $\rho(v)$ as a regression problem with a set of measurements that give out data set as $(v, \rho(v))$. The regression problem would be to determine a nonlinear regression function as shown in (8.7). But, we will show soon that a direct application of regression is not necessarily easy for this problem.

8.3.1.2 Extraction Algorithms

We treat each measurement of the parameter of interest F as a sampling of the quantity in (8.4). Given N samples of a chip and M number of measurement sites on each chip, we group the measured data $f_{k,i}$ by their chip locations as follows: $f_{k,\cdot}$=[$f_{k,1}, \ldots, f_{k,M}$] for $k = 1$ to N, or by their site locations as follows: $f_{\cdot,i} = [f_{1,i}, \ldots, f_{N,i}]$ for $i = 1$ to M. For better presentation, we denote the actual variance as σ^2 with an uppercase letter in subscript, like σ_G^2 for the global variation component, and denote the extracted variance as σ^2 with a lowercase letter in subscript, like σ_g^2 for the extracted global variation component.

Since the spatial variation is modeled as a *homogeneous and isotropic random field* in a two-dimensional space $\mathscr{R}^2$, for the parameter of interest at *arbitrary* two different points, their covariance is

$$\mathrm{cov}(F_i, F_j) = \mathrm{cov}(X_g, X_g) + \mathrm{cov}(X_{s,i}, X_{s,j}) = \sigma_G^2 + \rho(v)\sigma_S^2, \tag{8.9}$$

where $\rho(v)$ is the spatial correlation coefficient between two locations that are v distance apart. In other words, we can characterize the process variation by extracting the inter-chip global variation σ_G^2, intra-chip spatial variation σ_S^2, and the correlation function $\rho(v)$.

For the parameter of interest at two different locations with distance of v, the *overall process correlation* between them is thus given by:

$$\rho_v \equiv \frac{\mathrm{cov}(F_i, F_j)}{\sigma_{F_i}\sigma_{F_j}} = \frac{\sigma_G^2 + \rho(v)\sigma_S^2}{\sigma_G^2 + \sigma_S^2 + \sigma_R^2}. \tag{8.10}$$

Because the spatial correlation $\rho(v)$ is a function of the distance v, so is the overall process correlation ρ_v. As $\rho(v)$ is homogeneous and isotropic, so is ρ_v. Because of the one-to-one correspondence between spatial correlation $\rho(v)$ and the overall process correlation ρ_v, extracting the spatial correlation function $\rho(v)$ is equivalent to extracting the overall process correlation function ρ_v.

We approximate the overall chip variance σ_F^2 by computing the *unbiased sample variance* of $f_{k,i}$ as follows:

$$\sigma_F^2 \approx \sigma_f^2 = \frac{1}{MN-1}\left(\sum_i \sum_k f_{k,i}^2 - \frac{(\sum_i \sum_k f_{k,i})^2}{MN}\right). \tag{8.11}$$

For all samples of the parameter of interest F within a particular chip c, because the inter-chip global variation X_g changes the value of parameter for all samples with the same chip by the same amount, the *overall within-chip variance* is thus given by:

$$\sigma_{F_c}^2 = \sigma_S^2 + \sigma_R^2. \tag{8.12}$$

We estimate the overall within-chip variation by computing the *unbiased sample variance* of $f_{k,\cdot}$ as follows:

$$\sigma_{F_c}^2 \approx \sigma_{f_k}^2 = \frac{1}{M-1}\left(\sum_i f_{k,i}^2 - \frac{(\sum_i f_{k,i})^2}{M}\right). \tag{8.13}$$

For different $f_{k,\cdot}$, we may get different estimation of $\sigma_{F_c}^2$ caused by inconsistent measurement. To improve the accuracy, we estimate the overall within-chip variance by taking the average value of $\sigma_{f_k}^2$. We denote the resulting average value as $\sigma_{f_c}^2 \approx \sigma_{F_c}^2$.

Knowing the estimation of the overall chip variance σ_f^2 and the overall within-chip variance $\sigma_{f_c}^2$, we extract the inter-chip global variation by:

$$\sigma_G^2 = \sigma_F^2 - \sigma_{F_c}^2 \approx \sigma_g^2 = \sigma_f^2 - \sigma_{f_c}^2. \tag{8.14}$$

For any two different sets of $f_{\cdot,i}$ and $f_{\cdot,j}$ at two different sites that are v distance apart, we estimate the covariance of F_i and F_j by computing the *unbiased sample covariance* of $f_{\cdot,i}$ and $f_{\cdot,j}$ as follows:

$$\mathrm{cov}(F_i, F_j) \approx \mathrm{cov}(f_{\cdot,i}, f_{\cdot,j}) = \frac{\sum_k f_{k,i} f_{k,j}}{N-1} - \frac{\sum_k f_{k,i} \sum_k f_{k,j}}{N(N-1)}. \tag{8.15}$$

For simplicity, we also denote $\mathrm{cov}(f_{\cdot,i}, f_{\cdot,j})$ as $\mathrm{cov}(v)$ to show that it is a function of two points that are v distance apart.

According to (8.9) and (8.14), we estimate the product of spatial variation σ_S^2 and spatial correlation $\rho(v)$ as follows:

$$\sigma_S^2 \cdot \rho(v) = \mathrm{cov}(F_i, F_j) - \sigma_G^2 \approx \mathrm{cov}(v) - \sigma_g^2. \tag{8.16}$$

Because $\rho(v)$ is a function of v, we need to compute $\rho(v)$ for different pairs of sites with different distances in order to obtain the full description of $\rho(v)$. But, there are two challenges in doing that: (1) we do not know the exact value of spatial variation σ_S^2; (2) because of unavoidable measurement errors, the data set computed as above may not be consistent. Therefore, in the following, we propose a robust technique to find the spatial correlation function $\rho(v)$ and σ_S^2 accurately. Moreover, the resulting $\rho(v)$ is guaranteed to be a valid spatial correlation function.

Given the data set $(v, \mathrm{cov}(v))$ as computed from (8.15), we formulate the robust spatial variation extraction problem as the following regression-like optimization problem:

$$\begin{aligned} \min_{\Phi,\sigma_s^2} : \ & \| \sigma_s^2 \textstyle\int_0^\infty J_0(\omega v) d(\Phi(\omega)) - \mathrm{cov}(v) + \sigma_g^2 \|, \\ s.t. \quad & \sigma_s^2 \le \sigma_{f_c}^2, \\ & \textstyle\int_0^\infty \frac{d\Phi(\omega)}{(1+\omega^2)^p} < \infty. \end{aligned} \tag{8.17}$$

In other words, we find a valid spatial correlation function by solving a constrained nonlinear optimization problem, so that the resulting spatial correlation function minimizes the total error with respect to measurement data. After obtaining $\Phi(\omega)$, we plug it into (8.7) to obtain the valid spatial correlation function $\rho(v)$.

The above problem formulation is very general and applies to any real nondecreasing function $\Phi(\omega)$. For practical use, however, there is no need to enumerate all possible choices of $\Phi(\omega)$ in order to find the optimal $\rho(v)$.

Therefore, to make the above problem tractable, we can approximate the experimentally measured correlation function with a function selected from a family of functions that are proved to be valid spatial correlation functions. To serve such a purpose, it is sufficient to choose a family of functions $\Phi(\omega)$ so that the $\rho(v)$ obtained from (8.7) contains a rich set of functions for the purpose of modeling spatial correlation.

It has been shown in [22] that by choosing a proper family function of $\Phi(\omega)$, we obtain a very general family of spatial correlation functions:

$$\rho(v) = 2\left(\frac{bv}{2}\right)^{s-1} K_{s-1}(bv)\Gamma(s-1)^{-1}, \tag{8.18}$$

where K is the modified Bessel function of the second kind, Γ is the gamma function, and b and s are two real parameter numbers that regulate the shape of the function. By varying b and s, we obtain different spatial correlation functions.

Without loss of generality, in the following, Eq. (8.18) will be used as the candidate correlation function in (8.17). Moreover, 2-norm is used as a measure of the objective function in (8.17). Therefore, we rewrite the regression optimization problem as given in (8.17) as follows:

$$\begin{aligned} \min_{b,s,\sigma_s^2} &: \sum[2\sigma_s^2(\tfrac{bv}{2})^{s-1}K_{s-1}(bv)\Gamma(s-1)^{-1} - \mathrm{cov}(v) + \sigma_g^2]^2, \\ s.t. &\qquad \sigma_s^2 \le \sigma_{fc}^2. \end{aligned} \tag{8.19}$$

This is a constrained nonlinear least-square problem, and we can solve it efficiently via any nonlinear least-square technique [23]. Note that problem (8.19) is not a convex problem in general, hence we cannot guarantee to find a global optimal solution. But, as this kind of least-square minimization problem is well studied in the literature, good solvers are available to find a solution with high quality.

After solving the above problem, we obtain the estimated spatial variation component $\sigma_S^2 \approx \sigma_s^2$, and the parameter b and s. By plugging b and s into (8.18), we obtain the estimated spatial correlation function $\overline{\rho(v)} \approx \rho(v)$. Therefore, we have obtained all information about the spatial variation component: both the variance of spatial variation and the spatial correlation function.

The overall algorithm for characterizing the process variation is summarized as shown in Fig. 8.1:

Fig. 8.1 Algorithm for characterization of process variation

1	Extract global variation σ_g^2 by (8.14);
2	Solve (8.19) to obtain σ_s^2 and b and s;
3	Extract $\overline{\rho(v)}$ by plugging b and s into (8.18);
4	Extract random variation σ_r^2 by (8.20);
5	Extract overall process correlation by (8.10);

We first extract the global variation component σ_g^2 by using formula (8.14). We then solve the nonlinear least-square optimization problem as defined in (8.19) to obtain the spatial variation component σ_s^2, and the parameter of b and s that define the spatial correlation function for a homogeneous and isotropic random field as shown in (8.18). According to (8.5), we extract the random variation component by using the following formula:

$$\sigma_R^2 = \sigma_F^2 - \sigma_G^2 - \sigma_S^2 \approx \sigma_r^2 = \sigma_f^2 - \sigma_g^2 - \sigma_s^2. \tag{8.20}$$

By plugging all variation components into (8.10), we obtain the overall process correlation at any distance.

8.3.1.3 Experimental Results

Because of the sensitivity of manufacturing data, we employ a Monte Carlo model of measurement to verify the robustness and accuracy of our extraction algorithms in this section. One of the advantages of using Monte Carlo simulation is that it allows us to simulate different variation scenarios and measurement settings that are difficult to control in reality. By comparing the extracted variation components with the known variation components used in the Monte Carlo model, we can quantitatively examine how robust and how accurate our extraction algorithms are in the presence of different amount of measurement errors. Such a study is useful because it provides us the confidence in applying the algorithms to real wafer measurement.

In this experiment, the Monte Carlo model is based on a valid correlation function $\rho(v)$ that follows a homogeneous and isotropic random field, but with different variation amounts for the three variation components (σ_G^2, σ_S^2, and σ_R^2). We simulate the measurement process by generating a set of measurement data from N number of sample chips and M number of measurement sites on each chip. To model the reality due to measurement error, we add a Gaussian noise with different variation amounts during the Monte Carlo sampling.

We report the experiment results in Table 8.1. According to Table 8.1, we see that our algorithm is very accurate in extracting different variation components, yet very robust to different amount of random noise. For example, with $N = 2000$, $M = 60$, and Noise $= 10\%$, our extracted results have about 0.4% error for the global variation, 1.9% error for the spatial variation, and 2.0% error for the spatial correlation function. When the noise amount changes from 10% to 100%,

Table 8.1 Process variation extraction

N	M	Noise	err(σ_G^2)	err(σ_S^2)	err($\rho(v)$)
2000	60	10%	0.4%	-1.9%	2.0%
		50%	0.3%	-2.8%	2.7%
		100%	0.3%	-2.6%	3.7%
1500	60	10%	4.1%	2.5%	0.9%
		50%	3.9%	2.1%	1.0%
		100%	3.8%	2.0%	1.2%
1000	60	10%	7.5%	1.2%	1.0%
		50%	7.2%	1.0%	1.0%
		100%	6.9%	1.4%	1.0%
500	60	10%	17.8%	10.9%	6.6%
		50%	18.3%	6.1%	4.8%
		100%	18.6%	4.7%	3.1%
1000	50	10%	6.5%	0.8%	2.8%
		50%	5.7%	-0.4%	3.0%
		100%	5.1%	-3.0%	3.5%
	40	10%	8.6%	-4.1%	6.5%
		50%	8.7%	-3.9%	7.0%
		100%	8.9%	-2.3%	8.4%

the accuracy of our results almost does not change at all. This convincingly shows that our extraction algorithm is very resilient to the measurement noise.

We further test the robustness of our algorithm by reducing the number of chip samples N from 2000 to 1500, 1000, and 500. We see that when there are reasonable number of chip samples (1500 and 1000), our algorithm still gives quite accurate results, and the maximum error for the global variation is no more than 10%, and the maximum error in either the spatial variation or spatial correlation function is less than 5%. When the chip samples drop to 500, we start to see a larger error (but no more than 20%) in the extracted global variation. These observations are expected, because according to the statistical sampling theories, there is a lower bound on the number of samples in order to obtain reasonably accurate statistics.

Moreover, we observe that because of the optimization procedure used to extract the spatial variation and spatial correlation function as shown in (8.19), the extraction of those two parts is not as sensitive to the number of sample chips as the global variation extraction does.

We further fix the number of sample chips N to be 1000 and vary the number of measurement sites M on the chip from 60 to 50 and 40 to study how the accuracy of our algorithm changes. From Table 8.1, we see that our algorithm still gives quite accurate results. When M changes from 60 to 40, we only see slight increase of errors for all extracted variation components, and none of them has more than 10% error.

8.3.2 *Statistical Chip Testing and Yield Optimization*

Chip testing is an important step in the manufacturing of integrated circuits. The fundamental task of chip testing is to determine whether a chip should be accepted or discarded, i.e., chip disposition, by measuring various parameters of chips (called *surrogate metrics*) and setting up their corresponding screening criteria (also called *chip disposition criteria*). The accepted chip is assumed to have its *target metrics* meeting the customer specification. Typical target metrics are operational frequency, power consumption, and robustness, while the corresponding surrogate metrics may be frequency of ring oscillators, IDDQ, and process windows. Target metrics are usually difficult and too expensive to measure directly. If the decision is wrong, we risk shipping "bad" chips to customers. Hence, the quality of chip disposition criteria directly impacts both the manufacturing yield and contracted product quality loss (PQL). This is particularly so with the increasingly large process variation in nanoscale manufacturing.

Without considering variability, a circuit is typically optimized for a single combination of process parameters. As a result of manufacturing, however, we receive chips corresponding to various combinations of process parameters. Deterministic optimization cannot guarantee that the chip satisfies design requirements for all or most of these combinations. Statistical optimization would target to solve this problem. The goal of statistical optimization is to maximize yield while satisfying timing, area, power, and other design constraints. This goal can be achieved only by considering the whole space of process variations.

In this section, we briefly discuss a number of techniques on statistical chip testing and yield optimization resulting from the explicit modeling of process variability from measurement data.

8.3.2.1 Statistical Test Margin Computation

In a sequential circuit, data signals are first launched from launching flip-flops at the trigger of launching clocks, then propagated through combinational logic, and lastly are latched at the capturing flip-flops at the trigger of capturing clocks. Data signals can be latched by capturing flip-flops only if theirs arrival time T_A are less than the respective required times T_R. The difference $S = T_R - T_A$ between the required and actual arrival times is timing slack. Zero slack is the minimum value at which the circuit can operate correctly. The required time T_R can be expressed in terms of cycle time T_{clk} as $T_R = T_{\text{clk}} - \tau$, where τ accounts for such effects as clock skew, latch setup time, and so on. Thus, timing slack can be expressed as $S = T_R - T_A = T_{\text{clk}} - \tau - T_A$.

The timing slack of a collection of paths is defined as the minimum of their individual slacks. *chip slack* is the timing slack of all paths of the chip; and *test slack* is the timing slack of only those paths that are tested. *test margin* refers to the *additional slack required during testing*.

Because of process variation, chip and test slacks are modeled as correlated random variables. Parameterized statistical static timing analysis [24, 25] approximates them as linear forms:

$$S = S_0 + \sum_{i=1}^{n} a_i \Delta X_i + a_R \Delta R_a \tag{8.21}$$

where ΔX_i and ΔR_a are zero-mean unit Gaussians. Variables ΔX_i model globally correlated variations of process parameters (including spatial correlation) and ΔR_a models uncorrelated variation. S_0 is the mean or nominal value of the slack. Coefficients a_i and a_R are sensitivities to the corresponding variations. For techniques on how to obtain those coefficients from the extracted process variation (including spatial variation) as discussed above can be found in [26]. The benefit of this representation is that the correlation between two canonical forms can be immediately judged based on sensitivities to common process variables.

Given chip and test slacks S_C, S_T in the form of (8.21) and a maximum allowed shipped product quality loss (SPQL) q, compute a uniform test margin S_M as the solution of the optimization problem:

$$\max_{S_M} P(S_T \geq S_M) \tag{8.22}$$

$$\text{s.t.} \quad P(S_C \leq 0 | S_T \geq S_M) \leq q, \tag{8.23}$$

where $P(S_T \geq S_M)$ is the probability of shipping a chip and $P(S_C \leq 0|S_T \geq S_M)$ is the conditional probability that a shipped chip is deficient (i.e., SPQL).

By taking advantage of the measurement of per-chip surrogate metrics, we can also compute the test margin that is specific to each chip. Such a formulation will look like as follows.

Given chip, test, and surrogate measurement's slacks S_C, S_T, and S_P in the form of (8.21), for each value of the surrogate metric S_P for individual chip, compute the per-chip test margin $S_M(S_P)$ to maximize the fraction of shipped chips without exceeding the required SPQL q, that is:

$$\max_{S_M(S_P)} P\left(S_T \geq S_M\left(S_P\right)\right) \tag{8.24}$$

$$\text{s.t.} \quad P\left(S_C \leq 0 | S_T \geq S_M\left(S_P\right)\right) \leq q. \tag{8.25}$$

The objective function and constraint of this problem look similar to those of (8.22) and (8.23). The only difference is the dependence of test margin on measured PSRO slack. However, this difference dramatically changes the optimization problem. Instead of computing a single optimal value of S_M, we need to compute the function $S_M(S_P)$ that delivers the optimal solution for each chip. Now, both the objective function and constraint are functionals.

By going through some convolved mathematical manipulations (please refer to [27] for detailed derivations), we can show that the above two statistical optimization problems can all be reduced to some kind of root finding problems, to which many numerical techniques can be readily applied.

8.4 Hierarchical Multitask Learning for Wafer Quality Prediction

In this section, we present an optimization framework for hierarchical multitask learning (MTL) for wafer quality prediction in semiconductor manufacturing. The goal is to predict the wafer quality based on measured process variables.

8.4.1 Problem Formulation

Wafers are produced from multiple chambers, and each chamber has multiple sides. We build a unique task model for each side of a chamber to predict the quality of wafers produced therein based on the measured process variables. We also treat tasks from the same chamber as a group of tasks, and the hierarchical task relatedness is reflected in the fact that models of different sides of the same chamber are more similar to each other (task-level relatedness) than those from different chambers (group-level relatedness). Moreover, the measured process variables can be grouped into two categories: the independent process variables corresponding to controlled variables (such as pressure, power, temperature, and gas flows) that are subject to advanced process control (APC), and the dependent process variables that are heavily affected by those controlled variables in a complicate way (examples include impedance, electric bias, and throttle valve positions).

This level of understanding of the semiconductor manufacturing process results in some major differences between our formulation and many previous MTL formulations, such as [28–31]. First, most previous formulations aim to uncover the grouping of tasks, whereas the grouping of tasks is known in our setting , and the challenge is to model the hierarchical task relatedness. Second, previous formulations focus on various amount of task-level relatedness, and are lack of group-level relatedness, whereas we integrate both the task- level and the group-level relatedness. Third, in previous formulations, the input features are treated equally, whereas the input features are partitioned into two sets (independent variables and dependent variables) in our setting, which requires us imposing different constraints on them to accommodate the task-level and group-level relatedness.

More formally, suppose that we have C chambers, i.e., C groups of tasks, and the tth chamber, i.e., the tth group, has S_t sides (tasks). The process variables measured for each wafer can be further partitioned into two sets: the set of d_1 independent variables and the set of d_2 dependent variables. For the sth side of the tth chamber, let $n_{t,s}$ denote the number of wafers produced therein; $\boldsymbol{A}_{t,s} \in \mathbb{R}^{n_{t,s} \times d_1}$ denote the measured independent variables for all the wafers; $\boldsymbol{B}_{t,s} \in \mathbb{R}^{n_{t,s} \times d_2}$ denote the measured dependent variables; and $\boldsymbol{y}_{t,s} \in \mathbb{R}^{n_{t,s}}$ denote the quality of the wafers (such as the deposition thickness). Our goal is to leverage the relatedness among multiple sides (tasks) and multiple chambers (groups) to better predict the wafer quality of $\boldsymbol{y}_{t,s}$.

To predict the wafer quality $\boldsymbol{y}_{t,s}$, we adopt a linear model as is commonly used in virtual metrology [32, 33] as follows:

$$\hat{\boldsymbol{y}}_{t,s} = \boldsymbol{A}_{t,s}\boldsymbol{\alpha}_t + \boldsymbol{B}_{t,s}\boldsymbol{\beta}_{t,s},$$

where $\hat{\boldsymbol{y}}_{t,s}$ is the predicted wafer quality for the sth side of the tth chamber; $\boldsymbol{\alpha}_t \in \mathbb{R}^{d_1}$ and $\boldsymbol{\beta}_{t,s} \in \mathbb{R}^{d_2}$ are the coefficient vectors for the independent and dependent variables, respectively. Note that the coefficient vectors for independent variables are modeled the same for different sides of the same chamber because of the fact that independent variables are *controlled* parameters with specific APC set points during manufacturing.

Putting everything together, we have the following mathematical formulation of the proposed MTL problem:

$$\begin{aligned} \min \mathscr{Q}(\boldsymbol{\alpha}_t, \boldsymbol{\beta}_{t,s}, \boldsymbol{M}) = \min \sum_{t=1}^{T} \{ \sum_{s=1}^{S_t} [\mathscr{L}(\boldsymbol{y}_{t,s}, \boldsymbol{A}_{t,s}\boldsymbol{\alpha}_t, \boldsymbol{B}_{t,s}\boldsymbol{\beta}_{t,s}) \\ + \lambda_1 \|\boldsymbol{\beta}_{t,s} - \boldsymbol{M}\boldsymbol{\alpha}_t\|^p] \} + \lambda_2 \mathscr{R}(\boldsymbol{\alpha}_1, \ldots, \boldsymbol{\alpha}_T) \end{aligned} \tag{8.26}$$

where $\mathscr{L}(\cdot)$ denotes the loss function; $\boldsymbol{M} \in \mathbb{R}^{d_2 \times d_1}$ is the transformation matrix to be determined that models the complicate relationship between independent variables and dependent variables by connecting their coefficient vectors through this transformation matrix; $\mathscr{R}(\cdot)$ is a function of all the coefficient vectors for the independent variables; and p, λ_1, and λ_2 are positive parameters.

Intuitively, the first term of the objective function $\mathscr{Q}$ measures the prediction error; the second term measures the approximation error of the coefficient vector for the dependent variables using the transformed coefficient vector for the independent variables; and the last term imposes similarity on all the coefficient vectors for the independent variables. Notice that various tasks are coupled in two different ways, depending on whether they are in the same group or not, i.e., sharing the same group index t. If two tasks come from the same group t, they share the same coefficient vector $\boldsymbol{\alpha}_t$ for the independent variables, and their coefficient vectors for the dependent variables are both close to $\boldsymbol{M}\boldsymbol{\alpha}_t$; on the other hand, if two tasks come from different groups, they are related only by the coefficient vectors for the independent variables. By minimizing such an objective function, we leverage the hierarchical task relatedness to construct models for all the tasks.

Many existing single-task and multitask learning methods can be seen as special cases of the formulation as shown in (8.26). For example, if we only have the dependent variables, and $\boldsymbol{\alpha}_t$ is fixed at 0, all the tasks are decoupled, and solving (8.26) is equivalent to solving each task individually. If we only have the independent variables, $\boldsymbol{\beta}_{t,s}$ is fixed at 0, and $\boldsymbol{M}$ is a 0 matrix, all the tasks are coupled such that: (1) tasks from the same group share the same coefficient vector, and (2) tasks from different groups have similar coefficient vectors. In the case that each group has only one task, this is consistent with the formulation in [34, 35].

The objective function (8.26) also relates to convex multitask feature learning proposed in [29]. To see this, assume that we only have a set of dependent variables $\boldsymbol{B}_{t,s}$, and let $\lambda_1 \to \infty$. In this way, tasks within the same group share the same coefficient vector $\boldsymbol{M}\boldsymbol{\alpha}_t$, and different groups have similar $\boldsymbol{\alpha}_t$. Furthermore, if $S_t = 1$ ($t = 1, \ldots, T$), and $\mathscr{R}()$ is the (2,1)-norm of the matrix consisting of $\boldsymbol{\alpha}_1, \ldots, \boldsymbol{\alpha}_T$, we get the same optimization problem in [29].

8.4.2 HEAR Algorithm

In this section, we study an instantiation of the formulation (8.26) as follows:

$$\min \mathscr{Q}(\boldsymbol{\alpha}_t, \boldsymbol{\beta}_{t,s}, \boldsymbol{M}, \boldsymbol{\alpha}_0) = \min \sum_{t=1}^{T} \{ \sum_{s=1}^{S_t} [(\boldsymbol{y}_{t,s} - \boldsymbol{A}_{t,s}\boldsymbol{\alpha}_t - \boldsymbol{B}_{t,s}\boldsymbol{\beta}_{t,s})^2 + \lambda_1 \|\boldsymbol{\beta}_{t,s} - \boldsymbol{M}\boldsymbol{\alpha}_t\|^2] \} + \lambda_2 \sum_{t=1}^{T} (\boldsymbol{\alpha}_t - \boldsymbol{\alpha}_0)^2, \tag{8.27}$$

where $p = 2$ from (8.26) and $\boldsymbol{\alpha}_0 \in \mathbb{R}^{d_1}$ is a common vector to be determined for all $\boldsymbol{\alpha}_t$, reflecting the fact that the APC process tries to find common set points for independent variables across chambers.

Note that (8.27) is a nonconvex optimization problem. However, it is easy to see that given $\boldsymbol{\alpha}_t$ and $\boldsymbol{\beta}_{t,s}$ ($t = 1, \ldots, T$, $s = 1, \ldots, S_t$), $\mathscr{Q}$ is convex with respect to $\boldsymbol{M}$ and $\boldsymbol{\alpha}_0$; on the other hand, given $\boldsymbol{M}$ and $\boldsymbol{\alpha}_0$, $\mathscr{Q}$ is convex with respect to $\boldsymbol{\alpha}_t$ and $\boldsymbol{\beta}_{t,s}$ ($t = 1, \ldots, T$, $s = 1, \ldots, S_t$). Therefore, we propose a block coordinate descent-based algorithm to find the solution to (8.27), i.e., we repeatedly update $\boldsymbol{M}$ and $\boldsymbol{\alpha}_0$ based on the current $\boldsymbol{\alpha}_t$ and $\boldsymbol{\beta}_{t,s}$, and vice versa until convergence.

To do so, we first take the partial derivative of $\mathscr{Q}$ with respect to $\boldsymbol{M}$ and $\boldsymbol{\alpha}_0$, and set it to 0, obtaining:

$$\boldsymbol{M} = \sum_{t=1}^{T} \left(\sum_{s=1}^{S_t} \boldsymbol{\beta}_{t,s} \right) \boldsymbol{\alpha}_t' \left(\sum_{t=1}^{T} S_t \boldsymbol{\alpha}_t \boldsymbol{\alpha}_t' \right)^{-1} \tag{8.28}$$

$$\boldsymbol{\alpha}_0 = \frac{1}{T}\sum_{t=1}^{T}\boldsymbol{\alpha}_t \tag{8.29}$$

where $\boldsymbol{\alpha}'_t$ denotes the transpose of $\boldsymbol{\alpha}_t$. From (8.28) and (8.29), we see that both $\boldsymbol{M}$ and $\boldsymbol{\alpha}_0$ are used by different groups of tasks, hence their estimation leverages the information from all the tasks.

On the other hand, $\forall t = 1, \ldots, T$, let $\boldsymbol{\gamma}_t \in \mathbb{R}^{d_1+S_t d_2}$ denote $[\boldsymbol{\alpha}'_t, \boldsymbol{\beta}'_{t,1}, \ldots, \boldsymbol{\beta}'_{t,S_t}]'$, $\boldsymbol{\gamma}_0 \in \mathbb{R}^{d_1+S_t d_2}$ denote $[\boldsymbol{\alpha}_0, 0, \ldots, 0]'$, $\boldsymbol{y}_t \in \mathbb{R}^{\sum_{s=1}^{S_t} n_{t,s}}$ denote $[\boldsymbol{y}'_{t,1}, \ldots, \boldsymbol{y}'_{t,S_t}]'$, $\boldsymbol{X}_t \in \mathbb{R}^{\left(\sum_{s=1}^{S_t} n_{t,s}\right)\times\left(d_1+S_t d_2\right)}$ denote

$$\begin{bmatrix} \boldsymbol{A}_{t,1} & \boldsymbol{B}_{t,1} & \boldsymbol{0} & \cdots & \boldsymbol{0} \\ \boldsymbol{A}_{t,2} & \boldsymbol{0} & \boldsymbol{B}_{t,2} & \cdots & \boldsymbol{0} \\ \vdots & \vdots & \vdots & \ddots & \vdots \\ \boldsymbol{A}_{t,S_t} & \boldsymbol{0} & \cdots & \boldsymbol{0} & \boldsymbol{B}_{t,S_t} \end{bmatrix}$$

$\boldsymbol{I}_t \in \mathbb{R}^{(d_1+S_t d_2)\times(d_1+S_t d_2)}$ denote

$$\begin{bmatrix} \boldsymbol{I}_{d_1\times d_1} & 0 \cdots 0 \\ 0 & 0 \cdots 0 \\ \vdots & \vdots \ddots \vdots \\ 0 & 0 \cdots 0 \end{bmatrix}$$

where $\boldsymbol{I}_{d_1\times d_1}$ denotes $d_1 \times d_1$ identity matrix, and $\boldsymbol{M}_t \in \mathbb{R}^{(d_1+S_t d_2)\times(d_1+S_t d_2)}$ denote

$$\begin{bmatrix} \boldsymbol{M}'\boldsymbol{M} & -\boldsymbol{M}' & \cdots & -\boldsymbol{M}' \\ -\boldsymbol{M} & \boldsymbol{I}_{d_2\times d_2} & \cdots & 0 \\ \vdots & \vdots & \ddots & \vdots \\ -\boldsymbol{M} & 0 & \cdots & \boldsymbol{I}_{d_2\times d_2} \end{bmatrix}$$

Given $\boldsymbol{M}$ and $\boldsymbol{\alpha}_0$, $\mathcal{Q}$ can be written as a function of $\boldsymbol{\gamma}_t$ $(t = 1, \ldots, T)$:

$$\mathcal{Q}(\boldsymbol{\gamma}_t) = \sum_{t=1}^{T}\{\|\boldsymbol{y}_t - \boldsymbol{X}_t\boldsymbol{\gamma}_t\|^2 + \lambda_1\boldsymbol{\gamma}'_t\boldsymbol{M}_t\boldsymbol{\gamma}_t + \lambda_2(\boldsymbol{\gamma}_t - \boldsymbol{\gamma}_0)'\boldsymbol{I}_t(\boldsymbol{\gamma}_t - \boldsymbol{\gamma}_0)\}$$

Taking the partial derivative of $\mathcal{Q}$ with respect to $\boldsymbol{\gamma}_t$ and setting it to 0, we have

$$\boldsymbol{\gamma}_t = (\boldsymbol{X}'_t\boldsymbol{X}_t + \lambda_1\boldsymbol{M}_t + \lambda_2\boldsymbol{I}_t)^{-1}(\boldsymbol{X}'_t\boldsymbol{y}_t + \lambda_2\boldsymbol{\gamma}_0) \tag{8.30}$$

where we make use of the fact that $\boldsymbol{I}_t\boldsymbol{\gamma}_0 = \boldsymbol{\gamma}_0$.

From (8.30), we can see that if we fix $\boldsymbol{M}$ and $\boldsymbol{\alpha}_0$, different groups of tasks are decoupled. Therefore, the coefficient vector $\boldsymbol{\gamma}_t$ for the tth task can be individually estimated. However, the tasks within the same group are still coupled due to the sharing of the coefficient vector $\boldsymbol{\alpha}_t$ for the independent variables as well as the connection between the coefficient vectors $\boldsymbol{\alpha}_t$ and $\boldsymbol{\beta}_{t,s}$ for the two sets of variables via the transformation matrix $\boldsymbol{M}$.

Since (8.30) involves computing the inverse of a $(d_1+S_t d_2)\times(d_1+S_t d_2)$ matrix, it may be very time consuming when the number of tasks within each group is large. To address this problem, we notice that the matrix inversion involves two parts: $\boldsymbol{X}_t'\boldsymbol{X}+\lambda_2\boldsymbol{I}_t$ which is fixed for each task, and $\lambda_1\boldsymbol{M}_t$ which is updated in each iteration. Furthermore, $\boldsymbol{M}_t$ can be rewritten as follows:

$$\boldsymbol{M}_t = \boldsymbol{M}_{t1}\begin{bmatrix} \boldsymbol{I}_{d_2\times d_2} & -\boldsymbol{I}_{d_2\times d_2} \\ -\boldsymbol{I}_{d_2\times d_2} & \boldsymbol{0} \end{bmatrix}\boldsymbol{M}_{t1}' \tag{8.31}$$

where $\boldsymbol{M}_{t1}$ is defined as follows:

$$\boldsymbol{M}_{t1} = \begin{bmatrix} \boldsymbol{M}' & \boldsymbol{0} \\ \boldsymbol{0} & \boldsymbol{I}_{d_2\times d_2} \\ \vdots & \vdots \\ \boldsymbol{0} & \boldsymbol{I}_{d_2\times d_2} \end{bmatrix}$$

Let $\boldsymbol{C}_t \in \mathbb{R}^{(d_1+S_t d_2)\times(d_1+S_t d_2)}$ denote $(\boldsymbol{X}_t'\boldsymbol{X}_t+\lambda_2\boldsymbol{I}_t)^{-1}$. In the following analysis, when the context is clear, we omit the subscript of $\boldsymbol{C}_t$, and let $\boldsymbol{C}_{i,j}$ denote its block in the ith row and jth column, s.t., $\boldsymbol{C}_{1,1} \in \mathbb{R}^{d_1\times d_1}$, $\boldsymbol{C}_{i,1} \in \mathbb{R}^{d_2\times d_1}$, $\boldsymbol{C}_{1,j} \in \mathbb{R}^{d_1\times d_2}$, and $\boldsymbol{C}_{i,j} \in \mathbb{R}^{d_2\times d_2}$ $(i, j = 2, \ldots, S_t)$. Then using Woodbury formula, we have

$$(\boldsymbol{X}_t'\boldsymbol{X}_t + \lambda_1\boldsymbol{M}_t + \lambda_2\boldsymbol{I}_t)^{-1} = \boldsymbol{C} - \boldsymbol{C}\boldsymbol{M}_{t1}\boldsymbol{D}^{-1}\boldsymbol{M}_{t1}'\boldsymbol{C} \tag{8.32}$$

where $\boldsymbol{D}$ can be written as follows:

$$\boldsymbol{D} = \begin{bmatrix} \boldsymbol{M}\boldsymbol{C}_{1,1}\boldsymbol{M}' & \boldsymbol{M}\sum_{j=2}^{S_t+1}\boldsymbol{C}_{1,j} \\ \sum_{i=2}^{S_t+1}\boldsymbol{C}_{i,1}\boldsymbol{M}' & \sum_{i=2}^{S_t+1}\sum_{j=2}^{S_t+1}\boldsymbol{C}_{i,j} \end{bmatrix}$$

In this way, in each iteration, instead of computing the inverse of a $(d_1 + S_t d_2) \times (d_1 + S_t d_2)$ matrix, we only compute the inverse of a $(2d_2) \times (2d_2)$ matrix, which is independent of the number of tasks within each group.

The proposed HEAR algorithm is presented in Algorithm 6. It works as follows. In Step 1, we initialize the coefficient vectors to be vectors of all 1s. In Step 2, we compute matrix $\boldsymbol{C}_t$, which is the inverse of $\boldsymbol{X}_t'\boldsymbol{X}_t + \lambda_2\boldsymbol{I}_t$ $(t = 1, \ldots, T)$. It will be used to update the coefficient vectors via (8.32). Next, we repeatedly update $\boldsymbol{M}$, $\boldsymbol{\alpha}_0$, as well as all the coefficient vectors N times. Based on the coefficient vectors obtained via the proposed HEAR algorithm, given an unlabeled example from the

sth task of the tth group with independent variables $\boldsymbol{a}_{t,s} \in \mathbb{R}^{d_1}$ and dependent variables $\boldsymbol{b}_{t,s} \in \mathbb{R}^{d_2}$, we predict its output using $\boldsymbol{a}'_{t,s}\boldsymbol{\alpha}_t + \boldsymbol{b}'_{t,s}\boldsymbol{\beta}_{t,s}$.

Algorithm 6 HEAR Algorithm

1: **REQUIRE** $\boldsymbol{A}_{t,s}$, $\boldsymbol{B}_{t,s}$, $\boldsymbol{y}_{t,s}$, $(t = 1, \ldots, T,\ s = 1, \ldots, S_t)$, λ_1, λ_2, maximum number of iterations N
2: **ENSURE** $\boldsymbol{\alpha}_t$, $\boldsymbol{\beta}_{t,s}$ $(t = 1, \ldots, T, s = 1, \ldots, S_t)$, $\boldsymbol{M}$, $\boldsymbol{\alpha}_0$
3: Initialize $\boldsymbol{\alpha}_t$ and $\boldsymbol{\beta}_{t,s}$ $(t = 1, \ldots, T, s = 1, \ldots, S_t)$ to be vectors of all 1s
4: Compute $\boldsymbol{C}_t = (\boldsymbol{X}'_t\boldsymbol{X}_t + \lambda_2\boldsymbol{I}_t)^{-1}$ $(t = 1, \ldots, T)$
5: **for** $i = 1$ to N **do**
6: Update $\boldsymbol{M}$ using (8.28)
7: Update $\boldsymbol{\alpha}_0$ using (8.29)
8: Update $\boldsymbol{\alpha}_t$ and $\boldsymbol{\beta}_{t,s}$ using (8.30) and (8.32)

According to [36], the convergence of the proposed HEAR to a local optimum is guaranteed since the objective function (8.27) is convex with respect to each block (i.e., $\boldsymbol{M}$, $\boldsymbol{\alpha}_0$, and $\boldsymbol{\gamma}_t$). Notice that in each iteration of the HEAR algorithm, the processing time is linear with respect to the total number of training examples $\sum_{t=1}^{T}\sum_{s=1}^{S_t} n_{t,s}$, which is a direct result from (8.30) and (8.32).

8.4.3 Experimental Results

We demonstrate the performance of the proposed algorithm on real data from semiconductor manufacturing. This manufacturing process deposits dielectric materials as capping film on wafers, whose quality is mainly determined by the deposition thickness. For the purpose of this study, we used wafer data collected from four chambers with each having two sides. Each wafer is associated with measurements of 56 independent variables and 31 dependent variables. The drift and variation of both types of variables can cause wafer quality variation, and we choose wafer deposition thickness as the target metric in this study. Our data set consists of the measurements of 1651 wafers together with their deposition thickness over a 5-month period.

For comparison purpose, we also implemented two commonly used algorithms for multitask learning. The first one is CASO [30], which improves the ASO algorithm [37] by including an additional regularizer in the objective function and making use of alternating structure optimization to solve the convex relaxation problem. The second one is GMTL [28], which is based on a neural network model and uses gating of tasks to make some tasks more similar to each other than others.

The test errors are averaged over all the sides (tasks) for the three algorithms as shown in Fig. 8.2, where the x-axis is the number of labeled wafers used for training, and the y-axis is the root mean squared error (RMSE) average over all the tasks and multiple runs. This figure shows that the performance of our proposed algorithm is

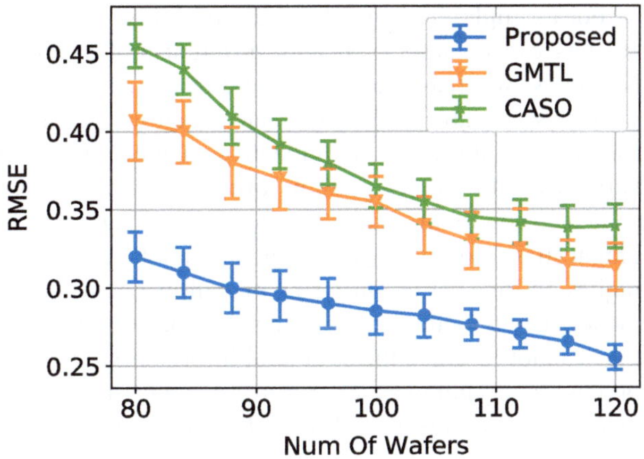

Fig. 8.2 Comparison results on multitask learning

consistently better than the other two under different training sets, especially when the number of labeled wafers is small. This improvement is due to the fact that our formulation leverages the grouping of the tasks and imposes hierarchical task relatedness accordingly, resulting better models for the relationship between input process variables and output wafer quality measurements.

8.5 Co-clustering Structural Temporal Data from Semiconductor Manufacturing

To manufacture a semiconductor device, multiple tools need to be deployed following a recipe process, and each tool has multiple chambers to carry out the task. The time series data associated with various process variables measured from all chambers can be organized as a two-dimensional array, where each row of the array corresponds to one chamber, each column corresponds to one process variable, and each element in this array corresponds to the measurements of the process variable over time. Such structural temporal data contain rich information about the manufacturing process, and thus can be exploited to help domain experts gain more insights into the recipe of manufacturing process.

8.5.1 Problem Formulation

Let $z^{i,j}$ denote the time series in the ith row and the jth column, $i = 1, \ldots, M$, $j = 1, \ldots, N$. Each time series $z^{i,j}$ has $T^{i,j}$ observations, i.e., $z^{i,j} = \{z_1^{i,j}, \ldots, z_{T^{i,j}}^{i,j}\}$. Given the above structured time series data, our goal is to simultaneously cluster the $M \times N$ time series into R row clusters and C column clusters, resulting in $R \times C$ clusters in total.

For example, in device fabrication of semiconductor manufacturing, the R row clusters can be associated with different tools, which contain various number of chambers, whereas the C column clusters can be associated with different types of process variables. The simultaneous clustering of rows (chambers) and columns (process variables) can help identify chambers with similar behaviors and process variables with similar patterns over time. Such information can be further used to detect abnormal chambers and process variables for quality control and fault diagnostics. However, this problem cannot be readily solved using existing techniques on time series clustering or multiway clustering. For time series clustering, most existing methods are designed for unstructured time series data [38, 39], and thus cannot leverage the structural information from the underlying manufacturing process. For multiway clustering, existing methods take as input one or more matrices of scalers [40, 41], and cannot be applied to matrices of time series. Therefore, we have to develop a new formulation of co-clustering of such structural temporal data.

Considering the manufacturing process, where the time series collected from chambers in the same tool exhibit similar patterns due to their physical proximity, and the time series associated with process variables of the same type are similar to each other due to their control patterns, we need to impose the following two constraints on the co-clustering problem: (1) row constraint, i.e., time series from the same row should be assigned to the same row cluster, and (2) column constraint, i.e., time series from the same column should be assigned to the same column cluster.

8.5.2 C-Struts Algorithm

Let $\hat{z}^{r,:}$ denote the rth row cluster for $r = 1, \ldots, R$; $\hat{z}^{:,c}$ denote the cth column cluster for $c = 1, \ldots, C$; Φ_1 denote the mapping from $\{z^{i,j}\}$ to $\{\hat{z}^{r,:}\}$ (row cluster); and Φ_2 denote the mapping from $\{z^{i,j}\}$ to $\{\hat{z}^{:,c}\}$ (column cluster). Considering the row and column constraints, we have $\Phi_1(z^{i,j}) = \Phi_1(z^{i,:}) = \hat{z}^{r,:}$, where $z^{i,:}$ denotes the time series on the ith row, and $\Phi_2(z^{i,j}) = \Phi_2(z^{:,j}) = \hat{z}^{:,c}$, where $z^{:,j}$ denotes the time series on the jth column.

For each time series $z^{i,j}$, we first extract a number of features $\boldsymbol{\beta}^{i,j}$ to represent the time series. We denote the set of $\boldsymbol{\beta}^{i,j}$ as $\boldsymbol{\beta}$.

For example, the time series can be represented via the auto-regressive (AR) model, i.e., assuming that its current value at time stamp t can be regressed on the past values up to a maximum lag L: $z_t^{i,j} = \sum_{l=1}^{L} \beta_l^{i,j} \cdot z_{t-l}^{i,j} + \epsilon_t^{i,j}$, where $\beta_l^{i,j}$ are the parameters, and $\epsilon_t^{i,j}$ are IID random variables. Such a problem can be solved via:

$$\boldsymbol{\beta}^{i,j} = \arg\min \sum_t (z_t^{i,j} - \sum_{l=1}^{L} \beta_l^{i,j} \cdot z_{t-l}^{i,j})^2 + \alpha \mathscr{R}(\boldsymbol{\beta}^{i,j}) \tag{8.33}$$

where $\mathscr{R}(\cdot)$ is a regularizer on the parameters, e.g., p-norm of $\boldsymbol{\beta}^{i,j}$, and α is a positive parameter that balances between the mean squared error and the regularizer. Such optimization problem can be solved using ridge regression if $\mathscr{R}(\boldsymbol{\beta}^{i,j}) = \|\boldsymbol{\beta}^{i,j}\|^2$, Lasso [42] if $\mathscr{R}(\boldsymbol{\beta}^{i,j}) = |\boldsymbol{\beta}^{i,j}|$, elastic net [43] if $\mathscr{R}(\cdot)$ is a linear combination of both the 2-norm and the 1-norm of $\boldsymbol{\beta}^{i,j}$, etc.

Alternatively, we can also extract $\boldsymbol{\beta}^{i,j}$ via the auto-regressive moving-average (ARMA) model, i.e., $z_t^{i,j} = c^{i,j} + \sum_{l=1}^{L} \beta_l^{i,j} \cdot z_{t-l}^{i,j} + \epsilon_t^{i,j} + \sum_{l'=1}^{L'} \gamma_{l'}^{i,j} \cdot \epsilon_{t-l'}^{i,j}$, where $c^{i,j}$ is the expectation of $\boldsymbol{z}^{i,j}$, and $\gamma_{l'}^{i,j}$ are additional parameters. If all the time series in the two-dimensional array are of the same length, we could also use PCA or DFT to extract the features, both of which have been used in time series clustering.

Similar to [41], we model all these time series features as random variables, and the joint probability of $\boldsymbol{\beta}$, $z^{i,:}$ and $z^{:,j}$ can be approximated by the following auxiliary probability distribution:

$$
\begin{aligned}
& p(\boldsymbol{\beta}, z^{i,:}, z^{:,j}) \approx q(\boldsymbol{\beta}, z^{i,:}, z^{:,j}) \\
& \doteq \mu^{i,j} p(\hat{z}^{r,:}, \hat{z}^{:,c}) p(z^{i,:}|\hat{z}^{r,:}) p(z^{:,j}|\hat{z}^{:,c}) p(\boldsymbol{\beta}|z^{i,:}) p(\boldsymbol{\beta}|z^{:,j})
\end{aligned}
$$

where the coefficient $\mu^{i,j}$ is to ensure that $q(\cdot)$ is a valid probability distribution, that is:

$$
\mu^{i,j} = \frac{1}{\int_{\boldsymbol{\beta}} p(\boldsymbol{\beta}|z^{i,:}) p(\boldsymbol{\beta}|z^{:,j})}
$$

Based on the auxiliary probability distribution $q(\cdot)$, the parameters $\boldsymbol{\beta}$ can be generated as follows. We first draw the row cluster $\hat{z}^{r,:}$ and the column cluster $\hat{z}^{:,c}$ from $p(\hat{z}^{r,:}, \hat{z}^{:,c})$; based on these clusters, we then draw each row $z^{i,:}$ according to $p(z^{i,:}|\hat{z}^{r,:})$, and each column $z^{:,j}$ according to $p(z^{:,j}|\hat{z}^{:,c})$; finally, we draw the parameters $\boldsymbol{\beta}$ based on both $z^{i,:}$ and $z^{:,j}$ according to $\mu^{i,j} p(\boldsymbol{\beta}|z^{i,:}) p(\boldsymbol{\beta}|z^{:,j})$.

It can be proven that $q(\cdot)$ has the following property.

Lemma 8.1 (Properties of $q(\cdot)$)

1. *The marginal probability of $\hat{z}^{r,:}$ ($\hat{z}^{:,c}$) is the same under $p(\cdot)$ and $q(\cdot)$. To be specific:*

$$
q(\hat{z}^{r,:}) = p(\hat{z}^{r,:}) \tag{8.34}
$$

$$
q(\hat{z}^{:,c}) = p(\hat{z}^{:,c}) \tag{8.35}
$$

2. *The conditional probability of $z^{i,:}$ ($z^{:,j}$) given $\hat{z}^{r,:}$ ($\hat{z}^{:,c}$) is the same under $p(\cdot)$ and $q(\cdot)$. To be specific:*

$$
q(z^{i,:}|\hat{z}^{r,:}) = p(z^{i,:}|\hat{z}^{r,:}) \tag{8.36}
$$

$$
q(z^{:,j}|\hat{z}^{:,c}) = p(z^{:,j}|\hat{z}^{:,c}) \tag{8.37}
$$

3. *$z^{i,:}$ is conditionally independent of the column cluster $\hat{z}^{:,c}$ given the row cluster $\hat{z}^{r,:}$; $z^{:,j}$ are conditionally independent of the row cluster $\hat{z}^{r,:}$ given the column cluster $\hat{z}^{:,c}$. To be specific:*

$$q(z^{i,:}|\hat{z}^{r,:}, \hat{z}^{:,c}) = q(z^{i,:}|\hat{z}^{r,:}) \tag{8.38}$$

$$q(z^{:,j}|\hat{z}^{r,:}, \hat{z}^{:,c}) = q(z^{:,j}|\hat{z}^{:,c}) \tag{8.39}$$

From the above properties, we can see that the approximation probability $q(\cdot)$ keeps the marginal probability of the row/column clusters, as well as the conditional probability of each row/column given a row/column cluster. Furthermore, the conditional independence in Eqs. (8.38) and (8.39) is consistent with both row constraints and column constraints.

Therefore, we propose to construct a prototype for each row/column cluster based on $q(\cdot)$. To this end, $\forall r = 1, \ldots, R$, we first compute the posterior distribution of $\boldsymbol{\beta}$ given $\hat{z}^{r,:}$ as follows:

$$\begin{aligned}
q(\boldsymbol{\beta}|\hat{z}^{r,:}) &= \frac{q(\boldsymbol{\beta}, \hat{z}^{r,:})}{q(\hat{z}^{r,:})} \\
&= \frac{\sum_{\hat{z}^{:,c}} \sum_{\substack{z^{i,:}:\Phi_1(z^{i,:})=\hat{z}^{r,:} \\ z^{:,j}:\Phi_2(z^{:,j})=\hat{z}^{:,c}}} q(\boldsymbol{\beta}, z^{i,:}, z^{:,j})}{p(\hat{z}^{r,:})} \\
&= \sum_{\hat{z}^{:,c}} \sum_{\substack{z^{i,:}:\Phi_1(z^{i,:})=\hat{z}^{r,:} \\ z^{:,j}:\Phi_2(z^{:,j})=\hat{z}^{:,c}}} \mu^{i,j} p(\hat{z}^{:,c}|\hat{z}^{r,:}) p(z^{i,:}|\hat{z}^{r,:}) p(z^{:,j}|\hat{z}^{:,c}) \\
&\quad \cdot p(\boldsymbol{\beta}|z^{i,:}) p(\boldsymbol{\beta}|z^{:,j})
\end{aligned}$$

Then for the rth row cluster, we define its prototype as the expected value of $\boldsymbol{\beta}$ given the row cluster, that is:

$$\hat{\boldsymbol{\beta}}^{r,:} = E_{q(\boldsymbol{\beta}|\hat{z}^{r,:})}(\boldsymbol{\beta})$$

If both $p(\boldsymbol{\beta}|z^{i,:})$ and $p(\boldsymbol{\beta}|z^{:,j})$ follow a Gaussian distribution such that $p(\boldsymbol{\beta}|z^{i,:}) \propto \exp(-\frac{1}{2\sigma_R^2}(\boldsymbol{\beta} - \boldsymbol{\beta}^{i,:})^T(\boldsymbol{\beta} - \boldsymbol{\beta}^{i,:}))$, and $p(\boldsymbol{\beta}|z^{:,j}) \propto \exp(-\frac{1}{2\sigma_C^2}(\boldsymbol{\beta} - \boldsymbol{\beta}^{:,j})^T(\boldsymbol{\beta} - \boldsymbol{\beta}^{:,j}))$, where $\boldsymbol{\beta}^{i,:}$ denotes the column average of $\boldsymbol{\beta}^{i,j}$, $\boldsymbol{\beta}^{:,j}$ denotes the row average, σ_R and σ_C are both positive parameters, that is:

$$\boldsymbol{\beta}^{i,:} = \frac{1}{N}\sum_{j=1}^{N} \boldsymbol{\beta}^{i,j} \tag{8.40}$$

$$\boldsymbol{\beta}^{:,j} = \frac{1}{M}\sum_{i=1}^{M} \boldsymbol{\beta}^{i,j} \tag{8.41}$$

In this case, the row cluster prototype can be derived as follows:

$$\begin{aligned}
\hat{\boldsymbol{\beta}}^{r,:} &= \sum_{\hat{z}^{:,c}} \sum_{\substack{z^{i,:}:\Phi_1(z^{i,:})=\hat{z}^{r,:} \\ z^{:,j}:\Phi_2(z^{:,j})=\hat{z}^{:,c}}} p(\hat{z}^{:,c}|\hat{z}^{r,:})p(z^{i,:}|\hat{z}^{r,:})p(z^{:,j}|\hat{z}^{:,c}) \\
&\quad \cdot E_{\mu^{i,j} p(\boldsymbol{\beta}|z^{i,:})p(\boldsymbol{\beta}|z^{:,j})}(\boldsymbol{\beta}) \\
&= \sum_{\hat{z}^{:,c}} \sum_{\substack{z^{i,:}:\Phi_1(z^{i,:})=\hat{z}^{r,:} \\ z^{:,j}:\Phi_2(z^{:,j})=\hat{z}^{:,c}}} p(\hat{z}^{:,c}|\hat{z}^{r,:})p(z^{i,:}|\hat{z}^{r,:})p(z^{:,j}|\hat{z}^{:,c}) \\
&\quad \cdot \frac{\sigma_C^2 \boldsymbol{\beta}^{i,:} + \sigma_R^2 \boldsymbol{\beta}^{:,j}}{\sigma_R^2 + \sigma_C^2}
\end{aligned} \tag{8.42}$$

Similarly, the column cluster prototype $\hat{\boldsymbol{\beta}}^{:,c}$ can be obtained from the following equation:

$$\begin{aligned}
\hat{\boldsymbol{\beta}}^{:,c} &= \sum_{\hat{z}^{r,:}} \sum_{\substack{z^{i,:}:\Phi_1(z^{i,:})=\hat{z}^{r,:} \\ z^{:,j}:\Phi_2(z^{:,j})=\hat{z}^{:,c}}} p(\hat{z}^{r,:}|\hat{z}^{:,c})p(z^{i,:}|\hat{z}^{r,:})p(z^{:,j}|\hat{z}^{:,c}) \\
&\quad \cdot \frac{\sigma_C^2 \boldsymbol{\beta}^{i,:} + \sigma_R^2 \boldsymbol{\beta}^{:,j}}{\sigma_R^2 + \sigma_C^2}
\end{aligned} \tag{8.43}$$

Notice that the row and column cluster prototypes in Eqs. (8.42) and (8.43) are not centroids. Instead, they are weighted combination of the row and column average of the parameters, where the weights are obtained via the approximation probability $q(\cdot)$.

For the conditional probability of each row/column given its row/column cluster, we propose to estimate its value as follows:

$$p(z^{i,:}|\hat{z}^{r,:}) \propto \exp\left(-\frac{1}{2\sigma_R^2}(\boldsymbol{\beta}^{i,:} - \hat{\boldsymbol{\beta}}^{r,:}\right)^T (\boldsymbol{\beta}^{i,:} - \hat{\boldsymbol{\beta}}^{r,:})) \tag{8.44}$$

$$p(z^{:,j}|\hat{z}^{:,c}) \propto \exp\left(-\frac{1}{2\sigma_C^2}(\boldsymbol{\beta}^{:,j} - \hat{\boldsymbol{\beta}}^{:,c}\right)^T (\boldsymbol{\beta}^{:,j} - \hat{\boldsymbol{\beta}}^{:,c})) \tag{8.45}$$

Together with the fact that $\sum_{z^{i,:}:\Phi_1(z^{i,:})=\hat{z}^{r,:}} p(z^{i,:}|\hat{z}^{r,:}) = 1$ and $\sum_{z^{:,j}:\Phi_1(z^{:,j})=\hat{z}^{:,c}} p(z^{:,j}|\hat{z}^{:,c}) = 1$, we can obtain the exact value of these conditional probabilities.

To estimate the joint probability of each row/column cluster pair, we use the empirical probability mass of the time series that have been mapped to the corresponding row/column cluster. To be specific:

$$p(\hat{z}^{r,:}, \hat{z}^{:,c}) = \frac{|z^{i,j}|\Phi_1(z^{i,j}) = \hat{z}^{r,:}, \Phi_2(z^{i,j}) = \hat{z}^{:,c}|}{M \times N} \tag{8.46}$$

where $|z^{i,j}|\Phi_1(z^{i,j}) = \hat{z}^{r,:}, and \Phi_2(z^{i,j}) = \hat{z}^{:,c}|$ denotes the number of times series that have been mapped to the rth row cluster and the cth column cluster.

Based on the above discussion, the optimal mapping functions Φ_1 and Φ_2 could be found by maximizing the following objective function:

$$\min_{\Phi_1,\Phi_2} E_p \left(\sum_{r=1}^{R} \sum_{\Phi_1(z^{i,:})=\hat{z}^{r,:}} \|\boldsymbol{\beta}^{i,:} - E_{q(\boldsymbol{\beta}|\hat{z}^{r,:})}(\boldsymbol{\beta})\|^2 \right.$$
$$\left. + \sum_{c=1}^{C} \sum_{\Phi_2(z^{:,j})=\hat{z}^{:,c}} \|\boldsymbol{\beta}^{:,j} - E_{q(\boldsymbol{\beta}|\hat{z}^{:,c})}(\boldsymbol{\beta})\|^2 \right)$$

where the outermost expectation is with respect to the true joint probability $p(\cdot)$, and the inner expectations are with respect to the auxiliary probability $q(\cdot)$.

The above optimization problem can be solved by repeatedly estimating the probability matrix of observed times series, which can then be used as the input of a variety of multiway clustering algorithms, such as information-theoretic co-clustering [41] if both R and C are known, or cross-association [40] if R and C are not known.

The proposed C-struts algorithm is summarized in Algorithm 7. We first compute the parameters $\boldsymbol{\beta}^{i,j}$ for each time series in the two-dimensional array; then, compute the row and column averages of the parameters $\boldsymbol{\beta}^{i,:}/\boldsymbol{\beta}^{:,j}$; and then, randomly assign the rows to the R row clusters, and assign the columns to the C column clusters. We start the iterations until convergence (not to exceed n_{iter} times) by first computing the row and column cluster prototypes $\hat{\boldsymbol{\beta}}^{r,:}/\hat{\boldsymbol{\beta}}^{:,c}$, and then reassigning each row/column to the closest row and column cluster prototype.

Note that we have so far focused on temporal data that fit into a two-dimensional array. But, the proposed C-Struts algorithm can be naturally extended to multidimensional arrays. For example, if the underlying structure is a three-dimensional array instead of a matrix, the auxiliary probability distribution can be defined as follows:

$$\begin{aligned} & q(\boldsymbol{\beta}, z^{i,:,:}, z^{:,j,:}, z^{:,:,k}) \\ & = \mu^{i,j,k} p(\hat{z}^{r,:,:}, \hat{z}^{:,c,:}, \hat{z}^{:,:,o}) p(z^{i,:,:}|z^{r,:,:}) p(z^{:,j,:}|z^{:,c,:}) \\ & \cdot p(z^{:,:,k}|z^{:,:,o}) p(\boldsymbol{\beta}|z^{i,:,:}) p(\boldsymbol{\beta}|z^{:,j,:}) p(\boldsymbol{\beta}|z^{:,:,k}) \end{aligned}$$

It consists of three parts: the first part is the joint probability of clusters on the three dimensions; the second part is the conditional probability of a single element (e.g., a row) given the cluster on a certain dimension; and the last part is the conditional probability of the parameters given the elements on different dimensions. Based on the above auxiliary probability distribution, we could modify the prototypes on

Algorithm 7 C-Struts Algorithm

REQUIRE $z^{i,j}, i = 1, \ldots, M, j = 1, \ldots, N, R, C, n_{iter}$
ENSURE Φ_1, Φ_2
for $i = 1$ to M **do**
    for $j = 1$ to N **do**
        Compute the parameters $\boldsymbol{\beta}^{i,j}$ by solving Equation (8.33) using ridge regression, Lasso, etc.;
for $i = 1$ to M **do**
    Compute the row average $\boldsymbol{\beta}^{i,:}$ using Equation (8.40);
for $j = 1$ to N **do**
    Compute the column average $\boldsymbol{\beta}^{:,:}$ using Equation (8.41);
Randomly initialize Φ_1 and Φ_2;
for $k = 1$ to n_{iter} **do**
    for $r = 1$ to R **do**
        Compute the row cluster prototype $\hat{\boldsymbol{\beta}}^{r,:}$ using Equation (8.42);
    for $c = 1$ to C **do**
        Compute the column cluster prototype $\hat{\boldsymbol{\beta}}^{:,c}$ using Equation (8.43);
    for $i = 1$ to M **do**
        Update $\Phi_1(z^{i,:}) \leftarrow \arg\min_{\hat{z}^{r,:}} \|\boldsymbol{\beta}^{i,:} - \hat{\boldsymbol{\beta}}^{r,:}\|^2$;
    for $j = 1$ to N **do**
        Update $\Phi_2(z^{:,j}) \leftarrow \arg\min_{\hat{z}^{:,c}} \|\boldsymbol{\beta}^{:,j} - \hat{\boldsymbol{\beta}}^{:,c}\|^2$;

the three dimensions accordingly, based on which we could repeatedly update the cluster membership using the same C-Struts Algorithm 7.

8.5.3 Experimental Results

We report the results on a data set collected from semiconductor manufacturing. The data set corresponds to an etching step with 19 process variables that forms three categories based on the process control practice: "gas and pressure," "power," and "others." The etching step lasts for 133 s, and two data points are collected during each second for all the process variables. It is concurrently running in five tools, each having six chambers. Here, the goal is to identify similar chamber and process variable behaviors.

Figure 8.3 presents the co-clustering results, where the various colors indicate the ground truth, and the rows and columns are rearranged so that rows and columns assigned to the same row and column cluster are grouped together. We see that 30 chambers have been correctly assigned to the row clusters corresponding to tools that the chambers belong to. This is consistent with the assumption that chambers from the same tool are more likely to be similar to each other than those from different tools. On the other hand, of the three column clusters generated by our method, the first cluster corresponds to the "other" category, and it mistakenly includes process variables from the "gas and pressure" category. This might be due

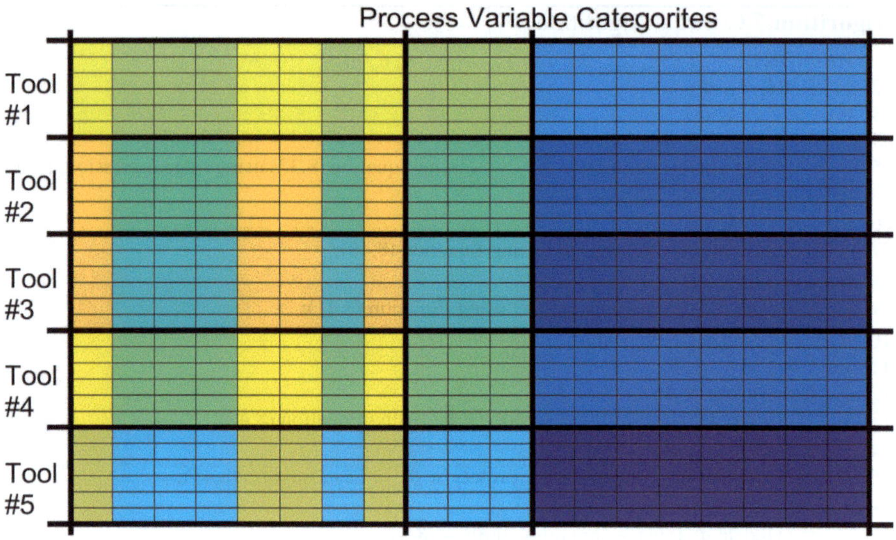

Fig. 8.3 Co-clustering results of tools and process variables

Fig. 8.4 Clustering results comparison

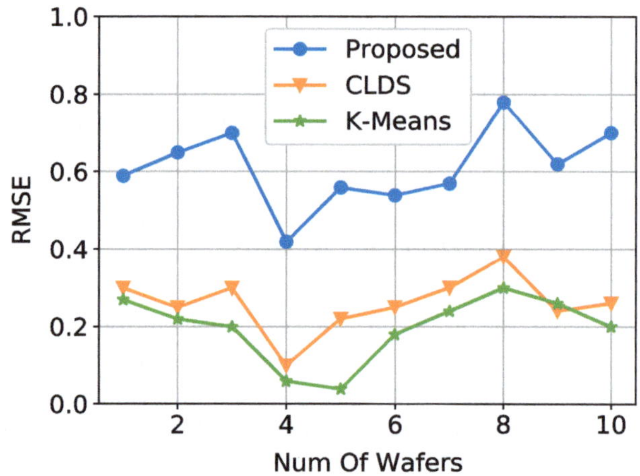

to the nature of data quality, i.e., the "other" category contains process variables of different types, and is less well defined as the "gas and pressure" or the "power" category.

We also compare the performance of the proposed approach with existing methods for time series clustering, including CLDS [14] and *k*-means. Different from the proposed method, neither CLDS nor k-means takes into consideration the structural information associated with the underlying two-dimensional array. Figure 8.4 shows the comparison results for ten wafers, and it clearly demonstrates the high-quality results of the proposed approach over existing methods.

8.6 Conclusions

We have shown in this paper, through a number of concrete use cases, how machine learning and big data analytics can be used to solve the semiconductor industry applications. The key message is that many of the existing machine learning algorithms are not necessarily well positioned to solve these challenging problems arising from the latest semiconductor manufacturing process, and novel techniques involving temporal, structural, and hierarchical properties of the manufacturing process need to be developed to solve these problems. Moreover, these so-developed techniques can be readily applied to many other machine learning and big data analytic applications, opening up endless opportunities for professionals working in the traditional semiconductor industry to contribute back to the machine learning and big data analytics communities.

References

1. J. Chen, Y. Chen, X. Du, C. Li, J. Lu, S. Zhao, X. Zhou, Big data challenge: a data management perspective. Front. Comp. Sci. **7**(2), 157–164 (2013)
2. D. Analytics, Analytics trends 2015: a below-the-surface look, in *White Paper* (2015)
3. G. Newell, N. Bekhazi, R. Morgan, Optimizing storage and I/O for distributed processing on enterprise and high performance compute (HPC) systems for mask data preparation software (CATS), Technical Report, Synopsys, Inc., 2007
4. D. Kurz, C.D. Luca, J. Pilz, Monitoring virtual metrology reliability in a sampling decision system, in *Conference on Automation Science and Engineering* (2013)
5. A. Johnson, S. McLoone, A dynamic sampling methodology for within product virtual metrology, in *29th International Manufacturing Conference* (2012)
6. J. Attenberg, K. Weinberger, A. Dasgupta, Collaborative email-spam filtering with the hashing-trick, in *Proceedings of the Sixth Annual Collaboration, Electronic messaging, Anti-Abuse and Spam Conference* (2009)
7. O. Chappelle, P. Shivaswamy, S. Vadrevu, Multi-task learning for boosting with application to web search ranking, in *Proceedings of the 16th ACM SIGKDD International Conference on Knowledge Discovery and Data Mining (KDD)* (2010), pp. 1189–1198
8. A. Torralba, K.P. Murphy, W.T. Freeman, Sharing features: efficient boosting procedures for multiclass object detection, in *Proceedings of the IEEE Conference on Computer Vision and Pattern Recognition (CVPR)* (2004), pp. 762–769
9. M. Aitkin, N. Longford, Statistical modelling issues in school effectiveness studies. J. R. Stat. Soc. A **149**, 1–43 (1986)
10. M. Daniels, C. Gatsonis, Hierarchical generalized linear models in the analysis of variations in health care utilization. J. Am. Stat. Assoc. **94**, 29–38 (1999)
11. Y. Chen, B. Hu, E.J. Keogh, G.E.A.P.A. Batista, Dtw-d: time series semi-supervised learning from a single example, in *KDD* (2013), pp. 383–391
12. B. Hu, Y. Chen, E.J. Keogh, Time series classification under more realistic assumptions, in *SDM* (2013), pp. 578–586
13. J. Zakaria, A. Mueen, E.J. Keogh, Clustering time series using unsupervised-shapelets, in *ICDM* (2012), pp. 785–794
14. L. Li, B.A. Prakash, Time series clustering: complex is simpler!, in *ICML* (2011), pp. 185–192
15. L. Li, B.A. Prakash, C. Faloutsos, Parsimonious linear fingerprinting for time series. J. Proc. VLDB Endow. **3**(1), 385–396 (2010)

16. T. Rakthanmanon, B.J.L. Campana, A. Mueen, G.E.A.P.A. Batista, M.B. Westover, Q. Zhu, J. Zakaria, E.J. Keogh, Searching and mining trillions of time series subsequences under dynamic time warping, in *KDD* (2012), pp. 262–270
17. L. Wei, E.J. Keogh, X. Xi, M. Yoder, Efficiently finding unusual shapes in large image databases. Data Min. Knowl. Disc. **17**(3), 343–376 (2008)
18. B.-K. Yi, N. Sidiropoulos, T. Johnson, H.V. Jagadish, C. Faloutsos, A. Biliris, Online data mining for co-evolving time sequences, in *ICDE* (2000), pp. 13–22
19. S. Papadimitriou, J. Sun, C. Faloutsos, Streaming pattern discovery in multiple time-series, in *VLDB* (2005), pp. 697–708
20. Y.-J. Chang, Y. Kang, C.-L. Hsu, C.-T. Chang, T.Y. Chan, Virtual metrology technique for semiconductor manufacturing, in *International Joint Conference on Neural Networks, 2006. IJCNN '06* (2006), pp. 5289–5293
21. A. Yaglom, Some classes of random fields in n-dimensional space, related to stationary random processes. Theory Probab. Appl. **2**, 273–320 (1957)
22. R.L. Bras, I. Rodriguez-Iturbe, *Random Functions and Hydrology* (Dover Publishers, Mineola, 1985)
23. T. Coleman, Y. Li, An interior, trust region approach for nonlinear minimization subject to bounds. SIAM J. Optim. **6**, 418–445 (1996)
24. C. Visweswariah, K. Ravindran, K. Kalafala, S.G. Walker, S. Narayan, First-order incremental block-based statistical timing analysis, in *DAC*, San Diego, CA, June 2004, pp. 331–336
25. H. Chang, S.S. Sapatnekar, Statistical timing analysis considering spatial correlations using a single PERT-like traversal, in *ICCAD*, San Jose, CA, November 2003, pp. 621–625
26. R. Chen, L. Zhang, V. Zolotov, C. Visweswariah, J. Xiong, Static timing: back to our roots, in *Asia and South Pacific Design Automation Conference*, Seoul, South Korea, March 2008, pp. 310–315
27. J. Xiong, V. Zolotov, C. Visweswariah, P. Habitz, Optimal margin computation for at-speed test, in *Conference on Design, Automation and Test in Europe*, Munich, Germany, March 2008, pp. 622–627
28. B. Bakker, T. Heskes, Task clustering and gating for Bayesian multitask learning. J. Mach. Learn. Res. **4**, 83–99 (2003)
29. A. Argyriou, T. Evgeniou, M. Pontil, Convex multi-task feature learning. Mach. Learn. **73**(3), 243–272 (2008)
30. J. Chen, L. Tang, J. Liu, J. Ye, A convex formulation for learning shared structures from multiple tasks, in *ICML* (2009), p. 18
31. J. Chen, J. Zhou, J. Ye, Integrating low-rank and group-sparse structures for robust multi-task learning, in *KDD* (2011), pp. 42–50
32. P. Kang, D. Kim, H.-J. Lee, S. Doh, S. Cho, Virtual metrology for run-to-run control in semiconductor manufacturing. Expert Syst. Appl. **38**, 2508–2522 (2011)
33. S. Lynn, J. Ringwood, E. Ragnoli, S. McLoone, N. MacGearailt, Virtual metrology for plasma etch using tool variables, in *Advanced Semiconductor Manufacturing Conference* (2009)
34. Y. Zhang, J.G. Schneider, Learning multiple tasks with a sparse matrix-normal penalty, in *NIPS* (2010), pp. 2550–2558
35. H. Liu, M. Palatucci, J. Zhang, Blockwise coordinate descent procedures for the multi-task lasso, with applications to neural semantic basis discovery, in *ICML* (2009), p. 82
36. D.G. Luenberger, *Linear and Nonlinear Programming*, 2nd edn. (Addison-Wesley, Massachusetts, 1973)
37. R.K. Ando, T. Zhang, A framework for learning predictive structures from multiple tasks and unlabeled data. J. Mach. Learn. Res. **6**, 1817–1853 (2005)
38. J. Zakaria, A. Mueen, E.J. Keogh, Clustering time series using unsupervised-shapelets, in *ICDM* (2012), pp. 785–794
39. T. Rakthanmanon, B.J.L. Campana, A. Mueen, G.E.A.P.A. Batista, M.B. Westover, Q. Zhu, J. Zakaria, E.J. Keogh, Searching and mining trillions of time series subsequences under dynamic time warping, in *KDD* (2012), pp. 262–270

40. D. Chakrabarti, S. Papadimitriou, D.S. Modha, C. Faloutsos, Fully automatic cross-associations, in *KDD* (2004), pp. 79–88
41. I.S. Dhillon, S. Mallela, D.S. Modha, Information-theoretic co-clustering, in *KDD* (2003), pp. 89–98
42. R. Tibshirani, Regression shrinkage and selection via the lasso. J. R. Stat. Soc. Ser. B **58**, 267–288 (1996)
43. H. Zou, T. Hastie, Regularization and variable selection via the elastic net. J. R. Stat. Soc. Ser. B (Stat Methodol.) **67**(2), 301–320 (2003)

Chapter 9
Machine Learning-Based Aging Analysis

Arunkumar Vijayan, Krishnendu Chakrabarty, and Mehdi B. Tahoori

9.1 Introduction

Bias temperature instability is one of the major aging mechanisms that can cause failures or degradation of digital circuits in their lifetime and hence, circuits are designed with a pessimistic guard band for ensuring resilience. Design-time solutions and guard bands for resilience are no longer sufficient for integrated circuits (ICs) fabricated at nanoscale technology nodes. This is due to large variations in the fabrication process, workload, and working conditions, which makes design-time solutions inefficient and impractical [13, 29]. Therefore, there is a need for runtime solutions based on real-time monitoring and adaptation. Chip manufacturers incorporate dynamic adaptation strategies such as voltage/frequency scaling, body biasing, and thermal management in response to slow-down due to aging, high temperature, current surge, process variations, etc., but the *adaptation policies* are *static*. The decisions taken in response to system behavior are "hard-coded," e.g., in lookup tables, boot ROM, firmware, etc.; hence, today's adaptation methods are more reactive than predictive, and there is no solution available to train the adaptation policy dynamically in response to changes in chip behavior.

Most existing delay monitoring schemes can only support reactive aging mitigation techniques [1, 4, 9]. A reactive mitigation approach can be too late to prevent the occurrence of a system failure. Hence, there is a need for proactive aging mitigation where suitable adaptation actions are adopted in advance. An ideal proactive delay monitoring technique should be capable of capturing fine-grained aging trends, i.e.,

---

A. Vijayan (✉) · M. B. Tahoori
Karlsruhe Institute of Technology, Karlsruhe, Germany
e-mail: arun.v@kit.edu; mehdi.tahoori@kit.edu

K. Chakrabarty
Duke University, Durham, NC, USA
e-mail: krish@duke.edu

I. M. Elfadel et al. (eds.), *Machine Learning in VLSI Computer-Aided Design*,
https://doi.org/10.1007/978-3-030-04666-8_9

monitoring aging trends in the scale of typical workload phases (in the order of milli/micro seconds). Some of the existing proactive techniques require the system to be physically aged in order to capture the aging trends [1, 37]. The time constant for aging is in the order of weeks or months. Hence, these methods cannot be used for fine-grained aging monitoring to support proactive aging mitigation.

The design of an effective aging-aware delay monitoring scheme for proactive aging mitigation is challenging for two reasons: (1) it is impractical to accurately measure the delay degradation of a circuit over a short time period. This is because traditional delay monitoring sensors track path delays, and the path-delay degradation for a short time period is too small for these sensors to capture, i.e., they can only track coarse-grained aging trends; (2) degradation rate depends on the currently running workload and working conditions. In other words, different workloads can degrade the timing of various circuit paths differently and the sensors placed during design time cannot capture this workload effect.

We present a method to perform low-cost and fine-grained workload-induced stress monitoring for accurate aging-induced delay prediction. The key idea here is to track recent circuit activity and extrapolate these trends to activate different mitigation measures, which include task migration [7], dynamic voltage and frequency scaling (DVFS) [5], adaptive body biasing (ABB) [31], and core power gating [5], based on the severity of the captured stress. This approach involves the ranking of workloads or workload phases based on their aging-stress severity. This ranking can then be used to activate aging mitigation techniques proactively. Note that the aging computation and prediction in this work are made on the basis of aging projection. The circuit is simulated with a representative workload for a short time period and the extracted behavior of the circuit is then extrapolated to estimate the aging trends. The workloads are then ranked based on these trends.

It has been shown that the impact of workload on aging can be computed on the basis of the gate-level *signal probabilities* (SPs) [26]. However, it is computationally impractical to simulate large designs at the gate-level on a cycle-by-cycle basis for realistic workloads. In addition, online monitoring of circuit node SPs of a large design with millions of gates is impractical due to area and power overhead constraints. We show that, to rank workloads in terms of severity of aging, we can also use the SPs of a small set of flip-flops as a surrogate measure of signal probability. These flip-flops are chosen at design time by carefully studying the correlation between aging and flip-flop SPs for representative workloads. We utilize support-vector machines (SVM) to develop a predictive model for estimating aging based on SPs. This model is subsequently utilized at runtime as a software thread to evaluate the online workload-induced aging-stress severity. Moreover, the monitoring of a small set of flip-flops reduces the hardware cost of monitoring and is computationally less burdensome since not all flip-flops in the design need to be monitored at runtime. Furthermore, we propose a technique to sample these selected flip-flops infrequently in order to reduce power overhead.

Our experiments on two embedded processors running realistic workloads show that the accuracy of the proposed aging prediction method is extremely high (the prediction is nearly perfect), even in the case when the prediction is based only on the SPs of 0.64% of the total number of flip-flops. The key benefits of the proposed aging prediction method are listed below: (1) **Proactive**: the aging trend is predicted before a measurable delay degradation happens, hence countermeasures can be considered in a timely manner (2) **Accurate**: the simulation results show that the correlation coefficient of actual aging trends and predicted aging trends is extremely high (higher than 0.9, where 1.0 indicates perfect correlation), (3) **Low overhead**: since the proposed technique is based on the SP values of a small set of flip-flops, it imposes minimal area and power overhead.

Aging trends can be predicted by monitoring all flip-flops in the design by reusing the on-chip design for test (DfT) infrastructure. However, this approach does not consider the overhead of accessing all flip-flops in the design for online monitoring. Hence, we propose a space-sampling approach that selects a small set of flip-flops offline based on their aging information; only these flip-flops are monitored during runtime. We introduce two different techniques to achieve space sampling: (1) correlation-based and (2) fan-in cone-based flip-flop selection. We advocate the fan-in cone-based method as the preferred approach because of its lower offline-characterization requirements.

9.2 Negative Bias Temperature Instability

Negative bias temperature instability (NBTI) is one of the major reliability threats in nanoscale technology nodes. This phenomenon causes increase in V_{th} of PMOS transistors along with a decrease in transconductance (g_m) and saturation current (I_{Dsat}) at negative gate voltages and elevated temperatures. The modelling of this phenomenon is complicated due to a recovery phase that involves a partial recovery of transistor degradation when a positive gate voltage is applied. The long-term effect of NBTI on a design is heavily dependent on the type and history of stress due to the workload under execution.

NBTI can be explained using two different mechanisms, (1) reaction–diffusion (RD) model, and (2) trapping/detrapping (TD) model. According to RD model, increase in threshold voltage occurs due to the breaking of covalent (Si–H) bonds at interface resulting in the generation of interface traps as illustrated in Fig. 9.1a. The increase in V_{th} is assumed to follow a power law relation with the time under stress and an exponential relation with the applied stress voltage.

The TD model proposes a fundamentally different phenomenon behind the effect of threshold voltage increase, i.e., the capture and emission of charge carriers by interface traps as illustrated in Fig. 9.1b. The capture time constant determines the probability of trapping and emission time constant determines the probability of detrapping. The threshold voltage increases gradually with the change in the number of occupied traps and hence, follows a logarithmic relation in V_{th} shift. TD model assumes a logarithmic relation with stress time and an exponential relation with temperature and stress voltage.

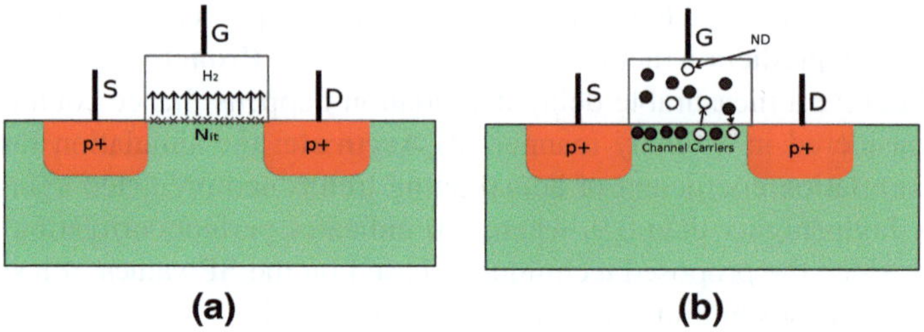

Fig. 9.1 NBTI aging models [41]. (**a**) Reaction–diffusion mechanism. (**b**) Trapping–detrapping mechanism

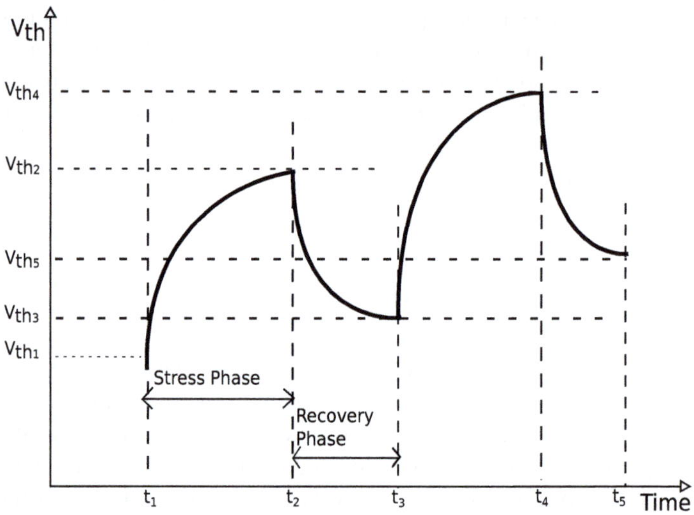

Fig. 9.2 Illustration of stress and recovery phases due to NBTI

In short, both aging mechanisms assume a stress phase and a recovery phase for a transistor under BTI as shown in Fig. 9.2 based on the type of bias applied. During a normal workload execution, the bias of a transistor alternates causing alternate stress and recovery phases. In long term, the cumulative effect of BTI heavily depends on the type of workload executed on the design.

9.3 Related Prior Work

Worst-case guard bands have been used in industry for many years. Designers use this conservative approach to ensure that circuits will operate under worst-case temperature, voltage, and workload conditions [20, 33]. However, these worst-case assumptions are too pessimistic and hence, performance of the circuits is significantly compromised [1, 42]. In addition, due to statistical behavior of aging mechanisms, similar devices may age differently even for the same environmental and workload conditions, which makes aging in the field even less predictable [21]. Finally, guard bands cannot keep up with aging challenges in newer technologies [3, 28].

An alternative approach for achieving resilience is referred to as online circuit failure prediction. This approach predicts the occurrence of a failure before errors actually occur [1]. Prediction requires information collection in real time on temperature, signal activity, signal delay, and analysis of the data. Such information is usually collected through ring oscillators, temperature sensors, delay sensors, and special circuit structures. Circuit failure prediction can be used to take actions to prevent the chip from failing, and these actions are collectively referred to as online self-healing [22, 44].

Dynamic reliability management (DRM) techniques were proposed in [15, 23].

In [38], the processor uses runtime adaptation to respond to changing application behavior to maintain its lifetime reliability target. In [39], architectural-level models were developed for lifetime-reliability-aware analysis of applications and architectures. In [40], selective redundancy is applied at the micro-architectural level. The method proposed in [43] slows aging through application scheduling and voltage changes at key moments. In [16], the processor relies on compute cycles in which computations get finished early under higher supply voltages, followed by idle cycles that provide relaxation to recover from aging.

Dynamic adaptation is often implemented through the use of embedded lookup tables (LUTs). The adaptation policies are encoded and stored in the LUTs, but they are predetermined at design time. A representative LUT-based adaptation policy was proposed in [6]. When a hardware unit changes its state from standby to active mode, the power-management unit fetches a codeword from the LUT, which is provided as input to the adaptive body-bias controller. Another LUT-based technique adjusts the power-supply voltage and body bias to compensate for aging [25]. In [45], a LUT stores, on a case-by-case basis, the optimum values of body bias, supply voltage, and clock frequency to compensate for droop and temperature variations. In [37], a built-in proactive tuning (BIPT) system was proposed, based on a canary circuit that generates predictive warning signals. In [30], the authors proposed control policies to achieve better energy efficiency and lower cost than worst-case guard-banding. Dynamic cooling was introduced as an additional tuning parameter.

A number of the above methods have been adopted for industrial circuits. In [14], the design of a Texas Instruments 3.5G baseband and multimedia applications processor is presented. This SoC consists of multiple independently controlled power domains that use dynamic voltage/frequency scaling and adaptive voltage scaling. In addition, it implements adaptive body biasing [45]. In [11], runtime adaptation techniques are described for Intel's Itanium architecture microprocessor. The core supply voltage and clock frequency are dynamically modulated in order to maximize performance within the power envelope [36]. In [27], the SmartReflex power-management techniques implemented on the OMAP3430 Mobile Multimedia Processor are presented. Active power reduction is achieved through aggressive voltage/frequency scaling and process compensation.

To enable adaptive aging mitigation, online monitoring of circuit degradation is required. The state-of-the-art monitoring methods include in-situ sensors [1, 9], tunable replica circuits [4], and representative critical reliability path-based monitoring [46]. However, these monitoring methods suffer from three problems: (1) additional hardware or design modifications, which are undesirable due to their associated area and power overhead, (2) intrinsic conflict between the accuracy of the method and hardware overhead, and (3) existing techniques cannot track short-term aging trends, since they are designed to capture an aggregated measure of the degradation when a significant delay increase takes place. Therefore, the mitigation techniques based on these monitoring systems are *reactive*. *Proactive* aging mitigation has been advocated as a promising approach that can be more effective than reactive techniques [1, 37].

The proposed method is conceptually different from any critical-path-based aging prediction method [10, 46]. During the online phase, critical-path monitoring methods rely on aging sensors. These sensors measure the actual path-delay increase, which typically occurs in longer time-scales. Even if high-resolution aging sensors [24] are used to capture the aging rate, the circuit delay needs to increase at least a few picoseconds in order to be detectable. The duration required for this minimum aging-induced delay increase (ΔD_{min}) is typically in the scale of days or weeks. In addition, circuit aging has a logarithmic relationship with time, i.e., the rate of circuit delay degradation decreases over time. Hence, the sensors cannot provide aging-rate information within an acceptable time for mitigation actions. Figure 9.3 shows ΔD_{min} at different points of time in a 3-year operation of the circuit. It implies that an aging sensor of resolution 1 ps requires around 44 days to capture a delay increase in a processor operating at 1 GHz after 3 years of operation. Since the circuit is closer to a timing failure after 3 years in comparison to its initial state, this significant increase in ΔD_{min} is unacceptable for any aging mitigation technique.

In contrast, our method evaluates the aging stress imposed by the workload-under-execution which does not require the physical aging of any of the critical

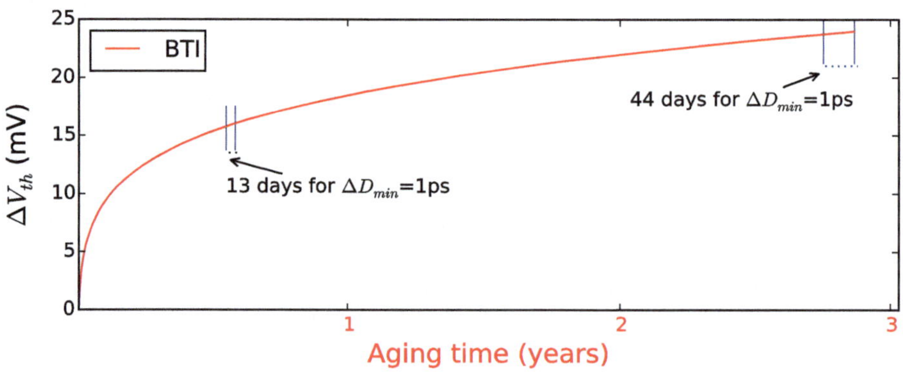

Fig. 9.3 The difference in ΔD_{min} at different time points in the lifetime of a circuit

paths. The flip-flops are tracked online to capture the characteristics of the workload, and we obviate the need to track the physical delay increase of the corresponding circuit path. As a result, we can achieve timely workload-stress estimations and aging rates that can be used in turn to take appropriate proactive and fine-grain mitigating actions and prevent the circuit from aging at higher rates.

9.4 Proposed Technique

The proposed aging prediction method is based on tracking the severity of the workload-induced runtime stress. This information can be used to guide fine-grained proactive aging mitigation policies. The flip-flops in the design that can represent the workload information are identified based on their behavior during offline workload characterization. A prediction model is constructed based on the correlation between the behavior of these selected flip-flops and the aging behavior of the circuit under different workload executions. These selected flip-flops are monitored during runtime using additional monitoring hardware in order to extract their online behavior. This extracted information is translated by the prediction model to corresponding aging trends. The output from the aging prediction model can be used to proactively actuate aging mitigation measures.

The overall flow of the method can be divided into two parts: (1) Offline correlation analysis, which consists of representative flip-flop selection, and prediction model generation, (2) Runtime stress monitoring. During offline correlation analysis, a small set of representative flip-flops are selected whose signal probabilities correlate with the aging trend of the entire circuit. Then, based on the SPs of these representative flip-flops, an aging-severity prediction model is generated by extracting workload information. This prediction model is deployed as a software in the system. During runtime, on-chip hardware is used to capture the features relevant for predicting aging-induced delay, which are then fed to the prediction model in real time. This hardware samples the state of these representative flip-flops at runtime to extract their signal probabilities, using a so-called *time sampling* scheme, which is then sent to the software-based prediction model, trained and developed at design time, to predict aging severity of the running workload.

The subsequent sections describe the details of the offline correlation analysis and runtime monitoring schemes, respectively.

9.5 Offline Correlation Analysis and Prediction Model Generation

During design time, the effect of workload on aging-induced delay is analyzed. A set of representative workloads is applied to the circuit and the SPs of all nodes are

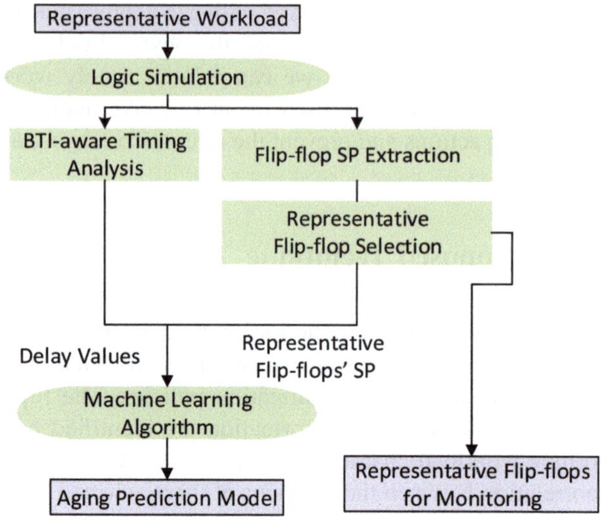

Fig. 9.4 Flowchart showing the steps involved in the offline-characterization phase

calculated for each workload. The amount of circuit delay is projected by assuming the same amount and type of workload over a fixed period of time. The delay of each gate is updated based on the SP values and then aging-aware static timing analysis is performed [10]. In this way, the aging-induced delay corresponding to each representative workload is obtained. We select a small set of representative flip-flops that can represent the aging behavior of the whole circuit as explained in Sect. 9.5.3. The SPs of these representative flip-flops together with the circuit delay values are used to train an SVM-based model. The goal here is to capture the impact of workload on circuit delay by constructing an analytical aging-prediction model.

The flowchart in Fig. 9.4 illustrates the procedure of construction of the aging prediction model. In summary, this phase involves: (1) estimation of BTI-induced delay degradation, flip-flop SP extraction and selection of representative flip-flops, and (2) construction of an aging prediction model using SVM.

9.5.1 Aging-Induced Delay Degradation and SP Extraction

Bias temperature instability (BTI)-induced threshold voltage degradation of the transistors in logic gates depends on their input SP values. For each representative workload, SPs of the primary inputs are propagated in order to determine the SPs of the internal nodes. This is achieved by annotating the SP values of the primary inputs and carrying out zero delay simulation using Synopsys Power Compiler. The resulting SPs of the internal nodes are extracted from the switching-activity-interchange-format (SAIF) file generated by Synopsys Power Compiler. SPs and switching activities are required for power-profile analysis, and commercial logic simulators have a built-in feature to dump such information for desired signals and time intervals in a SAIF file. The SP values at the inputs of each logic gate are then translated to the threshold-voltage degradation of the transistors within these gates.

During logic synthesis, the standard delay format (SDF) file of the circuit is generated, which contains the delay information about all the flip-flops and logic gates in the netlist. For each workload under consideration, these delay values of logic gates are updated to aging-induced delay values based on the SP values at the gate inputs. Static timing analysis with the updated gate delays yields an estimate of the aging-induced critical delay of the circuit. Note that STA is block-based and it implicitly considers the delay increase in all possible circuit paths.

Aging-aware timing analysis is carried out using the following steps:

1. Define the circuit netlist as a graph with logic gates as nodes and the wires as the edges.
2. Traverse through the graph, covering the forward cone of each flip-flop in the circuit until the next flip-flop in the path is reached.
3. For each node in the circuit, the propagation delay until that node is calculated by performing sum and max operations in a block-based analysis.
4. The maximum delay for the entire circuit is obtained in this way and is defined as the aging-induced circuit delay.

The only difference of this approach from conventional STA carried out by a commercial tool is that, we use aging-induced delay values of logic gates to compute circuit delay.

9.5.2 Predictor Training Using Support-Vector Machines

The next step in the offline phase is to train a predictor based on a training set consisting of the flip-flop SPs and aging-induced delay for several representative workloads. In this work, we use SVM, a supervised learning algorithm, to train the predictor. SVMs are very popular because of their resilience to over-fitting, robustness to outliers, and high prediction accuracy for a wide range of applications [8].

The delay shift due to BTI is a deterministic function only when the SP values of all nodes in the critical/near-critical paths are available in the online phase. Since it is expensive to track SPs of all of these nodes, we track a few selected flip-flops and deploy SVM to get accurate aging estimation at low cost. Note that we do not record and use the SPs of all the flip-flops in the design. Therefore, it is not feasible to use deterministic methods that rely on a large volume of data. SVM is efficient in this scenario since it utilizes a small subset of features after feature elimination.

An SVM-based model is trained based on the set of $(SP, Delay)$ pairs from the final training set. Before the training process, each set of SPs in the training set is converted to a vector form. We refer to this vector as the SP vector. For training an SVM-based predictor, the SP vectors are mapped into a higher-dimensional feature space and an optimal hyperplane (regression line) is constructed in this space. Let

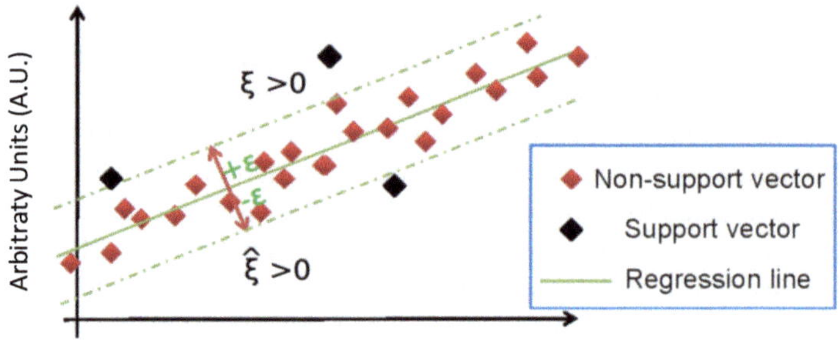

Fig. 9.5 Illustration of the support-vector regression model

Table 9.1 Hypothetical training set with four data samples

Workload	Set of SPs {SP1,SP2,SP3,SP4,SP5}	Normalized aging-induced delay
W1	0.1, 0.1, 0.4, 0.8, 0.2	0.4
W2	0.2, 0.6, 0.2, 0.1, 0.5	0.8
W3	0.1, 0.4, 0.1, 0.6, 0.5	0.1
W4	0.2, 0.1, 0.7, 0.4, 0.3	0.6

$(x_i, y_i)_{i=1}^{S}$ denote the training set, where $x_i \in \mathbb{R}^d$, and $y_i \in \mathbb{R}$. The training set consists of S SP vectors, $x_1, x_2, ..., x_S$, and each SP vector has d features and a corresponding target value (aging-induced delay), y_i, which is a real number. In this work, the ϵ-support-vector regression technique has been used. The objective of ϵ-support-vector regression is to find a regression line that fits most points with in ϵ-margin ($\epsilon > 0$), as shown in Fig. 9.5. The equation has the following form: $f(x) = \sum_{i=1}^{S} \beta_i k(x, x_i) + b$, where $b = \frac{1}{S}\sum_{i=1}^{S}(y_i - \frac{\beta_i}{|\beta_i|}\epsilon - (\sum_{j=1}^{S} \beta_i k(x_i, x_j)))$, $k(x_i, x_j)$ is the kernel function used to map the SP vectors to a high-dimensional feature space, and $(\alpha_i)_{i=1}^{S}$ are the Lagrange multipliers introduced during the process of derivation of this equation.

To illustrate the SVM-based regression methodology, consider a hypothetical scenario with four training workloads. In addition, assume that this system has five flip-flops. We form the training set as shown in Table 9.1, where the second column in each row corresponds to the set of SPs and the third column corresponds to the normalized aging-induced delay value. Using this training set and a linear kernel, we obtain the following model to predict aging-induced delay based on a given set of SPs ($s = \{s_1, s_2, s_3, s_4, s_5\}$):

$$f(\boldsymbol{x}) = 4.45 \cdot s_1 + 1.67 \cdot s_2 + 0.22 \cdot s_3 + 0.20 \cdot s_4 - 2.31 \cdot s_5 + 0.01 \quad (9.1)$$

Suppose we have a set of SPs {0.2 0.1 0.7 0.4 0.3}, which is the fourth row in Table 9.1, as the input to our model. The aging-induced delay is evaluated to be $y = 0.6$ using (9.1). Let us consider a second set of SPs {0.1 0.2 0.6 0.1 0.1}. In this case, the model evaluates to $y = 0.7$.

The runtime of the prediction model depends on the number of its variables, i.e., the number of monitored flip-flop SPs. Since a typical design has millions of flip-flops, the runtime can be considerably large. Moreover, it is extremely expensive to monitor all the flip-flops in every clock cycle. In the following sub-sections, we describe space and time sampling-based approaches to reduce these overheads.

9.5.3 Representative Flip-Flop Selection (Space Sampling)

We propose two methods to select a small subset of flip-flops whose SPs are highly correlated to aging-induced delay: (1) Correlation-based flip-flop selection and (2) fan-in cone-based flip-flop selection.

9.5.3.1 Correlation-Based Flip-Flop Selection

In correlation-based flip-flop selection, flip-flops are selected based on their behavior for different workloads under consideration. The overall flow of this method is illustrated in Fig. 9.6. During workload characterization, several workloads with varying characteristics are executed on the synthesized gate-level netlist of the chip. The SPs of flip-flop values under various workloads are extracted. For the same workloads, aging-induced delay values due to the workload-stress are also extracted using the aging-analysis framework. A feature-selection method is then employed to select significant flip-flops whose SPs are highly correlated with aging-induced circuit delay. The benefits of employing feature selection are: (1) it reduces over-fitting by eliminating redundant data, (2) it improves accuracy by eliminating irrelevant data, and (3) it reduces training time by reducing the size of the training set.

Let us assume that N *workload phases* are available at design time for training. The aging-correlation analysis flow, explained in Sect. 9.5, can be used to generate a training set $\{(SP_i, Delay_i)\}_{i=1}^{N}$ of size N, where the set SP contains the SPs of all M flip-flops $SP_i = (SP_{ij})_{j=1}^{M}$ and $Delay_i$ represents the aging-induced delay under each workload phase in our design. Our goal is to find a set of m flip-flops, $m \ll M$, whose SPs are highly correlated with aging-induced delay. This can be carried out using a univariate feature-selection method that takes the set of N $\{(SP_i, Delay_i)\}_{i=1}^{N}$ pairs and the parameter m as input and returns m features (flip-flops). These m features are selected using an m-best feature-selection algorithm in which the correlation between each individual feature and the circuit delay is evaluated, and the m best features are retained as the output [19].

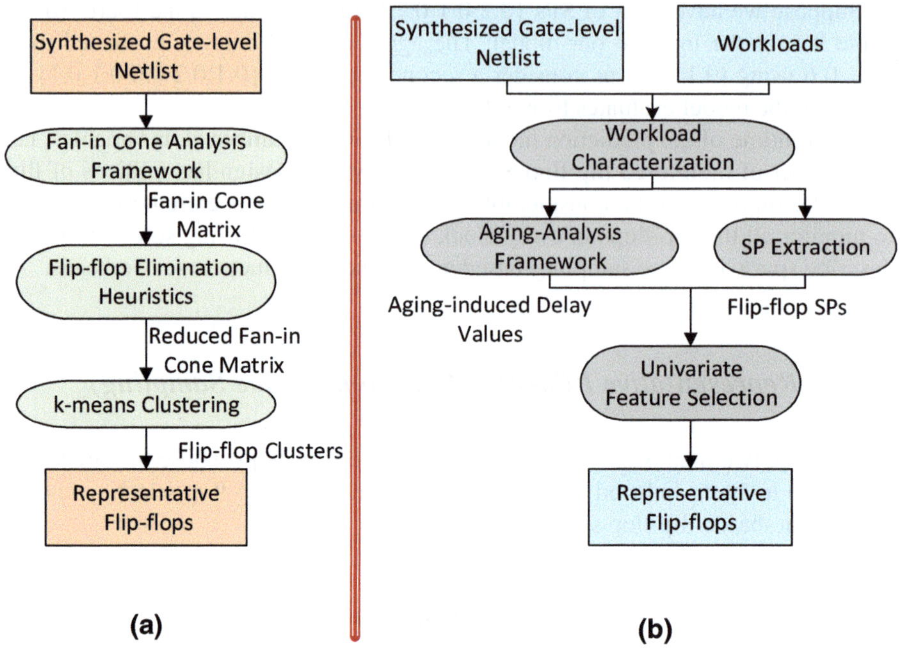

Fig. 9.6 Overall flow of space-sampling techniques to identify representative flip-flops. (**a**) Fan-in cone-based flip-flop selection. (**b**) Correlation-based flip-flop selection

To illustrate the correlation-based flip-flop selection method, consider the hypothetical scenario explained in Sect. 9.5.2. The Pearson correlation coefficients [2], which take values in $[-1, 1]$, are used to represent the correlation between the SP of flip-flop i, $1 \leq i \leq 4$, and the delay value. The Pearson correlation coefficient between two sets of values, $(x_i)_{i=1}^{n}$ and $(y_i)_{i=1}^{n}$, is evaluated as follows:

$$r = \sum_{i=1}^{n}(x_i - \bar{x})(y_i - \bar{y}) \Bigg/ \left(\sqrt{\sum_{i=1}^{n}(x_i - \bar{x})}\sqrt{\sum_{i=1}^{n}(y_i - \bar{y})}\right) \tag{9.2}$$

where $\bar{x} = \frac{1}{n}\sum_{i=1}^{n} x_i$ and $\bar{y} = \frac{1}{n}\sum_{i=1}^{n} y_i$. We obtain the correlation coefficients for the five $(SP, Delay)$ pairs to be 0.2, 0.5, 0.8, and 0.4, respectively. Therefore, Flip-flop 2 (FF2) and Flip-flop 3 (FF3) are selected as being the most effective for prediction. Similarly, if our objective is to select the best three flip-flops, then Flip-flop 2, Flip-flop 3, and Flip-flop 4 are selected.

9.5.3.2 Fan-In Cone-Based Flip-Flop Selection

Although the correlation-based method provides high accuracy, it demands very high characterization effort, especially for larger designs with large number of flip-flops. The accuracy of the correlation-based method depends on the number of data samples, i.e., the number of workloads characterized. Moreover, the simulation of a sufficient number of workloads for correlation analysis is extremely time-consuming. The details of runtime will be discussed in Sect. 9.7.5.3. Hence, we propose a more practical method that selects flip-flops independent of workload characteristics.

In fan-in cone-based sampling, flip-flops are characterized by the gates in their fan-in cones. We assume that flip-flops with a larger number of logic paths through them hold more information in terms of workload since the flow of logic values during workload execution occurs through flip-flops. Intuitively, the overlap in fan-in cones of flip-flops is assumed to represent the amount of redundancy in the information. If one flip-flop has a significant overlap in its fan-in cone with another flip-flop, only one of them needs to be selected for monitoring. For example, consider the scenario shown in Fig. 9.7. The gates present in the fan-in cones of flip-flops FF1, FF2, and FF3 are shown along with the overlap in their fan-in cones. Hence, from FF1, FF2, and FF3, only FF1 and FF2 are selected for monitoring.

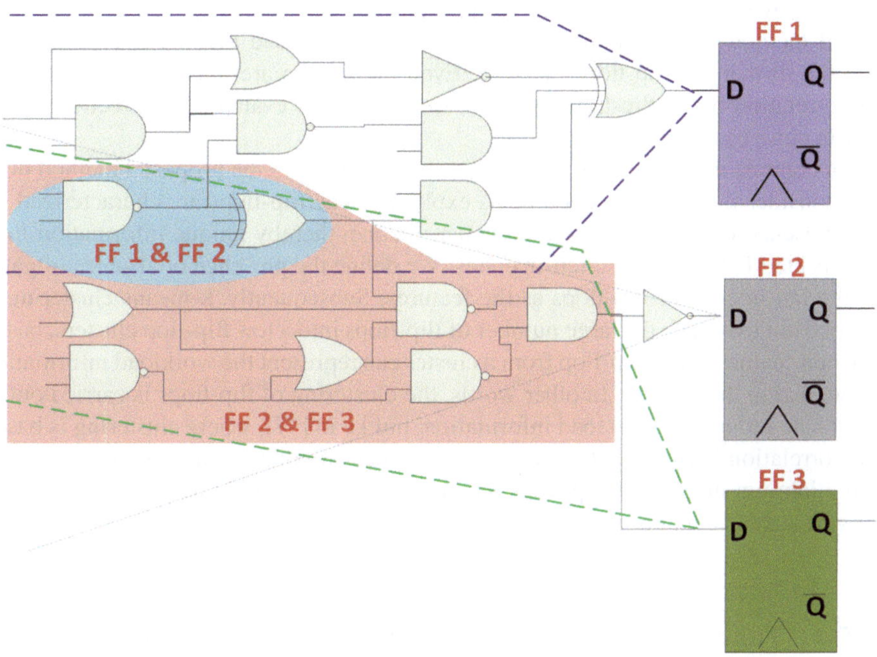

Fig. 9.7 Fan-in cone characteristics of flip-flops

The overall flow of fan-in cone-based flip-flop selection is shown in Fig. 9.6. The synthesized gate-level netlist of the design is required for fan-in cone analysis. A fan-in cone analysis framework captures the logic gates in the fan-in cone of each flip-flop. This involves a backward traversal of timing paths starting from the flip-flop input until we reach a primary input or another flip-flop's output. In this way, a fan-in cone matrix M is generated of size $n \times m$, where n and m refer to the number of flip-flops and gates in the design, respectively. Each element M_{ij} in the fan-in cone matrix can be either "0" or "1," representing the absence or presence of the gate G_j in the fan-in cone of flip-flop FF_i, respectively. For a realistic design, the size of the fan-in cone matrix would be huge for further computations. Hence, heuristics are used to eliminate a few rows or columns. This includes (1) elimination of flip-flops with only a few gates in their fan-in cone, and (2) elimination of flip-flops with exactly the same gates in their fan-in cone. Thus we obtain a reduced fan-in cone matrix for further processing.

The set of flip-flops with unique logic gates and minimum overlap in their fan-in cones is identified using k-means clustering with the reduced fan-in cone matrix as the input. The k-means clustering method, an unsupervised machine learning technique for feature selection [18], can generate k clusters with each cluster containing flip-flops having similar features, i.e., significant overlap in their fan-in cones. The representative flip-flops for monitoring can be obtained by selecting one flip-flop from each such cluster.

Since we observed (from our results) that the effectiveness of the fan-in cone-based method is comparable to that of correlation-based method, we propose to use the former for flip-flop selection. Even though we are not advocating the use of correlation-based flip-flop selection, we utilize it to evaluate the effectiveness of fan-in cone-based selection.

In contrast to [17], the fan-in cone-based algorithm in the proposed method does not consider the critically aged paths explicitly. The flip-flops are characterized by the presence of unique gates in their fan-in cone, thereby unique information flow is expected for the specific fan-in cones. We define the presence or absence of gates in the fan-in cone of flip-flops as the features. Subsequently, k-means clustering is carried out to reduce the large number of flip-flops into a few flip-flop clusters, based on their features. One flip-flop from a cluster can represent the workload information contained in that cluster. In other words, the clustering of flip-flops is carried out to eliminate redundant workload information, unlike in [17], where clustering is based on correlation between path delays or adjacency of nodes. Our objective is to find the minimum number of flip-flops having unique information to cover a maximum of gate states (signal probabilities).

9.5.4 *Time Complexity of Flip-Flop Selection Methods*

The execution time required for flip-flop selection methods is an important concern. The runtime for correlation-based flip-flop selection is much higher compared to fan-in cone-based selection due to the extensive workload characterization required to generate a sufficient number of training and testing samples.

We derive an expression for the runtime of fan-in cone-based flip-flop selection method in the worst-case. It is not possible to derive a simple closed-form expression for the runtime of correlation-based flip-flop selection method since it involves extensive workload characterization. Let the number of gates in the circuit be G, the number of flip-flops be F, and the number of signals (nets) be N. In order to generate the fan-in cone characteristics, a depth-first search is carried out starting from the flip-flop's input node until the traversal reaches the output of another flip-flop or a primary input. Therefore, for each flip-flop, the time required for finding the fan-in cone is $O(G + N)$. Since there are F flip-flops in our circuit, the runtime required for generating the fan-in cone matrix is $O(F(G + N))$.

The runtime of the heuristic used to eliminate flip-flops with the same gates in their fan-in cone is $O(F^2G)$ since the comparison of two vectors of length G takes $O(G)$ time. The runtime of the heuristic used to eliminate flip-flops whose fan-in cone has gates less than 5% of the number of gates in the largest fan-in cone is $O(F)$, if we assume that we can obtain the size of a vector in unit time. The runtime of k-means clustering algorithm is $O(nkdi)$ [35], where n is the number of input vectors (F in our problem), k is the number of clusters, d is the size of each vector (G in our problem), and i the number of iterations needed until convergence. The number of iterations until convergence is often small, and results only improve slightly after the first ten iterations. Therefore, the number of iterations until convergence is limited to ten in most implementations of k-means clustering algorithm. As a result, we can consider i as a constant in our analysis. We then obtain the runtime of the k-means clustering algorithm to select a fixed number of representative flip-flops as $O(FG)$. Since $G \gg F$ for large designs, we conclude that the runtime of the fan-in cone-based flip-flop selection method is $O(F(FG + N))$.

9.5.5 Time Sampling

The proposed aging-induced delay prediction method is based on the SPs of a small number of representative flip-flops. However, monitoring the representative flip-flops in every clock cycle to extract their SPs can be expensive in terms of total power consumption. This overhead in power consumption can be reduced by time-sampling the flip-flops instead of monitoring them in every clock cycle. In time-sampling, the flip-flops are sampled at regular intervals of time. For example, x% time-sampling implies that the flip-flops are sampled in x% clock cycles out of the total number of clock cycles in a workload segment. It is intuitive to see that the SP value of a flip-flop evaluated with a low sampling rate approximately represents the original SP value since SP is the average logic value of the flip-flop over the whole period under consideration.

Figure 9.8 shows the additional hardware required to perform time sampling. A shadow latch and a counter are added to monitor the SP of each representative

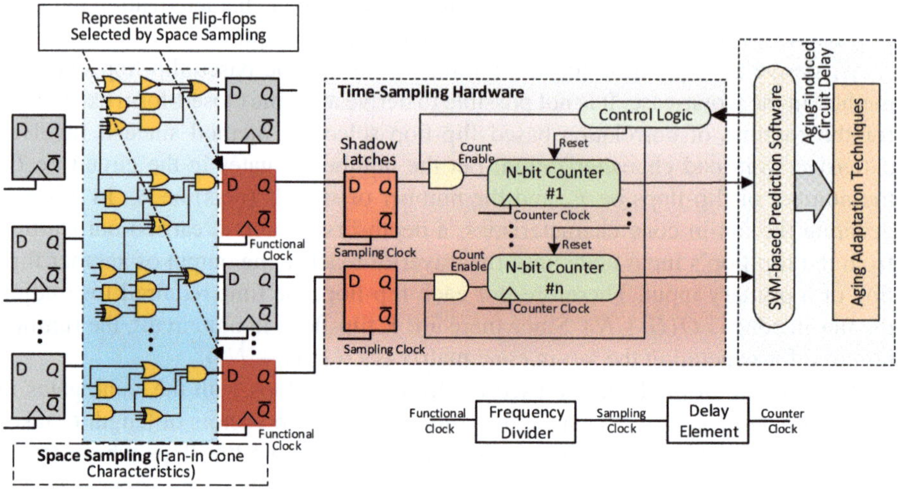

Fig. 9.8 Illustration of flip-flop SP monitoring methodology

flip-flop. We also require a frequency divider to generate the clock for the shadow latches from the functional clock. The frequency divider is designed based on the time-sampling rate. For example, if we choose 1% time-sampling, then a divide-by-100 frequency divider is needed to generate the clock for the shadow latches. The clock generated for the shadow latches cannot be used for the counters as we have to take into account the setup time of the flip-flops used to design the counter. We therefore use a delayed clock for the counters.

Our results in Sect. 9.7.1 show that, even with very low sampling rates, significantly high prediction accuracy can be achieved.

9.6 Runtime Stress Monitoring

The SPs of the flip-flops relevant to the aging prediction are monitored during runtime using a synchronous up-counter, as shown in Fig. 9.8. The counter is enabled whenever the output of the flip-flop being monitored takes logic value "1"; therefore, the counter tracks the number of clock cycles for which the flip-flop output takes logic value "1." The outputs of all the counters in the design are sampled at uniform time intervals, and these values are stored in a buffer. It should be clarified that the sampling frequency mentioned here is the frequency at which the counter values are sampled while time-sampling rate described in Sect. 9.5.5 specifies the frequency at which monitoring flip-flops are sampled. The sampling frequency to sample the counter values is decided at design time based on the characteristics of the representative workloads. The sampling frequency also determines the width of the counter that needs to be implemented. For example, if the output of a counter is sampled every 1000 clock cycles, then we need a counter that is at least 10 bits wide. Some additional control logic, as shown in Fig. 9.8, is also required to reset the counters after every sampling operation. The buffered outputs of all the counters are then transferred to the SVM-based prediction model, and then, this model translates these workload phase characteristics to aging-induced delay. The system can then take an appropriate protection measure based on the predicted aging trend.

There are two methods that can potentially be used to implement the prediction software. In the first method, the predictor is executed as a thread on any idle core on-chip. The role of this software thread is to collect flip-flop SP data from every core on-chip. These data, which are stored in a buffer on every core (Fig. 9.8), are transferred from one core to another based on a *handshake* mechanism. The core on which the thread is executing broadcasts a read signal to all the other cores. The core that is ready for this read operation sends back an acknowledgement (permission to read). Once the core executing the thread receives an acknowledgement from another core, it starts reading data from the buffer on that core and stores them in its buffer. In this way, data from all the cores are collected before the aging prediction process is started. This method of executing the prediction software does not require any additional hardware. However, an idle core may not always be available, in which case the system operation has to be interrupted to execute the prediction software. Moreover, migrating the predictor between different idle cores also involves overhead.

The other method is to execute the prediction software on a dedicated programmable microcontroller. The microcontroller communicates with every core on-chip to obtain flip-flop SP data. Therefore, it is necessary to define the interface between the microcontroller and the on-chip hardware. In [12], the communication between the energy management microcontroller and the processor core occurs through the industry-standard inter-integrated circuit (I^2C) interface. The microcontroller can read and write to the registers on every on-chip core through this interface. Therefore, the flip-flop SP data can be accessed by the microcontroller through the I^2C interface using buffer read operations. This method of implementation of the prediction software does not interrupt the processes running on any core and therefore, has minimal performance overhead. However, additional hardware cost is incurred to implement the dedicated microcontroller on-chip.

9.7 Results

For the experiments, we selected two open-source embedded processors implemented at RTL, namely Leon3 and OpenRISC 1200 (OR1200), such that we can run realistic application workloads. The effectiveness of the proposed approach was evaluated using these two processors. We implemented an SVM-based predictor, and the dependence of the prediction accuracy on the number of monitored flip-flops and on the time-sampling rate were studied for both processors. We also compared the accuracy of the two proposed flip-flop selection methods. Experiments were run on a 64 bit Linux machine with 16 GB of RAM and quad-core Intel Xeon processors running at 2.53 GHz.

9.7.1 Experimental Setup

OR1200 is five-stage pipeline embedded processor based on the 32-bit ORBIS32 instruction set architecture (ISA). Leon3 is a 32-bit processor based on the SPARC-V8 RISC ISA. The processor was synthesized using Synopsys Design Compiler with Nangate 45 nm library [32]. Our method is not limited to any particular technology node. The aging model used for evaluation can be easily updated to incorporate any newer technology node.

Six programs from MiBench, namely crc32, bitcount, qsort, susan, sha, stringsearch, and basicmath, were executed on the synthesized netlists of Leon3. Each workload was divided into several smaller workload segments of 10^3 clock cycles to collect the required number of workload samples for training and validation. Workload samples constituting the training set were used to construct a predictive model, and those constituting the test set were used to evaluate the accuracy of the predictive model. The aging-induced circuit delay values of Leon3 were computed using an NBTI-based aging analysis framework. The SPs of the flip-flops in Leon3 were obtained for each workload phase from a SAIF file generated by performing a post-synthesis simulation in ModelSim. The SVM algorithms used to train and validate the aging predictor were implemented using Scikit-learn [34] with the built-in LibSVM software package.

For aging analysis, the entire logic core of the processor (including all combinational and sequential elements), except memory blocks (such as caches and register files), is considered to be a single circuit. Aging-aware STA is carried out on this core logic to extract aging-induced circuit delay. In our analysis, we have not included any analog blocks in the circuit. However, analog blocks can be included in the timing paths as black boxes if their propagation delays are defined.

9.7.2 SVM Training and Validation

The training data set for each processor benchmark consisted of 2000 SP vectors and their corresponding aging-induced delay values. A non-overlapping set of 2000 SP vectors and their corresponding aging-induced delay values were used to test the prediction model. The best values for the radial basis function (RBF) kernel parameters used in the SVM model were determined using a five-fold cross-validation approach. In this approach, the training set was divided into five equal subsets and each subset was validated using a model trained on the remaining four subsets. A grid search was carried out on the parameters and the parameter values corresponding to the highest cross-validation accuracy were chosen. The prediction model was then trained using these best parameter values and the complete training set.

9.7.3 *Evaluation of Prediction Accuracy*

We have selected representative flip-flops based on the two different space-sampling techniques as mentioned in Sect. 9.5.3. The metric used to evaluate the accuracy of prediction is the Pearson correlation coefficient.

Furthermore, we applied time-sampling techniques to reduce the frequency at which the selected representative flip-flops are sampled.

9.7.3.1 Joint Space–Time Sampling

Figure 9.9 illustrates the results for joint space–time sampling on Leon3 and OR1200 processors. Joint time–space sampling involves selection of a set of flip-flops with space-sampling and further sampling of the logic values in these flip-flops at a pre-defined time-sampling rate. We obtained the results for four different time-sampling rates (1%, 0.1%, 0.01%, and 0.001%) and four sets of flip-flops with different sizes (8, 16, 32, and 64).

The prediction accuracy, evaluated as the Pearson correlation coefficient, under fan-in cone-based flip-flop selection and correlation-based flip-flop selection is above 0.85 for sampling rates greater than 0.001% and flip-flop selection size greater than 8. The overall trend shows a significant reduction in accuracy when time-sampling rate reduces to 0.001% from 0.01%. Moreover, the prediction accuracy increases with an increase in the number of selected flip-flops for correlation-based approach. However, it is not necessarily the case for the fan-in cone-based approach. The reason is that fan-in cone-based selection is a heuristic approach; therefore, it

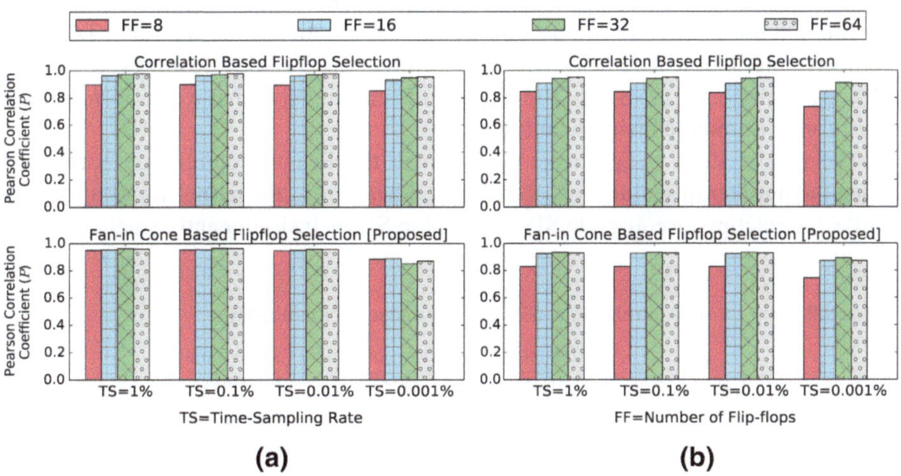

Fig. 9.9 Results obtained using joint time–space sampling of flip-flops. (**a**) Leon3. (**b**) OR1200

does not necessarily find optimal solutions. As a result, there are some cases in which the accuracy decreases even when more number of flip-flops are selected.

Since the correlation-based flip-flop selection method takes into account the nature of workload running on the chip, this method can provide better results for sufficiently large number of flip-flops. Hence, for a selection of 64 flip-flops, correlation-based flip-flop selection method provides a higher accuracy of 0.9784 compared to 0.9603 of fan-in cone-based flip-flop selection method in Leon3 for a 1% time-sampling rate. However, as the number of flip-flops selected is reduced to 8, fan-in cone-based sampling predicts with a better accuracy of 0.9514 compared to 0.8993 of the correlation-based method for Leon3. Hence, it is important to choose between these methods based on the feasibility and cost of flip-flop monitoring.

In summary, the key take-away messages from these results can be outlined as follows:

- A lower time-sampling rate can be adopted to reduce the power overhead of monitoring without compromising the accuracy of aging prediction.
- Hardware monitoring costs can be reduced by selectively monitoring a small number of flip-flops while maintaining high prediction accuracy.
- The proposed fan-in cone-based flip-flop selection heuristic, with much lower runtime and complexity, is as effective as correlation-based flip-flop selection method having large runtime requirements.

9.7.3.2 Step-by-Step Correlation

The increase in threshold voltage of a transistor due to BTI depends on the SP at the gate terminal of that transistor. For accurate aging calculation, we require SP values for all internal nodes of the circuit netlist. Instead of monitoring the SP values of the internal nodes directly, we use flip-flop SPs to predict the aging trend. Table 9.2 compares prediction accuracies with two different observables (feature vectors): (1) SPs of circuit nodes, (2) SPs of flip-flops in the circuit. Note that we perform feature selection to eliminate insignificant features.

The results show insignificant variation in the Pearson correlation coefficient when the number of features exceeds 16. In other words, the aging-information loss while compacting circuit-node SPs to flip-flop SPs is insignificantly small.

Table 9.2 Step-by-step correlation of SPs to aging-induced circuit delay

		Pearson correlation coefficient (SP → delay)			
		LEON3		OR1200	
		Circuit nodes	Flip-flops	Circuit nodes	Flip-flops
Number of features selected	64	0.9893	0.9865	0.9889	0.9835
	32	0.9889	0.9846	0.9862	0.9813
	16	0.9846	0.9786	0.9719	0.9713
	8	0.9767	0.9565	0.9589	0.9323

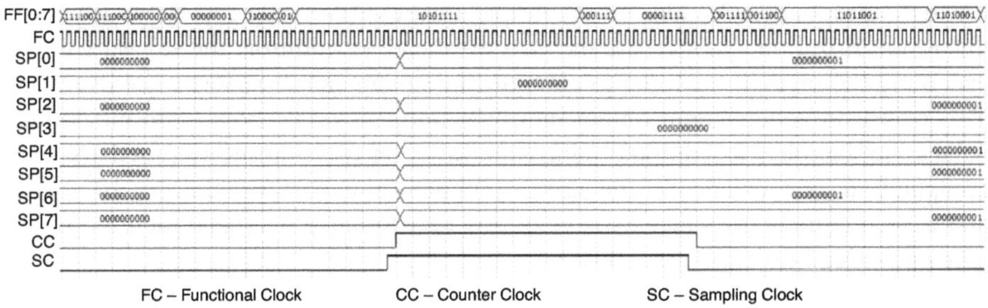

Fig. 9.10 Timing simulation results for the time-sampling hardware

9.7.4 Validation of Time-Sampling Hardware Design

The proposed method employs a time-sampling hardware as described in Sect. 9.5.5 that mainly consists of a shadow latch and a synchronous up-counter to monitor flip-flop SPs. We validated our design by carrying out timing simulations using its VHDL model. The model was implemented to sample eight flip-flops in 1% clock cycles out of the total number of clock cycles in a workload segment (assumed to be 10^5). Figure 9.10 shows timing simulation results for this design. The clock signals for the shadow latches (sampling clock) and counters (counter clock) in our design are generated from the functional clock. The flip-flop values are sampled by the shadow latches when the sampling clock is high, and the counters are updated on the rising edge of the counter clock.

9.7.5 Overheads

For area and power estimations, the processor logic cores along with their corresponding memory blocks were considered. The overheads associated with the time-sampling hardware are extracted by adding it to the original processor netlist and the updated netlist is re-synthesized. The size of the counter is estimated based on the size of a workload segment, i.e., the number of clock cycles constituting one workload segment and the time-sampling rate chosen. For instance, if the workload segment is of size 10^6 with a time-sampling rate of 1%, a 14-bit counter is required.

9.7.5.1 Performance Overhead

Static timing analysis (STA) is carried out on the circuits using Synopsys Primetime to estimate the performance overhead. The results show that the additional hardware does not affect the maximum circuit delay, hence no performance overhead. This can be attributed to the re-optimization carried out by the synthesis tool to compensate for the additional load added to the output of the monitoring flip-flops. However, this re-optimization can increase the area overhead. In addition, the software thread executing the predictive model to compute aging can cause performance overhead. This overhead is dependent on the sampling frequency, i.e., how often the counter values are sampled to compute aging, and also on the runtime required for each computation. For instance, if the aging computation is performed every 10^6 cycles and if each computation takes 40 cycles, the performance overhead incurred by the software-thread execution is 0.004%.

9.7.5.2 Area and Power Overhead

The area overhead incurred due to the additional hardware for eight flip-flops under monitoring is 0.42% for Leon3 and 0.38% for OR1200 with 0.001% time-sampling. The size of the counters required to track SPs decreases with a reduction in the time-sampling rate. Moreover, a part of this area increase is due to the optimization carried out by the synthesis tool in the form of resizing of gates. Therefore, the actual area overhead will be lower than the values that we obtained.

For Leon3 and OR1200, the power overheads for a time-sampling rate of 0.001% were found to be as low as 0.07% and 0.12%, respectively. The dynamic power overhead of the monitoring hardware is significantly reduced with lower time-sampling rates, and hence, the overall power overhead becomes lower. In other words, the leakage power of the monitoring hardware has the major contribution to overall power overhead.

9.7.5.3 Overhead at Design Time

The correlation-based flip-flop selection required an estimated CPU time of 9.5 h while fan-in cone-based flip-flop selection required only 21 min. Therefore, the runtime for correlation-based flip-flop selection is much higher compared to fan-in cone-based selection. Since the effectiveness of the fan-in cone-based method is comparable to that of correlation-based method, we advocate the use of the former for flip-flop selection. In addition, the runtime in our setup to train the SVM model offline with the selected 64 features (flip-flops obtained through fan-in cone-based flip-flop selection) was less than 1 min for both processors.

9.8 Conclusions

We have proposed a method to predict the aging-induced delay based on flip-flop SPs. Unlike existing delay-monitoring schemes based on hardware sensors, our method imposes minimal area and power overhead since we rely on the SPs of a small number of flip-flops, which can be obtained by attaching simple counters to the flip-flop outputs. This method also makes it possible to capture fine-grained aging trends that can support proactive aging mitigation techniques. Simulation results for two embedded processors demonstrate that the proposed method can accurately predict workload-induced aging trends.

References

1. M. Agarwal, B.C. Paul, M. Zhang, S. Mitra, Circuit failure prediction and its application to transistor aging, in *IEEE VLSI Test Symposium* (2007), pp. 277–286
2. J. Benesty, J. Chen, Y. Huang, I. Cohen, Pearson correlation coefficient, in *Noise Reduction in Speech Processing* (Springer, Berlin, 2009), pp. 1–4
3. S. Borkar, Designing reliable systems from unreliable components: the challenges of transistor variability and degradation. IEEE Micro **25**(6), 10–16 (2005)
4. K. Bowman et al., Circuit techniques for dynamic variation tolerance, in *ACM/IEEE Design Automation Conference* (2009), pp. 4–7
5. T.-B. Chan, J. Sartori, P. Gupta, R. Kumar, On the efficacy of NBTI mitigation techniques, in *Design, Automation and Test in Europe Conference and Exhibition* (IEEE, Piscataway, 2011), pp. 1–6
6. B. Choi, Y. Shin, Lookup table-based adaptive body biasing of multiple macros, in *International Symposium on Quality Electronic Design* (2007), pp. 533–538
7. S. Corbetta, W. Fornaciari, NBTI mitigation in microprocessor designs, in *Proceedings of the Great Lakes Symposium on VLSI* (ACM, New York, 2012)
8. C. Cortes, V. Vapnik, Support-vector networks. Mach. Learn. **20**(3), 273–297 (1995)
9. D. Ernst, N.S. Kim, S. Das, S. Pant, R. Rao, T. Pham, C. Ziesler, D. Blaauw, T. Austin, K. Flautner, T. Mudge, Razor: a low-power pipeline based on circuit-level timing speculation, in *IEEE/ACM International Symposium on Microarchitecture* (2003), pp. 7–18
10. F. Firouzi, F. Ye, K. Chakrabarty, M.B. Tahoori, Aging- and variation-aware delay monitoring using representative critical path selection. ACM Trans. Des. Autom. Electron. Syst. **20**(3), 39 (2015)
11. T. Fischer, J. Desai, B. Doyle, S. Naffziger, B. Patella, A 90-nm variable frequency clock system for a power-managed itanium architecture processor. J. Solid-State Circuits **41**(1), 218–228 (2006)
12. M. Floyd, M. Allen-Ware, K. Rajamani, B. Brock, C. Lefurgy, A.J. Drake, L. Pesantez, T. Gloekler, J.A. Tierno, P. Bose, A. Buyuktosunoglu, Introducing the adaptive energy management features of the POWER7 chip. IEEE/ACM Int. Symp. Microarchitecture **31**(2), 60–75 (2011)
13. M. Fojtik, D. Fick, Y. Kim, N. Pinckney, D. Harris, D. Blaauw, D. Sylvester, Bubble razor: an architecture-independent approach to timing-error detection and correction, in *IEEE International Solid-State Circuits Conference Digest of Technical Papers* (IEEE, Piscataway, 2012), pp. 488–490
14. G. Gammie, A. Wang, M. Chau, S. Gururajarao, R. Pitts, F. Jumel, S. Engel, P. Royannez, R. Lagerquist, H. Mair, J. Vaccani, G. Baldwin, K. Heragu, R. Mandal, M. Clinton, D. Arden, U. Ko, A 45 nm 3.5G baseband-and-multimedia application processor using adaptive body-bias and ultra-low-power techniques, in *International Solid-State Circuits Conference* (2008), pp. 258–611
15. B. Guenin, K.C. Gross, A. Gribok, A. Urmanov, A new sensor validation technique for the enhanced RAS of high end servers, in *International Conference on Machine Learning: Models, Technologies and Applications* (2004)
16. S. Gupta, S.S. Sapatnekar, Employing circadian rhythms to enhance power and reliability. ACM Trans. Des. Autom. Electron. Syst. **18**(3), 38 (2013)

17. S. Gupta, S.S. Sapatnekar, Variation-aware variable latency design. IEEE Trans. Very Large Scale Integration Syst. **22**(5), 1106–1117 (2014)
18. A.K. Jain, Data clustering: 50 years beyond K-means. Pattern Recogn. Lett. **31**(8), 651–666 (2010)
19. A. Jain, D. Zongker, Feature selection: evaluation, application, and small sample performance. IEEE Trans. Pattern Anal. Mach. Intell. **19**(2), 153–158 (1997)
20. K. Kang, H. Kufluoglu, M.A. Alam, K. Roy, Efficient transistor-level sizing technique under temporal performance degradation due to NBTI, in *IEEE International Conference on Computer Design* (2007), pp. 216–221
21. K. Kang, S.P. Park, K. Roy, M.A. Alam, Estimation of statistical variation in temporal NBTI degradation and its impact on lifetime circuit performance, in *International Conference on Computer-Aided Design* (2007)
22. B. Kapoor, S. Hemmady, S. Verma, K. Roy, M.A. D'Abreu, Impact of SoC power management techniques on verification and testing, in *International Symposium on Quality Electronic Design* (2009), pp. 692–695
23. E. Karl, D. Blaauw, D. Sylvester, T. Mudge, Multi-mechanism reliability modeling and management in dynamic systems. Trans. VLSI Syst. **16**(4), 476–487 (2008)
24. J. Keane, X. Wang, D. Persaud, C.H. Kim, An all-in-one silicon odometer for separately monitoring HCI, BTI, and TDDB. IEEE J. Solid-State Circuits **45**(4), 817–829 (2010)
25. S.V. Kumar, C.H. Kim, S.S. Sapatnekar, NBTI-aware synthesis of digital circuits, in *ACM/IEEE Design Automation Conference* (2007), pp. 370–375
26. D. Lorenz, G. Georgakos, U. Schlichtmann, Aging analysis of circuit timing considering NBTI and HCI, in *IEEE International On-Line Testing Symposium* (2009), pp. 3–8
27. H. Mair, A. Wang, G. Gammie, D. Scott, P. Royannez, S. Gururajarao, M. Chau, R. Lagerquist, L. Ho, M. Basude, N. Culp, A. Sadate, D. Wilson, F. Dahan, J. Song, B. Carlson, U. Ko, A 65-nm mobile multimedia applications processor with an adaptive power management scheme to compensate for variations, in *IEEE Symposium on VLSI Circuits* (2007)
28. J.W. McPherson, Reliability challenges for 45 nm and beyond, in *ACM/IEEE Design Automation Conference* (2006), pp. 176–181
29. E. Mintarno, J. Skaf, R. Zheng, J. Velamala, Y. Cao, S. Boyd, R.W. Dutton, S. Mitra, Optimized self-tuning for circuit aging, in *Design, Automation and Test in Europe Conference and Exhibition* (2010), pp. 586–591
30. E. Mintarno, J. Skaf, R. Zheng, J.B. Velamala, Y. Cao, S. Boyd, R.W. Dutton, S. Mitra, Self-tuning for maximized lifetime energy-efficiency in the presence of circuit aging. IEEE Trans. Comput. Aided Des. Integr. Circuits Syst. **30**(5), 760–773 (2011)
31. H. Mostafa, M. Anis, M. Elmasry, Adaptive body bias for reducing the impacts of NBTI and process variations on 6T SRAM cells. IEEE Trans. Circuits Syst. **58**(12), 2859–2871 (2011)
32. Nangate 45 nm open cell library v1.3. http://www.nangate.com
33. B.C. Paul, K. Kang, H. Kufluoglu, M.A. Alam, K. Roy, Temporal performance degradation under NBTI: estimation and design for improved reliability of nanoscale circuits, in *Design, Automation and Test in Europe Conference and Exhibition* (2006), pp. 780–785
34. F. Pedregosa, G. Varoquaux, A. Gramfort, V. Michel, B. Thirion, O. Grisel, M. Blondel, P. Prettenhofer, R. Weiss, V. Dubourg et al., Scikit-learn: machine learning in python. J. Mach. Learn. Res. **12**, 2825–2830 (2011)
35. D. Pelleg, A. Moore, Accelerating exact K-means algorithms with geometric reasoning, in *ACM International Conference on Knowledge Discovery and Data Mining* (1999), pp. 277–281
36. C. Poirier, R. McGowen, C. Bostak, S. Naffziger, Power and temperature control on a 90-nm itanium family processor. J. Solid-State Circuits **41**(1), 229–237 (2006)
37. N. Shah, R. Samanta, M. Zhang, J. Hu, D. Walker, Built-in proactive tuning system for circuit aging resilience, in *IEEE Defect and Fault Tolerance in VLSI Systems* (2008), pp. 96–104
38. J. Srinivasan, S.V. Adve, P. Bose, J.A. Rivers, The case for lifetime reliability-aware microprocessors, in *ACM International Symposium on Computer architecture*, vol. 32 (2004), p. 276

39. J. Srinivasan, S.V. Adve, P. Bose, J.A. Rivers, Lifetime reliability: toward an architectural solution. IEEE Micro **25**, 70–80 (2005)
40. J. Srinivasan, S.V. Adve, P. Bose, J.A. Rivers, Exploiting structural duplication for lifetime reliability enhancement, in *ACM International Symposium on Computer architecture*, June 2005, pp. 520–531
41. K. Sutaria, A. Ramkumar, R. Zhu, R. Rajveev, Y. Ma, Y. Cao, BTI-induced aging under random stress waveforms: modeling, simulation and silicon validation, in *Design Automation Conference (DAC)* (2014), pp. 1–6
42. D. Sylvester, D. Blaauw, E. Karl, Elastic: an adaptive self-healing architecture for unpredictable silicon. Des. Test **23**(6), 484–490 (2006)
43. A. Tiwari, J. Torrellas, Facelift: hiding and slowing down aging in multicores, in *2008 41st IEEE/ACM International Symposium on Microarchitecture* (2008)
44. J.W. Tschanz, J.T. Kao, S.G. Narendra, R. Nair, D.A. Antoniadis, A.P. Chandrakasan, V. De, Adaptive body bias for reducing impacts of die-to-die and within-die parameter variations on microprocessor frequency and leakage. J. Solid-State Circuits **37**(11), 1396–1402 (2002)
45. J. Tschanz, N.S. Kim, S. Dighe, J. Howard, G. Ruhl, S. Vangal, S. Narendra, Y. Hoskote, H. Wilson, C. Lam, M. Shuman, C. Tokunaga, D. Somasekhar, S. Tang, D. Finan, T. Karnik, N. Borkar, N. Kurd, V. De, Adaptive frequency and biasing techniques for tolerance to dynamic temperature-voltage variations and aging, in *International Solid-State Circuits Conference* (2007), pp. 292–604
46. S. Wang, J. Chen, M. Tehranipoor, Representative critical reliability paths for low-cost and accurate on-chip aging evaluation, in *IEEE/ACM International Conference on Computer-Aided Design* (2012), pp. 736–741

Part III
Machine Learning for Failure Modeling

I came to the conclusion that, in order to succeed, it would become necessary to advance the art of construction itself.

Charles Babbage

Chapter 10
Extreme Statistics in Memories

Amith Singhee

10.1 Cell Failure Probability: An Extreme Statistic

$F_{f,array}$, the chip or array failure probability, is the probability of *one or more* cells failing in any one array. Hence, if we manufacture 10 arrays, and 6 of them have a total of 8 faulty cells, our array failure probability is (approximately) $6/10$, and not $8/10$. Another, equivalent way to define $F_{f,array}$ is: the probability of the worst cell in an array being faulty. Hence, if we are measuring some performance metric y (e.g., static noise margin), we are interested in the statistics of the worst value of y from among N values, where N is the number of cells in our array. Let us define this worst value as M_N, and let us assume that by worst, we mean the maximum, i.e., large values of y are bad. If small values of y are bad, then we can use the maximum of $-y$ as the worst value.

$$M_N = max(Y_1, \ldots, Y_N) \tag{10.1}$$

where $Y_1, \ldots, Y_N$ are the measured performance metric values for the N cells in an array. Also suppose that the failure threshold for y is y_f, i.e., any cell with $y > y_f$ is defined as failing. Then, $F_{f,array}$ can be defined as

$$F_{f,array} = P(M_N > y_f) \tag{10.2}$$

that is, the probability of the worst case y in the array being greater than the failure threshold y_f. Now,

$$P(M_N > y_f) = 1 - P(M_N < y_f) \tag{10.3}$$

A. Singhee (✉)
IBM Research, Bangalore, India
e-mail: asinghee@in.ibm.com

I. M. Elfadel et al. (eds.), *Machine Learning in VLSI Computer-Aided Design*,
https://doi.org/10.1007/978-3-030-04666-8_10

where $P(M_N \leq y_f)$ is the probability of the worst case cell passing: this is the same as the probability of all the cells in the array passing; that is,

$$P(M_N \leq y_f) = P(Y_1 \leq y_f, \ldots, Y_N \leq y_f) \quad (10.4)$$

Now, we assume that the failure of any cell in the array is independent of the failure of any other cell in the array. This is reasonable because we are modeling failures due to random variations, which largely tend to be spatially independent. For instance, a well-designed CMOS SRAM bitcell would fail from random variation mainly because of random mismatch between its left and right halves induced primarily because of random dopant fluctuation. It is well known that random dopant fluctuation in one transistor is largely independent of the fluctuation in another transistor. Given this assumption of independence of cell failure events, we can write (10.4) as

$$P(M_N \leq y_f) = \prod_{i=1}^{N} P(Y_i \leq y_f) = [P(Y \leq y_f)]^N \quad (10.5)$$

Now,

$$P(Y \leq y_f) = 1 - P(Y \geq y_f) \quad (10.6)$$

where $P(Y_i > y_f)$ is the failure probability of a cell, i.e., $F_{f,array}$. Hence, combining all equations from (10.2) to (10.6), we have

$$c = 1 - [1 - F_{f,cell}]^N, or \quad (10.7)$$

$$F_{f,cell} = 1 - [1 - F_{f,array}]^{1/N} \quad (10.8)$$

10.1.1 Units of Failure Probability

There are some different units used to represent failure probability, especially in the context of extremely low probability. Some relevant ones are as follows:

- Raw probability (P): This is the most inconvenient representation, for example, a 32 Mb array with an array level requirement of $F_{f,array} < 1\%$ will need a cell failure probability $< 3 \times 10^{-10}$.
- Parts-per-million (ppm): This is the expected number of failure events (failing cells) in 1 million events (cells), and is given by $P \times 1$ million. So, a raw probability of 3×10^{-10} is the same as 3×10^{-4} ppm.
- Parts-per-billion (ppb): This is the expected number of failure events in 1 billion events and is given by $P \times$ 1 billion. Our example of 3×10^{-10} would be 0.3 ppb.

Table 10.1 Cell failure probability $F_{f,cell}$, in parts-per-billion (ppb), computed from Eq. (10.8) for a range of array sizes N and array yield specifications $F_{f,array}$

$N \downarrow$, $F_{f,array} \rightarrow$	<1%	<0.1%
1 Mb (2^{20})	9.59	0.95
2 Mb	4.79	0.48
4 Mb	2.4	**0.24**
8 Mb	1.2	**0.12**
16 Mb	0.6	**0.06**
32 Mb	0.3	**0.03**

Typical cases are highlighted in bold

- Equivalent sigma (σ): This is the $(1 - P)$-th quantile on the standard normal distribution, and is given by

$$m = \phi^{-1}(1 - P) = -\phi^{-1}(P) \tag{10.9}$$

where ϕ^{-1} is the inverse of the standard normal cumulative distribution function (CDF). So, if $P = 3 \times 10^{-10}$, we get $m = 6.19\sigma$. This is the most widely used measure of extreme failure probability in the domain of memory design.

10.1.2 An Example of Extreme Statistics in Memories

Table 10.1 shows the cell failure probability in parts-per-billion, calculated for several values of array yield specification and several values of array size. Looking at the highlighted typical values, we can see that the cell yield requirements are very stringent: only a few cells per 10 billion can fail. Apart from the significant difficulty in design such a robust memory cell, it is difficult to even estimate in simulation, the failure probability of a given cell design with acceptable accuracy.

10.1.3 Incorporating Redundancy

To increase the fault tolerance of the memory, designers may choose to include redundancy in the array. A common approach is to have redundant columns in the array that are used to replace defective columns in the chip, see, for example, [11]. As array sizes increase, the chances of having a failing cell in one array increase. In reality, this failure can be due to radiation-induced soft errors, manufacturing defects, or process variation. Redundancy allows increased tolerance to such failures. In the simplest case, if there is one redundant column, the chip can tolerate one failing cell. The use of redundancy, of course, changes the relation between the array failure probability and the cell failure probability; however, a direct relation still exists and can be derived [22]. If the array needs to be able to tolerate up to r faulty bits, the maximum acceptable bitcell failure probability can be estimated for a given array failure probability via:

$$F_{f,array} \approx 1 - \sum_{k=0}^{r} \frac{(F_{f,cell}N)^k e^{-F_{f,cell}N}}{k!} \tag{10.10}$$

From the preceding discussion, we now know how to compute the specification on the cell failure probability, given a specification on the array failure probability. Suppose now that we design a cell. How do we estimate its failure probability to check if it meets this criterion? This will be the focus of the following chapters.

10.2 Extremes: Tails and maxima

Suppose we want to model the rare event statistics of the write time of an SRAM cell. Figure 10.1 shows an example of the distribution of the write time.

We see that it is skewed to the right with a *heavy* right tail. A typical approach is to run a Monte Carlo with a small sample size (e.g., 1000) and fit a standard analytical distribution to the data, for example, a normal or a lognormal distribution. Such an approach can be accurate for fitting the "body" of the distribution, but will often be grossly inaccurate in the tail of the distribution: the skewness of the actual distribution or the heaviness of its tail will be difficult to match. As a result, any prediction of the statistics of rare events, lying far in the tail, will be inaccurate.

Let F denote the cumulative distribution function (CDF) of the write time y, and let us define a tail threshold t to mark the beginning of the tail (e.g., the 99-th percentile). Let z be the excess over the threshold t. We can write the conditional CDF of the tail as

$$F_t(Z) = P(Y - t \leq Z | Y > t) = \frac{F(Z + t) - F(t)}{1 - F(t)} \tag{10.11}$$

and the overall CDF as

$$F(Z + t) = (1 - F(t))F_t(Z) + F(t) \tag{10.12}$$

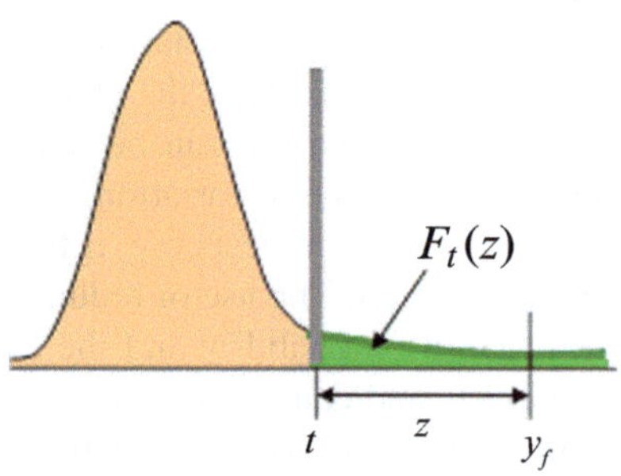

Fig. 10.1 A possible skewed distribution for some SRAM metric (e.g., write time)

If we know $F(t)$ and can estimate the conditional CDF of the tail $F_t(Z)$ accurately, we can accurately estimate rare event statistics. For example, the yield for some extreme threshold $y_f > t$ is given as

$$F(y_f) = (1 - F(t))F_t(y_f - t) + F(t) \tag{10.13}$$

and the corresponding failure probability $F_f(y_f) = 1 - F(y_f)$ is given as

$$F_f(y_f) = (1 - F(t))(1 - F_t(y_f - t)), \tag{10.14}$$

$F(t)$ can be accurately estimated using a few thousand simulations, since t is not too far out in the tail. Then, the problem here is to efficiently estimate the conditional tail CDF F_t as a simple analytical form, which can then be used to compute statistical metrics such as Eqs. (10.13) and (10.14) for rare events. Of course, here we assume that any threshold y_f of interest will be far into the tail, such that $y_f \gg t$. This is easily satisfied for any real memory design scenario, since memory sizes are typically >1 MB, requiring failure probabilities on individual bitcells ≪1 ppm in order to achieve overall memory yields in an acceptable range (typically >99%).[1] We also assume that the extreme values of interest lie only in the upper tail of the distribution. This is without any loss of generality, because any lower tail can be converted to the upper tail by replacing $y = -y$. This same approach of fitting a CDF to the exceedances over some threshold has been developed and widely applied by hydrologists under the name of the peaks over threshold (POT) method [3]. In their case though, the data is from historical record and not synthetically generated.

Now, suppose that $Y_1, Y_2, \ldots$ is a sequence of independent, identically distributed random variables from the CDF F. For any sample $Y_1, Y_2, \ldots, Y_N$ of size N, define the *sample maximum* as

$$M_N = max(Y_1, Y_2, \ldots, Y_N),\ N \geq 2 \tag{10.15}$$

The probability of $M_N \leq y$ is the probability of all of $Y_1, Y_2, \ldots, Y_N$ being $\leq y$, as in:

$$P(M_N \leq y) = P(Y_1 \leq y, \ldots, Y_N \leq y), = \prod_{i=1}^{N} P(Y_i \leq y) = F^N(y). \tag{10.16}$$

The maximum is a measure of the expected worst case value of any performance metric. Consequently, the statistics of the maximum can help us analyze the

[1] To put this in perspective, note that achieving 99% yield for even 1000 bitcells requires marginal yield of 99.999% on each cell. For a parameter with an unskewed Gaussian distribution, this equates to over 4.2 standard deviations beyond the mean, far above the 2.3 assumed by a 99th percentile value. With mega-bits of memory, it is clear that assuming thresholds far into the tail is reasonable.

statistical worst case behavior of a circuit and estimate accurate statistical design margins. We now look at some results from extreme value theory (EVT) that directly apply to these problems of estimating rare event statistics: the sample maxima and the conditional tail CDF.

10.2.1 Sample Maximum: Limiting Distributions

An important result from EVT addresses the question: What are the possible limiting distributions of M_N as $N \to \infty$? This result is stated in the following theorem by Fisher and Tippett [4].

Theorem 10.1 (Fisher–Tippett [4]) *If there exist normalizing constants a_N, b_N, and some non-degenerate CDF H, such that*

$$P\left(\frac{M_N - b_N}{a_N} \leq y\right) = F^N(a_N y + b_N) \to H(y) \text{ as } N \to \infty,\ y \in \Re \tag{10.17}$$

then H belongs to the type of one of the following three CDFs:

$$Fréchet : \Phi_a(y) = \begin{cases} 0,\ y \leq 0, \\ e^{-y^{-a}},\ y > 0 \end{cases},\ \alpha > 0$$

$$Weibull : \Psi_a(y) = \begin{cases} e^{-(-y)^a},\ y \leq 0, \\ 1,\ y > 0 \end{cases},\ \alpha > 0 \tag{10.18}$$

$$Gumbel : \Lambda(y) = e^{-e^{-y}},\ y \in \Re$$

This amazing result formed the foundation of estimation of rare event statistics. Roughly, it says that for a very large class of CDFs, we can model the distribution of the normalized sample maximum M_N as one of the three standard distributions: Fréchet, Weibull, and Gumbel. These three CDFs can be combined together into a generalized extreme value (GEV) distribution:

$$H_\zeta(y) = \begin{cases} e^{-(1-\zeta y)^{1/\zeta}},\ \zeta \neq 0 \\ e^{-e^{-y}},\ \zeta = 0 \end{cases},\ where 1 - \zeta y > 0. \tag{10.19}$$

The three CDFs are obtained as follows:

- $\zeta = -\alpha^{-1} < 0$ gives the Fréchet CDF Φ_a,

Table 10.2 Some common distributions lying in $MDA(H_\zeta)$

H_ζ	Distributions in $MDA(H_\zeta)$
$\Phi_{-1/\zeta}$	Cauchy, Pareto, Loggamma
$\Psi_{1/\zeta}$	Uniform, Beta
Λ	Normal, Lognormal, Gamma, Exponential

- $\zeta = -\alpha^{-1} > 0$ gives the Weibull CDF Ψ_a, and
- $\zeta = 0$ gives the Gumbel CDF Λ.

The condition equation (10.17) is commonly stated as F *lies in the maximum domain of attraction of* H, or $F \in MDA(H)$. Hence, for non-degenerate H, Theorem 2.1 can be stated succinctly as

$$F \in MHA(H) \implies H \text{ is of type } H_\zeta$$

The conditions for which $F \in MDA(H)$ for some non-degenerate H are quite general for most practical purposes, and known well. Gnedenko [5] provided the first rigorous proof for the Fisher–Tippett theorem, showing conditions on F required for the convergence to each of the three limiting CDFs. For details regarding these conditions, see [3, 17, 18]. Here, we only list some common distributions belonging to $MDA(H_\zeta)$, in Table 10.2, and immediately proceed to a similar result for the conditional CDF of the tail.

10.2.2 Distribution Tail: Limiting Distributions

We recall the definition of F_t as the conditional tail CDF for a tail threshold t, as in Eq. (10.11). Then, the following is true.

Theorem 10.2 (Balkema and de Haan [1] and Pickands [15]) *For every* $\zeta \in \Re$, $F \in MDA(H_\zeta)$ *if and only if*

$$\lim_{t\to\infty} \sup_{z\geq 0} |F_t(z) - G_{\zeta,\beta(t)}(z)| = 0 \tag{10.20}$$

for some positive function $\beta(t)$, *where* $G_{\zeta,\beta(Z)}$ *is the generalized Pareto distribution (GPD)*

$$G_{\zeta,\beta}zZ) = \begin{cases} 1-(1-\zeta\frac{z}{\beta})^{1/\zeta}, \zeta \neq 0, z \in D(\zeta,\beta) \\ 1-e^{-z/\beta}, \zeta = 0, z \geq 0 \end{cases} \quad , where \tag{10.21}$$

$$D(\zeta,\beta) = \begin{cases} [0,\infty], \zeta \leq 0 \\ [0,\beta/\zeta], \zeta > 0 \end{cases} .$$

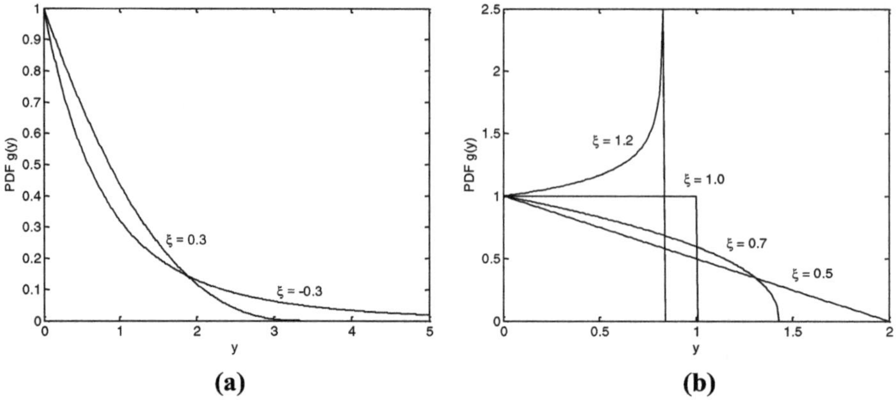

Fig. 10.2 The probability density function for a GPD with $\beta = 1$. We get long unbounded tails for $\zeta \leq 0$. (**a**) $\xi < 1/2$. (**b**) $\xi \geq 1/2$

In other words, for any distribution F in the maximum domain of attraction of the GEV distribution, the conditional tail distribution F_t converges to a GPD (Fig. 10.2) as we move further out in the tail. This is an extremely useful result: it implies that, if we can generate enough points in the tail of a distribution ($y \geq t$), in most practical cases, we can fit the simple, analytical GPD to the data and make predictions further out in the tail. This approach would be independent of the circuit or the performance metric being considered.

Of course, two important questions remain:

- How do we efficiently generate a large number of points in the tail ($y \geq t$)?
- How do we fit the GPD to the generated tail points?

We answer these in the following sections.

10.3 Analysis of Tails and Extreme Values

In this section we will look at some useful statistics and techniques for analyzing and understanding the behavior of extreme values.

10.3.1 Order Statistics and Quantiles

Monte Carlo based methods or statistical measurements yield samples of some quantity, for instance, the write time of an SRAM cell. Each sampled value is then a random variable. It is often instructive to study the quantiles and order statistics of the sample. For instance, the median is the 50% quantile (or 50-th percentile) of the distribution, and is often used as a robust measure of the most common case. The sample maximum M_N from Eq. (10.15) is a measure of the 100% quantile.

Order Statistics

Let us start with order statistics first. Once again, let $Y_1, Y_2, \ldots, Y_N$ be the i.i.d. random variables drawn from the CDF F. First we define the order statistics for such a sample. If we sort the sample in increasing order, as

$$Y_{1,N} \leq Y_{2,N} \leq \cdots \leq Y_{N,N},$$

to obtain the ordered sample, then Yk, N is the k-th order statistic. Note that $Y_{1,N} = min(Y_1, \ldots, Y_N)$ and $Y_{1,N} = max(Y_1, \ldots, Y_N)$ are the sample minimum and maximum, respectively. $Y_{k,N}$ is also referred to as the $(N-k+1)$-th upper order statistic. We can easily infer an empirical CDF as an approximation for F from the order statistics. For any real y, we write this empirical CDF, $F_N(y)$, as

$$F_N(y) = \frac{1}{N}|\{Y_{k,N} : Y_{k,N} \leq y, 1 \leq k \leq N\}|;$$

that is, the fraction of sample values that are less than y. Figure 10.3 shows such an empirical CDF computed using a sample from a lognormal distribution. The distribution of the k-th order statistic drawn from the CDF F can be easily derived as follows. We denote the CDF of the k-th order statistic by $F_{k,N}(y)$. Then, we can

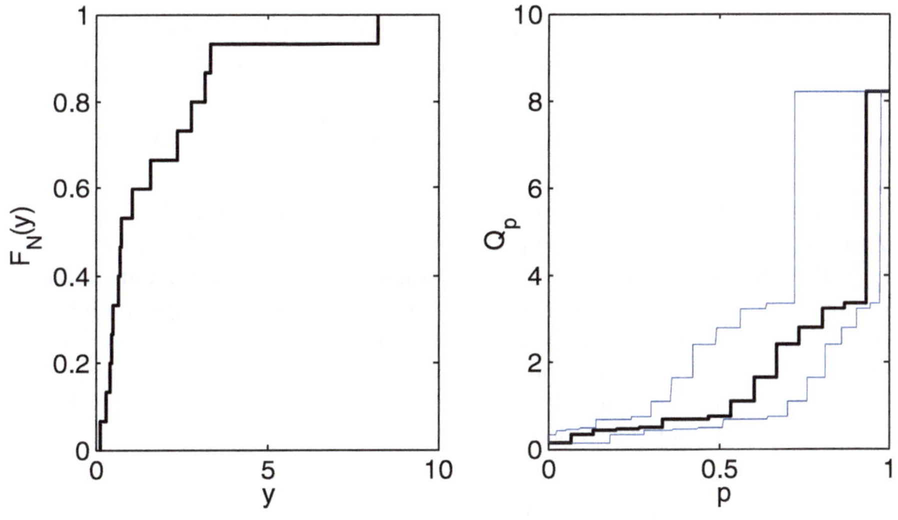

Fig. 10.3 Empirical CDF and sample quantiles estimated for a sample of 15 values drawn from a lognormal distribution. The quantile plot also shows the 90% confidence intervals, estimated using Eq. (10.23)

write

$$
\begin{aligned}
F_{k,N}(y) &= P(Y_{k,N} \le y) \\
&= P(Y_{k,N} \le y < Y_{k+1,N}) + P(Y_{k+1,N} \le y < Y_{k+2,N}) \\
&\quad + \cdots + P(Y_{N,N} \le y) \\
&= {}^{N}C_{N-k}(1 - F(y))^{N-k} F(y)^{k} + \cdots + {}^{N}C_{0}(1 - F(y))^{0} F(y)^{N} \\
&= \sum_{r=0}^{N-k} {}^{N}C_{r}(1 - F(y))^{r} F(y)^{N-r},
\end{aligned}
$$

Quantiles

The P-th percentile ($0 < P < 100$) is that value of y that has a cumulative probability of $P/100$. Hence, we divide the entire domain of y into 100 equiprobable bins and the percentiles are the bin boundaries. Quantiles are generalizations of this concept. Instead of 100 bins, if we make q equiprobable bins, then the resulting boundary values are the q-quantiles. For $0 < k < q$, and integer k and q, the k-th q-quantile or just k-th q-tile is the smallest value of y that has a cumulative probability no less than k/q. We can further generalize this idea to the p probability quantile, where $0 < p < 1$. The quantile for probability p is the smallest value y that has a cumulative probability no less than p. If we denote this quantity as Q_p, then we can define it in terms of the quantile function $F^{-1}(p)$:

$$
Q_p = F^{-1}(p) = \inf\{y \in \Re : F(y) \ge p\}, 0 < p < 1.
$$

Quantiles can be estimated from sampled data, using the order statistics. A common estimate of the *sample quantile* for probability p is given as

$$
Q_{p,N} = F^{-1}(p) = Y_{k,N} : \frac{k-1}{N} < p \le \frac{k}{N}, k = 1, \ldots, N. \tag{10.22}
$$

This expression lets us compute, for instance, the 99-th percentile of the write time of an SRAM cell from some sampling of the write time: $p = 0.99$. Figure 10.3 shows the empirical quantile function for a sample of lognormal random variables. An important question now is: *What is the distribution of the sample quantile?* If we know this distribution, we can estimate confidence intervals for any quantile estimate. We address this question next. Suppose we have estimated some quantile $Q_{p,N}$ using an ordered sample. An empirical distribution for Qp can be estimated using the ordered sample as follows:

$$
F_{Q_p,N}(y) = \begin{cases} P(Q_p \le Y_{k,N}) : k = \sup\{i : Y_{i,N} \le y\}, & y \ge Y_{1,N} \\ 0, \quad y < Y_{1,N} \end{cases}
$$

Now,

$$\begin{aligned} P(Q_p \le Y_{k,N}) &= P(Q_p \le Y_{1,N}) + \cdots + P(Y_{k-1,N} < Q_p \le Y_{k,N}) \\ &= {}^N C_0 p^0 (1-p)^{N-0} + \cdots +^N C_{k-1} p^{k-1} (1-p)^{N-k+1} \\ &= \sum_{r=0}^{k-1} [{}^N C_r p^r (1-p)^{N-r}] \\ &= B(k-1; N, p), \end{aligned}$$

where $B(.; N, p)$ is the binomial distribution with parameters N and p. This gives us a convenient estimate for the confidence interval of any quantile estimate. Suppose, we wish to compute $P\%$ confidence limits for the quantile estimate $Q_{p,N}$ computed using Eq. (10.22). We can use the following estimate:

$\lfloor Y_{l,N}, Y_{h,N} \rfloor$ where

$$l = B^{-1}(0.5 - p/200; N, p) + 1, h = B^{-1}(0.5{+}p/200; N, p) + 1. \tag{10.23}$$

Here, $Y_{l,N}$ and $Y_{h,N}$ are order statistics. The confidence interval estimates for our lognormal example are shown in Fig. 10.3. For large sample cases, the binomial distribution $B(y, N, p)$ is well approximated by a normal distribution with mean N_p and variance $N_p(1-p)$. Hence, if N is large, we can estimate the indices l and h in Eq. (10.23) as

$$l = Np - \Delta, h = Np + \Delta, where \Delta = \Phi^{-1}(0.5 + P/200)\sqrt{Np(1-p)}$$

where Φ^{-1} is the inverse standard normal CDF.

10.3.1.1 Mean Excess Plot

A common tool for graphical exploration of statistical data, particularly of distribution tails, is the *sample mean excess plot*, an idea introduced by Davison and Smith [2]. It is particularly useful for identifying the generalized Pareto distribution from sampled tails. The *mean excess function* for a given threshold t is defined as

$$e(t) = E(y - t | y > t);$$

that is, the mean of exceedances over t. Plotting $e(t)$ against t gives us the mean excess plot. The sample mean excess function is the sample version of $e(t)$. For a given sample $y_i : i = 1, \ldots, n$, it is defined as

$$e_n(t) = \frac{\sum_{i=1}^{n} (y_i - t)_+}{|y_i : y_i > t|}, \text{ where } (.)_+ = max(., 0);$$

that is, the sample mean of only the exceedances over t. A plot of $e_n(t)$ against t gives us the sample mean excess plot. The mean excess function of a GPD $G_{\zeta,\beta}$ can be shown (see [3]) to be a straight line given by

$$e(t) = \frac{\beta - \zeta t}{1 + \zeta}, \quad for\ t \in D(\zeta, \beta), \tag{10.24}$$

where $D(\zeta, \beta)$ is as defined in Theorem 2.3. Hence, if the sample mean excess function of any data sample starts to follow roughly a straight line from some threshold, then it is an indication that the exceedances over that threshold follow a GPD. Suppose, we have some sample data $Y_1, \ldots, Yn$, and we wish to identify if and at what value (tail threshold) the GPD convergence sets in. We can exploit the mean excess plot to test the GPD convergence of any tail threshold we choose. Let us look at an example to illustrate these ideas. Let us generate a large sample from the standard normal distribution, and fit GPDs at two tail thresholds: $t = 1$ and $t = 2.5$. We use the maximum likelihood estimation method, as described in the following sections, to obtain these fits. The resulting mean excess relations from Eq. (10.24) are then compared against the sample mean excess plot of the normal sample. This comparison is shown in Fig. 10.4. It is obvious that the GPD behavior has not set in at $t = 1$, since the sample mean excess plot does not follow the straight line mean excess relation of the GPD. However, going further out into the tail, at $t = 2.5$, it seems that the GPD behavior has indeed emerged.

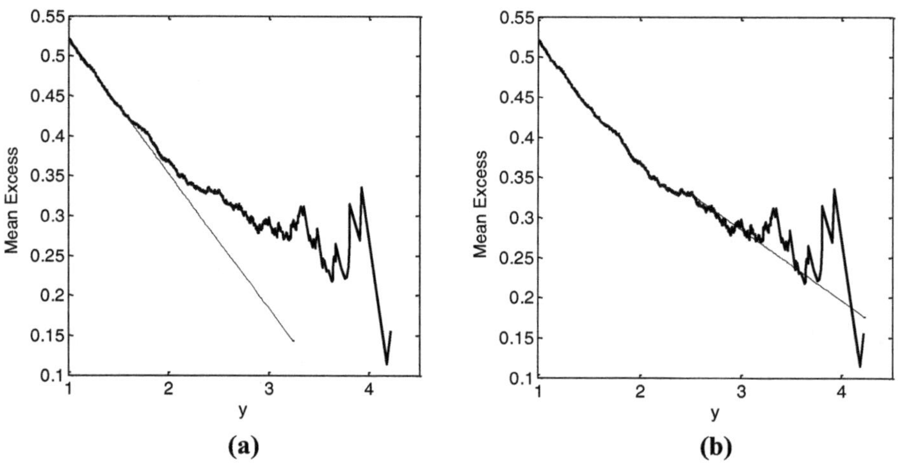

Fig. 10.4 Sample mean excess plot for a sample drawn from the standard normal. The GPD behavior of linear mean excess plot appears to have set in beyond a tail threshold of 2.5, as indicated by the GPD fits at two different thresholds (shown as straight lines). (**a**) Threshold $t = 1$. (**b**) Threshold $t = 2.5$

10.4 Estimating the Tail: Learning the GPD Parameters from Data

For now, let us suppose that we can generate a reasonably large number of points in the tail of our performance distribution. For this we might, theoretically, use standard Monte Carlo simulation with an extremely large sample size, or, more practically, the statistical blockade sampling method we will discuss in Sect. 10.5. Let this data be $Z = (Z_1, \ldots, Z_n)$, where each Z_i is the exceedance over the tail threshold t ($Z_i > 0$, for all i). All Z_i are i.i.d. random variables with common CDF F_t. Then we have the problem of estimating the optimal GPD parameters ζ, β from this tail data, so as to best fit the conditional tail CDF F_t. There are several options; we review three of the most popular ones here. In particular we focus on two methods that require no manual effort and can be completely optimized.

10.4.1 Maximum Likelihood Estimation

Maximum likelihood estimation (MLE) is a standard statistical estimation technique that tries to estimate those model parameter values (here ζ, β of the GPD) that maximize the "chances" of obtaining the data that we have observed. The probability density function of a GPD $G_{\zeta,\beta}$ is given as

$$g_{\zeta,\beta}(z) = \begin{cases} \frac{1}{\beta}(1 - \zeta\frac{z}{\beta})^{\frac{1}{\zeta}-1}, & \zeta \neq 0, z \in D(\zeta, \beta) \\ \frac{1}{\beta}e^{-z/\beta}, \zeta = 0, z \geq 0 \end{cases}, \tag{10.25}$$

where $D(\zeta, \beta)$ is defined in Theorem 2.3. Recall that all Z_i are i.i.d. random variables with common CDF F_t. We assume that F_t is of the form of a GPD. The likelihood ("chances") of having seen this data from an underlying GPD is the multivariate probability density associated with it, and is given as

$$\ell(\zeta, \beta|Z) = g_{\zeta,\beta}(Z_1, \ldots, Z_n) = \prod_{i=1}^{n} g_{\zeta,\beta}(Z_i).$$

Since $\ell(\zeta, \beta|Z)$ can be too small for accurate computation with finite accuracy, it is typical to use the *log-likelihood function*

$$ln(\ell(\zeta, \beta|Z)) = \sum_{i=1}^{n} ln(g_{\zeta,\beta}(Z_i)).$$

which increases monotonically with ℓ. MLE then computes (ζ, β) to maximize this log-likelihood, as

$$(\hat{\zeta}, \hat{\beta})_{mle} = \arg\max_{\zeta,\beta} \sum_{i=1}^{n} ln(g_{\zeta,\beta}(Z_i)) \tag{10.26}$$

Substitution of Eq. (10.25) in (10.26) and subsequent algebra allows for a simplification to a one dimensional search that can be exploited by a careful implementation of a Newton–Raphson algorithm. Details regarding such an implementation are shown in [6]. Smith [24] studies convergence when F_t is not exactly of GPD form, and provides limit results for the distributions of $(\hat{\zeta}, \hat{\beta})$ for each of the three cases, $F \in MDA(\Phi_{-1/\zeta})$, $F \in MDA(\Lambda)$, and $F \in MDA(\Psi_{1/\zeta})$. For $\zeta < 1/2$, the MLE estimates are asymptotically normal and efficient (bias = 0) under certain regularity assumptions on F. If (ζ, β) are the exact values to be estimated, then as the sample size $n \to \infty$, the variance of the MLE estimates is given as

$$\text{var}\begin{bmatrix}\hat{\zeta}\\ \hat{\beta}\end{bmatrix} \to \frac{1-\zeta}{n}\begin{bmatrix}1-\zeta & \beta\\ \beta & 2\beta^2\end{bmatrix}, \qquad \zeta < \frac{1}{2}.$$

When $\zeta \geq 1/2$, MLE convergence can be difficult and special techniques are needed, as discussed in [25]. However, $\zeta \geq 1/2$ is usually rare, since it corresponds to a finite tail with $g_{\zeta,\beta}(z) > 0$ at the endpoint (Fig. 10.2).

10.4.2 Probability-Weighted Moment Matching

Probability-weighted moments (PWMs) [9] of a continuous random variable Y with CDF F are generalizations of the standard moments, and are defined as

$$M_{p,r,s} = E[Y^p F^r(Y)(1-F(Y))^s].$$

The standard p-th moment is given by $M_{p,0,0}$. For the GPD, we have a convenient relationship between $M_{1,0,s}$ and (ζ, β), given by

$$m_s = M_{1,0,s} = \frac{\beta}{(1+s)(1+s+\zeta)}, \zeta > 0.$$

Then, we can write

$$\beta = \frac{2m_0 m_1}{m_0 - 2m_1}, \zeta = \frac{m_0}{m_0 - 2m_1} - 2.$$

We estimate these PWMs from the data sample, as

$$\hat{m}_s = \frac{1}{n}\sum_{i=1}^{n}(1-q)^s Y_{i,n},$$

where $Y_{1,n} \leq Y_{2,n} \leq \cdots \leq Y_{n,n}$ is the *ordered sample*, and

$$q_i = \frac{i+\gamma}{n+\delta}.$$

Here, γ and δ are fitting parameters. We can use $\gamma = -0.35$ and $\delta = 0$, as suggested for GPD fitting in [8]. The estimates converge to the exact values as $n \to \infty$, and are asymptotically normally distributed with covariance given by

$$\text{var}\begin{bmatrix}\hat{\xi}\\ \hat{\beta}\end{bmatrix} \to \frac{n^{-1}}{(1+2\xi)(3+2\xi)} \times$$

$$\begin{bmatrix}(1+\xi)(2+\xi)^2(1+\xi+2\xi^2) & \beta(2+\xi)(2+6\xi+7\xi^2+2\xi^3)\\ \beta(2+\xi)(2+6\xi+7\xi^2+2\xi^3) & \beta^2(7+18\xi+11\xi^2+2\xi^3)\end{bmatrix} \tag{10.27}$$

Based on an extensive simulation study, Hosking and Wallis [8] suggests that the PWM method often has lower bias than moment matching and MLE for sample sizes up to 500. Also, the MLE search Eq. (10.26) is shown to suffer from some convergence problems when ξ is estimated close to $1/2$. Finally, the study also suggests that PWM matching gives more reliable estimates of the variability of the estimated parameters, as per Eq. (10.27). Based on these reasons, we choose PWM matching here.

Once we have estimated a GPD model of the conditional CDF above a threshold t, we can estimate the failure probability for any value y_f by substituting the GPD in Eq. (10.14) as

$$P(Y > y_f) \approx (1 - F(t))(1 - G_{\xi,\beta}(y_f - t)). \tag{10.28}$$

The next section addresses the important remaining question: How do we efficiently generate a large number of points in the tail ($y \geq t$)?

10.5 Statistical Blockade: Sampling Rare Events

Let any circuit performance metric, or simply, output y be computed as

$$y = f_{sim}(x) \tag{10.29}$$

Here, x is a point in the statistical parameter (e.g., V_t, t_{ox}) space, or simply, the input space, and f_{sim} includes expensive SPICE simulation. We assume that y has

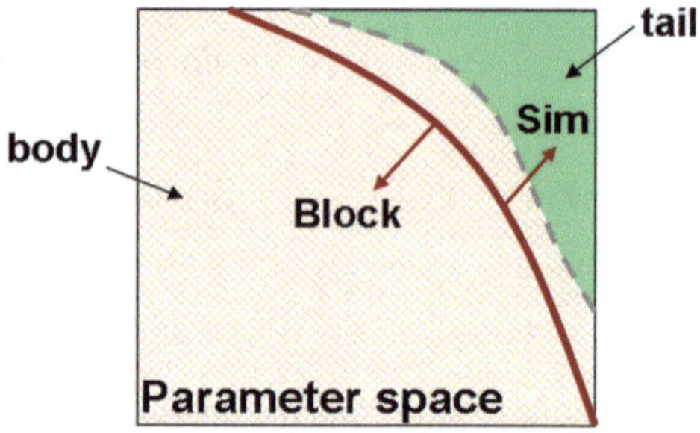

Fig. 10.5 The tail and body regions in the statistical parameter space. The dashed line is the exact tail region boundary for tail threshold t. The solid line is the relaxed boundary modeled by the classifier for a classification threshold $t_c < t$

some probability distribution F, with an extended tail. Suppose, we define a large tail threshold t for y, then from the developments in Sect. 10.2 we know that we can approximate the conditional tail CDF F_t by a generalized Pareto distribution $G_{\xi,\beta}$. Section 10.4 shows how we can estimate the GPD parameters (ξ, β) from data drawn from the tail distribution. We now discuss an efficient tail sampling strategy that will generate the tail points for fitting this GPD.

Corresponding to the tail of output distribution, we expect a "tail region" in the input space: any statistical parameter values drawn from this tail region will give an output value $y > t$. Figure 10.5 shows an example of such a tail region for two inputs. The rest of the input space is called the "body" region, corresponding to the body of the output distribution F. In Fig. 10.5 these two regions are separated by a dashed line.

The key idea behind the efficient sampling technique is to identify the tail region and simulate only those Monte Carlo points that are likely to lie in this tail region. Here, we exploit the common fact that *generating* the random values for a Monte Carlo sample point is very cheap compared to actually *simulating* the point as in Eq. (10.29). Hence, if we generate points as in standard Monte Carlo, but *block*—not simulate—those points that are unlikely to fall in the tail region, we can drastically cut down the total time spent. We use a classifier to distinguish between the tail and body regions, and to block out the body points. A classifier [7] is an indicator function that takes as input any point in the input space (the statistical parameter space) and predicts the membership of this point in one of the multiple classes (the "body" or "tail" classes). In the context of Fig. 10.5, it essentially builds a model of the boundary between the tail and body regions. Using this model of the boundary it can label points from the Monte Carlo sample set as either "tail" or "body." Only the "tail" points are then simulated. We also refer to this classifier as the blockade filter and its blocking activity as *blockade filtering*.

To build this model of the tail region boundary, the classifier can be trained with a small (e.g., 1000 points) training set of simulated Monte Carlo sample points. However, it is difficult, if not impossible to build an exact model of the boundary in general. Misclassifications, at least on points unseen during training, are unavoidable. Hence, we relax the accuracy requirement to allow for classification error. This is done by building the classification boundary at a classification threshold t_c that is less than the tail threshold t. Since we have assumed that only

Require: training sample size n_0 (e.g., 1,000)
total sample size n
percentages p_t (e.g., 99%), p_c (e.g., 97%)

$\mathbf{X} = \text{MonteCarlo}(n_0)$
$\mathbf{y} = f_{\text{sim}}(\mathbf{X})$
$t = \text{Percentile}(\mathbf{y}, p_t)$
$t_c = \text{Percentile}(\mathbf{y}, p_c)$
$C = \text{BuildClassifier}(\mathbf{X}, \mathbf{y}, t_c)$ // C is a classifier
$\mathbf{y} = f_{\text{sim}}(\text{Filter}(C, \text{MonteCarlo}(n)))$
$\mathbf{y}_{\text{tail}} = \{y_i \in \mathbf{y} : y_i > t\}$
$(\xi, \beta) = \text{FitGPD}(\mathbf{y}_{\text{tail}} - t)$

Fig. 10.6 Algorithm: the statistical blockade algorithm for efficiently sampling rare events and estimating their probability distribution

the upper (right) tail is relevant, the tail region corresponding to t will be a subset of the tail region corresponding to t_c, if $t_c < t$. This will help to ensure that, even if the classifier is imperfect, it is unlikely that it will misclassify points in the true tail region (for t). The relaxed boundary corresponding to such a t_c is shown as the solid line in Fig. 10.5. The statistical blockade algorithm is then as in Algorithm in Fig. 10.6.

The thresholds $t = p_t$-th percentile and $t_c = p_c$-th percentile are estimated from the small initial Monte Carlo run, which also gives the n_0 training points for the classifier. Typical values for these constants are shown in Algorithm in Fig. 10.6. The function *MonteCarlo(n)* generates n points in the statistical parameter space, which are stored in the $n \times s$ matrix X, where s is the input dimensionality. Each row of X is a point in s dimensions. y is a vector of output values computed from simulations. The function $BuildClassifier(X, y, t_c)$ trains and returns a classifier using the training set (X, y) and classification threshold t_c. The function $Filter(C, X)$ blocks the points in X classified as "body" by the classifier C, and returns only the points classified as "tail." $FitGPD(y_{tail} - t)$ computes the parameters (ζ, β) for the best GPD approximation $G_{\zeta,\beta}$ to the conditional CDF of the exceedances of the tail points in y_{tail} over t. We can then use this GPD model to compute statistical metrics for rare events, for example, the failure probability for some threshold y_f, as in Eq. (10.28). This sampling procedure is also illustrated in Fig. 10.7.

Note on Unbiasing the Classifier

An important technical point to note about the classifier construction is as follows. The training set will typically have many more body points than tail points. Hence, even if all or most of the tail points are misclassified, the training error will be low as long as most of the body points are correctly classified. This will result in a classifier that is biased to allow more misclassifications of points in the tail region. However, we need to minimize misclassification of tail points to avoid distorting the statistics of the simulated tail points. Hence, we need to reverse bias the classification error.

Using the technique proposed in [14], we penalize misclassifications of tail points more than misclassifications of body points. Let γ_t and γ_b be possibly different

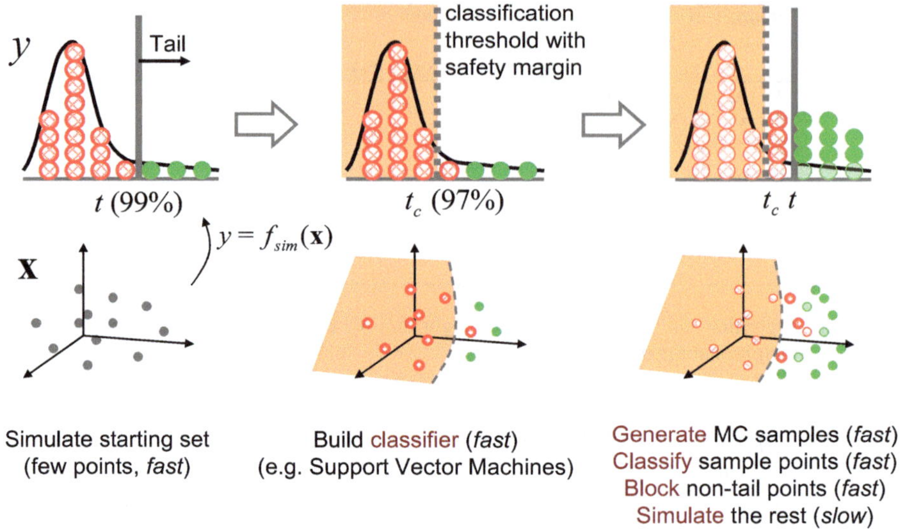

Fig. 10.7 The efficient tail (rare event) sampling method of statistical blockade

penalty factors for the "tail" and "body" classes: misclassifications of any training points in the tail (body) region are penalized by a factor of $\gamma_t (\gamma_b)$. We can obtain an unbiased classifier if we choose

$$\frac{\gamma_t}{\gamma_b} = \frac{\text{Number of 'body' points}}{\text{Number of 'tail' points}}$$

Note on Sampling and Commercially Available Device Models

It is common for commercially available device models to include Monte Carlo variants. Care must be taken when using these to ensure that tails are being modeled correctly and that the sample space includes correct tail distributions. For example, many such models use the "AGAUSS" function, which generates a set of sample points within a set of bounds, typically set to $+/- 3\sigma$. Such an approach can produce some rare events (e.g., two independent parameters each at $+3\sigma$ is an event of about 2 ppm probability) but miss others that are much more likely (e.g., one parameter being above $+3.5\sigma$ while the other is less than 1σ occurs at a rate of 200 ppm). For this reason, when working with EVT, it is usually desirable to be very explicit about how sample points are generated and how they are incorporated into circuit simulations. The mechanics of doing this as part of a simulation flow vary considerably among commercially available device models, even those from a single foundry. A good approach for unencrypted models is to substitute externally generated sample points for those used by the model, taking care to consider underlying physical realities (avoiding negative V_t for NFETs, for example). For encrypted models, an approximation can be made through the use

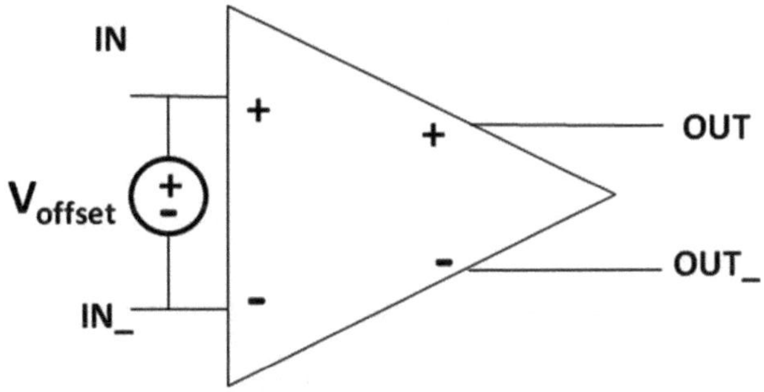

Fig. 10.8 Adding an offset to a sense amplifier to compensate for V_t shift on input pass transistors

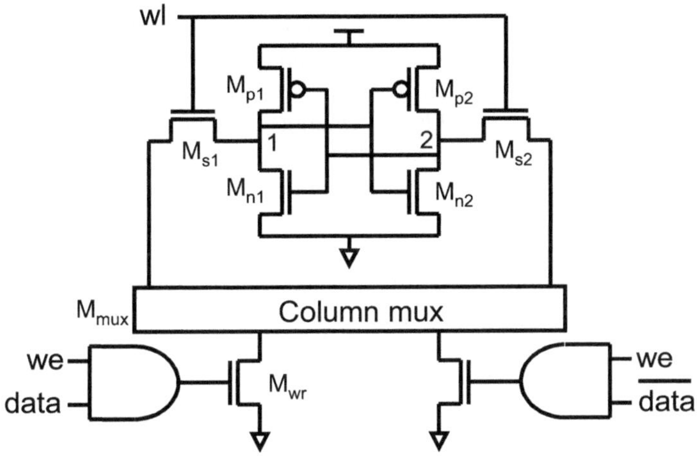

Fig. 10.9 A 6-*transistor* SRAM cell with write driver and column mux

of offsets. For example, Fig. 10.8 shows a method of using a parameterized voltage source providing a voltage offset to mimic threshold voltage variation in an NFET.

Example: 6T SRAM Cell

Let us now apply the statistical blockade method to a 6T SRAM cell to illustrate its use. The circuit is shown in Fig. 10.9. Here, we use the Cadence 90*nm* Generic PDK library for the *BSIM3V3* [13] device models, and model RDF on every transistor as independent, normally distributed threshold voltage (V_t) variation per transistor. The V_t standard deviation is taken to be $\sigma(Vt) = 5(WL)^{-0.5} V$, where W and L are the transistor width and length, respectively, in *nm*. We also include a global gate oxide thickness variation, also normally distributed and with a standard deviation of 2% of the nominal thickness. This gives us a total of 9 statistical parameters. The metric we are measuring here is the write time τ_w: the time between the wordline going high, to the non-driven cell node (node 2) transitioning. Here, "going high" and "transitioning" imply crossing 50% of the full voltage change. The rare event statistical metric we will compute is the failure probability $F_f(y_f)$ for any failure threshold y_f. We will represent this failure probability as the equivalent quantile y_σ on the standard normal distribution:

$$y_\sigma = \phi^{-1}(1 - F_f(y_f)) = \phi^{-1}(F(y_f)),$$

where ϕ is the standard normal CDF. For example, a failure probability of $F_f = 0.00135$ implies a cumulative probability of $F = 1 - F_f = 0.99865$. The equivalent point on a standard normal, having the same cumulative probability, is $y_\sigma = 3$. In other words, any y_f with a failure probability of 0.00135 is a "3σ" point.

Let us now compute y_σ for different failure thresholds using statistical blockade. The blockade filter is a support vector machine classifier [10], built at a classification threshold of $t_c = 97$-th percentile, using a training sample set of $n_0 = 1000$ points. The tail threshold t is defined as the 99-th percentile. We use Eq. (10.28), but the points used to estimate (ζ, β) are obtained from blockade filtering a sample of 100,000 points. This gives us 4379 tail candidates, which on simulation yield 978 true tail points ($\tau_w > t$), to which the GPD is fit. We compare these results with a 1 million point Monte Carlo run, from which we estimate y_σ empirically.

Table 10.3 shows a comparison of the y_σ values estimated by the two different methods, and Fig. 10.10 compares the conditional tail CDFs computed from the empirical method and from statistical blockade, both showing a good match.

Some observations highlighting the efficiency of statistical blockade can be made immediately.

- The empirical method fails beyond 2.6 FO4, corresponding to about 1 ppm circuit failure probability, because there are no points generated by the Monte Carlo run so far out in the tail.
- Fitting a GPD model to the tail points allows us to make predictions far out in the tail, even though we have no points that far out.

Table 10.3 Prediction of failure probability as y_σ by methods I, II, and III, for a 6T SRAM cell

$\tau_w(y_f)(FO4)$	(I) Standard Monte Carlo	(III) Statistical blockade
2.4	3.404	3.379
2.5	3.886	3.868
2.6	4.526	4.352
2.7	∞	4.845
2.8	∞	5.356
2.9	∞	5.899
3.0	∞	6.493
Num. Sims.	1,000,000	5379

The number of simulations for statistical blockade includes the 1000 training points. The write time values are in "fanout of 4 delay" units

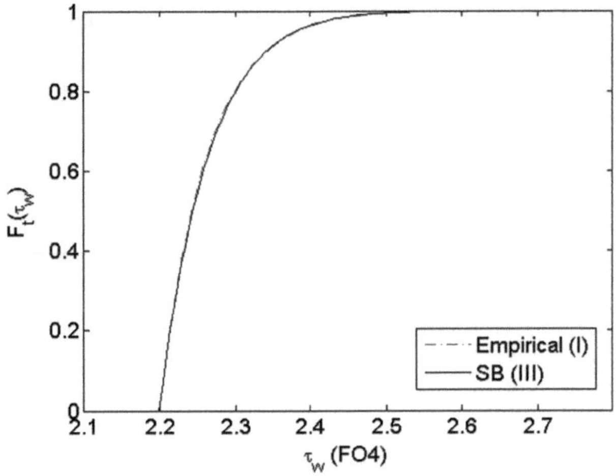

Fig. 10.10 Comparison of GPD tail model from statistical blockade (5379 simulations) and the empirical tail CDF (1 million simulations) for the write time of the 6T SRAM cell

- Using blockade filtering, coupled with the GPD tail model, we can drastically reduce the number of simulations (from 1 million to 5379) with very small change to the tail model.

However, the tail model cannot be relied on too far out from the available data, because of the decrease in statistical confidence as we move further out in the tail. We further discuss and attack this problem in Sect. 10.5.2.

10.5.1 Conditionals and Disjoint Tail Regions

SRAM performance metrics are often computed for two states of the SRAM cell: while storing a 1, and while storing a 0. The final metric value is then a maximum or a minimum of the vales for these two states. The presence of such *conditionals* (max, min) can result in *disjoint* tail regions in the statistical parameter space, making it difficult to use a single classifier to define the boundary of the tail region. Let us look at an example to illustrate this problem.

Consider the 6T SRAM cell. With technology scaling reaching nanometer feature sizes, sub-threshold and gate leakage become very significant. In particular, for the large memory blocks seen today the standby power consumption due to leakage can be intolerably high. Supply voltage (V_{dd}) scaling [12] is a powerful technique to reduce this leakage, whereby the supply voltage is reduced when the memory bank is not being accessed. However, lowering V_{dd} also makes the cell unstable, ultimately resulting in data loss at some threshold value of V_{dd}, known as the *data retention voltage* or DRV. Hence, the DRV of an SRAM cell is the lowest supply voltage that still preserves the data stored in the cell. DRV can be computed as follows:

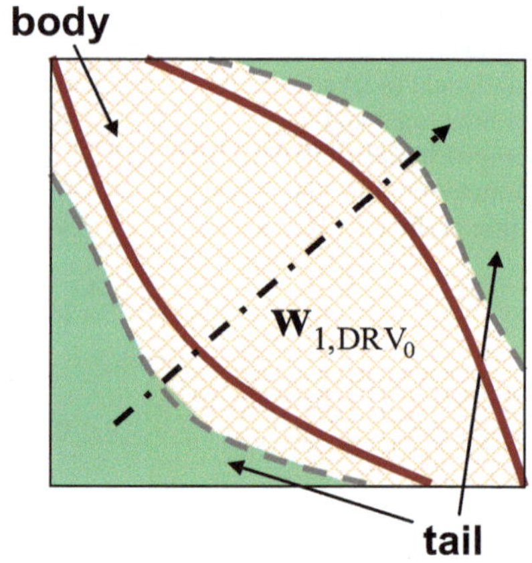

Fig. 10.11 A circuit metric (e.g., DRV) with two disjoint tail regions. The tail regions are separated from the body region by dashed lines. w_{1,DRV_0} is the direction of maximum variation of the circuit metric

$$DRV = max(DRV_0, DRV_1), \tag{10.30}$$

where DRV_0 is the DRV when the cell is storing a 0, and DRV_1 it the DRV when it is storing a 1. If the cell is balanced (symmetric), with identical left and right halves, then $DRV_0 = DRV_1$. However, if there is any mismatch due to process variations, they become unequal. This creates the situation where the standard statistical blockade classification technique would fail because of the presence of disjoint tail regions.

Suppose we run a 1000 point Monte Carlo, varying all the mismatch parameters in the SRAM cell according to their statistical distributions. This would give us distributions of values for DRV_0, DRV_1, and DRV. In certain parts of the mismatch parameter space $DRV_0 > DRV_1$, and in other parts, $DRV_0 < DRV_1$. This is clearly illustrated in Fig. 10.12: let us see how. First, we extract the direction in the parameter space that has maximum influence on DRV_0. This direction is shown as w_{1,DRV_0} in the two dimensional parameter space example of Fig. 10.11. As we move along this vector in the statistical parameter space, the corresponding DRV_0 increases from low to high (from bad to good).[2] Other directions would show changes in DRV_0 that are no greater than the change along this direction. The figure plots the simulated DRV_0 and DRV_1 values from a 1000 point Monte Carlo run, along this direction. It is clear that the two DRV measures are inversely related: one decreases as the other increases.

Now, let us take the maximum as in Eq. (10.30), and choose the classification threshold tc equal to the 97-th percentile. Then we pick out the worst 3% points from the classifier training data and plot them against the same latent variable in Fig. 10.12, as squares. Note that we have not trained the classifier yet, we are just looking at the points that the classifier would have to classify as being in the tail. We

[2]Here we extract this vector using the SiLVR tool described in [19, 21], so that w_{1,DRV_0} is essentially the projection vector of the first latent variable of DRV_0.

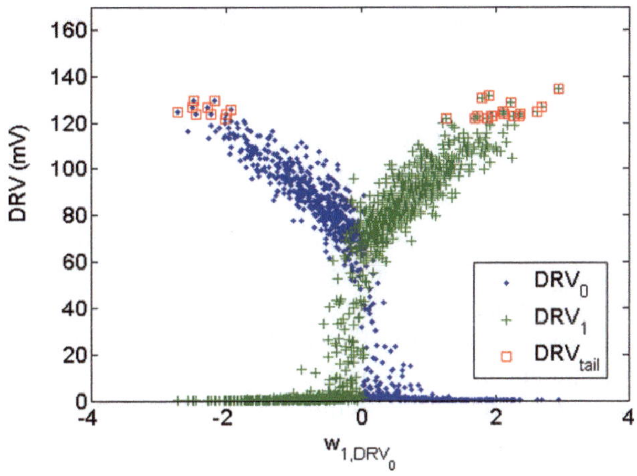

Fig. 10.12 Behavior of DRV_0 and DRV_1 along the direction of maximum variation in DRV_0. The worst 3% DRV values are plotted as squares, showing the disjoint tail regions (along this direction in the parameter space)

can clearly see that these points (squares) lie in two disjoint parts of the parameter space.

Since the true tail region defined by the tail threshold $t > t_c$ will be a subset of the classifier tail region (defined by t_c), it is obvious that the true tail region consists of two disjoint regions of the parameter space. This is illustrated in our two dimensional example in Fig. 10.11. The dark tail regions on the top-right and bottom-left corners of the parameter space correspond to the large DRV values shown as squares in Fig. 10.12.

The Solution

Instead of building a single classifier for the tail in Eq. (10.30), let us build two separate classifiers, one for the 97-th percentile (t_c, DRV_0) of DRV_0, and another for the 97-th percentile (t_c, DRV_1) of DRV_1. The generated Monte Carlo samples can then be filtered through both these classifiers: points classified as "body" by *both* the classifiers will be blocked, and the rest will be simulated. In the general case for arbitrary number of arguments in the conditional, let the circuit metric be given as

$$y = max(y_0, y_1, \ldots) \tag{10.31}$$

Note that a $min()$ operator can easily be converted to $max()$ by negating the arguments. The resulting general algorithm is then as follows:

1. Perform initial sampling to generate training data to build the classifiers, and estimate tail and classification thresholds, t_i and $t_{c,i}$, respectively, for each y_i, $i = 0, 1, \ldots,$. Also estimate the tail threshold t for y.
2. For each argument, $y_i, i = 0, 1, \ldots,$ of the conditional Eq. (10.31), build a classifier C_i at a classification threshold $t_{c,i}$ that is less than the corresponding tail threshold ti.

3. Generate more points using Monte Carlo, but block the points classified as "body" by all the classifiers. Simulate the rest and compute y for the simulated points.

Hence, in the case of Fig. 10.11, we build a separate classifier for each of the two boundaries. The resulting classification boundaries are shown as solid lines. From the resulting simulated points, those with $y > t$ are chosen as tail points for further analysis, e.g., for computing a GPD model for the tail distribution of y.

Note that this same algorithm can also be used for the case of multiple circuit metrics. Each metric would have its own thresholds and its own classifier, just like each argument in Eq. (10.31), the only difference being that we would not be computing any conditional. Note also that adding additional classifiers increases the number of points that require simulation: if two non-overlapping classifiers each identify 3% of points, then overall 6% of points will need simulation.

10.5.2 Extremely Rare Events and Their Statistics

The GPD tail model can be used to make predictions regarding rare events that are farther out in the tail than any of the data we used to compute the GPD model. Indeed, this is the compelling reason for adopting the GPD model. However, as suggest by common intuition and the estimate variance relations of Sect. 10.4, we expect the statistical confidence in the estimates to decrease as we predict farther out in the tail. Let us run an experiment to test this expectation.

Let us go back to our SRAM cell write time example. Suppose we run 50 runs of Monte Carlo with $n_{MC} = 100{,}000$ points each and compute a GPD tail model from each run, using tail threshold $t =$ the 99-th percentile write time. This gives us 50 slightly different pairs of the GPD parameters (ξ, β), one for each of 50 GPD models so computed, and 50 different estimates of the $m\sigma$ point, where $m \in [3, 6]$. These estimates are shown in Fig. 10.13.

As expected, the spread of the estimates increases as we extrapolate further with the GPD model. The general inference is that we should not rely on the GPD tail model too far out from our data. How then do we compute a more reliable GPD model further out in the tail?

A solution is to sample further out in the tail and use a higher tail threshold for building the GPD model of the tail. This is, of course, "easier said than done." Suppose we wish to support our GPD model with data up to the 6σ point. The failure probability of a 6σ value is roughly 1 parts-per-*billion*, corresponding to a 99% chip yield requirement for a 10 Mb cache.

This is definitely not an impractical requirement. However, for a 99% tail threshold, even a perfect classifier ($t_c = t$) will only reduce the number of simulations to an extremely large 10 million. If we decide to use a 99.9999% threshold, the number of simulations will be reduced to a more practical 1000 tail points (with a perfect classifier). However, we will need to simulate an extremely

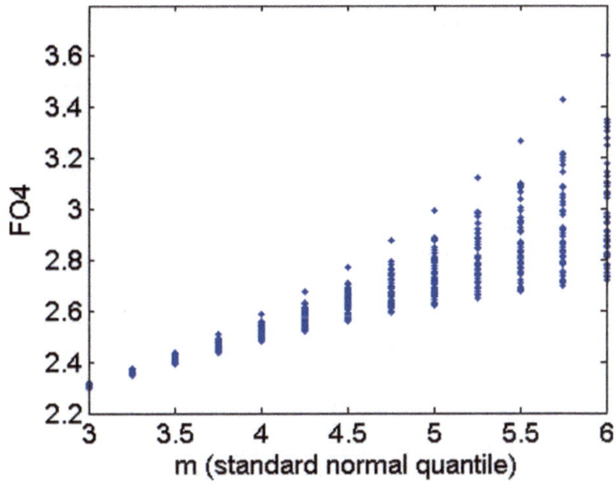

Fig. 10.13 The spread of $m\sigma$ point estimates across 50 runs of statistical blockade

large number of points (≥1 million) to generate a classifier training set with at least one point in the tail region. In both cases, the circuit simulation counts are too high. We now describe a recursive formulation of statistical blockade [23, 27] that reduces this count drastically.

10.5.3 A Recursive Formulation of Statistical Blockade

Let us first assume that there are no conditionals. For a tail threshold equal to the a-th percentile, let us represent it as t_a, and the corresponding classification threshold as t_c^a. For this threshold, build a classifier C^a and generate sufficient points beyond the tail threshold, $y > t^a$, so that a higher percentile ($t^b, t_c^b, b > a$) can be estimated. For this new, higher threshold (t_c^b), a new classifier C^b is trained and a new set of tail points ($y > t^b$) is generated. This new classifier will block many more points than C^a, significantly reducing the number of simulations. This procedure is repeated to push the threshold out more until the tail region of interest is reached. The complete algorithm, applied to general conditionals, is shown in Algorithm in Fig. 10.14. This pseudocode uses a conditional tail threshold of the 99-th percentile at each recursion stage: $(ta, tca) = (99\%, 97\%)$ points, $(t_b, t_c^b) = (99.99\%, 99.97\%)$ points, and so on. Consequently, the total sample size (without filtering) n is restricted to some power of 100, times 1000:

$$n = 100^j.1000, j = 0, 1, \ldots \quad (10.32)$$

These threshold values are an extension of the values chosen in the original statistical blockade paper [20], where the authors relied on empirical evidence. A general version of the algorithm for arbitrary thresholds is presented in [18, 28].

```
Require: initial sample size n0 (e.g., 1,000);
         total sample size n; tail sample size nt;
         performance metric function y = max(y0, y1,...)

X = MonteCarlo(n0)
n' = n0
Y = fsim(X) // Simulate the initial Monte Carlo sample set
ytail,i = Y•,i, i = 0, 1,... // The i-th column of Y contains values for yi in y = max(y0, y1,...)
Xtail,i = X, i = 0, 1,...
while n' < n
    Δn = 99n' // Number of points to filter in this recursion step
    n' = n' + Δn // Total number of points filtered by the end of this recursion stage
    X = MonteCarloNext(Δn) // The next Δn points in the Monte Carlo sequence
    forall i : yi is an argument in y = max(y0, y1,...)
        (Xtail,i, ytail,i) = GetWorst(1,000, Xtail,i, ytail,i) // Get the 1,000 worst points
        t = Percentile(ytail,i, 99)
        tc = Percentile(ytail,i, 97)
        C^i = BuildClassifier(Xtail,i, ytail,i, tc)
        (Xtail,i, ytail,i) = GetGreaterThan(t, Xtail,i, ytail,i) // Get the points with yi > t
        Xcand,i = Filter(C^i, X) // Candidate tail points for yi
    endfor
    X = [X^T_cand,0  X^T_cand,1  ...]^T // Union of all candidate tail points
    Y = fsim(X) // Simulate all candidate tail points
    ycand,i = {Yj,i : Xj,• ∈ Xcand,i}, i = 0, 1,... // Extract the tail points for yi
    ytail,i = [y^T_tail,i  y^T_cand,i]^T, Xtail,i = [X^T_tail,i  X^T_cand,i]^T, i = 0, 1,... // All tail points till now
endwhile
ytail = MaxOverRows([ytail,0  ytail,1 ...]) // compute the conditional
ytail = GetWorst(nt, ytail)
(ξ, β) = FitGPD(ytail - min(ytail))
```

Fig. 10.14 Algorithm: the recursive statistical blockade algorithm with fixed sequences for the tail and classification thresholds: $t = 99\%, 99.99\%, 99.9999\%, \ldots$ points, and $t_c = 97\%, 99.97\%, 99.9997\%, \ldots$ points. The total sample size is given by Eq. (10.32)

Practical implementation of this general algorithm is, however, difficult and is a topic for further research.

The functions that appear also in Algorithm 10.6 do the same work here, hence, we do not reiterate their description. f_{sim} now returns multiple outputs: it computes the values of all the arguments of the conditional in $y = max(y_0, y_1, \ldots)$. For example, in the case of DRV, it will return the values of DRV_0 and DRV_1. These values, for any one Monte Carlo point, are stored in one row of the result matrix Y. The function *MonteCarloNext* (Δn) returns the *next* Δn points in the sequence of points generated till now. The function $GetWorst(n, X, y)$ returns the n worst values in the vector y and the corresponding rows of the matrix X. This functionality naturally extends to the two argument $GetWorst(n, y)$. $GetGreaterThan(t, X, y)$ returns the elements of y that are greater than t, along with the corresponding rows of X.

The algorithms presented here are in iterative form, rather than recursive form. To see how the recursion works, suppose we want to estimate the 99.9999% tail. To generate points at and beyond this threshold, we first estimate the 99.99% point

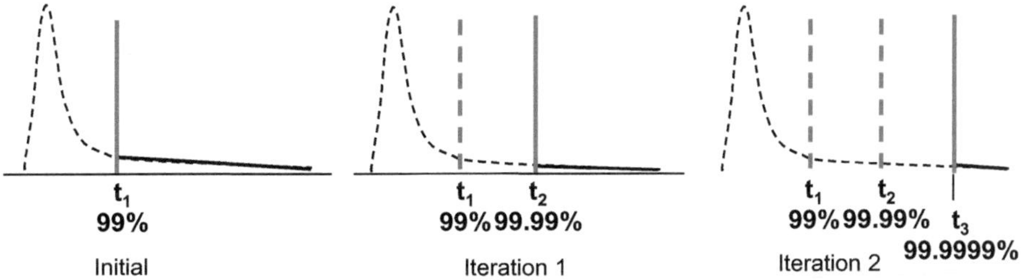

Fig. 10.15 Recursive formulation of statistical blockade as in algorithm in Fig. 10.14

and use a classifier at the 99.97% point to generate these points efficiently. To build this classifier in turn, we first estimate the 99% point and use a classifier at the 97% point. Figure 10.15 illustrates this recursion on the PDF of any one argument in the conditional equation (10.31).

An Experiment with DRV

We now test the recursive statistical blockade method on another SRAM cell test case, where we compute the data retention voltage (DRV) as in Eq. (10.30). In this case the SRAM cell is implemented in an industrial 90 nm process. Wang et al. [26] develop an analytical model for predicting the CDF of the DRV that uses not more than 5000 Monte Carlo points. The CDF is given as

$$F(y) = 1 - \mathrm{erfc}(y_0) + \frac{1}{4}\mathrm{erfc}^2(y_0), \quad where\ y_0 = \frac{\mu_0 = k(y - V_0)}{\sqrt{2}\sigma_0} \tag{10.33}$$

where y is the DRV value and erfc() is the complementary error function [16]. k is the sensitivity of the static noise margin (SNM) of the SRAM cell to the supply voltage, computed using a DC sweep. μ_0 and σ_0 are the mean and standard deviation of the SNM (SNM_0), for a user-defined supply voltage V_0. SNM_0 is the SNM of the cell while storing a 0. These statistics are computed using a short Monte Carlo run of 1500–5000 sample points. We direct the reader to [26] for complete details regarding this analytical model of the DRV distribution. The q-th quantile can be estimated as

$$DRV(q) = \frac{1}{k}(\sqrt{2}\sigma_0 \mathrm{erfc}^{-1}(2 - 2\sqrt{q}) - \mu_0) + V_0. \tag{10.34}$$

Here $DRV(q)$ is the supply voltage $V_d d$ such that $P(DRV(q)) \leq Vdd = q$. Let us now compute the DRV quantiles as $m\sigma$ points, such that q is the cumulative probability for the value m from a standard normal distribution. We will use five different methods to estimate the DRV quantiles for $m \in [3, 8]$:

1. Analytical: Use Eq. (10.34).
2. Recursive statistical blockade without the GPD model: Algorithm in Fig. 10.14 is run for $n = 1$ billion. This results in three recursion stages, corresponding to

total sample sizes of $n' = 100{,}000$, 10 million, and 1 billion Monte Carlo points, respectively. The worst DRV value for these three recursion stages are estimates of the 4.26σ, 5.2σ, and 6σ points, respectively.
3. GPD model from recursive statistical blockade: The 1000 tail points from the last recursion stage of the recursive statistical blockade run are used to fit a GPD model, which is then used to predict the DRV quantiles.
4. Normal: A normal distribution is fit to data from a 1000 point Monte Carlo run, and used to predict the DRV quantiles.
5. Lognormal: A lognormal distribution is fit to the same set of 1000 Monte Carlo points, and used for the predictions.

The results are shown in Fig. 10.16. From the plots in the figure, we can immediately see that the recursive statistical blockade estimates are very close to the estimates from the analytical model. This shows the efficiency of the recursive formulation in reducing the error in predictions for events far out in the tail.

Table 10.4 shows the number of circuit simulations performed at each recursion stage. The total number of circuit simulations is 41,721. This is not small, but in comparison to standard Monte Carlo (1 billion simulations), and basic, non-recursive statistical blockade (approximately, 30 million with $t_c = 97$-th percentile) it is extremely fast. 41,721 simulations for DRV computation of a 6T SRAM cell can be completed in several hours on a single computer today. With the advent

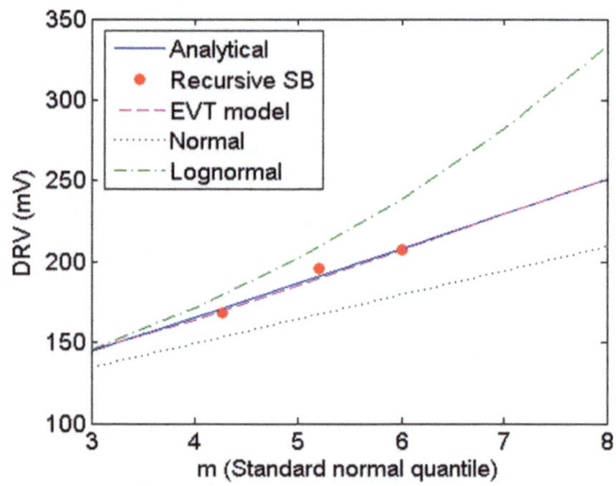

Fig. 10.16 Estimates of DRV quantiles from five estimation methods. The GPD model closely fits the analytical model Eq. (10.33). The solid circles show the worst DRV values from the three recursion stages of statistical blockade sampling. The normal and lognormal models are quite inaccurate

Table 10.4 Number of circuit simulation needed by recursive statistical blockade to generate a 6σ point

Recursion stage	Number of simulations
Initial	1000
1	11,032
2	14,184
3	15,505
Total	41,721
Speedup over Monte Carlo	23,969x
Speedup over statistical blockade	719x

of multi-core processors, the total simulation time can be drastically reduced with proper implementation.

Note that we can extend the prediction power to 8σ with the GPD model, without any additional simulations. Standard Monte Carlo would need over 1.5 quadrillion circuit simulations to generate a single 8σ point. For this case, the speedup over standard Monte Carlo is extremely large. As expected, the normal and lognormal fits show large errors. The normal fit is unable to capture the skewness of the DRV distribution. On the other hand, the lognormal distribution has a heavier tail than the DRV distribution.

10.6 Conclusions

We discussed an approach to rapidly estimate rare event statistics of memory circuit. Statistical blockade, in its recursive formulation, makes estimation of rare event statistics practical. It combines a classification, Monte Carlo simulation, and extreme value theory to achieve several orders of magnitude speedup over traditional Monte Carlo simulation. This capability can be of immense use to designers of high capacity memories: SRAM, DRAM, non-volatile memories. Since it exploits rigorous limit theorems from extreme value theory, it has the unique capability of estimating the entire distribution of rare events, even with limited data.

Acknowledgements This work was supported by the MARCO/DARPA Focus Research Center for Circuit and System Solutions (C2S2) and the Semiconductor Research Corporation.

References

1. A.A. Balkema, L. de Haan, Residual life time at great age. Ann. Probab. **2**(5), 792–804 (1974)
2. A.C. Davison, R.L. Smith, Models for exceedances over high thresholds (with discussion). J. R. Stat. Soc. Ser. B Methodol. **52**, 393–442 (1990)
3. P. Embrechts, C. Klüppelberg, T. Mikosch, *Modelling Extremal Events for Insurance and Finance*, 4th edn. (Springer, Berlin, 2003)
4. R.A. Fisher, L.H.C. Tippett, Limiting forms of the frequency distribution of the largest or smallest member of a sample. Proc. Camb. Philol. Soc. **24**, 180–190 (1928)
5. B. Gnedenko, Sur la distribution limite du terme maximum d'une aleatoire. Ann. Math. **44**(3), 423–453 (1943)
6. S.D. Grimshaw, Computing maximum likelihood estimates for the generalized Pareto distribution. Technometrics **35**(2), 185–191 (1993)
7. T. Hastie, R. Tibshirani, J. Friedman, *The Elements of Statistical Learning: Data Mining, Inference, and Prediction* (Springer, New York, 2001)
8. J.R.M. Hosking, J.R. Wallis, Parameter and quantile estimation for the generalized Pareto distribution. Technometrics **29**(3), 339–349 (1987)
9. J.R.M. Hosking, The theory of probability weighted moments, IBM Research Report, RC12210, 1986

10. T. Joachims, Making large-scale SVM learning practical, in *Advances in Kernel Methods - Support Vector Learning*, ed. by B. Schölkopf, C. Burges, A. Smola (MIT Press, Cambridge, 1999)
11. B. Joshi, R.K. Anand, C. Berg, J. Cruz-Rios, A. Krishnamurthi, N. Nettleton, S. Ngu-yen, J. Reaves, J. Reed, A. Rogers, S. Rusu, C. Tucker, C. Wang, M. Wong, D. Yee, J.-H. Chang, A BiCMOS 50MHz cache controller for a superscalar microprocessor, in *International Solid-State Circuits Conference* (1992)
12. R.K. Krishnamurthy, A. Alvandpour, V. De, S. Borkar, High-performance and low-power challenges for sub-70nm microprocessor circuits, in *Proceedings of Custom Integrated Circuits Conference* (2002)
13. W. Liu, X. Jin, J. Chen, M.-C. Jeng, Z. Liu, Y. Cheng, K. Chen, M. Chan, K. Hui, J. Huang, R. Tu, P. Ko, C. Hu, BSIM 3v3.2 Mosfet Model Users' Manual, Tech. Report No. UCB/ERL M98/51, University of California, Berkeley, 1988
14. K. Morik, P. Brockhausen, T. Joachims, Combining statistical learning with a knowledge-based approach - a case study in intensive care monitoring, in *Proceedings of 16th International Conference on Machine Learning* (1999)
15. J. Pickands III, Statistical inference using extreme order statistics. Ann. Stat. **3**(1), 119–131 (1975)
16. W.H. Press, B.P. Flannery, A.A. Teukolsky, W.T. Vetterling, *Numerical Recipes in C: The Art of Scientific Computing*, 2nd edn. (Cambridge University Press, Cambridge, 1992)
17. S.I. Resnick, *Extreme Values, Regular Variation and Point Processes* (Springer, New York, 1987)
18. A. Singhee, Novel algorithms for fast statistical analysis of scaled circuits. PhD Thesis, Electrical and Computer Engineering, Carnegie Mellon University (2007)
19. A. Singhee, SiLVR: projection pursuit for response surface modeling, in *Machine Learning in VLSI Computer Aided Design*, ed. by I.M. Elfadel, D. Boning, X. Li (Springer, Berlin, 2018)
20. A. Singhee, R.A. Rutenbar, Statistical Blockade: a novel method for very fast Monte Carlo simulation of rare circuit events, and its application, in *Proceedings of Design Automation & Test in Europe* (2007)
21. A. Singhee, R.A. Rutenbar, Beyond low-order statistical response surfaces: latent variable regression for efficient, highly nonlinear fitting, in Proceedings of IEEE/ACM Design Automation Conference (2007)
22. A. Singhee, R. Rutenbar, *Extreme Statistics in Nanoscale Memory Design* (Springer, New York, 2010)
23. A. Singhee, J. Wang, B.H. Calhoun, R.A. Rutenbar, Recursive Statistical Blockade: an enhanced technique for rare event simulation with application to SRAM circuit design, in *Proceeding of International Conference on VLSI Design* (2008)
24. R.L. Smith, Estimating tails of probability distributions. Ann. Stat. **15**(3), 1174–1207 (1987)
25. R.L. Smith, Maximum likelihood estimation in a class of non-regular cases. Biometrika **72**, 67–92 (1985)
26. J. Wang, A. Singhee, R.A. Rutenbar, B.H. Calhoun, Modeling the minimum standby supply voltage of a full SRAM array, in *Proceedings of European Solid-State Circuits Conference* (2007)
27. J. Wang, A. Singhee, R.A. Rutenbar, B.H. Calhoun, Two fast methods for estimating the minimum standby supply voltage for large SRAMs. IEEE Trans. Comput. Aided Des. **29**(12), 1908–1920 (2010)
28. K. Zhang, *Embedded Memories for Nanoscale VLSIs* (Springer, New York, 2009)

Chapter 11
Fast Statistical Analysis Using Machine Learning

Rouwaida Kanj, Rajiv V. Joshi, Lama Shaer, Ali Chehab, and Maria Malik

A model is a simplification or approximation of reality and hence will not reflect all of reality

Kenneth P. Burnham
David R. Anderson

11.1 Introduction: Logistic Regression-Based Importance Sampling Methodology for Statistical Analysis of Memory Design

With the advent of technology scaling, process variations pose a serious challenge to the manufacturing and reliability of integrated circuit designs. Memory designs use the smallest devices on the chip and tend to suffer most in terms of functionality and yield. With tight constraints of less than one failing part per million, traditional statistical analysis techniques like the Monte Carlo analysis become impractical, and hence arises the need for robust statistical methodologies for rare event estimation [1]. Several methodologies have been proposed to speed up the statistical analysis of memory designs. The authors in [2] proposed statistical models for the SRAM static noise margin during Read and Write. In [3], the authors propose probability distribution models for the SRAM cell dynamic behavior in terms of Read and Write

R. Kanj (✉) · L. Shaer · A. Chehab
Maroun Semaan Faculty of Engineering and Architecture, American University of Beirut, Beirut, Lebanon
e-mail: rk105@aub.edu.lb; las24@aub.edu.lb; chehab@aub.edu.lb

R. V. Joshi
IBM TJ Watson Labs, Yorktown Heights, NY, USA
e-mail: rvjoshi@us.ibm.com

M. Malik
George Mason University, Fairfax, VA, USA
e-mail: mmalik9@gmu.edu

I. M. Elfadel et al. (eds.), *Machine Learning in VLSI Computer-Aided Design*,
https://doi.org/10.1007/978-3-030-04666-8_11

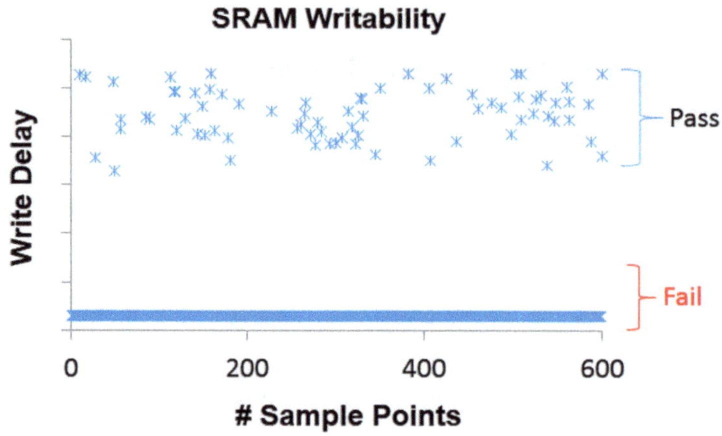

Fig. 11.1 SRAM writability simulations. Subject to random process variations, the SRAM cell may fail to write [8]

access times. The authors in [4] propose the statistical blockade methodology to predict the fail probability by building models for the distribution tail regions.

A mixture importance sampling (MixIS)-based yield analysis methodology was proposed in [5]. The methodology is circuit-simulation based and allows for the analysis of the SRAM cell dynamic margins. It achieves several orders of magnitude speedup compared to traditional Monte Carlo by distorting the natural distribution towards the fail region. To further speed up the circuit-simulation-based analysis, the authors in [6] proposed efficient on-demand spline-based table lookup transistor models for the SPICE simulations. In [7], sparse regression-based models were proposed to speed up the circuit simulations. However, often functional fails translate into discontinuities in the performance metric space (Fig. 11.1). Hence, they rely on a thresholding technique for proper sample point pass/fail prediction.

We propose a logistic regression-based prediction framework for speeding up the importance sampling methodology [8]. We rely on logistic regression to model the chances of a sample point fail and hence avoid dealing with the SRAM cell functional discontinuities. For regularization, we rely on a cross-validation-based ordered feature selection technique. The proposed approach is efficient and results in highly accurate models with less than 4.5% false prediction rates. The chapter is organized as follows. Section 11.2 presents an overview of Monte Carlo and variance reduction methods. Section 11.3 provides a background review of linear regression and logistic regression techniques. Section 11.4 presents the fast importance sampling methodology, and finally, Sect. 11.5 presents the application to a 14 nm industrial FinFET SRAM design.

11.2 Monte Carlo, Importance Sampling, and Variance Reduction Methods

Statistical inference is the act of inferring information from random samples to generalize that of the population [9]. For memory designs, we are mostly interested in statistical inference for proportions. A memory cell can be viewed as a Bernoulli trial with a predesigned probability of fail, P_f and the size of the memory array represents the number of trials, n. Accurately estimating P_f helps the designer predict the overall chip yield, according to

$$Yield = (1 - P_f)^n. \tag{11.1}$$

Traditional statistical methodologies for estimating the probability of fail include standard Monte Carlo. However, with millions of memory cells in the array and criteria of less than one part per million fails, designers must ensure very low cell fail probabilities. This renders Monte Carlo very slow, and raises the need for variance reduction techniques. In this section, we provide a review of Monte Carlo followed by an overview of mixture importance sampling as an efficient variance reduction method for rare fail estimation of memory designs.

11.2.1 Monte Carlo

One of the early Monte Carlo experiments dates back to Buffon's needle experiment in 1700s to estimate the value of pi [10]. In the 1940s Ulam used Monte Carlo to study the diffusion of the neutrons [11]. Nowadays, Monte Carlo has a wide range of applications in the different fields of physical sciences, engineering, and applied sciences. Monte Carlo methods are often used for finding approximate yet accurate numerical estimation to otherwise complex problems. We list below few examples.

1. Numerical integration: $Int = \int_a^b f(x)dx$
2. Statistical inference of a mean $\mu_x = \int x.pdf(x)dx$ of the variable x with a probability density function $pdf(x)$.
3. Statistical inference of proportions, e.g., estimating a probability of fail $P_f = \int_R pdf(x)dx$ over a region of fail R. The integral evaluation can be complex as it requires specific knowledge of the fail region.

We focus on the latter for purposes of our application.

11.2.1.1 Estimating the Fail Probability of a Circuit Design

Given a circuit design subject to a set of process variation parameters $\mathbf{x_p}$. Also given a set of performance metrics $f_i(\mathbf{x_p})$, such as the circuit delay, power, and gain, along with the corresponding set of performance metric fail criteria $\{f_i^0\}$, we define the indicator function $I(x)$ for a given performance metric to be:

$$I(x) = \begin{cases} 0 & \text{if } f_i(\mathbf{x_p}) \leq f_i^0, \\ 1 & \text{otherwise} \end{cases} \quad (11.2)$$

The probability of fail, P_f, can then be defined as

$$P_f = \int_R pdf(x)dx = \int I(x)pdf(x)dx = E(I(x)) \quad (11.3)$$

Monte Carlo method approximates the expectation function in (11.3) by computing the average value of the indicator function over a set of random sample points. Hence, given a random sample of size N, P_f is computed as the ratio of the number of failing sample points to the total number of sample points (11.4).

$$\hat{P_f} = \frac{1}{N}\sum_{i=1}^{N} I(x) \quad (11.4)$$

The outcome of Monte Carlo is not deterministic and varies from one random sample set to another. The variance of the estimate, $\sigma^2_{P_f}$, for a sample set of size N can be determined according to (11.5).

$$\sigma^2_{P_f} = \frac{(1 - P_f) * P_f}{N} \quad (11.5)$$

Therefore, to reduce the variance one needs to increase the number of sample points. For small probabilities, the number of sample points and hence the number of circuit simulations becomes prohibitively large (see Table 11.1 for an example). Hence, it is not efficient to rely on traditional Monte Carlo techniques for accurately estimating such rare probabilities. More robust and efficient statistical analysis

Table 11.1 The number of Monte Carlo sample points needed to achieve a standard error to the probability of fail ratio $\frac{\sigma_{P_f}}{P_f} = 10\%$ for different P_f values

Probability of fail P_f	Number of Monte Carlo Sample points N
10^{-3}	10^5
10^{-4}	10^6
10^{-5}	10^7
10^{-6}	10^8

methodologies such as variance reduction methods are necessary to enhance speed of the statistical analysis methodology and the accuracy of the estimate.

11.2.2 Variance Reduction

For rare fail events, variance reduction methods improve the accuracy of the estimate compared to standard Monte Carlo. Common variance reduction methods include the following [12].

1. Control variates method: Standard Monte Carlo method typically aims to estimate the expected value of a variable Y, μ_y, according to (11.6). The control variates method introduces a variable Z correlated to Y, such that while the variance of Y may be large, the variance of $(Y - cZ)$ is small. The variable Z is often derived as a model or approximation to Y. The optimal value of c is determined as the ratio of the covariance of Y and Z to the variance of Z. Accordingly, the control variates method estimates μ_{CV}, the expected value of the variable $(Y - cZ)$, as a more efficient and an unbiased estimator of μ_y.

$$\mu_y = \frac{1}{N}\sum_{i=1}^{N} Y_i \tag{11.6}$$

$$c = \frac{Cov(Y, Z)}{Var(z)} \tag{11.7}$$

$$\mu_{CV} = \frac{1}{N}\sum_{i=1}^{N}(Y_i - c(Z_i - \mu_Z)) \tag{11.8}$$

2. Antithetic variables: For every sample path, the antithetic variables method samples the antithetic path, i.e., for every direction, it samples the opposite direction. This is shown to reduce the variance of the estimate and reduce the number of sample points to be generated.
3. Stratified sampling: The population is divided into strata. The sample points are evenly divided among the different strata. Simple random sampling is performed within each stratum. This allows for better coverage of the variability space.
4. Importance sampling: The natural distribution is distorted to place more emphasis on sampling the critical fail region. The true metric is sampled, and the estimate is unbiased by correcting for the ratio of the true to distorted probability density function. The methodology does not require knowledge of functional approximations as is the case for the control variates method. Furthermore, it is more sophisticated than stratified sampling in that it zooms in to critical regions. Henceforth, we will focus on importance sampling methods.

11.2.3 Importance Sampling

Importance sampling is a variance reduction method that enhances the efficiency of standard Monte Carlo by distorting the (natural) distribution, $pdf(x)$, to produce more sample points in the important region(s) [13]. It can be best explained by Eq. (11.9), where $pdf_{IS}(x)$ is the importance sampling distribution. Figure 11.2 presents an example where a uniform distribution is used to generate more sample points in the tail region compared to the natural Gaussian distribution. More efficient importance sampling distributions entail shifting the center of the natural distribution towards the important region, such that $pdf_{IS}(x) = pdf(x - x_{IS})$. It has been shown that determining the optimal shift center requires the knowledge of the estimate itself. In [5], the authors rely on a first stage of uniform sampling to determine the shift center for the mixture importance sampling phase.

$$\int f(x)pdf(x)dx = \int f(x)\frac{pdf(x)}{pdf_{IS}(x)}pdf_{IS}(x)dx \tag{11.9}$$

There are three forms of importance sampling.

1. Integrated importance sampling.
2. Ratioed importance sampling.
3. Regression importance sampling.

To unbias the estimate, the indicator function is adjusted as illustrated in (11.10) to account for the distortion of the natural sampling distribution. The probability of fail estimate, $\hat{P}_{f_{IS}}$, and its variance, $\sigma^2_{Pf:IS}$, can then be derived according to (11.11) and (11.12) for the integrated importance sampling case.

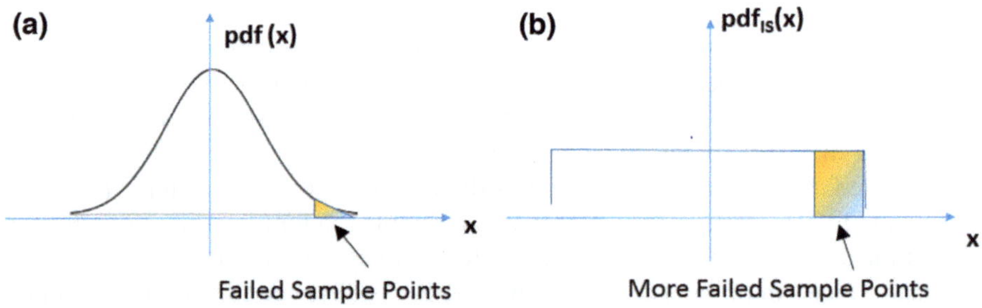

Fig. 11.2 (**a**) The natural distribution provides a small number of sample points in the failure region. (**b**) The importance sampling distribution $pdf_{IS}(x)$ provides more sample points in the failure region

$$Y(x_i) = I(x_i) * \frac{pdf(x_i)}{pdf_{IS}(x_i)} \tag{11.10}$$

$$\hat{P}_{f_{IS}} = \frac{1}{N} \sum_{i=1}^{N} Y(x_i) \tag{11.11}$$

$$\sigma^2_{Pf:IS} = \frac{1}{N(N-1)} \sum_{i=1}^{N} (Y(x_i) - (\bar{Y})) \tag{11.12}$$

11.3 Logistic Regression

Statisticians rely on regression to model or interpret the behavior of the response variable as function of the independent or explanatory variables. In the event of continuous response, they rely on linear regression to model the outcome. On the other hand, when the outcome is categorical, logistic regression is the most popular type of regression to model such data [14]. Logistic regression is used in many fields and spans a wide range of applications in engineering, medical sciences, and finance [15, 16]. As an example of a semiconductor manufacturing application, the authors in [17] relied on logistic regression to model "when the test chip will or will not have bit failures" based on specific features of the test chip.

In this section, we provide an overview of simple linear regression, followed by a review of logistic regression methodologies.

11.3.1 Simple Linear Regression Review

In traditional linear regression, a function $f(x)$ is modeled as a linear combination of the independent variables vector **x** according to the following equation [18]:

$$f(\mathbf{x}) \approx h_\theta(\mathbf{X}) = \sum_{i=0}^{D} \theta_i * x_i \tag{11.13}$$

where $\{\theta_i, i = 1, 2, 3, \ldots, D\}$ are the model coefficients, $\{\mathbf{x} = x_i : i = 1, 2, 3, \ldots, D\}$ represents the D-dimensional feature vector, θ_0 represents the constant term, x_0 represents the intercept term, and the hypothesis function $h(\boldsymbol{\theta})$ is the approximate linear regression model to $f(x)$.

Given a training set of N sample points, $N > D$, we define

$$\mathbf{X} = \begin{pmatrix} x_1^{(1)} & x_1^{(2)} & \dots & x_1^{(N)} \\ x_2^{(1)} & x_2^{(2)} & \dots & x_2^{(N)} \\ \vdots & \vdots & \dots & \vdots \\ x_D^{(1)} & x_D^{(2)} & \dots & x_D^{(N)} \end{pmatrix} \tag{11.14}$$

and $\mathbf{F} = [f^{(1)} f^{(1)} \dots f^{(N)}]^T$, where $f^{(j)}$ and $\mathbf{x}^{(j)}$ represent the value of $f(\mathbf{x})$ and $\mathbf{x}$ at the jth sample point, respectively. Usually $N > D$, and the system of equations to find the best θ based on the training data is overdetermined. We typically solve for $\boldsymbol{\theta}$ by minimizing the cost function in (11.15) which represents the error between $h_{\boldsymbol{\theta}}$ and f [19].

$$J(\mathbf{x}, \boldsymbol{\theta}) = \frac{1}{2N} \sum_{i=1}^{N} (h_\theta(\mathbf{x^{(i)}}) - f^{(i)})^2 \tag{11.15}$$

Table 11.2 presents the batch gradient descent approach, based on the least mean squares update rule, to solve for θ; α represents the learning rate, and θ^0 represents the initial guess for $\boldsymbol{\theta}$. $\boldsymbol{\theta}^k$ represents the value of $\boldsymbol{\theta}$ at the kth iteration.

It is worth noting that for linear regression, a closed form solution for $\boldsymbol{\theta}$ that minimizes J_θ exists and can be derived according to (11.16).

$$\boldsymbol{\theta} = (\mathbf{X}^T\mathbf{X})^{-1}\mathbf{X}^T\mathbf{F} \tag{11.16}$$

To avoid overfitting and improve accuracy, regularized linear regression solves for a penalized form of the cost function (11.17) , where $l_p = 0, 1$, or 2 [20]. Thus, there are three types of regularization: the $L_{0\text{-}norm}$, $L_{1\text{-}norm}$, and $L_{2\text{-}norm}$ regularizations. The $L_{0\text{-}norm}$ guarantees sparseness by identifying the optimal subset of independent variables; however, it is NP-hard. The $L_{1\text{-}norm}$ solution enforces sparseness and can be approximated. The $L_{2\text{-}norm}$ is simpler, minimizes the magnitude of the parameters, however, it does not enforce sparseness.

$$J(\mathbf{x}, \boldsymbol{\theta}) = \frac{1}{2N} \sum_{i=1}^{N} (h_\theta(\mathbf{x^{(i)}}) - f^{(i)})^2 + \lambda||\boldsymbol{\theta}||_{l_p} \tag{11.17}$$

Table 11.2 Pseudo-code for the batch gradient descent method

Line #	Code
1.	Initialize $\boldsymbol{\theta} = \boldsymbol{\theta}^0$
2.	$\boldsymbol{\theta}_j^{k+1} = \boldsymbol{\theta}_j^k + \alpha \sum_{i=0}^{N} (h_\theta(\mathbf{x^{(i)}}) - f^{(i)})x_j^{(i)}$

11.3.2 Logistic Regression Overview

Logistic regression belongs to the family of machine learning tools developed to model categorical data including fault prediction. Similar to support vector machines, it is used to develop a model for a categorical response as function of a set of independent random variables. Logistic regression can be used to model outcomes that are binomial, multinomial, or ordinal [21]. For purposes of our applications, we focus on the case where the outcome, y, is binary. Thus, we define the hypothesis function, $h_{\boldsymbol{\theta}}(x)$, as follows:

$$h_{\boldsymbol{\theta}}(x) = g(\boldsymbol{\theta}^T \mathbf{x}) = \frac{1}{1 + exp^{-\boldsymbol{\theta}^T \mathbf{x}}} \tag{11.18}$$

where the sigmoid function $g(z)$ returns a probability value between 0 and 1.

$$g(z) = \frac{1}{1 + exp^{-z}} \tag{11.19}$$

The hypothesis function computes a probability value that determines whether the expected outcome is "0" or "1" [22].

$$y = \begin{cases} 1 & \text{if } h_\theta(x) > 0.5, \\ 0 & \text{otherwise} \end{cases} \tag{11.20}$$

Hence, the probability of a sample point, $p(y|\mathbf{x}; \boldsymbol{\theta})$, can be computed using the hypothesis function according to (11.21) [19]. The likelihood function for a set of sample points is the product of the individual sample point probabilities, and the logistic regression parameter vector $\boldsymbol{\theta}$ can be determined by maximizing the log-likelihood function in (11.22). This corresponds to minimizing the cost function: $J(\mathbf{x}, \boldsymbol{\theta}) = -l(\boldsymbol{\theta})$.

$$p(y|\mathbf{x}; \boldsymbol{\theta}) = (h_{\boldsymbol{\theta}}(\mathbf{x}))^y (1 - h_{\boldsymbol{\theta}}(\mathbf{x}))^{1-y} \tag{11.21}$$

$$l(\boldsymbol{\theta}) = \sum_{i=1}^{N} log(h_\theta(\mathbf{x^{(i)}})).y^{(i)} + log(1 - h_\theta(\mathbf{x^{(i)}})).(1 - y^{(i)}) \tag{11.22}$$

One may rely on the stochastic gradient ascent rule (11.23) [19] to solve for $\boldsymbol{\theta}$. This is similar to the batch gradient (descent) rule derived earlier for the linear regression. However, for the stochastic approach, the updates are performed per sample point, and for the logistic regression the function $h_{\boldsymbol{\theta}}$ is nonlinear.

$$\boldsymbol{\theta}_j^{k+1} = \boldsymbol{\theta}_j^k + \alpha(y^{(i)} - h_\theta(\mathbf{x^{(i)}}))x_j^{(i)} \tag{11.23}$$

The authors in [23] compare several numerical techniques for solving for $\boldsymbol{\theta}$. These include: Newton's method, coordinate ascent method, conjugate gradient ascent method, fixed Hessian Newton method, quasi-Newton method, and the iterative scaling method. In terms of the computational complexity, they found the conjugate gradient ascent and quasi-Newton methods to be the fastest. In their work, they rely on an alternative representation for the logistic regression as will be discussed in the following subsection.

11.3.2.1 Logistic Regression: Numerical Optimization Techniques

When the observed outcome y takes values ± 1, the logistic regression model can be represented by the probability function (11.24) [23].

$$p(y = \pm 1|\mathbf{x}, \boldsymbol{\theta}) = g(y\boldsymbol{\theta}^T\mathbf{x}) = \frac{1}{1 + exp(-y\boldsymbol{\theta}^T\mathbf{x})} \tag{11.24}$$

In this case, the log-likelihood function in (11.22) can be written as:

$$l(\boldsymbol{\theta}) = \sum_{1}^{N} log(1 + exp(-y^{(i)}\boldsymbol{\theta}^T\mathbf{x}^{(i)})) \tag{11.25}$$

The two forms in (11.22) and (11.25) are equivalent due to the inherent property of the sigmoid function that $g(-z) = 1 - g(z)$. Numerical optimization techniques to solve for optimal θ can rely on either formulation for the optimization. Caution must be taken, however, based on the choice of the values of y. Particularly, (11.25) is not valid when $y = 0$.

As stated earlier, following the formulations in (11.24) and (11.25), the authors in [23] compared several techniques for optimizing the maximum a-posteriori parameter estimate. The different optimization methodologies rely on the following three key computational elements.

1. The diagonal matrix A, such that the diagonal element a_{ii} is given by

$$a_{ii} = g(\boldsymbol{\theta}^T\mathbf{x}^{(i)})(1 - g(\boldsymbol{\theta}^T\mathbf{x}^{(i)})) \tag{11.26}$$

2. The Hessian matrix, **H** defined as

$$\mathbf{H}(\boldsymbol{\theta}) = -\mathbf{X}A\mathbf{X}^T \tag{11.27}$$

3. The gradient, **grad**

$$\mathbf{grad}(\boldsymbol{\theta}) = \nabla_\theta l(\boldsymbol{\theta}) = \sum_{i=1}^{N}(1 - g(y^{(i)}\boldsymbol{\theta}^T\mathbf{x}^{(i)}))y^{(i)}\mathbf{x}^{(i)} \tag{11.28}$$

For purposes of illustration, we present here their implementation of Newton's method for parameter optimization. We refer the reader to [23] for the details of the other techniques. A typical Newton's method step calculates the new $\boldsymbol{\theta}$ from its old value by relying on the Hessian and the gradient according to (11.29).

$$\boldsymbol{\theta}^{k+1} = \boldsymbol{\theta}^{k} + \mathbf{H}(\boldsymbol{\theta}^{\mathbf{k}})^{-1}\mathbf{grad}(\boldsymbol{\theta}^{\mathbf{k}}) \tag{11.29}$$

By defining $\mathbf{z}$ according to (11.30), Eq. (11.29) can be simplified to the form presented in (11.31) where both $\mathbf{z}$ and $\mathbf{A}$ are evaluated using $\boldsymbol{\theta}^{k}$.

$$z_i = \mathbf{x}^T\boldsymbol{\theta}_j + \frac{(1 - g(y_i\boldsymbol{\theta}^T\mathbf{x_i}))y_i}{a_{ii}} \tag{11.30}$$

$$\boldsymbol{\theta}^{\mathbf{k+1}} = (\mathbf{XAX^T})^{-1}\mathbf{XAz} \tag{11.31}$$

Hence, the solution for the Newton step corresponds to that of a weighted least squares problem, and this implementation is often referred to as the iteratively reweighted least squares method.

11.4 Proposed Methodology

In this section, we present a logistic regression-based fast statistical analysis methodology. At the core of its engine is a mixture importance sampling technique which comprises two sampling stages: the uniform sampling stage and the importance sampling stage. We rely on the uniform sampling stage data to determine the shift center of the importance sampling distribution, and to develop a logistic regression model for predicting the design functionality of the importance sampling points. We employ ordered feature selection along with cross-validation in order to minimize the model error and avoid overfitting. We describe the methodology overview in Sect. 11.4.1.

11.4.1 Methodology Overview

11.4.1.1 Terminology and Definitions

Let the vector $\mathbf{x_p} = \{x_{p_i}, i = 1, 2, 3, \ldots, D\}$ represent the D-dimensional process variation parameter vector. Given N sample points, we generate a feature vector $\mathbf{x}$ using linear, quadratic, and interaction terms derived from $\mathbf{x_p}$ along with the constant term.

$$\mathbf{X} = \begin{pmatrix} x_{p1}^{(1)} & x_{p1}^{(2)} & \dots & x_{p1}^{(N)} \\ x_{p2}^{(1)} & x_{p2}^{(2)} & \dots & x_{p2}^{(N)} \\ \vdots & \vdots & \dots & \vdots \\ (x_{p1}^{(1)})^2 & (x_{p1}^{(2)})^2 & \dots & (x_{p1}^{(N)})^2 \\ (x_{p2}^{(1)})^2 & (x_{p2}^{(2)})^2 & \dots & (x_{p2}^{(N)})^2 \\ \vdots & \vdots & \dots & \vdots \\ x_{p1}^{(1)}x_{p2}^{(1)} & x_{p1}^{(2)}x_{p2}^{(2)} & \dots & x_{p1}^{(N)}x_{p2}^{(N)} \\ \vdots & \vdots & \dots & \vdots \\ 1 & 1 & \dots & 1 \end{pmatrix} \tag{11.32}$$

Let $\mathbf{F} = [f^{(1)} f^{(1)} \dots f^{(N)}]$ be the response vector representing the SRAM cell functionality. The outcome vector, $\mathbf{Y}$, is derived from $f^{(i)}$, such that $y_i = 0$ for a functional cell, and $y_i = 1$ for a failing cell.

$$y_i = \begin{cases} 1 & \text{if } f^{(i)} \text{ is fail,} \\ 0 & \text{otherwise} \end{cases} \tag{11.33}$$

11.4.1.2 Methodology Flow Diagram

Figure 11.3 presents an overview of the proposed logistic regression-based MixIS methodology (LRegSim). It comprises two major phases corresponding to the two stages of the mixture importance methodology [5].

- In the first phase, we generate uniform random sample points, $\mathbf{X_{p_{uni}}}$, and we rely on circuit simulations to generate the corresponding performance metric indicator values $\mathbf{Y_{uni}}$. We use these sample points to generate the quadratic features vector, $\mathbf{X_{uni}}$ from $\mathbf{X_{p_{uni}}}$, and we normalize it to build the corresponding regularized logistic regression model and determine the corresponding model $\boldsymbol{\theta}$, referred to hereon as $\boldsymbol{\theta}_{\mathbf{model}}$.
- In the second phase, we generate the importance sample points, $\mathbf{X_{p_{MixIS}}}$ and rely on the logistic regression model developed in Phase 1 to predict the failing sample points and hence estimate the design yield. It is worth noting that Phase 2 data is typically a subspace of Phase 1 data, since the latter varies all the parameters uniformly between $\pm 6\sigma$.

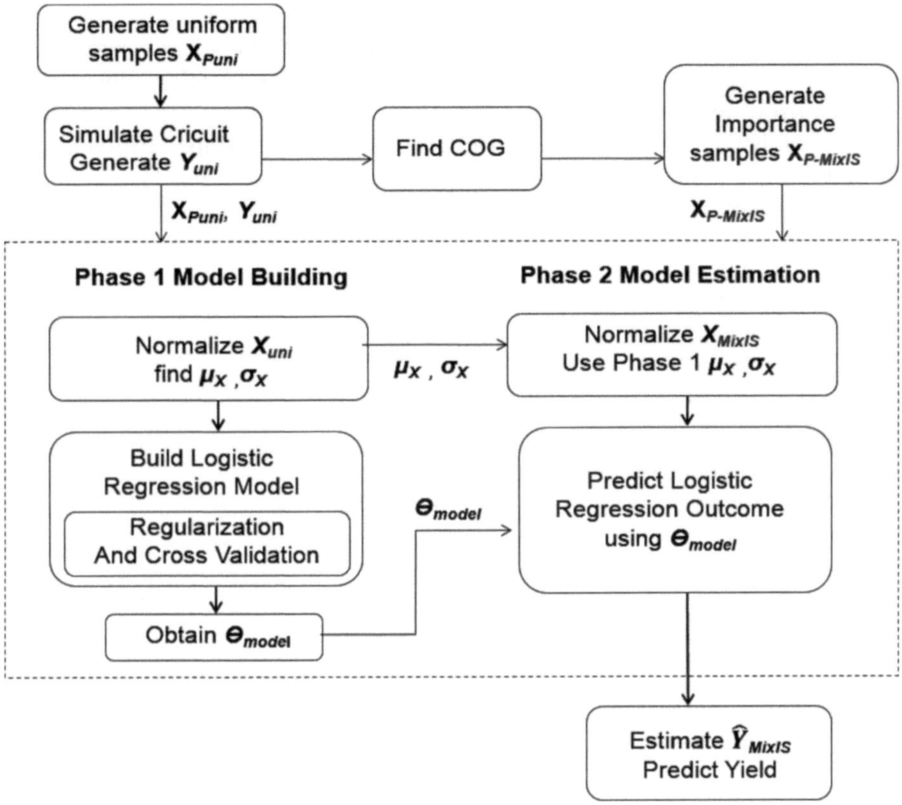

Fig. 11.3 An overview of the logistic regression-based yield analysis methodology. The methodology relies on Phase 1 simulations to build a model that predicts the response of Phase 2 importance sample points [8]

11.4.1.3 Regularization Framework

In [24], the authors convert the $L_{1\text{-}norm}$ regularized logistic regression problem into an L_1-constrained iterative least squares problem. The authors in [25] rely on local derivatives to determine critical features that lead to cost reduction that outweighs the introduced feature weight. For our implementation [8], we propose a two-step heuristic approach for $L_{0\text{-}norm}$ feature selection.

1. **Initialization and ordering:** First, we build the initial logistic regression model, and order the feature criticality based on their resultant $\boldsymbol{\theta}$ values.
2. **Cross-validation-based search:** For each iteration j, we find the average cross-validation error corresponding to the top j critical features. The best model corresponds to the model with minimal cross-validation error. For more details on cross-validation, we refer the reader to Sect. 11.4.1.4.

Table 11.3 presents the pseudo-code for the proposed methodology. In the initialization phase, using Phase 1 data, the quadratic features are generated and then normalized. An initial modeling step involving all features allows obtaining an ordered list of the features based on the initial model θ values. The search for the best model is performed by adding one feature at a time from the ordered list and identifying the model with the smallest cross-validation error. For the importance sampling phase, the quadratic features are generated for the importance samples,

Table 11.3 Pseudo-code for the proposed methodology [8]

Logistic Regression-Based Yield Analysis Methodology
Parameters:
Given process variations and response vectors $\mathbf{X}_{\mathbf{P}_{\mathbf{uni}}}, \mathbf{F}_{\mathbf{uni}}, \mathbf{X}_{\mathbf{pMixIS}}$
Initialization:
1. Generate $\mathbf{Y}_{\mathbf{uni}}, \mathbf{X}_{\mathbf{uni}} \leftarrow$ quadMap($\mathbf{X}_{\mathbf{P}_{\mathbf{uni}}}$) using (32),and (33)
2. Generate mean $\boldsymbol{\mu}_{\mathbf{X}_{\mathbf{uni}}}$ and standard deviation$\boldsymbol{\sigma}_{\mathbf{X}_{\mathbf{uni}}}$ for $\mathbf{X}_{\mathbf{uni}}$
3. Normalize Uniform feature $\mathbf{X}_{\mathbf{n}_{\mathbf{uni}}} = \frac{\mathbf{X}_{\mathbf{uni}} - \mu_{\mathbf{X}_{\mathbf{uni}}}}{\sigma_{\mathbf{X}_{\mathbf{uni}}}}$
4. Generate $\boldsymbol{\theta}_{\mathbf{init}} =$ buildLogistic($\mathbf{X}_{\mathbf{n}_{\mathbf{uni}}}, \mathbf{Y}$)
5. $(\boldsymbol{\theta}_{\mathbf{sort}}, \mathbf{I}_{\mathbf{dx}}) = \mathrm{sort}(\boldsymbol{\theta}_{\mathbf{init}}), \lvert\boldsymbol{\theta}_{\mathbf{sort(i)}} = \boldsymbol{\theta}_{\mathbf{init}}(\mathbf{I}_{\mathbf{dx,i}})$
Model Building (Phase 1 data):
For: j = 1:size($\boldsymbol{\theta}_{\mathbf{sort}}$)
Perform Cross-Validation over j first critical vars
1. $\mathbf{X}_{\mathbf{crit}} \leftarrow \{\mathbf{X}_{\mathbf{n\_uni}}(\mathbf{I}_{\mathbf{dx}}(1:j))\}$
2.**For** K = 1:4
Datasets:$\{\mathbf{X}_{\mathbf{crit}_{\mathbf{test}}}, \mathbf{Y}_{\mathbf{test}}\}_K, \{\mathbf{X}_{\mathbf{crit}_{train}}, \mathbf{Y}_{train}\}_K$
$\boldsymbol{\theta}_{\mathbf{K,j}} \leftarrow buildLogistic(\{\mathbf{X}_{\mathbf{crit}_{train}}, \mathbf{Y}_{train}\}_K)$
$\mathbf{Error}_{\mathbf{K,j}} \leftarrow predictLogistic(\mathbf{X}_{\mathbf{crit}_{test}}, \mathbf{Y}_{\mathbf{test}}, \boldsymbol{\theta}_{\mathbf{K,j}})$
end
3.$\mathbf{Error}_{\mathbf{j}} \Leftarrow \mathrm{avg}(\mathbf{Error}_{\mathbf{K,J}})$
4.**If** ($\mathbf{Error}_{\mathbf{j}} < \mathbf{min}$)
$\mathbf{min} = \mathbf{Error}_{\mathbf{j}}$; $\mathbf{count} = 0$;
$\mathbf{critical}_{\mathbf{vars}} \leftarrow \mathbf{j}, \mathbf{X}_{\mathbf{crit}}, \mathbf{Y}, \boldsymbol{\lambda} = \mathbf{0}$)
$\boldsymbol{\theta}_{\mathbf{model}} \leftarrow$ buildLogistic($\mathbf{X}_{\mathbf{Crit}}, \mathbf{Y}$)
Else
$\mathbf{count} = \mathbf{count} + \mathbf{1}$
if(count > **4)** break; **endif**
endif
end
Importance Sample and Yield Prediction (Phase 2 data)
1.Generate $\mathbf{X}_{\mathbf{MixIS}} \leftarrow quadMap(\mathbf{X}_{\mathbf{P}_{\mathbf{MixIS}}})$ based on (1), (2)
2.Normalize features $\mathbf{X}_{\mathbf{n}_{\mathbf{MixIS}}} = \frac{\mathbf{X}_{\mathbf{MixIS}} - \mu_{\mathbf{X}_{\mathbf{uni}}}}{\sigma_{\mathbf{X}_{\mathbf{uni}}}}$
3.$\widehat{\mathbf{Y}}_{\mathbf{MixIS}} =$ predictLogistic($\mathbf{X}_{\mathbf{n}_{\mathbf{MixIS}}}, \boldsymbol{\theta}_{\mathbf{crit}}$)
4.$\widehat{Yield} \leftarrow Yield(\mathbf{X}_{\mathbf{P}_{\mathbf{MixIS}}}, \widehat{\mathbf{Y}}_{\mathbf{MixIS}})$

and the features are normalized using the same mean and standard deviation values obtained in Phase 1. The developed model is then used to predict the outcomes which are in turn used to predict the yield. Note the following function definitions.

1. *quadMap* corresponds to the mapping of Eq. (11.32).
2. *buildLogistic* solves for $\boldsymbol{\theta}$ by maximizing (11.22).
3. *predictLogistic* predicts y according to (11.20).

Next, we provide a brief overview of cross-validation.

11.4.1.4 Cross-validation

In K-folds cross-validation [26], the dataset is divided into k folds. $k - 1$ folds are used for the training and one set is used for testing the model error. The process is repeated k-times each time selecting a different fold for the test set. The average error obtained from the k runs is used to determine the cross-validation error. There are trade-offs in the choice of k. Larger k values are associated with reduced bias in the model error at the expense of higher runtime.

For our purposes, our objective is to determine the model with the lowest number of features corresponding to the minimum cross-validation error. Hence, without loss of generality, and to maintain a low runtime overhead, we divide the sample points into 4 mutually exclusive folds, and run the cross-validation process 4 times as illustrated in Fig. 11.4.

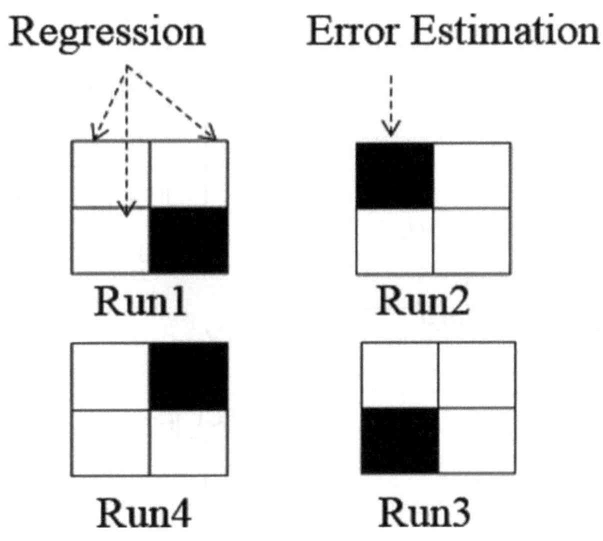

Fig. 11.4 Example of fourfold cross-validation [8]

11.5 Application to State-of-the-Art FinFET SRAM Design

In this section, we apply the methodology to analyze a state-of-the-art industrial 14 nm FinFET SRAM design with write-assist circuitry. We compare the results to those of a fully circuit-simulation-based approach.

11.5.1 Selective Boosting and Write-Assist Circuitry

Figure 11.5 presents the sketch of a 6-transistor SRAM cell. In planar CMOS technology, designers optimize the SRAM cell β and γ ratios in terms of the pulldown-to-passgate (PD/PG) and pullup-to-passgate (PU/PG) device ratios. This helps achieve robust cell designs while addressing the different design considerations in terms of compactness, functionality, and performance. FinFET SRAM designs bring forth challenges in terms of maintaining proper device sizing due to quantization. A minimum-sized FinFET SRAM cell has the PU, PD, and PG devices set to a single fin each, and hence, the β and γ ratios are set to one. This degrades functionality at low voltage operation, and raises the need for robust write-assist circuitry and selective boosting techniques [27]. The selective boosting can be applied to selective paths including the write drivers, wordline, and the cell. This helps enhance the cell yield while maintaining operation at low Vdd. Figure 11.6

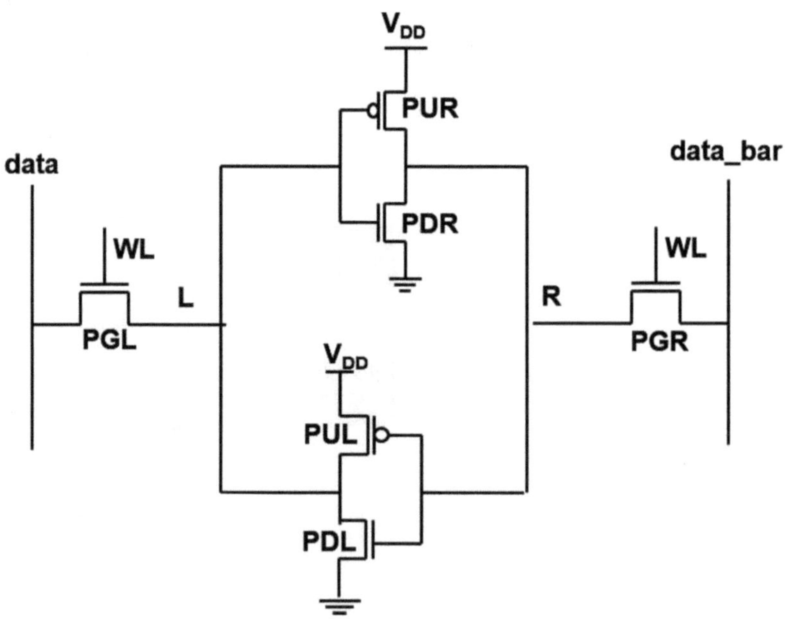

Fig. 11.5 Schematic of a 6-transistor SRAM cell

presents the boosting circuitry schematic consisting of an nFET and pFET device in parallel. When the gate switches, it couples the Vddv to a value above Vdd. This boosts the virtual supply node which feeds selected memory paths by more than 100 mV. The rest of the circuit operates at Vdd.

Write-assist circuitry further improves the writability yield by negatively boosting the bitlines during write as illustrated in Fig. 11.7. The negative boost lowers the bitline voltage below zero and creates a larger voltage swing between the true

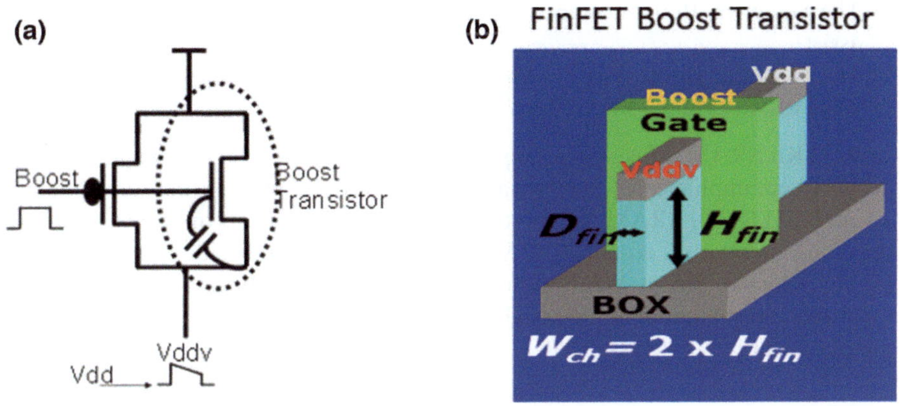

Fig. 11.6 (**a**) Capacitive coupling between the gate and the source nodes boosts the source voltage Vddv above Vdd when the "Boost" signal switches high [7] (**b**) Sketch of the FinFET Boost Transistor

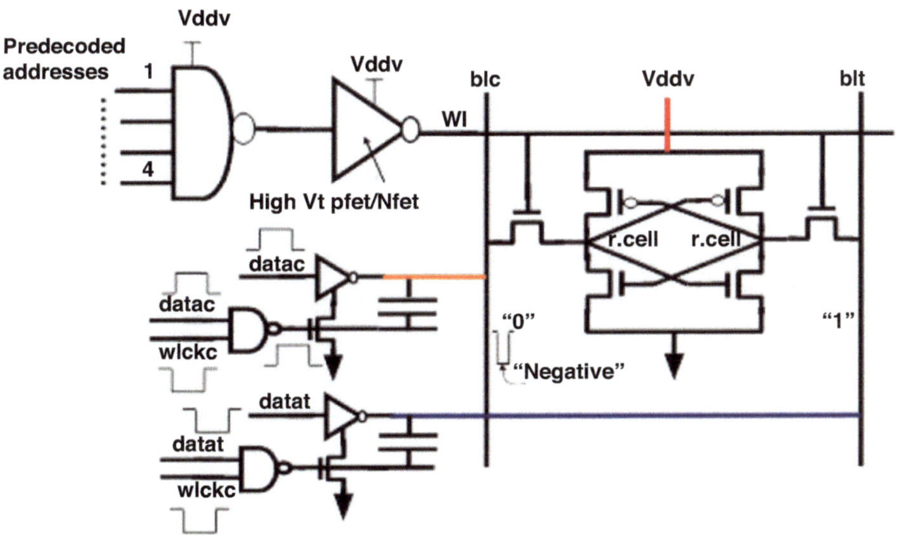

Fig. 11.7 Negative boost write assist is applied to the bitlines during write operation. Cell, write driver, and wordline receive boosted supply voltage (Vddv) [7]

and complement data values to be written to the cell. This in turn helps flip the cell data more easily and hence improves writability. The selective boost and write-assist circuitry are especially necessary for near threshold voltage operation where the devices are weak. For slightly higher voltages, selective boost can be turned on only for Read operation. Programmable boost designs [28] can also be considered (Fig. 11.8).

To study the benefits of the write-assist circuitry, we focus on the following two design styles summarized in Table 11.4.

1. The Selective_Boost design refers to the selective boost design with no write assist.
2. The Selective_Boost_WA design refers to the selective boost paired design with write assist.

A 72 Kb SRAM array was arranged in columns of 16 cells/bitlines and the two designs were implemented in hardware [27]. Figure 11.9 presents the corresponding 14 nm FinFET SOI SRAM die photo. Hardware measurements indicate a minimum operating voltage of around 0.40 V for the Selective_Boost design, and 0.35 V for the Selective_Boost_WA design. These results were validated by the statistical yield analysis methodology [5], and the results corroborate well with the hardware measurements.

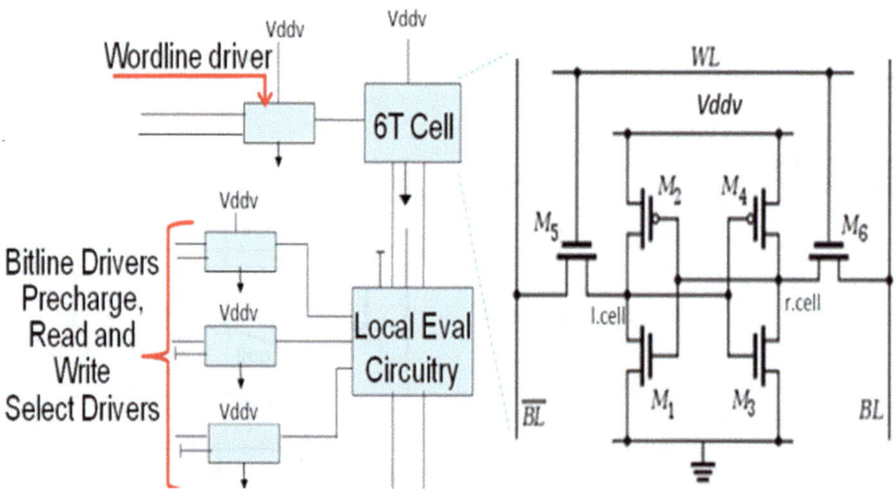

Fig. 11.8 Sketch of the SRAM cell cross-section including wordline, bitline, and write drivers[7]

Table 11.4 Summary of the designs under study

Design	Write assist	Label
Selective boost only	None	Selective_Boost
Selective boost with write assist	Negative bitline boost	Selective_Boost_WA

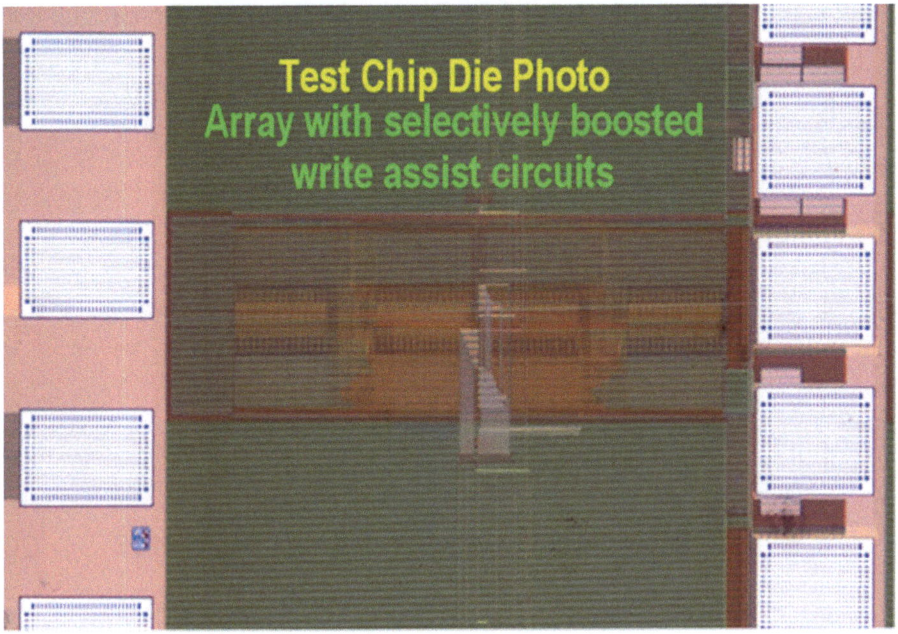

Fig. 11.9 Snapshot of the 14 nm FinFET SOI technology die [7]

11.5.2 Experimental Setup

We test the efficacy of the methodology by analyzing the 14 nm FinFET SRAM designs presented in Table 11.4. For accurate analysis and for purposes of model to hardware corroboration, our simulations include the peripheral logic along with the SRAM cell as illustrated in Fig. 11.8. Thus, our cross-section includes the write driver, bitline driver, wordline drivers, and the local evaluation circuitry. We study the SRAM cell functionality in terms of its dynamic write margin [29] defined as the ability to write a "0" or "1" onto the cell internal nodes. In the presence of process variations, it is possible that the cell pullup device (e.g., PUL) becomes strong, and the cell pass gate device (PGL) becomes weak thereby making it difficult or in some rare cases not possible to write a "0" to the left node of the SRAM. For purposes of our simulations, we apply process variations to the SRAM cell transistors and critical transistors of the local evaluation circuitry. For each device, we lump the different sources of variability in terms of metal gate granularity, line edge roughness, fin height, and random dopant fluctuations into an effective σ_{Vt}.The device threshold voltage is then subject to independent random process variations δ_{Vt} $N(0, \sigma_{Vt}^2)$ resulting in 45 features total. Our objective is to validate V_{min} improvement due to the write-assist circuitry. Accordingly, we rely on the proposed methodology to study the yield of the Selective_Boost design for Vdd = [0.40–0.43 V] and the Selective_Boost_WA design yield for Vdd = [0.35–0.40 V].

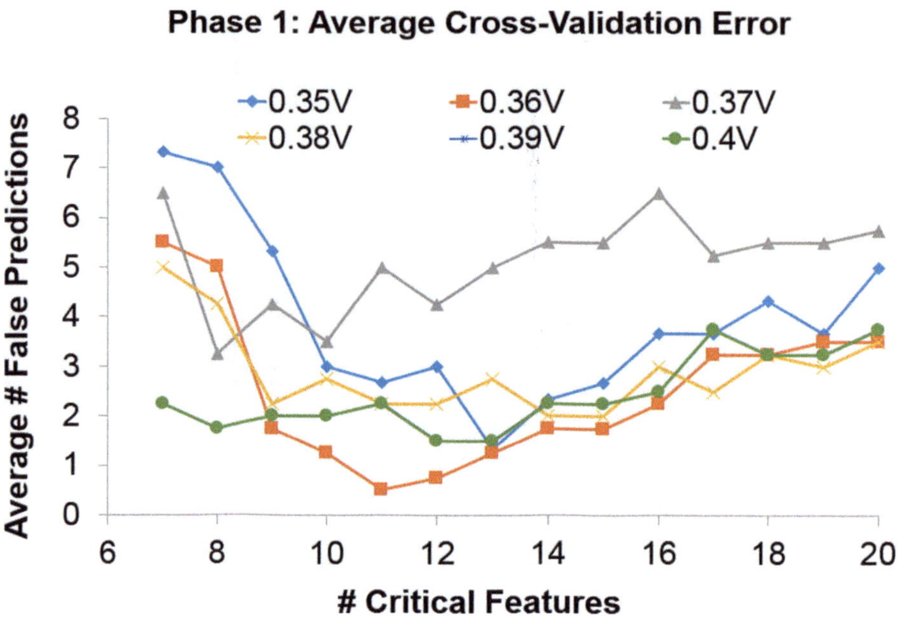

Fig. 11.10 Average number of false predictions versus the number of critical features [8]

11.5.3 *Analysis and Results*

We demonstrate the accuracy of the proposed methodology (LRegSim) compared to that of a pure circuit-simulation-based mixture importance sampling approach (CktSim). We report the results for the model building, model estimation, and importance sampling-based yield prediction.

11.5.3.1 Phase 1 Model Building: Uniform Sampling Stage

In this section, we present the results for the model building phase. Figure 11.10 presents the cross-validation error for the Selective_Boost_WA design as function of the number of critical features. The cross-validation error is reported as the average number of false predicts, which is computed as the total number of false positives and false negatives for the test set for the different iterations of the *k*-fold cross-validation. Figure 11.11 illustrates the optimal number of critical features for both designs. Experiments 1–6 correspond to the Selective_Boost_WA design simulations with Vdd = [0.35-0.4 V], and experiments 7–10 correspond to the Selective_Boost design simulations with Vdd = [0.4-0.43 V]. The optimal number of critical features is found to be in the range between 7 and 12 features for all the performed experiments. As illustrated in Fig. 11.12, the corresponding false predict rate is reported to be less than 2%.

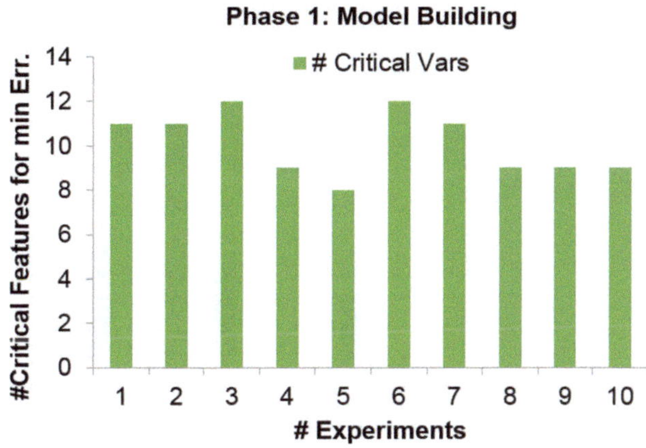

Fig. 11.11 Optimal number of critical features. Experiments 1 through 6 represent Selective_Boost_WA design for Vdd range of [0.35–0.40 V]. Experiments 7 to 10 represent the Selective_Boost design for Vdd range of [0.40–0.43 V] [8]

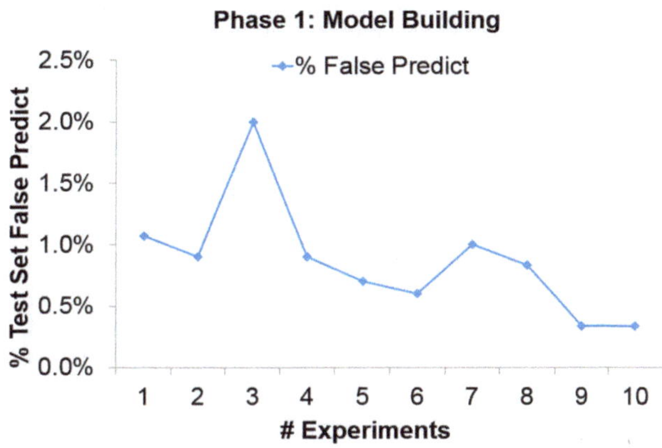

Fig. 11.12 Test set prediction error corresponding to the model based on the critical number of variables in Fig. 11.11. Experiments 1 through 6 represent Selective_Boost_WA design for Vdd range of [0.35–0.40 V]. Experiments 7 to 10 represent the Selective_Boost design for Vdd range of [0.40–0.43 V] [8]

11.5.3.2 Phase 2 Model Prediction: Importance Sampling Stage

For the importance sample points generated in Phase 2, we rely on the model developed in Phase 1 to predict the design writability. Figures 11.13 and 11.14 illustrate the results obtained for both designs; each experiment is associated with a different Vdd value as discussed earlier. It is obvious that the number of false predictions is very low given that the maximum recorded value is 22 false predictions out of 600 sample points for the Selective_Boost_WA design at $0.4V$. The average false prediction rate is reported to be around 4.5% for all the conducted experiments. This proves the efficiency of the proposed methodology. Note that no false negatives were observed for the Selective_Boost design experiments.

11.5.3.3 Yield Estimation and Convergence Analysis

The importance sample points response predicted in the previous section is used to estimate the yield. Importance sample weights are calculated according to Eq. (11.10) to predict the unbiased yield estimates. Figure 11.15 presents the

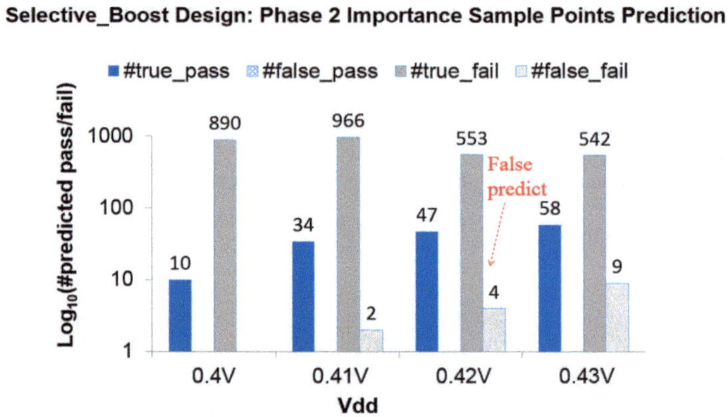

Fig. 11.13 Selective_Boost design Phase 2 pass/fail prediction for the importance sample points. No false passing (false negatives) points were reported for these experiments [8]

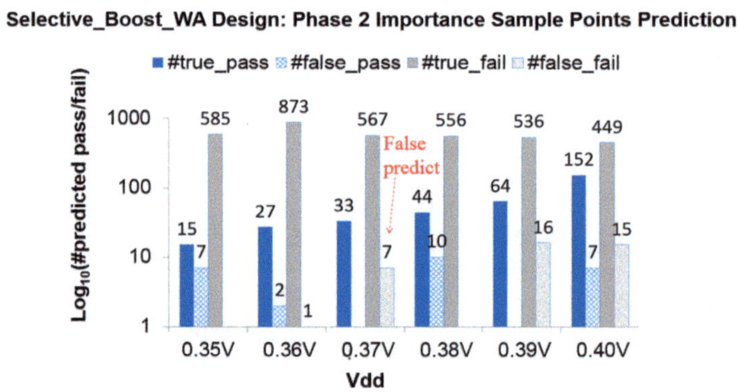

Fig. 11.14 Selective_Boost_WA design Phase 2 pass/fail prediction for the importance sample points [8]

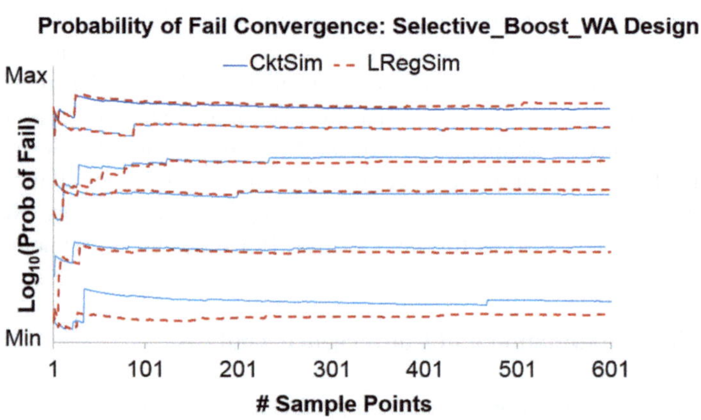

Fig. 11.15 Selective_Boost_WA design probability of fail convergence [8]

fail probability convergence for the Selective_Boost_WA design, and Fig. 11.16 presents the log plots for the different experiments for the proposed versus fully circuit-simulation-based approach. The results show excellent matching for the different ranges of estimated probability values of interest, including probabilities as low as 1e-12. This is further established in the yield plots of Figs. 11.17 and 11.18 where we note around 3% difference in the sigma yield estimates compared to the pure CktSim methodology. We therefore report a 50 mV improvement for the Selective_Boost_WA compared to the Selective_Boost methodology. These results corroborate well with the hardware data.

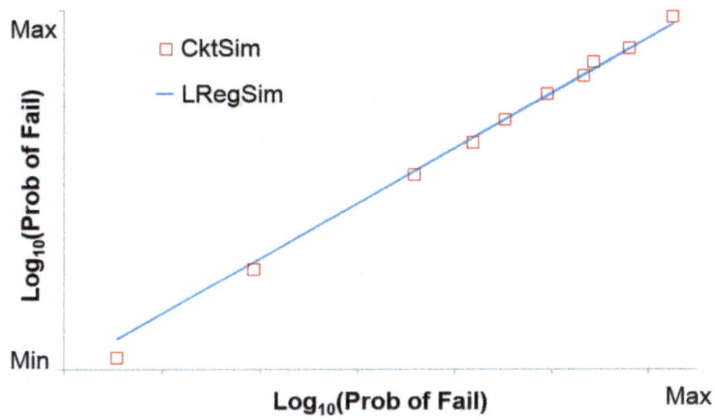

Fig. 11.16 Estimated probability of fail values is accurate for values as low as 1e−12. Results are normalized for max and min probabilities for all the experiments [8]

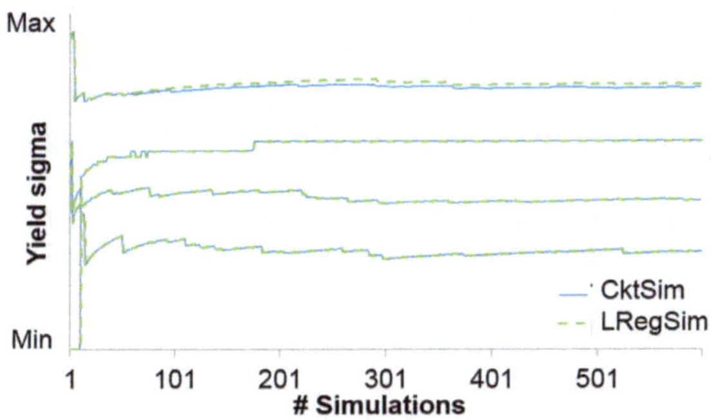

ΔYield (sigma)= True Yield - Estimated Yield	-0.005	0.03	0.04	0.06

Fig. 11.17 Selective_Boost design yield prediction. Results are normalized for max and min yield. True yield refers to CktSim approach, and estimated yield to LRegSim [8]

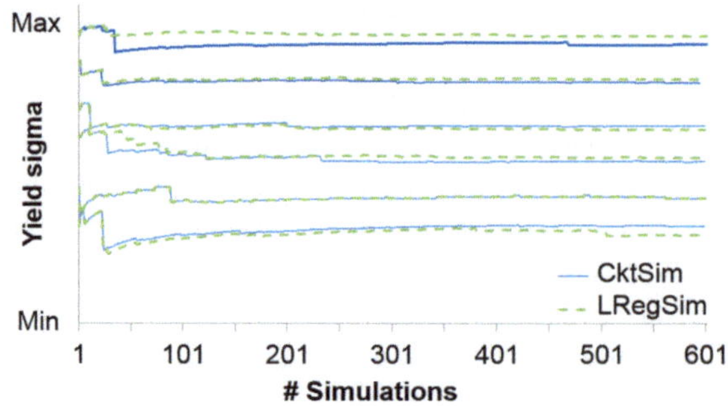

ΔYield (sigma)= True Yield - Estimated Yield	-0.01	-0.22	0.11	0.12

Fig. 11.18 Selective_Boost_WA design yield prediction. Results are normalized for max and min yield. True yield refers to CktSim approach, and estimated yield to LRegSim [8]

11.6 Conclusions

In this chapter, a logistic regression-based fast statistical analysis methodology is proposed for low fail probability estimation. It relies on a two-step importance sampling methodology to speed up simulations compared to traditional Monte Carlo. To further improve the simulation runtime, a regularized logistic regression model developed using the first stage sample points is used to predict the response of second stage sample points. The methodology is comprehensive and computationally efficient and can be applied to a wide variety of applications. We demonstrate the accuracy of the methodology on 14 nm FinFET SRAM designs.

Acknowledgements The authors would like to thank the Maroun Semaan Faculty of Engineering and Architecture at the American University of Beirut for supporting Ph.D. student Miss Lama Shaer.

References

1. X. Li, H. Liu, Statistical regression for efficient high-dimensional modeling of analog and mixed-signal performance variations, in *Design Automation Conference, 2008. DAC 2008. 45th ACM/IEEE* (IEEE, New York, 2008), pp. 38–43
2. K. Agarwal, S, Nassif, Statistical analysis of SRAM cell stability, in *Proceedings of the 43rd annual Design Automation Conference* (ACM, New York, 2006), pp. 57–62
3. S. Mukhopadhyay, H. Mahmoodi, K. Roy, Statistical design and optimization of SRAM cell for yield enhancement, in *Proceedings of the 2004 IEEE/ACM International Conference on Computer-Aided Design* (IEEE Computer Society, Washington, DC, 2004), pp. 10–13
4. A. Singhee, R.A. Rutenbar, Statistical blockade: a novel method for very fast Monte Carlo simulation of rare circuit events, and its application, in *Design, Automation, and Test in Europe* (Springer, New York, 2008), pp. 235–251

5. R. Kanj, R. Joshi, S. Nassif, Mixture importance sampling and its application to the analysis of SRAM designs in the presence of rare failure events, in *Design Automation Conference, 2006 43rd ACM/IEEE* (IEEE, New York, 2006), pp. 69–72
6. R. Kanj, T. Li, R. Joshi, K. Agarwal, A. Sadigh, D. Winston, S. Nassif, Accelerated statistical simulation via on-demand Hermite spline interpolations, in *Proceedings of the International Conference on Computer-Aided Design* (IEEE Press, New York, 2011), pp. 353–360
7. M. Malik, R.V. Joshi, R. Kanj, S. Sun, H. Homayoun, T. Li, Sparse regression driven mixture importance sampling for memory design, in *IEEE Transactions on Very Large Scale Integration (VLSI) Systems* (2017)
8. L. Shaer, R. Kanj, R. Joshi, M. Malik, A. Chehab, Regularized logistic regression for fast importance sampling based SRAM yield analysis, in *2017 18th International Symposium on Quality Electronic Design (ISQED)* (IEEE, New York, 2017), pp. 119–124
9. G.E.P. Box, J. Stuart Hunter, W. Gordon Hunter, *Statistics for Experimenters: Design, Innovation, and Discovery*, vol. 2 (Wiley, New York, 2005)
10. J.F. Ramaley, Buffon's noodle problem. Am. Math. Mon. **76**(8), 916–918 (1969)
11. N. Metropolis, S. Ulam, The Monte Carlo method. J. Am. Stat. Assoc. **44**(247), 335–341 (1949)
12. C.P. Robert. *Monte Carlo Methods* (Wiley Online Library, 2004)
13. T.C. Hesterberg, Advances in importance sampling, PhD Thesis, 1988
14. D.W Hosmer Jr., S. Lemeshow, R.X. Sturdivant, *Applied Logistic Regression*, vol. 398 (Wiley, New York, 2013)
15. A. Mojsilovic, A logistic regression model for small sample classification problems with hidden variables and non-linear relationships: an application in business analytics, in *IEEE International Conference on Acoustics, Speech, and Signal Processing, 2005. Proceedings (ICASSP'05)*, vol. 5 (IEEE, New York, 2005), pp. v–329
16. S. Ahmad, N.M. Ramli, H. Midi, Outlier detection in logistic regression and its application in medical data analysis, in *2012 IEEE Colloquium on Humanities, Science and Engineering (CHUSER)* (IEEE, New York, 2012), pp. 503–507
17. R.S. Collica, A logistic regression yield model for SRAM bit fail patterns, in *The IEEE International Workshop on Defect and Fault Tolerance in VLSI Systems, 1993* (IEEE, New York, 1993), pp. 127–135
18. X. Li, H. Liu, Statistical regression for efficient high-dimensional modeling of analog and mixed-signal performance variations, in *Design Automation Conference, 2008. DAC 2008. 45th ACM/IEEE* (IEEE, New York, 2008), pp. 38–43
19. A. Ng, Cs229 lecture notes (2017). Retrieved from http://cs229.stanford.edu/notes/cs229-notes-all/
20. X. Li, Finding deterministic solution from underdetermined equation: large-scale performance variability modeling of analog/RF circuits. IEEE Trans. Comput. Aided Des. Integr. Circuits Syst. **29**(11), 1661–1668 (2010)
21. D. Böhning, Multinomial logistic regression algorithm. Ann. Inst. Stat. Math. **44**(1), 197–200 (1992)
22. D.T, Larose, *Data Mining Methods & Models* (Wiley, New York, 2006)
23. T.P. Minka, A comparison of numerical optimizers for logistic regression, Unpublished draft (2003)
24. S.-I. Lee, H. Lee, P. Abbeel, A.Y. Ng, Efficient L1 regularized logistic regression, in *AAAI*, vol. 6 (2006), pp. 401–408
25. S. Perkins, J. Theiler, Online feature selection using grafting, in *Proceedings of the 20th International Conference on Machine Learning (ICML-03)* (2003), pp. 592–599
26. R. Kohavi et al., A study of cross-validation and bootstrap for accuracy estimation and model selection. in *Proceedings of International Joint Conference on Artificial Intelligence, IJCAI*, Stanford, CA, vol. 14 (1995), pp. 1137–1145
27. R.V. Joshi, M. Ziegler, H. Wetter, C. Wandel, H, Ainspan, 14nm finfet based supply voltage boosting techniques for extreme low v min operation, in *2015 Symposium on VLSI Circuits (VLSI Circuits)* (IEEE, Piscataway, 2015), pp. C268–C269

28. R.V. Joshi, M.M. Ziegler, Programmable supply boosting techniques for near threshold and wide operating voltage SRAM, in *2017 IEEE Custom Integrated Circuits Conference (CICC)* (IEEE, Piscataway, 2017), pp. 1–4
29. R. Joshi, R. Kanj, S. Nassif, D. Plass, Y. Chan et al., Statistical exploration of the dual supply voltage space of a 65nm PD/SOI CMOS SRAM cell, in *Proceeding of the 36th European Solid-State Device Research Conference, 2006. ESSDERC 2006* (IEEE, Piscataway, 2006), pp. 315–318

Chapter 12
Fast Statistical Analysis of Rare Circuit Failure Events

Jun Tao, Shupeng Sun, Xin Li, Hongzhou Liu, Kangsheng Luo, Ben Gu, and Xuan Zeng

12.1 Introduction

As integrated circuit (IC) technology advances, the ever increasing process variation has become a growing concern [5]. A complex IC, containing numerous memory components, is required to meet the design specification not only at the nominal process corner, but also under large-scale process variations. To achieve sufficiently high yield, the failure rate of each individual memory component must be extremely small. For instance, the failure rate of an SRAM bit-cell must be less than

J. Tao (✉) · X. Zeng (✉)
State Key Laboratory of ASIC and System, School of Microelectronics, Fudan University, Shanghai, China
e-mail: taojun@fudan.edu.cn; xzeng@fudan.edu.cn

S. Sun
Department of Electrical and Computer Engineering, Carnegie Mellon University, Pittsburgh, PA, USA
e-mail: shupengs@ece.cmu.edu

X. Li (✉)
Department of Electrical and Computer Engineering, Duke University, Durham, NC, USA
e-mail: xinli.ece@duke.edu

H. Liu · K. Luo
Cadence Design Systems, Inc., Pittsburgh, PA, USA
e-mail: hliu@cadence.com; ksluo@cadence.com

B. Gu
Cadence Design Systems, Inc., Austin, TX, USA
e-mail: gxin@cadence.com

I. M. Elfadel et al. (eds.), *Machine Learning in VLSI Computer-Aided Design*,
https://doi.org/10.1007/978-3-030-04666-8_12

$10^{-8}{\sim}10^{-6}$ for a typical SRAM design [2, 12]. Due to this reason, efficiently analyzing the rare failure event for the individual memory component becomes an important task for the IC design community.

The simple way to estimate the failure probability is to apply the well-known crude Monte Carlo (CMC) technique [3]. CMC directly draws random samples from the probability density function (PDF) that models device-level variations, and performs a transistor-level simulation to evaluate the performance value for each sample. When CMC is applied to estimate an extremely small failure rate (e.g., $10^{-8}{\sim}10^{-6}$), most random samples do not fall into the failure region. Hence, a large number of (e.g., $10^7{\sim}10^9$) samples are needed to accurately estimate the small failure probability, which implies that CMC can be extremely expensive for our application of rare failure rate estimation.

To improve the sampling efficiency, importance sampling (IS) methods have been proposed in the literature [7, 13, 15, 17, 20]. Instead of sampling the original PDF, IS samples a distorted PDF to get more samples in the important failure region. The efficiency achieved by IS highly depends on the choice of the distorted PDF. The traditional IS methods apply several heuristics to construct a distorted PDF that can capture the most important failure region in the variation space. Such a goal, though easy to achieve in a low-dimensional variation space, is extremely difficult to fulfill when a large number of random variables are used to model process variations.

Another approach to improving the sampling efficiency, referred to as statistical blockade, has recently been proposed [18]. This approach first builds a classifier with a number of transistor-level simulations, and then draws random samples from the original PDF. Unlike CMC where all the samples are evaluated by transistor-level simulations, statistical blockade only simulates the samples that are likely to fall into the failure region or close to the failure boundary based on the classifier. The efficiency achieved by this approach highly depends on the accuracy of the classifier. If the variation space is high-dimensional, a large number of transistor-level simulations are needed to build an accurate classifier, which makes the statistical blockade method quickly intractable.

In addition to the aforementioned statistical methods, several deterministic approaches have also been proposed to efficiently estimate the rare failure probability [10, 14]. These methods first find the failure boundary, and then calculate the failure probability by integrating the PDF over the failure region in the variation space. Though efficient in a low-dimensional variation space, it is often computationally expensive to accurately determine the failure boundary in a high-dimensional space especially if the boundary has a complicated shape (e.g., non-convex or even discontinuous).

Most of these traditional methods [7, 9, 10, 13–15, 17, 18, 20, 22, 23] have been successfully applied to SRAM bit-cells to estimate their rare failure rates where only a small number of (e.g., $6{\sim}20$) independent random variables are used to model process variations and, hence, the corresponding variation space is low-dimensional. It has been demonstrated in the literature that estimating the rare failure probability in a high-dimensional space (e.g., hundreds of independent random variables to model the device-level variations for SRAM) becomes increasingly important [21]. Unfortunately, such a high-dimensional problem cannot be efficiently handled by most traditional methods. It, in turn, poses an immediate need of developing a new CAD tool to accurately capture the rare failure events in a high-dimensional variation space with low computational cost.

To address this technical challenge, we first describe a novel subset simulation (SUS) technique. The key idea of SUS, borrowed from the statistics community [1, 6, 11], is to express the rare failure probability as the product of several large conditional probabilities by introducing a number of intermediate failure events. As such, the original problem of rare failure probability estimation is cast to an equivalent problem of estimating a sequence of conditional probabilities via multiple phases. Since these conditional probabilities are relatively large, they are substantially easier to estimate than the original rare failure rate.

When implementing the SUS method, it is difficult, if not impossible, to directly draw random samples from the conditional PDFs and estimate the conditional probabilities, since these conditional PDFs are unknown in advance. To address this issue, a modified Metropolis (MM) algorithm is adopted from the literature [1] to generate random samples by constructing a number of Markov chains. The conditional probabilities of interest are then estimated from these random samples. Unlike most traditional techniques [7, 9, 10, 13–15, 17, 18, 20, 22, 23] that suffer from the dimensionality issue, SUS can be efficiently applied to high-dimensional problems, which will be demonstrated by the experimental results in Sect. 12.2.

To define the intermediate failure events required by SUS, the performance of interest (PoI) must be continuous. In other words, SUS can only analyze a continuous PoI. For many rare failure events, however, PoIs are discrete (e.g., the output of a voltage-mode sense amplifier). Realizing this limitation, we further describe a scaled-sigma sampling (SSS) approach to efficiently estimate the rare failure rates for discrete PoIs in a high-dimensional space. SSS is particularly developed to address the following two fundamental questions: (1) how to efficiently draw random samples from the rare failure region, and (2) how to estimate the rare failure rate based on these random samples. Unlike CMC that directly samples the variation space and therefore only few samples fall into the failure region, SSS draws random samples from a distorted PDF for which the standard deviation (i.e., sigma) is scaled up. Conceptually, it is equivalent to increasing the magnitude of process variations. As a result, a large number of samples can now fall into the failure region. Once the distorted random samples are generated, an analytical model derived from the theorem of "soft maximum" is optimally fitted by applying maximum likelihood estimation (MLE). Next, the failure rate can be efficiently estimated from the fitted model.

The remainder of this chapter is organized as follows. In Sect. 12.2, we will summarize the SUS approach and, next, the SSS approach will be presented in Sect. 12.3. Finally, we conclude in Sect. 12.4.

12.2 Subset Simulation

Suppose that the vector

$$\mathbf{x} = \begin{bmatrix} x_1 & x_2 & \cdots & x_M \end{bmatrix}^T \tag{12.1}$$

is an M-dimensional random variable modeling device-level process variations. In a process design kit, the random variables $\{x_m; m = 1, 2, \ldots, M\}$ in (12.1) are typically modeled as a jointly Normal distribution [7, 9, 10, 13–15, 17, 18, 20, 22, 23]. Without loss of generality, we further assume that $\{x_m; m = 1, 2, \ldots, M\}$ are mutually independent and standard Normal (i.e., with zero mean and unit variance) and its joint PDF is

$$f(\mathbf{x}) = \prod_{m=1}^{M} p_m(x_m) = \prod_{m=1}^{M} \left[\frac{1}{\sqrt{2\pi}} \cdot \exp\left(-\frac{x_m^2}{2} \right) \right] = \frac{\exp\left(- \|\mathbf{x}\|_2^2 / 2 \right)}{\left(\sqrt{2\pi} \right)^M}, \tag{12.2}$$

where $p_m(x_m)$ is the 1-D PDF for x_m, and $\| \bullet \|_2$ denotes the L_2-norm of a vector. Any correlated random variables that are jointly Normal can be transformed to the independent random variables $\{x_m; m = 1, 2, \ldots, M\}$ by principal component analysis [3]. Then, the failure rate of a circuit can be mathematically represented as:

$$P_F = \Pr(\mathbf{x} \in \Omega) = \int_{\mathbf{x} \in \Omega} f(\mathbf{x}) \cdot d\mathbf{x}, \tag{12.3}$$

where Ω denotes the failure region, i.e., the subset of the variation space where the PoI does not meet the specification.

Instead of directly estimating the rare failure probability P_F, SUS expresses P_F as the product of several large conditional probabilities by introducing a number of intermediate failure events in the variation space. Without loss of generality, we define K intermediate failure events $\{\Omega_k; k = 1, 2, \ldots, K\}$ as:

$$\Omega_1 \supset \Omega_2 \supset \cdots \supset \Omega_{K-1} \supset \Omega_K = \Omega. \tag{12.4}$$

Based on (12.4), we can express P_F in (12.3) as:

$$P_F = \Pr(\mathbf{x} \in \Omega) = \Pr\left(\mathbf{x} \in \Omega_K, \mathbf{x} \in \Omega_{K-1}\right). \tag{12.5}$$

Equation (12.5) can be re-written as:

$$P_F = \Pr\left(\mathbf{x} \in \Omega_K \,\middle|\, \mathbf{x} \in \Omega_{K-1}\right) \cdot \Pr\left(\mathbf{x} \in \Omega_{K-1}\right). \tag{12.6}$$

Similarly, we can express $\Pr\left(\mathbf{x} \in \Omega_{K-1}\right)$ as:

$$\Pr\left(\mathbf{x} \in \Omega_{K-1}\right) = \Pr\left(\mathbf{x} \in \Omega_{K-1} \left| \mathbf{x} \in \Omega_{K-2}\right.\right) \cdot \Pr\left(\mathbf{x} \in \Omega_{K-2}\right). \tag{12.7}$$

From (12.4), (12.6), and (12.7), we can easily derive:

$$P_F = \Pr\left(\mathbf{x} \in \Omega_1\right) \cdot \prod_{k=2}^{K} \Pr\left(\mathbf{x} \in \Omega_k \left| \mathbf{x} \in \Omega_{k-1}\right.\right) = \prod_{k=1}^{K} P_k, \tag{12.8}$$

where

$$P_1 = \Pr\left(\mathbf{x} \in \Omega_1\right), \tag{12.9}$$

$$P_k = \Pr\left(\mathbf{x} \in \Omega_k \left| \mathbf{x} \in \Omega_{k-1}\right.\right) \quad (k = 2, 3, \ldots, K). \tag{12.10}$$

If $\{\Omega_k; k = 1, 2, \ldots, K\}$ are properly chosen, all the probabilities $\{P_k; k = 1, 2, \ldots, K\}$ are large and can be efficiently estimated. Once $\{P_k; k = 1, 2, \ldots, K\}$ are known, the rare failure probability P_F can be easily calculated by (12.8).

Note that the failure events $\{\Omega_k; k = 1, 2, \ldots, K\}$ are extremely difficult to specify in a high-dimensional variation space. For this reason, we do not directly define $\{\Omega_k; k = 1, 2, \ldots, K\}$ in the variation space. Instead, we utilize their corresponding subsets $\{F_k; k = 1, 2, \ldots, K\}$ in the performance space:

$$F_k = \left\{y(\mathbf{x}); \mathbf{x} \in \Omega_k\right\} \quad (k = 1, 2, \ldots, K), \tag{12.11}$$

where $y(\mathbf{x})$ denotes the PoI as a function of $\mathbf{x}$. Since $y(\mathbf{x})$ is typically a scalar, $\{F_k; k = 1, 2, \ldots, K\}$ are just one-dimensional subsets of $\mathbb{R}$ and, therefore, easy to be specified. Once $\{F_k; k = 1, 2, \ldots, K\}$ are determined, $\{\Omega_k; k = 1, 2, \ldots, K\}$ are implicitly known. For instance, to know whether a given $\mathbf{x}$ belongs to Ω_k, we first run a transistor-level simulation to evaluate $y(\mathbf{x})$. If $y(\mathbf{x})$ belongs to F_k, $\mathbf{x}$ is inside Ω_k. Otherwise, $\mathbf{x}$ is outside Ω_k.

In what follows, we will use a simple 2-D example to intuitively illustrate the basic flow of SUS. Figure 12.1 shows this 2-D example where two random variables $\mathbf{x} = [x_1\ x_2]^T$ are used to model the device-level process variations, and Ω_1 and Ω_2 denote the first two subsets in (12.4). Note that Ω_1 and Ω_2 are depicted for illustration purposes in this example. In practice, we do not need to explicitly know Ω_1 and Ω_2, as previously explained.

Our objective is to estimate the probabilities $\{P_k; k = 1, 2, \ldots, K\}$ via multiple phases. Starting from the 1st phase, we simply draw L_1 independent random samples $\{\mathbf{x}^{(1,l)}; l = 1, 2, \ldots, L_1\}$ from the PDF $f(\mathbf{x})$ to estimate P_1. Here, the superscript "1" of the symbol $\mathbf{x}^{(1,l)}$ refers to the 1st phase. Among these L_1 samples, we identify a subset of samples $\{\mathbf{x}_F^{(1,t)}; t = 1, 2, \ldots, T_1\}$ that fall into Ω_1, where T_1 denotes the total number of samples in this subset. As shown in Fig. 12.1(a), the red

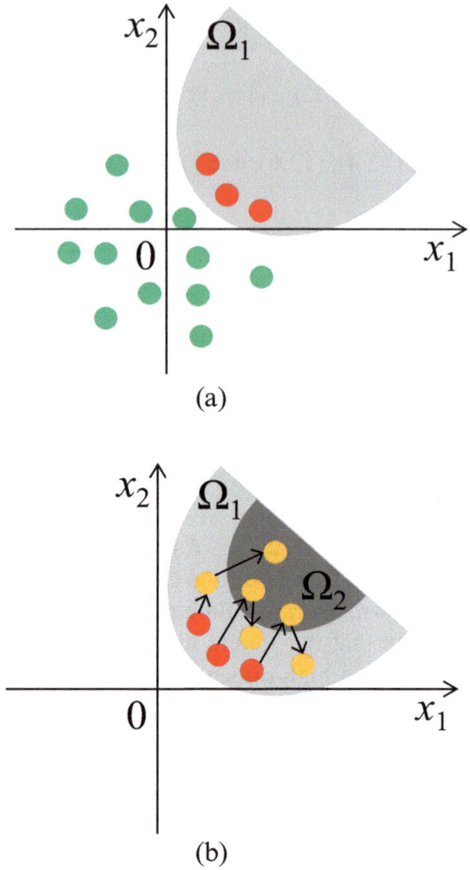

Fig. 12.1 A 2-D example is used to illustrate the procedure of probability estimation via multiple phases by using SUS: (**a**) generating MC samples and estimating P_1 in the 1st phase, and (**b**) generating MCMC samples and estimating P_2 in the 2nd phase

points represent the samples that belong to Ω_1 and the green points represent the samples that are out of Ω_1. In this case, P_1 can be estimated as:

$$P_1^{\text{SUS}} = \frac{1}{L_1} \cdot \sum_{l=1}^{L_1} I_{\Omega_1}\left[\mathbf{x}^{(1,l)}\right] = \frac{T_1}{L_1}, \tag{12.12}$$

where P_1^{SUS} denotes the estimated value of P_1, and $I_{\Omega_1}(\mathbf{x})$ represents the indicator function

$$I_{\Omega_1}(\mathbf{x}) = \begin{cases} 1 & \mathbf{x} \in \Omega_1 \\ 0 & \mathbf{x} \notin \Omega_1 \end{cases}. \tag{12.13}$$

If P_1 is large, it can be accurately estimated with a small number of random samples (e.g., L_1 is around $10^2 \sim 10^3$).

Next, in the 2nd phase, we need to estimate the conditional probability $P_2 = \Pr(\mathbf{x} \in \Omega_2 | \mathbf{x} \in \Omega_1)$. Towards this goal, one simple idea is to directly draw random samples from the conditional PDF $f(\mathbf{x}|\mathbf{x} \in \Omega_1)$ and then compute the mean of the indicator function $I_{\Omega_2}(\mathbf{x})$

$$I_{\Omega_2}(\mathbf{x}) = \begin{cases} 1 & \mathbf{x} \in \Omega_2 \\ 0 & \mathbf{x} \notin \Omega_2 \end{cases}. \tag{12.14}$$

This approach, however, is practically infeasible since $f(\mathbf{x}|\mathbf{x} \in \Omega_1)$ is unknown in advance. To address this issue, we apply a modified Metropolis (MM) algorithm [1] to generate a set of random samples that follow the conditional PDF $f(\mathbf{x}|\mathbf{x} \in \Omega_1)$.

MM is a Markov chain Monte Carlo (MCMC) technique [3]. Starting from each of the samples $\{\mathbf{x}_F^{(1,t)}; t = 1, 2, \ldots, T_1\}$ that fall into Ω_1 in the 1st phase, MM generates a sequence of samples that form a Markov chain. In other words, there are T_1 independently generated Markov chains in total and $\mathbf{x}_F^{(1,t)}$ is the 1st sample of the t-th Markov chain. To clearly explain the MM algorithm, we define the symbol $\mathbf{x}^{(2,t,1)} = \mathbf{x}_F^{(1,t)}$, where $t \in \{1, 2, \ldots, T_1\}$. The superscripts "2" and "1" of $\mathbf{x}^{(2,t,1)}$ refer to the 2nd phase and the 1st sample of the Markov chain, respectively.

For our 2-D example, we start from $\mathbf{x}^{(2,1,1)} = [x_1^{(2,1,1)}\ x_2^{(2,1,1)}]^T$ to form the 1st Markov chain. To generate the 2nd sample $\mathbf{x}^{(2,1,2)}$ from $\mathbf{x}^{(2,1,1)}$, we first randomly sample a new value x_1^{NEW} from a 1-D transition PDF $q_1[x_1^{NEW}|x_1^{(2,1,1)}]$ that must satisfy the following condition [1]:

$$q_1\left[x_1^{NEW}\middle|x_1^{(2,1,1)}\right] = q_1\left[x_1^{(2,1,1)}\middle|x_1^{NEW}\right] \tag{12.15}$$

There are many possible ways to define $q_1[x_1^{NEW}|x_1^{(2,1,1)}]$ in (12.15) [1]. For example, a 1-D Normal PDF can be used

$$q_1\left[x_1^{NEW}\middle|x_1^{(2,1,1)}\right] = \frac{1}{\sqrt{2\pi}\cdot\sigma_1}\cdot\exp\left\{-\frac{\left[x_1^{NEW} - x_1^{(2,1,1)}\right]^2}{2\cdot\sigma_1^2}\right\}. \tag{12.16}$$

where $x_1^{(2,1,1)}$ and σ_1 are the mean and standard deviation of the distribution, respectively. Here, σ_1 is a parameter that usually be empirically chosen[19].

Next, we compute the ratio

$$r = \frac{p_1\left(x_1^{NEW}\right)}{p_1\left(x_1^{(2,1,1)}\right)}, \tag{12.17}$$

where $p_1(x_1)$ is the original PDF of the random variable x_1 shown in (12.2). A random sample u is then drawn from a 1-D uniform distribution with the following PDF:

$$f(u) = \begin{cases} 1 & 0 \le u \le 1 \\ 0 & \text{Otherwise} \end{cases}, \tag{12.18}$$

and the value of $x_1^{(2,1,2)}$ is set as

$$x_1^{(2,1,2)} = \begin{cases} x_1^{NEW} & u \le \min(1, r) \\ x_1^{(2,1,1)} & u > \min(1, r) \end{cases}. \tag{12.19}$$

A similar procedure is applied to generate $x_2^{(2,1,2)}$. Once $x_1^{(2,1,2)}$ and $x_2^{(2,1,2)}$ are determined, we form a candidate $\mathbf{x}^{NEW} = [x_1^{(2,1,2)}\ x_2^{(2,1,2)}]^T$ and use it to create the sample $\mathbf{x}^{(2,1,2)}$

$$\mathbf{x}^{(2,1,2)} = \begin{cases} \mathbf{x}^{NEW} & \mathbf{x}^{NEW} \in \Omega_1 \\ \mathbf{x}^{(2,1,1)} & \mathbf{x}^{NEW} \notin \Omega_1 \end{cases}. \tag{12.20}$$

By repeating the aforementioned steps, we can create other samples to complete the Markov chain $\{\mathbf{x}^{(2,1,l)}; l = 1, 2, \ldots, L_2\}$, where L_2 denotes the length of the Markov chain in the 2nd phase. In addition, all other Markov chains can be similarly formed. Since there are T_1 Markov chains and each Markov chain contains L_2 samples, the total number of the MCMC samples is $T_1 \cdot L_2$ for the 2nd phase. Figure 12.1(b) shows the sampling results in the 2nd phase for our 2-D example. In Fig. 12.1(b), the red points represent the initial samples $\{\mathbf{x}^{(2,t,1)}; t = 1, 2, \ldots, T_1\}$ of the Markov chains and they are obtained from the 1st phase. The yellow points represent the MCMC samples created via the MM algorithm in the 2nd phase. It has been proved in [1] that all these MCMC samples $\{\mathbf{x}^{(2,t,l)}; t = 1, 2, \ldots, T_1; l = 1, 2, \ldots, L_2\}$ in Fig. 12.1(b) approximately follow $f(\mathbf{x}|\mathbf{x} \in \Omega_1)$. In other words, we have successfully generated a number of random samples that follow our desired distribution for the 2nd phase.

Among all the MCMC samples $\{\mathbf{x}^{(2,t,l)}; t = 1, 2, \ldots, T_1; l = 1, 2, \ldots, L_2\}$, we further identify a subset of samples $\{\mathbf{x}_F^{(2,t)}; t = 1, 2, \ldots, T_2\}$ that fall into Ω_2, where T_2 denotes the total number of the samples in this subset. The conditional probability P_2 can be estimated as:

$$P_2^{\text{SUS}} = \frac{1}{T_1 \cdot L_2} \cdot \sum_{t=1}^{T_1} \sum_{l=1}^{L_2} I_{\Omega_2}\left[\mathbf{x}^{(2,t,l)}\right] = \frac{T_2}{T_1 \cdot L_2}, \tag{12.21}$$

where P_2^{SUS} denotes the estimated value of P_2.

By following the aforementioned idea, we can estimate all the probabilities $\{P_k; k = 1, 2, \ldots, K\}$. Once the values of $\{P_k; k = 1, 2, \ldots, K\}$ are estimated, the rare failure rate P_F is calculated by

$$P_F^{SUS} = \prod_{k=1}^{K} P_k^{SUS}, \tag{12.22}$$

where P_F^{SUS} represents the estimated value of P_F by using SUS. If we have more than two random variables, estimating the probabilities $\{P_k; k = 1, 2, \ldots, K\}$ can be pursued in a similar way [19].

To efficiently apply SUS, we must carefully choose the subset $\{F_k; k = 1, 2, \ldots, K\}$ so that the probability P_k will be close to 0.1, where $k \in \{1, 2, \ldots, K\}$. In this case, even if the failure rate P_F is extremely small (e.g., $10^{-8}{\sim}10^{-6}$), SUS only needs a small number of (e.g., $K = 6{\sim}8$) phases to complete. Furthermore, it only requires a few hundred samples during each phase to accurately estimate the probability P_k.

In addition, to quantitatively assess the accuracy of the proposed SUS estimator, we must estimate its confidence interval (CI). To this end, we need to know the distribution of P_F^{SUS}. Since P_F^{SUS} is equal to the multiplication of $\{P_k^{SUS}; k = 1, 2, \ldots, K\}$, we must carefully study the statistical property of P_k^{SUS} in order to derive the distribution for P_F^{SUS}.

To be specific, P_1^{SUS} is calculated by using (12.12) with L_1 independent and identically distributed (i.i.d.) samples drawn from $f(\mathbf{x})$. Hence, according to the central limit theorem (CLT) [16], P_1^{SUS} approximately follows a Normal distribution

$$P_1^{SUS} \sim N(P_1, v_1), \tag{12.23}$$

where the mean value P_1 is defined in (12.9) and the variance value v_1 can be approximated as [16]

$$v_1 \approx \frac{1}{L_1} \cdot P_1^{SUS} \cdot \left(1 - P_1^{SUS}\right). \tag{12.24}$$

On the other hand, the conditional probability P_k^{SUS}, where $k \in \{2, 3, \ldots, K\}$, can be estimated by using the MCMC samples $\{\mathbf{x}^{(k,t,l)}; t = 1, 2, \ldots, T_{k-1}; l = 1, 2, \ldots, L_k\}$ created by MM:

$$P_k^{SUS} = \frac{1}{T_{k-1} \cdot L_k} \cdot \sum_{t=1}^{T_{k-1}} \sum_{l=1}^{L_k} I_{\Omega_k}\left[\mathbf{x}^{(k,t,l)}\right], \tag{12.25}$$

where $I_{\Omega_k}[\mathbf{x}]$ represents the indicator function

$$I_{\Omega_k}(\mathbf{x}) = \begin{cases} 1 & \mathbf{x} \in \Omega_k \\ 0 & \mathbf{x} \notin \Omega_k \end{cases}. \tag{12.26}$$

Since the MCMC samples $\{\mathbf{x}^{(k,t,l)}; l = 1, 2, \ldots, L_k\}$, where $t \in \{1, 2, \ldots, T_{k-1}\}$, are strongly correlated, they cannot be considered as i.i.d. samples. For this reason, we cannot directly apply CLT to derive the distribution for the estimator P_k^{SUS} in (12.25).

To address this issue, we define a set of new random variables

$$s^{(k,t)} = \frac{1}{L_k} \cdot \sum_{l=1}^{L_k} I_{\Omega_k}\left[\mathbf{x}^{(k,t,l)}\right], \tag{12.27}$$

where $t \in \{1, 2, \ldots, T_{k-1}\}$. Studying (12.27) reveals two important observations. First, $s^{(k,t)}$ only depends on the t-th Markov chain $\{\mathbf{x}^{(k,t,l)}; l = 1, 2, \ldots, L_k\}$. Since different Markov chains are created from different initial samples $\{\mathbf{x}^{(k,t,1)}; t = 1, 2, \ldots, T_{k-1}\}$, the random variables $\{s^{(k,t)}; t = 1, 2, \ldots, T_{k-1}\}$ are almost statistically independent. Second, since all initial samples $\{\mathbf{x}^{(k,t,1)}; t = 1, 2, \ldots, T_{k-1}\}$ follow the same conditional PDF $p(\mathbf{x} \left| \mathbf{x} \in \Omega_{k-1}\right.)$ and all the Markov chains are generated by following the same procedure, the random variables $\{s^{(k,t)}; t = 1, 2, \ldots, T_{k-1}\}$ must be identically distributed. For these reasons, we can consider $\{s^{(k,t)}; t = 1, 2, \ldots, T_{k-1}\}$ as a set of i.i.d. random variables.

Based on (12.27), P_k^{SUS} in (12.25), where $k \in \{2, 3, \ldots, K\}$, can be re-written as

$$P_k^{\mathrm{SUS}} = \frac{1}{T_{k-1}} \cdot \sum_{t=1}^{T_{k-1}} s^{(k,t)} \tag{12.28}$$

and, as a result, approximately follows a Normal distribution according to CLT:

$$P_k^{\mathrm{SUS}} \sim N\left(P_k, v_k\right), \tag{12.29}$$

where P_k is defined in (12.10) and

$$v_k \approx \frac{1}{\left(T_{k-1} - 1\right) \cdot T_{k-1}} \cdot \sum_{t=1}^{T_{k-1}} \left[s^{(k,t)} - P_k^{\mathrm{SUS}}\right]^2. \tag{12.30}$$

To further derive the distribution for P_F^{SUS} in (12.22) based on (12.23) and (12.29), we take logarithm on both sides of (12.22) because it is much easier to handle summation than multiplication

$$\log\left(P_F^{\mathrm{SUS}}\right) = \sum_{k=1}^{K} \log\left(P_k^{\mathrm{SUS}}\right). \tag{12.31}$$

To derive the distribution of $\{log(P_k^{SUS}); k = 1, 2, \ldots, K\}$, we approximate the nonlinear function $\log(\bullet)$ by the first-order Taylor expansion around the mean value

P_k of the random variable P_k^{SUS}:

$$\log\left(P_k^{\mathrm{SUS}}\right) \approx \log\left(P_k\right) + \frac{P_k^{\mathrm{SUS}} - P_k}{P_k} \approx \log\left(P_k\right) + \frac{P_k^{\mathrm{SUS}} - P_k}{P_k^{\mathrm{SUS}}}. \tag{12.32}$$

According to the linear approximation in (12.32), $\log(P_k^{\mathrm{SUS}})$ follows a Normal distribution

$$\log\left(P_k^{\mathrm{SUS}}\right) \sim N\left[\log\left(P_k\right), v_{\log,k}\right], \tag{12.33}$$

where

$$v_{\log,k} = \frac{v_k}{\left(P_k^{\mathrm{SUS}}\right)^2}, \tag{12.34}$$

and $k \in \{1, 2, \ldots, K\}$.

Since $\log(P_F^{\mathrm{SUS}})$ is the summation of several "approximately" Normal random variables $\{\log(P_k^{\mathrm{SUS}}); k = 1, 2, \ldots, K\}$, $\log(P_F^{\mathrm{SUS}})$ also approximately follows a Normal distribution [16]

$$\log\left(P_F^{\mathrm{SUS}}\right) \sim N\left\{\mathrm{MEAN}\left[\log\left(P_F^{\mathrm{SUS}}\right)\right], \mathrm{VAR}\left[\log\left(P_F^{\mathrm{SUS}}\right)\right]\right\} \tag{12.35}$$

Based on (12.8), (12.31), and (12.33), $\mathrm{MEAN}[\log(P_F^{\mathrm{SUS}})]$ can be expressed as

$$\mathrm{MEAN}\left[\log\left(P_F^{\mathrm{SUS}}\right)\right] = \sum_{k=1}^{K} \log\left(P_k\right) = \log\left(\prod_{k=1}^{K} P_k\right) = \log\left(P_F\right), \tag{12.36}$$

and $\mathrm{VAR}[\log(P_F^{\mathrm{SUS}})]$ can be calculated as

$$\begin{aligned} \mathrm{VAR}\left[\log\left(P_F^{\mathrm{SUS}}\right)\right] &= \mathrm{VAR}\left[\sum_{k=1}^{K} \log\left(P_k^{\mathrm{SUS}}\right)\right] \\ &= \sum_{k=1}^{K} v_{\log,k} + 2 \cdot \sum_{i=1}^{K-1} \sum_{j=i+1}^{K} \mathrm{COV}\left[\log\left(P_i^{\mathrm{SUS}}\right), \log\left(P_j^{\mathrm{SUS}}\right)\right] \end{aligned}, \tag{12.37}$$

where $\mathrm{COV}(\bullet, \bullet)$ denotes the covariance of two random variables.

When applying MCMC, we often observe that an MCMC sample is strongly correlated to its adjacent sample. However, the correlation quickly decreases as the distance between two MCMC samples increases. Therefore, we can assume that the samples used to estimate $\log(P_i^{\text{SUS}})$ are weakly correlated to the samples used to estimate $\log(P_j^{\text{SUS}})$, if the distance between i and j is greater than 1 (i.e., $|i - j| > 1$). Based on this assumption, (12.37) can be approximated as

$$\text{VAR}\left[\log\left(P_F^{\text{SUS}}\right)\right] \approx \sum_{k=1}^{K} v_{\log,k} + 2 \cdot \sum_{k=1}^{K-1} \text{COV}\left[\log\left(P_k^{\text{SUS}}\right), \log\left(P_{k+1}^{\text{SUS}}\right)\right]. \tag{12.38}$$

Accurately estimating the covariance between $\log(P_k^{\text{SUS}})$ and $\log(P_{k+1}^{\text{SUS}})$ is nontrivial. Here, we derive an upper bound for $\text{COV}[\log(P_k^{\text{SUS}}), \log(P_{k+1}^{\text{SUS}})]$ [16]:

$$\text{COV}\left[\log\left(P_k^{\text{SUS}}\right), \log\left(P_{k+1}^{\text{SUS}}\right)\right] \leq \sqrt{v_{\log,k} \cdot v_{\log,k+1}}, \tag{12.39}$$

where $k \in \{1, 2, \ldots, K - 1\}$. Substituting (12.39) into (12.38) yields

$$\text{VAR}\left[\log\left(P_F^{\text{SUS}}\right)\right] \leq \sum_{k=1}^{K} v_{\log,k} + 2 \cdot \sum_{k=1}^{K-1} \sqrt{v_{\log,k} \cdot v_{\log,k+1}} = v_{\log,\text{SUS}}. \tag{12.40}$$

In this chapter, we approximate $\text{VAR}[\log(P_F^{\text{SUS}})]$ by its upper bound $v_{\log,\text{SUS}}$ defined in (12.40) to provide a conservative estimation for the CI. Based on (12.36) and (12.40), (12.35) can be re-written as

$$\log\left(P_F^{\text{SUS}}\right) \sim N\left[\log\left(P_F\right), v_{\log,\text{SUS}}\right]. \tag{12.41}$$

According to (12.41), we can derive the CI for any given confidence level. For instance, the 95% CI is expressed as

$$\left[\exp\left(\log(P_F^{\text{SUS}}) - 1.96 \cdot \sqrt{v_{\log,\text{SUS}}}\right), \exp\left(\log(P_F^{\text{SUS}}) + 1.96 \cdot \sqrt{v_{\log,\text{SUS}}}\right)\right]. \tag{12.42}$$

To demonstrate the efficacy of SUS, we consider an SRAM column example designed in a 45 nm CMOS process, as shown in Fig. 12.2. In this example, our PoI is the read current I_{READ}, which is defined as the difference between the bit-line currents I_{BL} and $I_{\text{BL\_}}$ (i.e., $I_{\text{READ}} = I_{\text{BL}} - I_{\text{BL\_}}$) when we start to read CELL <0>. If I_{READ} is greater than a pre-defined specification, we consider the SRAM circuit as "PASS". For process variation modeling, the local V_{TH} mismatch of each transistor is considered as an independent Normal random variable. In total, we have 384 independent random variables (i.e., 64 bit-cells × 6 transistors per bit-cell = 384).

We first run CMC with 10^9 random samples, and the estimated failure rate is 1.1×10^{-6}, which is considered as the "golden" failure rate in this example. Next, we compare SUS with the traditional importance sampling technique: MNIS [17], where 2000 simulations are used to construct the distorted PDF. We repeatedly run

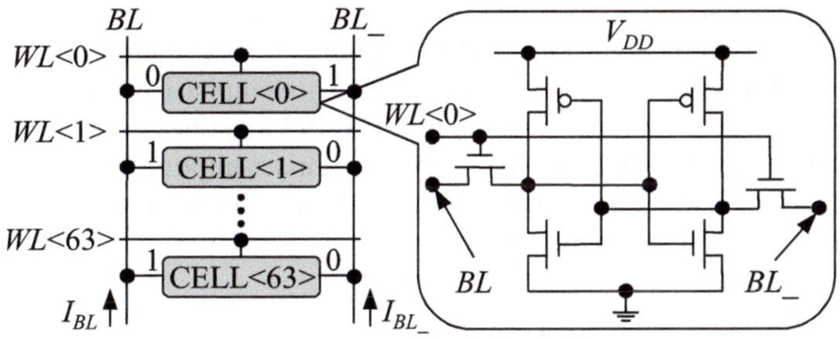

Fig. 12.2 The simplified schematic is shown for an SRAM column consisting of 64 bit-cells designed in a 45 nm CMOS process

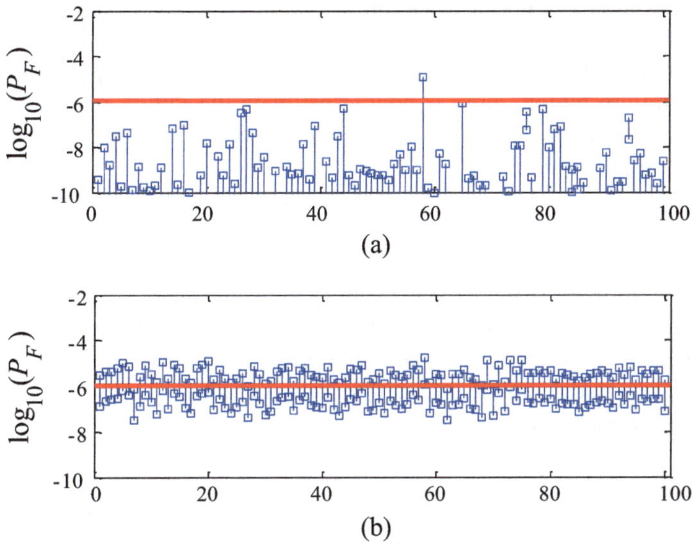

Fig. 12.3 The 95% confidence intervals (blue bars) of the SRAM read current example are estimated from 100 repeated runs with 6000 transistor-level simulations in each run for: (**a**) MNIS and (**b**) SUS. The red line represents the "golden" failure rate

MNIS and SUS for 100 times with 6000 transistor-level simulations in each run. Figure 12.3 shows the 100 estimated 95% CIs for each method, where each blue bar represents the CI of a single run, and the red line represents the "golden" failure rate.

In this example, only a single CI estimated from 100 repeated runs by MNIS can cover the "golden" failure rate, implying that MNIS fails to estimate the CIs accurately. This is an important limitation of MNIS, and generally most of the importance sampling techniques, since the user cannot reliably know the actual "confidence" of the estimator in practice. For the SUS approach, however, there are 95 CIs out of 100 runs that cover the "golden" failure rate. More importantly, the CIs estimated by SUS are relatively tight, which implies that SUS achieves substantially better accuracy than the traditional MNIS approach in this example.

Before ending this section, we would like to emphasize that to define the subsets $\{F_k; k = 1, 2, \ldots, K\}$ required by SUS, PoI must be continuous. Realizing this limitation, we further describe a scaled-sigma sampling (SSS) approach to efficiently estimate the rare failure rates for discrete PoIs in a high-dimensional space, which will be presented in the next section.

12.3 Scaled-Sigma Sampling

Unlike the traditional importance sampling methods that must explicitly identify the high-probability failure region, SSS takes a completely different strategy to address the following questions: (1) how to efficiently draw random samples from the high-probability failure region, and (2) how to estimate the failure rate based on these random samples. In what follows, we will derive the mathematical formulation of SSS and highlight its novelties.

For the application of rare failure rate estimation, a failure event often occurs at the tail of the PDF $f(\mathbf{x})$. Given (12.2), it implies that the failure region Ω is far away from the origin $\mathbf{x} = \mathbf{0}$, as shown in Fig. 12.4(a). Since the failure rate is extremely small, the traditional CMC analysis cannot efficiently draw random samples from the failure region. Namely, many samples cannot reach the tail of $f(\mathbf{x})$.

To address the aforementioned sampling issue, SSS applies a simple idea. Given $f(\mathbf{x})$ in (12.2), we scale up the standard deviation of $\mathbf{x}$ by *a scaling factor s* $(s > 1)$, yielding the following distribution:

$$g(\mathbf{x}) = \prod_{m=1}^{M} \left[\frac{\exp\left(-x_m^2 / 2s^2\right)}{\sqrt{2\pi} s} \right] = \frac{\exp\left(-\|\mathbf{x}\|_2^2 / 2s^2\right)}{\left(\sqrt{2\pi} \cdot s\right)^M}. \tag{12.43}$$

Once the standard deviation of $\mathbf{x}$ is increased by a factor of s, we conceptually increase the magnitude of process variations. Hence, the PDF $g(\mathbf{x})$ widely spreads over a large region and the probability for a random sample to reach the far-away failure region increases, as shown in Fig. 12.4(b).

It is important to note that the mean of the scaled PDF $g(\mathbf{x})$ remains $\mathbf{0}$, which is identical to the mean of the original PDF $f(\mathbf{x})$. Hence, for a given sampling location $\mathbf{x}$, the likelihood defined by the scaled PDF $g(\mathbf{x})$ remains inversely proportional to the length of the vector $\mathbf{x}$ (i.e., $\|\mathbf{x}\|_2$). Namely, it is more (or less) likely to reach the sampling location $\mathbf{x}$, if the distance between the location $\mathbf{x}$ and the origin $\mathbf{0}$ is smaller (or larger). It, in turn, implies that the high-probability failure region associated with the original PDF $f(\mathbf{x})$ remains the high-probability failure region after the PDF is scaled to $g(\mathbf{x})$, as shown in Fig. 12.4(a) and (b). Scaling the PDF from $f(\mathbf{x})$ to $g(\mathbf{x})$

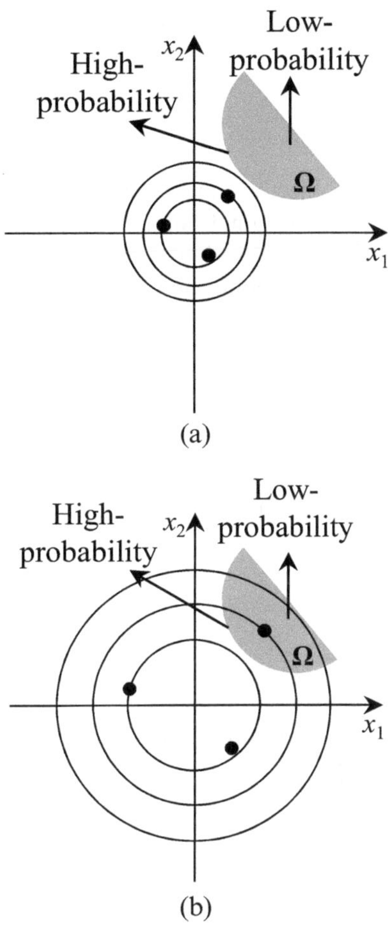

Fig. 12.4 The proposed SSS is illustrated by a 2-D example where the grey area Ω denotes the failure region and the circles represent the contour lines of the PDF. (**a**) Rare failure events occur at the tail of the original PDF $f(\mathbf{x})$ and the failure region is far away from the origin $\mathbf{x} = \mathbf{0}$. (**b**) The scaled PDF $g(\mathbf{x})$ widely spreads over a large region and the scaled samples are likely to reach the far-away failure region

does not change the location of the high-probability failure region; instead, it only makes the failure region easy to sample.

Once the scaled random samples are drawn from $g(\mathbf{x})$ in (12.43), we need to further estimate the failure rate P_F defined in (12.3). To this end, one straightforward way is to apply the importance sampling method [3]. Such a simple approach, however, has been proved to be intractable when the dimensionality (i.e., M) of the variation space is high [21]. Namely, it does not fit the need of high-dimensional failure rate estimation in this chapter.

Instead of relying on the theory of importance sampling, SSS attempts to estimate the failure rate P_F from a completely different avenue. We first take a look at the "scaled" failure rate P_G corresponding to $g(\mathbf{x})$:

$$P_G = \int_{\mathbf{x}\in\Omega} g(\mathbf{x}) \cdot d\mathbf{x} = \int_{-\infty}^{+\infty} I_\Omega(\mathbf{x}) \cdot g(\mathbf{x}) \cdot d\mathbf{x}, \tag{12.44}$$

where $I_\Omega(\mathbf{x})$ represents the indicator function:

$$I_\Omega(\mathbf{x}) = \begin{cases} 1 & \mathbf{x} \in \Omega \\ 0 & \mathbf{x} \notin \Omega \end{cases}. \tag{12.45}$$

Our objective is to study the relation between the scaled failure rate P_G in (12.44) and the original failure rate P_F in (12.3). Towards this goal, we partition the M-dimensional variation space into a large number of identical hyper-rectangles with the same volume and the scaled failure rate P_G in (12.44) can be approximated as:

$$P_G \approx \sum_k I_\Omega\left[\mathbf{x}^{(k)}\right] \cdot g\left[\mathbf{x}^{(k)}\right] \cdot \Delta\mathbf{x}, \tag{12.46}$$

where $\Delta\mathbf{x}$ denotes the volume of a hyper-rectangle. The approximation in (12.46) is accurate, if each hyper-rectangle is sufficiently small. Given the definition of $I_\Omega(\mathbf{x})$ in (12.45), Eq. (12.46) can be re-written as:

$$P_G \approx \sum_{k \in \Omega} g\left[\mathbf{x}^{(k)}\right] \cdot \Delta\mathbf{x}, \tag{12.47}$$

where $\{k; k \in \Omega\}$ represents the set of all hyper-rectangles that fall into the failure region.

Substituting (12.43) into (12.47), we have

$$P_G \approx \frac{\Delta\mathbf{x}}{\left(\sqrt{2\pi} \cdot s\right)^M} \cdot \sum_{k \in \Omega} \exp\left[-\frac{\left\|\mathbf{x}^{(k)}\right\|_2^2}{2s^2}\right]. \tag{12.48}$$

Taking the logarithm on both sides of (12.48) yields:

$$\log P_G \approx \log \frac{\Delta\mathbf{x}}{(2\pi)^{M/2}} - M \cdot \log s + \operatorname*{lse}_{k \in \Omega}\left[-\frac{\left\|\mathbf{x}^{(k)}\right\|_2^2}{2s^2}\right], \tag{12.49}$$

where

$$\operatorname*{lse}_{k \in \Omega}\left[\frac{-\left\|\mathbf{x}^{(k)}\right\|_2^2}{2s^2}\right] = \log\left\{\sum_{k \in \Omega} \exp\left[-\frac{\left\|\mathbf{x}^{(k)}\right\|_2^2}{2s^2}\right]\right\} \tag{12.50}$$

stands for the log-sum-exp function. The function lse(•) in (12.50) is also known as the "soft maximum" from the mathematics [4]. It can be bounded by

$$\max_{k\in\Omega}\left[-\frac{\left\|\mathbf{x}^{(k)}\right\|_2^2}{2s^2}\right]+\log(T)\geq \operatorname*{lse}_{k\in\Omega}\left[-\frac{\left\|\mathbf{x}^{(k)}\right\|_2^2}{2s^2}\right]\geq \max_{k\in\Omega}\left[-\frac{\left\|\mathbf{x}^{(k)}\right\|_2^2}{2s^2}\right], \tag{12.51}$$

where T denotes the total number of hyper-rectangles in Ω.

In general, there exist a number of (say, T_0) dominant hyper-rectangles that are much closer to the origin $\mathbf{0}$ than other hyper-rectangles in the set $\{\mathbf{x}^{(k)}; k \in \Omega\}$. Without loss of generality, we assume that the first T_0 hyper-rectangles $\{\mathbf{x}^{(k)}; k = 1, 2, \ldots, T_0\}$ are dominant. Hence, we can approximate the function lse(•) in (12.50) as

$$\operatorname*{lse}_{k\in\Omega}\left[-\frac{\left\|\mathbf{x}^{(k)}\right\|_2^2}{2s^2}\right]\approx \log\left\{\sum_{k=1}^{T_0}\exp\left[-\frac{\left\|\mathbf{x}^{(k)}\right\|_2^2}{2s^2}\right]\right\}. \tag{12.52}$$

We further assume that these dominant hyper-rectangles $\{\mathbf{x}^{(k)}; k = 1, 2, \ldots, T_0\}$ have similar distances to the origin $\mathbf{0}$. Thus, Eq. (12.52) can be approximated by

$$\operatorname*{lse}_{k\in\Omega}\left[\frac{-\left\|\mathbf{x}^{(k)}\right\|_2^2}{2s^2}\right]\approx \max_{k\in\Omega}\left[-\frac{\left\|\mathbf{x}^{(k)}\right\|_2^2}{2s^2}\right]+\log(T_0). \tag{12.53}$$

Substituting (12.53) into (12.49) yields

$$\log P_G \approx \alpha + \beta\cdot\log s + \frac{\gamma}{s^2}, \tag{12.54}$$

where

$$\begin{aligned}
\alpha &= \log\frac{\Delta\mathbf{x}}{(2\pi)^{M/2}} + \log(T_0)\\
\beta &= -M\\
\gamma &= \max_{k\in\Omega}\left[-\frac{\left\|\mathbf{x}^{(k)}\right\|_2^2}{2}\right]
\end{aligned}. \tag{12.55}$$

Equation (12.54) reveals the important relation between the scaled failure rate P_G and the scaling factor s. The approximation in (12.54) does not rely on any specific assumption of the failure region. It is valid, even if the failure region is non-convex or discontinuous.

While (12.55) shows the theoretical definition of the model coefficients α, β and γ, finding their exact values is not trivial. For instance, the coefficient γ is determined by the hyper-rectangle that falls into the failure region Ω and is closest to the origin $\mathbf{x} = \mathbf{0}$. In practice, without knowing the failure region Ω, we cannot directly find out the value of γ. For this reason, we fit the analytical model in (12.54) by linear regression. Namely, we first estimate the scaled failure rates $\{P_{G,q}; q = 1, 2, \ldots, Q\}$ by setting the scaling factor s to a number of different values $\{s_q; q = 1, 2, \ldots, Q\}$. As long as the scaling factors $\{s_q; q = 1, 2, \ldots, Q\}$ are sufficiently large, the scaled failure rates $\{P_{G,q}; q = 1, 2, \ldots, Q\}$ are large and can be accurately estimated with a small number of random samples. Next, the model coefficients α, β, and γ are fitted by linear regression based on the values of $\{(s_q, P_{G,q}); q = 1, 2, \ldots, Q\}$. Once α, β, and γ are known, the original failure rate P_F in (12.3) can be predicted by extrapolation. Namely, we substitute $s = 1$ into the analytical model in (12.54):

$$\log P_F^{\mathrm{SSS}} = \alpha + \gamma, \tag{12.56}$$

where P_F^{SSS} denotes the estimated value of P_F by SSS. Apply the exponential function to both sides of (12.56) and we have

$$P_F^{\mathrm{SSS}} = \exp\left(\alpha + \gamma\right). \tag{12.57}$$

To make the SSS method of practical utility, maximum likelihood estimation is applied to fit the model coefficients in (12.54). The MLE solution can be solved from an optimization problem and it is considered to be statistically optimal for a given set of random samples.

Without loss of generality, we assume that N_q scaled random samples $\{\mathbf{x}^{(n)}; n = 1, 2, \ldots, N_q\}$ are collected for the scaling factor s_q. The scaled failure rate $P_{G,q}$ can be estimated by MC

$$P_{G,q}^{\mathrm{MC}} = \frac{1}{N_q} \cdot \sum_{n=1}^{N_q} I_\Omega\left(\mathbf{x}^{(n)}\right), \tag{12.58}$$

where $I_\Omega(\mathbf{x})$ is the indicator function defined in (12.45). The variance of the estimator $P_{G,q}^{\mathrm{MC}}$ in (12.58) can be approximated as [16]

$$v_{G,q}^{\mathrm{MC}} = P_{G,q}^{\mathrm{MC}} \cdot \frac{1 - P_{G,p}^{\mathrm{MC}}}{N_q}. \tag{12.59}$$

If the number of samples N_q is sufficiently large, the estimator $P_{G,q}^{\mathrm{MC}}$ in (12.58) follows a Gaussian distribution according to CLT [16]

$$P_{G,q}^{\mathrm{MC}} \sim \mathrm{Gauss}\left(P_{G,q}, v_{G,p}^{\mathrm{MC}}\right), \tag{12.60}$$

where $P_{G,q}$ denotes the actual failure rate corresponding to the scaling factor s_q.

Note that the model template in (12.54) is expressed for $\log P_G$, instead of P_G. To further derive the probability distribution for $\log P_{G,q}^{\mathrm{MC}}$, we adopt the first-order delta method from the statistics community [16]. Namely, we approximate the nonlinear function $\log(\bullet)$ by the first-order Taylor expansion around the mean value $\log P_{G,q}$ of the random variable $\log P_{G,q}^{\mathrm{MC}}$

$$\log P_{G,q}^{\mathrm{MC}} \approx \log P_{G,q} + \frac{P_{G,q}^{\mathrm{MC}} - P_{G,q}}{P_{G,q}} \approx \log P_{G,q} + \frac{P_{G,q}^{\mathrm{MC}} - P_{G,q}}{P_{G,q}^{\mathrm{MC}}}. \tag{12.61}$$

Based on the linear approximation in (12.61), $\log P_{G,q}^{\mathrm{MC}}$ follows the Gaussian distribution

$$\log P_{G,q}^{\mathrm{MC}} \sim \mathrm{Gauss}\left[\log P_{G,q}, \frac{v_{G,q}^{\mathrm{MC}}}{\left(P_{G,q}^{\mathrm{MC}}\right)^2}\right]. \tag{12.62}$$

Equation (12.62) is valid for all scaling factors $\{s_q; q = 1, 2, \ldots, Q\}$. In addition, since the scaled failure rates corresponding to different scaling factors are estimated by independent Monte Carlo simulations, the estimated failure rates $\{P_{G,q}^{\mathrm{MC}}; q = 1, 2, \ldots, Q\}$ are mutually independent. Therefore, the Q-dimensional random variable

$$\log \mathbf{P}_G^{\mathrm{MC}} = \left[\log P_{G,1}^{\mathrm{MC}} \quad \log P_{G,2}^{\mathrm{MC}} \quad \cdots \quad \log P_{G,Q}^{\mathrm{MC}}\right]^T \tag{12.63}$$

satisfies the following jointly Gaussian distribution:

$$\log \mathbf{P}_G^{\mathrm{MC}} \sim \mathrm{Gauss}\left(\boldsymbol{\mu}_G, \boldsymbol{\Sigma}_G\right), \tag{12.64}$$

where the mean vector $\boldsymbol{\mu}_G$ and the covariance matrix $\boldsymbol{\Sigma}_G$ are equal to

$$\boldsymbol{\mu}_G = \left[\log P_{G,1} \quad \log P_{G,2} \quad \cdots \quad \log P_{G,Q}\right]^T \tag{12.65}$$

$$\boldsymbol{\Sigma}_G = \mathrm{diag}\left[\frac{v_{G,1}^{\mathrm{MC}}}{\left(P_{G,1}^{\mathrm{MC}}\right)^2}, \frac{v_{G,2}^{\mathrm{MC}}}{\left(P_{G,2}^{\mathrm{MC}}\right)^2}, \ldots, \frac{v_{G,Q}^{\mathrm{MC}}}{\left(P_{G,Q}^{\mathrm{MC}}\right)^2}\right], \tag{12.66}$$

where $\mathrm{diag}(\bullet)$ denotes a diagonal matrix.

The diagonal elements of the covariance matrix $\boldsymbol{\Sigma}_G$ in (12.66) can be substantially different. In other words, the accuracy of $\{\log P_{G,q}^{\mathrm{MC}}; q = 1, 2, \ldots, Q\}$ associated with different scaling factors $\{s_q; q = 1, 2, \ldots, Q\}$ can be different, because the scaled failure rates $\{P_{G,q}; q = 1, 2, \ldots, Q\}$ strongly depend on the scaling factors. In general, we can expect that if the scaling factor s_q is small, the scaled failure rate $P_{G,q}$ is small and, hence, it is difficult to accurately estimate $\log P_{G,q}$ from a small number of random samples. For this reason, instead of equally "trusting" the estimators $\{\log P_{G,q}^{\mathrm{MC}}; q = 1, 2, \ldots, Q\}$, we must carefully model the "confidence" for each estimator $\log P_{G,q}^{\mathrm{MC}}$, as encoded by the covariance matrix $\mathbf{\Sigma}_G$ in (12.66). Such "confidence" information will be fully exploited by the MLE framework to fit a statistically optimal model.

Since the scaled failure rates $\{P_{G,q}; q = 1, 2, \ldots, Q\}$ follow the analytical model in (12.54), the mean vector $\boldsymbol{\mu}_G$ in (12.65) can be re-written as

$$\boldsymbol{\mu}_G = \alpha + \beta \cdot \begin{bmatrix} \log s_1 \\ \log s_2 \\ \vdots \\ \log s_Q \end{bmatrix} + \gamma \cdot \begin{bmatrix} s_1^{-2} \\ s_2^{-2} \\ \vdots \\ s_Q^{-2} \end{bmatrix} = \mathbf{A} \cdot \mathbf{\Theta}, \tag{12.67}$$

where

$$\mathbf{A} = \begin{bmatrix} 1 & \log s_1 & s_1^{-2} \\ 1 & \log s_2 & s_2^{-2} \\ \vdots & \vdots & \vdots \\ 1 & \log s_Q & s_Q^{-2} \end{bmatrix} \tag{12.68}$$

$$\mathbf{\Theta} = \begin{bmatrix} \alpha & \beta & \gamma \end{bmatrix}^T. \tag{12.69}$$

Equation (12.68) implies that the mean value of the Q-dimensional random variable $\log \mathbf{P}_G^{\mathrm{MC}}$ depends on the model coefficients α, β, and γ. Given $\{P_{G,q}^{\mathrm{MC}}; q = 1, 2, \ldots, Q\}$, the key idea of MLE is to find the optimal values of α, β, and γ so that the likelihood of observing $\{P_{G,q}^{\mathrm{MC}}; q = 1, 2, \ldots, Q\}$ is maximized.

Because the random variable $\log \mathbf{P}_G^{\mathrm{MC}}$ follows the jointly Gaussian distribution in (12.64), the likelihood associated with the estimated failure rates $\{P_{G,q}^{\mathrm{MC}}; q = 1, 2, \ldots, Q\}$ is proportional to

$$L \sim \exp\left[-\frac{1}{2}\left(\log \mathbf{P}_G^{\mathrm{MC}} - \boldsymbol{\mu}_G\right)^T \cdot \mathbf{\Sigma}_G^{-1} \cdot \left(\log \mathbf{P}_G^{\mathrm{MC}} - \boldsymbol{\mu}_G\right)\right]. \tag{12.70}$$

Taking the logarithm for (12.70) yields

$$\log L \sim -\left(\log \mathbf{P}_G^{\mathrm{MC}} - \boldsymbol{\mu}_G\right)^T \cdot \mathbf{\Sigma}_G^{-1} \cdot \left(\log \mathbf{P}_G^{\mathrm{MC}} - \boldsymbol{\mu}_G\right). \tag{12.71}$$

Substitute (12.67) into (12.71), and we have

$$\log L \sim -\left(\log \mathbf{P}_G^{\mathrm{MC}} - \mathbf{A} \cdot \mathbf{\Theta}\right)^T \cdot \mathbf{\Sigma}_G^{-1} \cdot \left(\log \mathbf{P}_G^{\mathrm{MC}} - \mathbf{A} \cdot \mathbf{\Theta}\right). \tag{12.72}$$

Note that the log-likelihood $\log L$ in (12.72) depends on the model coefficients α, β, and γ, because the vector $\mathbf{\Theta}$ is composed of these coefficients as shown in (12.69). Therefore, the MLE solution of α, β, and γ can be determined by maximizing the log-likelihood function

$$\max_{\mathbf{\Theta}} \quad -\left(\log \mathbf{P}_G^{\mathrm{MC}} - \mathbf{A}\cdot\mathbf{\Theta}\right)^T \cdot \mathbf{\Sigma}_G^{-1} \cdot \left(\log \mathbf{P}_G^{\mathrm{MC}} - \mathbf{A}\cdot\mathbf{\Theta}\right). \tag{12.73}$$

Since the covariance matrix $\mathbf{\Sigma}_G$ is positive definite, the optimization in (12.73) is convex. In addition, since the log-likelihood $\log L$ is simply a quadratic function of $\mathbf{\Theta}$, the unconstrained optimization in (12.73) can be directly solved by inspecting the first-order optimality condition [4]

$$\begin{aligned} &\frac{\partial}{\partial \mathbf{\Theta}}\left[-\left(\log \mathbf{P}_G^{\mathrm{MC}} - \mathbf{A}\cdot\mathbf{\Theta}\right)^T \cdot \mathbf{\Sigma}_G^{-1} \cdot \left(\log \mathbf{P}_G^{\mathrm{MC}} - \mathbf{A}\cdot\mathbf{\Theta}\right)\right] \\ &= 2\cdot\mathbf{A}^T\cdot\mathbf{\Sigma}_G^{-1}\cdot\left(\log \mathbf{P}_G^{\mathrm{MC}} - \mathbf{A}\cdot\mathbf{\Theta}\right) = \mathbf{0} \end{aligned}. \tag{12.74}$$

Based on the linear equation in (12.74), the optimal value of $\mathbf{\Theta}$ can be determined by

$$\mathbf{\Theta} = \left(\mathbf{A}^T\cdot\mathbf{\Sigma}_G^{-1}\cdot\mathbf{A}\right)^{-1}\cdot\mathbf{A}^T\cdot\mathbf{\Sigma}_G^{-1}\cdot\log \mathbf{P}_G^{\mathrm{MC}}. \tag{12.75}$$

Studying (12.75) reveals an important fact that the estimators $\{\log P_{G,q}^{\mathrm{MC}}; q = 1, 2, \ldots, Q\}$ are weighted by the inverse of the covariance matrix $\mathbf{\Sigma}_G$. Namely, if the variance of the estimator $\log P_{G,q}^{\mathrm{MC}}$ is large, $\log P_{G,q}^{\mathrm{MC}}$ becomes non-critical when determining the optimal values of α, β, and γ. In other words, the MLE framework has optimally weighted the importance of $\{\log P_{G,q}^{\mathrm{MC}}; q = 1, 2, \ldots, Q\}$ based on the "confidence" level of these estimators. Once α, β, and γ are solved by MLE, the original failure rate P_F can be estimated by (12.57).

To apply MLE, we need a set of pre-selected scaling factors $\{s_q; q = 1, 2, \ldots, Q\}$. In practice, appropriately choosing these scaling factors is a critical task due to several reasons. First, if these scaling factors are too large, the estimator P_F^{SSS} based on extrapolation in (12.57) would not be accurate, since the extrapolation point $s = 1$ is far away from the selected scaling factors. Second, if the scaling factors are too small, the scaled failure rates $\{P_{G,q}; q = 1, 2, \ldots, Q\}$ are extremely small and they cannot be accurately estimated from a small number of scaled random samples. Third, the failure rates for different performances and/or specifications can be quite different. To estimate them both accurately and efficiently, we should choose small scaling factors for the performance metrics with large failure rates, but large scaling factors for the performance metrics with small failure rates. Hence, finding an appropriate set of scaling factors for all performances and/or specifications can be extremely challenging.

In this chapter, a number of evenly distributed scaling factors covering a relatively large range are empirically selected. For the performance metrics with large failure rates, the scaled failure rates corresponding to a number of small scaling factors can be used to fit the model template in (12.54). On the other hand, the scaled failure rates corresponding to a number of large scaling factors can be used for the performance metrics with small failure rates. As such, a broad range of performances and/or specifications can be accurately analyzed by the SSS method.

While the MLE algorithm is able to optimally estimate the model coefficients α, β, and γ and then predict the failure rate P_F, it remains an open question how we can quantitatively assess the accuracy of our SSS method. Since SSS is based upon Monte Carlo simulation, a natural way for accuracy assessment is to calculate the confidence interval of the estimator P_F^{SSS}. However, unlike the traditional estimator where a statistical metric is estimated by the average of multiple random samples and, hence, the confidence interval can be derived as a closed-form expression, our proposed estimator P_F^{SSS} is calculated by linear regression with nonlinear exponential/logarithmic transformation. Accurately estimating the confidence interval of P_F^{SSS} is not a trivial task.

To address the aforementioned challenge, a bootstrapping based technique [8] is developed to accurately estimate the CI of the SSS estimator. The key idea of bootstrap is to re-generate a large number of random samples based on a statistical model without running additional transistor-level simulations. These random samples are then used to repeatedly calculate the value of P_F^{SSS} in (12.57) for multiple times. Based on these repeated runs, the statistics (hence, the confidence interval) of the estimator P_F^{SSS} can be accurately estimated.

In particular, we start from the estimated failure rates $\{P_{G,q}^{\text{MC}}; q = 1, 2, \ldots, Q\}$. Each estimator $P_{G,q}^{\text{MC}}$ follows the Gaussian distribution in (12.60). The actual mean $P_{G,q}$ in (12.60) is unknown; however, we can approximate its value by the estimated failure rate $P_{G,q}^{\text{MC}}$. Once we know the statistical distribution of $P_{G,q}^{\text{MC}}$, we can re-sample its distribution and generate N_{RS} sampled values $\{P_{G,q}^{MC(n)}; n = 1, 2, \ldots, N_{\text{RS}}\}$. This re-sampling idea is applied to all scaling factors $\{s_q; q = 1, 2, \ldots, Q\}$, thereby resulting in a large data set $\{P_{G,q}^{MC(n)}; q = 1, 2, \ldots, Q; n = 1, 2, \ldots, N_{\text{RS}}\}$. Next, we repeatedly run SSS for N_{RS} times and get N_{RS} different failure rates $\{P_F^{SSS(n)}; n = 1, 2, \ldots, N_{\text{RS}}\}$. The confidence interval of P_F^{SSS} can then be estimated from the statistics of these failure rate values.

Note that to apply SSS, we only need a set of scaling factors and their corresponding scaled failure rates: $\{(s_q, P_{G,q}); q = 1, 2, \ldots, Q\}$. As long as $\{s_q; q = 1, 2, \ldots, Q\}$ are sufficiently large, $\{P_{G,q}; q = 1, 2, \ldots, Q\}$ are not small probability values and, therefore, can be efficiently estimated by CMC. When applying CMC, we only need to determine whether the random samples belong to the failure region. Namely, the PoI does not have to be continuous. Due to this reason, SSS can be applied to estimate the rare failure rates for both continuous and discrete PoIs. However, since SUS explores additional information from the continuous performance values, SUS is often preferred over SSS when we handle continuous PoIs.

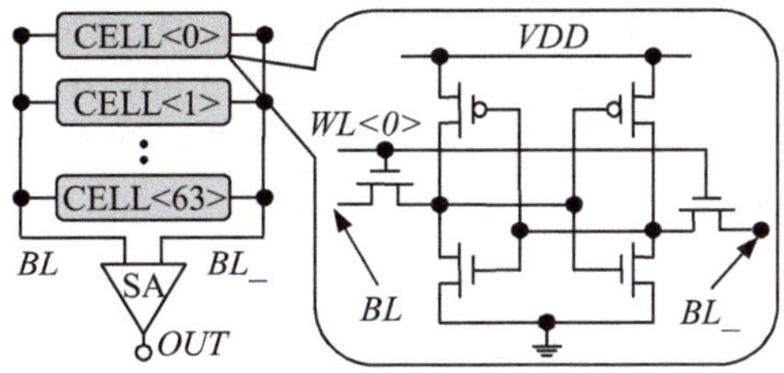

Fig. 12.5 The simplified schematic is shown for an SRAM column consisting of 64 bit-cells and a sense amplifier (SA) designed in a 45 nm CMOS process

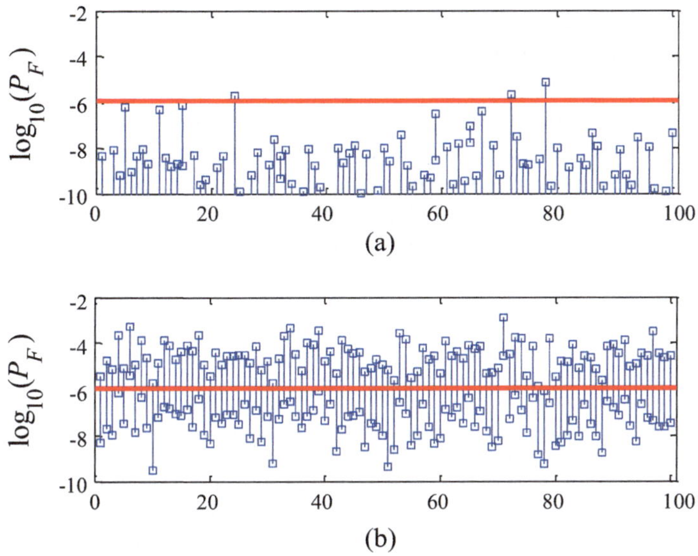

Fig. 12.6 The 95% confidence intervals (blue bars) of the SRAM example are estimated from 100 repeated runs with 6000 transistor-level simulations in each run for: (**a**) MNIS and (**b**) SSS. The red line represents the "golden" failure rate

To demonstrate the efficacy of SSS, we consider an SRAM column consisting of 64 bit-cells and a sense amplifier (SA) designed in a 45 nm CMOS process. Figure 12.5 shows the simplified circuit schematic of this SRAM column example. Similar to the SRAM read current example shown in Fig. 12.2, we consider the local V_{TH} mismatch of each transistor as an independent Normal random variable. In total, we have 384 independent random variables. In this example, the output of SA is considered as the PoI. If the output is correct, we consider the circuit as "PASS". Hence, the PoI is binary, and we cannot apply SUS in this example. For comparison purposes, we run MNIS [17] and SSS for 100 times with 6000 transistor-level simulations in each run. As shown in Fig. 12.6, there are 3 and 97 CIs out of 100 runs that cover the "golden" failure rate for MNIS and SSS, respectively. Here, the "golden" failure rate is estimated by CMC with 10^9 random samples. MNIS, again, fails to accurately estimate the corresponding CIs. SSS, however, successfully estimates the CIs. These results demonstrate that SSS is superior to the traditional MNIS method in this SRAM example, where the dimensionality of the variation space is more than a few hundred.

12.4 Conclusions

Rare failure event analysis in a high-dimensional variation space has attracted more and more attention due to aggressive technology scaling. To address this technical challenge, we summarize two novel approaches: SUS and SSS. Several SRAM examples are used to demonstrate the efficacy of SUS and SSS. More experimental results of SUS and SSS can be found in the recent publications [19, 21]. Both SUS and SSS are based upon solid mathematical background and do not pose any specific assumption on the failure region. Hence, they can be generally applied to estimate the rare failure rates of a broad range of other circuits, e.g., DFF.

References

1. S. Au, J. Beck, Estimation of small failure probabilities in high dimensions by subset simulation. Probab. Eng. Mech. **16**(4), 263–277 (2001)
2. A. Bhavnagarwala, X. Tang, J. Meindl, The impact of intrinsic device fluctuations on CMOS SRAM cell stability. IEEE J. Solid-State Circuits **36**(4), 658–665 (2001)
3. C. Bishop, *Pattern Recognition and Machine Learning* (Prentice Hall, Upper Saddle River, 2007)
4. S. Boyd, L. Vandenberghe, *Convex Optimization* (Cambridge University Press, Cambridge, 2009)
5. B. Calhoun, Y. Cao, X. Li, K. Mai, L. Pileggi, R. Rutenbar, K. Shepard, Digital circuit design challenges and opportunities in the era of nanoscale CMOS. Proc. IEEE **96**(2), 343–365 (2008)
6. F. Cérou, P. Moral, T. Furon, A. Guyader, Sequential Monte Carlo for rare event estimation. Stat. Comput. **22**(3), 795–808 (2012)
7. L. Dolecek, M. Qazi, D. Shah, A. Chandrakasan, Breaking the simulation barrier: SRAM evaluation through norm minimization, in *International Conference on Computer-Aided Design* (2008), pp. 322–329
8. B. Efron, R. Tibshirnani, *An Introduction to the Bootstrap* (Chapman & Hall/CRC, London, 1993)
9. R. Fonseca, L. Dilillo, A. Bosio, P. Girard, S. Pravossoudovitch, A. Virazel, N. Badereddine, A statistical simulation method for reliability analysis of SRAM core-cells, in *Design Automation Conference* (2010), pp. 853–856
10. C. Gu, J. Roychowdhury, An efficient, fully nonlinear, variability aware non-Monte-Carlo yield estimation procedure with applications to SRAM cells and ring oscillators, in *IEEE Asia and South Pacific Design Automation Conference* (2008), pp. 754–761
11. A. Guyader, N. Hengartner, E. Matzner-Løber, Simulation and estimation of extreme quantiles and extreme probabilities. Appl. Math. Optim. **64**(2), 171–196 (2011)
12. R. Heald, P. Wang, Variability in sub-100nm SRAM designs, in *International Conference on Computer-Aided Design* (2004), pp. 347–352

13. R. Kanj, R. Joshi, S. Nassif, Mixture importance sampling and its application to the analysis of SRAM designs in the presence of rare failure events, in *Design Automation Conference* (2006), pp. 69–72
14. R. Kanj, R. Joshi, Z. Li, J. Hayes, S. Nassif, Yield estimation via multi-cones, in *Design Automation Conference* (2012), pp. 1107–1112
15. K. Katayama, S. Hagiwara, H. Tsutsui, H. Ochi, T. Sato, Sequential importance sampling for low-probability and high-dimensional SRAM yield analysis, in *International Conference on Computer-Aided Design* (2010), pp. 703–708
16. A. Papoulis, S. Pillai, *Probability, Random Variables and Stochastic Process* (McGraw-Hill, New York, 2001)
17. M. Qazi, M. Tikekar, L. Dolecek, D. Shah, A. Chandrakasan, Loop flattening and spherical sampling: highly efficient model reduction techniques for SRAM yield analysis, in *Design, Automation & Test in Europe* (2010), pp. 801–806
18. A. Singhee, R. Rutenbar, Statistical blockade: very fast statistical simulation and modeling of rare circuit events, and its application to memory design. IEEE Trans. Comput.-Aided Des. Integr. Circuits Syst. **28**(8), 1176–1189 (2009)
19. S. Sun, X. Li, Fast statistical analysis of rare circuit failure events via subset simulation in high-dimensional variation space, in *International Conference on Computer-Aided Design* (2014), pp. 324–331
20. S. Sun, Y. Feng, C. Dong, X. Li, Efficient SRAM failure rate prediction via Gibbs sampling. IEEE Trans. Comput.-Aided Des. Integr. Circuits Syst. **31**(12), 1831–1844 (2012)
21. S. Sun, X. Li, H. Liu, K. Luo, B. Gu, Fast statistical analysis of rare circuit failure events via scaled-sigma sampling for high-dimensional variation space, in *International Conference on Computer-Aided Design* (2013), pp. 478–485
22. R. Topaloglu, Early, accurate and fast yield estimation through Monte Carlo-alternative probabilistic behavioral analog system simulations, in *IEEE VLSI Test Symposium* (2006), pp. 137–142
23. J. Wang, S. Yaldiz, X. Li, L. Pileggi, SRAM parametric failure analysis, in *Design Automation Conference* (2009), pp. 496–501

Chapter 13
Learning from Limited Data in VLSI CAD

Li-C. Wang

13.1 Introduction

Applying machine learning tools to analyze data collected from design and test processes can encounter different types of learning problems. One common type of analytics is based on dividing data samples into two classes. For example, the analytic begins with m samples where there are m_p *positive samples* and the rest are *negative samples*. A set of n features $f_1, \ldots, f_n$ are used to describe each sample. The goal is to learn a model based on those features to differentiate one or more of the positive samples from the negative ones. It is often that m_p is very small and in some cases, even $m_p = 0$. In addition, m is limited as well. A recent paper [1] discusses several applications that involve this type of analytics.

For example, Fig. 13.1 illustrates an application in functional verification [2, 3]. With a functional verification environment, a *testbench* is instantiated through constrained random test generation (CRTG) into a set of *functional tests*. The design under verification can be a SoC (System-on-Chip) and each test can be a C program (or a sequence of instructions). Simulating the tests results in *simulation traces* which in this case are the data to be analyzed. The task might concern a *coverage point* CP in the design and the goal is to help improve the coverage of CP.

To run an analytic tool, a dataset is formatted as that illustrated in Fig. 13.2. First, we need to define what a sample represents. In the context of analyzing simulation traces, a sample can be defined as the activities observed in a simulation cycle. Next, we need to describe the activities with a set of features. Each feature can be defined based on a design signal. Then, each sample can be represented as a vector of digital values. For example, these values can be 0, 1, a rising transition, and a

L.-C. Wang (✉)
Department of Electrical and Computer Engineering, University of California, Santa Barbara, Santa Barbara, CA, USA
e-mail: licwang@ece.ucsb.edu

I. M. Elfadel et al. (eds.), *Machine Learning in VLSI Computer-Aided Design*,
https://doi.org/10.1007/978-3-030-04666-8_13

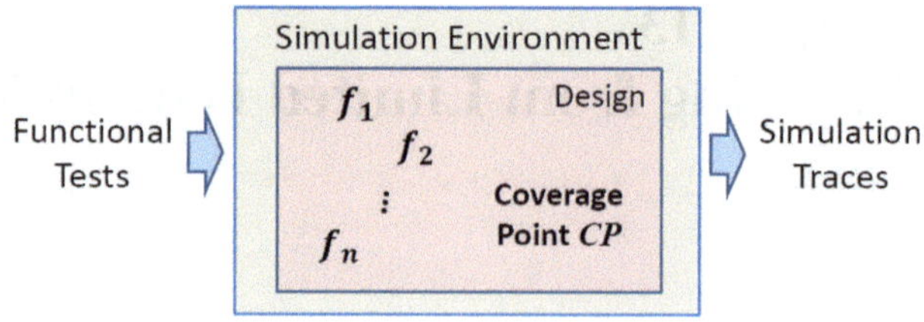

Fig. 13.1 An application in functional verification

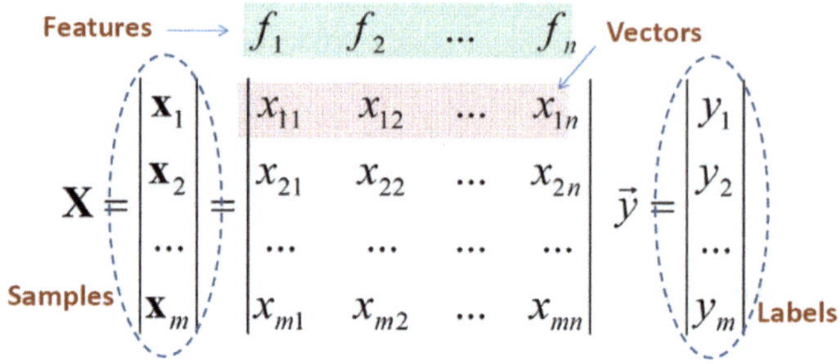

Fig. 13.2 The format of a dataset to be analyzed

falling transition. Each sample ($\mathbf{x}$) is also associated with a label (y) based on the coverage status of CP. The coverage status can be from the same cycle or in the next few cycles. If CP is covered, the sample is labeled as positive ($+1$). Otherwise, it is labeled as negative (-1).

Usually, the coverage point CP is of concern because it receives very few or no coverage in the simulation. This means that in the dataset, there are very few or no positive samples. The goal of the analytics can be for deriving a *rule* such that satisfying the rule is likely to achieve coverage of CP. Such a rule can be a combination of some feature values. After a rule is reported by an analytic tool, the testbench is modified to produce more tests that satisfy the rule. These tests are simulated and the quality of the rule is observed in the simulation result [2, 3].

In addition to functional verification, the paper [1] discusses several other applications that involve this type of analytics. For example, in physical verification, a given layout can be scanned into a sequence of *snippets* [4], i.e. a small window of layout image, as illustrated in Fig. 13.3. Each snippet can be described with a set of features, such as attributes related to shape, spacing, materials, etc. A positive sample can be defined as a snippet that potentially causes an issue (i.e., a defect-prone spot). Then, the rest are negative samples. Our interest is often in predicting the positive samples. Such an application can also encounter a dataset where there are many more negative samples than positive ones. Moreover, because different positive samples may be due to different reasons, it is likely that for a particular analysis, we are interested in modeling only one or few selected positive samples.

Another example is timing verification. In this case, samples can be defined as paths (e.g., a path from one flip-flop to another flip-flop). For example, Fig. 13.4 shows critical paths collected from a silicon experiment [5] and their predicted

Fig. 13.3 Raster scan to extract layout snippets

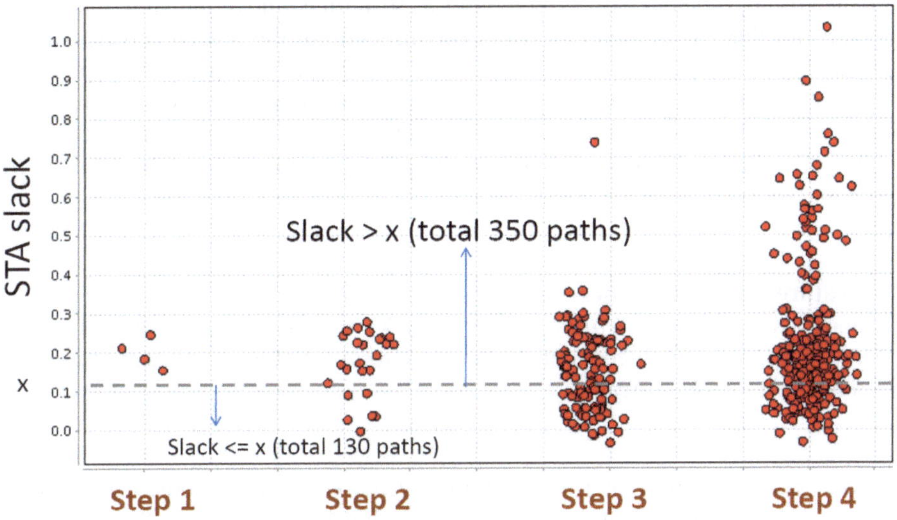

Fig. 13.4 350 critical paths not predicted by STA

timing slacks from a static timing analysis (STA) tool. For a path, its actual delays are measured on a set of process cores. The measurement is carried out by *frequency stepping* where step 1 has the lowest frequency. Each dot shown in the plot represents a path. For example, at step 1 there are four silicon-critical paths.

For this particular design, the assumption is that if the timing slack is less than a selected value (denoted as x), then the path is reported as an STA-critical path. In the plot, the (normalized) value of x is shown and it can be observed that 350 out of the 480 paths shown in the plot are not reported as STA-critical paths. In fact, based on the given x value, the STA reports 21,589 STA-critical paths and only 130 of those paths show up in this plot as a silicon-critical path [5].

To understand why a silicon-critical path is not an STA-critical path, one can start the analysis with the first four silicon-critical paths shown on the left of the plot. In this case, the positive samples are those 4 paths. The negative samples could be the remaining 21,459 (= 21,589 − 130) STA-critical paths which do not show up as a silicon-critical path [5]. To enable the analysis, a set of design-related features are developed for describing a path. These features can be based on, for example, usage of the cells, layout properties, path location, and so on [5].

Limited Number of Samples For the applications discussed above, a given dataset can have very few or no positive samples. Moreover, the number of total samples can also be limited. For example, in functional verification simulation cost is usually a main concern. Hence, the number of cycles simulated in each run is limited. In the timing verification application, the number of paths with measured timing on silicon chips is limited.

In those applications, there can be many choices to derive a feature set. Hence, the underlying problem can be thought of as searching among the large number of features for those few important ones. The difficulty of this search depends on the availability of the positive samples, the total samples, and the number of potential features to search on. For example, to search on a large number of potential features, one might need a large number of samples and at least some positive samples. If obtaining sufficient samples or positive samples is practically prohibited, then the problem can become challenging.

Feature-Based Analytics Analytics involved in the applications discussed above can be called *feature-based analytics* which means the underlying problem is to search for a small combination of features or feature values among a large set of features.

This search is different from traditional *feature selection* studied in machine learning. A traditional feature selection algorithm is entirely data-driven. On the other hand, the core problem faced in feature-based analytics is that the data can be insufficient to determine the importance of some features when they are included in a dataset. If such a dataset is run with a traditional feature selection algorithm the result can be misleading. This is because a typical learning algorithm usually reports some optimized result based on the data but does not report if the data is sufficient or not for learning the result.

Because of the data limitation, feature-based analytics in practice can be more intuitively captured as an iterative feature search process [1]. In each iteration, a subset of features are selected to run an analytic tool with the samples in hand. The result is evaluated until the outcome is satisfied.

Learning About the Features In general, learning can be seen as achieving two tasks, learning about the features and learning to construct a model based on the features. Traditionally, these two tasks are separated. A learning algorithm focuses on model building and a feature selection algorithm focuses on feature evaluation. Modern machine learning approach such as *deep learning* [6] puts much more emphasis on learning about the features. Similar to that, the emphasis in the applications discussed in this chapter can also be seen as more about learning the features.

In addition to the applications mentioned above, several other analytic tasks involving post-silicon test data can also be seen as learning about the features. For example, one application is to build a model to predict the Fmax (functional maximum frequency) of a chip [7]. In this application, a feature can be based on a flip-flop, a test pattern, a ring oscillator, or a selected critical path. The feature value is a delay structurally measured. The term "structurally" means to measure the delay using some scan structure. The main challenge for this analytic is to decide what features to use. If an effective set of features are used, an accurate model can be built rather easily. Otherwise, it can be quite difficult to learn a good model.

Another example is production yield optimization [8, 9]. In this context, a feature can be an *e-test* characterizing a process parameter on each wafer. A feature can also be based on a measurable property of a production tool. The analytic is to uncover a feature or a combination of features to explain a yield issue. Then, the goal is to improve the yield by adjusting those feature values accordingly [8]. In this context, the learning problem is almost entirely about learning the features.

13.2 Iterative Feature Search

The iterative feature search process is illustrated in Fig. 13.5. First, it is common that a tool from the machine learning toolbox (e.g., [10]) expects a dataset that is formatted as depicted in Fig. 13.2. To produce such a dataset, three steps are performed: sample selection, feature selection, and dataset construction.

The sample selection step defines what a sample is and selects a set of m samples $\mathbf{x}_1, \ldots, \mathbf{x}_m$ for the analysis. For example, in functional verification a sample consists of signal activities in a simulation cycle. In physical verification, a sample is a layout snippet. In timing verification, a sample is a path. For Fmax prediction, a sample is a chip. In yield optimization, a sample can be a chip, a wafer, or a lot. If the analysis is based on *supervised learning*, a label value y_i is calculated for each sample $\mathbf{x}_i$. If it is for *unsupervised learning*, no label is required.

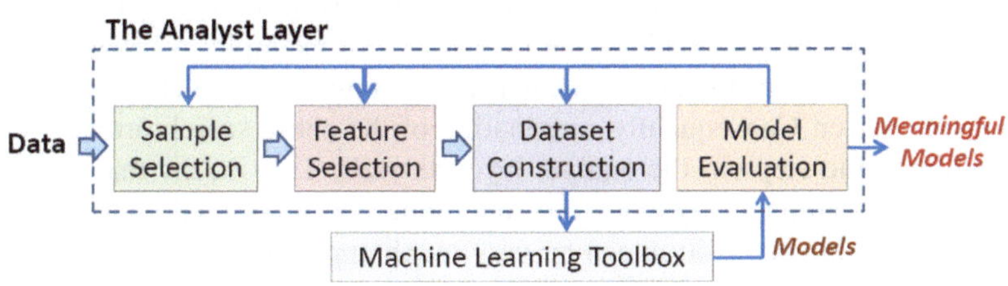

Fig. 13.5 Iterative feature search process

Calculating the label for a sample might need a separate analysis itself. For example, in yield analysis a wafer can be classified as good or bad and deciding this binary label can be based on outlier analysis [8] (i.e., a bad wafer has a yield number that is classified as an outlier). Furthermore, it is possible that a particular analysis does not use all samples. For example, after some initial analysis, it might be decided that a subset of samples require a more focused analysis.

An analytic task begins with an initial set of features. This feature set is often developed by consulting a domain expert. Then, in each iteration a subset of features $f_1, \ldots, f_n$ are selected. After this subset is determined, in the dataset construction step for each sample $\mathbf{x}_i$ its feature values $(x_{i1}, \ldots, x_{in})$ are computed. This computation might require running a machine learning tool also. For example, the original values of a feature might be divided into ranges and this division can be based on a clustering algorithm [11].

After a dataset is constructed, running a machine learning tool on the dataset can result in one or more models. For example, a tool can allow setting some parameter values to affect the optimization objective of the learning algorithm. With different parameter values, different models can be obtained. The models then go through a model evaluation step. In this step, the meaningfulness of a model is assessed in the context of the particular application. If model evaluation cannot determine a meaningful model, then a different dataset is required. This invokes a new iteration that can involve redoing one or more of the previous three steps.

13.2.1 The Need for Domain Knowledge

In Fig. 13.5, domain knowledge is involved in the analyst layer for the dataset preparation and model evaluation. And because of this, learning in view of the figure can be seen as using the data to enhance one's domain knowledge. In other words, Knowledge + Data $\Rightarrow$ Learning [1]. In theory, learning from data would not be possible without any prior knowledge [12]. Hence, the view in Fig. 13.5 is not entirely new. However, the view emphasizes the need for domain knowledge in practice where there is a tradeoff between the knowledge requirement and the data requirement in an analytic task, even though such a tradeoff might not be easily quantifiable.

In view of Fig. 13.5, it is obvious that the effectiveness of the learning is not determined solely by the machine learning tool. For example, if relevant features are missing in a dataset, then no tool can produce a model completely capturing the underlying answer. Consequently, automation of the entire search process requires automation of the steps in the analyst layer as well. This means that automation requires implementing a way to acquire and model the analyst's domain knowledge and this knowledge acquisition can be very much application dependent (see, e.g., [9, 13–15]).

13.2.2 The Model Evaluation Step

In practice, the model evaluation step can be expensive and/or time consuming. For example, in functional verification, the step might involve modifying the testbench according to a learned rule and generating some new tests. Then, these new tests are simulated to decide if the rule is meaningful. In other applications, the evaluation might involve meetings with the design team. In the context of production yield optimization, the meetings could involve engineers from a foundry outside the company. Consequently, each search iteration can be delayed and this bottleneck is outside the machine learning toolbox.

13.2.3 The Tool Requirement

To speed up the search process, ideally the search desires a machine learning tool that reports not only a learned model but also a quality measure for the model. This quality measure can help decide in the model evaluation step whether or not to go through an expensive evaluation process.

In traditional machine learning, a tool is designed to output an "optimized" model based on a given dataset where the optimization objective depends on the learning algorithm. Then, the quality of the model is evaluated through *cross-validation* separated from the learning algorithm. In cross-validation, there are two datasets: a *training* dataset and a *validation* dataset. The model is learned with the training dataset and its accuracy is calculated by applying the model to the validation dataset. The accuracy seen on the validation dataset is supposed to represent how the model will perform if it is applied on future unseen samples.

While cross-validation is a common practice, the no-free-lunch (NFL) theorem in machine learning [12] warns about its misuse in practice. Unless one can ensure that the validation dataset is somewhat a complete representation for the future unseen samples (which is often not the case in practice), cross-validation may provide little assurance in practice for an application.

More importantly, for the applications mentioned before, in practice cross-validation might not be a viable option to assess the quality of a model. This is due to the limitation on the positive samples and/or on the total samples as discussed above. This means that in Fig. 13.5, the quality of a model has to be assessed in a separate model evaluation step which can be expensive. The only possible assurance a learning tool can provide on its output models is that they are "optimized" with respect to some optimization objective. However, whether such an optimization objective is meaningful with respect to the particular application context can be quite difficult to assess.

In view of Fig. 13.5, ideally, the search desires a machine learning tool that by itself can provide some quality assurance for its output model. This can alleviate the need for cross-validation. This requirement leads to the main question for the machine learning toolbox: What additional quality assurance a learning algorithm can provide?

In summary, there are two areas of concern in an iterative search process: (1) How to provide more quality assurance for the models entering the model evaluation step? (2) How to effectively incorporate the domain knowledge in the iterative search process? In the rest of the chapter, the discussion will focus more on the first question. Then, in Sect. 13.7 we briefly review several recent works [9, 13–15] related to the second question.

13.3 Assumptions in Machine Learning

To address the model quality concern, we need to understand why cross-validation is needed in the first place, i.e. why a machine learning algorithm does not provide an assurance for its model quality and demands cross-validation to evaluate its model? To facilitate the discussion, Fig. 13.6 illustrates a theoretical setup in the context of *supervised learning*.

In this setup, a hypothesis space H is assumed. H is a set of functions (hypotheses) and one of them f is the target function (true answer) to be learned. A sample generator G produces a set of m samples $\mathbf{x}_1, \ldots, \mathbf{x}_m$ according to an unknown but fixed distribution D. For each sample $\mathbf{x}_i$, its label y_i is calculated as $f(\mathbf{x}_i)$. Then, the dataset comprising the m pairs $(\mathbf{x}_i, y_i)$ is given to a learning algorithm L to learn. The algorithm L outputs its answer h. Ideally, if the answer is correct, we would have $\forall \mathbf{x}$ generated from G, $f(\mathbf{x}) = h(\mathbf{x})$.

In theory, f has to be *learnable* [16, 17] in order for a learning algorithm to achieve some sort of learning. To ensure learnability, some assumptions need to be made in view of the setup. There are five areas to make an assumption, as marked in Fig. 13.6.

The first assumption concerns the hypothesis space H. It is intuitive that learnability depends on the complexity of H, i.e. the more complex the H is, the more difficult the learning is (hence less learnable). If H is finite and enumerable, then its complexity can be measured more easily. For example, if H is the set of all Boolean functions based on n variables, then H contains 2^{2^n} distinct functions.

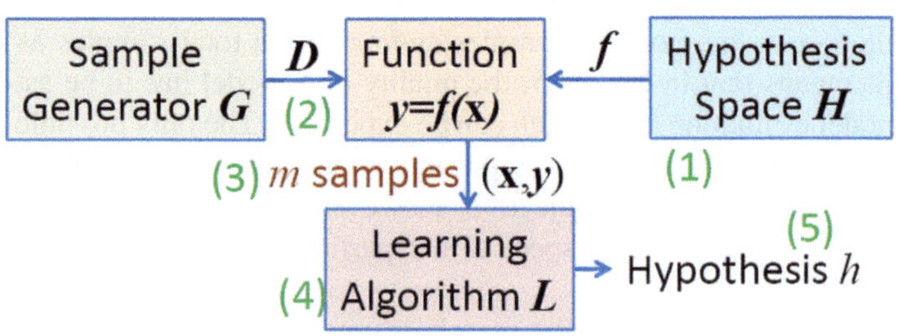

Fig. 13.6 Five areas to make an assumption in order to enable learning

The difficulty is when H is infinite and/or uncountable. In this case, one cannot rely on counting to define its complexity. One theory to measure the complexity of H is based on its ability to fit the data. This concept is called the *capacity* of H which can be characterized as the *VC dimension* [18]. The VC dimension (VC-D) also represents the minimum number of samples required to identify an f randomly chosen from H. To learn, one needs to make an assumption on the VC-D, for example VC-D should be on the order of $poly(n)$ (polynomial in n, the number of features). Otherwise, the number of required samples can be too large for the learning to be practical.

The second assumption concerns the sample generator G. The common assumption is that G produces its samples by drawing a sample randomly according to a fixed distribution D. Hence, as far as the learning concerns, all future samples are generated according to the same distribution.

The third assumption concerns the number of samples (m) available to the learning algorithm. This m has to be at least as large as the VC-D. Then, the fourth assumption concerns the complexity of the learning algorithm. Even though m is sufficiently large, learning the function f can still be computationally hard [17].

The computational hardness can be characterized in terms of the traditional NP-Hardness notion [17] or the hardness to break a cryptography function [19]. For example, learning a 3-term DNF (disjunctive normal form) formula using the DNF representation is hard [17]. In fact, except for a few special cases, learning based on a Boolean functional class is usually hard [17]. Moreover, learning based on a simple neural network is hard [20]. The computational hardness implies that in practice for most of the interesting learning problems, the learning algorithm can only be a heuristic. Consequently, its performance cannot be guaranteed on all problem instances.

The last assumption concerns how the answer h is evaluated. In the discussion of the other assumptions above, we implicitly assume that the "closeness" of h to f is evaluated through an error function $Err()$, for example $Err(h, f) = Prob(h(\mathbf{x}) \neq f(\mathbf{x}))$ for a randomly drawn $\mathbf{x}$. Notice that with such an $Err()$, an acceptable answer does not require $h = f$. As far as the learning concerns, as long as their outputs are the same with a high probability, h is an acceptable answer for f. This is because the purpose of the learning is for prediction and the goal is to have a predictor whose accuracy is high enough. However, for many applications in design and test processes, the use of a learning model is not for prediction but for interpretation. For those applications, adopting such an error function can be misleading.

13.4 Traditional Machine Learning

When applying a machine learning algorithm, a practitioner is often instructed to pay attention to the issue of model *overfitting*. In practice, one way to observe overfitting is through cross-validation. Let D_T and D_V denote the training and validation datasets. Let $EmErr(h, D)$ be an error function to report an *empirical error rate* by applying a model h onto the dataset D. Let the learning error rate be $e_T = EmErr(h, D_T)$ and validation error rate be $e_V = EmErr(h, D_V)$. In learning, a learning algorithm has only D_T to work on. Hence, the algorithm can try to improve on e_T but does not know what the resulting e_V might look like.

Overfitting means that the learning algorithm continues to improve on e_T, but e_T and e_V deviate from each other and hence, the improvement does not translate to e_V. In contrast, *underfitting* means that e_T is high and hence there is still room for improvement. Figure 13.7 illustrates these concepts where the x-axis can be thought of as a scale to reduce e_T based on employing a more complex model.

Refer back to Fig. 13.6. A learning algorithm usually has no knowledge regarding the actual hypothesis space H where the function f is drawn. To learn, the learning algorithm assumes a hypothesis space H_L to begin with. This is usually done by assuming a model representation, for example, such as a particular neural network design. When the assumption of H_L is over-constrained, its capacity is smaller than the capacity of H. Then, it is possible that H_L does not contain a hypothesis h that is close enough to f. As a result, it is not possible for $Err(h, f)$ to approach zero. This can be considered as another perspective to understand the concept of underfitting.

In practice, to avoid underfitting the learning begins with an assumed H_L that is as less constrained as possible, i.e. with a capacity as large as possible. In practice, this assumption is constrained by the computational resources for the learning. This is because assuming a more complex H_L usually implies more computational overhead in the learning. Also, assuming a more complex H_L means more samples needed to cover the hypothesis space. More importantly, with a H_L whose capacity is larger than the capacity of H, obtaining sufficient data samples to achieve a complete coverage on H_L might not be practically possible.

For example, Fig. 13.8 depicts such a situation. After a learning algorithm checks the hypotheses in a hypothesis space H_L against all samples in the training dataset D_T, the space can be divided into two subsets. The first includes all hypotheses h' that are inconsistent with the samples, for example $EmErr(h', D_T) \neq 0$. Then, the rest R (it is called the *version space*) includes all *consistent* hypothesis h where

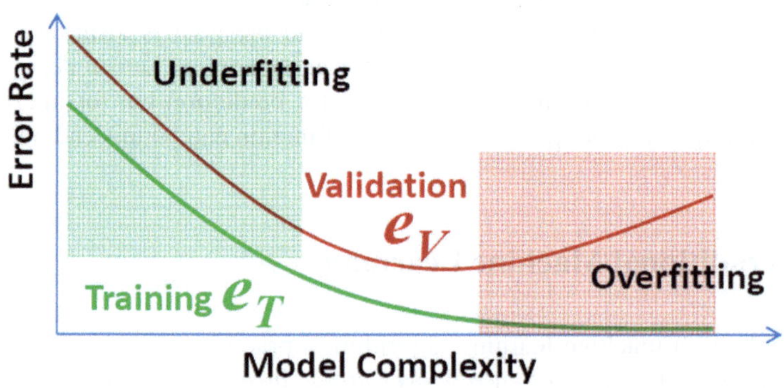

Fig. 13.7 Underfitting vs. overfitting

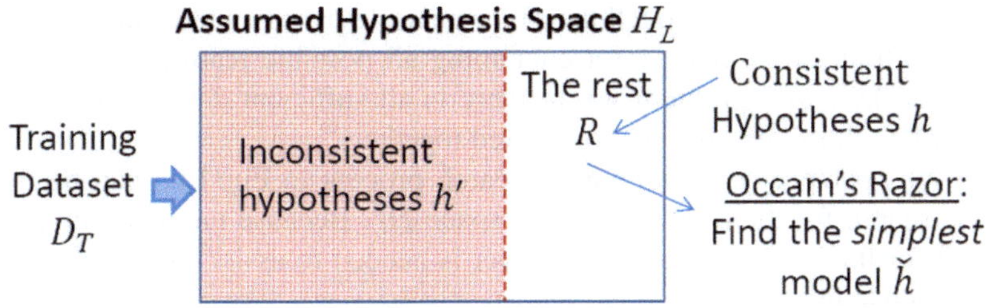

Fig. 13.8 Occam's Razor in machine learning

$EmErr(h, D_T) = 0$. In overfitting, *R* contains two or more distinct answers. Here "distinct" means existing one **x** where $\mathbf{x} \notin D_T$ and the two hypotheses h_1, h_2 result in two different values, i.e. $h_1(\mathbf{x}) \neq h_2(\mathbf{x})$. In other words, there exists a sample to differentiate h_1 and h_2 but this sample is not in D_T. In practice *R* can contain a large number of distinct hypotheses after learning on D_T, for example due to insufficient samples used in the learning.

13.4.1 Occam's Razor

In machine learning, a common strategy to pick a model in the version space *R* is based on the Occam's Razor principle, i.e. picking the "simplest" model as the answer. Applying the Occam's Razor principle requires a measure for model simplicity. Suppose this measure is defined. Then picking the model in *R* becomes an optimization problem, i.e. optimizing the model according to the simplicity measure. Because such an optimization problem can be computationally hard, a heuristic is usually developed to tackle the problem and such a heuristic does not guarantee always finding the optimal model.

In theory, there is also some subtlety to apply the Occam's Razor principle in learning [21]. Furthermore, the definition of a simplicity measure might or might not have a physical meaning in an application context. For all those reasons, an "optimized" model reported by a learning tool might provide little assurance on e_V. Consequently, cross-validation is required to evaluate the model even though it is considered as an optimized model by a learning algorithm.

13.4.2 Avoid Overfitting

Ideally, one would like to assume a H_L whose capacity is the same as H. However, this can be extremely difficult to accomplish in practice. In the context of feature-based analytics discussed before, underfitting can mean a required feature is not included in the initial feature set. Overfitting can mean there are features included where no data are available to tell their relevance. To avoid underfitting, one desires to begin with an initial feature set containing all possible features. However, this strategy can lead to a problem where there is no sufficient data to find the exact answer (when those features are considered together).

From this perspective, the iterative search process in Fig. 13.5 can be thought of as the search for the right hypothesis space H_L. Because the data might not be sufficient for the search, the Analyst Layer is needed to assist the search based on domain knowledge.

13.5 An Adjusted Machine Learning View

As discussed in Sect. 13.4, a traditional machine learning algorithm is designed to find an optimized model that fits a dataset based on a given hypothesis space assumption. In practice, such an approach may provide little quality assurance on its answer and hence, increases the burden of the model evaluation step in Fig. 13.5. In addition, cross-validation might not be a viable option for the applications considered due to limitation on the available samples for the learning.

For feature-based analytics, the essence of the problem is finding the right hypothesis space assumption. From this perspective, it is desirable to design a machine learning tool that can automatically evaluate a hypothesis space assumption before finding a fitting model. With this in mind, an adjusted view for the machine learning algorithm can be stated as the following: To search for a *hypothesis space assumption* where there is exactly one hypothesis that fits all samples in a given dataset.

In other words, the adjusted machine learning view adds an additional constraint for judging the quality of the learning—the resulting hypothesis has to be unique in terms of the assumed hypothesis space.

It is interesting to note that the adjusted machine learning view is compatible with the Occam's Razor principle and with the structural risk minimization (SRM) proposed in [18]. The difference is the addition of the uniqueness requirement. In fact, an early work to justify the Occam's Razor principle in machine learning [22] already suggested that the simplest model is better because the model is more likely to be unique, i.e. it is harder to find another answer with the same complexity which can fit the data. In a sense, the adjusted machine learning view makes the uniqueness an explicit requirement for learning, which is already implicitly hinted by adopting the Occam's Razor principle.

13.5.1 Search for a Hypothesis Space Assumption

To implement the adjusted machine learning view, we need a way to determine the uniqueness for a given hypothesis space assumption. In addition, we also need a way to define a set of hypothesis space assumptions. To facilitate the search we may desire to order those hypothesis spaces with increasing capacity as H_1, H_2, ..., $H_i, H_{i+1}, \ldots$ where H_p is less complex than H_q for all $p < q$. This ordering enables one to evaluate a set of hypothesis spaces incrementally by following the Occam's Razor principle (and the SRM in [18]). Figure 13.9 illustrates the idea.

In this figure, a version space R_i represents the set of consistent hypotheses (see Fig. 13.8) in H_i based on a given dataset. In the figure, $R_1, \ldots, R_{i-1}$ are all empty, meaning that no hypothesis in those hypothesis spaces can be found to fit all the samples in the dataset (i.e., these hypothesis spaces underfits the dataset). H_i is the first hypothesis space where R_i is not empty. Depending on how stringent the uniqueness requirement is applied, the search may continue into H_{i+1}. For example, uniqueness may be defined as $|R| = 1$, i.e. containing exactly one consistent hypotheses. Alternatively, the requirement might be relaxed to be such as $0 < |R| \leq 5$. If the requirement is satisfied by H_i, then the consistent hypothesis (or hypotheses) is reported.

A machine learning tool implementing the idea in Fig. 13.9 can provide two advantages. First, the output includes a hypothesis space assumption used to obtain the answer(s). This provides additional information for its user to judge the meaningfulness of a reported model. Second, it is possible that the tool results in a situation where no hypothesis space satisfies the uniqueness requirement. In this case, the learning can fail. This failure can immediately triggers an adjustment to the feature set in use or the set of the hypothesis spaces assumed. This means to start another iteration in the search process in Fig. 13.5, without involving the expensive model evaluation step.

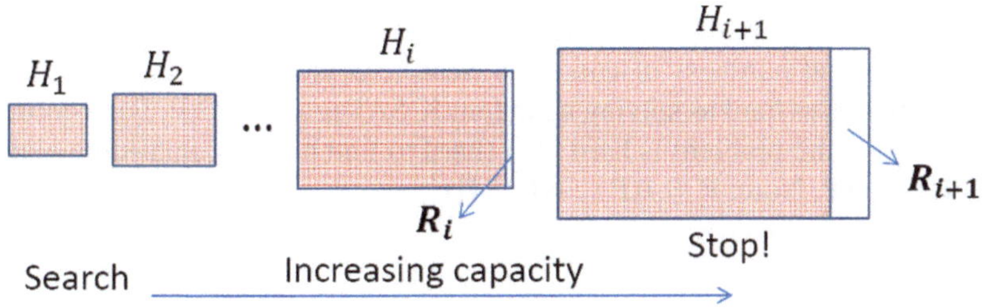

Fig. 13.9 Search for a hypothesis space assumption

13.6 A SAT-Based Implementation

One major challenge for implementing a learning tool following the idea presented in Fig. 13.9 is to obtain an ordered sequence of hypothesis spaces. This ordering should be based on measuring the capacities of the hypothesis spaces. If a hypothesis space contains an infinite number of hypotheses, it is difficult to measure its capacity. In theory the capacity can be measured in terms of its VC dimension [18]. However, in practice the *effective capacity* of a hypothesis space is also limited by the learning algorithm [6], making its estimation quite difficult.

For applications discussed in Sect. 13.1, however, the number of features is limited. Moreover, the number of values a feature can take can be limited as well. Hence, for those applications we can assume that each feature has only a limited number of possible values. If we use a pseudo feature to represent a particular feature value, then we can further assume that all the features in use are binary, i.e. yes for the occurrence of the feature value and no otherwise. Therefore, from a theoretical perspective the underlying learning problem can be treated as a Boolean learning problem.

13.6.1 Boolean Learning

Given n features, the Boolean space contains 2^{2^n} Boolean functions. A hypothesis space is a set of Boolean functions in this space. Usually, a hypothesis space is specified with a representation. For example, a k-term DNF restricts the functions to those representable with k product terms. Each product term (also called a *monomial*) can have up to n literals (features in positive or negative forms).

For a given Boolean hypothesis space, its capacity can be measured in terms of the number of Boolean functions contained in the space. Hence, assuming that the underlying learning problem is Boolean learning avoids the difficulty for measuring the capacity of an infinite hypothesis space. However, it does not avoid the computational hardness discussed in Sect. 13.3 before. For example, learning DNF formulas is as hard as solving a random K-SAT problem [23]. For k-term DNF, even for $k = 3$ the problem is hard [17] (hard to find a polynomial-time algorithm unless **RP** = **NP**. Note: **P** $\subseteq$ **RP** $\subseteq$ **NP** [24]).

13.6.2 Monomial Learning

The simplest case for k-term DNF learning is $k = 1$. In this case, the problem is essentially to learn a *monomial*. From the perspective of computational learning theory (CLT), monomials are efficiently learnable [17]. Even though monomial learning is considered as an easy problem in CLT, it can still be a hard problem in view of Fig. 13.9. In Fig. 13.6, consider that D is a uniform distribution. Suppose the true answer f is the monomial $f_1 \cdots f_j$ for a large j. It is likely that no positive sample is generated even for a large m, i.e. for all samples $\mathbf{x}$ produced by G we have $f(\mathbf{x}) = 0$. Consequently, the output model is simply the constant 0. From the CLT perspective, the constant 0 is a good answer because the error probability for $f(\mathbf{x}) \neq 0$ on a randomly drawn $\mathbf{x}$ is extremely low. However, in practice a constant 0 most often means the learning has failed.

With the adjusted learning view (Fig. 13.9), the concept of learnable is not based on an error function $Err()$. Instead, learnable can be viewed as achieving a version space with a very small $|R_i|$ (e.g., $|R_i| = 1$). With this change, monomial learning can be hard when there lacks a positive sample. In fact, if there is only one positive sample and many negative samples, finding the shortest monomial is NP-hard [25].

13.6.3 Hypothesis Space Pruning

Take monomial learning as an example. The ordered sequence of hypothesis spaces in Fig. 13.9 can be defined as $H_1, \ldots, H_n$ where n is the number of features. Each H_l comprises the monomials of the same length l (i.e., with l literals). For example, H_1 comprises the monomials of length 1, i.e. $\{f_1, f_1', f_2, f_2', \ldots, f_n, f_n'\}$. H_2 comprises the monomials of length 2, which has $2^2\binom{n}{2}$ monomials: For every pair of features f_i, f_j, where $i \neq j$, we have 2^2 monomials $\{f_i f_j, f_i' f_j, f_i f_j', f_i' f_j'\}$. In general, H_l comprises $2^l\binom{n}{l}$ monomials.

For a given H_l and a dataset, our goal is to compute the version space R_l. This computation can be based on removing the monomials that are inconsistent with the samples. For example, suppose $n = 3$ and $l = 2$. A negative sample "001" removes the following hypotheses: $f_1' f_2'$, $f_1' f_3$, $f_2' f_3$. A positive sample "100" removes all hypotheses except for: $f_1 f_2'$, $f_1 f_3'$, $f_2' f_3'$. In general, a negative sample removes $\binom{n}{l}$ hypotheses while a positive sample removes all but the $\binom{n}{l}$ hypotheses. Therefore, the pruning power of a positive sample is larger than a negative sample.

The ordered sequence of hypothesis spaces can be generalized to include k-term DNF learning for up to a small k, for example $k \leq 3$. In this case, there are two variables to define an ordering of the hypothesis spaces, l as the total number of literals in a hypothesis and k as the total number of terms. For example, the hypothesis spaces can be ordered in such a way that H_{l_1,k_1} is before H_{l_2,k_2} in the sequence if $l_1 < l_2$ or $l_1 = l_2 \wedge k_1 < k_2$.

In pruning a hypothesis space H, each sample can be seen as a constraint that removes a subset of hypotheses in the space. Suppose our uniqueness requirement is $|R| = 1$, i.e. the version space after the pruning contains exactly one hypothesis. Then the pruning can be formulated as a Boolean satisfiability (SAT) problem. A satisfiable assignment represents a hypothesis that is consistent with all samples. To check for the uniqueness requirement we need to invoke SAT twice. The first time is to find a satisfiable assignment A. Then, A is converted into a constraint to block itself. After this constraint is added to the SAT formula, if the result is unsatisfiable then we know A is the only satisfiable assignment. If not, we know $|R| > 1$.

13.6.4 SAT Encoding

To convert the pruning problem into a SAT problem, we need three groups of CNF (conjunctive normal form) clauses: (1) clauses to constrain the hypothesis space based on given n, l, k, where n is the number of features, l is the number of literals in a hypothesis, and k is the number of terms; (2) clauses to constrain the space based on positive samples (if any), and (3) clauses to constrain the space based on negative samples. Let m_p be the number of positive samples and m_n be the number of negative samples. Then, the encoding method described below results in a CNF formula with $\Theta(nkl + km_p)$ symbols and $\Theta(nkl + nkm_p + km_n)$ clauses.

The key idea for the encoding is that each feature can appear in positive, or in negative, or does not appear at all in a term. Hence, three variables are used to represent these three cases for a feature:

- $X^j_{i,1}$ is True iff the i-th feature in the j-th term appears in negative form
- $X^j_{i,2}$ is True iff the i-th feature in the j-th term appears in positive form
- $X^j_{i,3}$ is True iff the i-th feature does not appear in the j-th term

Since exactly one of the three cases is true, one-hot constraints are required to enforce the requirement:

$$\Pi^k_{j=1}\Pi^n_{i=1}\left(X^j_{i,1} + X^j_{i,2} + X^j_{i,3}\right)\left(\neg X^j_{i,1} + \neg X^j_{i,2}\right)$$
$$\left(\neg X^j_{i,1} + \neg X^j_{i,3}\right)\left(\neg X^j_{i,2} + \neg X^j_{i,3}\right).$$

For a given (l, k), we need to constrain the space to those hypotheses containing only l literals. This involves a cardinality constraint. The performance of different encoding methods for a cardinality constraint can be found in [26]. In our implementation, we choose the sequential counter method [27] because its performance is comparable to other encoding methods and it has the unit propagation property[26]. The encoding for the cardinality constraint requires additional $l(nk - 1)$ new symbols and $\Theta(nkl)$ clauses. Further detail can be found in [28].

To illustrate the conversion from a positive samples into clauses, consider a positive sample $s = 101$. For a single term to be evaluated as true, feature 1 and feature 3 must not appear in negative form and feature 2 must not appear in positive form. Then, at least one term must be evaluated as true. A naive encoding leads to n^k clauses, which is not feasible. To overcome this challenge, additional k symbols,

$A^1, A^2, \ldots, A^k$ are used such that A^j is true if and only if the j-th term is evaluated as true. The number of clauses reduces to $(n + 1)k + 1$. The requirement of at least one term is evaluated as true is encoded by a single clause:

$$\left(\Sigma_{j=1}^{k} A^j\right),$$

and for each j, the relation of A^j and $X_{i,\delta}^j$ is maintained by

$$\Pi_{i=1}^{n}\left(\neg X_{i,2-s[i]}^{j} + \neg A^j\right), \text{ and}$$

$$\left(\Sigma_{i=1}^{n} X_{i,2-s[i]}^{j} + A^j\right).$$

Similarly, suppose $s = 101$ is a negative sample. For a single term to be evaluated as false, at least one of feature 1 and feature 3 must appear in negative form or feature 2 appear in positive form. Besides, all the terms must be evaluated as false. For each sample, k clauses are required. The overall encoding is

$$\Pi_{j=1}^{k}\left(\Sigma_{i=1}^{n} X_{i,2-s[i]}^{j}\right).$$

13.6.5 *Effect of the Uniqueness Requirement*

To illustrate the effect of including the uniqueness requirement in learning, in this section the performance of the SAT-based learning tool is compared to a popular decision-tree learning tool, the CART tool from the Python machine learning library [10]. The discussion focuses on why uniqueness can be a desirable property in learning. Detail of our learning tool is described in [28] which is named VeSC-CoL (version space cardinality based concept learning) and uses the Lingeling solver [29] for SAT.

In the experiment, we assume the number of features $n = 100$. We further assume the length of the true answer is 6 which can be a k-term DNF formula for $k = 1, 2, 3$. In each case, a k is randomly picked and the true answer is also randomly picked from all the k-term DNF hypotheses. The dataset contains exactly k positive samples which is randomly generated. Then, negative samples are also randomly generated.

Figure 13.10 shows a comparison result for VeSC-CoL and CART. The experiment includes 20 cases. The x-axis shows the number of negative samples used in the learning up to that particular point. In the experiment, the k positive samples are always used first, before any negative samples are used.

For CART, the figure shows the number of cases the CART tool correctly reports the true answer at each x value. For VeSC-CoL, the figure shows two numbers at each x value. The first is the number of cases where VeSC-CoL reports a unique hypothesis as its answer. This is marked as a red dot. The second is the number of

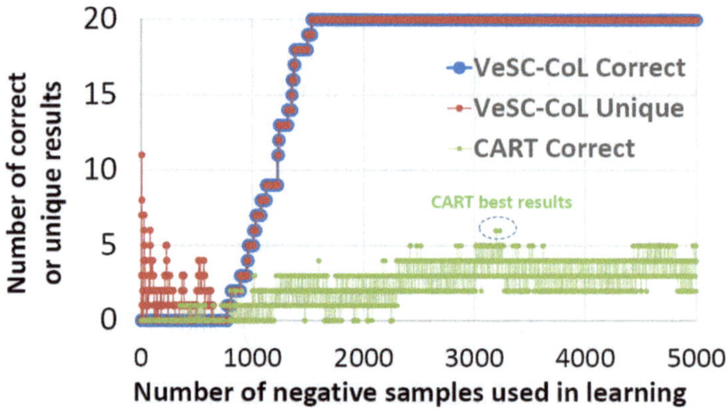

Fig. 13.10 A comparison between VeSC-CoL and CART

cases where VeSC-CoL correctly finds the true answer. This is marked as a blue dot. When these two numbers coincide, the figure shows an overlap of red dots circled by blue edge.

For CART, the best scenario is that it finds the true answer in 6 out of the 20 cases. This happens occasionally on some particular x values after 3000. For the range of the x values shown, CART performs slightly better as more samples are provided. However, in addition to being poor, its performance fluctuates quite frequently as more samples are added.

For VeSC-CoL, notice that a unique hypothesis found by the tool does not always guarantee it is the correct answer. However, this happens only when the x value is still relatively small. After about $x = 650$, a unique hypothesis is always the correct answer. More interestingly from $x = 650$ and up to about $x = 1500$, as more samples are added to the learning, the VeSC-CoL result always improves. After $x = 1500$, VeSC-CoL finds all the 20 true answers and the result does not change with more samples added.

Figure 13.11 presents the result of VeSC-CoL from a different perspective and focuses on x value up to 800. For each of the y label from 1 to 20, the figure marks the x values where VeSC-CoL finds a unique hypothesis but it is not the true answer. We can call each range of such x values a *mistake window*. Figure 13.11 shows where such mistake windows occur as the number of samples increases.

It is interesting to observe that most of the mistake windows occur when the x value is less than 200. The largest window size is 70. Most importantly, a mistake window occurs only for $x \leq 647$. After more than 647 samples are used in the learning, no mistake window occurs (also see Fig. 13.10 above).

To see what a mistake window means, refer back to the hypothesis space search process depicted in Fig. 13.9 before. Suppose H_i is the hypothesis space that contains the true answer. When the samples used in the learning are not sufficient, the version space R_i can still include many unfiltered hypotheses. However, it is possible that the samples are enough to filter out most of the hypotheses in a simpler hypothesis space, say H_2. By chance, this filtering might result in one hypothesis left in H_2. Then, because of its uniqueness, VeSC-CoL would report the hypothesis as its answer. However, as more samples are added to the learning, this hypothesis will be filtered out and the wrong answer is removed. In other words, the largest window size 70 shown in Fig. 13.11 means that such a mistake made by VeSC-CoL is corrected after 70 additional samples are used. From this perspective, Fig. 13.11 shows that in all cases when VeSC-CoL makes a mistake, the mistake is corrected after up to 70 new samples added in the learning.

Note that each mistake window shown in Fig. 13.11 corresponds to a hypothesis space that is simpler to the hypothesis space containing the true answer. Because the true answer is in a hypothesis space with $l = 6$ and k randomly selected between 1 and 3, there are many hypothesis spaces before the true answer as the hypothesis spaces are arranged according to the ordering discussed in Sect. 13.6.3.

Figure 13.12 shows the result by redoing the experiment with $l = 3$ and k randomly selected between 1 and 2. In other words, the result shows what happens if we make the true answer easier to learn. For VeSC-CoL, its performance is similar to that shown in Fig. 13.10 before. After $x > 124$, VeSC-CoL finds the true answer in all 20 cases. Recall that this number is about 1500 in Fig. 13.10 before.

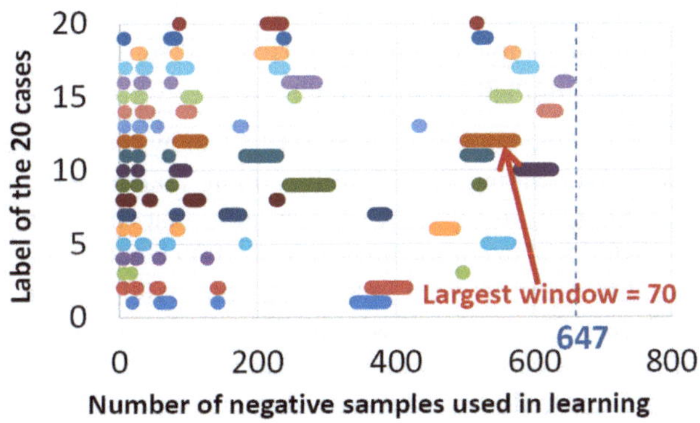

Fig. 13.11 The mistake windows in VeSC-CoL learning when the x value is small

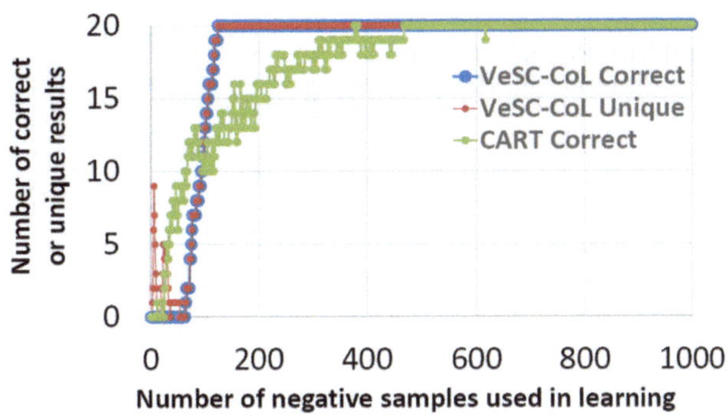

Fig. 13.12 Result by redoing the experiment with $l = 3$ and $k = 1, 2$

The performance of CART is substantially better than that shown in Fig. 13.10. It is interesting to observe that in some cases (only occur when $x < 90$), CART can actually perform better than VeSC-CoL, i.e. finding true answers for more cases. However, CART result can still fluctuate as more samples are used. But eventually (after about $x > 500$), CART can also find all true answers.

The two results presented above can be summarized as follows:

- Finding a unique answer in a hypothesis space does not guarantee it is the correct answer. However, if there are enough samples to filter out all other hypotheses in the hypothesis space containing the true answer, the true answer is guaranteed to be found.
- After some initial samples, the performance of VeSC-CoL becomes *consistent*, i.e. more samples always leading to a better result. This is not the case for CART.
- If the true answer is easy to learn, CART can be better than VeSC-CoL, i.e. can find the true answer with fewer samples. However, the difference in the number of samples required to learn between CART and VeSC-CoL is not that significant. In contrast, if the true answer is hard to learn, VeSC-CoL can significantly outperform CART.
- For a hard-to-learn answer, VeSC-CoL requires much less samples to learn than CART. However, because VeSC-CoL involves a SAT solver, it is computationally more expensive. In a sense, VeSC-CoL enables its user to trade computational cost for sample requirement. Hence, if samples are limited, VeSC-CoL can be a useful alternative for learning.

To conclude this section, Fig. 13.13 shows another result with the same setting as that for generating Fig. 13.10 by adding 100 more cases. As seen, the performance of VeSC-CoL in Fig. 13.13 is similar to that shown in Fig. 13.10. The performance of CART is also comparable, at $x = 5000$ CART finds the true answer for 22 cases. The difference between VeSC-CoL and CART remains significant.

Figure 13.14 then shows what happens to the CART's performance if the number of negative samples is increased up to 50,000. Close to the right end in the figure, CART correctly finds the true answer for 45 cases. For additional 8 cases, CART

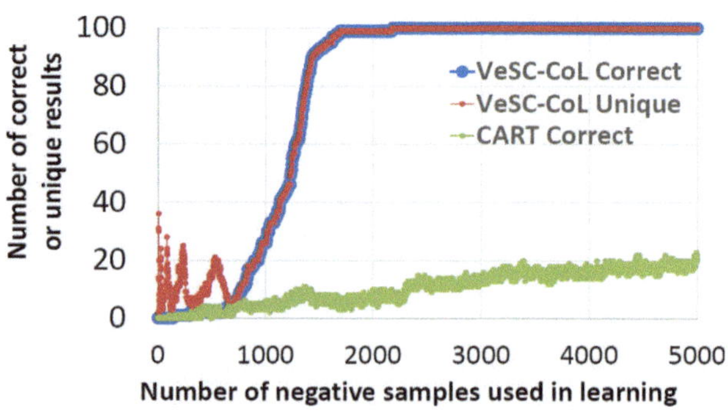

Fig. 13.13 Result by redoing the experiment with 100 cases

finds the true answer occasionally at some particular x values but the result is not stable. As a result, the number of correctly found cases fluctuates between 47 and 53 on the right side of the figure.

Table 13.1 then shows what types of the cases CART can find the true answer and what types of the cases CART cannot. Recall that the length of each true answer is fixed at 6. A true answer can be a 1-term, 2-term, or 3-term DNF formula. If it is a 1-term DNF, it is a length-6 monomial. If it is a 2-term DNF, the lengths of the two terms are represented as (l_1, l_2) where $l_1 + l_2 = 6$. Similarly, for a 3-term DNF, the lengths are represented as (l_1, l_2, l_3) where $l_1 + l_2 + l_3 = 6$.

Observe from Table 13.1 that if the true answer is a 1-term monomial, CART can find the true answer for all of them. If it is a 2-term DNF with lengths (1, 5), CART can also find the true answer. For those (1, 1, 4) cases, CART's result fluctuates between 2 and 8. For other cases, CART does not find the true answer at any x value even with up to 50,000 negative samples in use.

Suppose we only consider those 53 cases CART can find the true answer at least once. Figure 13.14 shows that it still requires about 20,000 negative samples for CART to find all those answers. This is compared to the sample requirement for VeSC-CoL in Fig. 13.13, where after about 1700 samples VeSC-CoL correctly finds all 100 true answers.

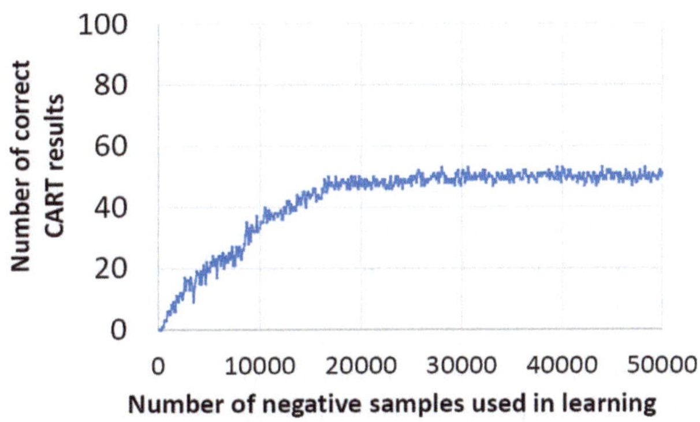

Fig. 13.14 CART result by adding up to 50,000 negative samples in learning

Table 13.1 Detail of the CART result for the 100 cases in Fig. 13.14

Formula	Lengths	# cases	# Correctly found
1-term	6	35	35
2-term	(1, 5)	10	10
2-term	(2, 4)	10	0
2-term	(3, 3)	15	0
3-term	(1, 1, 4)	10	2 to 8*
3-term	(1, 2, 3)	10	0
3-term	(2, 2, 2)	10	0

*This number fluctuates between 2 and 8

From the results presented in this section, observe that the performance of VeSC-CoL is quite consistent. In contrast, the performance of CART is not. In Figs. 13.10, 13.13, and 13.14 CART's performance does not look satisfactory. In Fig. 13.12, CART seems to be doing fine. This reveals that the search strategy of Fig. 13.9 with the uniqueness requirement can result in a learning process more consistent, as comparing to a traditional learning method such as CART whose performance can be largely affected by how easy or difficult to learn the true answer.

13.7 Incorporating Domain Knowledge

As discussed in Sect. 13.2.1, the analyst layer in Fig. 13.5 is driven mostly by domain knowledge. Hence, to automate the entirety of the iterative search process, one has to consider automation of the steps in the analyst layer. The analyst layer essentially comprises two major components, the dataset preparation component including the sample selection, feature selection, and dataset construction, and the model (or result) evaluation component.

For example, in the context of functional verification the dataset preparation involves two major tasks: generation of the tests to run the simulation and the selection of signals to encode a simulation trace. The data quality, in this case the quality of the simulation traces largely depends on the tests used in the simulation. To automate the test generation, the work in [14] presents a *constrained process discovery* approach that learns a test generation model based on example tests (e.g., C programs) written by a verification engineer. The test generation model can then be executed to produce tests automatically. For signal selection, often a verification engineer relies on reading the specification document to select the important and relevant signals. To automate this task, the work in [15] develops a text-mining approach to extract signals from a specification document.

In the context of production yield optimization, the dataset preparation involves making several choices to decide what data to analyze and what type of analytic to run [8]. In practice, these choices are made by the analyst. The work in [9] presents a way to learn how an analyst makes such choices in sequence for performing a yield data analytic task. The learning result is captured in a *process model* that looks like a flowchart where each node in the model is a software script. The model can then be executed automatically as if the analyst would perform the analytics for resolving a yield issue. For automating the result evaluation component, recently the work in [13] presents a learning approach to construct a plot recognizer based on example plots instructed by an analyst. Such a plot recognizer can be used to automatically recognize the meaningfulness of an analytic result when it is presented as a plot. The implementation is based on the recently proposed generative adversarial networks (GAN) learning approach [30, 31].

In general, where to apply learning and what type of learning to apply for automating the analyst layer is largely application dependent. For example, in the context of functional verification, learning is applied to automate two different tasks in the dataset preparation and the learning approaches involved are fundamentally different, i.e. constrained process discovery [14] vs. text mining [15]. In the context of production yield optimization, learning is applied to automate both the dataset preparation and result evaluation, and the learning approaches are also different, i.e. process mining [9] vs. GAN [13]. These examples illustrate the diversity of the problem for automating the analyst layer. Overall, whether the analyst layer for a given application can be fully automated remains an open question to be explored with future research.

13.8 Conclusions

In this chapter, the discussion focuses on a particular type of learning encountered in some design and test applications where the data samples are limited, which is given the name *feature-based analytics*. Because of the data limitation, in practice it is more intuitive to view feature-based analytics in terms of an iterative feature search process as depicted in Fig. 13.5. The effectiveness of this search depends on the steps conducted in the analyst layer as well as on the machine learning tool in use. For the machine learning tool, we explain the challenges to adopt a traditional machine learning problem formulation view. Instead, an adjusted machine learning view is presented, and illustrated in Fig. 13.9. In the adjusted view, uniqueness of an answer is included as an additional requirement for learning and the focus of the learning is shifted from finding a model to finding a hypothesis space. A SAT-based approach to realize this adjusted learning view in the context of learning a k-term DNF formula is presented and its benefits are illustrated with several experiment results. Finally, for automating the tasks in the analyst layer, several recent works are briefly discussed as examples to illustrate the diversity of the problem. More future research is required in order to assess if the analyst layer can be fully automated and such automation can also be very much application dependent.

Acknowledgements This work is supported in part by National Science Foundation under grant No. 1618118 and in part by Semiconductor Research Corporation with project 2016-CT-2706.

The author would also like to thank his doctoral student Kuo-Kai Hsieh especially for his help on Sects. 13.6.4 and 13.6.5.

This chapter is an extension from author's prior work "Machine Learning for Feature-Based Analytics," in Proceeding 2018 International Symposium on Physical Design, pp. 74-81 ©2018 Association for Computing Machinery, Inc. http://doi.acm.org/10.1145/3177540.3177555. Reprinted by permission.

References

1. L.-C. Wang, Experience of data analytics in EDA and test - principles, promises, and challenges. IEEE Trans. CAD **36**(6), 885–898 (2017)
2. W. Chen, L.-C. Wang, J. Bhadra, Simulation knowledge extraction and reuse in constrained random processor verification, in *ACM/IEEE Design Automation Conference* (2013)
3. L.-C. Wang, Data mining in functional test content optimization, in *ACM/IEEE Asian South Pacific Design Automation Conference* (2015)
4. G. Drmanac, F. Liu, L.-C. Wang, Predicting variability in nanoscale lithography processes, in *ACM/IEEE Design Automation Conference* (2009)
5. J. Chen, B. Bolin, L.-C. Wang, J. Zeng, D. (Gagi) Drmanac, M. Mateja, Mining AC delay measurements for understanding speed-limiting paths, in *IEEE International Test Conference* (2010)
6. I. Goodfellow, Y. Benjio, A. Courville, *Deep Learning* (The MIT Press, Cambridge, 2016)
7. J. Chen, L.-C. Wang, P.-H. Chang, J. Zeng, S. Yu, M. Metaja, Data learning techniques and methodology for Fmax prediction, in *IEEE International Test Conference* (2009)
8. J. Tikkanen, S. Siatkowski, N. Sumikawa, L.-C. Wang, M.S. Abadir, Yield optimization using advanced statistical correlation methods, in *IEEE International Test Conference* (2014)
9. S. Siatkowski, L.-C. Wang, N. Sumikawa, L. Winemberg, Learning the process for correlation analysis, in *IEEE VLSI Test Symposium* (2017)
10. F. Pedregosa et al., Scikit-learn: machine learning in python. J. Mach. Learn. Res. **12**, 2825–2830 (2010)
11. N. Callegari, D. (Gagi) Drmanac, L.-C. Wang, M.S. Abadir, Classification rule learning using subgroup discovery of cross-domain attributes responsible for design-silicon mismatch, in *ACM/IEEE Design Automation Conference*, 374–379 (2010)
12. D.H. Wolpert, The lack of a priori distinctions between learning algorithms. Neural Comput. **8**(7), 1341–1390 (1996)
13. M. Nero, J. Shan, L.-C. Wang, N. Sumikawa, Concept recognition in production yield data analytics, in *IEEE International Test Conference* (2018)
14. K. Hsieh, L.-C. Wang, W. Chen, J. Bhadra, Learning to produce direct tests for security verification using constrained process discovery, in *Design Automation Conference* (2017)
15. K-K. Hsieh, S. Siatkowski, L.-C. Wang, W. Chen, J. Bhadra, Feature extraction from design documents to enable rule learning for improving assertion coverage, in *ACM/IEEE Asia South Pacific Design Automation Conference* (2017)
16. L.G. Valiant, A theory of learnable. Commun. ACM **27**(11), 1134–1142 (1984)
17. M.J. Kearns, U. Vazirani, *An Introduction to Computational Learning Theory* (The MIT Press, Cambridge, 1994)
18. V. Vapnik, *The Nature of Statistical Learning Theory* (Springer, New York, 2000)
19. M.J. Kearns, U. Vazirani, Cryptographic limitations on learning Boolean formulae and finite automata. J. ACM **14**(1), 67–95 (1994)
20. A. Daniely, N. Linial, S. Shaleve-Shwartz, From average case complexity to improper learning complexity, in *ACM Symposium on Theory of Computing* (2014), pp. 441–448.
21. D.H. Wolpert, The relationship between Occam's Razor and convergent guessing. Complex Syst. **4**, 319–368 (1990)
22. J. Pearl, On the connection between the complexity and credibility of inferred models. Int. J. General Syst. **4**, 255–264 (1978)
23. A. Daniely, S. Shalev-Shwartz, Complexity theoretic limitations on learning DNF's, in *PMLR: Proceeding of Machine Learning Research*, vol. 49 (2016), pp. 815–830
24. R. Motwani, P. Raghavani, *Randomized Algorithms* (Cambridge University Press, Cambridge, 1995)
25. D. Haussler, Quantifying inductive bias: AI learning algorithms and valiant's learning framework. Artif. Intell. **36**, 177–221 (1998)

26. Y. Ben-Haim, A. Ivrii, O. Margalit, A. Matsliah, Perfect hashing and CNF encodings of cardinality constraints, in *International Conference on Theory and Applications of Satisfiability Testing* (Springer, Berlin, 2012), pp. 397–409
27. C. Sinz, Towards an optimal CNF encoding of boolean cardinality constraints, in *International Conference on Principles and Practice of Constraint Programming* (Springer, Berlin, 2005), pp. 827–831
28. K. Hsieh, L.-C. Wang, A concept learning tool based on calculating version space cardinality (2018). arXiv:1803.08625v1
29. A. Biere, Lingeling, Plingeling and Treengeling entering the SAT competition, in *Proceedings of SAT Competition* (2013)
30. I. Goodfellow, J. Pouget-Abadie, M. Mirza, B. Xu, D. Warde-Farley, S. Ozair, A. Courville, J. Bengio, Generative adversarial networks (2014). arXiv:1406.2661
31. A. Radford, L. Metz, S. Chintala, Unsupervised representation learning with deep convolutional generative adversarial networks (2016). arXiv:1511.06434v2

Part IV
Machine Learning for Analog Design

The successful construction of all machinery depends on the perfection of the tools employed, and whoever is a master in the art of tool-making possesses the key to the construction of all machines.

Charles Babbage

Chapter 14
Large-Scale Circuit Performance Modeling by Bayesian Model Fusion

Jun Tao, Fa Wang, Paolo Cachecho, Wangyang Zhang, Shupeng Sun, Xin Li, Rouwaida Kanj, Chenjie Gu, and Xuan Zeng

14.1 Introduction

The aggressive technology scaling has made it more difficult than ever to design and manufacture high-performance analog and mixed-signal (AMS) circuits (e.g., RF front-end, high-speed I/O link, etc.) that are robust to large-scale process variations [1–3]. Most AMS performance metrics (e.g., random offset) are extremely sensitive to the inter-die and/or intra-die variations associated with today's nanoscale manufacturing technology. For this reason, AMS circuits are not as scalable as digital

J. Tao (✉) · X. Zeng (✉)
State Key Laboratory of ASIC and System, School of Microelectronics, Fudan University, Shanghai, China
e-mail: taojun@fudan.edu.cn; xzeng@fudan.edu.cn

F. Wang · S. Sun
Department of Electrical and Computer Engineering, Carnegie Mellon University, Pittsburgh, PA, USA
e-mail: fwang1@ece.cmu.edu; shupengs@ece.cmu.edu

P. Cachecho · R. Kanj
Department of Electrical and Computer Engineering, American University of Beirut, Beirut, Lebanon

W. Zhang
Cadence Design Systems, Inc., Pittsburgh, PA, USA
e-mail: wzhang@cadence.com

X. Li (✉)
Department of Electrical and Computer Engineering, Duke University, Durham, NC, USA
e-mail: xinli.ece@duke.edu

C. Gu
Strategic CAD Labs, Intel Corporation, Hillsboro, OR, USA
e-mail: chenjie.gu@intel.com

I. M. Elfadel et al. (eds.), *Machine Learning in VLSI Computer-Aided Design*,
https://doi.org/10.1007/978-3-030-04666-8_14

circuits and they have been considered as the major bottleneck for future scaling of integrated circuit (IC) technology [4].

To address these challenges associated with AMS circuits, adaptive post-silicon tuning has been recently proposed [5–7] to facilitate the continuous scaling of AMS circuits. The adoption of post-silicon tuning, however, introduces tremendous new design challenges. Each AMS circuit now becomes a large-scale, complex system that can adaptively vary over time. It, in turn, brings up enormous technical problems related to computer-aided design of tunable AMS circuits.

- **Challenges for pre-silicon validation:** A tunable AMS circuit must contain a number of control knobs (e.g., switches, tunable bias voltage/current, etc.) for post-silicon reconfiguration. In some cases, these circuits may even include on-chip sensors and control blocks for "self-healing". Hence, the complexity of these AMS circuits substantially increases and the corresponding simulation cost for pre-silicon validation becomes increasingly large. This cost issue is especially critical for highly complex AMS circuits, such as phase-locked loop and high-speed I/O link, where a single transistor-level simulation may take a few days or even a few weeks to finish. In this case, it is extremely expensive to repeatedly run a large number of simulations over all process variations and environmental corners to validate a given design.
- **Challenges for post-silicon tuning:** Post-silicon tuning requires us to first measure the circuit performance by either on-chip sensors and/or off-chip equipment and then determine the optimal setup for all control knobs based on the measurement results. Towards this goal, a *tuning policy* must be developed to map the measured performance to the knob setup. In order to learn such a tuning policy over all possible process and environmental conditions, a lot of measurement data must be collected [8–11]. Furthermore, the tuning policy must be periodically re-calibrated to accommodate the process shift over time, thereby requiring additional measurement data. Collecting all these silicon data can be expensive, or even infeasible, as silicon measurement has become extremely time-consuming today.

In the literature, several data mining techniques [12–14] have been studied to reduce the cost (i.e., the number of required measurement data) of functional verification and test. However, for AMS pre-silicon validation and post-silicon tuning, most CAD methodologies today still need to collect a large amount of simulation and measurement data for pre-silicon validation and post-silicon tuning, resulting in prohibitively high cost. The challenging issue here is how to make AMS validation and tuning affordable by *reducing the amount of required data*. This fundamental issue has not been appropriately addressed by the state-of-the-art CAD tools. It, in turn, poses an immediate need to develop a radically new CAD framework for efficient validation and tuning of today's AMS circuits in order to facilitate the continuous scaling of ICs.

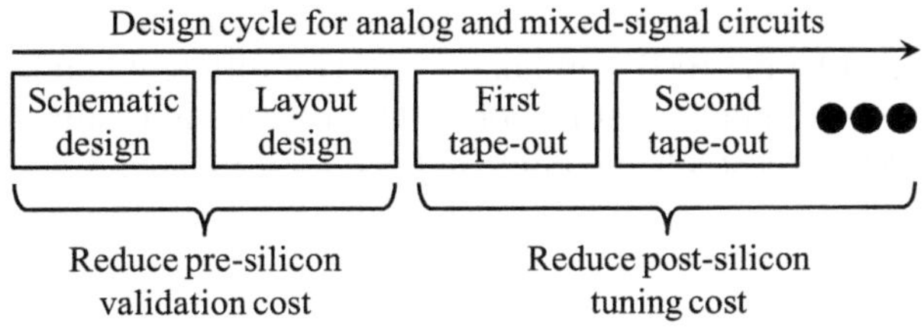

Fig. 14.1 Bayesian Model Fusion (BMF) reuses the early-stage data to facilitate efficient validation and tuning of AMS circuits at a late stage

In this chapter, we describe a novel idea of *Bayesian Model Fusion* (BMF) to address the aforementioned challenges [15–18]. BMF is motivated by the fact that today's AMS design cycle typically spans multiple stages, as shown in Fig. 14.1. At each stage, simulation and/or measurement data are collected to validate and/or tune the AMS circuit, before moving to the next stage. Most conventional AMS validation and tuning methods rely on the data collected at a single stage only and they completely ignore the data that are *already* generated at the previous stages. The key idea of BMF, however, is to *reuse* the early-stage data to facilitate efficient validation and tuning of AMS circuits at a late stage. As such, the amount of required data can be substantially reduced at the late stage. For instance, as shown in Fig. 14.1, we can borrow the simulation data from the schematic design to efficiently estimate the performance distributions of the post-layout simulation and, as a result, reduce pre-silicon validation cost.

Mathematically, BMF is derived from the theory of *Bayesian inference* [19, 20]. Starting from a set of early-stage data, it first "learns" the prior knowledge from these existing data and statistically "encodes" them as a *prior distribution*. Next, the prior distribution is combined with very few late-stage data to efficiently solve the pre-silicon validation and/or post-silicon tuning problem via Bayesian inference. By *fusing* the early-stage and late-stage data together through Bayesian inference, the simulation and/or measurement cost can be significantly reduced.

The BMF framework is generally applicable to a broad range of practical problems related to pre-silicon validation and post-silicon tuning of AMS circuits. The following are a few representative examples.

- **Example problem for pre-silicon validation:** How do we estimate the pre-silicon performance distribution and parametric yield of an AMS circuit based on *10* post-layout Monte Carlo simulation samples only?
- **Example problem for post-silicon tuning:** How do we learn the post-silicon tuning policy for a tunable AMS circuit based on the measurement data collected from *2* silicon chips only?

These problems seem extremely challenging, if not impossible, to solve. However, we will show in this chapter that BMF can successfully address these fundamental-yet-challenging problems and makes them tractable, as will be demonstrated by the simulation and measurement data of several industrial circuit examples.

The remainder of this chapter is organized as follows. In Sect. 14.2, we first describe the BMF formulation for two different applications of pre-silicon validation: (1) moment estimation and (2) distribution estimation. Next, we further extend BMF to post-silicon tuning in Sect. 14.3. Finally, we conclude in Sect. 14.4.

14.2 Pre-silicon Validation

In this section, we describe the theoretical framework of BMF for pre-silicon validation. While pre-silicon validation is a broad area involving many different research topics, our focus is to accurately estimate the statistics of a given AMS circuit performance (e.g., gain of an amplifier, phase noise of an oscillator, etc.). These statistical metrics include moments (e.g., mean and variance) and distributions (e.g., probability density function and cumulative distribution function) that are particularly useful for estimating the parametric yield of the AMS circuit.

14.2.1 Moment Estimation

Moment estimation is an important task to predict the performance distribution and, consequently, the parametric yield. In particular, if the performance distribution is Gaussian, it is fully characterized by the first two moments (i.e., mean and variance) [21]. Even if the performance distribution is non-Gaussian, it can be accurately estimated by finding the high-order moments [22].

Without loss of generality, we consider a given AMS performance of interest x. Most conventional statistical algorithms require a lot of data to estimate the moments of x [21–23]. However, in practice, pre-silicon simulation (e.g., post-layout simulation) is time-consuming. Especially for large-scale, complex AMS circuits such as phase-locked loop and high-speed I/O link, only a limited amount of (e.g., less than 5) simulation runs can be afforded. The key idea of BMF is to reuse the early-stage (e.g., schematic-level) data to accurately estimate the late-stage (e.g., post-layout) moments with very few late-stage data [16].

In this sub-section, we consider mean estimation as an example to illustrate the mathematical formulation of BMF. It should be noted that the BMF methodology can be generally extended to estimate other high-order moments as well [16]. Assume that the circuit performance x follows a Gaussian distribution; however, the early-stage performance distribution $pdf_E(x)$ and the late-stage performance distribution $pdf_L(x)$ have different mean and variance values:

$$pdf_E(x) = \frac{1}{\sqrt{2\pi} \cdot \sigma_E} \cdot \exp\left[-\frac{(x - \mu_E)^2}{2 \cdot \sigma_E^2} \right] \sim Gauss\left(\mu_E, \sigma_E^2\right), \tag{14.1}$$

$$pdf_L(x) = \frac{1}{\sqrt{2\pi} \cdot \sigma_L} \cdot \exp\left[-\frac{(x - \mu_L)^2}{2 \cdot \sigma_L^2} \right] \sim Gauss\left(\mu_L, \sigma_L^2\right), \tag{14.2}$$

where μ_E and σ_E are the mean and standard deviation of the early-stage distribution, respectively, and μ_L and σ_L are the mean and standard deviation of the late-stage distribution, respectively.

The early-stage mean μ_E can be estimated from the early-stage data. In most practical applications, the early-stage data are collected to validate the early-stage design, before we move to the late stage. For this reason, we should already know the early-stage mean μ_E, before estimating the late-stage mean μ_L. Namely, we assume that the early-stage mean μ_E is provided as the input to the BMF framework. Our goal is to find an accurate estimation of the late-stage mean μ_L based on very few late-stage data.

Unlike the conventional moment estimation techniques that completely ignore the correlation between the early-stage mean μ_E and the late-stage mean μ_L, BMF fully exploits such correlation information to reduce the amount of required data at the late stage. It consists of two major steps: (1) statistically encoding the prior knowledge of the early-stage data as a prior distribution and (2) optimally determining the late-stage mean μ_L by combining the prior distribution and very few late-stage data. In what follows, we will describe the mathematical formulations of these two steps, respectively.

Given the early-stage mean μ_E, we need to extract the prior knowledge that can be used to efficiently estimate the late-stage mean μ_L. On one hand, we expect that the early-stage mean μ_E and the late-stage mean μ_L are similar. In other words, the difference between μ_E and μ_L is small. On the other hand, the early-stage mean μ_E is not exactly identical to the late-stage mean μ_L due to multiple reasons. For example, comparing the schematic-level (i.e., early-stage) performance distribution with the post-layout (i.e., late-stage) performance distribution, the mean values of these two distributions can be different because the post-layout simulation of an AMS circuit includes the device and interconnect parasitics that are not available during the schematic-level simulation of the same circuit.

To statistically encode the "common" information between the early-stage mean μ_E and the late-stage mean μ_L, we conceptually assume that the "uncertainty" of the late-stage mean μ_L follows a prior distribution:

$$pdf(\mu_L) = \frac{1}{\sqrt{2\pi} \cdot \sigma_\mu} \cdot \exp\left[-\frac{(\mu_L - \mu_E)^2}{2 \cdot \sigma_\mu^2} \right] \sim Gauss\left(\mu_E, \sigma_\mu^2\right), \tag{14.3}$$

where $pdf(\mu_L)$ represents a Gaussian distribution with the mean μ_E and the standard deviation σ_μ. In practice, the value of the standard deviation σ_μ is unknown; however, it can be estimated by the cross-validation method developed by the machine learning community [20], as will be further discussed later in this sub-section.

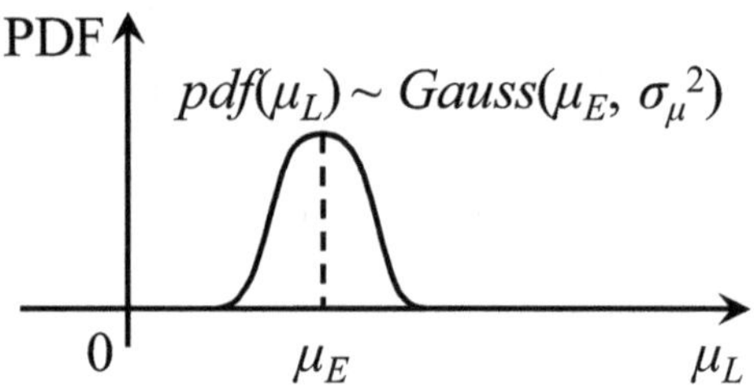

Fig. 14.2 A simple example is shown to illustrate the approach of extracting the prior knowledge from the early-stage data where the late-stage mean μ_L is close to the early-stage mean μ_E as encoded by the prior distribution $pdf(\mu_L)$

Figure 14.2 shows a simple example of the prior distribution for the late-stage mean μ_L. The prior distribution defined in (14.3) has a twofold meaning. First, the Gaussian distribution $pdf(\mu_L)$ is peaked at its mean value $\mu_L = \mu_E$, implying that the early-stage mean μ_E and the late-stage mean μ_L are likely to be similar. In other words, since the Gaussian distribution $pdf(\mu_L)$ exponentially decays with $(\mu_L - \mu_E)^2$, it is unlikely to observe a late-stage mean μ_L that is completely different from the early-stage mean μ_E. Second, the standard deviation σ_μ in (14.3) encodes our "confidence" of the prior knowledge. If the standard deviation σ_μ is small, the prior distribution is narrowly peaked around its mean value μ_E, implying that the late-stage mean is possibly close to the early-stage mean μ_E. Otherwise, if the standard deviation σ_μ is large, the prior distribution $pdf(\mu_L)$ widely spreads over a large range and the late-stage mean μ_L can possibly take a value that is far away from the early-stage mean μ_E.

Given the prior distribution $pdf(\mu_L)$ in (14.3), we further combine $pdf(\mu_L)$ with a few late-stage random samples $\{x_{L,n}; n = 1, 2, \ldots, N\}$ to accurately estimate the late-stage mean μ_L. These late-stage samples can be generated by running Monte Carlo simulation for the AMS circuit. Once the late-stage data are available, they can tell us additional information about the late-stage mean μ_L and, hence, help us to accurately estimate μ_L.

Based on Bayes' theorem [19, 20], the uncertainties of the late-stage mean μ_L after knowing the data $\{x_{L,n}; n = 1, 2, \ldots, N\}$ can be mathematically described by the following *posterior distribution*:

$$pdf(\mu_L|\mathbf{x}_L) \propto pdf(\mu_L) \cdot pdf(\mathbf{x}_L|\mu_L), \tag{14.4}$$

where $\mathbf{x}_L$ is a vector containing the late-stage random samples $\{x_{L,n}; n = 1, 2, \ldots, N\}$. In (14.4), the prior distribution $pdf(\mu_L)$ is defined by (14.3). The conditional distribution $pdf(\mathbf{x}_L|\mu_L)$ is referred to as the likelihood function. It measures the probability of observing the late-stage data $\{x_{L,n}; n = 1, 2, \ldots, N\}$ associated with the performance distribution $pdf_L(x)$:

$$pdf(\mathbf{x}_L|\mu_L) = \prod_{n=1}^{N} pdf_L(x_{L,n}|\mu_L). \tag{14.5}$$

Substituting (14.2) into (14.5) yields:

$$pdf(\mathbf{x}_L|\mu_L) = \prod_{n=1}^{N} \frac{1}{\sqrt{2\pi} \cdot \sigma_L} \cdot \exp\left[-\frac{(x_{L,n} - \mu_L)^2}{2 \cdot \sigma_L^2}\right]. \tag{14.6}$$

As shown in (14.6), the likelihood function $pdf(\mathbf{x}_L|\mu_L)$ depends on the late-stage mean μ_L that we aim to solve. The mean value μ_L controls the location of the probability density function $pdf_L(x)$ and, therefore, directly influences the likelihood function in (14.6).

Even after the late-stage data $\{x_{L,n}; n = 1, 2, \ldots, N\}$ are available, the late-stage mean μ_L is not deterministic. They must be modeled by the probability density function $pdf(\mu_L|\mathbf{x}_L)$ (i.e., the posterior distribution) in (14.4). Depending on the shape of the posterior distribution $pdf(\mu_L|\mathbf{x}_L)$, the late-stage mean μ_L does not take all possible values with equal probability. If the posterior distribution $pdf(\mu_L|\mathbf{x}_L)$ reaches its maximum value at μ_L^*, the value μ_L^* is considered as the optimal estimation of the late-stage mean, since it is the mean value that is most likely to occur. Such a method is referred to as the *maximum-a-posteriori* (MAP) estimation in the literature [19, 20].

The aforementioned MAP estimation can be mathematically formulated as an optimization problem:

$$\max_{\mu_L} \quad pdf(\mu_L) \cdot pdf(\mathbf{x}_L|\mu_L). \tag{14.7}$$

Substituting (14.3) and (14.6) into (14.7) and taking the logarithm for the merit function, we get:

$$\begin{aligned} \max_{\mu_L} \quad & -\frac{(\mu_L - \mu_E)^2}{2 \cdot \sigma_\mu^2} - \sum_{n=1}^{N} \frac{(x_{L,n} - \mu_L)^2}{2 \cdot \sigma_L^2} - \log\left(\sqrt{2\pi} \cdot \sigma_\mu\right) \\ & - N \cdot \log\left(\sqrt{2\pi} \cdot \sigma_L\right) \end{aligned}. \tag{14.8}$$

Based on the first-order optimality condition [24], we have:

$$-\frac{\mu_L - \mu_E}{\sigma_\mu^2} - \sum_{n=1}^{N} \frac{\mu_L - x_{L,n}}{\sigma_L^2} = 0. \tag{14.9}$$

Solving (14.9) results in the optimal estimation of the late-stage mean μ_L:

$$\mu_L = \frac{\sigma_L^2}{N \cdot \sigma_\mu^2 + \sigma_L^2} \cdot \mu_E + \frac{\sigma_\mu^2}{N \cdot \sigma_\mu^2 + \sigma_L^2} \cdot \sum_{n=1}^{N} x_{L,n}. \tag{14.10}$$

Defining the symbol:

$$\rho = \frac{\sigma_\mu}{\sigma_L}, \tag{14.11}$$

Equation (14.10) can be re-written as:

$$\mu_L = \frac{1}{N \cdot \rho^2 + 1} \cdot \mu_E + \frac{\rho^2}{N \cdot \rho^2 + 1} \cdot \sum_{n=1}^{N} x_{L,n}. \tag{14.12}$$

Studying (14.12) reveals two important observations. First, the estimated late-stage mean μ_L is a function of the early-stage mean μ_E. If the early-stage mean μ_E appropriately carries the prior information of the late-stage mean μ_L, it can help us to accurately estimate the late-stage mean μ_L via Bayesian inference. Second, the optimal estimation of the late-stage mean μ_L depends on both the standard deviation σ_μ of the prior distribution $pdf(\mu_L)$ in (14.3) and the standard deviation σ_L of the late-stage performance distribution $pdf_L(x)$ in (14.2). In practice, both σ_μ and σ_L are unknown. However, they can be estimated by cross-validation [20]. Importantly, since μ_L is a function of the ratio ρ in (14.11), we only need to estimate the ratio ρ, instead of the individual values of σ_μ and σ_L, by cross-validation.

To demonstrate the efficacy of BMF for mean estimation, we consider an industrial example of high-speed I/O link. One of its critical performance metrics is time margin. It measures the eye width at the receiver and is directly related to the bit error rate (BER) of the link. In this example, our objective is to estimate the mean of time margin for 8 different configurations where one configuration refers to a specific environmental corner (i.e., supply voltage and temperature) and a specific testing setup (i.e., channel for data transmission). Because accurately estimating the time margin requires us to pass a large number of random bits through the link and, hence, is extremely time-consuming, we cannot afford to simulate or measure a large number of dies (i.e., generate a large number of random samples) for mean estimation. For test and comparison purposes, two different estimators are implemented: (1) the conventional direct estimation based on sample mean [21] and (2) the BMF technique described in this sub-section. When applying BMF, we borrow the statistical information from one configuration and consider it as our prior knowledge to estimate the mean values of other configurations.

Figure 14.3 shows the relative error of mean estimation over 8 different configurations. Studying Fig. 14.3 reveals an important fact that given the same number of random samples, BMF consistently achieves 1.5× error reduction over the conventional direct estimation based on sample mean. On the other hand, to achieve the same accuracy, BMF can reduce the number of required random samples and, hence, the simulation and/or measurement cost by 2.25× over the conventional approach.

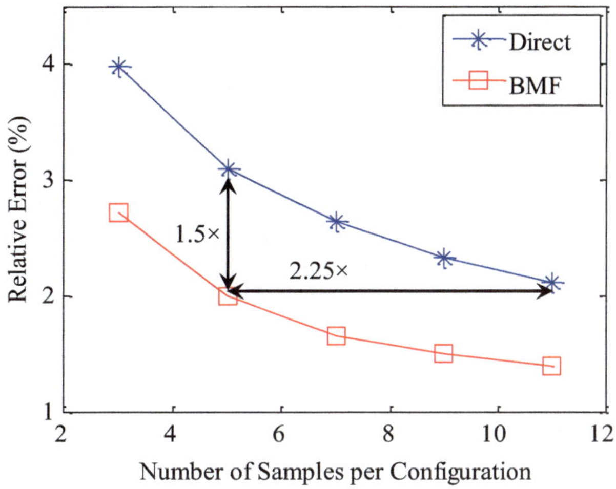

Fig. 14.3 Relative error of mean estimation over 8 different configurations is shown for the conventional direct estimation and the BMF method

14.2.2 Distribution Estimation

In addition to moment estimation, BMF can also be applied to estimate the probability distribution of a given AMS performance metric x. Unlike most conventional algorithms that require a lot of data to determine the distribution $pdf(x)$ [21–23], BMF attempts to reuse the early-stage data so that the late-stage performance distribution can be accurately estimated with very few late-stage data [15].

Towards this goal, instead of assuming that the early-stage and late-stage probability density functions $pdf_E(x)$ and $pdf_L(x)$ follow the Gaussian distributions in (14.1)–(14.2), we approximate $pdf_E(x)$ and $pdf_L(x)$ as the linear combinations of a set of basis functions:

$$pdf_E(x) \approx \sum_{m=1}^{M} \alpha_{E,m} \cdot b_m(x), \tag{14.13}$$

$$pdf_L(x) \approx \sum_{m=1}^{M} \alpha_{L,m} \cdot b_m(x), \tag{14.14}$$

where $\{b_m(x); m = 1, 2, \ldots, M\}$ contains the basis functions (e.g., orthogonal polynomials [25], wavelet basis functions [26], etc.), $\{\alpha_{E,m}; m = 1, 2, \ldots, M\}$ and $\{\alpha_{L,m}; m = 1, 2, \ldots, M\}$ contain the early-stage and late-stage coefficients, respectively, and M represents the total number of basis functions.

For a given set of basis functions $\{b_m(x); m = 1, 2, \ldots, M\}$, the performance distributions $pdf_E(x)$ and $pdf_L(x)$ are uniquely determined by the coefficients $\{\alpha_{E,m}; m = 1, 2, \ldots, M\}$ and $\{\alpha_{L,m}; m = 1, 2, \ldots, M\}$. Here, we assume that the early-stage coefficients $\{\alpha_{E,m}; m = 1, 2, \ldots, M\}$ are already known and the late-stage coefficients $\{\alpha_{L,m}; m = 1, 2, \ldots, M\}$ should be estimated to determine the late-stage performance distribution $pdf_L(x)$. To statistically define the prior knowledge for the late-stage performance distribution $pdf_L(x)$, we consider the following two possible approaches.

- **Encoding the prior information based on coefficient magnitude:** Several previous works in the literature have demonstrated that when approximating a nonlinear function such as $pdf_E(x)$ or $pdf_L(x)$ in (14.13)–(14.14) by a set of appropriately selected basis functions (e.g., wavelet basis), many of the resulting coefficients are close to zero [26]. In other words, these coefficients carry a unique *sparse* pattern. To statistically encode such a sparse pattern as our prior knowledge, we exploit the fact that if the early-stage coefficient $\alpha_{E,m}$ has a large (or small) magnitude, it is likely that the late-stage coefficient $\alpha_{L,m}$ also has a large (or small) magnitude. In particular, we model each late-stage coefficient as a zero-mean Gaussian distribution:

$$\begin{aligned} pdf(\alpha_{L,m}) &= \frac{1}{\sqrt{2\pi} \cdot \lambda \cdot |\alpha_{E,m}|} \cdot \exp\left(-\frac{\alpha_{L,m}^2}{2 \cdot \lambda^2 \cdot \alpha_{E,m}^2}\right), \\ &\sim Gauss\left(0, \lambda^2 \cdot \alpha_{E,m}^2\right) \quad (m = 1, 2, \ldots, M) \end{aligned} \tag{14.15}$$

  where the standard deviation of the Gaussian distribution is equal to $\lambda \cdot |\alpha_{E,m}|$. In (14.15), the parameter λ is positive and it controls the variance of the distribution. The appropriate value of λ can be determined by cross-validation [20].

  The standard deviation $\lambda \cdot |\alpha_{E,m}|$ in (14.15) encodes the magnitude information of the late-stage coefficient $\alpha_{L,m}$. If the magnitude of the early-stage coefficient $|\alpha_{E,m}|$ is small and, hence, the standard deviation $\lambda \cdot |\alpha_{E,m}|$ is small, the prior distribution $pdf(\alpha_{L,m})$ is narrowly peaked around zero, implying that the late-stage coefficient $\alpha_{L,m}$ is possibly close to zero. Otherwise, if the magnitude of the early-stage coefficient $|\alpha_{E,m}|$ is large and, hence, the standard deviation $\lambda \cdot |\alpha_{E,m}|$ is large, the prior distribution $pdf(\alpha_{L,m})$ widely spreads over a large range and the late-stage coefficient $\alpha_{L,m}$ can possibly take a value that is far away from zero. Figure 14.4(a) shows a simple example of the prior distribution for two late-stage coefficients $\alpha_{L,1}$ and $\alpha_{L,2}$ where $|\alpha_{E,1}|$ is small and $|\alpha_{E,2}|$ is large.
- **Encoding the prior information based on coefficient value:** If the values of the early-stage and late-stage coefficients, in addition to their magnitudes, are close, we can define a prior distribution that is different from (14.15):

$$\begin{aligned} pdf(\alpha_{L,m}) &= \frac{1}{\sqrt{2\pi} \cdot \lambda \cdot |\alpha_{E,m}|} \cdot \exp\left[-\frac{(\alpha_{L,m} - \alpha_{E,m})^2}{2 \cdot \lambda^2 \cdot \alpha_{E,m}^2}\right], \\ &\sim Gauss\left(\alpha_{E,m}, \lambda^2 \cdot \alpha_{E,m}^2\right) \quad (m = 1, 2, \ldots, M) \end{aligned} \tag{14.16}$$

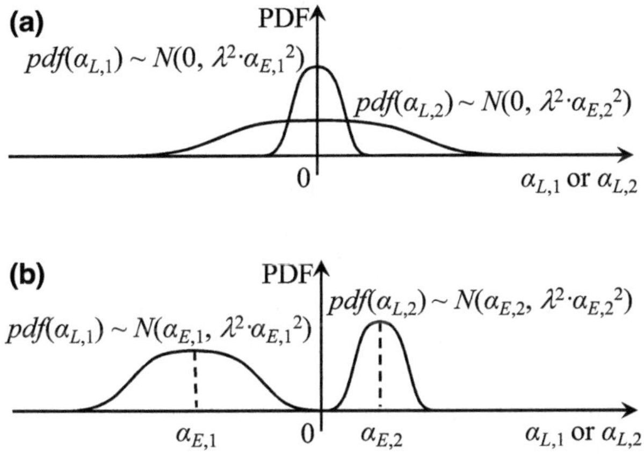

Fig. 14.4 Two simple examples are shown to illustrate the approaches of extracting the prior knowledge from the early-stage data. (**a**) Encode the prior information based on coefficient magnitude where the late-stage coefficient $\alpha_{L,1}$ is possibly close to zero and the late-stage coefficient $\alpha_{L,2}$ can possibly be far away from zero. (**b**) Encode the prior information based on coefficient value where the late-stage coefficients $\alpha_{L,1}$ and $\alpha_{L,2}$ are close to the early-stage coefficients $\alpha_{E,1}$ and $\alpha_{E,2}$, respectively, and both $\alpha_{L,1}$ and $\alpha_{L,2}$ are provided with a relatively equal opportunity to deviate from the corresponding early-stage coefficients $\alpha_{E,1}$ and $\alpha_{E,2}$

where $pdf(\alpha_{L,m})$ represents a Gaussian distribution with the mean value $\alpha_{E,m}$ and the standard deviation $\lambda \cdot |\alpha_{E,m}|$. Similar to (14.15), the parameter λ in (14.16) is positive and its value can be determined by cross-validation [20].

The prior distribution in (14.16) has a twofold meaning. First, the Gaussian distribution $pdf(\alpha_{L,m})$ is peaked at its mean value $\alpha_{L,m} = \alpha_{E,m}$, implying that the early-stage coefficient $\alpha_{E,m}$ and the late-stage coefficient $\alpha_{L,m}$ are likely to be similar. Second, the standard deviation of the prior distribution $pdf(\alpha_{L,m})$ is proportional to $|\alpha_{E,m}|$. It means that the absolute difference between the late-stage coefficient $\alpha_{L,m}$ and the early-stage coefficient $\alpha_{E,m}$ can be large (or small), if the magnitude of the early-stage coefficient $|\alpha_{E,m}|$ is large (or small). Restating in words, each late-stage coefficient $\alpha_{L,m}$ has been provided with a relatively equal opportunity to deviate from the corresponding early-stage coefficient $\alpha_{E,m}$. Figure 14.4b shows a simple example of the aforementioned prior distribution for two late-stage coefficients $\alpha_{L,1}$ and $\alpha_{L,2}$ where $\alpha_{E,1}$ is negative and $\alpha_{E,2}$ is positive.

Comparing the two prior distributions in (14.15) and (14.16), we notice that $pdf(\alpha_{L,m})$ in (14.16) carries the sign information (i.e., positive or negative) of the late-stage coefficient, while $pdf(\alpha_{L,m})$ in (14.15) does not. In practice, the efficacy of different prior definitions is case-dependent. How to choose the optimal prior distribution for a given application case remains an open question and should be considered as an important topic for future research.

To complete the definition of the prior distribution for all late-stage coefficients $\{\alpha_{L,m}; m = 1, 2, \ldots, M\}$, we further assume that these coefficients are statistically independent and their joint distribution is represented as:

$$pdf(\boldsymbol{\alpha}_L) = \prod_{m=1}^{M} pdf(\alpha_{L,m}), \tag{14.17}$$

where $\boldsymbol{\alpha}_L$ is a vector containing all late-stage coefficients $\{\alpha_{L,m}; m = 1, 2, \ldots, M\}$ and the prior distribution $pdf(\alpha_{L,m})$ for the mth late-stage coefficient $\alpha_{L,m}$ can be taken from either (14.15) or (14.16), depending on the specific application case. The independence assumption in (14.17) simply implies that we do not know the correlation information among these coefficients as our prior knowledge. The correlation information will be learned from the late-stage data, when the posterior distribution is calculated by Bayesian inference.

Given the prior distribution $pdf(\boldsymbol{\alpha}_L)$ defined in (14.17), we further combine $pdf(\boldsymbol{\alpha}_L)$ with a few late-stage random samples $\{x_{L,n}; n = 1, 2, \ldots, N\}$ to calculate the posterior distribution:

$$pdf(\boldsymbol{\alpha}_L|\mathbf{x}_L) \propto pdf(\boldsymbol{\alpha}_L) \cdot pdf(\mathbf{x}_L|\boldsymbol{\alpha}_L), \tag{14.18}$$

where $\mathbf{x}_L$ is a vector containing the late-stage random samples $\{x_{L,n}; n = 1, 2, \ldots, N\}$. In (14.18), the likelihood function $pdf(\mathbf{x}_L|\boldsymbol{\alpha}_L)$ measures the probability of observing the late-stage data $\{x_{L,n}; n = 1, 2, \ldots, N\}$ associated with the performance distribution $pdf_L(x)$ in (14.14):

$$pdf(\mathbf{x}_L|\boldsymbol{\alpha}_L) = \prod_{n=1}^{N} \sum_{m=1}^{M} \alpha_{L,m} \cdot b_m(x_{L,n}). \tag{14.19}$$

To determine the late-stage coefficients $\{\alpha_{L,m}; m = 1, 2, \ldots, M\}$ by MAP estimation, we formulate the following optimization:

$$\max_{\boldsymbol{\alpha}_L} \quad pdf(\boldsymbol{\alpha}_L) \cdot pdf(\mathbf{x}_L|\boldsymbol{\alpha}_L). \tag{14.20}$$

Substituting (14.17) and (14.19) into (14.20) and taking the logarithm for the merit function, we have:

$$\max_{\boldsymbol{\alpha}_L} \quad \sum_{m=1}^{M} \log[pdf(\alpha_{L,m})] + \sum_{n=1}^{N} \log\left[\sum_{m=1}^{M} \alpha_{L,m} \cdot b_m(x_{L,n})\right] \tag{14.21}$$

In addition, since the integral of any probability density function must be equal to 1, we need to further consider the following constraint when solving the late-stage coefficients $\{\alpha_{L,m}; m = 1, 2, \ldots, M\}$:

$$\int_{-\infty}^{+\infty} pdf_L(x) \cdot dx = \sum_{m=1}^{M} \alpha_{L,m} \cdot \int_{-\infty}^{+\infty} b_m(x) \cdot dx = 1 \tag{14.22}$$

Combining (14.21) and (14.22) results in the following constrained optimization problem:

$$\begin{aligned} \max_{\boldsymbol{\alpha}_L} \quad & \sum_{m=1}^{M} \log[pdf(\alpha_{L,m})] + \sum_{n=1}^{N} \log\left[\sum_{m=1}^{M} \alpha_{L,m} \cdot b_m(x_{L,n})\right] \\ s.t. \quad & \sum_{m=1}^{M} \alpha_{L,m} \cdot \int_{-\infty}^{+\infty} b_m(x) \cdot dx = 1 \end{aligned} \tag{14.23}$$

It can be proven that the merit function in (14.23) is concave [15], if the prior distribution $pdf(\alpha_{L,m})$ is taken from either (14.15) or (14.16). Furthermore, the constraint function is simply linear with respect to the problem unknowns $\{\alpha_{L,m}; m = 1, 2, \ldots, M\}$, implying that the constraint set is convex. For these reasons, the optimization in (14.23) is a convex programming problem and can be solved both efficiently (i.e., with low computational cost) and robustly (i.e., with guaranteed global optimum) [24]. Once the late-stage coefficients $\{\alpha_{L,m}; m = 1, 2, \ldots, M\}$ are found from (14.23), the late-stage performance distribution $pdf_L(x)$ in (14.14) is determined.

As an example to demonstrate the efficacy of BMF for distribution estimation, Fig. 14.5(a) shows the simplified circuit schematic of an SRAM (static random-access memory) read path designed in a commercial 32 nm CMOS process. In this example, our objective is to reuse the schematic-level (i.e. early-stage) simulation data to efficiently estimate the post-layout (i.e., late-stage) probability distribution of the read path delay from the word line (*WL*) to the sense amplifier output (*Out*). For testing and comparison purposes, 1000 random samples are generated by both schematic-level and post-layout simulations. The runtime to generate a single schematic-level simulation sample is 26.44 s, and it is 104.58 s for post-layout simulation. Two different techniques are implemented: (1) the conventional kernel estimation based on Gaussian kernel with optimal bandwidth [27] and (2) the BMF technique based on DCT (discrete cosine transform) basis functions [26]. Figure 14.5(b) shows the estimation error for the cumulative distribution function (i.e., the integral of the probability density function). Note that the conventional kernel estimation requires 10× more simulation samples than BMF to achieve the same accuracy. In other words, BMF achieves 10× runtime speedup over kernel estimation in this example. It reduces the total post-layout simulation time from 2.91 h (i.e., 100 samples) to 17.43 min (i.e., 10 samples).

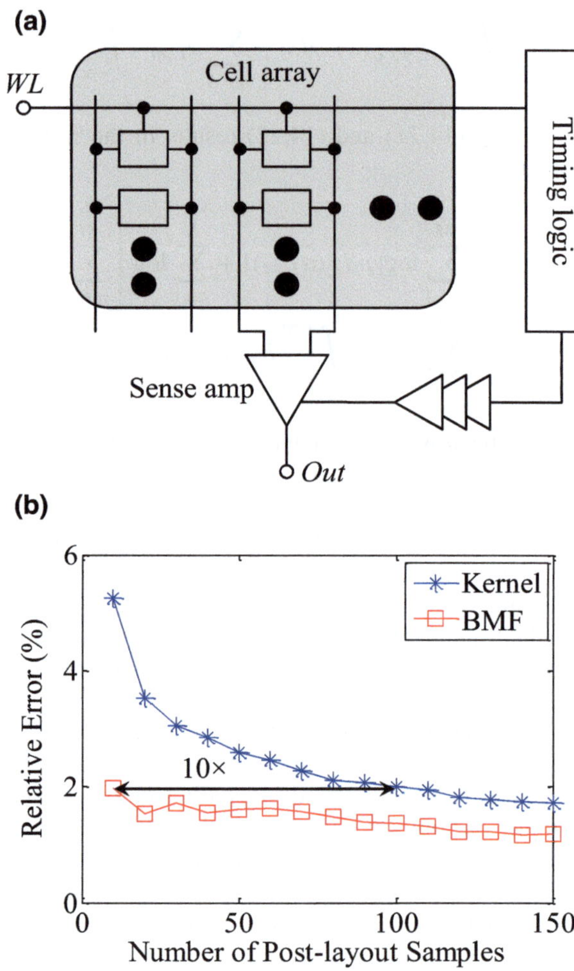

Fig. 14.5 (**a**) A simplified circuit schematic is shown for an SRAM read path designed in a commercial 32 nm CMOS process. (**b**) Estimation error of the cumulative distribution function for the post-layout read path delay is shown for the conventional kernel estimation and the BMF technique

14.3 Post-silicon Tuning

In this section, we further extend BMF to the application of post-silicon tuning. In particular, we focus on the problem of tuning policy generation with consideration of large-scale process shift over time. Namely, we aim to solve a regression modeling problem to find the mapping between the measured AMS performance and the control knob setup:

$$f(\mathbf{x}) \approx \sum_{m=1}^{M} \alpha_m \cdot b_m(\mathbf{x}), \tag{14.24}$$

where $\mathbf{x}$ is a vector containing the measurement results of circuit performance, f represents the optimal value of a control knob, $\{\alpha_m; m = 1, 2, \ldots, M\}$ contains the model coefficients, $\{b_m(\mathbf{x}); m = 1, 2, \ldots, M\}$ contains the basis functions (e.g., polynomials), and M is the total number of basis functions. Note that the regression modeling problem in (14.24) is substantially *different* from the distribution estimation problem in (14.13)–(14.14), even though both problems involve the linear combination of M basis functions.

There have been many regression modeling algorithms that were developed in the literature [28–37]. Most of these existing methods require a lot of data to fit the model, especially if the function $f(\mathbf{x})$ is strongly nonlinear and/or high-dimensional. BMF, however, reuses the early-stage data so that $f(\mathbf{x})$ can be accurately approximated with very few late-stage data [17, 18]. In what follows, we will describe the BMF formulation for the application of post-silicon tuning.

Similar to (14.13)–(14.14), we consider two different regression models: the early-stage model $f_E(\mathbf{x})$ and the late-stage model $f_L(\mathbf{x})$:

$$f_E(\mathbf{x}) \approx \sum_{m=1}^{M} \alpha_{E,m} \cdot b_m(\mathbf{x}) \tag{14.25}$$

$$f_L(\mathbf{x}) \approx \sum_{m=1}^{M} \alpha_{L,m} \cdot b_m(\mathbf{x}), \tag{14.26}$$

where $\{\alpha_{E,m}; m = 1, 2, \ldots, M\}$ and $\{\alpha_{L,m}; m = 1, 2, \ldots, M\}$ represent the early-stage and late-stage coefficients, respectively. The early-stage model $f_E(\mathbf{x})$ can be fitted by either the pre-silicon data collected from numerical simulation or the post-silicon data measured from previous tape-out. We assume that the early-stage model $f_E(\mathbf{x})$ is already available. Given the early-stage coefficients $\{\alpha_{E,m}; m = 1, 2, \ldots, M\}$, we can define the prior distribution $pdf(\boldsymbol{\alpha}_L)$ based on (14.15)–(14.17), similar to the case of pre-silicon validation described in Sect. 14.2.2.

Once the prior distribution $pdf(\boldsymbol{\alpha}_L)$ is defined, we collect a few late-stage data $\{(\mathbf{x}^{(n)}, f_L^{(n)}); n = 1, 2, \ldots, N\}$, where $\mathbf{x}^{(n)}$ and $f_L^{(n)}$ are the values of $\mathbf{x}$ and $f_L(\mathbf{x})$ at the nth data point, respectively, by measuring several chips from the new tape-out. Next, we combine the prior distribution $pdf(\boldsymbol{\alpha}_L)$ with the late-stage data $\{(\mathbf{x}^{(n)}, f_L^{(n)}); n = 1, 2, \ldots, N\}$ to solve the late-stage coefficients $\{\alpha_{L,m}; m = 1, 2, \ldots, M\}$ by MAP estimation. As such, a new tuning policy is efficiently generated to accommodate the process shift. To this end, we formulate the following posterior distribution:

$$pdf(\boldsymbol{\alpha}_L|\mathbf{f}_L) \propto pdf(\boldsymbol{\alpha}_L) \cdot pdf(\mathbf{f}_L|\boldsymbol{\alpha}_L), \tag{14.27}$$

where $\boldsymbol{\alpha}_L$ is a vector containing all late-stage coefficients $\{\alpha_{L,m}; m = 1, 2, \ldots, M\}$ and $\mathbf{f}_L$ is a vector containing the late-stage data $\{f_L^{(n)}; n = 1, 2, \ldots, N\}$.

To derive the likelihood function $pdf(\mathbf{f}_L|\boldsymbol{\alpha}_L)$, we further assume that the error for the late-stage model $f_L(\mathbf{x})$ can be represented as a random variable that follows a zero-mean Gaussian distribution and, hence, the approximate equality in (14.26) can be re-written as:

$$f_L(\mathbf{x}) = \sum_{m=1}^{M} \alpha_{L,m} \cdot b_m(\mathbf{x}) + \varepsilon_L, \tag{14.28}$$

where ε_L denotes the modeling error:

$$pdf\ (\varepsilon_L) = \frac{1}{\sqrt{2\pi} \cdot \sigma_0} \cdot \exp\left(-\frac{\varepsilon_L^2}{2 \cdot \sigma_0^2}\right) \sim Gauss\left(0, \sigma_0^2\right). \tag{14.29}$$

In (14.29), the standard deviation σ_0 indicates the magnitude of the modeling error. Its value can be determined by cross-validation [20]. Given (14.28)–(14.29), since the modeling error associated with the nth data point $(\mathbf{x}^{(n)}, f_L^{(n)})$ is simply one sample of the random variable ε_L, it follows the Gaussian distribution:

$$f_L^{(n)} - \sum_{m=1}^{M} \alpha_{L,m} \cdot b_m\left(\mathbf{x}^{(n)}\right) \sim Gauss\left(0, \sigma_0^2\right). \tag{14.30}$$

Therefore, the probability of observing the nth data point is:

$$\begin{aligned} pdf\left(f_L^{(n)}|\boldsymbol{\alpha}_L\right) = & \frac{1}{\sqrt{2\pi} \cdot \sigma_0} \cdot \exp\left\{-\frac{1}{2 \cdot \sigma_0^2} \cdot\right. \\ & \left.\left[f_L^{(n)} - \sum_{m=1}^{M} \alpha_{L,m} \cdot b_m\left(\mathbf{x}^{(n)}\right)\right]^2\right\} \end{aligned} \tag{14.31}$$

Assuming that all sampling points are independently generated, we can write the likelihood function $pdf(\mathbf{f}_L|\boldsymbol{\alpha}_L)$ as:

$$pdf(\mathbf{f}_L|\boldsymbol{\alpha}_L) = \prod_{n=1}^{N} pdf\left(f_L^{(n)}|\boldsymbol{\alpha}_L\right). \tag{14.32}$$

Combining (14.17), (14.27), and (14.32) results in the following optimization formulation for MAP estimation:

$$\max_{\boldsymbol{\alpha}_L} \quad \sum_{m=1}^{M} \log[pdf(\alpha_{L,m})] - \frac{1}{2 \cdot \sigma_0^2} \cdot \sum_{n=1}^{N} \left[f_L^{(n)} - \sum_{m=1}^{M} \alpha_{L,m} \cdot b_m\left(\mathbf{x}^{(n)}\right)\right]^2. \tag{14.33}$$

Note that the MAP formulation in (14.33) is completely different from the formulation in (14.23) that was previously developed for distribution estimation of pre-silicon validation. It can be proven that the unconstrained optimization in (14.33) is convex [17, 18] and, hence, can be solved both efficiently and robustly [24]. Once the late-stage coefficients $\{\alpha_{L,m}; m = 1, 2, \ldots, M\}$ are found from (14.33), the new tuning policy (i.e., the late-stage model) $f_L(\mathbf{x})$ in (14.26) is determined.

As an example for demonstration, we apply BMF to generate the tuning policy for a VCO (voltage-controlled oscillator) designed in a commercial 32 nm CMOS process, as shown in Fig. 14.6(a). In this example, the objective of post-silicon tuning is to appropriately set the bias voltage V_{Bias} controlled by a DAC so that the phase noise of the VCO is minimized. To this end, we measure a number of impor-

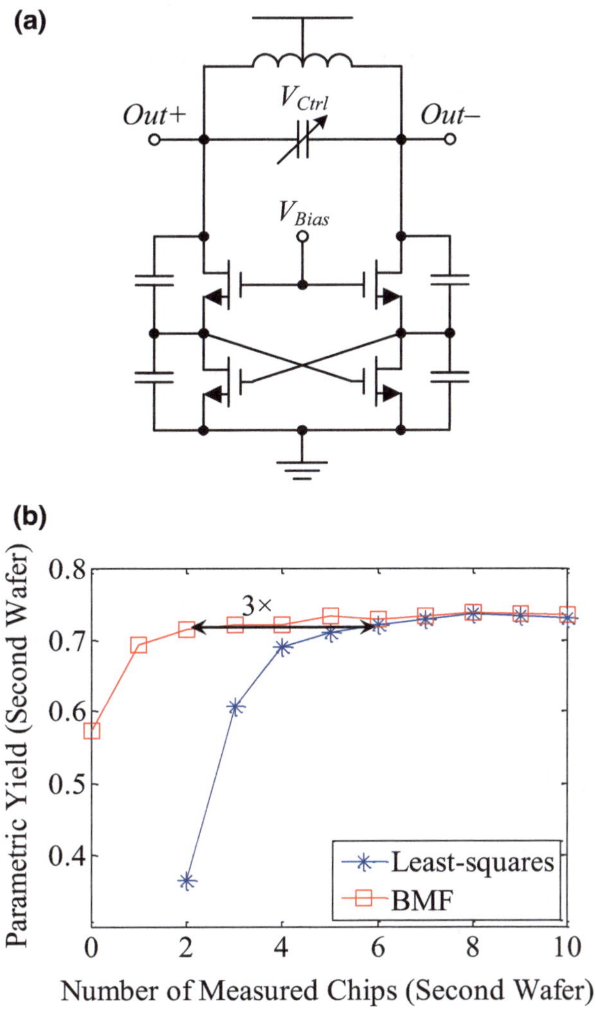

Fig. 14.6 (**a**) A simplified circuit schematic is shown for a VCO designed in a commercial 32 nm CMOS process where 116 chips from two wafers are measured. (**b**) Parametric yield achieved by post-silicon tuning for the second wafer is shown for two different tuning policies generated by the conventional least-squares fitting and the BMF technique, respectively

tant performance metrics (i.e., oscillation frequency, oscillation amplitude, bias current, etc.) under the nominal bias voltage and then determine the optimal value of the control knob V_{Bias} based on these measurement results. Based on these measurement results, a quadratic polynomial model $f(\mathbf{x})$ can be fitted to estimate the phase noise. Next, V_{Bias} corresponding to the minimum phase noise is chosen as the optimal solution by using a brute-force search algorithm based on the model $f(\mathbf{x})$.

For this VCO example, two wafers with 116 VCO chips in total are produced by an industrial partner. These two wafers are not manufactured at the same time and a significant process shift is observed. Our goal is to reuse the measurement data from the first wafer (i.e., the early-stage data) to efficiently generate the tuning policy for the second wafer in order to accommodate the large-scale process shift. For testing and comparison purposes, two different techniques are implemented: (1) the conventional least-squares fitting [28] and (2) the BMF technique described in this section. Figure 14.6(b) plots the parametric yield achieved by post-silicon tuning for the second wafer as a function of the number of measured chips. Note that BMF only needs to measure 2 chips from the second wafer in order to generate the new tuning policy with the yield higher than 70%. To achieve the same parametric yield, the conventional least-squares fitting must measure 6 chips (3× more) in this example.

14.4 Conclusions

In this chapter, a Bayesian Model Fusion (BMF) framework is described to reduce the simulation and/or measurement cost for both pre-silicon validation and post-silicon tuning. The key idea of BMF is to reuse the data that was previously generated at an early stage. It statistically extracts the prior knowledge from the early-stage data to facilitate efficient late-stage validation and/or tuning with minimal simulation and/or measurement cost. BMF is mathematically derived from the theory of Bayesian theory and its efficacy is demonstrated for several industrial circuit examples in this chapter. We envision that BMF can be generally applied to a broad range of circuit applications and a large number of follow-up research works are needed in this emerging area.

References

1. S. Nassif, Modeling and analysis of manufacturing variations, in *IEEE Conference on Custom Integrated Circuits* (2001), pp. 223–228
2. X. Li, J. Le, L. Pileggi, *Statistical Performance Modeling and Optimization* (Now Publishers, Hanover, 2007)
3. Semiconductor Industry Associate, *International Technology Roadmap for Semiconductors* (Semiconductor Industry Association, Washington, 2011)

4. R. Rutenbar, G. Georges, J. Roychowdhury, Hierarchical modeling, optimization and synthesis for system-level analog and RF designs. Proc. IEEE **95**(3), 640–669 (2007)
5. A. Tang, F. Hsiao, D. Murphy, I. Ku, J. Liu, S. D'Souza, N. Wang, H. Wu, Y. Wang, M. Tang, G. Virbila, M. Pham, D. Yang, Q. Gu, Y. Wu, Y. Kuan, C. Chien, M. Chang, A low-overhead self-healing embedded system for ensuring high yield and long-term sustainability of 60GHz 4Gb/s radio-on-a-chip, in *IEEE International Solid-State Circuits Conference* (2012), pp. 316–318
6. S. Bowers, K. Sengupta, K. Dasgupta, A. Hajimiri, A fully-integrated self-healing power amplifier, in *IEEE Radio Frequency Integrated Circuits Symposium* (2012)
7. J. Plouchart, M. Ferriss, A. Natarajan, A. Valdes-Garcia, B. Sadhu, A. Rylyakov, B. Parker, M. Beakes, A. Babakani, S. Yaldiz, L. Pileggi, R. Harjani, S. Reynolds, J. Tierno, D. Friedman, A 23.5GHz PLL with an adaptively biased VCO in 32nm SOI-CMOS, in *IEEE Custom Integrated Circuits Conference* (2012)
8. N. Kupp, H. Huang, P. Drineas, Y. Makris, Post-production performance calibration in analog/RF devices, in *IEEE International Test Conference* (2010), pp. 245–254
9. D. Han, B. Kim, A. Chatterjee, DSP-driven self-tuning of RF circuits for process-induced performance variability. IEEE Trans. Very Large Scale Integr. Syst. **18**(2), 305–314 (2010)
10. E. Acar, S. Ozev, Low cost MIMO testing for RF integrated circuits. IEEE Trans. Very Large Scale Integr. Syst. **18**(9), 1348–1356 (2010)
11. S. Yaldiz, V. Calayir, X. Li, L. Pileggi, A. Natarajan, M. Ferriss, J. Tierno, Indirect phase noise sensing for self-healing voltage controlled oscillators, in *IEEE Custom Integrated Circuits Conference* (2011)
12. L. Wang, Data mining in functional test content optimization, in *IEEE Asia and South Pacific Design Automation Conference* (2015), pp. 308–315
13. L. Wang, Experience of data analytics in EDA and test-principles, promises, and challenges. IEEE Trans. Comput. Aided Des. Integr. Circuits Syst. **36**(6), 885–898 (2017)
14. W. Chen, L. Wang, J. Bhadra, M. Abadir, Simulation knowledge extraction and reuse in constrained random processor verification, in *Design Automation Conference* (2013)
15. X. Li, W. Zhang, F. Wang, S. Sun, C. Gu, Efficient parametric yield estimation of analog/mixed-signal circuits via Bayesian model fusion, in *International Conference on Computer-Aided Design* (2012), pp. 627–634
16. C. Gu, E. Chiprout, X. Li, Efficient moment estimation with extremely small sample size via Bayesian inference for analog/mixed-signal validation, in *Design Automation Conference* (2013)
17. F. Wang, W. Zhang, S. Sun, X. Li, C. Gu, Bayesian model fusion: large-scale performance modeling of analog and mixed-signal circuits by reusing early-stage data, in *Design Automation Conference* (2013)
18. S. Sun, F. Wang, S. Yaldiz, X. Li, L. Pileggi, A. Natarajan, M. Ferriss, J. Plouchart, B. Sadhu, B. Parker, A. Valdes-Garcia, M. Sanduleanu, J. Tierno, D. Friedman, Indirect performance sensing for on-chip analog self-healing via Bayesian model fusion, in *IEEE Custom Integrated Circuits Conference* (2013)
19. G. Casella, R. Berger, *Statistical Inference* (Duxbury Press, Pacific Grove, 2001)
20. C. Bishop, *Pattern Recognition and Machine Learning* (Prentice Hall, Englewood Cliffs, 2007)
21. A. Papoulis, U. Pillai, *Probability, Random Variables and Stochastic Processes* (McGraw-Hill, New York, 2001)
22. X. Li, J. Le, P. Gopalakrishnan, L. Pileggi, Asymptotic probability extraction for nonnormal performance distributions. IEEE Trans. Comput. Aided Des. Integr. Circuits Syst. **26**(1), 16–37 (2007)
23. A. Graupner, W. Schwarz, R. Schufny, Statistical analysis of analog structures through variance calculation. IEEE Trans. Circuits Syst. I Fundam. Theory Appl. **49**(8), 1071–1078 (2002)
24. S. Boyd, L. Vandenberghe, *Convex Optimization* (Cambridge University Press, Cambridge, 2004)
25. P. Beckmann, *Orthogonal Polynomials for Engineers and Physicists* (Golem Press, Boulder, 1973)

26. R. Gonzalez, R. Woods, *Digital Image Processing* (Prentice Hall, Upper Saddle River, 2007)
27. B. Silverman, *Density Estimation for Statistics and Data Analysis* (Chapman & Hall/CRC, London, 1986)
28. G. Box, N. Draper, *Empirical Model-Building and Response Surfaces* (Wiley, New York, 1987)
29. J. Swidzinski, M. Styblinski, G. Xu, Statistical behavioral modeling of integrated circuits. IEEE Int. Symp. Circuits Syst. **6**, 98–101 (1998)
30. W. Daems, G. Gielen, W. Sansen, An efficient optimization-based technique to generate posynomial performance models for analog integrated circuits, in *Design Automation Conference* (2002), pp. 431–436
31. H. Liu, A. Singhee, R. Rutenbar, L. Carley, Remembrance of circuits past: macromodeling by data mining in large analog design spaces, in *Design Automation Conference* (2002), pp. 437–442
32. X. Li, J. Le, L. Pileggi, A. Strojwas, Projection-based performance modeling for inter/intra-die variations, in *International Conference on Computer-Aided Design* (2005), pp. 721–727
33. Z. Feng et al., Performance-oriented statistical parameter reduction of parameterized systems via reduced rank regression, in *International Conference on Computer-Aided Design* (2006), pp. 868–875
34. A. Singhee, R. Rutenbar, Beyond low-order statistical response surfaces: latent variable regression for efficient, highly nonlinear fitting, in *Design Automation Conference* (2007), pp. 256–261
35. A. Mitev, M. Marefat, D. Ma, J. Wang, Principle Hessian direction based parameter reduction with process variation, in *International Conference on Computer-Aided Design* (2007), pp. 632–637
36. T. McConaghy, G. Gielen, Template-free symbolic performance modeling of analog circuits via canonical-form functions and genetic programming. IEEE Trans. Comput. Aided Des. Integr. Circuits Syst. **28**(8), 1162–1175 (2009)
37. X. Li, Finding deterministic solution from underdetermined equation: large-scale performance modeling of analog/RF circuits. IEEE Trans. Comput. Aided Des. Integr. Circuits Syst. **29**(11), 1661–1668 (2010)

Chapter 15
Sparse Relevance Kernel Machine-Based Performance Dependency Analysis of Analog and Mixed-Signal Circuits

Honghuang Lin, Asad Khan, and Peng Li

15.1 Introduction

As the complexity of *analog/mixed-signal* (AMS) circuits keeps increasing at a rapid pace, the tasks of design, verification, and test have become significant challenges. Nevertheless, it is essential to characterize the dependencies of circuit performances/specifications on various circuit and device parameters or test signatures for purposes such as design, verification, and test optimization. However, doing so is not trivial since the targeted dependencies are usually complex and nonlinear with deep-rooted correlations, making it arduous to reliably quantify the importance of numerous parameters.

For characterizing sophisticated circuit systems, machine learning techniques based on circuit simulations or measurements have been proven to be effective and produced promising outcomes. For example, *support vector machines* (SVMs) [34] are used as nonlinear classifiers in [19] to capture the mapping from input parameters to circuit performance. A regression extension to SVM is employed in [1] to rank circuit parameters based on their correlations with unexpected timing deviations. Additionally, Bayesian inference is often used to build statistical circuit models. For instance, a co-learning Bayesian model is proposed in [36] to efficiently model the performance of AMS circuits.

However, building machine learning-based circuit models faces two key challenges: (1) the availability of training data is limited, since circuit simulations or silicon measurements are usually expensive for sophisticated AMS systems; and (2) to describe an AMS system, it needs a huge number of analog signals and design/device parameters, leading to an extremely high dimensional problem. Since

H. Lin · A. Khan · P. Li (✉)
Texas Instruments, Dallas, TX, USA
e-mail: h-lin10@ti.com; a-khan1@ti.com; pli@tamu.edu

I. M. Elfadel et al. (eds.), *Machine Learning in VLSI Computer-Aided Design*,
https://doi.org/10.1007/978-3-030-04666-8_15

the sensitivities of circuit performances to various parameters may vary vastly, it is instrumental to reliably analyze circuit parameter criticality during the extraction of accurate circuit models. While traditional feature selection or importance ranking techniques may help to identify and select some important parameters out of a large parameter set, building models only with the selected parameters usually degrades the model performance and few of those techniques can guide the model to achieve higher accuracy. These difficulties present important roadblocks to analog/mixed-signal circuit characterization and performance dependency analysis with machine learning techniques.

In the machine learning domain, traditionally, feature selection [3, 5, 9] may be performed by combinatorial search [27, 35] to incrementally add or remove features from the selected subset, which is evaluated by the performance of its resulting predictor. Another kind of method [2, 24, 29] trims the feature space with regularization-based methods, like introducing new regularization terms into the cost function, to perform the feature selection. All these techniques are considered as *linear* feature selection methods since they handle or formulate the features in a linear manner, such as combinatorial search or L_1 norm regularization. To capture the nonlinear dependencies among features and their nonlinear "relevancy" to the targets, some other methods [6, 10, 17] switch the roles of features and samples in their learning models and apply the kernel method to the features instead of samples. These kinds of methods also belong to the category of regularization-based methods since their optimization models using Euclidean inner product as their kernel functions are equivalent to Lasso regression [29] based on the conclusion provided by Li et al. [17]. A drawback of such methods is that their results usually only improve the regularization of the learning model but not the accuracy, since they work independently as preceding filters of the training process.

Moreover, in complex scenarios, especially in complex circuit applications, where features' relevancy may vary in a large range, proper weighting of the features may improve the learning quality by balancing the impact of features on target. In this sense, feature selection is just a 0–1 binary weighting scheme, whose capability is limited when it comes to these scenarios. The task of assigning weights directly to the features and embed the weighted samples into a learning model is extremely computationally challenging, since most commonly used learning models are nonlinear, making weights difficult to manipulate and costly to optimize in the training process.

To address the above challenges, this work proposes a novel Bayesian learning framework for characterizing analog circuits with sparse statistical regression/classification models. The proposed framework is named *sparse relevance kernel machine* (SRKM) and can be considered as a significant extension to the SVM [34] and *relevance vector machine* (RVM) [30]. Instead of directly manipulating the features in a traditional way, the original kernel function is "atomized" into bilinear terms with weighting factors that obliquely reflect the relevancy of the features, leading to the newly defined feature kernel described in the next section. Then, the training model of the SRKM is developed following the RVM framework to achieve a sparse model for both regression and classification. The

SRKM simultaneously seeks relevant training samples (i.e., vectors) and parameters (i.e., features) to derive a sparse model in both the vector and parameter spaces. As a result, the SRKM not only produces accurate models learned from a moderate amount of simulation or measurement data, but also computes a probabilistically inferred weighting factor quantifying the criticality of each parameter as part of the overall learning framework, hence offering a powerful enabler for variability modeling, failure diagnosis, and test development. In addition, an iterative algorithm is developed for efficient training of the proposed SRKM.

The proposed SRKM is capable of solving both classification and regression problems. Compared to other popular kernel-based learning techniques, the SRKM produces more accurate models, requires less amount of training data, and extracts more reliable parametric ranking. The effectiveness of SRKM is demonstrated in examples including statistical variability modeling of a *low-dropout regulator* (LDO), *built-in self-test* (BIST) development of a charge-pump *phase-locked loop* (PLL), and applications of building statistical variability models for a commercial automotive interface design.

This chapter is organized as follows: In Sect. 15.2, we propose the feature kernel weighting scheme and learning model based on the kernel methods adopted in other SVM-related techniques. Then, in Sect. 15.3, the learning model of the SRKM is developed following the sparse Bayesian learning framework. An iterative efficient algorithm is developed to remedy the additional complexity that stems from the inclusion of more variables. Lastly, to illustrate the aforementioned advantages of the SRKM, details and results of three experiments are provided in Sect. 15.4.

15.2 Feature Kernel Weighting

Learning models of circuits are usually defined as: assuming that there are F circuit parameters of interest with which the circuit is described by a parameter (feature) vector $\mathbf{x}$, a sample of the circuit is defined by a pair $\{\mathbf{x}_i, t_i\}$ where t_i is the circuit performance under the configuration $\mathbf{x}_i$. By collecting a number of N samples, the objective of the learning task is to capture the mapping $\Psi : \mathbf{x} \rightarrow t$ with a function $y(\mathbf{x})$ whose output can be used as a prediction of the performance t. If the performance t is a quantified value, this falls into the regression category in machine learning. If t is a binary label, for example, *pass* or *fail* in the context of verification or test, then the learning is a binary classification task. Both regression and classification can be solved by kernel machines such as support vector machines.

15.2.1 Kernel Methods

Support vector machines (SVMs) [25, 33] have been widely used in the EDA domain as a powerful supervised learning toolbox solving classification and regression problems. According to a recent experiment conducted by Fernández-Delgado et al. [8], the comparison of 179 classifiers evaluated on 121 data sets shows that SVM is still among the top learning methods.

The excellence of SVM mainly relies on two portions of its model: the cost function and the famous "kernel trick." Similar to some other learning methods, the cost function of SVM is composed of a fitting loss term and a smoothing penalty term. By using different formulas for losses and penalties, several variants of SVM [12, 13, 26, 28] are derived from the original quadratic optimization problem. Such composition of cost functions provides exceptional robustness and regularization [39]. On the other hand, kernel trick or kernel method has shown great success in handling nonlinear problems.

By using the Lagrange multipliers, the exploration of the optimal separating hyperplane in the mapped higher dimensional space can be expressed as the following optimization problem:

$$\begin{aligned} \underset{\boldsymbol{\alpha}}{\text{minimize}} \quad & f_{\text{SVM}}(\boldsymbol{\alpha}) = \frac{1}{2}\sum_{i=1}^{N}\sum_{j=1}^{N} y_i y_j \alpha_i \alpha_j \langle \hat{\mathbf{x}}_{\mathbf{i}}, \hat{\mathbf{x}}_{\mathbf{j}} \rangle - \sum_{i=1}^{N} \alpha_i, \\ \text{subject to} \quad & \sum_{i=1}^{N} y_i \alpha_i = 0, \\ & 0 \le \alpha_i \le C, \forall i. \end{aligned} \tag{15.1}$$

where y_i denotes the binary label of the training sample $\mathbf{x_i}$, and $\hat{\mathbf{x}}_{\mathbf{i}}$ represents a projection of $\mathbf{x_i}$ in a higher dimensional space.

In SVM, the idea of solving nonlinear problems is to map the original input vectors from the *input space*, which are not linearly solvable, into a higher dimensional space and explore a linear solution in that space. It is clearly shown in (15.1) that the optimization problem can be solved by merely knowing the inner product of any pair of the mapped input vectors $\langle \hat{\mathbf{x}}_{\mathbf{i}}, \hat{\mathbf{x}}_{\mathbf{j}} \rangle$, without explicitly defining the mapping $\mathbf{x} \rightarrow \hat{\mathbf{x}}$. Therefore, defining a kernel function K that satisfies or represents $K(\mathbf{x_i}, \mathbf{x_j}) = \langle \hat{\mathbf{x}}_{\mathbf{i}}, \hat{\mathbf{x}}_{\mathbf{j}} \rangle$ and substituting it into the optimization problem should achieve the same results as what is produced by making the mapping $\mathbf{x} \rightarrow \hat{\mathbf{x}}$. As long as the kernel function K satisfies Mercer's condition [33], there exists a feature space where the inner product is generated by K. In this light, a kernel function can be considered as an implicit definition of a certain mapping.

15.2.2 Weighting via Atomized Kernel

Although the implicit mapping defined by kernel methods is very powerful in solving nonlinear problems, similar to other learning techniques, it may easily be confused by irrelevant or redundant features, which are commonly seen in circuit

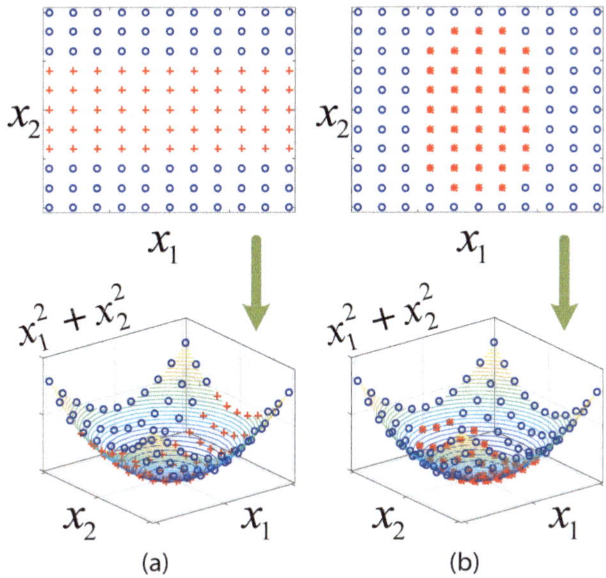

Fig. 15.1 Examples that cannot be linearly solved by the kernel function corresponding to the mapping $(x_1, x_2) \rightarrow (x_1, x_2, x_1^2 + x_2^2)$, where blue circles and red stars denote two different classes, respectively. (**a**) x_2 is the only relevant feature while x_1 is purely irrelevant. (**b**) x_1 is the dominant feature while x_2 provides subtle information to reflect the classes

applications. For example, in Fig. 15.1, if we only sample the center and the four corners, all the samples can be linearly separated by using the mapping $(x_1, x_2) \rightarrow (x_1, x_2, x_1^2 + x_2^2)$, which is equivalent to using kernel function $K(\mathbf{x}_a, \mathbf{x}_b) = \mathbf{x}_a^T \mathbf{x}_b + \mathbf{x}_a^T \mathbf{x}_a \mathbf{x}_b^T \mathbf{x}_b$. But if we keep sampling more evenly in the 2-D input space to include those small circles and asterisks in the training set, it's possible that, as Fig. 15.1 shows, one of the two features is actually irrelevant to the target, or features may have quite different relevancy. In both scenarios, by using the same kernel function or its equivalent mapping, the mapped input vectors can no longer be linearly solved in the 3-D feature space.

From another perspective, kernel function $K(\mathbf{x}_i, \mathbf{x}_j)$ is usually viewed as a representation of the covariance or similarity between $\mathbf{x}_i$ and $\mathbf{x}_j$. If $\mathbf{x}_i$ and $\mathbf{x}_j$ contain notable amount of noise or redundant features, the kernel function may not be able to correctly reflect the covariance or similarity. For example, Gaussian kernel, also known as *radial basis function* (RBF) and defined as $K(\mathbf{x}_i, \mathbf{x}_j) = \exp(-\gamma \|\mathbf{x}_i - \mathbf{x}_j\|^2)$, is a commonly used kernel function. Any noisy redundant feature included in $\mathbf{x}_i$ and $\mathbf{x}_j$ will directly affect the term $\|\mathbf{x}_i - \mathbf{x}_j\|^2$, leading to a kernel function value that cannot truly reflect the similarity of the two samples. Such vulnerability is critical and a flexible weighting scheme is needed especially when there are only limited number of samples.

To approach such problems, based on the implicit mapping mechanism, we propose to construct a new kernel by atomizing the existing kernel functions to achieve implicit feature weighting. Considering the examples shown in Fig. 15.1, to get rid of the interference from the redundant or much less relevant feature, the mapping can be further extended into a 6-dimensional space:

$$\hat{\mathbf{x}} = (x_1, \epsilon x_1, x_2, \epsilon x_2, x_1^2 + \epsilon x_2^2, \epsilon x_1^2 + x_2^2)^T,$$

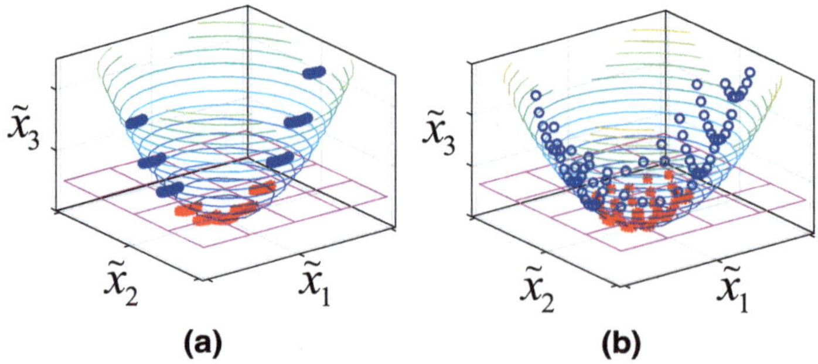

Fig. 15.2 3-D projections of the new 6-D mapping where the two examples in Fig. 15.1 are linearly solved by the purple planes: (**a**) $\tilde{x}_1 = x_2 + \epsilon x_1$, $\tilde{x}_2 = 0$, and $\tilde{x}_3 = x_2^2 + \epsilon^2 x_1^2$; (**b**) $\tilde{x}_1 = x_1 + \epsilon x_2$, $\tilde{x}_2 = \upsilon(x_2 + \epsilon x_1)$, and $\tilde{x}_3 = (x_1^2 + \epsilon^2 x_2^2) + \upsilon(x_2^2 + \epsilon^2 x_1^2)$, where υ is a small positive weight

where a small ϵ can weaken the interference of one feature to the other in the last two dimensions of $\hat{\mathbf{x}}$. As shown in Fig. 15.2a and b respectively corresponding to Fig. 15.1a and b, the 6-D space is linearly projected to 3-D spaces for the purpose of illustration, and the mapped samples are linearly solvable in the new 3-D spaces, meaning linear solutions exist in spaces which are linearly transformed from the 6-D space defined by $\hat{\mathbf{x}}$.

Since implicit mappings defined by kernel functions are more favorable, for the exact mapping $\mathbf{x} \to \tilde{\mathbf{x}}$ defined above, its corresponding kernel function is:

$$\begin{aligned}\tilde{K}(\mathbf{x_a}, \mathbf{x_b}) =& (\mathbf{x_a}^{(1)})^T \mathbf{x_b}^{(1)} + (\mathbf{x_a}^{(1)})^T \mathbf{x_a}^{(1)} (\mathbf{x_b}^{(1)})^T \mathbf{x_b}^{(1)} \\ &+ (\mathbf{x_a}^{(2)})^T \mathbf{x_b}^{(2)} + (\mathbf{x_a}^{(2)})^T \mathbf{x_a}^{(2)} (\mathbf{x_b}^{(2)})^T \mathbf{x_b}^{(2)},\end{aligned}$$

where $\mathbf{x}^{(1)} = diag(1, \epsilon) \cdot \mathbf{x}$ and $\mathbf{x}^{(2)} = diag(\epsilon, 1) \cdot \mathbf{x}$. By substituting the original kernel function K into the new kernel, it can be re-written as:

$$\begin{aligned}\tilde{K}(\mathbf{x_a}, \mathbf{x_b}) =& K(diag(1, \epsilon) \cdot \mathbf{x_a}, diag(1, \epsilon) \cdot \mathbf{x_b}) \\ &+ K(diag(\epsilon, 1) \cdot \mathbf{x_a}, diag(\epsilon, 1) \cdot \mathbf{x_b}),\end{aligned}$$

where the first term and second term are original kernels with vectors scaled by $diag(1, \epsilon)$ and $diag(\epsilon, 1)$, respectively. In other words, each terms scale down one feature with a small ϵ and map $\mathbf{x}$ to a portion of the dimensions in $\tilde{\mathbf{x}}$ which are insensitive to that feature.

More generally, we define a scaling diagonal matrix $\mathbf{S_i}(\epsilon)$ with:

$$\mathbf{S_i}(\epsilon) = diag(\mathbf{s_i}) \tag{15.2}$$

and

$$\mathbf{s_i}(j) = \begin{cases} 1, & j = i, \\ \epsilon, & j \neq i, \end{cases}$$

where $\epsilon \in [0, 1]$. For any existing kernel function $K(\mathbf{x_a}, \mathbf{x_b})$, we define a new *feature kernel* function as:

$$K_i(\mathbf{x_a}, \mathbf{x_b}; \epsilon) = K(\mathbf{S_i}(\epsilon) \cdot \mathbf{x_a}, \mathbf{S_i}(\epsilon) \cdot \mathbf{x_b}). \tag{15.3}$$

Such feature kernel mainly maintains the sensitivity to the i-th feature in both $\mathbf{x_a}$ and $\mathbf{x_b}$, with other features scaled by ϵ. If $\epsilon = 1$, such feature kernel is identical to the original kernel. If $\epsilon = 0$, the i-th feature is completely isolated since all the other features are zero out from the original kernel.

Assuming that kernel K maps samples from an F-dimensional input space to a d-dimensional space, which may be vulnerable to irrelevant or nonlinearly relevant features, we now atomize K into the sum of F weighted feature kernels by assigning one for each feature:

$$\tilde{K}(\mathbf{x_a}, \mathbf{x_b}; \epsilon) = \sum_{i=1}^{F} v_i K_i(\mathbf{x_a}, \mathbf{x_b}; \epsilon). \tag{15.4}$$

It will result in a mapping from the F-dimensional input space to another $(d \cdot F)$-dimensional space. In this much higher dimensional space, we are expecting that the inner product of any pair of vectors can be expressed or approximated by a linear combination of the feature kernels.

In addition, for the i-th feature kernel K_i, it mainly represents the information provided by the i-th feature since the influence of other features in the corresponding "axes" of the mapped space is scaled down if $\epsilon < 1$ or completely removed if $\epsilon = 0$. Therefore, we propose to perform feature weighting via the weighting parameters v_i as demonstrated in (15.4). Larger $|v_i|$ means the i-th kernel is more important in the kernel model. For example, by using $\epsilon = 0.1$ as the scaling parameter, in Fig. 15.2a, the kernel function corresponding to the projection is $\tilde{K} = 0 \cdot K_1 + K_2$ while in Fig. 15.2b, it is $\tilde{K} = K_1 + v \cdot K_2$ where $v = 3/8$. These two atomized kernel functions clearly reflect the relevancy of the two features.

One of the advantages of this weighting scheme is that the weighting parameters are much easier to manipulate compared to the schemes that directly apply weighting parameters to the input vectors. Secondly, the parameter ϵ makes the model more flexible by smoothly morphing from regular kernels ($\epsilon = 1$) to feature selection ($\epsilon = 0$). Moreover, similar to the original kernel method, this weighting scheme actually avoids defining explicit feature weighting by instead weighting the linear combination of the feature kernels, which can be considered as an implicitly defined nonlinear weighting scheme.

For an existing kernel function K that satisfies Mercer's condition [33], the newly defined feature kernel K_i should also satisfy Mercer's condition. As a result, after $\mathbf{x}$ is normalized, for all square integrable $g(\mathbf{x})$ we have:

$$\int_{\chi\times\chi} \tilde{K}(\mathbf{x_a}, \mathbf{x_b}; \epsilon) g(\mathbf{x_a}) g(\mathbf{x_b}) d\mathbf{x_a} d\mathbf{x_a}$$
$$= \sum_{i=1}^{F} \int_{\chi\times\chi} v_i K_i(\mathbf{x_a}, \mathbf{x_b}; \epsilon) g(\mathbf{x_a}) g(\mathbf{x_b}) d\mathbf{x_a} d\mathbf{x_a} \geq 0,$$

for all $\mathbf{x_a}, \mathbf{x_b} \in \chi$ as long as $v_i \geq 0, \forall i$.

15.2.3 Learning Model with Feature Kernels

In SVM, the training model is often solved in its dual form and, by leveraging *Karush–Kuhn–Tucker* (KKT) conditions, the prediction model is based upon the following decision function:

$$y(\mathbf{x}; \mathbf{w}) = \sum_{i=1}^{N} w_i K(\mathbf{x}, \mathbf{x_i}), \tag{15.5}$$

where $\{\mathbf{x_i}\}_{i=1}^{N}$ are the training examples and w_i are actually the Lagrange multipliers referred to as α_i in (15.1). The training process of learning methods using (15.5) as their decision function is to infer all the parameters w_i in (15.5) given the corresponding targets $\{t_i\}_{i=1}^{N}$ of the training examples $\{\mathbf{x_i}\}_{i=1}^{N}$.

Taking the error e_i between t_i and $y(\mathbf{x_i}; \mathbf{w})$ into consideration, the kernel-based machine can be written as:

$$\mathbf{t} = \Phi_w \cdot \mathbf{w} + \mathbf{e}, \tag{15.6}$$

where Φ_w is an $N \times N$ matrix defined by $\Phi_w(i, j) = K(\mathbf{x_i}, \mathbf{x_j})$, $\mathbf{t}$ is simply the target vector with $\mathbf{t}(i) = t_i$, and $\mathbf{e}$ is the error vector for the N training samples. Learning methods like SVM aim at minimizing $\mathbf{e}$ with fitting loss term and regularizing $\mathbf{w}$ with smoothing regularization term simultaneously.

As previously mentioned, most popular kernel functions such as Gaussian kernel (also known as *radial basis function*) and polynomial kernel are vulnerable to features with complicated relevancy. For training data with F features, we define a new learning model by embedding the sum of weighted feature kernels as a new kernel function into (15.6) to handle feature relevancy:

$$\mathbf{t} = {\Phi_{wv}}^{(\epsilon)} \cdot (\mathbf{w} \otimes \mathbf{v}) + \mathbf{e}, \tag{15.7}$$

where ${\Phi_{wv}}^{(\epsilon)}$ is an $N \times (NF)$ matrix defined by ${\Phi_{wv}}^{(\epsilon)}(i, (j-1)F + k) = K_k(\mathbf{x_i}, \mathbf{x_j}; \epsilon)$ with $i, j \in [1, N]$ and $k \in [1, F]$. Besides, $\mathbf{w} \otimes \mathbf{v}$ in (15.7) is the tensor product of vector $\mathbf{w}$ and $\mathbf{v}$ which yields an $(NF) \times 1$ row vector with the definition $(\mathbf{w} \otimes \mathbf{v})((j-1)F + k) = w_j v_k$ where $j \in [1, N]$ and $k \in [1, F]$.

A useful property of this new kernel machine is that the roles of $\mathbf{w}$ and $\mathbf{v}$ in the model (15.7) are relatively symmetric and transposable. Model (15.6) can be derived from model (15.7) by moving $\mathbf{v}$ from the tensor product to the design matrix defined by ${\Phi_w}^{(\epsilon)}(i, j) = \sum_{k=1}^{F} v_k K_k(\mathbf{x_i}, \mathbf{x_j}; \epsilon)$ with $i, j \in [1, N]$. Similarly, the following model can be derived from (15.7) by instead moving $\mathbf{w}$ to the design matrix:

$$\mathbf{t} = {\Phi_v}^{(\epsilon)} \cdot \mathbf{v} + \mathbf{e}, \tag{15.8}$$

where ${\Phi_v}^{(\epsilon)}$ is an $N \times F$ matrix defined by ${\Phi_v}^{(\epsilon)}(i, k) = \sum_{j=1}^{N} w_j K_k(\mathbf{x_i}, \mathbf{x_j}; \epsilon)$ with $i \in [1, N]$ and $k \in [1, F]$. This property indicates that the exploration process of $\mathbf{w}$ and $\mathbf{v}$ may be unified, which will be discussed in detail in the next section.

15.3 Sparse Relevance Kernel Machine

While the new kernel machine is capable of solving feature weighting during the training process, the searching space of the solution is inflated by the number of features. A sparse treatment is highly appealing since the sparsity may help to reduce the complexity. And in the scenario of feature selection, that is, in the cases of $\epsilon = 0$, it may help to filter out irrelevant or redundant features as much as possible if $\mathbf{v}$ tends to be sparse. Since Bayesian learning frameworks with sparse prior [30, 37, 38] are capable of producing highly sparse models, we develop a sparse relevance kernel machine under the Bayesian learning framework in this section.

15.3.1 Relevance Vector Machine

The *relevance vector machine* (RVM)[30] is a sparse Bayesian model providing a viable probabilistic framework, whose Bayesian network model is shown in Fig. 15.3.

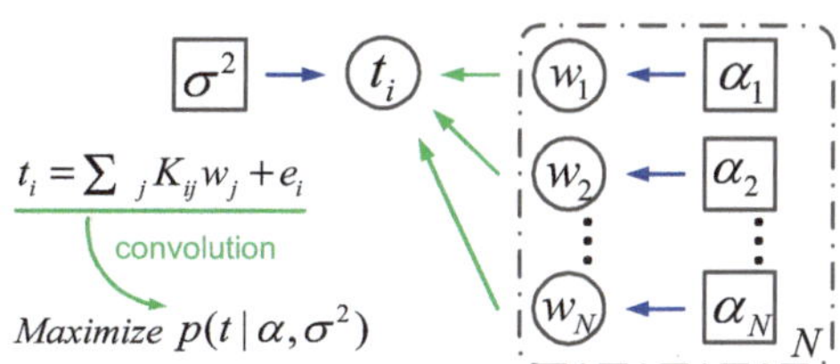

Fig. 15.3 The network model of the RVM where $K_{ij} = K(x_i, x_j)$. Circles denote random variables and squares denote deterministic model parameters

Given the input vectors $\{\mathbf{x_n}\}_{n=1}^N$ and their corresponding targets $\{t_n\}_{n=1}^N$, the RVM is used to probabilistically determine the model (15.6). The given target values are modeled as independent Bernoulli random variables with the following probability distribution:

$$P(\mathbf{t}|\mathbf{w}) = \prod_{i=1}^{N} \sigma[y(\mathbf{x_i}; \mathbf{w})]^{t_i} \cdot \sigma[1 - y(\mathbf{x_i}; \mathbf{w})]^{1-t_i}, \tag{15.9}$$

where σ is the sigmoid link function $\sigma(y) = (1 + e^{-y})^{-1}$.

Different from deterministic learning models (e.g., the SVM) that compute $\mathbf{w}$ directly, the RVM defines the prior distribution of $\mathbf{w}$ as independent zero-mean Gaussian random variables with variance $\boldsymbol{\alpha}$, and compute $\boldsymbol{\alpha}$ instead of $\mathbf{w}$ in the training process:

$$p(\mathbf{w}|\boldsymbol{\alpha}) = \prod_{n=1}^{N} \mathcal{N}(0, \alpha_n^{-1}). \tag{15.10}$$

If $\alpha_i < \infty$, w_i is called a *relevance vector* since it has a variance greater than zero, allowing x_i making contributions to the decision function. Note that the RVM performs prediction with the posterior probability of the internal variables (i.e., the weights). Via convolution of Gaussian distributions, the covariance and mean of the posterior $p(\mathbf{w}|\mathbf{t}, \boldsymbol{\alpha})$ can be shown to be respectively:

$$\Sigma = (\Phi^T \Gamma \Phi + \mathbf{A})^{-1} \tag{15.11}$$

$$\boldsymbol{\mu} = \Sigma \Phi^T \Gamma \mathbf{t}, \tag{15.12}$$

where $\mathbf{A} = diag(\boldsymbol{\alpha})$, $\Gamma = diag(\boldsymbol{\gamma})$, and $\boldsymbol{\gamma}(i) = \sigma[y(\mathbf{x_i}; \mathbf{w})] \cdot \sigma[1 - y(\mathbf{x_i}; \mathbf{w})]$.

The objective of the Bayesian network is to find the most probable model parameters with the given training samples, i.e., to maximize the posterior probability $p(\mathbf{w}, \boldsymbol{\alpha}|\mathbf{t})$. Based on the Bayes' rule, such objective is equivalent to:

$$\arg\max_{\boldsymbol{\alpha}} p(\mathbf{t}|\boldsymbol{\alpha}). \tag{15.13}$$

A training algorithm is proposed in [32] to compute the optimal $\boldsymbol{\alpha}$. After the training, for any $\hat{\mathbf{x}}$, its predicted probability of being labeled 1 is:

$$\sigma(\hat{y}) = \sigma(\boldsymbol{\mu}^T \boldsymbol{\phi}(\hat{\mathbf{x}})) \tag{15.14}$$

where $\boldsymbol{\phi}(\hat{\mathbf{x}})$ is a vector of size N whose i-th entry is defined by $\boldsymbol{\phi}(\hat{\mathbf{x}})(i) = K(\hat{\mathbf{x}}, \mathbf{x_i})$.

15.3.2 Bayesian Learning Model for SRKM

The RVM is a learning model based on the decision function (15.5) focusing on producing a sparse $\mathbf{w}$. Under the same framework, a feature selection technique called *relevance feature vector machine* (RFVM) [6] is proposed to exchange the roles of samples and features by defining the "feature vector" $\mathbf{f}_i = (\mathbf{x}_1(i), \mathbf{x}_2(i), \ldots, \mathbf{x}_F(i))^T$ and a new design matrix $\Phi'(i, j) = K(\mathbf{f}_i, \mathbf{f}_j)$. This is a regularization-based feature selection method, which focuses on building a filter method [5] by ranking the features.

To achieve high model accuracy and high quality feature weighting with the new kernel machine (15.7), we propose a *sparse relevance kernel machine* (SRKM) whose conceptual structure is shown in Fig. 15.4.

Without using the new kernel machine model (15.7), learning method with embedded feature weighting is often realized by directly assigning weights $\mathbf{v}$ to the parameters. However, as we discussed previously, since most kernel functions are nonlinear, it is extremely difficult to develop the models in the Bayesian learning contexts. For example, by assigning Gaussian prior to $\mathbf{v}$ in a similar way $p(\mathbf{v}|\boldsymbol{\beta}) = \prod_{n=1}^{F} \mathcal{N}(0, \beta_n^{-1})$, the Bayesian network described in Fig. 15.3 is extended to derive the new model on the left of Fig. 15.5. Assuming Gaussian kernel with linear direct feature weighting

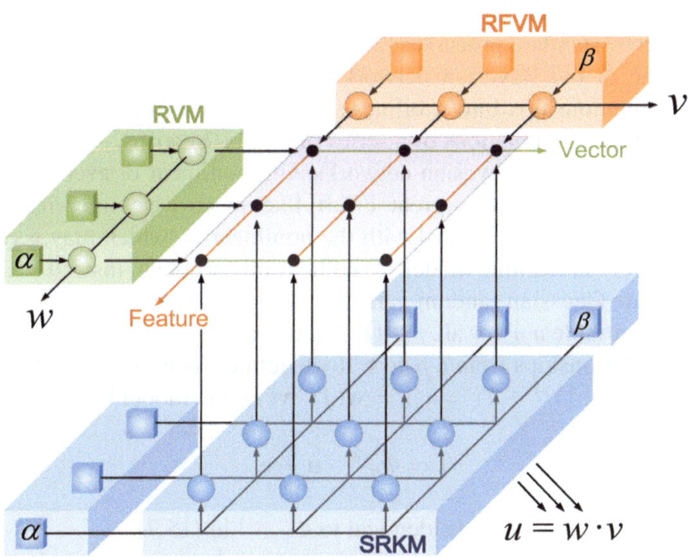

Fig. 15.4 The conceptual SRKM model. Small black dots denote training data, with each row representing a vector and each column representing a feature. Circles denote random variables and squares denote deterministic model parameters

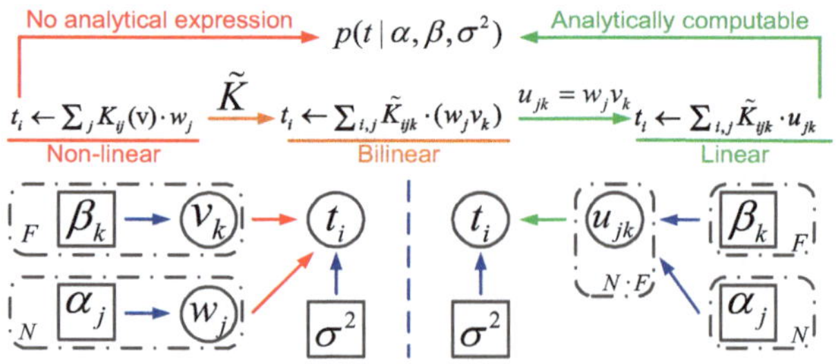

Fig. 15.5 Development of SRKM Bayesian network where $K_{ijk} = K_k(\mathbf{x}_i, \mathbf{x}_j; \epsilon)$. Circles denote random variables and squares denote deterministic model parameters

$$K(\mathbf{x_i}, \mathbf{x_j}) = e^{-\gamma \|diag(\mathbf{v}) \cdot (\mathbf{x_i} - \mathbf{x_j})\|^2},$$

is employed, the deterministic relationship from $\mathbf{w}$ and $\mathbf{v}$ to t is highly nonlinear. As a result, the following optimization objective of the Bayesian training process is not analytically computable and hence hinders the optimization-based training process:

$$p(\mathbf{t}|\boldsymbol{\alpha}, \boldsymbol{\beta}, \sigma^2) = \iint p(\mathbf{t}|\mathbf{w}, \mathbf{v}, \sigma^2) \cdot p(\mathbf{w}|\boldsymbol{\alpha}) \cdot p(\mathbf{v}|\boldsymbol{\beta}) d\mathbf{w} d\mathbf{v}. \quad (15.15)$$

As the first step towards developing the SRKM, we embed the new kernel machine (15.7) into the Bayesian learning framework to handle the feature weighting problem. Here we treat ϵ in the feature kernel as a kernel parameter pre-defined before the training process, similar to the γ parameter in the Gaussian kernel, which can be selected by standard processes like cross validation. Leveraging the relevance kernel machine in the Bayesian network helps to simplify the highly nonlinear deterministic relationships into a bilinear form. As described in (15.7), $\mathbf{t}$ can be expressed as linear combinations of a series of $\mathbf{w}(j)\mathbf{v}(k)$ terms.

However, inference in Bayesian networks with nonlinear or even bilinear deterministic relationships requires great effort like piece-wise linearization [7] or dynamic discretization [23] to deal with the nonlinearity, which may greatly boost the complexity of the learning model. To address this problem, instead of defining $\mathbf{w}$ and $\mathbf{v}$ as separate Gaussian random variables, we replace the term $\mathbf{w}(j)\mathbf{v}(k)$ with a single random variable u_{jk} for all $j \in [1, N]$ and $k \in [1, F]$, which results in linear deterministic relationships from u_{jk} to $\mathbf{t}$. If we define a new vector $\mathbf{u}$ of size $(N \cdot F)$ whose entry $\mathbf{u}((j-1)N + k) = u_{jk} = \mathbf{w}(j) \cdot \mathbf{v}(j)$, the model becomes:

$$\mathbf{t} = \Phi_u{}^{(\epsilon)} \cdot \mathbf{u} + \mathbf{e}, \quad (15.16)$$

where the design matrix $\Phi_u{}^{(\epsilon)}$ is identical to $\Phi_{wv}{}^{(\epsilon)}$ in (15.7).

In the RVM, the zero-mean Gaussian prior distribution of $\mathbf{w}$ tends to help the model converge to a sparse $\mathbf{w}$ since the resulting marginal prior distribution over $\mathbf{w}$ is the product of *Student-t* distributions. Similarly, to achieve sparsity in $\mathbf{u}$, we also define their prior distributions as independent zero-mean Gaussian distributions. Considering the nature of $\mathbf{u}$, if $u_{j,k}$ is irrelevant, meaning the distribution of $u_{j,k}$

is infinitely peaked at zero, either the i-th sample ($\{u_{j,k}\}_{k=1}^{F}$) or the j-th parameter ($\{u_{j,k}\}_{j=1}^{N}$) should be irrelevant as well. To reflect this, we define a proper prior for $\mathbf{u}$ as:

$$p(\mathbf{u}|\boldsymbol{\alpha}, \boldsymbol{\beta}) = \prod_{j=1}^{N} \prod_{k=1}^{F} \mathcal{N}(0, \alpha_j^{-1} \beta_k^{-1}), \tag{15.17}$$

which leads to our proposed computable linear Bayesian network shown on the right of Fig. 15.5 and our conceptual model described in Fig. 15.4. An infinite $\alpha_j \beta_k = \infty$ means $u_{jk} = 0$ and the corresponding feature kernel is irrelevant to the final decision function. In addition, if $\alpha_j \to \infty$, all the $\{u_{j,k}\}_{k=1}^{F}$ are zero, meaning the j-th sample is discarded from the set of relevance vectors. Likewise, if $\beta_k < \infty$, the k-th parameter is relevant and there should be at least one non-zero $u_{i,j}$ for $i \in [1, N]$.

Under the same Bayesian inference framework, the posterior covariance and mean of $p(\mathbf{u}|\mathbf{t}, \boldsymbol{\alpha}, \boldsymbol{\beta}, \sigma^2)$ in the proposed Bayesian network are found to be:

$$\Sigma_u = (\sigma^{-2} ({\Phi_u}^{(\epsilon)})^T {\Phi_u}^{(\epsilon)} + \mathbf{A_u})^{-1}, \tag{15.18}$$

$$\boldsymbol{\mu}_\mathbf{u} = \sigma^{-2} \Sigma_u ({\Phi_u}^{(\epsilon)})^T \mathbf{t}, \tag{15.19}$$

where $\mathbf{A_u} = diag(\alpha_1\beta_1, \alpha_1\beta_2, \ldots, \alpha_N\beta_F)$ and ${\Phi_u}^{(\epsilon)}$ is the new design matrix defined in (15.16). The formulas (15.18) and (15.19) are in the same form as the posterior covariance and mean of w in the RVM, and consequently solvable with the existing RVM algorithms.

The SRKM classification model behaves analogously to the regression model but using a Bernoulli likelihood instead of Gaussian for the target with the following sigmoid link function:

$$\varrho(y) = \frac{1}{1 + e^{-y}}. \tag{15.20}$$

As a result, the Bernoulli likelihood is:

$$P(\mathbf{t}|\mathbf{u}) = \prod_{i=1}^{N} \varrho[y(\mathbf{x_i}; \mathbf{u})]^{t_i} \cdot \varrho[1 - y(\mathbf{x_i}; \mathbf{u})]^{1-t_i}, \tag{15.21}$$

where the targets $t_i \in \{0, 1\}$ are for binary classifications. With likelihood in the form of (15.21), there is no closed-form expressions for $\mathbf{u}$ and hence the Laplace approximation procedure [21, 32] should be utilized and nested in each iteration of the training.

15.3.3 Efficient Training Algorithm

The marginal likelihood maximization [32] required in training the RVM model is solved in an iterative process similar to the well-known *expectation maximization* (EM) algorithm. Due to the required matrix operations, the worst case computational complexity in each iteration is $O(N^2F + N^2M)$ [14] if there are M relevance vectors in that iteration and F features in total. By performing one-time pre-computation of the full $N \times N$ design matrix Φ with the complexity of $O(N^2F)$ and pre-computing $(\Phi^T\Phi)$ (or normalizing Φ as the implementation of [31]), the complexity of each iteration can be reduced to $O(NM^2)$ by using $O(N^2)$ memory instead of $O(NM)$.

For the SRKM, by solving (15.16) with this algorithm, since the size of the vector $\mathbf{u}$ is $(N \cdot F)$, a large number of features F will blow the worst case computational complexity for each iteration from $O(NM^2)$ to $O(NFM^2E^2)$ if there are M relevance vectors and E relevance features in that iteration.

To address this computational challenge, our proposed efficient algorithm leverages the property that $\mathbf{w}$ and $\mathbf{v}$ are interchangeable in the bilinear Bayesian network (15.7), and that either vectors can be merged into the design matrix to reduce our model to (15.6) or (15.8). As Fig. 15.6 shows, fixing $\boldsymbol{\alpha}$ and moving the resulting expectation of $\mathbf{w}$ into the design matrix (i.e., converting $\Phi_u^{(\epsilon)}$ in (15.16) to $\Phi_v^{(\epsilon)}$ in (15.8)) will reduce every column in Fig. 15.6 to a single weight v_j with its prior β_j. Similarly, row-wise reduction by fixing $\boldsymbol{\beta}$ converts the proposed network to another RVM network with $\mathbf{w}$ and $\boldsymbol{\alpha}$.

The above discussion suggests an efficient two-level iterative training process. In each iteration of the top level, we reduce the model either row-wise or column-wise, and update $\boldsymbol{\alpha}$ or $\boldsymbol{\beta}$ subsequently. In the second level, the original algorithm [32] can be employed to solve either Model (15.6) or Model (15.8). The complexity in each iteration is now reduced to either $O(NM^2)$ or $O(FE^2)$.

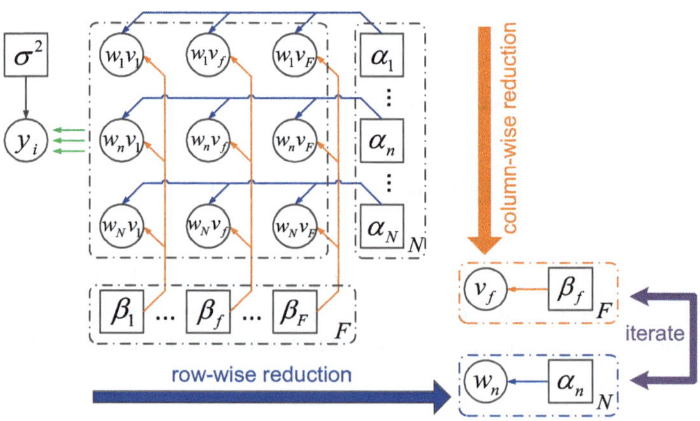

Fig. 15.6 Efficient SRKM model with network reduction. Circles denote random variables and squares denote deterministic model parameters

15.4 Experiments

To demonstrate the superiority of the proposed SRKM, we compare its performance with popular learning-based techniques including the SVM [34] and the RVM [30]. We also compare the SRKM with the RFVM [6] in terms of parameter (feature) ranking. For the purposes of parameter ranking and selection, we use feature kernel with $\epsilon = 0$ in all the experiments.

15.4.1 *Variability Analysis of an LDO*

Building an accurate regression model for a given analog performance and performing feature ranking among all sorts of process parameters are key to the understanding of the impacts of process variabilities on analog circuits. Since simulations or measurements are usually expensive, it is of great significance to build an accurate regression model and obtain reliable parameter weighting with a moderate amount of samples, which turns out to be a task well handled by the proposed method.

We investigate the process variations in a realistic *low-dropout regulator* (LDO) design (Fig. 15.7) proposed in [16]. We build SRKMs to analyze the impact of process variations on LDO specifications including its quiescent current, undershoot of the output voltage V_{out}, and load regulation. Channel length variations of all transistors in the LDO are modeled at the SPICE level using a commercial 90 nm CMOS technology design kit. We use various numbers of simulation samples to build a regression model relating the model process parameters with each targeted specification and test the accuracies of these models using a testing set of 1000 simulation samples. The results are shown in Fig. 15.8.

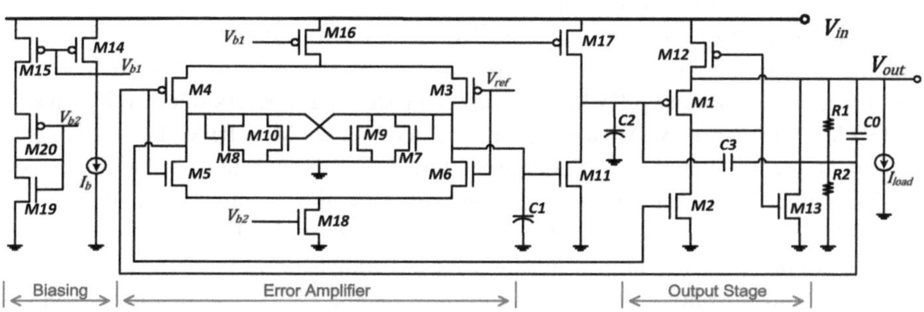

Fig. 15.7 Low-dropout regulator (LDO)

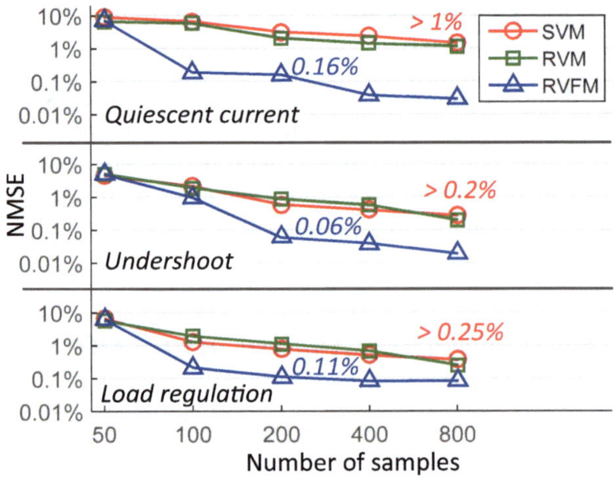

Fig. 15.8 Regression performance comparison

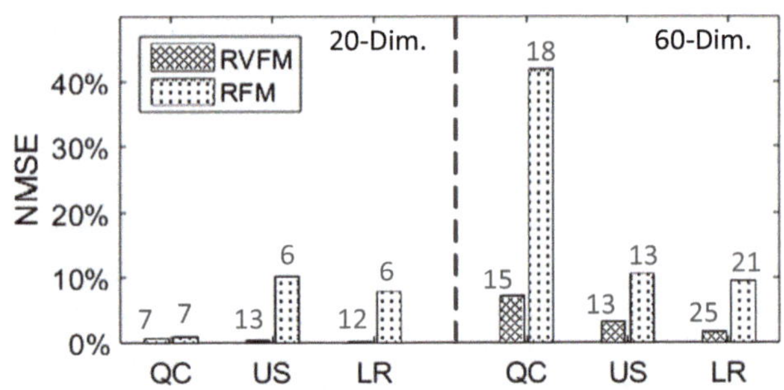

Fig. 15.9 Ranking quality comparison with the number of selected parameters (QC quiescent current, US undershoot, LR load regulation)

In this experiment, *normalized mean square error* (NMSE) is used as the metric to evaluate the performance of the predictors trained with different techniques. As Fig. 15.8 shows, the SRKM out-performs the popular SVM and RVM in all cases by achieving one order of magnitude lower NMSEs.

We compare the ranking produced by the SRKM and the RFVM on feature ranking in Fig. 15.10. To evaluate the quality of the ranking, for each design specification, we train two RVMs only in the process parameters selected by SRKM and RFVM, respectively. A parameter is selected by SRKM or RFVM if its expected weight is greater than 0.01. Such procedure is firstly applied to the regression model with 20 channel length variations, i.e., the three columns on the left of Fig. 15.9. Then, the same procedure is applied to an expanded full parameter set of 60 parameters involving variations of each transistor's channel length, oxide thickness, and threshold voltage (on the right of Fig. 15.9). The resulting NMSEs and the numbers of parameters selected indicate that the SRKM produces more reliable parameter weighting and reaches similar sparsity compared to the RFVM.

We use design knowledge to provide further insights and validation for the parameter rankings of the 20 channel length variations computed by the SRKM. For example, based on the analysis in [15] a majority portion of the multi-loop LDO's

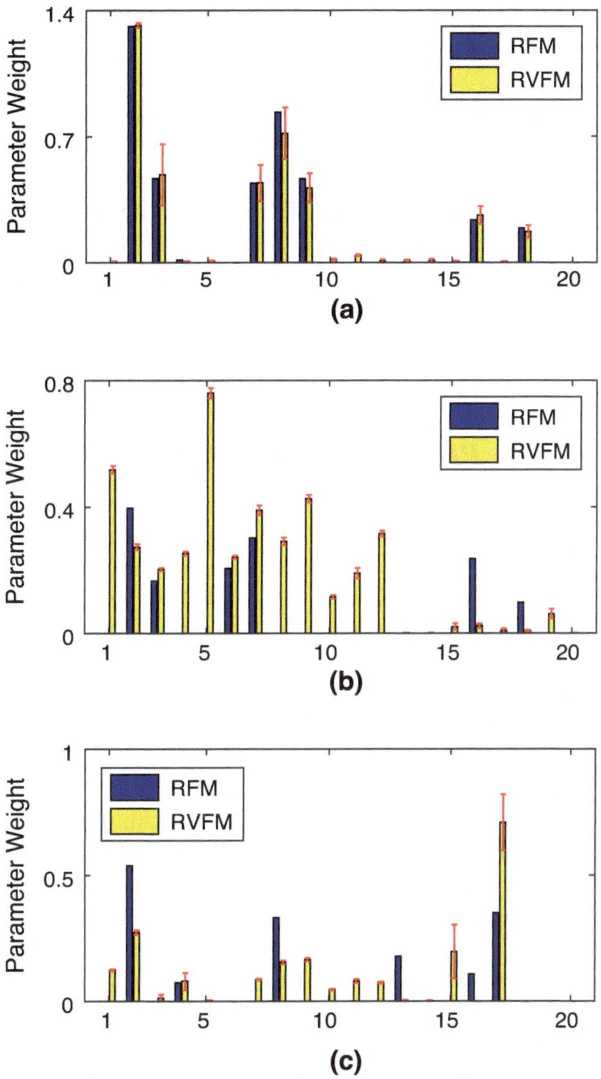

Fig. 15.10 Weights of transistor's channel length variation in the model of: (**a**) quiescent current; (**b**) undershoot; (**c**) load regulation. Red lines represent the 95% confidence intervals estimated by SRKM

quiescent current is consumed by the fastest two loops in the output stage and hence the variation on $M2$ has significant impact on the quiescent current. Moreover, variations on $M3$, $M7$, $M8$, and $M9$ may lead to mismatches and considerable changes at the two output nodes of the error amplifier, one of which is V_g of $M2$. This analysis matches the ranking shown in Fig. 15.10a.

The undershoot of the LDO is mainly determined by the load capacitor and the loop bandwidth, which is further determined by the error amplifier (involving $M3 \sim M10$), the fast loop in the output stage ($M12$), and the in-band zeros locations defined as:

$$\omega_{LCZ} \approx \sqrt{\frac{g_{m1} g_{m11} g_a}{g_{m2} C_{C2}(C_{C1} + C_{C3})}}, \tag{15.22}$$

where g_a is the output admittance of the error amplifier defined by the g_m of $M7 \sim M10$. The ranking of the SRKM in Fig. 15.10b is reliable since it captures all these relevant variations.

Load regulation of the LDO is mainly determined by the DC loop gain, which is the product of the gains of all stages in the loop. The gain of the EA stage is inversely proportional to the g_m of $M7 \sim M10$ and the second stage is comprised of $M17$ and $M11$. Again, the ranking of the SRKM as shown in Fig. 15.10c successfully identifies all these important variations.

15.4.2 PLL BIST Scheme Optimization

Built-in-self-test (BIST) is very effective in detecting operational failures of deployed analog/mixed-signal circuits. Based on the concept of alternative test, efficient BIST solutions can be formed by collecting low-cost test signatures and relating the signatures to targeted performance specifications using statistical prediction models. The effectiveness of BIST heavily depends on the quality of the selected signatures and the tradeoffs between accuracy, overhead, and test time. We apply the RFVM to the BIST of a charge-pump PLL targeting three key specifications: *lock-time* (LT), *frequency overshoot* (OVS), and *jitter* (JT) (Table 15.1).

Figure 15.11 shows the PLL along with three BIST schemes using various test signatures. Jitter, frequency overshoot, and lock-time are important specifications but cannot be easily measured directly on the chip. To capture failures in those specifications, the first candidate BIST scheme [40] collects the readouts of the counter in the divider as its test signature, while the second scheme [11] collects the accumulated *up* and *dn* phase detector outputs via integrators and *time-to-digital converters* (TDCs). The third scheme is an example of IDDQ testing, measuring the quiescent currents of the *charge-pump* (CP) and the *voltage control oscillator* (VCO) as test signatures similar to the approach of [22].

The first two schemes operate in a special test mode which instead of feeding back the divider output, it first feeds the one-buffer delayed reference clock to the phase detector for 8 reference cycles with a cycle time of 0.1 μs. Then, the reference

Table 15.1 BIST scheme synthesis

Spec.	Original best accuracy	Synthesized signatures			Test time (μs)			Test time reduction	Optimized accuracy
		Sch.1	Sch.2	Sch.3	Sch.1&2	Sch.3	Total		
JT	97.22%	1–3	1	VCO	0.3	0.6	0.9	43.75%	99.98%
OVS	98.00%	1–3	1–7	CP	0.4	0.6	1.0	37.50%	99.88%
LT	96.40%	1–4	1–2	VCO	0.4	0.6	1.0	37.50%	99.95%

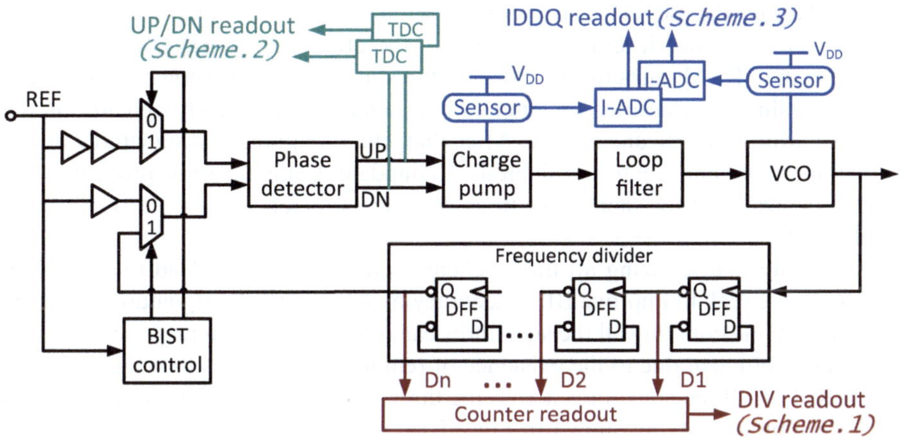

Fig. 15.11 A PLL and three BIST schemes

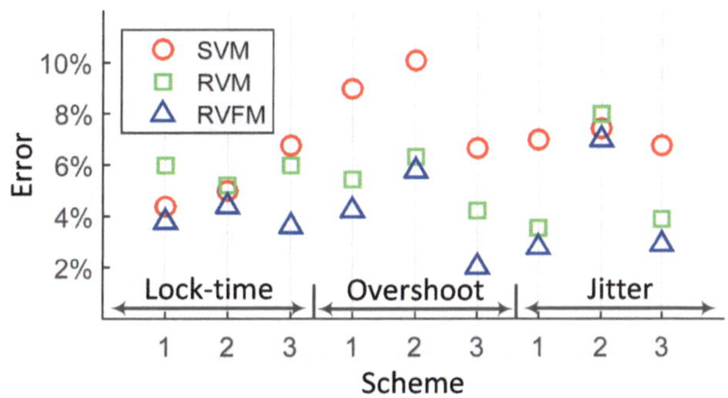

Fig. 15.12 Classifier performance comparison

clock input is replaced by the double delayed reference clock for another 8 cycles. Each cycle generates one signature for Scheme 1 and two for Scheme 2, making a total of 16 and 32 signatures for Scheme 1 and 2, respectively. Scheme 3 reads out two signatures, i.e., the CP and VCO quiescent currents, in the quiescent mode.

Recently, learning-based classifiers like the SVM have been trained to perform the failure detection in BIST [4, 40]. To make better usage of the collected test signatures, we apply the proposed SRKM in each scheme. We fit the target specification into a sigmoid function before we employ the SRKM as a classier for failure detection. Three classification techniques, the SVM, RVM, and SRKM, are trained with 200 simulation samples and tested with 4000 samples. The classifying errors are compared in Fig. 15.12 which shows the superior BIST classifier accuracy of the proposed SRKM.

In addition, the SRKM also produces reliable ranking among test signatures, which can be further leveraged to improve the efficiency of BIST schemes. For example, the SRKM ranks the 16 test signatures in Scheme 1 as shown in Fig. 15.13 when building the classifier for jitter failure detection. The tenth signature is the last one with a significant weight. After that, the remaining six signatures are of little importance and can be considered as redundant. Using the same procedure, we reduce the test time for each of the three specifications for Scheme 1 as reported in Table 15.2.

Assuming that realizing all three schemes on-chip does not lead to significant overhead, we seek to improve BIST accuracy by leveraging the signatures of all the schemes. While combing all the signatures can offer the best accuracy, it may not be completely efficient due to the existence of redundant signatures. For this, we train an SRKM on all the signatures across the three schemes to predict the jitter. Based on the signature ranking shown in Fig. 15.14, we collect the first three signatures in Scheme 1 and the first signature in Scheme 2. Although the third last signature in Scheme 2 also possesses a notable weight, collecting such signature is not cost-effective in terms of test time, and thus it is discarded. For Scheme 3, only the quiescent current of VCO is selected, which can be measured in 0.6 μs according to [22].

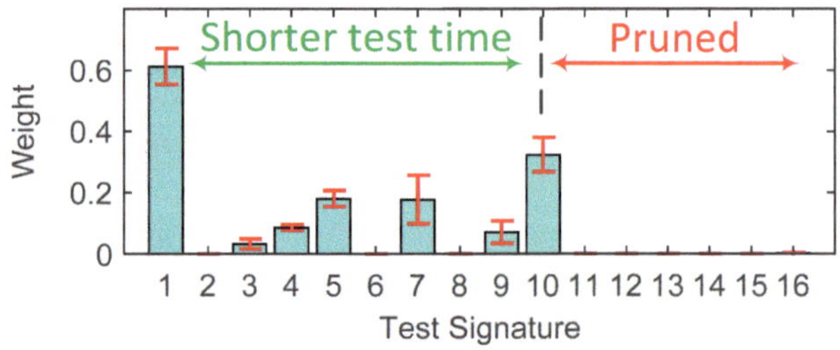

Fig. 15.13 Signature ranking for jitter prediction with Scheme 1

Table 15.2 Test time optimization of Scheme 1

Spec.	Original accuracy	Selected readouts	Resulting accuracy	Test time reduction
JT	97.22%	1–10	96.20%	37.5%
OVS	95.78%	1–12	94.89%	25.0%
LT	96.20%	1–6	97.00%	62.5%

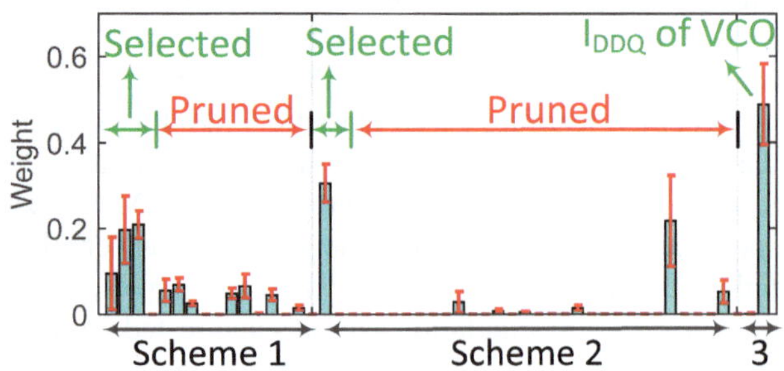

Fig. 15.14 Signature ranking for jitter prediction with all three schemes

Based on these five selected signatures, we synthesize an optimized combined BIST scheme for each specification and show the results in Table 15.1. As can be been, by using the proposed SRKM, the BIST accuracy can be boosted to over 99.88% with a test time reduction of about 40%.

15.4.3 Binary Classification for Functional Check

The above two examples illustrate the effectiveness of the regression version of SRKM. This experiment demonstrates the superiority of the proposed SRKM algorithm for classification by applying it to analyzing a commercial dual-lane data communication AMS system shown in Fig. 15.15. As a product designed for automotive applications, functional safety is a key requirement. One of the safety features, the *thermal shutdown* (TSD) function, is investigated in this experiment to analyze the impact of process variations on this feature.

Figure 15.15 shows the functional blocks that are electrically related to the TSD feature. The BG block is a bandgap reference that provides temperature independent voltage reference to the IREF block. The IREF block provides reference currents to multiple blocks, including the TX, RX, and TSD blocks. The Lane 1 and Lane 2 in the system share the same design, including the TSD implementation. The shutdown temperature threshold is designed to be 190 °C, meaning both Lane 1 and Lane 2 should turn off their driver to the bus when the temperature exceeds 190 °C.

We collected results from 1000 Monte Carlo simulations, and label them as 1 or 0 according to whether the lane can turn off the driver (TX block) at 190 °C or not. We use half of the simulations as the training data to build an SRKM classification model for each lane, and use the other half to test the performance of the trained classifier. Based on the *process design kit* (PDK) we use, there are 15,015 process parameters involved in each simulation. In other words, our goal is to build 15,015-dimensional classifiers with merely 500 samples.

As shown in Table. 15.3, the widely used SVM can only produce a classifier with an accuracy of about 60%, while the proposed SRKM can achieve an accuracy of around 85%. For the trained SVM model, all 500 samples become the support vectors, indicating that the SVM cannot find the regularity from the given training

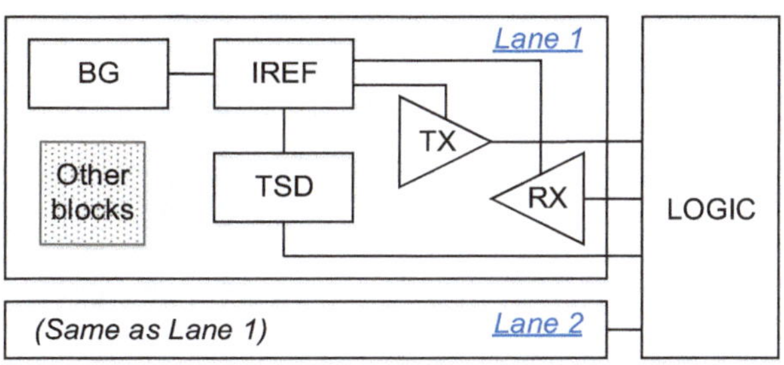

Fig. 15.15 Block diagram of a dual-lane data communication AMS system

data. On the contrary, the SRKM successfully achieves sparsity in the vector space (2 relevance vectors for Lane 1 and 23 for Lane 2 out of 500 samples) and the feature space (38 relevance features for Lane 1 and 45 relevance features for Lane 2 out of 15,015 process parameters).

In Fig. 15.16, we sum up the produced SRKM feature weights in each functional blocks and lanes to reflect the block-wise impact of process variations. For the SRKM trained for the Lane 1 TSD, relatively small feature weights have been assigned to the process parameters from Lane 2, which matches the structure of the system, where Lane 2 is indeed not directly connected to Lane 1 and hence should have minimal impact on the Lane 1 TSD. Similarly, in the SRKM trained for Lane 2 TSD, the impact of process parameters from Lane 1 is also minimal.

Based on the block-wise break-down of the sum of feature weights, it implies that the process parameters from the BG and IREF blocks, which determine the reference current of the TSD block, are more critical than the process parameters from the TSD block, which is the implementation of the TSD function. This also matches with the design intuition since the TSD block mainly involves components such as comparators whose performances are more resilient with respect to process variations, while currents are usually more vulnerable to process variations. Therefore, it can be anticipated that the feature weighting results are reliable.

Table 15.3 Learning model performance comparison (*SV/RV* support vectors/relevance vectors; *RF* relevance features)

	Model	Accuracy	# SV/RV	# RF
Lane 1	SVM	56.4%	500	–
	SRKM	85.8%	2	38
Lane 2	SVM	61.0%	500	–
	SRKM	84.6%	23	45

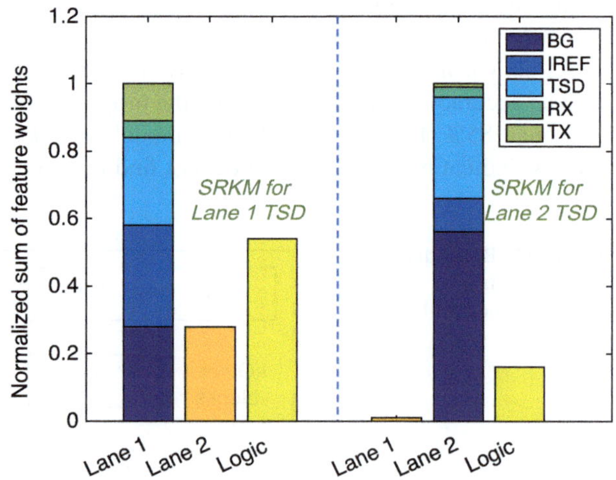

Fig. 15.16 Process variation impacts on TSD of various blocks

15.5 Conclusions

This paper proposes a novel sparse Bayesian learning framework named sparse relevance kernel machine to capture circuit characteristics and analyze circuit performance dependencies on assorted parameters or signatures via a statistical model. The advantages of the proposed framework are demonstrated in examples including statistical variability modeling of an LDO, a BIST scheme optimization of a charge-pump PLL, and building statistical variability models for a commercial automotive interface design.

The framework of the learning model was originally developed in our earlier works [18, 20]. Since then, we have extended the framework to handle both regression and classification problems, and have explored potentials of the new learning model in practical circuit applications. One major limitation of the current learning model is the computational complexity. Although we developed an iterative algorithm to remedy the complexity increased by the feature kernel weighting mechanism, based on the discussion in Sect. 15.3.3, the computational complexity of SRKM is one order higher than the widely used SVM in each iteration. And the overall convergence is also slower since SRKM has two levels of iterations. As a result, to solve applications with huge amount of data, the training algorithm needs to be further optimized for better efficiency.

Moreover, the feature kernel weighting mechanism proposed in Sect. 15.2 is very flexible. Since sparsity is very useful in our applications, we developed the learning model of SRKM in the Bayesian learning framework, but the feature kernel weighting mechanism can definitely fit into other kernel-based learning framework such as sequential minimal optimization (SMO), which poses potentials in other application scenarios where sparsity is not in need.

Acknowledgements This material is based upon work supported by the Semiconductor Research Corporation (SRC) through Texas Analog Center of Excellence at the University of Texas at Dallas (Task ID:2712.004).

References

1. P. Bastani, N. Callegari, L.C. Wang, M.S. Abadir, Statistical diagnosis of unmodeled systematic timing effects, in *Proceedings of the 45th Annual Design Automation Conference* (ACM, New York, 2008), pp. 355–360
2. C.M. Bishop, *Neural Networks for Pattern Recognition* (Oxford University Press, Oxford, 1995)
3. A.L. Blum, P. Langley, Selection of relevant features and examples in machine learning. Artif. Intell. **97**(1), 245–271 (1997)
4. A. Bounceur, B. Brahmi, K. Beznia, R. Euler, Accurate analog/RF BIST evaluation based on SVM classification of the process parameters, in *2014 9th International Design & Test Symposium (IDT)* (IEEE, Piscataway, 2014), pp. 55–60
5. G. Chandrashekar, F. Sahin, A survey on feature selection methods. Comput. Electr. Eng. **40**(1), 16–28 (2014)

6. H. Cheng, H. Chen, G. Jiang, K. Yoshihira, Nonlinear feature selection by relevance feature vector machine, in *Machine Learning and Data Mining in Pattern Recognition* (Springer, Berlin, 2007), pp. 144–159
7. B.R. Cobb, P.P. Shenoy, Nonlinear deterministic relationships in Bayesian networks, in *Symbolic and Quantitative Approaches to Reasoning with Uncertainty* (Springer, Berlin, 2005), pp. 27–38
8. M. Fernández-Delgado, E. Cernadas, S. Barro, D. Amorim, Do we need hundreds of classifiers to solve real world classification problems? J. Mach. Learn. Res. **15**(1), 3133–3181 (2014)
9. I. Guyon, S. Gunn, M. Nikravesh, L.A. Zadeh, *Feature Extraction: Foundations and Applications*, vol. 207 (Springer, Berlin, 2008)
10. S. Hochreiter, K. Obermayer, Nonlinear feature selection with the potential support vector machine, in *Feature Extraction* (Springer, Berlin, 2006), pp. 419–438
11. S.W. Hsiao, N. Tzou, A. Chatterjee, A programmable BIST design for PLL static phase offset estimation and clock duty cycle detection, in *2013 IEEE 31st VLSI Test Symposium (VTS)* (IEEE, Piscataway, 2013), pp. 1–6
12. X. Huang, L. Shi, J.A. Suykens, Ramp loss linear programming support vector machine. J. Mach. Learn. Res. **15**(1), 2185–2211 (2014)
13. S.S. Keerthi, O. Chapelle, D. DeCoste, Building support vector machines with reduced classifier complexity. J. Mach. Learn. Res. **7**, 1493–1515 (2006)
14. D.E. King, Dlib-ml: a machine learning toolkit. J. Mach. Learn. Res. **10**, 1755–1758 (2009)
15. S. Lai, Modeling, design and optimization of IC power delivery with on-chip regulation. Doctoral dissertation, Texas A&M University, 2014
16. S. Lai, P. Li, A fully on-chip area-efficient CMOS low-dropout regulator with fast load regulation. Analog Integr. Circuits Signal Process. **72**(2), 433–450 (2012)
17. F. Li, Y. Yang, E.P. Xing, From lasso regression to feature vector machine, in *Advances in Neural Information Processing Systems* (2005), pp. 779–786
18. H. Lin, Algorithms for verification of analog and mixed-signal integrated circuits. Doctoral dissertation, Texas A&M University, 2016
19. H. Lin, P. Li, Circuit performance classification with active learning guided sampling for support vector machines. IEEE Trans. Comput. Aided Des. Integr. Circuits Syst. **34**(9), 1467–1480 (2015). https://doi.org/10.1109/TCAD.2015.2413840
20. H. Lin, P. Li, Relevance vector and feature machine for statistical analog circuit characterization and built-in self-test optimization, in *Proceedings of the 53rd Annual Design Automation Conference* (2016), pp. 11–16
21. D.J. MacKay, The evidence framework applied to classification networks. Neural Comput. **4**(5), 720–736 (1992)
22. S. Maltabas, O.K. Ekekon, K. Kulovic, A. Meixner, M. Margala, An IDDQ BIST approach to characterize phase-locked loop parameters, in *2013 IEEE 31st VLSI Test Symposium (VTS)* (IEEE, Piscataway, 2013), pp. 1–6
23. M. Neil, M. Tailor, D. Marquez, Inference in hybrid Bayesian networks using dynamic discretization. Stat. Comput. **17**(3), 219–233 (2007)
24. J. Neumann, C. Schnörr, G. Steidl, Combined SVM-based feature selection and classification. Mach. Learn. **61**(1–3), 129–150 (2005)
25. B. Schölkopf, A.J. Smola, *Learning with Kernels: Support Vector Machines, Regularization, Optimization, and Beyond* (MIT Press, Cambridge, 2002)
26. A. Smola, B. Scholkopf, G. Ratsch, Linear programs for automatic accuracy control in regression, in *Ninth International Conference on (Conf. Publ. No. 470) Artificial Neural Networks, 1999 (ICANN 99)*, vol. 2 (IET, Stevenage, 1999), pp. 575–580
27. P. Somol, P. Pudil, J. Novovičová, P. Paclık, Adaptive floating search methods in feature selection. Pattern Recogn. Lett. **20**(11), 1157–1163 (1999)
28. J.A. Suykens, J. Vandewalle, Least squares support vector machine classifiers. Neural Process. Lett. **9**(3), 293–300 (1999)
29. R. Tibshirani, Regression shrinkage and selection via the lasso. J. R. Stat. Soc. Ser. B Methodol. **58**, 267–288 (1996)

30. M.E. Tipping, Sparse Bayesian learning and the relevance vector machine. J. Mach. Learn. Res. **1**, 211–244 (2001)
31. M.E. Tipping, An efficient matlab implementation of the sparse Bayesian modelling algorithm (version 2.0). Vector Anomaly, March 2009
32. M.E. Tipping, A.C. Faul et al., Fast marginal likelihood maximisation for sparse Bayesian models, in *Proceedings of the Ninth International Workshop on Artificial Intelligence and Statistics* (2003)
33. V.N. Vapnik, *Statistical Learning Theory* (Wiley, New York, 1998)
34. V. Vapnik, S. Golowich, A. Smola, Support vector method for function approximation, regression estimation, and signal processing, in *Advances in Neural Information Processing Systems* (1997), pp. 281–287
35. D. Ververidis, C. Kotropoulos, Fast and accurate sequential floating forward feature selection with the Bayes classifier applied to speech emotion recognition. Signal Process. **88**(12), 2956–2970 (2008)
36. F. Wang, M., Zaheer, X. Li, J.O. Plouchart, A. Valdes-Garcia, Co-learning Bayesian model fusion: efficient performance modeling of analog and mixed-signal circuits using side information, in *Proceedings of the IEEE/ACM International Conference on Computer-Aided Design* (IEEE Press, Piscataway, 2015), pp. 575–582
37. P.M. Williams, Bayesian regularization and pruning using a Laplace prior. Neural Comput. **7**(1), 117–143 (1995)
38. D.P. Wipf, B.D. Rao, An empirical Bayesian strategy for solving the simultaneous sparse approximation problem. IEEE Trans. Signal Process. **55**(7), 3704–3716 (2007)
39. H. Xu, C. Caramanis, S. Mannor, Robustness and regularization of support vector machines. J. Mach. Learn. Res. **10**, 1485–1510 (2009)
40. G. Yu, P. Li, A methodology for systematic built-in self-test of phase-locked loops targeting at parametric failures, in *IEEE International Test Conference, 2007 (ITC 2007)* (IEEE, Piscataway, 2007), pp. 1–10

Chapter 16
SiLVR: Projection Pursuit for Response Surface Modeling

Amith Singhee

16.1 Motivation

In many situations it is desirable to have available an inexpensive model for predicting circuit performance, given the values of various statistical parameters in the circuit (e.g., V_t for the different devices in the circuit). Examples of such situations are 1) in a circuit optimization loop where quick estimates of yield might be necessary to drive the solution towards a high-yield design in reasonable run time, and 2) during manual design, a simple analytical model can provide insight into circuit operation using metrics such as sensitivities or using quick visualization, thus helping the designer to understand and tune the circuit. As an example, consider an operational amplifier (opamp) circuit, which has several performance metrics of interest, such as unity gain frequency or DC offset voltage. Statistical parameters that impact these performance metrics can include the threshold voltage values of the circuits in the opamp and the electrical properties of any passive components. A detailed analysis of an opamp and other examples, using the methods described in this chapter, is provided in Sect. 16.7. Director et al. [1] provides a good overview of general statistical design approaches. Even though the paper is not very recent, much of the literature on statistical design (yield optimization) over the last couple of decades proposes techniques that fall under the general types discussed therein. Such performance models in the statistical parameter space are commonly referred to as *response surface models*: we abbreviate this as RSM here. Initial approaches employed linear regression to model circuit performance metrics, as in [2]. Soon, the linear models were found to be inadequate for modeling nonlinear behavior and quadratic models were proposed in [3, 4] to reduce the modeling error.

A. Singhee (✉)
IBM Research, Bangalore, India
e-mail: asinghee@in.ibm.com

I. M. Elfadel et al. (eds.), *Machine Learning in VLSI Computer-Aided Design*,
https://doi.org/10.1007/978-3-030-04666-8_16

These low-order models worked sufficiently well for the technologies of yesteryears, but face fundamental difficulties going forward. Any solution now must address three large challenges:

- **Dimensionality**: The number of sources of variations in the circuit can be large. Even for a simple flip-flop, there can be over 50 sources, e.g., random dopant fluctuation (RDF), line edge roughness (LER), random poly crystal orientation (RCO) [5], and gate oxide thickness variation. Although many such sources can be absorbed into a few device-level parameters, for larger analog cells the dimensionality can still easily be in the hundreds. The number of variables, s, in a model determines the number of unknown model parameters that need to be estimated during model fitting. The number of SPICE-simulated points needed then is at least the number of unknown parameters.
- **Large variations**: The relative effect of every variation source is becoming very large. Just considering RDF, predictions indicate that the standard deviation of V_t can be 10% of the nominal V_t at the 70 nm node [6], growing to 21% for a 25 nm device [7], with 0.3 V V_t. If other variations (LER, RCO) are considered, the deviation is even higher [5].
- **Nonlinearity**: Not all performance/variable relationships are simple. A good example is the relationship between device V_t in a flip-flop and the flip-flop delay. Such nonlinearity is even more pronounced in the case of analog circuits.

Linear models are able to handle the dimensionality well, since the number of unknown model parameters is a slowly increasing $s + 1$, where s is the number of input variables. These models, however, fail to capture nonlinear behaviors, for which higher-order models are needed. Higher-order models can, however, have a very large number of parameters: a polynomial of degree d in s dimensions has ${}^{s+d}C_d$ terms. Hence, even a quadratic model in 100 dimensions can have 5151 parameters, requiring 5151 initial SPICE simulations to generate training points for the model! Recent attempts at reducing the number of unknowns in the quadratic model have resulted in very efficient techniques, namely PROBE [8] and kernel reduced-rank regression (RRR) [9]. Both these methods essentially reduce the rank of the quadratic model, the former doing it in a more natural near-optimal manner. We will look at PROBE in more detail in Sect. 16.2.3. However, these methods still suffer from the severe restriction of quadratic (which includes linear) behavior. In the presence of large variations, the nonlinearity in the circuit behavior is significant enough to make these models unusable, as we shall see in later sections.

In this chapter, we review the *latent variable regression* (LVR) [10] and *projection pursuit regression* (PPR) [11] strategies and show why they can be attractive in these scenarios. Roughly speaking, these techniques iteratively extract the next statistically most important variable (latent variable or LV), and minimize the error in fitting the remainder of the unexplained performance variation. Hence, they directly reduce the problem dimensionality. Further, these techniques can be accompanied with flexible, but compact, functional forms for the model, thus reducing *a priori* assumptions about the magnitude of variations and the behavior modeled. Using these ideas this chapter will develop an RSM strategy for silicon design problems—SiLVR—and show its superior performance in comparison to PROBE, in the context of the three challenges mentioned above. We will also see how the "designer's insight" can be obtained naturally from the structure of the SiLVR model, in the form of some quantitative measures and insightful visualization. Such insights into the circuit behavior can help the designer to better understand the behavior of the circuit during manual design, and guide the optimizer better during automatic sizing. SiLVR was first introduced by us in [12].

LVR methods have a long and interesting track record and encompass a variety of different techniques that follow the same philosophy to meet slightly different objectives, for example, partial least squares (PLS) [13], canonical correlation regression (CCR) [10], and reduced-rank regression [14]. A good survey and comparison is provided in [10]. These LVR techniques have found wide application, much of it outside the realm of silicon application, in areas ranging from chemometrics [15] to statistics [16], to bioinformatics [17]. Many LVR methods still assume a linear relationship, or use a low-order nonlinear kernel to explain the assumed nonlinear relationships. Thus, our own interest is in LVR methods that support a more flexible nonlinear framework. Here, Baffi et al. [18], and Malthouse et al. [19] are noteworthy. In addition to the single variable iterative extraction philosophy, these show how to use a neural network [20] to capture significant nonlinear behaviors. However, [18] suffers from unreliable and slow convergence during training and [19] uses an excessively complex model that can be prone to computational burdens and overfitting. SiLVR, although similar in flavor to these methods, uses a more compact model in a new unified training framework to remove these issues.

Although SiLVR derives its name from LVR, its philosophy finds a closer fit with projection pursuit [11, 21]. Both LVR and PP are very similar in the way they operate, but their theory and applications seem to have developed more or less independently: LVR in the world of chemometrics (PLS) and statistics (CCR, RRR), while PP in the world of statistics, approximation theory, and machine learning. Theoretical foundations for PP appear to be better developed, more so for nonlinear regression and the particular case of the SiLVR model (PP using sigmoidal functions). We will review relevant results from these as we move towards developing the SiLVR model architecture.

In the rest of this chapter, we briefly review linear and quadratic models, including PROBE, a low-rank quadratic model, after which we review the LVR and PP techniques, along with relevant theoretical results from approximation theory. Finally, we develop the SiLVR model, covering relevant details regarding model training, and show experimental results.

16.2 Prevailing Response Surface Models

Before we review linear and quadratic models, let us first concretely define the RSM problem. Let $\mathscr{X} = \mathbb{R}^s$ be the statistical parameter space and $\mathscr{Y} = \mathbb{R}^{s_\mathscr{Y}}$ be the circuit performance-metric or output space: $s_\mathscr{Y}$ is the number of outputs. For a given

$\mathbf{x} \in \mathscr{X}$, $\mathbf{y} = \mathbf{f}_{\text{sim}}(\mathbf{x}) \in \mathscr{Y}$ is evaluated using a SPICE-level circuit simulation. We want to find an approximation

$$\hat{\mathbf{y}} = \mathbf{f}_m(\mathbf{x}) \in \mathscr{Y} : \min_{\mathbf{f}_m} E(\|\mathbf{y} - \hat{\mathbf{y}}\|^2), \tag{16.1}$$

such that the function $\mathbf{f}_m$ is much cheaper to evaluate than $\mathbf{f}_{\text{sim}}$ in terms of computational cost. In this chapter, unless specifically mentioned, we will now consider only any one output y_i at a time, from the vector $\mathbf{y}$. This is for the sake of clarity of explanation, and we will drop the subscript i and use only y. Then, for the output y, we can write (16.1) as

$$\hat{y} = f_m(\mathbf{x}) \in \mathbb{R} : \min_{f_m} E(\|y - \hat{y}\|^2). \tag{16.2}$$

If we use the L_2 norm in (16.2), we achieve the least squared error fit. To obtain this model in practice, some n sample points $\{\mathbf{x}_i, y_i = f_{\text{sim}}(\mathbf{x})\}_{i=1}^{n}$ are generated using SPICE simulations and the following optimization problem is solved:

$$\min_{f_m} \sum_{i=1}^{n} |y_i - \hat{y}_i|^2 \quad \text{where} \quad \hat{y}_i = f_m(\mathbf{x}). \tag{16.3}$$

16.2.1 Linear Model

Linear models, such as the one used in [2], model the response y as a linear function of the parameters $\mathbf{x}$. Hence, a linear model can be written as

$$\hat{y} = \mathbf{a}^T \mathbf{x} + c, \tag{16.4}$$

where $\mathbf{a}$ is a vector of s unknown model parameters, $\mathbf{a} \in \mathbb{R}^s$ and c is an unknown real scalar. The total number of distinct, unknown model parameters is $n_p = s+1 = O(s)$. Given $n \geq n_p$ training sample points, we can estimate $\mathbf{a}$ and c, using the least squares form of (16.3), as

$$\left[\mathbf{a}^T \; c\right]^T = [\mathbf{X}\;\mathbf{1}]^+ \mathbf{Y}, \quad \text{where} \quad \mathbf{A}^+ = \left(\mathbf{A}^T \mathbf{A}\right)^{-1} \mathbf{A}^T, \tag{16.5}$$

where $\mathbf{X}$ is an $n \times s$ matrix with the i-th row being the i-th input sample point $\mathbf{x}_i^T$ and $\mathbf{Y}$ is an n-vector with the i-th element being the i-th output sample point. $[\mathbf{X}\;\mathbf{1}]$ means $\mathbf{X}$ is augmented by a column of ones. Figure 16.1a shows an example of a linear model with $s = 2$: we can immediately see that the linear model cannot capture the nonlinear relationship in the data and the errors are very large.

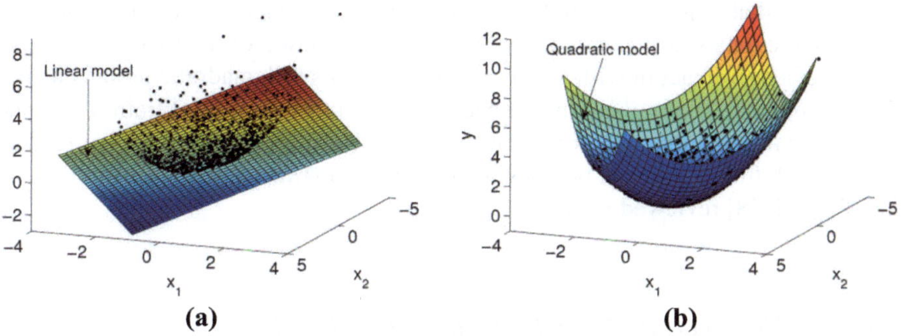

Fig. 16.1 A linear RSM cannot capture the quadratic behavior, while the quadratic RSM succeeds. (**a**) Linear model. (**b**) Quadratic model

16.2.2 Quadratic Model

Quadratic RSMs were proposed in [3, 4] to model nonlinearities when the linear model fails. The quadratic model can be written as

$$\hat{y} = \mathbf{x}^T \mathbf{A}\mathbf{x} + \mathbf{b}^T \mathbf{x} + c, \tag{16.6}$$

where $\mathbf{A}$ is a symmetric $s \times s$ matrix of unknowns, $\mathbf{b}$ is a vector of s unknowns, and c is an unknown scalar. The total number of distinct, unknown model parameters is $n_p = {}^{s+2}C_2 = (s+1)(s+2)/2 = O(s^2)$. Hence, the number of parameters grows quadratically with the number of dimensions. If we let $\mathbf{A}_i$ be the i-th row vector of $\mathbf{A}$ (written as a column vector), and the Kronecker product

$$\mathbf{x} \otimes \mathbf{x} = \begin{bmatrix} x_1^2 & x_1x_2 & \dots & x_1x_s & x_2x_1 & \dots & x_s^2 \end{bmatrix}^T,$$

we can write (16.6) as

$$\hat{y} = \mathbf{a}_e^T \mathbf{x}_e, \quad \text{where} \quad \mathbf{a}_e = \begin{bmatrix} \mathbf{A}_1^T & \dots & \mathbf{A}_s^T & \mathbf{b}^T & c \end{bmatrix}^T, \mathbf{x}_e = \begin{bmatrix} (\mathbf{x} \otimes \mathbf{x})^T & \mathbf{x}^T & 1 \end{bmatrix}^T, \tag{16.7}$$

which is similar in form to the linear model (16.4). Then, given $n \geq n_p$ training points, the least squared error estimate for the unknowns in (16.6) can be computed as

$$\mathbf{a}_e = \mathbf{X}_e^+ \mathbf{Y}, \tag{16.8}$$

where the i-th row of the matrix $\mathbf{X}_e$ is the vector $\mathbf{x}_e$ computed for the i-th input training point, and $\mathbf{X}_e^+$ is its pseudoinverse as in (16.5). In practice, the common (repeated) terms in $\mathbf{x} \otimes \mathbf{x}$ are combined, for example, x_1x_3 and x_3x_1. From Fig. 16.1b we can see that a quadratic model fits the data much better than a linear model. In this case the data was generated from a quadratic function of the two variables, and so we get a near-exact fit with a quadratic model. This full quadratic model can have a large number of unknowns and, hence, require a large number of training points n for proper fitting. Also, the computations in (16.8) can be very expensive for large s. This high fitting cost can be alleviated by using a *reduced-rank* quadratic model, like PROBE [8] reviewed next.

16.2.3 PROjection-Based Extraction (PROBE): A Reduced-Rank Quadratic Model

A reduced-rank quadratic RSM was proposed by Li et al. [8] to overcome the dimensionality problems of the full quadratic model. The matrix $\mathbf{A}$ in (16.6) is replaced by a low-rank approximation $\mathbf{A_L}$, given by

$$\mathbf{A}_L = \sum_{i=1}^{r} \lambda_i \mathbf{p}_i \mathbf{p}_i^T, \quad r < s, \tag{16.9}$$

where λ_i is the i-th dominant (largest) eigenvalue of $\mathbf{A}$ and $\mathbf{p}_i$ is the corresponding normalized eigenvector. It is known that this approximation minimizes the Frobenius-norm error for a given r: it is the optimal rank-r approximation [22]. Then the reduced quadratic model can be written as

$$\hat{y} = \sum_{i=1}^{r} \mathbf{x}^T \lambda_i \mathbf{p}_i \mathbf{p}_i^T \mathbf{x} + \mathbf{b}_i \mathbf{x} + c_i. \tag{16.10}$$

Thus, the model is a combination of r simpler quadratic models, where the i-th quadratic part varies along the projection along the i-th eigenvector $\mathbf{p}_i$. This is similar in flavor to the concept of projection pursuit that is used in a more powerful and flexible form for the SiLVR model proposed here.

Since the matrix $\mathbf{A}$ is not already known, an implicit method that does not need it is used in [8] to estimate its eigenvectors. The overall algorithm is shown in Algorithm 16.1. Once the i-th component in (16.10) is extracted, the predicted $g_i(\mathbf{x})$ values for all the sample points are subtracted out, so that the $(i + 1)$-th component fits the residual $y = y - g_i(\mathbf{x})$. Algorithm 16.2 extracts the i-th component using an implicit power iteration method, and constitutes the function `getRankOneQuadratic()` in Algorithm 16.1. The vector $\mathbf{q}_k \to \sqrt{\lambda_i}\mathbf{p}_i$ with $k \to \infty$ in Algorithm 16.2, for the i-th call to `getRankOneQuadratic()` in Algorithm 16.1. For a detailed explanation of the technique please refer to [8].

A rank-r quadratic model is effective in reducing the number of unknown model parameters and scales well with the number of dimensions s, if $r \ll s$: the number

Algorithm 16.1 The PROBE algorithm

Given training sample points $\{\mathbf{x}_j, y_j\}_{j=1}^n$
for i = 1 to r do
  $g_i(\mathbf{x}) \leftarrow$ getRankOneQuadratic($\{\mathbf{x}_j, y_j\}_{j=1}^n$)
  for all sample points $\{\mathbf{x}_j, y_j\}$
    $y_j \leftarrow y_j - g_i(\mathbf{x}_j)$
The rank-r model is $\hat{y} = \sum_{i=1}^r g_i(\mathbf{x})$

Algorithm 16.2 PROBE: `getRankOneQuadratic`($\{\mathbf{x}_j, y_j\}_{j=1}^n$) function to extract the rank-1 estimate

1: Given ϵ, a predefined tolerance
2: Randomly select $\mathbf{q}_0 \in \mathbb{R}^s$
3: $k = 0$, $\psi_0 = \infty$
4: repeat
5:   $k = k + 1$
6:   $\mathbf{q}_{k-1} = \mathbf{q}_{k-1}/\|\mathbf{q}_{k-1}\|_2$
7:   Solve (least squares error) $\min_{\mathbf{q}_k, \mathbf{b}_k, c} \psi_k$, where $\psi_k = \sum_{j=1}^n [y_j - (\mathbf{x}_j^T \mathbf{q}_k \mathbf{q}_{k-1}^T \mathbf{x}_j + \mathbf{b}_k \mathbf{x}_j + c_k)]^2$
8: until $|\psi_k - \psi_{k-1}| < \epsilon$
9: Return rank-1 estimate $g(\mathbf{x}) = \mathbf{x}^T \mathbf{q}_k \mathbf{q}_{k-1}^T \mathbf{x} + \mathbf{b}_k \mathbf{x} + c_k$

of model parameters is $n_p = 2r(s + 1) = O(rs)$, which increases linearly with s. The authors of [8] show that r is very small for the performance metrics of some commonly seen circuits: even a rank-1 model can suffice. However, the model still suffers from a quadratic behavior assumption. We will now review some techniques that, in the general case, make no assumption regarding the modeled behavior, and then show how we can maintain much of this generality using the proposed SiLVR model.

16.3 Latent Variables and Ridge Functions

For the rest of this chapter let us assume that all the training sample points have been normalized—scaled and translated to mean 0 and variance 1—in both the input and output spaces. This is for the sake of clear development of the following concepts, without any loss of generality.

16.3.1 Latent Variable Regression

With the assumption of normalized training points, the standard linear model for the $s_{\mathcal{Y}}$-vector of outputs $\mathbf{y}$ can be written as

$$\hat{\mathbf{y}} = \mathbf{A}\mathbf{x}, \tag{16.11}$$

where $\mathbf{A}$ is an $s_{\mathcal{Y}} \times s$ matrix of regression coefficients. Classically, latent variable regression (LVR) has been used to modify this linear model into a *reduced* linear model as

$$\hat{\mathbf{y}} = \mathbf{Z}\mathbf{W}_r\mathbf{x}. \tag{16.12}$$

Here $\mathbf{W}_r$ is an $r \times s$ matrix that projects the s-dimensional vector $\mathbf{x}$ to an r-dimensional space, where $r < s$, and $\mathbf{Z}$ is an $s_{\mathcal{Y}} \times r$-vector of regression coefficients over this reduced r-dimensional space. If we denote the i-th row of $\mathbf{W}_r$ by $\mathbf{w}_i$, then we can interpret $\mathbf{w}_i^T\mathbf{x}$ as the i-th coordinate in the reduced r-dimensional space. We will refer to $\mathbf{w}_i$ as the i-th *projection vector*, and the new variable $\mathbf{w}_i^T\mathbf{x}$ as the i-th *latent variable* t_i. Each coordinate w_{ij} of $\mathbf{w}_i$ will be referred to as the j-th *projection weight* of the i-th projection vector.

$$t_i = \mathbf{w}_i^T\mathbf{x}. \tag{16.13}$$

$\mathbf{W}_r$ is, then, the *projection matrix.*

The unknown parameters (the projection vectors $\mathbf{w}_i$ and the regression coefficients in $\mathbf{Z}$) can be chosen to satisfy a variety of criteria, each yielding a different LVR method (e.g., RRR, PLS, CCR) as shown in [10]. The relevant method here is reduced-rank regression (RRR), which solves the least squared error problem

$$\min_{\mathbf{W}_r,\mathbf{Z}} \|\mathbf{Y} - \mathbf{X}\mathbf{W}_r^T\mathbf{Z}^T\|^2, \tag{16.14}$$

where $\mathbf{X}, \mathbf{Y}$ are matrices of n sample points: each row is one sample point. From the discussion until now, the important idea to remember is that we are extracting the r statistically most important LVs ($\{t_1, \ldots, t_r\}$), such that the expected squared error is minimized, as in (16.14).

The problem of modeling nonlinear behavior, however, remains unsolved by these classical LVR techniques. Kernel-based methods try to address this issue by using the well-known "kernel trick": map the inputs ($\mathbf{x}$), using *fixed* nonlinear kernels ($f_K(\mathbf{x})$, e.g., a quadratic as in [9]), to a higher dimensional space, and then create a reduced linear model from this higher dimensional space to the output y [23]. This has severe limitations: it increases the problem dimensionality before reducing it, and, more importantly, assumes a known nonlinear relationship between $\mathbf{x}$ and y. Baffi [18] proposes adapting LVR to use a more flexible neural network [20] formulation, but the model fitting is very slow (a two-step process that iterates between model fitting and LV estimation) and unreliable (due to weak convergence of this two-step iteration). Malthouse [19] takes this further, but produces a very complex neural network model that can cause undesirable overfitting, especially for small training datasets, and has a large number of unknowns to fit. Also, both these methods solve a problem different from minimizing the least squared error as in (16.3). As we saw in Sect. 16.2.3, the PROBE method also uses a projection-based approach, but is restricted to a quadratic form.

The advantages of a flexible nonlinear LVR method are multiple and significant:

- It inherently reduces the dimensionality of the problem by extracting the LVs.
- The LVs are the "hidden variables" in the input space that impact the output in decreasing order of importance. Having this information can be of much use to the designer, as we shall see in the next few sections.
- The model would not be restricted to a small class of nonlinear behaviors.

All these features are very useful for addressing the problems mentioned in Sect. 16.1, and we will construct the SiLVR model to exploit all of them. First, though, we review the idea of projection pursuit, which bears close resemblance to LVR, and provides some theoretical foundation for the SiLVR model.

16.3.2 Ridge Functions and Projection Pursuit Regression

Projection pursuit regression (PPR) is a class of curve fitting algorithms, formally introduced first by Friedman and Stuetzle in [21] that approximate the output y as

$$\hat{y} = \sum_{i=1}^{r} g_i(\mathbf{w}_i^T \mathbf{x}), \tag{16.15}$$

where $\mathbf{w}_i$ is the i-th projection vector, similar to LVR, and $g_i : \mathbb{R} \rightarrow \mathbb{R}$ are unknown functions that might be parameterized functional forms (e.g., quadratic) or some nonparametric function, as in [21]. Hence, y is represented as the sum of nonlinear, univariate functions g_i, each varying along a different direction $\mathbf{w}_i$ in the input space. Each g_i function is called a *ridge function* [24] because for $s = 2$ it defines a 2-dimensional surface that is constant along one direction in the input space $\mathbb{R}^2$ (orthogonal to $\mathbf{w}_i$), leading to "ridges" in the topology. An example is shown in Fig. 16.2. In higher dimensions, a ridge function g_i is constant along the hyperplanes $\mathbf{w}_i^T \mathbf{x} = c$. Ridge functions have also been referred to as *plane waves* [25] historically, particularly in the field of partial differential equations [26] The representation in (16.15) is computed so as to minimize the modeling error as in (16.3). Given n training sample points, we can write this criterion as

$$\min_{r, \{g_i, \mathbf{w}_i\}_1^r} \sum_{j=1}^{n} \|y_j - \sum_{i=1}^{r} g_i(\mathbf{w}_i^T \mathbf{x}_j)\|^2. \tag{16.16}$$

From (16.15), we can see the similarity to LVR, where we are also trying to extract r best directions to predict the output. In fact, a *nonlinear* version of LVR optimizing (16.3) will accomplish precisely the same thing as PPR.

The representation of (16.15) is also a general form of a feedforward neural network with one hidden layer. Artificial neural networks were introduced first by McCullough and Pitts [27] to model the behavior of neurons in the nervous system.

We will refer to them as simply neural networks. Since then, neural networks have been the focus of much theoretical and applied research, and have been proposed in a large variety of forms [20, 28]. Here we refer to the simple feedforward form with one hidden layer of r nodes, which can be written mathematically as

$$\sum_{i=1}^{r} \alpha_i \sigma(\mathbf{w}_i^T \mathbf{x} + \beta_i), \ \alpha_i, \beta_i \in \mathbb{R}, \tag{16.17}$$

where $\sigma : \mathbb{R} \to \mathbb{R}$ is a fixed univariate function called the *activation function*. One such network is shown in Fig. 16.3. We will refer to such networks as 3-layer perceptrons (3LP) as per [29]: the first layer is just the layer of input nodes, layer two consists of the activation function nodes, and layer three consists of the output node. From (16.17), we can immediately see the similarity with the PPR model of (16.15). Hence, a 3LP is a special case of a PPR model. We will revisit the 3LP when we develop the SiLVR model, where we use it in a somewhat different manner.

Before we proceed further, let us look at a couple of simple examples to clarify the concept of PPR. Consider the functions

$$y_1 = (x_1 + 2x_2)^3, \qquad y_2 = x_1 x_2, \tag{16.18}$$

of which the second appears also in [30] and [11]. We can represent the first function as

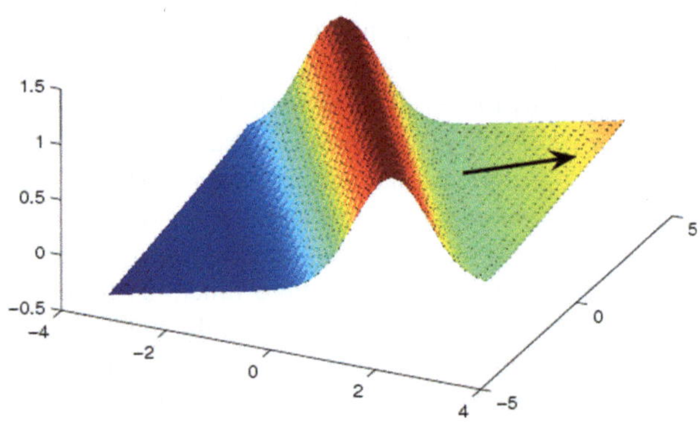

Fig. 16.2 Example of a ridge function. The arrow indicates the projection vector

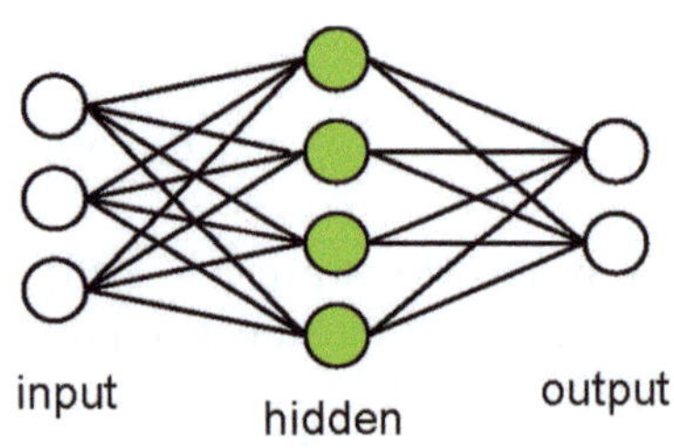

Fig. 16.3 A feedforward neural network with one hidden layer: a 3-layer perceptron

$$y_1 = t_1^3, \quad \text{where} \quad t_1 = (1\ 2)\begin{pmatrix} x_1 \\ x_2 \end{pmatrix}. \tag{16.19}$$

In this case projection along only one direction $\mathbf{w}_1 = (1, 2)$ is enough to model the entire function exactly. This is because the function varies only along that one direction. Hence, we have reduced the dimensionality of the input space to one. t_1 is the first LV, following the nomenclature from latent variable regression. On first glance, the second function could seem unfriendly to such linear projection-based decomposition. However, we can write y_2 as

$$y_2 = x_1 x_2 = 0.25(x_1 + x_2)^2 - 0.25(x_1 - x_2)^2, \tag{16.20}$$

which is the sum of two univariate ridge functions (quadratics) along the directions $\mathbf{w}_1 = (1, 1)$ and $\mathbf{w}_2 = (1, -1)$, in the form of (16.15). The functions are shown in Fig. 16.4.

It is interesting to note that the Fourier series representation of a function,

$$f(\mathbf{x}) = \sum_{k=1}^{r} a_k e^{i\bar{\omega}_k^T \mathbf{x}}, \tag{16.21}$$

is also a ridge function representation, where the projection vectors $\bar{\omega}_k^T$ are points in the s-dimensional Fourier domain. Note that the i in this equation is the imaginary unit, and not an index. Section 16.4 discusses a theorem from [30] that deals with representations similar to this.

Of course, for some unknown function f we would need to automatically extract the optimal projection directions and the corresponding ridge function. This "pursuit" of the optimal projections leads to the name projection pursuit. Before we discuss the algorithmic details of PPR, let us review some relevant results from approximation theory that establish a theoretical foundation for approximation using ridge functions. The reader who is more interested in the algorithmic considerations may skip forward to Sect. 16.5.

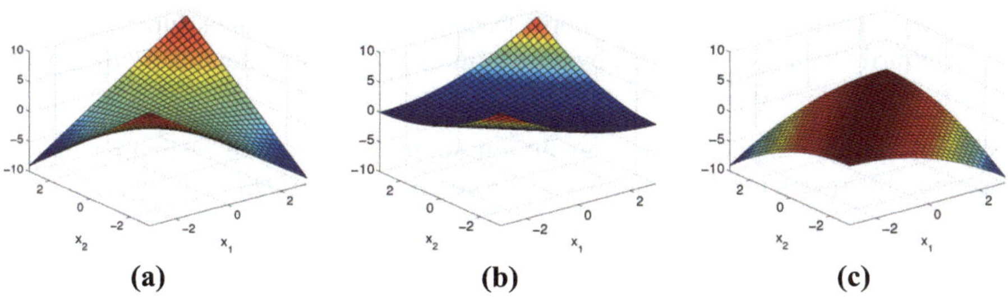

Fig. 16.4 The function of (16.20) and its component ridge functions. (**a**) $y = x_1 x_2$. (**b**) $0.25(x_1 + x_2)^2$. (**c**) $-0.25(x_1 - x_2)^2$

16.4 Approximation Using Ridge Functions: Density and Degree of Approximation

Before we can begin to develop algorithms for PPR, some more fundamental questions regarding ridge functions deserve attention. What can we approximate using ridge functions? How well can we approximate? To address these questions, let us first review some basic terminology from topology.

- **C**(X)—For some space or set X, $C(X)$ is the set of all continuous functions defined on X, $f(\mathbf{x}|\mathbf{x} \in X)$.
- ***p*-norm**—Given some function f over some space X, we define the p-norm as

$$\|f\|_p = \begin{cases} \left(\int_X |f(\mathbf{x})| d\mathbf{x}\right)^{1/p} & , 0 < p < \infty \\ \sup\limits_{\mathbf{x} \in X} |f(\mathbf{x})| & , p = \infty \end{cases} . \tag{16.22}$$

  A norm taken over a set $D \subset X$ will involve integrating or taking the supremum, as relevant, over only D. The norm is then denoted by $\|f\|_{p,D}$, unless obvious.
- $\mathbf{L}_p(X)$—For some space or set X, $L_p(X)$ is the set of continuous functions defined over X that have a finite p-norm over X, equivalently

$$L_p(X) = \{f : \|f\|_{p,X} < \infty\}. \tag{16.23}$$

- **Compact set**—A compact set in Euclidean space $\mathbb{R}^s$ is any subset of $\mathbb{R}^s$ that is closed and bounded. A more general definition for any space is as follows. A set D is compact if for every collection of open sets $\mathscr{U} = \{U_i\}$ such that $D \subset \cup_i U_i$, there is a *finite* subset $\{U_{i_j} : j = 1, \ldots, m\} \subset \mathscr{U}$ such that $D \subset \cup_{j=1,\ldots,m} U_{i_j}$. For example, the *closed* unit ball $\{\mathbf{x} : \|\mathbf{x}\|_2 \leq 1, \mathbf{x} \in \mathbb{R}^s\}$ is compact, while the *open* unit ball $\{\mathbf{x} : \|\mathbf{x}\|_2 < 1, \mathbf{x} \in \mathbb{R}^s\}$ is not compact.
- **Dense set**—Let $V_1 \subset V_2$ be two subsets of some space V. Then, V_1 is dense in V_2 if for any $v \in V_2$ and any $\epsilon > 0$, there is a $u \in V_1$ such that $\|v - u\| < \epsilon$ under the p-norm specified or assumed without confusion. For example, if V_2 is $C[0, 1]$ and V_1 is the space of all polynomials over $[a, b]$, then V_1 is dense in V_2 over $[a, b]$ under the ∞-norm, because every continuous function can be arbitrarily well approximated by polynomials, over some intervals $[a, b]$. This is the well-known Weierstrass approximation theorem [31].

We now provide some answers to the questions posed at the beginning of this section, by reviewing relevant results from the approximation theory of ridge functions.

16.4.1 Density: What Can Ridge Functions Approximate?

Theorem 16.1 (Diaconis and Shahshahani [30]) *Functions of the form* $\sum \alpha_i e^{\mathbf{w}_i^T \mathbf{x}}$ *with* $\alpha_i \in \mathbb{R}$, $\mathbf{w}_i \in \mathbb{Z}_+$ *are dense in* $C[0, 1]^s$ *under the* ∞*-norm.*

This theorem says that any continuous function over the unit cube in s dimensions can be arbitrarily well approximated by ridge functions of the exponential form. Even though this theorem restricts itself to exponential ridge functions, it does prove that there exists a ridge function representation (linear combination of exponentials) for any continuous function over the unit cube. Note that the unit cube domain can be relaxed to any compact set D in s dimensions by including all required (continuous) transformations in the function to be approximated.

We will present a proof here, since it is simple enough for the non-mathematician to follow, while at the same time it provides some good insight and is an interesting read. The proof will require the following, very well-known, Stone–Weierstrass theorem, a generalization of the Weierstrass theorem. For a proof of the Stone–Weierstrass theorem please refer to standard textbooks on analysis, e.g., [31]. Here we state a less general version of the theorem that suffices for our purposes.

Theorem 16.2 (Stone–Weierstrass) *Let* $D \subset \mathbb{R}^s$ *be a compact set, and let* V *be a subspace of* $C(D)$*, the space of continuous functions on* D*, such that*

(a) V *contains all constant functions,*
(b) $u, v \in V \Rightarrow uv \in V$*, and*
(c) For every pair $\mathbf{x}, \mathbf{y} \in D,\ \mathbf{x} \neq \mathbf{y},\ \exists v \in V$ *such that* $v(\mathbf{x}) \neq v(\mathbf{y})$.

Then, V *is dense in* $C(D)$*, i.e.,* $\forall v \in C(D),\ \epsilon > 0,\ \exists u \in V$ *such that* $\|u - v\|_\infty$

The Stone–Weierstrass theorem gives conditions such that V, the linear span of a given set of functions, is able to approximate any continuous function arbitrarily well. Using this, we can now prove Theorem 16.1.

Proof (Proof of Theorem 16.1) Here V is the space of functions of the form $\sum \alpha_i e^{\mathbf{w}_i^T \mathbf{x}}$ with $\alpha_i \in \mathbb{R}$, $\mathbf{w}_i \in \mathbb{Z}_+$, and it satisfies the three conditions in the Stone–Weierstrass theorem, for the compact set $D = [0, 1]^s$:

(a) $\alpha_i e^{\mathbf{0}^T \mathbf{x}}$ are all the constant functions, where $\mathbf{0}$ is the vector of zeros.
(b) $\sum_i \alpha_i e^{\mathbf{w}_i^T \mathbf{x}} \cdot \sum_j \beta_j e^{\mathbf{v}_j^T \mathbf{x}} = \sum_i \sum_j \alpha_i \beta_j e^{(\mathbf{w}_i + \mathbf{v}_j)^T \mathbf{x}}$, which lies in V.
(c) For any pair $\mathbf{x}, \mathbf{y} \in [0, 1]^p$ such that $\mathbf{x} \neq \mathbf{y}$, we must have $x_i \neq y_i$ for at least one $i \in 1, \ldots, s$. Then choose $\mathbf{w} = \mathbf{e}_i$, where the vector $\mathbf{e}_i$ is the unit vector along coordinate i. Then $e^{\mathbf{w}^T \mathbf{x}} = e^{x_i} \neq e^{\mathbf{w}^T \mathbf{y}} = e^{y_i}$.

Hence, the Stone–Weierstrass theorem applies and V is dense in $C[0, 1]^s$.

More general results regarding the density of ridge functions have been developed by several authors, notably Vostrecov and Kreines [25], Sun and Cheney [32], and Lin and Pinkus [33]. Let $\mathscr{W} \subseteq \mathbb{R}^s$ be the set of possible projection vectors, and define

$$\mathscr{R}_{\mathscr{W}} = \text{span}\left\{g(\mathbf{w}^T\mathbf{x}) : \mathbf{w} \in \mathscr{W}, g \in C(\mathbb{R}), \mathbf{x} \in \mathbb{R}^s\right\} \tag{16.24}$$

as the linear span of all possible ridge functions using univariate continuous functions along directions defined by the vectors in $\mathscr{W}$. Now we state two results that specify conditions on $\mathscr{W}$ such that $\mathscr{R}_{\mathscr{W}}$ is dense in $C(\mathbb{R}^s)$.

Theorem 16.3 (Vostrecov and Kreines [25]) *$\mathscr{R}_{\mathscr{W}}$ is dense on $C(\mathbb{R}^s)$ under the ∞-norm over compact subsets of $\mathbb{R}^s$ if and only if the only homogeneous polynomial of s variables that vanishes on $\mathscr{W}$ is the zero polynomial.*

A homogeneous polynomial is a polynomial whose terms all have the same degree. For example, $x_1^5 + x_1^2 x_2^3$ is a homogeneous polynomial of degree 5, while $x_1^5 + x_1^2$ is not. This theorem states that elements from $\mathscr{R}_{\mathscr{W}}$ can approximate any continuous function over any compact subset of $\mathbb{R}^s$ if and only if there is no nonzero homogeneous polynomial of s variables that has zeros at every point in $\mathscr{W}$. If we are allowed to choose any projection vector from $\mathbb{R}^s$, i.e., $\mathscr{W} = \mathbb{R}^s$, this is certainly true—the only homogeneous polynomial that is zero everywhere on $\mathbb{R}^s$ is the zero polynomial. Sun and Cheney state a similar, possibly simpler to visualize result:

Theorem 16.4 (Sun and Cheney [32]) *Let $s \geq 2$ and let $A_1, A_2, \ldots, A_s$ be subsets of $\mathbb{R}$. Put $\mathscr{W} = A_1 \times A_2 \times \cdots \times A_s$. $\mathscr{R}_{\mathscr{W}}$ is dense on $C(\mathbb{R}^s)$ under the ∞-norm over compact subsets of $\mathbb{R}^s$ if and only if at most one of the sets A_i is finite, and this finite set, if any, contains a nonzero element.*

Once again, if $\mathscr{W} = \mathbb{R}^s$, this condition is obviously met. These two theorems state necessary and sufficient conditions for the same outcome. Hence, the conditions must be equivalent. In fact, it is easy to see that the condition in Theorem 16.4 is sufficient for the condition in Theorem 16.3. If all sets A_i are infinite, then no nonzero homogeneous polynomial of s variables can vanish everywhere on $\mathscr{W} = \prod_i A_i$, since it would need to have an infinite number of roots. Now, consider the case that one set A_i is finite, with at least one nonzero element. Any nonzero homogeneous polynomial h_s of s variables with no x_i term would be a homogeneous polynomial h_{s-1} of $s-1$ variables. This h_{s-1} would then not vanish over $\prod_{j \neq i} A_j$, by the same argument, and, so neither would h_s vanish over $\mathscr{W}$. The only case that remains is when h_s does contain a term with x_i. If h_s now vanished everywhere on $\mathscr{W}$, it would vanish also at all points with x_i equal to a nonzero value from A_i. Replacing this value for x_i in h_s again gives us a homogeneous polynomial h_{s-1} in $s-1$ variables. Hence, by the same argument as before any nonzero h_s cannot vanish over $\mathscr{W}$.

These theorems answer the first question we asked at the beginning of this section: what can we approximate using ridge functions? The answer is essentially, any nonlinear function we are likely to encounter in practice. Now, we look at some results that try to answer the second question: how well can we approximate?

16.4.2 Degree of Approximation: How Good Are Ridge Functions?

One way to address the question, "how well can ridge functions approximate?," is to study the convergence of approximation using ridge functions—how does the error decrease as we increase the number of ridge functions in the model, or in other words, the model complexity? This is a difficult question for general ridge functions and there are some partial results here, notably [34–38]. Many of these results exploit constraints on the ridge functions to show the convergence behavior of the approximation.

From these, we state a general result, by Maiorov. Let $B^s = \{\mathbf{x} \in \mathbb{R}^s : \|\mathbf{x}\|_2 \leq 1\}$ be the closed unit ball. Let $W_2^{k,s}$ be a Sobolev class [39] of functions from $L_2(B^s)$. This is the class of functions $f \in L_2(B^s)$, for which all partial derivatives $\nabla_{\mathbf{x}}^{\mathbf{v}} f$ of order smaller than or equal to k ($\sum_{i=1}^{s} v_i \leq k$, where $\mathbf{v} = \{v_1, \ldots, v_s\}$), satisfy $\|\nabla_{\mathbf{x}}^{\mathbf{v}} f\|_{2,B^s} \leq 1$. These partial derivatives are taken in the weak sense [39]. Define

$$\mathscr{R}_r = \left\{ \sum_{i=1}^{j} g_i \left(\mathbf{w}_i^T \mathbf{x} \right) : j \leq r, \mathbf{w}_i \in \mathbb{R}, g \in L(\mathbb{R}) \right\}, \tag{16.25}$$

where by $L(\mathbb{R})$, we mean the space of all functions integrable on any compact subset of $\mathbb{R}$, or equivalently, there is some compact set $D \in \mathbb{R}$ such that $g \in L_1(D)$. For any two sets of functions, U, V, define the distance of U from V as

$$dist(U, V) = \sup_{u \in U} \inf_{v \in V} \|u - v\|_2. \tag{16.26}$$

In words, for any given $u \in U$ find the distance to the closest approximation v from V using the 2-norm. Then find the maximum of this distance over all possible $u \in U$. Hence, if U is the target set of functions to be approximated and V is the set of possible approximations, this metric computes the maximum error, using the best possible approximations from V. It then follows that $dist(W_2^{k,s}, \mathscr{R}_r)$ is the maximum error while approximating functions in $W_2^{k,s}$ using best fitting approximations from $\mathscr{R}_r$. Now, we are equipped to state the following.

Theorem 16.5 (Maiorov [35]) *For $k > 0$, $s \geq 2$, the following asymptotic relation holds:*

$$dist\left(W_2^{k,s}, \mathscr{R}_r\right) = \Theta\left(r^{-k/(s-1)}\right). \tag{16.27}$$

Here, Θ is the tight bound notation [40]. Hence, the maximum approximation error using r ridge functions decreases as $r^{-1/(s-1)}$, for a class of functions that satisfy a given smoothness criterion ($f \in W_2^{k,s}$).

Algorithm 16.3 The projection pursuit regression algorithm of Friedman and Stuetzle [21]

1: Given normalized training samples $\{\mathbf{x}_j, y_j\}_{j=1}^n$
2: $e_j \leftarrow y_j$, $j = 1, \ldots, n$ and $r = 0$
3: find $\mathbf{w}_{r+1}$ to maximize the fraction of variance explained by g_{r+1}:

$$I \leftarrow \max_{\mathbf{w}_{r+1} \in \mathscr{S}^{s-1}} 1 - \frac{\sum_{j=1}^n (e_j - g_{r+1}(\mathbf{w}_{r+1}^T \mathbf{x}_j))^2}{\sum_{j=1}^n e_j^2}. \tag{16.28}$$

g_{r+1} is the smooth along the direction $\mathbf{w}_{r+1}$. Rosenbrock's method [44] was used for the search.
4: if $I < \epsilon$
5: return $\{g_i, \mathbf{w}_i\}_{i=1}^r$
6: else
7: $e_j \leftarrow e_j - g_{r+1}(\mathbf{w}_{r+1}^T \mathbf{x}_j)$, $j = 1, \ldots, n$
8: $r \leftarrow r + 1$
9: go to step 3.
10: endif

All the results stated in this section provide us with some confidence that a ridge function-based approximation is theoretically feasible. Light [41] surveys some methods of constructing the approximation $\hat{y}$ if the original function y is known. However, all these results deal with functions and not with finite sample sets. In a practical response surface model generation scenario we would not know anything about the behavior of the function we are trying to approximate, but we would have a finite set of points from which we have to estimate the "best" projection vectors and functions in the RSM (16.16). The projection pursuit regression technique strives to accomplish precisely this with a statistical perspective. The next section reviews the original projection pursuit algorithm and some relevant convergence results.

16.5 Projection Pursuit Regression

The PPR algorithm, as proposed by Friedman and Stuetzle [21], takes a nonparametric approach to solve for the functions g_i and projection vectors $\mathbf{w}_i$ in (16.16). Each g_i is approximated using a smoothing over the training data. Let $\{t_j, y_j\}_{j=1}^n$ be our training data projected along some projection vector $\mathbf{w}$. In general, a smoothing-based estimate uses some sort of local averaging:

$$g(t) = \mathrm{AVE}_{t_j \in [t-h, t+h]}(y_j). \tag{16.29}$$

Here AVE can denote the mean, median, any weighted mean, or any other ways of averaging (e.g., nonparametric estimators in [42]). The parameter h defines the *bandwidth* or the *smoothing window*. We call the function g a *smooth*. Specific details of the smoothing method used by Friedman and Stuetzle can be found in [21, 43].

Their overall PPR algorithm is shown in Algorithm 16.3. We remind the reader that all training data has been normalized to mean 0 and variance 1, and denote the surface of the unit sphere in $\mathbb{R}^s$ as $\mathscr{S}^{s-1}$. Hence, $\mathscr{S}^{s-1}$ is the set of all s-vectors of magnitude 1.

We can see that the algorithm is iterative. At each iteration, it tries to extract the best direction $\mathbf{w}_{r+1}$ and the corresponding ridge function g_{r+1} so as to best approximate the residue values $\{e_j\}$ at that iteration. We can clearly see the similarity with latent variable regression. The i-th latent variable in this case is the displacement along the i-th projection vector $t_i = \mathbf{w}_i^T\mathbf{x}$. This iterative approach simplifies the problem of extracting all the required projections and ridge functions, by handling only one component at a time. This has the advantage of scoping down the problem to a one-dimensional curve fitting problem, from a very difficult high-dimensional curve fitting problem. Furthermore, since each component is extracted to maximally model the residue at that iteration, the latent variable associated with the i-th projection vector can be interpreted as the i-th most important variable for explaining the output behavior. This can be very useful for extracting some deep insight into the behavior of a circuit, when PPR is used for RSM building. We will revisit this observation and elaborate further on it when we explain the SiLVR model.

16.5.1 Smoothing and the Bias-Variance Trade-off

There is a subtle, but critical, observation we will make here regarding the ridge function that is extracted in any one iteration. This is best introduced using an illustration: we refer back to our example from (16.20), and reproduce it here in a slightly different form for the reader's convenience:

$$y_2 = x_1x_2 = 0.25(\{1, 1\}\cdot\mathbf{x})^2 - 0.25(\{1, -1\}\cdot\mathbf{x})^2. \tag{16.30}$$

Suppose the ridge function g_1 was unconstrained with regard to any smoothness requirement and was free to take up any shape. Then, given n training points, a perfect, zero-error interpolation could be performed along *any* direction $\mathbf{w}_1$. Figure 16.5 illustrates this. From (16.30) we know that $\mathbf{w}_1 = \{1, 1\}$ or $\{1, -1\}$ are two good candidates for the first projection vector. In fact, any $\{a, b\}$ such that $ab \neq 0$ is a good candidate because we can write

$$x_1x_2 = (4ab)^{-1}\left[(ax_1 + bx_2)^2 - (ax_1 - bx_2)^2\right]. \tag{16.31}$$

Therefore, $\{1, 0\}$ is a bad projection vector. Figure 16.5 shows 100 training points as (blue) dots, projected along the projection vectors $\{1, 0\}$ (Fig. 16.5a) and $\{1, 1\}$ (Fig. 16.5b). With unrestricted g_1 we can find perfect interpolations along both directions, shown as solid lines joining the projected training points. In both cases,

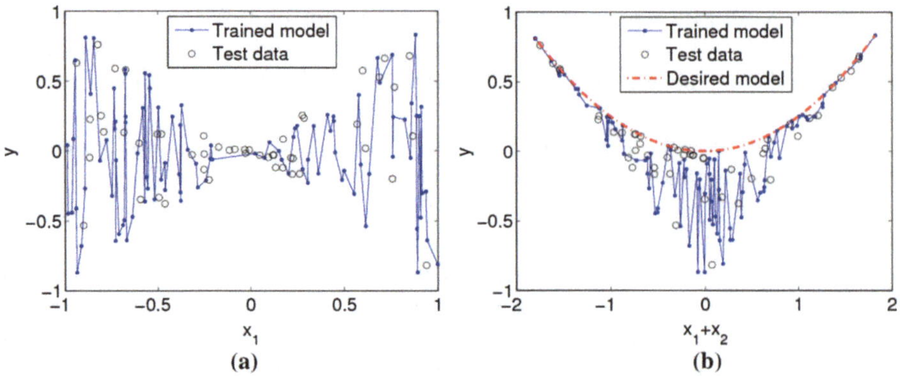

Fig. 16.5 Overfitting of training data along two different projection vectors ($\mathbf{w}_1$) for $y = x_1x_2$. **(a)** $\mathbf{w}_1 = 1, 0$. **(b)** $\mathbf{w}_1 = 1, 1$

the metric I in step 3 of Algorithm 16.3 is maximized to 1 and the algorithm has no way of determining which is the better direction. In fact, with such a flexible class of functions for g_1, all directions will have $I = 1$. Also, once the first ridge function is extracted, the algorithm will stop because all the variance in the training data will have been explained and I would be 0 for the second iteration, resulting in a final model with only one component ridge function model. The solid lines shown in Fig. 16.5a, b are, in fact, the final models. However, choosing the wrong projection vector $\mathbf{w}_1 = \{1, 0\}$ in Fig. 16.5a results in large errors on unseen test data, shown as black circles. This is, of course, as expected because the direction of projection is incorrect in the first place. However, even with the correct projection in Fig. 16.5b, we get large errors on unseen test data.

The problem here is the unrestricted flexibility in the function g_1. A more desired g_1 along $\mathbf{w}_1 = \{1, 1\}$ is actually a very smooth function in this case, shown as a (red) dash-dot line in Fig. 16.5b. This is the first term in expansion in (16.30). Note that this ridge function has large errors on the training data and does not try to exactly fit the training points along the projection. However, it lets the algorithm perform a second iteration, in which the second projection vector $\{1, -1\}$ is chosen and the second ridge function in (16.30) is extracted, giving us a near-exact two-component ridge function model. Such a class of smooth univariate functions will have a larger error along the incorrect direction of Fig. 16.5a and the algorithm will easily reject it. This illustrates the classic bias-variance trade-off in statistical learning [23]. If we minimize the bias in our estimated model by exactly fitting the training data, we will get a completely different approximation for a different set of training points, resulting in high variance. This choice also results in large errors on unseen points. If we minimize the variance, by estimating nearly the same model for different sets of training data, we need to reconcile with a larger training error. In the extreme version of this choice, any training sample will result in the same estimate of the model, meaning that we are not even using any information from the training data. Such extremes will also result in large errors on unseen data. Hence, we must find a balance such that we keep the error low on both the training data and on unseen test data. This is the classic problem of *generalization*.

This issue is particularly critical for the case of PPR. When we project the training data onto a single direction $\mathbf{w}_i$, there can be a lot of noise or variation in the output values because of smooth dependence on other directions orthogonal to $\mathbf{w}_i$, as in Fig. 16.5b for $\mathbf{w}_1$. If the function g_i is allowed too much flexibility, it will undesirably *overfit* the training data by fitting this orthogonal contribution to the behavior of the function. Hence, it is critical that any PPR algorithm employs some technique to avoid overfitting and improve the generalizability of the model. Friedman and Stuetzle used variable bandwidth smoothing to achieve this: the parameter h in (16.29) is adaptively changed to be larger in those parts of the projected input space where the function variation is estimated to be high, since this high variation is probably because of higher dependence on orthogonal directions in that region. Minimizing overfitting will a prime objective when we develop the proposed SiLVR model.

16.5.2 Convergence of Projection Pursuit Regression

PPR was proposed in [21] relying on intuitive arguments regarding why it should work and its advantages, as mentioned in the beginning of this section (Sect. 16.5). Unfortunately, the theoretical results developed for approximation using ridge functions (Sect. 16.4) do not directly apply to PPR because of at least two reasons. First, PPR uses a finite set of training points and does not have knowledge of the original function to be modeled. Second, PPR extracts each projection iteratively. Hence, it cannot rely on exact interpolation techniques, and must use statistical estimation. This was discussed in the context of the bias-variance trade-off and smoothing in Sect. 16.5.1. Also, this iterative scheme is a "greedy" approach, where at every step only the next best decision is taken—to select the next best projection and ridge function. The best decision at any given iteration might not be the best decision in the global sense. It might be better sometimes to not choose the ridge function that seems to be the best for the current iteration. In fact, later in this section, we will show an example where the choice made by PPR does not match the best choice suggested by analysis. Given this greedy nature, does the algorithm still converge to a good solution (to an accurate RSM)? Researchers in statistics have recognized these issues and questions, and there are some theoretical results showing convergence of PPR under different conditions [11, 45–47]. In this section we review some of these results.

Any set of training points will be drawn from some underlying probability distribution defined over the sampling space $\mathscr{X}$. We denote this distribution by P, and the probability density is denoted by p. This scenario is reasonable for our applications, since any statistical parameter (e.g., V_t) or design variable will follow some probability distribution (e.g., normal distribution) or lie uniformly in some bounded range. A bounded domain $D \in \mathbb{R}^s$ can be represented as a uniform distribution P that is nonzero for subsets in D and zero for subsets outside D. Any expectation computation will then be performed over the relevant probability distribution, unless differently specified. For example, the expectation (mean) of a circuit performance $y = f(\mathbf{x})$ will be computed as

$$E(y) = \int_P f(\mathbf{x}) dP = \int_{\mathscr{X}} f(\mathbf{x}) p(\mathbf{x}) d\mathbf{x}. \tag{16.32}$$

In general terms, P is the probability measure over the sample space $\mathscr{X}$ [48]. For our circuit applications $\mathscr{X}$ is typically $\mathbb{R}^s$.

In practice, the PPR algorithm has to deal with at least three nonidealities:

1. No exact knowledge of the original function $y = f(\mathbf{x})$—we have only a finite number of training points n.
2. Imperfect approximation technique for estimating the best univariate function g along any direction $\mathbf{w}$.
3. Imperfect search algorithm to search for the best $\mathbf{w}$ in any iteration.

To the best of our knowledge, there is no theoretical result establishing the convergence properties of PPR in the most general case allowing for all these nonidealities. However, there are results that make ideality assumptions for one or more of the three points mentioned above, but still provide insight into the general working of PPR.

Let us assume that we have a perfect version of PPR, free of the three nonidealities mentioned above. Then we ask the question,

What are the *best* projection vector $\mathbf{w}$
and the *best* univariate function g?

By *best* we mean the pair $(\mathbf{w}, g)$ that gives the best approximation; that is, minimizes the mean squared error. If we are in the i-th iteration, then we can define the residue e_{i-1} as

$$e_{i-1}(\mathbf{x}) = f(\mathbf{x}) - \sum_{j=1}^{i-1} g_j \left(\mathbf{w}^T \mathbf{x}\right). \tag{16.33}$$

Following (16.2), the best $(\mathbf{w}_i, g_i)$ will satisfy

$$(\mathbf{w}_i, g_i) = \arg\min_{\mathbf{w},g} E\left[\left(e_{i-1}(\mathbf{x}) - g(\mathbf{w}^T \mathbf{x})\right)^2\right]. \tag{16.34}$$

Let us first assume some candidate $\mathbf{w}$, and ask,

For any given projection vector $\mathbf{w}$,
what is the best univariate function g?

From (16.34), we know that the best g_i will minimize the error in approximating the residue

$$g_i = \arg\min_g E\left[\left(e_{i-1}(\mathbf{x}) - g(\mathbf{w}^T\mathbf{x})\right)^2\right]. \tag{16.35}$$

For every g, since $g(\mathbf{w}^T\mathbf{x})$ is constant ($= g(t)$) for all $\mathbf{w}^T\mathbf{x} = t$, we can write this criterion as follows. The best g_i will minimize the error in approximating the residue projected *along* $\mathbf{w}$:

$$g_i(t) = \arg\min_{g_t} E\left[\left(e_{i-1}(\mathbf{x}) - g_t\right)^2 |\mathbf{w}^T\mathbf{x} = t\right], \forall t, \tag{16.36}$$

where g_t is some scalar value. For any displacement t along the direction $\mathbf{w}$, we expect to see a distribution of values for the residue e_{i-1}, since multiple $\mathbf{x}$ will map to the same t. The best value of the new ridge function at t, $g_i(t)$, minimizes the mean squared error between the residue and g_i at t. The expectation here is taken over the marginal distribution of $\mathbf{x}$ in the hyperplane $\mathbf{w}^T\mathbf{x} = t$, which is a hyperplane normal to $\mathbf{w}$. This same criterion is applied for all t to obtain the complete function $g_i(t)$ for all values of t. Then, for any t, we can write

$$\begin{aligned} E\left[\left(e_{i-1}(\mathbf{x}) - g_t\right)^2 |\mathbf{w}^T\mathbf{x} = t\right] &= E\left[e_{i-1}^2(\mathbf{x}) - 2e_{i-1}(\mathbf{x})g_t + g_t^2|\mathbf{w}^T\mathbf{x} = t\right] \\ &= E\left[e_{i-1}^2(\mathbf{x})|\mathbf{w}^T\mathbf{x} = t\right] \\ &\quad -2g_t E\left[e_{i-1}(\mathbf{x})|\mathbf{w}^T\mathbf{x} = t\right] + g_t^2 \end{aligned} \tag{16.37}$$

since g_t is a constant for a given t. Then the optimal $g_i(t)$ for a given t is

$$\begin{aligned} g_i(t) &= g_t : \frac{d}{dg} E\left[\left(e_{i-1}(\mathbf{x}) - g_t\right)^2 |\mathbf{w}^T\mathbf{x} = t\right] = 0 \\ &\Rightarrow -2E\left(e_{i-1}(\mathbf{x})|\mathbf{w}^T\mathbf{x} = t\right) + 2g_i(t) = 0 \\ &\Rightarrow g_i(t) = E\left(e_{i-1}(\mathbf{x})|\mathbf{w}^T\mathbf{x} = t\right). \end{aligned} \tag{16.38}$$

Thus, the best value of $g_i(t)$ is the expectation of the residual $e_{i-1}(\mathbf{x})$. We have, thus, proved the following theorem that appears in [11]:

Theorem 16.6 *For any given projection vector* $\mathbf{w}$*, the best function* $g_i(t)$ *defined by (16.35) is given by*

$$g_i(t) = E\left(e_{i-1}(\mathbf{x})|\mathbf{w}^T\mathbf{x} = t\right). \tag{16.39}$$

This is an interesting result. The solution from this result can be quite different from what standard approximation theory would suggest. This is easily illustrated with our friendly example from (16.20) that is reproduced here for convenience:

$$y = x_1 x_2 = 0.25\left(\{1,1\}\cdot\mathbf{x}\right)^2 - 0.25\left(\{1,-1\}\cdot\mathbf{x}\right)^2. \tag{16.40}$$

Say we are considering one of the optimal directions $\mathbf{w}_1 = \{1, 1\}$. To achieve an exact approximation, as per (16.40), the best $g_1(t)$ is

$$g_1(t) = 0.25t^2, \tag{16.41}$$

which is just the first term on the right-hand side of (16.40) mapped on to the latent variable t. This function is shown as the (red) dash-dot line in Fig. 16.6, and also previously in Fig. 16.5b. It is indicated by "Best." However, the best $g_1(t)$ for PPR, as per Theorem 16.6, is the expectation of y taken over the hyperplane $\mathbf{w}_1^T\mathbf{x} = t$. Assuming that $\mathbf{x} \in [-1, 1]^2$, we can analytically compute this best g_1 function. This best g_1 is shown as the (black) solid line in Fig. 16.5b and is indicated by "PPR." We can clearly see that the two ridge functions are different. This difference is a result of PPR performing a greedy search by looking at only one projection at a time unlike the analysis in (16.40) which looks at the function as a whole over all the dimensions.

Given this optimal choice of g_i, we now ask

What then is the best projection vector $\mathbf{w}_i$ that will satisfy (16.34)?

From (16.34) and (16.39), we know that such a $\mathbf{w}_i$ must satisfy

$$\begin{aligned} \mathbf{w}_i &= \arg\min_{\mathbf{w}} E\left[\left(e_{i-1}(\mathbf{x}) - g_i\left(\mathbf{w}_i^T\mathbf{x}\right)\right)^2\right], \\ \text{where} \quad g_i(t) &= E\left(e_{i-1}(\mathbf{x})|\mathbf{w}^T\mathbf{x} = t\right). \end{aligned} \tag{16.42}$$

Expanding the first expectation we get

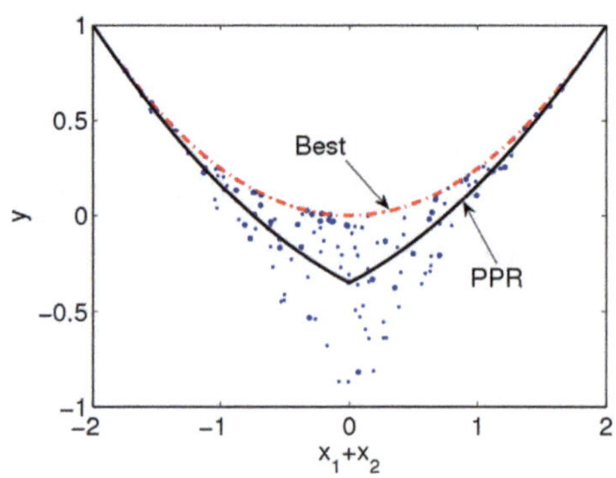

Fig. 16.6 Optimal ridge functions from analysis (red dash-dot) and PPR (black solid) can differ. This example is for $y = x_1 x_2$, along the projection vector $\{1, 1\}$

$$E\left[\left(e_{i-1}(\mathbf{x}) - g_i\left(\mathbf{w}_i^T\mathbf{x}\right)\right)^2\right] = E\left[e_{i-1}^2(\mathbf{x})\right] - 2E\left[e_{i-1}(\mathbf{x})g_i\left(\mathbf{w}_i^T\mathbf{x}\right)\right] \tag{16.43}$$

$$+ E\left[g_i^2\left(\mathbf{w}^T\mathbf{x}\right)\right].$$

Since $g_i(\mathbf{w}_i^T\mathbf{x})$ is a constant for all $\mathbf{w}_i^T\mathbf{x}$ equal to some constant t (it is a ridge function along $\mathbf{w}_i$), we have

$$E\left[g_i^2\left(\mathbf{w}^T\mathbf{x}\right)\right] = E\left[g_i^2(t)\right] \tag{16.44}$$

Let us now expand out the second expectation term on the right-hand side of (16.43) as

$$E\left[e_{i-1}(\mathbf{x})g_i\left(\mathbf{w}_i^T\mathbf{x}\right)\right] = \int_{\mathscr{X}} e_{i-1}(\mathbf{x})g_i\left(\mathbf{w}_i^T\mathbf{x}\right)p(\mathbf{x})d\mathbf{x} \tag{16.45}$$

Let us denote the marginal probability density of any t along $\mathbf{w}_i$ as $p_{\mathbf{w}_i}(t)$. Also let $\mathscr{X}_{\mathbf{w}_i}$ denote the range of $t = \mathbf{w}_i^T\mathbf{x}$ for $\mathbf{x} \in \mathscr{X}$. For our circuit applications, typically $\mathscr{X} = \mathbb{R}^s$ so that $\mathscr{X}_{\mathbf{w}_i} = \mathbb{R}$. Given these definitions, we can rewrite (16.45) as

$$E\left[e_{i-1}(\mathbf{x})g_i\left(\mathbf{w}_i^T\mathbf{x}\right)\right]$$
$$= \int_{t\in\mathscr{X}_{\mathbf{w}_i}}\left[\int_{\mathbf{x}:\mathbf{w}_i^T\mathbf{x}=t} e_{i-1}(\mathbf{x})g_i\left(\mathbf{w}_i^T\mathbf{x}\right)p\left(\mathbf{x}|\mathbf{w}_i^T\mathbf{x}=t\right)d\mathbf{x}\right]p_{\mathbf{w}_i}(t)dt \tag{16.46}$$

Since $g_i(\mathbf{w}_i^T\mathbf{x} = t)$ is a constant for a given t, we can take it out of the inner integral, giving

$$E\left[e_{i-1}(\mathbf{x})g_i\left(\mathbf{w}_i^T\mathbf{x}\right)\right]$$
$$= \int_{t\in\mathscr{X}_{\mathbf{w}_i}} g_i(t)\left[\int_{\mathbf{x}:\mathbf{w}_i^T\mathbf{x}=t} e_{i-1}(\mathbf{x})p\left(\mathbf{x}|\mathbf{w}_i^T\mathbf{x}=t\right)d\mathbf{x}\right]p_{\mathbf{w}_i}(t)dt \tag{16.47}$$

Now, the inner integral is nothing but $E(e_{i-1}(\mathbf{x})|\mathbf{w}^T\mathbf{x} = t)$, which is the same as $g_i(t)$ according to Theorem 16.6. Hence, we get

$$E\left[e_{i-1}(\mathbf{x})g_i\left(\mathbf{w}_i^T\mathbf{x}\right)\right] = \int_{t\in\mathscr{X}_{\mathbf{w}_i}} g_i^2(t)p_{\mathbf{w}_i}(t)dt = E\left[g_i^2(t)\right] \tag{16.48}$$

Substituting this and (16.44) in (16.43), we get

$$E\left[\left(e_{i-1}(\mathbf{x}) - g_i\left(\mathbf{w}_i^T\mathbf{x}\right)\right)^2\right] = E\left[e_{i-1}^2(\mathbf{x})\right] - E\left[g_i^2(t)\right], \text{ where } t = \mathbf{w}_i^T\mathbf{x} \tag{16.49}$$

Hence, we have proved the following, which appears in [11] without a proof:

Theorem 16.7 *The optimal* $\mathbf{w}_i$ *of (16.34) is the one that maximizes the variance of the function* g_i*, where* g_i *is chosen as in Theorem 16.6.*

One would expect any approximation of a function f to maximally explain the variance of f, and this result shows that PPR tries to achieve precisely this.

In Sect. 16.4.1 we saw that a ridge function approximation, like in (16.86), converges to the approximated function, but does the "greedy" and statistical PPR method converge? Jones addresses this question in [46] and proves strong convergence of PPR, as stated by the following theorem. Here, we assume ideality for conditions (1) and (2)—we have infinite number of points to exactly compute expectations, and we can compute the exact best functions g_i along any given direction, respectively. However, we do allow for error in estimating the optimal projection vector.

Theorem 16.8 (Jones [46]) *Let* $f(\mathbf{x}) \in L_2(P)$*, where* P *is the probability measure (distribution) for* $\mathbf{x} \in \mathbb{R}^s$*. Let PPR choose any possibly sub-optimal* $\mathbf{w}_r$ *such that* $E[g_r(\mathbf{w}_r^T\mathbf{x})^2] > \rho \cdot \sup_{\|\mathbf{b}\|_2=1, \mathbf{b}\in\mathbb{R}^s} E(g_r(\mathbf{b}^T\mathbf{x})^2]$ *for some fixed* $\rho, 0 < \rho < 1$*. Then,* $e_r(\mathbf{x}) \to 0$*, as* $r \to \infty$*.*

Hall, in [47], proves a convergence result for a scenario closer to practical PPR, accounting for many nonidealities. The only ideality assumption about the algorithm is that the search for the optimal projection vector is perfect, within the constraint of a finite number of training points n. This means that the sub-optimal projection may seem optimal because of the incomplete information from finite number of points, but the search algorithm will find this seemingly optimal projection. Also, the results in the paper are for the classical PPR technique [21] that employs some sort of smoothing (16.29) using a kernel function with window or bandwidth h, to estimate the function g along some direction $\mathbf{w}$. If K is the kernel function used and the training dataset is $\{\mathbf{x}_j, y_j\}_{j=1}^n$, then the estimate is given as

$$\hat{g}(t) = \frac{\sum_{j=1}^n y_j K\left[\left(t - \mathbf{w}^T\mathbf{x}_j\right)/h\right]}{\sum_{j=1}^N K\left[\left(t - \mathbf{w}^T\mathbf{x}_j\right)/h\right]}. \tag{16.50}$$

Hence, there is one kernel instance centered at the projection of every training point onto the vector $\mathbf{w}$ and the function value at any location along $\mathbf{w}$ is the weighted sum of contributions by each of these n kernels, the weights being the y_j output values associated with each kernel center. The bandwidth h determines the range of influence of each kernel center. Higher values of h lead to smoother estimates resulting in low variance error, but increase the bias error if h is too large. The denominator performs the appropriate normalization. This is the estimate used for prediction. A slightly different form of (16.50) is used for driving the search for the optimal projection vector. Please refer to [47] for details. The kernel is taken to satisfy the condition

$$\int_{-\infty}^{\infty} t^i K(t)dt = \begin{cases} 1 \, , i = 0 \\ 0 \, , 1 \leq i \leq k-1 \end{cases} , \tag{16.51}$$

and the first $k + 1$ directional derivatives of $p(\mathbf{x})$ and $f(\mathbf{x})$ exist and are continuous in $\mathbb{R}$. Under some more loose conditions on $p(\mathbf{x})$ and K, the following holds.

Theorem 16.9 (Hall [47]) *Let* $\mathbf{w}$ *and* g *be the optimal projection vector and ridge function for any PPR iteration, and* $\hat{\mathbf{w}}$ *and* $\hat{g}$ *be the sub-optimal estimates resulting from* n *training points and the imperfect kernel-based approximation of (16.50). Then, the error between* $\hat{g}(\hat{\mathbf{w}}^T\mathbf{x})$ *and* $g(\mathbf{w}^T\mathbf{x})$ *decreases as* $O(n^{-k/(2k+1)})$ *for appropriately chosen* h.

Implications of this result are discussed in [47] and are not immediately relevant here. However, it shows that convergence can be achieved even with significant nonidealities in the PPR algorithm, nonidealities that are unavoidable in any practical implementation. We are now well-equipped to develop the proposed SiLVR model with some confidence.

16.6 SiLVR

In this section we describe the SiLVR model and its features in detail.

16.6.1 *The Model*

SiLVR implements PPR, but uses building blocks and training algorithms that are different from the classical PPR method of [21]. The SiLVR model can be represented mathematically as a standard PPR model:

$$\hat{y} = \sum_{i=1}^{r} g_i \left(\mathbf{w}_i^T \mathbf{x} \right) . \tag{16.52}$$

However, the functions g_i are not purely nonparametric. In this case we use a linear combination of *sigmoidal* functions ($\sigma(\cdot)$) to represent g_i, as

$$g_i(t) = \sum_{j=1}^{q} a_{ij}\sigma(b_{ij}t + c_{ij}) : a_{ij}, b_{ij}, c_{ij} \in \mathbb{R},\ i \in \{1, 2, \ldots\},\ j \in \{1, \ldots, q\}, \tag{16.53}$$

where q is the number of sigmoids used for the approximation of one ridge function. The complete model can then be written as

$$\hat{y} = f_{\mathrm{SiLVR}}(\mathbf{x}) = \sum_{i=1}^{r}\sum_{j=1}^{q} a_{ij}\sigma\left(b_{ij}\mathbf{w}_i^T\mathbf{x} + c_{ij}\right). \tag{16.54}$$

A sigmoidal function or sigmoid is typically defined as a continuous, monotonic function $\sigma(t)$ such that $\lim_{t\to\infty}\sigma(t) = 1$ and $\lim_{t\to-\infty}\sigma(t) = 0$. Any such function taken through scaling and translation is also a sigmoid. Standard examples of sigmoidal functions are the logistic function

$$\sigma_l(t) = \frac{1}{1+e^{-t}}, \tag{16.55}$$

and the hyperbolic tangent function

$$\sigma_h(t) = \tanh(t) = \frac{e^{2t}-1}{e^{2t}+1}. \tag{16.56}$$

Both are shown in Fig. 16.7. In fact, these two sigmoids are equivalent in terms of their nonlinear approximation power using linear combinations because

$$\sigma_h(t) = 2\sigma_l(2t) - 1. \tag{16.57}$$

These functions have a very desirable property that their derivative is easily computed:

$$\frac{d\sigma_h}{dt} = 1 - \sigma_h^2. \tag{16.58}$$

This is useful for efficient training, which typically involves gradient computations within some optimizer. Section 16.6.4.2 shows how we exploit this property for SiLVR. Also, using a parametric model with few basis functions is much more efficient computationally than using data-dependent nonparametric methods like the PPR method of Friedman and Stuetzle. We will use the tanh function and refer to it simply as σ.

The i-th component of the model of (16.54) is shown graphically in Fig. 16.8. Along the lines of PPR or a nonlinear form of LVR, the model consists of two parts:

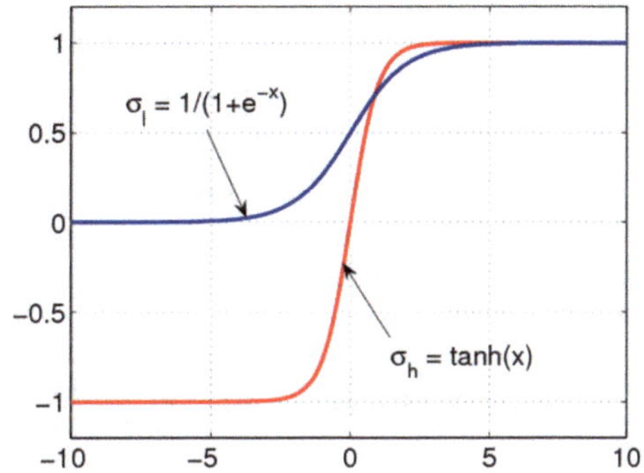

Fig. 16.7 Examples of sigmoidal functions

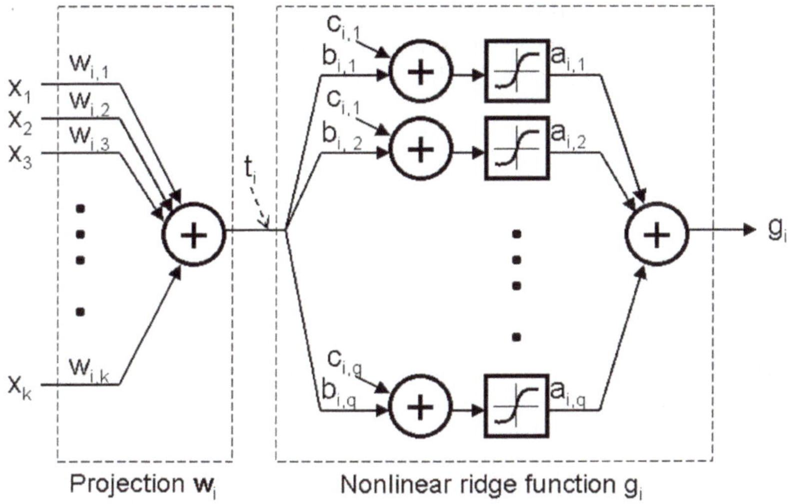

Fig. 16.8 The network corresponding to the i-th component in the SiLVR model (16.54)

1. A *linear* projection $t_i = \mathbf{w}_i^T \mathbf{x} = \sum_{j=1}^{s} w_{ij} x_j$ from the s-dimensional input space to one-dimensional latent variable (LV) t_i lying along the projection vector $\mathbf{w}_i$. The projection vector is chosen to be the one that is most important for explaining the behavior in the modeled output.
2. A *nonlinear* function $g_i(t_i)$ defined over this one-dimensional LV. This nonlinear function is a combination of q sigmoids, as in (16.53). This is essentially a 3-layer perceptron with only one input, one hidden layer with q sigmoidal nodes, and one output g_i.

Together, these two components define one ridge function. This representation of the ridge function allows us to interpret it as a neural network. Hence, we can draw upon the theory and algorithms from the domain of statistical inference using neural networks to compute the best model. Let us define a sampling version of the residue for the i-th LV, similar to (16.33).

$$e_{i,j} = y_j - \sum_{k=1}^{i} g_k \left(\mathbf{w}_k^T \mathbf{x}_j \right) = e_{i-1,j} - g_i \left(\mathbf{w}_i^T \mathbf{x}_j \right). \tag{16.59}$$

This is the value of the residue for the j-th sample point, after extracting the i-th ridge function. Then, we define our model fitting criterion for the i-th ridge function, similar to (16.34), as follows:

$$\left(\mathbf{w}_i, g_i \right) = \arg \min_{\mathbf{w},g} \sum_{j=1}^{n} \left| e_{i-1,j} - g \left(\mathbf{w}^T \mathbf{x}_j \right) \right|^2 . \tag{16.60}$$

Thus, the i-th ridge function is chosen so as to minimize the least squared error in fitting the residue at that iteration across the training set. More specifically to SiLVR, using (16.53), we can write this objective as

$$(\mathbf{w}_i, \mathbf{a}_i, \mathbf{b}_i, \mathbf{c}_i) = \arg \min_{\mathbf{w},\mathbf{a},\mathbf{b},\mathbf{c}} \sum_{j=1}^{n} \left[e_{i-1,j} - \sum_{k=1}^{q} a_k \sigma \left(b_k \mathbf{w}^T \mathbf{x}_j + c_k \right) \right]^2 . \tag{16.61}$$

We know from Theorem 16.6 that the best g_i along any given $\mathbf{w}_i$ is the expectation of the residue at that iteration, along $\mathbf{w}_i$. However, it is not easy to compute this expectation using arbitrary training points. Hence, we do not explicitly state this constraint in the optimization formulation and assume that a good optimizer will converge close to this optimum. We make a similarly reasonable assumption for the best $\mathbf{w}_i$ (given by Theorem 16.7).

Algorithm 16.4 The top-level SiLVR training algorithm

Given some fixed q, the number of sigmoids per LV
normalize the training points $\mathbf{x}_j, y_j{}_{j=1}^{n}$ to mean 0 and variance 1
$e_j \leftarrow y_j, j = 1, \ldots, n$
for $i = 1$ to r
 find the "best" ridge function of the form (16.53) to approximate e_j across all training points. This involves solving the objective function (16.61) appended with some penalty function to reduce overfitting (e.g., see (16.81)).
 $e_j \leftarrow e_j - \sum_{j=1}^{q} a_{ij} \sigma (b_{ij} t + c_{ij}), j = 1, \ldots, n$
the r-LV model is (16.54)

The basic training algorithm for SiLVR is shown in Algorithm 16.4. As expected, there is close similarity in the basic steps of the algorithm with the original PPR algorithm (Algorithm 16.3). The primary differences are in the representation of the ridge function, the formulation of the objective function of the "best" ridge function, and the search algorithm for the "best" ridge function. The straightforward least squares formulation of (16.61) can lead to overfitting issues of the type discussed in Sect. 16.5.1, and we absolutely must avoid such problems to achieve a well-behaved interpretable model from SiLVR. Hence, step 5 in Algorithm 16.4 introduced a modified objective to reduce overfitting. We discuss these issues in more detail in Sect. 16.6.4.

16.6.1.1 Model Complexity

Note that the number of model parameters to solve for is $s + 3q$ per LV, where q is the number of nodes in the sigmoid layer. Hence, if r is the total number of ridge functions that we use in the complete model, the number of model parameters is

$$n_p = r(s + 3q) = O(s), \tag{16.62}$$

which increases linearly with the dimensionality, assuming that the number sigmoids per ridge function is independent of the dimensionality. This is a reasonable assumption, since q is determined by the anticipated nonlinearity of the functions along a single direction. This one-dimensional curve fitting depends on the modeled behavior and not on the dimensionality s. Also, q can be kept very small—12 for our experiments across various circuit examples. The number of LVs, r, is also usually small in most cases—within 2 for our experiments. Hence, SiLVR can lead to very compact and yet, extremely flexible RSMs.

In the next section we review some theoretical results supporting this model formulation. For further algorithmic details on training the model and interpreting it in the context of RSM for circuits, the reader may skip forward to Sect. 16.6.4.

16.6.2 On the Convergence of SiLVR

In Sect. 16.5.2 we saw some results establishing the convergence of the PPR approach under some conditions that retain much practical relevance. These results do apply to SiLVR, but there is one extra consideration—the use of a *finite* number of sigmoids for estimating the ridge function g_i—that is not covered by them. We review some results in this section that help "plug this last hole."

Cybenko showed in [49] that any function continuous over the unit cube in s dimensions can be approximated arbitrarily well by a 3-layer perceptron (3LP). This is stated concretely in the following theorem.

Theorem 16.10 (Cybenko [49]) *Let σ be any continuous sigmoidal function. Then finite sums of the form $\sum_{j=1}^{q} a_j \sigma(\mathbf{b}_j^T \mathbf{x} + c_j)$ are dense in $C[0, 1]^s$. In other words, given any $f \in C[0, 1]^s$ and $\epsilon > 0$, there is a sum of this form, for which*

$$\left| \sum_{j=1}^{q} a_j \sigma\left(\mathbf{b}_j^T \mathbf{x} + c_j\right) - f(\mathbf{x}) \right| < \epsilon, \forall \mathbf{x} \in [0, 1]^s. \tag{16.63}$$

In the case of SiLVR, we only need the one-dimensional case of this theorem. It says that there exists a $g_i(t)$ of the form (16.53) that can approximate any continuous univariate function over any bounded region of the real line. Hornik et al. [50] showed similar density results for unbounded regions, with a finite probability distribution for the input space. This result gives us the confidence to use linear combinations of sigmoids as the ridge function approximators for SiLVR. Also, combined with Theorem 16.9 by Hall [47] they suggest convergence for a practical implementation of SiLVR. In fact, Chui and Li [51] have proved density results that can be directly applied to the complete model of SiLVR as in (16.54). We restate this here, relating it explicitly to SiLVR:

Theorem 16.11 (Chui and Li [51]) *Assume that the set of possible projection vectors* $\mathbf{w} \in \mathscr{W}$ *satisfies the condition of Theorem 16.4. Then, for any function* f *continuous over any compact set* $D \in \mathbb{R}^s$*, and any* $\epsilon > 0$*, there exists a SiLVR model as in (16.54) such that*

$$|f(\mathbf{x}) - f_{SiLVR}(\mathbf{x})| < \epsilon, \forall \mathbf{x} \in D. \tag{16.64}$$

This can be extended to handle the case of probability distributions of $\mathbf{x}$ over all of $\mathbb{R}^s$ using the arguments in Hornik et al. [50].

Barron [38] established bounds on the error of approximation using linear combinations of any fixed sigmoidal function. Here we only state the one-dimensional version. For a large class of functions f over some bounded set $B \in \mathbb{R}$, whose Fourier transform satisfies a finite-moment criterion (refer [38]), the following holds:

$$\int_B \left(f(t) - g(t)\right)^2 dP = O(1/q), \tag{16.65}$$

where $g(t)$ is a q-sigmoid approximation, as in (16.53). The $1/q$ behavior extends to s dimensions; that is, it is independent of the dimensionality. This says that for any given projection vector $\mathbf{w}$, the error in nonlinear function part of one component of the SiLVR model (Fig. 16.8) converges as $1/\sqrt{q}$, as long as the Fourier transform is bounded in the sense of [38]. The finite-moment criterion essentially restricts the spread of the Fourier transform of f. This translates to restricting the "sharpness" and discontinuity in the function f. A counter-example is the Dirac delta function, which has a uniform Fourier transform and is understandably very difficult to model with any accuracy using smooth sigmoids.

This result suggests that the more the number of sigmoids, the better the approximation we could achieve. However, in a sampling context, where we have only partial information because of a finite number of sampling points, this high model flexibility (complexity) can lead to overfitting problems. This overfitting problem is significantly exacerbated in the context of a PPR model like SiLVR, as discussed in Sect. 16.5.1. One way to counter overfitting is to reduce the model complexity by reducing the number of sigmoids q in the univariate approximation, so that the model is incapable of fitting the training sample exactly. Hence, there is this trade-off between high accuracy and less overfitting, i.e., between variance and bias in the model. A more detailed discussion of this issue in the context of PPR can be found in Sect. 16.5.1.

The interested reader can refer to several other results in the literature studying the density, convergence, and construction of neural networks under different conditions [29, 41, 52–55]. For now, we proceed on to discuss how we can interpret the SiLVR model in the context of response surface modeling for circuit.

16.6.3 *Interpreting the SiLVR Model*

The concept of latent variables behind SiLVR allows interpretations of the RSM that lead to useful insights in the context of circuits. We will look at two quantitative measures of these "designer's insight" that we can immediately extract from a 1-LV SiLVR model (i.e., $r = 1$).

16.6.3.1 Relative Global Sensitivity

Consider the function $y = f(x_1, x_2)$ of two variables shown in Fig. 16.9. The primary variation of f is along the shown direction $\{1, 2\}$. Assuming that SiLVR can extract this feature well, the first projection vector will be given by $\mathbf{w}_1 = \{1, 2\}$. The corresponding SiLVR model will be

$$\hat{y} = \sum_{j=1}^{q} a_{1j}\sigma\left(b_{1j}(1 \cdot x_1 + 2 \cdot x_2) + c_{1j}\right). \tag{16.66}$$

Hence, we can interpret from this that changes in x_2 have twice the impact on $\hat{y}$ as similar changes in x_1. In other words, $\hat{y}$ is twice as "sensitive" in a *global* sense to x_2 than to x_1. Then, if we normalize the projection vector to be of unit length

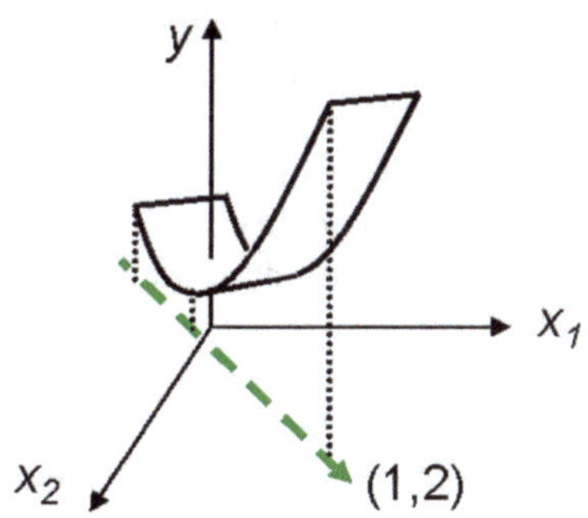

Fig. 16.9 A function of two variables with dominant latent variable along $\{1, 2\}$

$(\mathbf{w}_1/\|\mathbf{w}_1\|_2)$, we can interpret the normalized projection *weights* $(w_{1j}/\|\mathbf{w}_1\|_2)$ as estimates of *relative* global sensitivities of the output y to the inputs x_j. We can then define the relative global sensitivity to the j-th input variable as

$$S_j = w_{1j}/\|\mathbf{w}_1\|_2. \tag{16.67}$$

This measure of global sensitivity captures the designer's insight regarding which are the "important" variables or components in the circuit that have the most impact on the relevant circuit performance metric.

Note that these measures of global sensitivity are different from the standard measure of sensitivity $\partial f(\mathbf{x})/\partial x_j$ that models the *linear* relationship between y and the inputs x_j in a small neighborhood around a given point $\mathbf{x}$. S_j, however, takes a global view, not specific to any neighborhood around any point, but over the entire sampled input domain: it captures the overall contribution of the variable x_j to the variation in the output y. For a similar, more general interpretation of global sensitivity based on the analysis of variance (ANOVA), please refer to [56]. Of course, S_j can be believed to be a good estimate of the global sensitivity only if a 1-LV SiLVR model explains the behavior of the circuit performance sufficiently well. In general, the accuracy of these sensitivity estimates decreases with increasing error in the 1-LV model. However, as we shall see in the results Sect. 16.7, a 1-LV model can extract much of the circuit behavior for some commonly used circuits. A more general definition of sensitivity using a multi-LV SiLVR model may be possible using analysis of variance, on the lines of [56]. This, however, is not addressed here, and can be a potential component of future work.

16.6.3.2 Input-Referred Correlation

Suppose we have 1-LV SiLVR models for two different circuit performance metrics. We can then use the global sensitivity estimates of the previous section to define a measure of correlation between the two outputs that is robust to the presence of strong nonlinearities in the relationship between them. Let us first qualitatively define the idea of "nonlinear correlation." Two variables y_1 and y_2 are strongly correlated in a nonlinear sense if they have similar *causal dependencies*. This means that the perturbations that cause changes in y_1 also cause changes in y_2.

In a circuit design context, let us consider a standard two-stage opamp. If changing the widths of the input pair of transistors causes significant changes in both the DC gain and the DC offset of the opamp, we say that the DC gain and DC offset share this causal dependence on the width of the input devices. Extending this idea, if any design change made to impact the DC gain also impacts the DC offset and vice versa, we say that the two metrics have similar causal dependencies. Note that here we are not placing any conditions on the actual relationship between the two variables. For example, Pearson's linear correlation extracts the strength of the *linear* relationship and Spearman's rank correlation (Sect. 16.6.4.1) extracts the strength of any *monotonic* relationship. Here we relax such constraints and allow any, possibly *nonlinear*, relationship. A side-effect of not assuming monotonicity is that the sign of the relationship loses meaning. For correlation measures relying on monotonicity, positive correlation means that y_1 and y_2 increase together and negative correlation means that one decreases when the other increases. However if we do not have monotonicity, then both behaviors might be seen for the same pair of variables (e.g., y_1 is linear while y_2 is quadratic, but both have similar causal dependencies).

1-LV SiLVR models for y_1 and y_2 allow us to extract this measure of "nonlinear correlation" using *input-referred correlation* or IRC, defined as follows:

$$R(y_1, y_2) = \mathbf{S}^{(1)} \cdot \mathbf{S}^{(2)} = \frac{\mathbf{w}_1^{(1)} \cdot \mathbf{w}_1^{(2)}}{\|\mathbf{w}_1^{(1)}\| \|\mathbf{w}_1^{(2)}\|}, \tag{16.68}$$

where $\mathbf{S}^{(i)}$ is the vector of s relative global sensitivities (16.67) for y_i, and $\mathbf{w}_1^{(i)}$ is the corresponding first projection vector. Thus, IRC between y_1 and y_2 is the dot product of the relative global sensitivity vectors, or the normalized first projection vectors, of y_1 and y_2. Qualitatively, the IRC value is high if y_1 and y_2 are similarly sensitive to the same set of input variables. IRC can be useful for circuit design since it quantitatively captures the designer's insight regarding the dependencies between different performances in the circuit. Such insight can help guide designers to make well-informed design decisions that do not ignore significant trade-offs.

16.6.4 Training SiLVR

The last, but arguably the most important, piece of the SiLVR RSM methodology that we have not discussed yet is the training algorithm: how do we compute all the model parameters efficiently to achieve a near-optimal model, given a finite set of training points? We hinted at the relevant issues in Sect. 16.6.1. We now discuss these in detail. The search algorithm used to find the "best" ridge function in the SiLVR training algorithm (Algorithm 16.4) has to satisfy the following three important requirements.

1. **Good generalizability**: This means that the search should strive to minimize overfitting the training points and the influence from directions orthogonal to the candidate projection vector, as discussed in detail in Sect. 16.5.1. This is accomplished by using a variable bandwidth smoothing kernel in the original PPR algorithm [21], as in general nonparametric methods [42]. However, these techniques are not directly applicable to parametric methods like neural networks.
2. **Robust convergence**: The search should consistently settle on the same, or almost the same, model every time it is run. The desired property here is that we should be able to run the training algorithm just once and rely on the result,

knowing that it is very unlikely that the search will settle in some deeply inferior local minimum.

3. **Fast convergence**: While ensuring the previous two requirements, the search should not sacrifice too much in terms of speed and the training time should be reasonable (e.g., up to several seconds).

The following four techniques are used during step 5 of Algorithm 16.4 to ensure one or more of these requirements:

1. Initialization of projection vectors using **Spearman's rank correlation** [57]. This helps start the search closer to the optimal projection vector than just a random initialization, helping achieve robust and fast convergence (requirements 2 and 3).
2. The **Levenberg–Marquardt** algorithm [58] is used as the search algorithm. This algorithm blends the fast Gauss–Newton method with the robust steepest descent method to achieve fast convergence (requirement 3).
3. **Bayesian regularization** [59] is used to reduce model complexity by restricting the values of the model parameters. This helps reduce overfitting and meet requirement 1.
4. A modified fivefold **cross-validation** method is used to achieve a robust model that does not overfit the training data and, thus, helps satisfy requirements 1 and 2.

We will now discuss each of these techniques in some detail.

16.6.4.1 Initialization Using Spearman's Rank Correlation

We saw in Sect. 16.6.3.1 that the normalized projection weights of the first LV can be interpreted as relative global sensitivities of the output to the inputs. We can extend this same interpretation to the i-th LV. The normalized i-th projection weights $\{w_{ij}\}_{j=1}^{s}$ can be interpreted as relative global sensitivities of the residue being modeled by the i-th ridge function. Hence, if we can initialize the projection weights with some simple estimates of the relative global sensitivities, we can start the search closer to the global optimum, at least in the subspace of projection weights (the entire search space has all the model parameters $(\mathbf{a}_i, \mathbf{b}_i, \mathbf{c}_i, \mathbf{w}_i)$ as dimensions). The simple estimates we use here are the Spearman's rank correlation coefficients between the output and the different inputs. Spearman's rank correlation [57] between two variables x, y, given the sample set $\{x_j, y_j\}_{j=1}^{n}$, is given by

$$\rho_S(x, y) = \frac{\sum_{j=1}^{n}(P_j - \bar{P})(Q_j - \bar{Q})}{\sqrt{\sum_{j=1}^{n}(P_j - \bar{P})^2}\sqrt{\sum_{j=1}^{n}(Q_j - \bar{Q})^2}}, \tag{16.69}$$

where P_j and Q_j are the *ranks* of x_j and y_j in the sample set, as shown by the example in Table 16.1. To compute the rank of, say x_j, we sort all the x values in

Table 16.1 Example illustrating the concept of ranks for Spearman's rank correlation

x	P	y	Q
0.1	2	101	2
−0.1	1	89	1
0.89	4	130	3
0.76	3	132	4

The rank of a value is its position in a sorted list of its class, for example, 0.76 is third in the list of x values sorted in the increasing order

the increasing order and take the position of x_j in this sorted list as its rank. $\bar{P}$ and $\bar{Q}$ denote the means of the ranks. Hence, ρ_S is just Pearson's linear correlation on the ranks. However, this measure of correlation does not assume linearity like the latter, and, hence, gives better estimates of the sensitivities. It does assume a monotonic relationship between x and y. $\mathbf{w}_i$ is then initialized as the normalized vector of rank correlations between the inputs and the current residue.

$$\mathbf{w}_i = \frac{\{\rho_S\}}{\|\{\rho_S\}\|}, \; \{\rho_S\} = \{\rho_S(x_1, e), \ldots, \rho_S(x_s, e)\}. \tag{16.70}$$

If the actual relationship is non-monotonic, in the worst case the rank correlation will not capture it and the initialization will be similar to starting at the origin.

16.6.4.2 The Levenberg–Marquardt Algorithm

We use the Levenberg–Marquardt [58] algorithm to search for the best ridge function in step 5 of Algorithm 16.4. We refer to it as simply LM. LM has been found to be especially well suited for training neural networks with a least squared error formulation, as in [60]. It employs a blend of the fast, but sensitive Gauss–Newton method and the robust, but slower steepest descent. Steepest descent takes steps along the direction of maximum slope of the objective function. Gauss–Newton uses a quadratic approximation of the local region around the current point in the search space to estimate the minimum point. The same procedure is repeated from this new point in the next iteration. Gauss–Newton is a simplified version of Newton's method, and, like Newton's method, shows very desirable quadratic convergence close to the minimum point. However, for nonconvex surfaces, Gauss–Newton can get lost far from the global minimum. In such a situation steepest descent is a better choice.

We now delve briefly into the mathematical details of LM. Let us denote any point in our n_p-dimensional model parameter search space as $\mathbf{p}$. Also denote the objective function to be minimized by f for this discussion on LM. The steepest descent method is an iterative procedure that traces a sequence of points $\mathbf{p}_i$, ideally towards the desired minimum point $\mathbf{p}_{\min}$. The move from a current point $\mathbf{p}_{i-1}$ to the next point $\mathbf{p}_i$ is called a *step*. Steepest descent takes steps of the form

$$\mathbf{p}_i = \mathbf{p}_{i-1} - \delta \nabla f(\mathbf{p}_{i-1}), \tag{16.71}$$

where $\nabla f(\mathbf{p})$ is the gradient vector of f at the point $\mathbf{p}$, composed of the partial derivatives of f with respect to the model parameters,

$$\nabla f = \left\{ \frac{\partial f}{\partial p_1}, \ldots, \frac{\partial f}{\partial p_{n_p}} \right\}^T, \tag{16.72}$$

and δ is a step-size parameter. Hence, with each step, the search moves in the direction of decreasing f. As indicated by the step equation, the step length becomes smaller with smaller gradients, as happens close to a minimum point ($\nabla f(\mathbf{p}) = \mathbf{0}$). As a result, steepest descent performs well far from the optimum, but is a bad choice when the search is close to the optimum. The asymptotic convergence of steepest descent is linear.

Standard Newton's method speeds up the convergence by also using second-order information. The Newton step is given by

$$\mathbf{p}_i = \mathbf{p}_{i-1} - \left[\nabla^2 f(\mathbf{p}_{i-1}) \right]^{-1} \nabla f(\mathbf{p}_{i-1}), \tag{16.73}$$

where $\nabla^2 f(\mathbf{p}_{i-1})$ is the Hessian matrix composed of second-order partial derivatives of f with respect to the model parameters p_j. $\mathbf{p}_i$ here is basically the minimum point of a quadratic model of f around the previous iteration point $\mathbf{p}_{i-1}$.

Now, suppose that the function f is a least squares objective function,

$$f(\mathbf{p}) = \sum_{j=1}^{n} \epsilon_j^2(\mathbf{p}), \tag{16.74}$$

where $\epsilon_j(\mathbf{p})$ is the error for the j-th sample. An example is the SiLVR objective function in (16.61). By simple differentiation we get

$$\nabla f(\mathbf{p}) = J^T(\mathbf{p})\mathbf{e}(\mathbf{p}), \tag{16.75}$$

$$\nabla^2 f(\mathbf{p}) = J^T(\mathbf{p})J(\mathbf{p}) + \sum_{j=1}^{n} \epsilon_j(\mathbf{p}) \nabla^2 \epsilon_j(\mathbf{p}), \tag{16.76}$$

where J is the Jacobian matrix function

$$J = \begin{bmatrix} \frac{\partial \epsilon_1}{\partial p_1} & \frac{\partial \epsilon_1}{\partial p_2} & \cdots & \frac{\partial \epsilon_1}{\partial p_{n_p}} \\ \frac{\partial \epsilon_2}{\partial p_1} & \frac{\partial \epsilon_2}{\partial p_2} & \cdots & \frac{\partial \epsilon_2}{\partial p_{n_p}} \\ \vdots & \vdots & \ddots & \vdots \\ \frac{\partial \epsilon_n}{\partial p_1} & \frac{\partial \epsilon_n}{\partial p_2} & \cdots & \frac{\partial \epsilon_n}{\partial p_{n_p}} \end{bmatrix}. \tag{16.77}$$

The last term in (16.76) is very small near the solution, since $\epsilon_j(\mathbf{p})$ is very small there by definition, and it can be assumed to be ≈ 0. Then, we can write the Newton step of (16.73) as

$$\mathbf{p}_i = \mathbf{p}_{i-1} - \left[J^T(\mathbf{p})J(\mathbf{p})\right]^{-1} J^T(\mathbf{p})\mathbf{e}(\mathbf{p}). \tag{16.78}$$

This is the *Gauss–Newton* method. Note that this modification of Newton's method requires no explicit computation of second-order derivatives. Newton's method has quadratic convergence near the solution because it uses the second-order information in the Hessian, and the Gauss–Newton method also shows this convergence behavior. This is significantly faster than the steepest descent method. Hence, the Gauss–Newton method is preferable once the search is close to the solution. However, far from the solution, the Hessian can be ill-conditioned, or the quadratic model might be a bad approximation of the surface, causing the search to get "lost" and move away from the actual minimum. Here, the steepest descent method, which is insensitive to the second-order behavior of the surface, is more robust and is preferable.

Recognizing this, the LM step is a intuitive blend of the two

$$\mathbf{p}_i = \mathbf{p}_{i-1} - \left[J^T(\mathbf{p})J(\mathbf{p}) + \mu\mathbf{I}\right]^{-1} J^T(\mathbf{p})\mathbf{e}(\mathbf{p}), \tag{16.79}$$

where μ is an adaptive parameter: larger μ causes steepest descent steps, while smaller μ causes Gauss–Newton steps. Note that larger μ effectively improves the conditioning of the Hessian approximation $[J^T(\mathbf{p})J(\mathbf{p}) + \mu\mathbf{I}]$ by imposing diagonal dominance [22]. μ is multiplied by some factor $\beta \gg 1$ (increased) when a step results in an increase in f, and divided by β (decreased) when the step reduces f. In the former case, the steepest descent part of (16.79) is increased and in the latter, the Gauss–Newton part is increased. For further details, please refer to [60]. We can see the flavor of typical model-trust region methods [61] where good solutions from the quadratic model lead to increasing the belief in the quadratic model and bad solutions lead to decreasing the belief. In fact, the LM method can be developed as a model-trust region method [61]. Significant improvement over steepest descent or Gauss–Newton has been seen while using LM for neural network training, as shown in [60].

Apart from these desirable features of LM, we also note that computing the partial derivatives for J in (16.77) is very simple in the specific case of SiLVR, because of the easy derivate calculation for the tanh sigmoid, shown in (16.58). Using (16.61),

$$\frac{\partial \epsilon_j}{\partial p_i} = \frac{\partial}{\partial p_i}\left[e_j - \sum_{k=1}^{q} a_k\sigma\left(b_k\mathbf{w}^T\mathbf{x}_j + c_k\right)\right] = \sum_{k=1}^{q}\frac{\partial}{\partial p_i}\left[a_k\sigma\left(b_k\mathbf{w}^T\mathbf{x}_j + c_k\right)\right], \tag{16.80}$$

where p_i is one of a_k, b_k, c_k for some $k \in \{1, \ldots, q\}$. Note that we have dropped the subscript for the LV here. Hence, the only derivative we need to compute is for σ, which is easily done using (16.58).

16.6.4.3 Bayesian Regularization

Optimizing the objective function in (16.61) will drive the search towards a ridge function that exactly fits the sample points along the projection vector. As discussed in Sect. 16.5.1, this is not desirable for achieving a generalizable PPR model with low overfitting. *Regularization* is a standard technique used to constrain the model complexity and reduce this overfitting behavior and involves adding a penalty term to the standard least squared error objective. Roughly speaking, the penalty term models the model complexity, using the fitting parameters themselves. If we denote the pure data-driven standard objective of (16.61) by E_D, regularization augments it as follows:

$$\min_{\mathbf{p}} E_R, \quad E_R = \beta E_D + \alpha E_p, \tag{16.81}$$

where E_p is the sum of squares of the network parameters,

$$E_p = \mathbf{p}^T\mathbf{p} = \|\mathbf{w}_i\|_2^2 + \|\mathbf{a}_i\|_2^2 + \|\mathbf{b}_i\|_2^2 + \|\mathbf{c}_i\|_2^2, \tag{16.82}$$

and α and β determine the trade-off between accuracy and generalizability, or variance and bias, respectively. Restricting the values of the network parameters reduces the flexibility in the model and increases the smoothness of the response. This is analogous to increasing the bandwidth in kernel smoothing methods [42], as in the original PPR algorithm of [21]. Such a penalty is also known as a *roughness penalty*, and has been studied by several authors, for example, [23, 55, 59, 62].

A typical problem is estimating the proper values for α and β. A Bayesian formulation of this problem allows elegant, adaptive computation of these weights, as shown by MacKay in [59]. The argument for this formulation is as follows. We recognize that the optimal values for α and β are determined by the specific neural network (or any other model) structure M, and the available training data D. Given D and M, the posterior probability of some α, β is given by Bayes' rule as

$$P(\alpha, \beta|M, D) = \frac{P(D|\alpha, \beta, M)P(\alpha, \beta|M)}{P(D|M)}. \tag{16.83}$$

Under a Bayesian framework we want to use those values for α, β that maximize this probability. If we assume a prior density $P(\alpha, \beta|M)$, this can be achieved by maximizing $P(D|\alpha, \beta, M)$. $P(D|\alpha, \beta, M)$ is the likelihood of seeing the training data D, given M and some α, β. Let $\mathbf{p}^*$ denote the best choice of parameters that minimizes (16.81). Under assumptions of Gaussian prior distributions for noise in the training set, and for the network parameters, it can be shown [63] that the optimum values of α, β at $\mathbf{p}^*$ are

$$\alpha_0 = \frac{\gamma}{2E_p(\mathbf{p}^*)}, \quad \beta_0 = \frac{n-\gamma}{2E_D(\mathbf{p}^*)}, \quad \gamma = n_p - \frac{2\alpha_0}{\text{trace}(\nabla^2 E_R^*)}, \tag{16.84}$$

where $\nabla^2 E_R^*$ is the Hessian of the regularization objective function (16.81) at $\mathbf{p}^*$. γ is called the effective number of parameters and is a measure of the number of model parameters actually used for reducing the error. Foresee and Hagan [63] showed how this elegant formulation fits with the same elegance in the LM framework. The resulting algorithm for LM is shown in Algorithm 16.5, where the Hessian approximation from LM is used for $\nabla^2 E_R$.

Algorithm 16.5 Bayesian regularization in the Levenberg–Marquardt framework

1: initialize the network parameters normally, and set $\alpha = 0$ and $\beta = 1$
2: take one LM step (16.79) to minimize E_R
3: use the Gauss–Newton approximation for the Hessian, from LM

$$\nabla^2 E_R \approx 2\beta J^T J + 2\alpha \mathbf{I}_{n_p}, \tag{16.85}$$

where J is as in (16.77) and $\mathbf{I}_{n_p}$ is the $n_p \times n_p$ identity matrix, to compute $\gamma = n_p - 2\alpha/tr(\nabla^2 E_R)$
4: compute new estimates of α, β using the current point $\mathbf{p}_i$, and E_R in (16.84).
5: ifd converged as per LM criterion
6: **return**
7: else
8: go to step 2

16.6.4.4 Modified Fivefold Cross-Validation

Even with this regularization technique, we cannot be completely confident of the accuracy of the resulting model on unseen test data, since we are only optimizing for the training data. Also, the surface of the objective function in (16.81) can be nonconvex because it is defined by a sum of sigmoids. Hence, there are chances that the search may settle on a local optimum that is much worse than the desired global optimum (or, at least, a very good local optimum). A popular technique used

Algorithm 16.6 Modified fivefold cross-validation used to reduce overfitting and avoid local optima

```
Given a training set D of n points
divide D into 5 random, nonoverlapping sets {D_1, ..., D_5} which n/5 points each—D = ∪D_i
E* = ∞
for i = 1 to 5
    M ← 1-LV SiLVR model trained on D \ D_i
    E ← sum of squared error of M_i on D_i
    E < E*
        E* = E, M* = M
return M*
```

for selecting a model that is generalizable is *k-fold cross-validation*. For example, cross-validation may be used for selecting the number of sigmoids to be used in a neural network to achieve the best generalizability. For details on how it is used for model selection, the reader may refer to [23]. In our case, however, the model structure is fixed. We can still exploit cross-validation to address the two issues mentioned above by choosing good model parameter values.

The modified fivefold cross-validation that we use is shown in Algorithm 16.6. The algorithm trains five different SiLVR models, each time excluding one of the subsets D_i, then computes the testing error for each model on its unseen D_i, and finally, picks the model that has the lowest testing error. Hence, it ensures that testing error is used as the criterion for parameter selection, rather than only training error (with regularization). Furthermore, it runs five different training runs, significantly increasing the chances of finding a model close to the global optimum. Note that the cross-validation Algorithm 16.6 is run once for each LV—it is part of step 5 in Algorithm 16.4.

16.7 Experimental Results

SiLVR was implemented in Matlab. We now present some experimental results to demonstrate the performance of this implementation. We first test it on our example of (16.20), reproduced (once more) here in its general form, for the convenience of the reader.

$$\begin{aligned} y_2 = x_1 x_2 &= 0.25(x_1 + x_2)^2 - 0.25(x_1 - x_2)^2 \\ &= (4ab)^{-1}\left[(ax_1 + bx_2)^2 - (ax_1 - bx_2)^2\right]. \end{aligned} \tag{16.86}$$

We sampled 1000 values of x_1 and x_2 from a standard normal distribution $\mathcal{N}(0, 1)$, and trained SiLVR on the resulting set, using all the techniques of Sect. 16.6.4. The results are shown in Fig. 16.10. Figure 16.10 shows the training points and the surface extracted by a 2-LV SiLVR model. Comparing with Fig. 16.4 we see

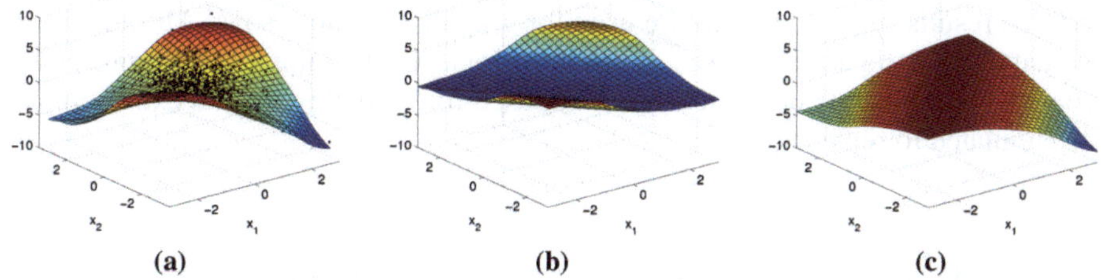

Fig. 16.10 A 2-LV SiLVR model for $y = x_1x_2$ for $(x_1, x_2) \sim \mathcal{N}(\mathbf{0}, \mathbf{I}_2)$. (**a**) The surface extracted by a 2-LV SiLVR model, and the training samples. (**b**) The first ridge function g_1 extracted by SiLVR, along $\mathbf{w}_1 = (0.7285, 0.6851)$. (**c**) The second ridge function g_2 extracted by SiLVR, along $\mathbf{w}_2 = (0.7296, -0.6838)$

that SiLVR does extract a reasonable approximation of the underlying assumption from the training set it is provided. We do see some artifacts in the under-sampled regions, but that is because of the lack of sufficient data there, and the heavy smoothing imposed on the training algorithm. Typically the under-sampled regions are less important because events occur rarely there, and errors in the model can be tolerated there. The well-sampled regions, however, must be modeled well and SiLVR does meet this criterion. Figure 16.10b, c shows the first and second ridge functions extracted, respectively (along the first and second projection vectors, respectively). Again, we see agreement in shape and alignment with the solutions in Fig. 16.4. From (16.86), we know that two candidate projection vectors are $\mathbf{w}_1 = (a, b)$ and $\mathbf{w}_2 = (a, -b)$ for $ab \neq 0$. The vectors that SiLVR extracts are $\mathbf{w}_1 = (0.7285, 0.6851)$ and $\mathbf{w}_2 = (0.7296, -0.6838)$, which are very close to the expected results. This implies that even if there are errors in the model in under-sampled regions, we should still obtain good estimates of the dominant projection vectors, and hence of the relative global sensitivities and IRC, when applicable. The average absolute error on the training set is 2.23%.

Now we show results for three realistic circuit test cases, each representing a different family of circuit behavior:

1. Master–slave flip-flop with the scan chain component,
2. Two-stage RC-compensated opamp, and
3. Sub-1 V bandgap voltage reference in CMOS.

The number of process parameters ranges from 13 to 122 (including one inter-die parameter). SiLVR is able to extract good estimates of the LVs, along with the accompanying ridge function model, using 1000 training samples for each case. The training points are generated using standard Monte Carlo sampling, following the probability distributions for the statistical parameters. The circuit simulator used is Spectre® [64] by Cadence Design Systems. We also compare SiLVR with a straightforward Matlab® implementation of a near-optimal, reduced quadratic model, built using the PROBE [8] algorithm, discussed in Sect. 16.2.3. The best PROBE results (up to rank 10) are used for graphical comparisons. All models are evaluated on a separate test set of 10,000 Monte Carlo samples. Training points where the circuit does not function are not used for modeling, but no extra samples are simulated to replace them.

16.7.1 Master–Slave Flip-Flop with Scan Chain

The first test case is a commonly seen master–slave flip-flop with scan chain shown in Fig. 16.11, which we refer to as simply MSFF. The circuit has been implemented using the 45 nm CMOS Predictive Technology Models of [65]. The variations considered are random dopant fluctuation (RDF) for all transistors and one global gate oxide thickness (t_{ox}) variation. The RDF is modeled as normally distributed independent threshold voltage (V_t) variation:

$$\delta V_t \sim \mathcal{N}\left(0, \left(\frac{13.5 V_{t0}}{\sqrt{WL}}\right)^2\right), \tag{16.87}$$

where W, L are the transistor width and length in nm, and V_{t0} is the nominal threshold voltage. This results in about 30% standard deviation for a minimum-sized transistor. This is large for current CMOS technologies, but we want to make sure that SiLVR is powerful enough for future technologies too, where large variations will be inevitable. The standard deviation for t_{ox} is taken as only 2% of the nominal value, since t_{ox} is typically better controlled than RDF. The number of statistical parameters, or input dimensionality of the model, is 31. We are modeling one output: the clock-output delay of the flip-flop (τ_{cq}). The setup time of the flip-flop is such that the variations result in the onset of some metastable behavior, resulting in strongly nonlinear behavior in some parts of the sampled region of the statistical parameter space. This realistic situation makes the modeling problem a harder test case.

Figure 16.12a shows the projection vector $\mathbf{w}_1$ for the first extracted latent variable t_1, and Fig. 16.12b plots the simulated and predicted delay values against t_1. The latter shows the predictions from SiLVR and also from the best reduced quadratic model. We can clearly observe two things: (1) only 6–8 out of 31 input dimensions (corresponding to transistors in the circuit) affect the output, and (2) SiLVR performs much better than a quadratic model. Also, note that just using one LV we are able to explain the general behavior of this test case. Figure 16.21 (a few pages later) shows error bars comparing the error of SiLVR against the error of PROBE, and we can immediately see the improvement in modeling accuracy. These are errors for the best SiLVR model and the best PROBE model: both PROBE and

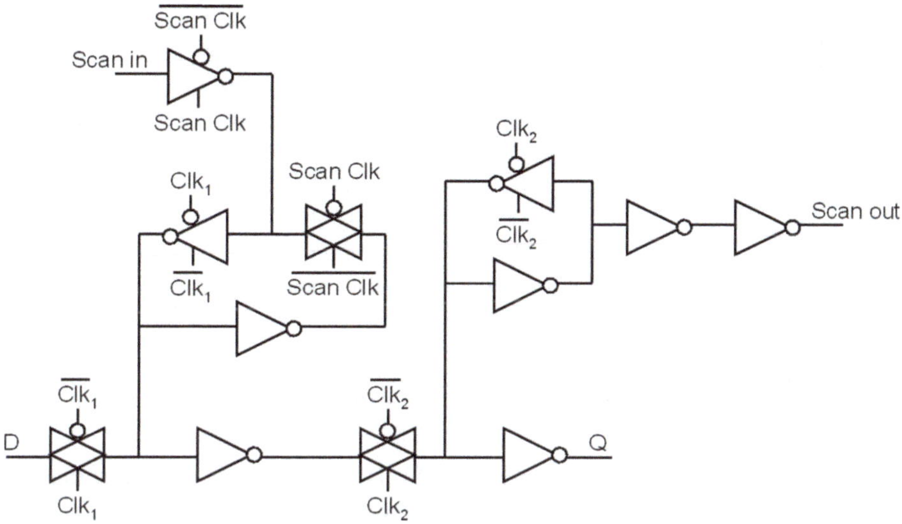

Fig. 16.11 A master–slave flip-flop with the scan chain component

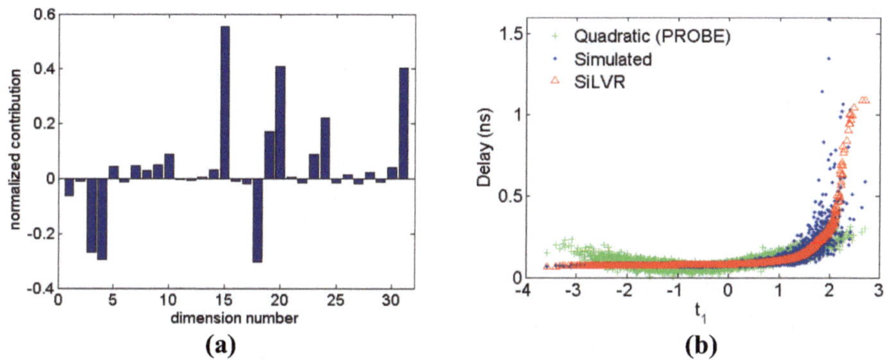

Fig. 16.12 Performance of SiLVR on the MSFF test case. (**a**) The normalized projection weights for the first LV. (**b**) Simulated, SiLVR-predicted (red triangle), and PROBE-predicted (green plus) delay values, plotted against the first LV

SiLVR perform best for rank-one and a one-LV model, respectively. Beyond this, we see only overfitting of the training data and larger errors on the test data. Table 16.2 shows this in more detail, as it compares the errors quantitatively with increasing number of LVs/rank: the best average error is reduced by 2.5×: from 16.3% for PROBE to 6.4%. The results for rank (number of LVs) greater than 6 do not provide any relevant insight and are excluded to avoid clutter.

Table 16.2 Average percentage error on a test set of 10,000 Monte Carlo samples, for MSFF and the voltage reference

	MSFF delay		Bandgap V_{ref}		V_{do}	
r	SiLVR	PROBE	SiLVR	PROBE	SiLVR	PROBE
1	6.41	16.3	0.278	0.639	11.1	35.4
2	7.24	19.7	0.341	0.707	10.4	39.9
3	8.15	21.3	0.374	0.726	11.9	41.9
4	8.17	22.2	0.375	0.736	11.9	42.3
5	8.35	22.9	0.374	0.739	11.9	42.4
6	7.79	23.2	0.374	0.738	11.9	42.4

r is the rank of the quadratic model or the number of LVs used in the SiLVR model, as applicable

16.7.2 Two-Stage RC-Compensated Opamp

This next test case [66], shown in Fig. 16.13, is representative of a large class of circuits in the *analog* domain: amplifiers. We test SiLVR on the DC, AC, and transient characteristics of the opamp. The opamp has been implemented using models from the Cadence 90 nm Generic PDK library. Once again, we model RDF on all transistors as independent variation on the threshold voltage V_t. We also include a global t_{ox} variation, and variations on the passives (resistors, capacitors) and the current source. All variations are assumed to be normally distributed. The V_t standard deviation is about 18% of nominal V_t:

$$\delta V_t \sim \mathcal{N}\left(0, \left(\frac{5mV}{\sqrt{WL}}\right)^2\right), \tag{16.88}$$

where W, L are the transistor width and length in μm. The standard deviation for t_{ox} is taken as 2% of the nominal value. Each passive and current source component has its own normally distributed variation with a standard deviation of 5%. The resulting input dimensionality for the RSM is 13, and we are modeling five performance metrics (outputs):

1. DC gain
2. Unity gain frequency (UGF)
3. Phase margin (PM)
4. Settling time (ST)
5. DC input offset

For details regarding these metrics and opamp operation, please refer to any standard textbook on circuit design, e.g., [66].

Figure 16.14 shows the first projection vector $\mathbf{w}_1$ for each of the outputs. Figure 16.15 plots the simulated and predicted values for each output against the respective first LV t_1, on the test points. The following observations are obtained immediately from these figures:

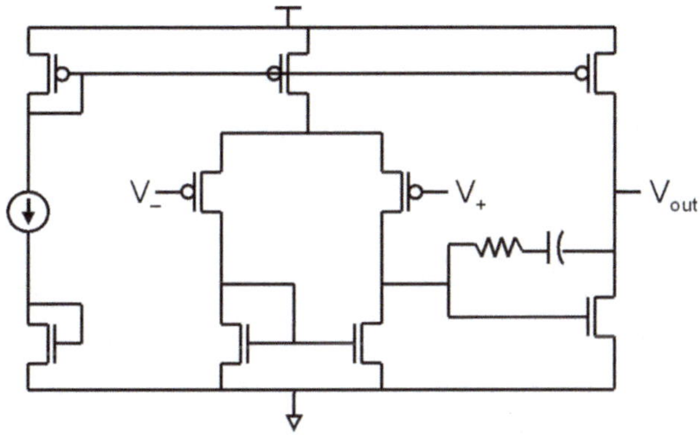

Fig. 16.13 A 2-stage RC-compensated operational amplifier

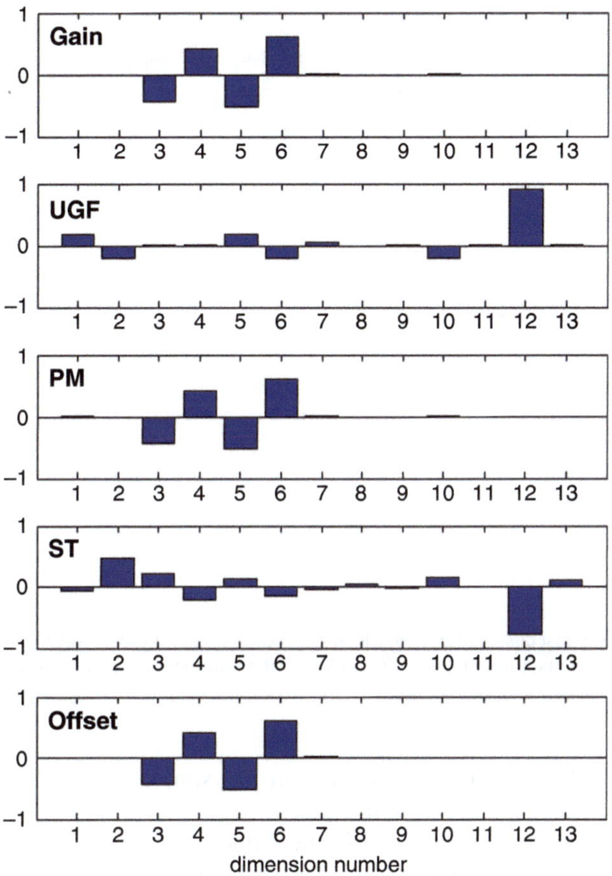

Fig. 16.14 Opamp test case: normalized projection vectors for the first LVs of the opamp metrics: we can see the strong relationship between gain, PM, and offset

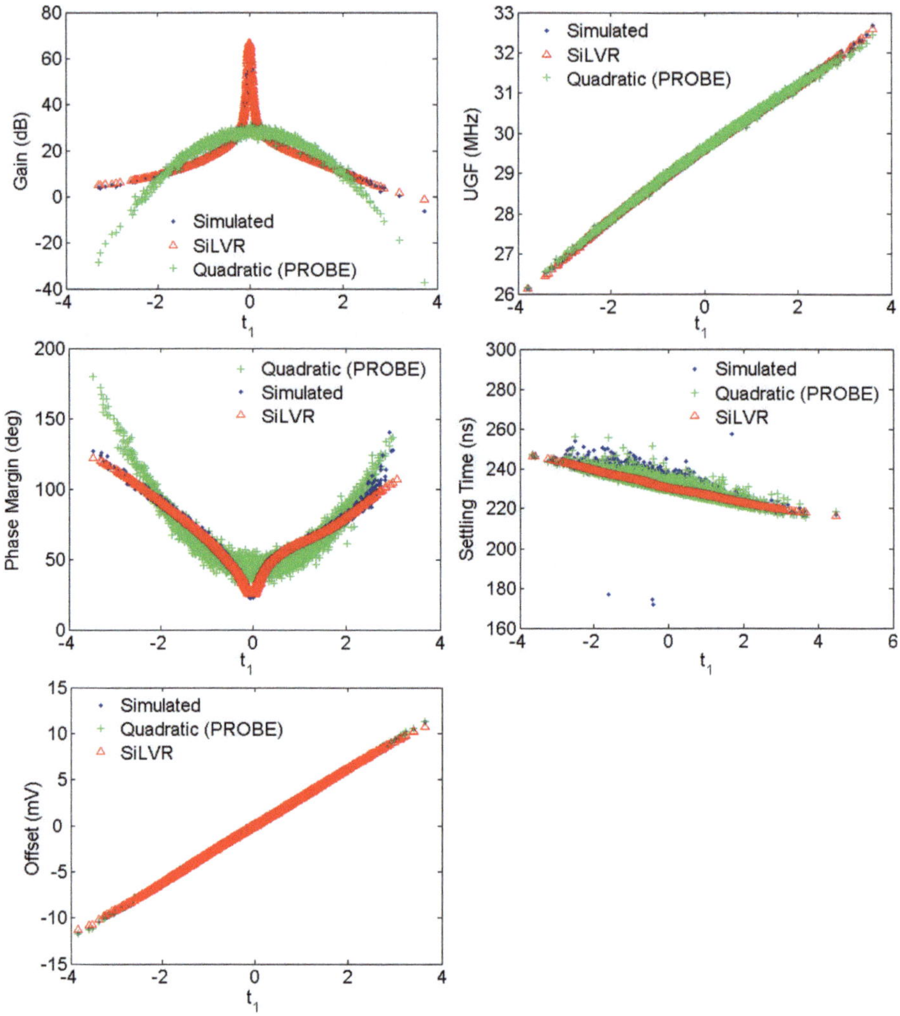

Fig. 16.15 Opamp test case: simulated, PROBE-predicted (green plus), and SiLVR-predicted (red triangle) opamp outputs, plotted against first LV. For the nonlinear cases, the simulated and SiLVR-predicted graphs coincide in many places

- Strongly nonlinear behavior exists even for simple circuits like the 2-stage opamp: for DC gain and PM in this case.
- A quadratic model performs well for near-linear behaviors, as expected, but has large errors for these strongly nonlinear behaviors.
- One (the first) LV is able to explain much of the behavior for some circuits: both the MSFF and the opamp till now.
- SiLVR is able to model even the strongly nonlinear behaviors reasonably well.

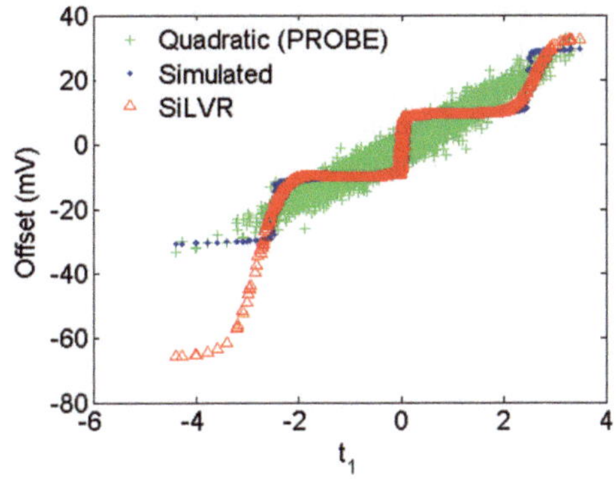

Fig. 16.16 Staircase behavior for incorrectly simulated input offset modeled by SiLVR

Apart from these obvious ones, we make some more subtle, but important observations. Having the explicit projections, as a result of the projection pursuit approach, provides deep insight into the circuit behavior. First, we can actually see the behavior clearly, removing any need for guesswork. We found a direct application of this advantage during the course of performing these experiments. Our initial results for the input offset showed a surprising (and, as it turned out, erroneous) step-shaped behavior when the offset values were plotted against the first LV extracted by SiLVR. This is shown in Fig. 16.16. This result can also be found in our initial publication [12]. However, this step behavior was unexpected for the offset of the opamp: we expected a near-linear behavior as described in [66]. This led to further investigation, resulting in the discovery of the cause of this anomaly—a tolerance parameter in the circuit simulator was too loose. On correcting the parameter, and re-running SiLVR training, we obtained the expected near-linear behavior shown in Fig. 16.15. This example shows just one of many possible scenarios where the better modeling flexibility, visualization potential, and interpretive power that SiLVR provides can be practically useful. The step behavior could be observed easily because we could reduce the dimensionality to the most important one and visualize the behavior easily.

SiLVR also provides us some quantitative measures to better understand circuit behavior: relative global sensitivities and input-referred correlation (Sect. 16.6.3). If we look at the projection vectors for gain, PM, and offset in Fig. 16.14, we can immediately see that these outputs depend almost identically on the same parameter subset (parameters 3–6): these are the driver and load devices in the input differential amplifier. Hence, they have similar causal dependencies, as defined in Sect. 16.6.3.2, and are strongly correlated in a nonlinear sense. This is confirmed by plotting all three simulated metrics against the first LV of *gain*, as in Fig. 16.17. This is where the power of SiLVR is really evident. Table 16.3 compares the rank correlation and linear correlation among these metrics, with the IRC. We see that the rank correlation performs better than linear correlation, but both completely fail to capture the strength of the relationship between gain and offset, and gain and PM. At the same time, IRC succeeds nicely.

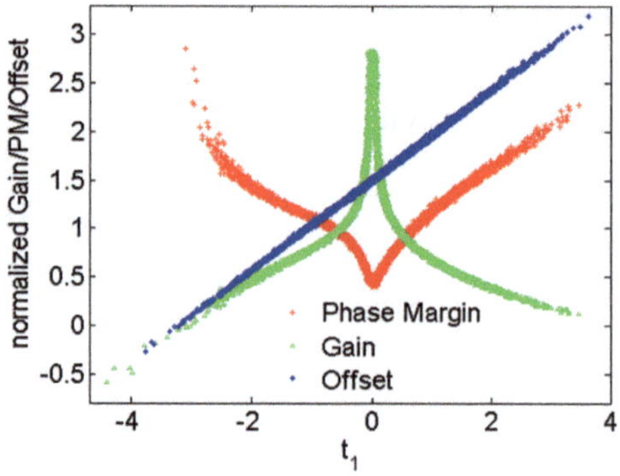

Fig. 16.17 Simulated opamp gain, phase margin, and offset plotted against the LV for gain, showing strong correlation among the three

Table 16.3 Rank and linear correlation compared with IRC as a measure of correlation between strongly correlated opamp metrics

Output pair	|Linear corr.|	|Rank corr.|	IRC
Gain-PM	0.871	0.986	1.000
PM-offset	0.119	0.161	1.000
Gain-offset	0.054	0.099	1.000

Table 16.4 Average percentage error on a test set of 10,000 Monte Carlo samples, for the opamp

	Opamp gain		UGF		PM		Settling time		Input offset	
r	SiLVR	PROBE	SiLVR	PROBE	SiLVR	PROBE	SiLVR	PROBE	SiLVR	PROBE
1	1.69	35.7	0.061	0.042	1.90	16.7	0.507	0.248	1.62	0.58
2	1.74	36.4	0.063	0.022	1.97	16.8	0.517	0.240	1.80	0.12
3	1.73	36.9	0.066	0.013	1.94	16.9	0.528	0.243	2.11	0.10
4	1.66	37.2	0.068	0.010	1.99	16.9	0.531	0.244	2.38	0.09
5	1.69	37.1	0.070	0.009	2.01	16.9	0.540	0.244	2.41	0.09
6	1.72	37.0	0.072	0.009	2.03	16.9	0.552	0.245	2.59	0.09

r is the rank of the quadratic model or the number of LVs used in the SiLVR model, as applicable

Figure 16.21 compares the average absolute percentage error for SiLVR and PROBE on this test case. Quantitative comparisons of the errors are provided in Table 16.4, for increasing number of LVs and rank of the quadratic model. The results for rank (number of LVs) greater than 6 do not provide any relevant insight and are excluded to avoid clutter. As expected, a quadratic model has large errors for the nonlinear, nonquadratic behavior of gain and PM. SiLVR can model these well and reduce the error significantly—by up to an absolute improvement of 34% for gain. The errors for the near-linear outputs worsen a little, but are still within reasonable limits. For this test case too, a 1-LV SiLVR model has the lowest testing error among all SiLVR models, as we can see from the table.

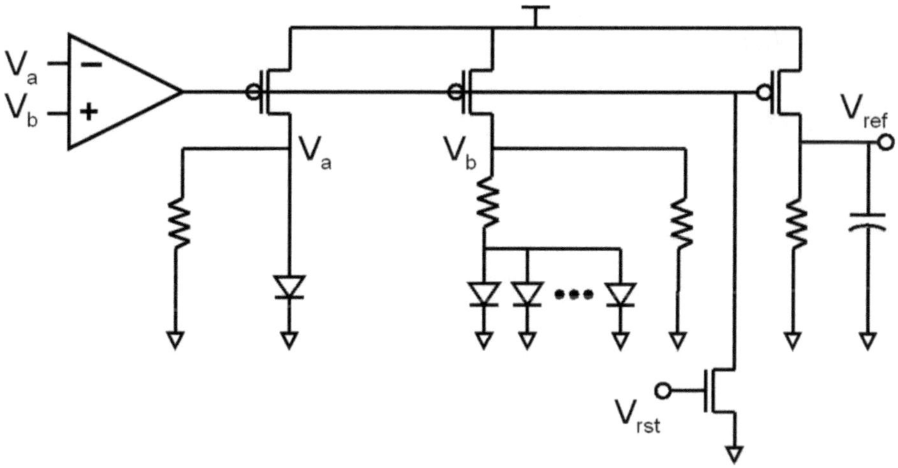

Fig. 16.18 Low-voltage CMOS bandgap voltage reference circuit from [67], with a parameter space of 122 dimensions

16.7.3 Sub-1 V CMOS Bandgap Voltage Reference

Figure 16.18 shows a low-voltage CMOS bandgap voltage reference circuit, proposed in [67]. The circuit is able to provide reference voltages that are less than 1 V, and is built using standard CMOS technology. It was chosen for its relevance in today's and tomorrow's low-voltage designs, and also because the related RSM problem has a high input dimensionality of 122 and strong nonlinear behavior. The opamp in the circuit is the same as in Sect. 16.7.2. The circuit has 101 diodes. The transistor device and variation models are the same 90 nm CMOS as the opamp. Variations in each diode are modeled as a normally distributed variation on the saturation current, with standard deviation of 10%. Each resistor and capacitor has its own normally distributed variation source, with a standard deviation of 5%. There are a total of 121 local variation parameters and one global t_{ox} variation. In this case, we measure two metrics: (1) the output voltage V_{ref}, and (2) the dropout voltage V_{do}. V_{do} is the difference between the supply voltage and V_{ref}, when V_{ref} falls by 1% of its nominal value: lower V_{do} implies a circuit more robust to variations in the supply voltage. The nominal V_{ref} we designed for is 600 mV.

Figure 16.19a shows the 122-dimensional projection vector for the first LVs of the bandgap performance metrics. Figure 16.20 plots the simulated and predicted outputs against their corresponding LVs. Here we see that PROBE performs well for the linearly behaved V_{ref}, but completely breaks down for the nonlinearly behaved V_{do}. SiLVR, however, *is* able to extract a good estimate of this strong nonlinear behavior, as evidenced also in Table 16.2. For the case of V_{do}, we can actually improve the fit further by using a 2-LV model, as shown both by Fig. 16.19b and by the last column in Table 16.2. The former plots the simulated and SiLVR-predicted

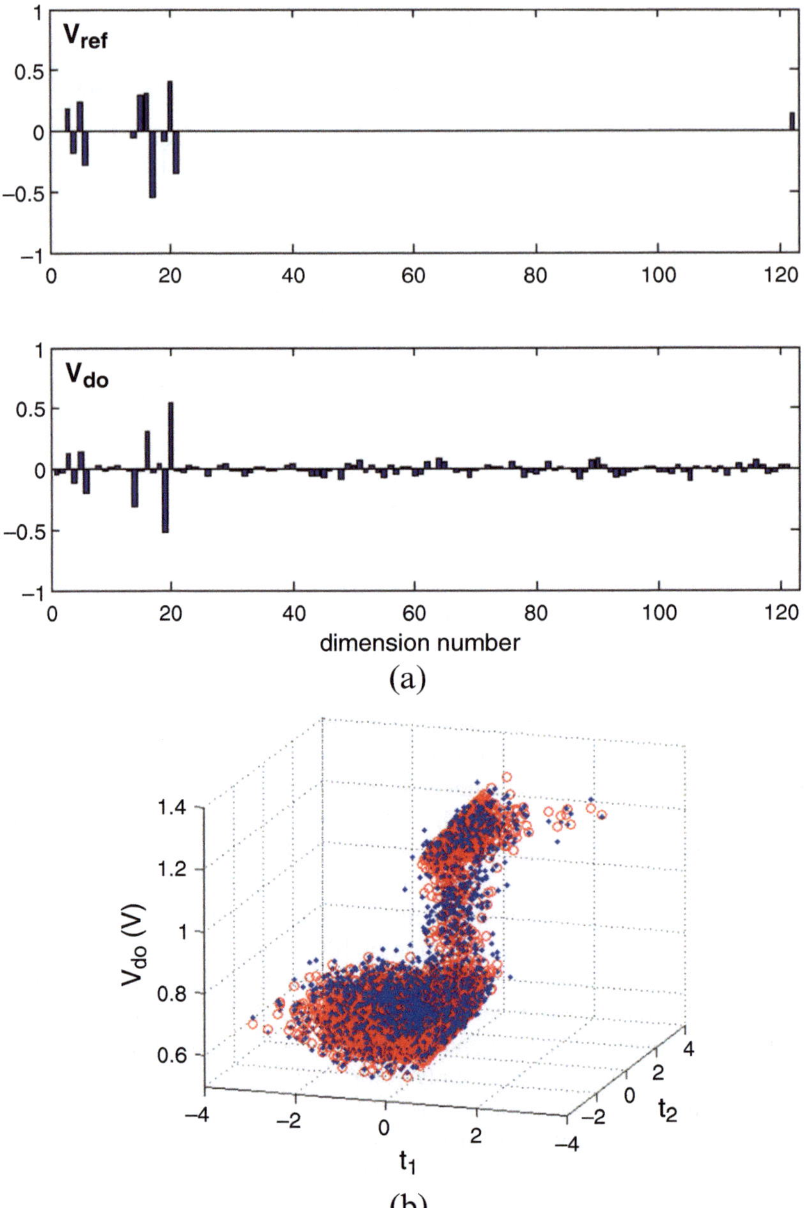

Fig. 16.19 Performance of SiLVR on the sub-1 V CMOS voltage reference circuit test case. (**a**) Normalized projection vector for the first LVs of the voltage reference circuit metrics. (**b**) Simulated and SiLVR-predicted V_{do} against the first two LVs

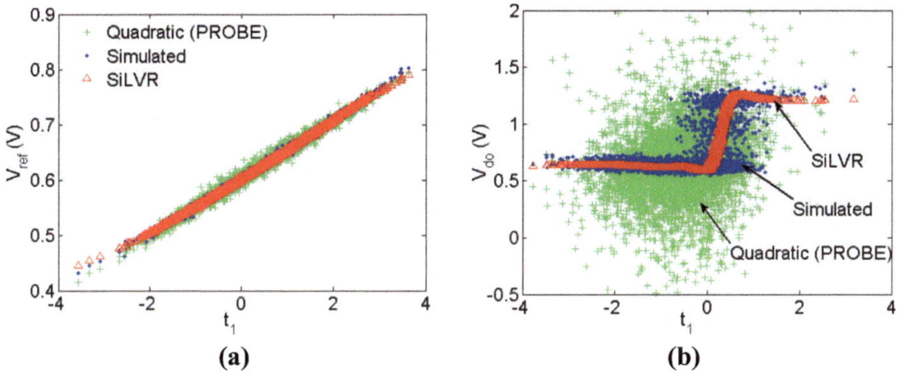

Fig. 16.20 Simulated, PROBE-predicted (green plus), and SiLVR-predicted (red triangle) outputs, plotted against the first LV, for the sub-1 V CMOS voltage reference circuit test case. (**a**) Comparison for V_{ref}. (**b**) Comparison for V_{do}—the quadratic model breaks down

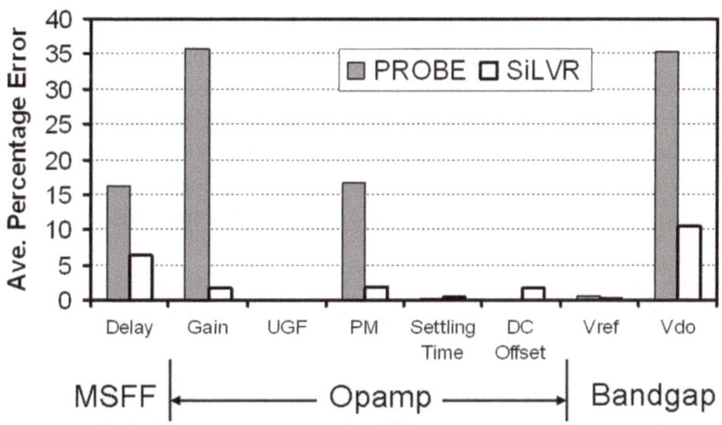

Fig. 16.21 Best SiLVR errors compared with best PROBE errors: PROBE shows large errors for the nonlinear performances. SiLVR significantly reduces those errors and maintains low errors for the near-linear cases (UGF, Settling Time, DC offset, V_{ref})

values of the dropout voltage for the test set points. Only for this figure, the SiLVR model was trained using a (separate) training set of 10,000 points to achieve a fit that is visually obvious in three dimensions. However, the results in Table 16.2 and in Fig. 16.21 are for a model trained using the standard sample size of 1000 points.

Hence, even though we started with a large dimensionality of 122, only 2 LVs can still explain most of the behavior. Also, the normalized inner product of the first two projection vectors $\frac{\mathbf{w}_1^T \mathbf{w}_2}{\|\mathbf{w}_1\| \|\mathbf{w}_2\|}$ is only 1.2e–3, meaning that they are almost orthogonal. This implies that SiLVR can extract almost all the information from the first LV before looking at the second LV. We saw similar results supporting this inference for the example $y = x_1 x_2$ at the beginning of Sect. 16.7.

16.7.3.1 Training Time

The training run times to build each LV are quite reasonable, even with the complex cross-validation strategy to improve neural network robustness: each LV requires 13–24 CPU seconds of Matlab® computation. This is especially attractive for higher dimensional cases like the voltage reference, where even the simple quadratic model of PROBE can be relatively expensive to train.

16.8 Conclusions

SiLVR possesses some very desirable features as an RSM technique. It elegantly handles nonlinear surfaces, enables performance-oriented dimensionality reduction, provides useful quantitative measures to understand the circuit design problem (relative global sensitivity and IRC), and enables insightful visualization of performance behavior in much reduced dimensions. Even before its application to automatic optimization, these features can find good use in the manual design process. For example, after running an increasingly popular Monte Carlo run for yield analysis, the designer can obtain a SiLVR model and all its by-product features in a few seconds without running any more simulations. This is a very simple use-mode that is minimally intrusive to most circuit design flows, and still provides a useful new design tool to the designer. The practical, real examples from the case of the opamp in Sect. 16.7.2 illustrate this usefulness.

Given this, much can still be done to extend the power and usefulness of SiLVR. Here we briefly introduce some possible directions of further research, targeting various aspects of this RSM strategy.

- Researchers in statistics and data mining have developed techniques to refine the PPR model. One relevant technique is the so-called *backfitting* [23]. In this procedure, after the i-th projection in the PPR model is extracted, the previously extracted projections are re-optimized, to better model the residue after removing the behavior modeled by this i-th projection.
- The SiLVR training algorithm emphasizes the reduction of overfitting, by employing a minimal number of sigmoids and using regularization and cross-validation techniques. It is not very clear if it currently sits at the best trade-off between accuracy and generalizability; that is, between variance and bias. This issue deserves further investigation to determine a near-optimal trade-off for the SiLVR model.
- Finally, a goal of most RSM strategies is to be employed as circuit performance models in automatic yield-aware optimization. Such is also the case for SiLVR. How can SiLVR be best incorporated in yield-aware circuit synthesis? This is a "loaded" question and any inquiry into its answer will require answering several other questions: How can SiLVR be adapted to work across both the statistical parameter and design variable spaces to allow larger model-trust regions in the

design space? How can the information available from SiLVR be best used for guiding the search algorithm? Which search algorithms fit best with SiLVR?

References

1. S.W. Director, P. Feldmann, K. Krishna, Statistical integrated circuit design. IEEE J. Solid-State Circuits **28**(3), 193–202 (1993)
2. P. Cox, P. Yang, S.S. Mahant-Shetti, P. Chatterjee, Statistical modeling for efficient parametric yield estimation of MOS VLSI circuits. IEEE Trans. Electron Devices **32**(2), 471–478 (1985)
3. T.-K. Yu, S.M. Kang, I.N. Hajj, T.N. Trick, Statistical performance modeling and parametric yield estimation of MOS VLSI. IEEE Trans. Comput. Aided Des. **6**(6), 1013–1022 (1987)
4. P. Feldmann, S.W. Director, Integrated circuit quality optimization using surface integrals. IEEE Trans. Comput. Aided Des. **12**(12), 1868–1879 (1993)
5. M. Hane, T. Ikezawa, T. Ezaki, Atomistic 3d process/device simulation considering gate line-edge roughness and poly-Si random crystal orientation effects, in *Proceedings of IEEE International Electron Devices Meeting* (2003)
6. T. Ezaki, T. Izekawa, M. Hane, Investigation of random dopant fluctuation induced device characteristics variation for sub-100 nm CMOS by using atomistic 3d process/device simulator, in *Proceedings of IEEE International Electron Devices Meeting* (2002)
7. D.J. Frank, Y. Taur, M. Ieong, H.-S.P. Wong, Monte Carlo modeling of threshold variation due to dopant fluctuation, in *Proceedings of International Symposium on VLSI Technology* (1999)
8. X. Li, J. Le, L.T. Pileggi, A. Stojwas, Projection-based performance modeling for inter/intra-die variations, in *Proceedings of IEEE/ACM International Conference on CAD* (2005)
9. Z. Feng, P. Li, Performance-oriented statistical parameter reduction of parameterized systems via reduced rank regression, in *Proceedings of IEEE/ACM International Conference on CAD* (2006)
10. A.J. Burnham, R. Viveros, J.F. MacGregor, Frameworks for latent variable multivariate regression. J. Chemom. **20**, 31–45 (1996)
11. P.J. Huber, Projection pursuit. Ann. Stat. **13**(2), 435–475 (1985)
12. A. Singhee, R.A. Rutenbar, Beyond low-order statistical response surfaces: latent variable regression for efficient, highly nonlinear fitting, in *Proceedings of IEEE/ACM Design Automation Conference* (2007)
13. S. Wold, A. Ruhe, H. Wold, W.J. Dunn III, The collinearity problem in linear regression. The partial least squares (PLS) approach to generalized inverses. J. Sci. Stat. Comput. **5**(3), 735–743 (1984)
14. G. Reinsel, R. Velu, *Multivariate Reduced-Rank Regression, Theory and Applications* (Springer, Berlin, 1998)
15. S. Wold, M. Sjstrm, L. Eriksson, PLS-regression: a basic tool of chemometrics. Chemom. Intell. Lab. Syst. **58**, 109–130 (2001)
16. P.T. Davies, M.K.-S. Tso, Procedures for reduced-rank regression. Appl. Stat. **31**(3), 244–255 (1982)
17. A-L. Boulesteix, K. Strimmer, Partial least squares: a versatile tool for the analysis of high-dimensional genomic data. Brief. Bioinform. **8**(1), 32–44 (2006)
18. G. Baffi, E.B. Martin, A.J. Morris, Non-linear projection to latent structures revisited (the neural network PLS algorithm). Comput. Chem. Eng. **23**(9), 1293–1307 (1999)
19. C. Malthouse, A.C. Tamhane, R.S.H. Mah, Nonlinear partial least squares. Comput. Chem. Eng. **21**(8), 875–890 (1997)
20. B.D. Ripley, *Pattern Recognition and Neural Networks* (Cambridge University Press, Cambridge, 1996)

21. J.H. Friedman, W. Stuetzle, Projection pursuit regression. J. Am. Stat. Assoc. **76**(376), 817–823 (1981)
22. G. Golub, C. Loan, *Matrix Computations* (JHU Press, Baltimore, 1996)
23. T. Hastie, R. Tibshirani, J. Friedman, *The Elements of Statistical Learning: Data Mining, Inference, and Prediction* (Springer, Berlin, 2001)
24. B.F. Logan, L.A. Shepp, Optimal reconstruction of a function from its projections. Duke Math. J. **42**, 645–659 (1975)
25. B.A. Vostrecov, M.A. Kreines, Approximation of continuous functions by superpositions of plane waves. Soviet Math. Dokl. **2**, 1326–1329 (1961)
26. F. John, *Plane Waves and Spherical Means Applied to Partial Differential Equations* (Interscience Publishers, New York, 1955)
27. W.S. McCullough, W. Pitts, A logical calculus of the ideas immanent in nervous activity. Null. Math. Biophys. **5**, 115–133 (1943)
28. I. Goodfellow, Y. Bengio, A. Courville, *Deep Learning* (MIT Press, Cambridge, 2016)
29. B. Irie, S. Miyake, Capabilities of three-layered perceptrons, in *International Conference on Neural Networks* (1988)
30. P. Diaconis, M. Shahshahani, On nonlinear functions of linear combinations. SIAM J. Sci. Stat. Comput. **5**(1), 175–191 (1984)
31. A.M. Bruckner, J.B. Bruckner, B.S. Thompson, *Real Analysis* (Prentice-Hall, New Jersey, 1997)
32. X. Sun, E.W. Cheney, The fundamentality of sets of ridge functions. Aequationes Math. **44**, 226–235 (1992)
33. V.Y. Lin, A. Pinkus, Fundamentality of ridge functions. J. Approx. Theory **75**, 295–311 (1993)
34. P.P. Petrushev, Approximation by ridge functions and neural networks. SIAM J. Math. Anal. **30**(1), 155–189 (1998)
35. V.E. Maiorov, On best approximation by ridge functions. J. Approx. Theory **99**, 68–94 (1999)
36. M. Burger, A. Neubauer, Error bounds for approximation with neural networks. J. Approx. Theory **112**, 235–250 (2001)
37. H.N. Mhaskar, Approximation by superposition of sigmoidal and radial basis functions. Adv. App. Math. **13**, 350–373 (1992)
38. A.R. Barron, Universal approximation bounds for superpositions of a sigmoidal function. IEEE Trans. Inf. Theory **39**(3), 930–945 (1993)
39. R.A. Adams, *Sobolev Spaces* (Academic Press, New York, 1975)
40. T.H. Cormen, C.E. Leiserson, R.L. Rivest, *Introduction to Algorithms*, 2nd edn. (MIT Press, Cambridge, 2001)
41. W. Light, Ridge functions, sigmoidal functions and neural networks, in *Approximation Theory*, ed. by E.W. Cheney, C.K. Chui, L.L. Schumaker, vol. VII (Academic Press, New York, 1992)
42. B.L.S. Prakasa Rao, *Nonparametric Functional Estimation* (Academic Press, New York, 1983)
43. J.H. Friedman, A variable span smoother, Department of Statistics Tech. Report LCS 05, Stanford University, 1984
44. H.H. Rosenbrock, An automatic method for finding the greatest or least value of a function. Comput. J. **3**, 175–184 (1960)
45. D. Donoho, I. Johnstone, P. Rousseeuw, W. Stahel, Projection pursuit (discussion). Ann. Stat. **13**(2), 496–500 (1985)
46. L.K. Jones, On a conjecture of Huber concerning the convergence of projection pursuit regression. Ann. Stat. **15**(2), 880–882 (1987)
47. P. Hall, On projection pursuit regression. Ann. Stat. **17**(2), 573–588 (1989)
48. M. Loéve, *Probability Theory I & II*, 4th edn. (Springer, Berlin, 1977)
49. G. Cybenko, Approximation by superpositions of sigmoidal functions. Math. Control Signals Syst. **2**, 303–314 (1989)
50. K. Hornik, M. Stinchcombe, H. White, Multilayer feedforward networks are universal approximators. Neural Netw. **2**, 359–366 (1989)
51. C.K. Chui, X. Li, Approximation by ridge functions and neural networks with one hidden layer. J. Approx. Theory **70**, 131–141 (1992)

52. K. Funahashi, On the approximate realization of continuous mappings by neural networks. Neural Netw. **2**, 183–192 (1989)
53. H.N. Mhaskar, Neural networks for optimal approximation of smooth and analytic functions. Neural Comput. **8**, 164–177 (1996)
54. C.K. Chui, X. Li, H.N. Mhaskar, Limitations of the approximation capabilities of neural networks with one hidden layer. Adv. Comput. Math. **5**, 233–243 (1996)
55. A.R. Barron, Statistical properties of artificial neural networks, in *Proceedings of 28th Conference on Decision and Control* (1989)
56. I.M. Sobol', S.S. Kucherenko, Global sensitivity indices for nonlinear mathematical models. Rev. Wilmott Mag. **2**, 2–7 (2005)
57. W.H. Press, B.P. Flannery, A.A. Teukolsky, W.T. Vetterling, *Numerical Recipes in C: The Art of Scientific Computing*, 2nd edn. (Cambridge University Press, Cambridge, 1992)
58. D. Marquardt, An algorithm for least squares estimation of non-linear parameters. J. Soc. Ind. Appl. Math. **11**, 431–441 (1963)
59. D.J.C. MacKay, A practical Bayesian framework for backpropagation networks. Neural Comput. **4**(3), 448–472 (1992)
60. M.T. Hagan, M.B. Menhaj, Training feedforward networks with the Marquardt algorithm. IEEE Trans. Neural Netw. **5**(6), 989–993 (1994)
61. J.E. Dennis, Jr., R.B. Schnabel, *Numerical Methods for Unconstrained Optimization and Nonlinear Equations* (SIAM, Philadelphia, 1996)
62. F. Girosi, M. Jones, T. Poggio, Regularization theory and neural network architectures. Neural Comput. **7**(2), 219–269 (1995)
63. F.D. Foresee, M.T. Hagan, Gauss–Newton approximation to Bayesian learning, in *Proceedings of International Conference on Neural Networks* (1997)
64. K. Kundert, *The Designer's Guide to SPICE and Spectre®* (Springer, Berlin, 1995)
65. W. Zhao, Y. Cao, New generation of predictive technology model for sub-45 nm early design exploration. IEEE Trans. Electron Devices **53**(11), 2816–2823 (2006)
66. P.R. Gray, P.J. Jurst, S.H. Lewis, R.G. Meyer, *Analysis and Design of Analog Integrated Circuits*, 4th edn. (Wiley, Hoboken, 2001)
67. H. Banba, H. Shiga, A. Umezawa, T. Miyaba, A CMOS bandgap reference circuit with sub-1-v operation. IEEE J. Solid-State Circuits **34**(5), 670–674 (1999)

Chapter 17
Machine Learning-Based System Optimization and Uncertainty Quantification for Integrated Systems

Hakki M. Torun, Mourad Larbi, and Madhavan Swaminathan

17.1 Introduction

Emerging trend in modern high-performance systems is leading to new challenges in efficient voltage regulation architectures. Some of these include delivering high current at a low voltage for increased processing power at very high switching frequencies to support dynamic voltage and frequency scaling (DVFS) for energy efficient multi-core architectures [1, 2]. The regulation efficiency of conventional standalone DC–DC converters is limited under such operating conditions by either the power delivery network (PDN) or AC/DC losses introduced by the inductor. The recent trend to address these issues is towards integrating IVRs either on-chip or in the package. Furthermore, IVRs provide a great opportunity for system miniaturization when combined with embedded inductors. In this chapter, our focus is on embedding solenoidal inductors in the package with a magnetic core [3].

Due to high levels of integration and increased switching frequency, IVRs have highly non-linear response to their control parameters which include number of phases, switching frequency, and the characteristics of embedded inductor including type and dimensions of the magnetic material. This creates unique challenges in deriving an accurate model that captures all the coupling between components comprising the system along with non-linearities.

Ultimately, the goal of the model is to perform design space exploration and optimization to determine the control parameters that will give the best performance. In this chapter, we present a new algorithm by building upon the importance sampling technique presented in [4], called *two-stage Bayesian optimization* (TSBO),

H. M. Torun · M. Larbi · M. Swaminathan (✉)
Georgia Institute of Technology, School of Electrical & Computer Engineering, Atlanta, GA, USA
e-mail: htorun3@gatech.edu; mourad.larbi@ece.gatech.edu; madhavan.swaminathan@ece.gatech.edu

I. M. Elfadel et al. (eds.), *Machine Learning in VLSI Computer-Aided Design*,
https://doi.org/10.1007/978-3-030-04666-8_17

that utilizes machine learning (ML)-based Bayesian optimization (BO) to perform tuning and optimization of the IVR over a large design space. The demonstrated algorithm is generic in the sense that it can be applied to other types of systems to handle design complexities caused by high dimensionality and multi-physics interactions. For instance, TSBO has been applied to 3D ICs and systems [5, 6], wireless power transfer systems [7], and high-speed channels [8].

Once the optimal design parameters have been found, uncertainties related to the fabrication process, such as intrinsic characteristics of materials, geometrical properties, and temperature variations, may cause a large variability of output response. Hence, it is essential to carry out uncertainty quantification (UQ) in order to ensure that the fabricated design satisfies all the electrical requirements. In this context, the brute force Monte Carlo (MC) method is not appropriate since it will require a high number of model evaluations to accurately estimate the variability of the output response. For this reason, a surrogate model, which aims at replacing the numerical model by an analytic approximation of the response, needs to be considered. In this chapter, we build up, from a non-intrusive technique, a surrogate model based on polynomial chaos (PC) [9, 10] to predict the variability of the IVR efficiency. The surrogate model is then used to estimate statistical quantities such as moments, probability density functions (PDF), and sensitivity analysis of the output.

The rest of this chapter is structured as follows: Sect. 17.2 presents a new design flow enabled by the TSBO algorithm; Sect. 17.3 provides background on black-box optimization and BO; Sect. 17.4 presents the TSBO algorithm with empirical analysis; Sect. 17.5 shows the application and performance of the algorithm presented in co-optimization of IVR and embedded inductor with the objective of maximizing IVR efficiency while minimizing the inductor area; Sect. 17.6 discusses background on UQ using sparse polynomial chaos followed by its application to IVR in Sect. 17.7 and conclusion in Sect. 17.8.

17.2 Optimization Oriented Design Flow

In practice, the use of optimization algorithms in EDA is mainly limited to fine-tuning an already good design. This is due to lack of a method capable of finding the optimum set of control parameters from the beginning. Conventional methods often depend on initial point selection, are limited to a few design parameters, and can lead to slow convergence to the desired solution. The majority of these methods are gradient based which can result in convergence to local extrema and require excessive sampling to calculate or approximate the gradient. As a result, designers are left with no choice but to hand-tune the parameters to meet design specifications, which requires decades of experience especially for complex structures.

Moreover, converging to a good design requires going through a rigorous, CPU extensive process that involves human intervention at multiple stages of the design flow, as shown in Fig. 17.1. The first step of a design cycle usually involves design

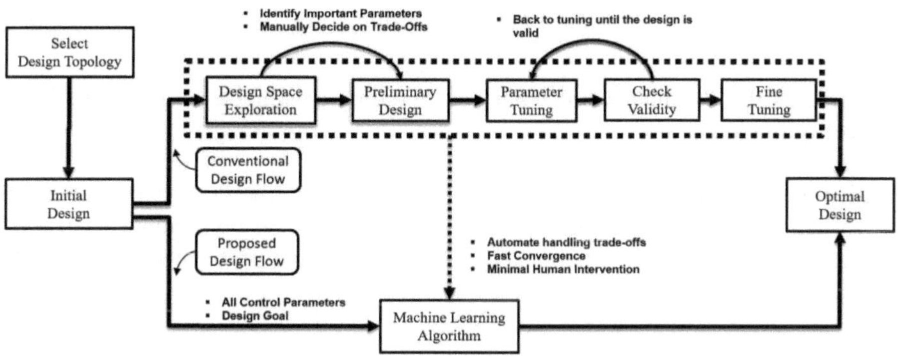

Fig. 17.1 The proposed optimization oriented design flow

space exploration to identify which parameters significantly affect the response of the system. This identification requires designers to manually decide on component-level trade-offs, which can result in suboptimal system-level performance due to high levels of coupling between different components comprising the system. Once a preliminary design has been developed, a burdensome parameter tuning process needs to be performed while checking the validity of the design during each cycle and finally, the winning parameters are selected for final implementation.

The proposed method, TSBO, is developed with the purpose to realizing the optimization oriented design flow in Fig. 17.1, which enables designers spend more time on innovation and less time on tuning the parameters to arrive at an optimized design. In order to achieve this, TSBO uses three new techniques to enhance ML-based BO, namely (1) incorporation of fast exploration and pure exploitation stages, (2) learning acquisition functions, and (3) intelligent sampling. The goal of each technique is to address aforementioned drawbacks of conventional methods, while ensuring that the proposed design flow remains practical in terms of minimizing CPU time and substantially reducing human intervention. For this flow, inputs to TSBO consist of control parameters of the system along with the design goal. The design goal can consist of a weighted sum of multiple objectives, i.e., prioritizing minimization of clock skew over reducing maximum temperature for 3D ICs, as an example [11]. By letting the designer determine the design goal, we allow domain expertise to couple with the ML algorithm for enabling faster convergence.

17.3 Black-Box Optimization

Non-convex optimization attracts attention in many fields of engineering problems since a major portion of realistic systems do not tend to have linear responses to their control parameters. The problem can be posed from the perspective of optimization as:

$$\max_{x \in R^d} f(x) \quad \text{or} \quad \min_{x \in R^d} f(x) \tag{17.1}$$

where x is the d dimensional input vector (control parameters) and $f(x)$ is the objective function.

In the domain of advanced electronics design, $f(x)$ can represent a high-dimensional, highly coupled, multi-uncertainty, and multi-scale system such as 3D ICs, IVRs, or high-speed channels. As it is not trivial to develop accurate and precise models of such systems, a necessity of approaching the problem in (17.1) in the *black-box* setting arises.

17.3.1 Why Machine Learning for Optimization?

While operating in the black-box setting, the only way to get information from the underlying system is to query the design under evaluation for certain parameters. In microelectronics applications, this is often obtained through multi-physics simulations which can be CPU extensive to achieve high accuracy.

Conventional optimization techniques take a step-by-step approach, meaning the search for next parameter only involves analyzing the current point. In order to do this in the black-box setting, simulations are performed in proximity of the current best point to identify patterns or approximate gradients. This can lead to excessive simulations or inaccurate approximations in high-dimensional systems and can result in convergence to local extrema.

In contrast, machine learning techniques leverage experience to make inference. In the context of optimization, this means that the next set of parameters is determined by analyzing all the previous observations, which enables capability of differentiating between local and global optima as well as using intelligent parameter elimination and selection to avoid excessive search.

BO based on Gaussian process is a ML technique that uses this concept to ensure convergence to global optima with a reduced number of simulations as compared to non-ML methods. Park et al. [11] provide a comparison of BO with other mostly used methods for EDA in Fig. 17.2 that show significant reduction in the required number of simulations.

The BO algorithm shown in Fig. 17.2, IMGPO [12], is a general purpose method, which does not necessarily address the challenges encountered in the EDA domain such as the ability to handle higher dimensions and focusing on reducing the number of simulations required instead of minimizing the algorithm progression time. The TSBO algorithm presented in this work is specifically developed to be more EDA oriented than general purpose BO algorithms by prioritizing reduction in total computational overhead through significantly reducing the number of simulations required for optimization.

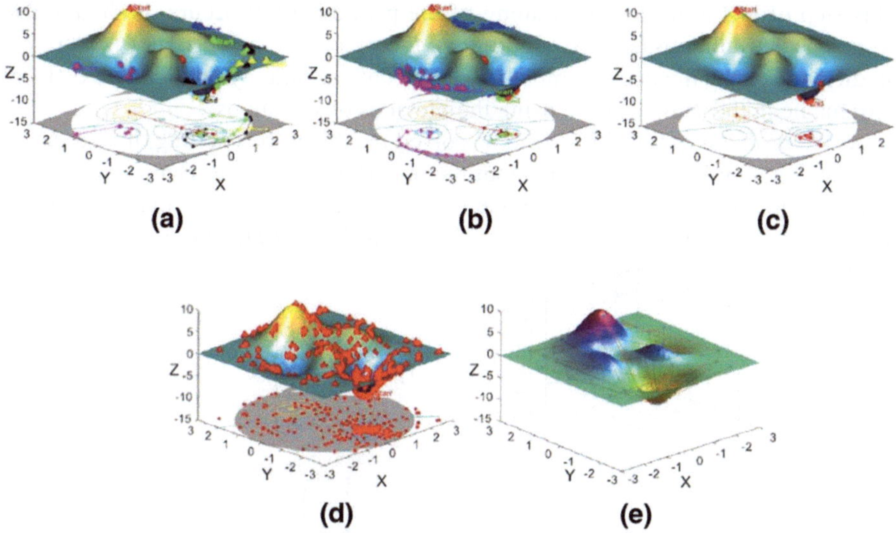

Fig. 17.2 Comparison of BO to other mostly used methods in EDA [11]. (**a**) Multistart (function counts = 345). (**b**) Global search (function counts = 273). (**c**) Pattern search (function counts = 272). (**d**) Genetic algorithm (function counts = 2650). (**e**) BO (function counts = 100)

17.3.2 Bayesian Optimization Based on Gaussian Process

In majority of black-box systems, including IVRs, it is not possible to directly access gradient information of $f(x)$ at an arbitrary point, x. Using BO based on Gaussian processes (GP), gradient information is not required; hence, it becomes a suitable and promising candidate for black-box optimization. BO is a well-known method in the machine learning community and has been mainly used for hyperparameter tuning for various machine learning algorithms. It is as an *active learning* algorithm, as in the process of optimization, BO selects the next point to maximize reward towards finding the solution of Eq. (17.1). The approach is based on Bayes' theorem:

$$P(f|D_{1:t}) \propto P(D_{1:t}|f)P(f) \tag{17.2}$$

where $D_{1:t} = \left[(x_1, y_1), (x_2, y_2), \ldots, (x_t, y_t)\right]$ represent observations made so far; $P(f)$, $P(f|D_{1:t})$, and $P(D_{1:t}|f)$ are probabilities of prior, posterior, and likelihood, respectively.

In BO, the underlying function to be optimized, $f(x)$, is represented as a joint, multi-dimensional GP with a mean (μ) and co-variance (K), given by:

$$f_{1:t} = \mathcal{N}(\mu(x_{1:t}), K(x_{1:t})) \tag{17.3}$$

This step of BO is where the training of the predictive GP occurs. It should also be noted that unlike other ML algorithms, the objective in BO is not to derive a predictor covering the entire sample space, but only accurately predict where global extrema lie in the sample space using the previous observations.

As the problem in Eq. (17.1) is to be solved in a black-box setting, it is assumed that there is no prior information about the underlying function. Hence, we choose a zero mean GP for $P(f)$ and use one of the popular kernel functions used in the literature, namely *automatic relevance determination (ARD) Matern 5/2 function*, given as:

$$K(x) = \begin{bmatrix} k(x_1, x_1) & \dots & k(x_1, x_t) \\ \vdots & \ddots & \vdots \\ k(x_t, x_1) & \dots & k(x_t, x_t) \end{bmatrix} \tag{17.4}$$

$$k(x_i, x_j) = \sigma_f^2 \left(1 + \sqrt{5}r + \frac{5}{3}r^2\right) e^{-\sqrt{5}r} \tag{17.5}$$

where $r(x_i, x_j) = \left(\sum_{d=1}^{D} \frac{(x_{i,d} - x_{j,d})^2}{\sigma_d^2}\right)^{1/2}$; σ_f and σ_d are hyperparameters of $K(x)$ which are updated during the training process to minimize the negative log marginal likelihood of the GP, given by:

$$\log p(y_{1:t} \mid x_{1:t}) = -\frac{1}{2} y_{1:t}^T \left(K + \sigma_n^2 I\right)^{-1} y_{1:t} - \frac{1}{2} \log |K + \sigma_n^2 I| - \frac{t}{2} \log 2\pi \tag{17.6}$$

In addition, using *ARD* type kernel has several advantages. As each input variable has its own scaling parameter to be determined in *ARD* type kernels, they enable interpretability of the underlying problem and act as an implicit sensitivity analysis. For instance, after the training of predictive GP is completed, if σ_d of one parameter is sufficiently higher compared to other parameters, it can be said that a change in this parameter has lower sensitivity on the predictions. Another possible use of σ_d parameters in Eq. (17.5) is transferring problem specific knowledge during the training of GP. If, for example, a certain parameter is believed to have higher effect on the output, the training process can be modified to force σ_d of that parameter to be smaller compared to others. However, in high-dimensional problems encountered in EDA, this can be very dangerous as predicting the importance of each parameter before having any prior training data can result in suboptimal designs.

The traditional approach of BO is obtaining an acquisition function, $u(x)$, based on a pre-defined strategy and uses auxiliary optimization on this acquisition function to find the new query point, x_{t+1}. Note that optimizing $u(x)$ does not require any additional queries, but uses the knowledge of previous samples to get a prediction at candidate points using:

$$\mu(x^*) = k^T \left(K + \sigma_n^2 I\right)^{-1} f_{1:t} \tag{17.7}$$

$$\sigma^2(x^*) = k^* - k^T \left(K + \sigma_n^2 I\right)^{-1} k \tag{17.8}$$

where $k^* = k(x^*, x^*)$ and $k = k(x^*, x_{1:t})$ are as in (17.5); K is given by (17.4), and σ_n is determined from the training. In the literature of BO, most prominent acquisition functions are *probability of improvement (PI), expected improvement (EI)*, and *upper confidence bound (UCB)* given as [13]:

$$u_{\mathrm{PI}} = \Phi\left((\mu(x) - \widetilde{f}^* - \zeta)/\sigma(x)\right) \tag{17.9}$$

$$u_{\mathrm{EI}} = (\mu(x) - \widetilde{f}^* - \zeta)\Phi(Z) + \sigma(x)\phi(Z) \tag{17.10}$$

$$u_{\mathrm{UCB}} = \mu(x) + S\sigma(x), \quad S = \sqrt{2ln(2\pi M^2/(12\eta))} \tag{17.11}$$

where $\widetilde{f}^*$ is the best point observed so far, ζ is a hyperparameter for u_{PI} and u_{EI}, M is the number of calls made to UCB, $(1 - \eta)$ is the probability of zero regret for GP-UCB [14], $Z = (\mu(x) - \widetilde{f}^* - \zeta)/\sigma(x)$, $\Phi(.)$ and $\phi(.)$ are the CDF and PDF of normal distribution, respectively.

A common situation encountered in the optimization of electronic systems is handling of design constraints. For instance, in the IVR example discussed, the range in the dimension of the physical parameters of the inductor could be constrained due to the process while area and efficiency constraints need to be satisfied in the final design. In the context of BO, there are several ways to introduce such design constraints in the optimization framework. One method is to penalize the objective function when the set of parameters lies in such unfeasible regions to prevent that point from being labelled as optimal. Another method is to modify the acquisition function so that the parameters in the unfeasible regions are less likely to be the next simulation point [15]. We refer readers to [13] for a more detailed theoretical background on Bayesian optimization and Gaussian processes.

17.4 Two-Stage Bayesian Optimization

In multi-armed bandit problems, the path to achieve optimal result goes through a trade-off between exploration and exploitation. For the case of aforementioned acquisition functions, u_{PI} and u_{EI} in Eqs. (17.9) and (17.10), the trade-off is made with the introduction of the hyperparameter ζ. IMGPO handles the trade-off with the assumption of existence of a tighter bound than UCB [12] and BamSOO takes the approach of eliminating regions where with high probability, the region does not contain global optima [16].

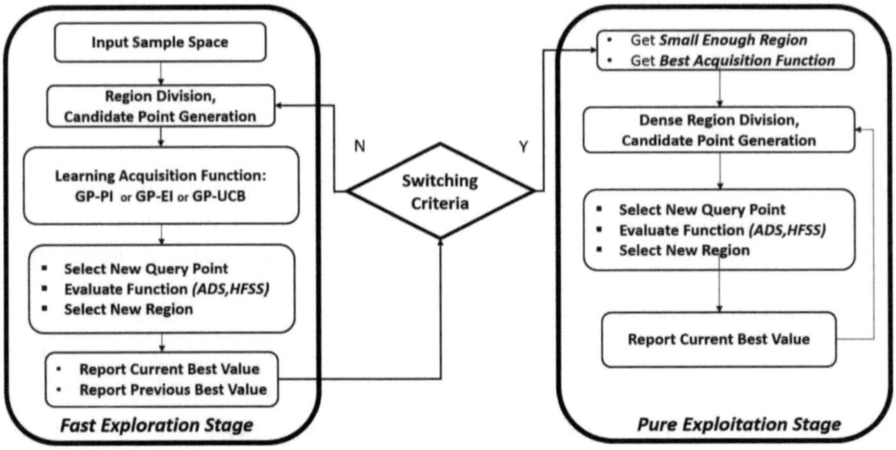

Fig. 17.3 Flowchart of the proposed algorithm

In the new algorithm we present, *two-stage Bayesian optimization (TSBO)*, we approach this trade-off in two stages, namely fast exploration stage and pure exploitation stage, which can also be interpreted as coarse and fine-tuning, respectively. In order to address EDA related challenges, we present two additional new techniques that increase convergence rate and extend the applicability of the algorithm to a variety of electronics design problems. These consist of (1) sub-learning process using acquisition functions and (2) a distinctive hierarchical partitioning tree construction scheme. The flowchart of TSBO is given in Fig. 17.3.

17.4.1 Fast Exploration and Pure Exploitation Stages

The first stage of TSBO, *fast exploration stage*, can be considered as coarse tuning, as the purpose of this stage is to rapidly find the region, A^d, in the sample space where the global optimum, $x^* = \arg\max_x f(x)$, is contained. When the optimization goal is minimization, $f(x)$ can be negated without any changes to the TSBO algorithm. For the purpose of finding a tight A^d rapidly, we divide the sample space X into 2^d regions of hyper rectangles, H_t, and generate candidate points, $c_{t,j}$, to determine the next sampling point, x_{t+1}. These candidate points, as illustrated in Fig. 17.4b, are chosen as the center of each region, given as:

$$c_{t,j} = \frac{H_{t,j,\min} + H_{t,j,\max}}{2}, \quad j = 1, 2, \ldots, t2^d \tag{17.12}$$

where $c_{t,j}$ is the *j*th candidate point of a total of $t2^d$ points at iteration t; $H_{t,j}$ is the region that $c_{t,j}$ belongs to, and $H_{t,j,\min}$ and $H_{t,j,\max}$ are corresponding lower

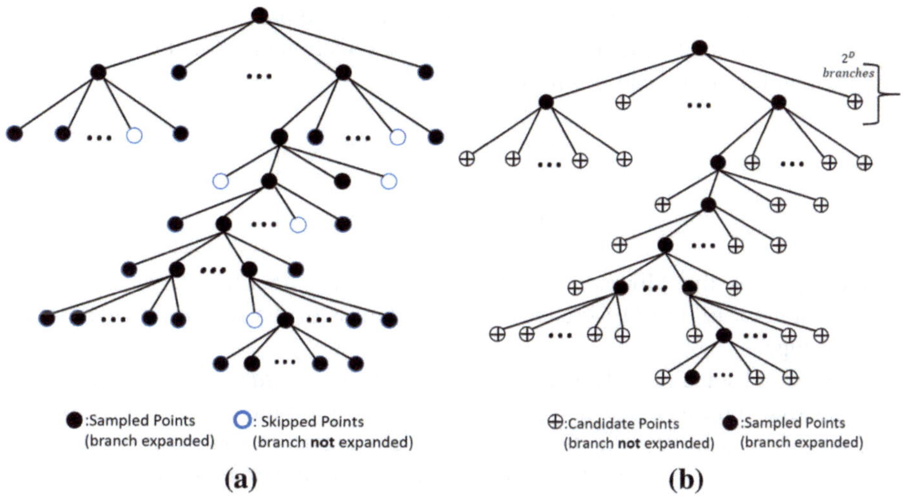

Fig. 17.4 Examples of hierarchical partitioning trees constructed by different algorithms. (**a**) BamSOO. (**b**) TSBO

and upper boundaries of each d dimensional region. In order to avoid relying on auxiliary optimization to select x_{t+1}, $u(x)$ is not optimized but evaluated at each candidate point as follows:

$$c_{(t,j^*)} = \arg\max_{c \in C} u^{(i)}(c) \tag{17.13}$$

$$x_{t+1} = c_{(t,j^*)} \tag{17.14}$$

where j^* denotes the selected candidate; C represents the finite set of candidate points, and $u^{(i)}(x)$ is $\mathrm{u_{UCB}}$, $\mathrm{u_{EI}}$, or $\mathrm{u_{PI}}$, selected sequentially as explained later in the subsection on *learning acquisition functions*. After querying the function at x_{t+1}, the new set of regions are generated using:

$$H_n = \bigcup_{i=1}^{2^d} h_i, \quad h_i \subset H_{(t,j^*)} \tag{17.15}$$

$$H_{t+1} = H_n \cup H_t \tag{17.16}$$

where H_n is the union of new regions, h_i, acquired by dividing H_{t,j^*} into 2^d hyper rectangles.

The first stage remains until it explores a small enough region ($x^* \in A^d$) such that the Euclidean distance between *previous best* and *current best* input points is negligible. This is illustrated in Fig. 17.3 as the switching criteria, given as:

$$n = \begin{cases} n+1, & \text{if } \|x_{\max} - x_{\mathrm{pmax}}\| < 10^{-3} \\ 0, & \text{otherwise} \end{cases} \tag{17.17}$$

where $x_{\max}$ and x_{pmax} are the input vector of current and previous best observations, respectively and switching occurs if $n > N$. The selection of A^d is illustrated by the rectangle in Fig. 17.6c and given as:

$$A^d = \begin{bmatrix} (1-\alpha)x_{\max,1} & (1+\alpha)x_{\max,1} \\ \vdots & \vdots \\ (1-\alpha)x_{\max,d} & (1+\alpha)x_{\max,d} \end{bmatrix} \tag{17.18}$$

where $x_{\max,i}$ is the Mourad: i-th dimension of $x_{\max}$ in Eq. (17.17) and α is a hyperparameter of TSBO for choosing the tightness of the region provided to the second stage.

The second stage of TSBO, namely *pure exploitation stage*, takes A^d and $u^*(x)$ from the first stage as inputs and performs fine-tuning for the optimization problem, i.e., increases accuracy. At this stage, the tight region A^d is divided into three new regions along its longest dimension at each iteration and candidate points are generated at each region in the same fashion as the first stage, but using the learned acquisition function of $u^*(x)$.

17.4.2 Hierarchical Partitioning Scheme

TSBO uses a distinctive hierarchical partitioning scheme that differentiates it from general purpose BO algorithms and makes it more EDA oriented. Typically, a hierarchical partitioning tree is constructed by fully committing to and expanding the selected branch, i.e., sampling each child node [17]. For example, BamSOO uses a region shrinking technique and the tree is expanded only in non-discarded regions [16]. However, the child node of the selected branch is fully expanded if it belongs to a non-discarded region. In TSBO, the selected branch is expanded to generate 2^d candidate points, i.e., child nodes, but the sampling only occurs at the most promising child node; hence, only one function query occurs per iteration. This overcomes the limitation in the number of branches that can be generated as in BamSOO and allows for rapid coverage of the sample space. In other words, with the cost of a few seconds of algorithm run time due to storing $t2^d$ points at each iteration, TSBO substantially reduces the number of required simulations needed to find x^*. An example partitioning tree constructed using TSBO is shown and compared to a tree generated by BamSOO in Fig. 17.4.

17.4.3 Learning Acquisition Functions

As the acquisition function is a pre-determined strategy, there is no guarantee that a single acquisition function will outperform others for every problem. GP-Hedge is a BO algorithm that uses popular acquisition functions defined in Eqs. (17.9)–(17.11)

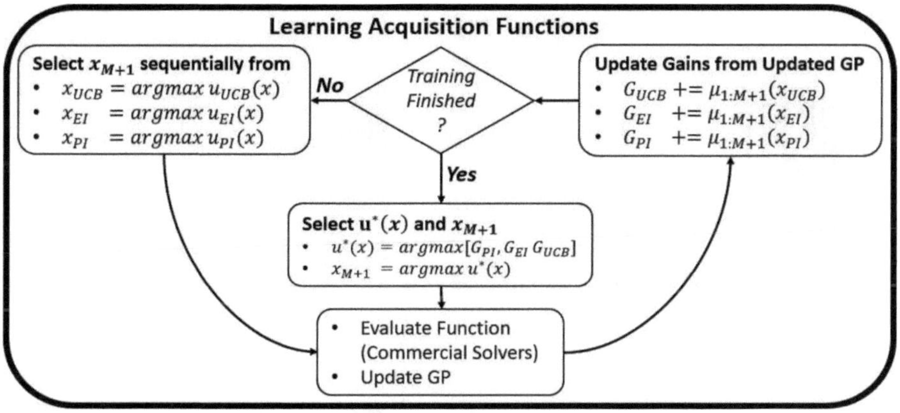

Fig. 17.5 Progression of learning acquisition function block in Fig. 17.3

and uses a probability-based gain strategy to determine which function best fits the problem [18]. We adapt this idea proposed in [18], however, we alter it and call it *learning acquisition functions*. Instead of choosing acquisition functions from a distribution, TSBO observes how each function behaves using the gain strategy given by:

$$G = \mu_{1:(t+1)}(\hat{x}_{1:t}) \tag{17.19}$$

where G is the accumulated gain, $(\hat{x}_{1:t})$ is the point suggested by each acquisition function after t observations, and $(\mu_{1:(t+1)})$ is the posterior mean of the GP model after $(t + 1)$ observations.

As summarized in Fig. 17.5, this sub-learning process of acquisition functions uses PI, EI, and UCB criteria defined in Eqs. (17.9)–(17.11) sequentially during training data collection and calculates gain of each strategy using Eq. (17.19). After M observations have been made, the algorithm quits the sequential strategy and starts to select $u^*(x)$ that has the largest gain so far. The gain functions for each $u(x)$ are continued to be updated at each iteration so that $u^*(x)$ is dynamically updated as simulations are performed.

Compared to GP-Hedge, BamSOO, and GP-UCB, this makes the algorithm deterministic along with the automatic initial point selection at the center of the sample space. The deterministic nature of TSBO makes it more robust.

17.4.4 Experiments on Challenge Functions

The progression of the proposed algorithm is provided in Fig. 17.6 and the objective is to maximize *peaks* function available in MATLAB, as an example. Here,

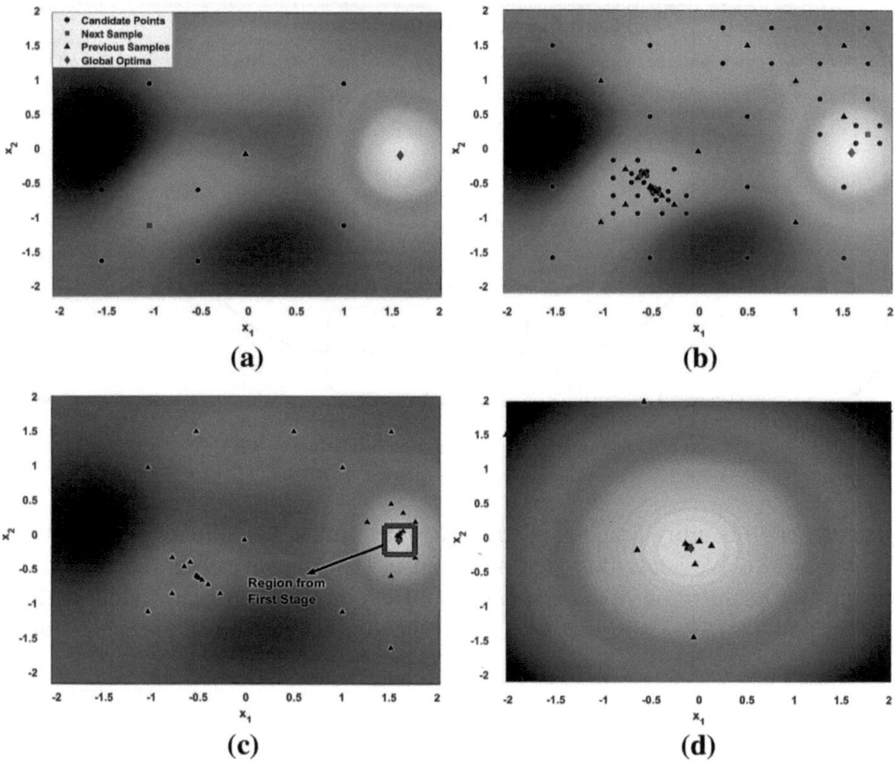

Fig. 17.6 Progression of TSBO for finding global optima of two dimensional Peaks function. (**a**) Starting point, $t = 1$. (**b**) $t = 20$. (**c**) End of first stage, $t = 42$. (**d**) End of first stage, $t = 42$

candidate points cover the entire sample space, but sampling is concentrated on only promising regions. The region captured by the first stage (rectangle in Fig. 17.6c) contains the global maximum, which is then used by the second stage to increase the accuracy up to 10^{-6}.

In order to test the performance of TSBO, we have considered four different challenge functions with different dimensions, which are common benchmarking problems in optimization [19]. Performance evaluation criteria in these experiments shown in Fig. 17.7 is *log distance to optima*, i.e., $\log_{10}(|f^* - \widetilde{f}^*|)$ where f^* is true optimum and $\widetilde{f}^*$ is the best value found after optimization. Kawaguchi showed that IMGPO outperforms BamSOO, GP-EI, and GP-PI on the same challenge functions used in this work [12]; hence, we compare TSBO with IMGPO and GP-UCB using MATLAB's *patternsearch* for auxiliary optimization of GP-UCB. Since the initial point is randomly selected in GP-UCB, we repeat the experiments 50 times and report the mean and standard deviation. As shown in Fig. 17.7, TSBO outperforms IMGPO and GP-UCB in these benchmark problems in terms of requiring less function evaluations for converging to the global optimum.

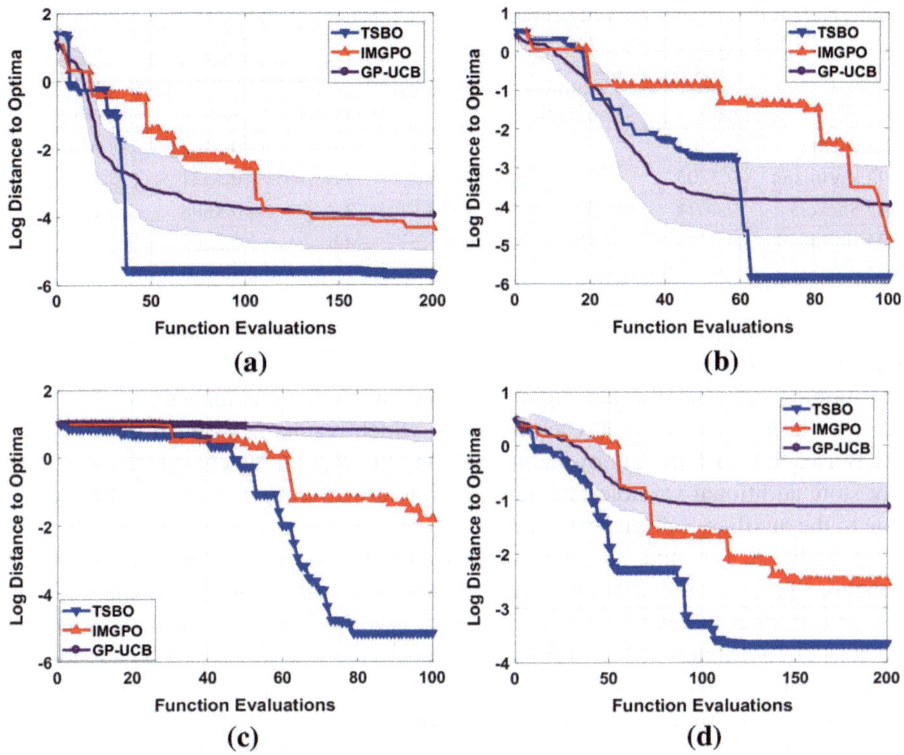

Fig. 17.7 Performance comparison for proposed algorithm, TSBO, on optimization challenge functions used in the literature. Functions are available in [19]. (**a**) 2D Branin function. (**b**) 3D Hartmann function. (**c**) 4D Shekel5 function. (**d**) 6D Hartmann function

Table 17.1 Algorithm progression times and accuracy within 100 iterations

	GP-UCB		IMGPO		TSBO	
	Run time (Var. memory)	Accuracy	Run time (Var. memory)	Accuracy	Run time (Var. memory)	Accuracy
2D Branin	10.38 s (0.99 Mb)	$10^{-3.8}$	1.28 s (1.65 Mb)	$10^{-2.5}$	6.15 s (2.85 Mb)	$10^{-5.6}$
3D Hartmann	10.52 s (0.10 Mb)	10^{-4}	1.35 s (1.66 Mb)	$10^{-4.8}$	4.35 s (2.01 Mb)	$10^{-5.8}$
4D Shekel5	11.35 s (0.10 Mb)	$10^{0.8}$	1.83 s (1.68 Mb)	$10^{-1.8}$	6.79 s (3.17 Mb)	$10^{-5.2}$
6D Hartmann	12.23 s (0.11 Mb)	$10^{-1.2}$	1.91 s (1.70 Mb)	$10^{-2.5}$	6.51 s (2.19 Mb)	$10^{-3.3}$

Furthermore, algorithm progression times (excluding function evaluation time) and total variable memory stored for 100 iterations are measured and compared in Table 17.1. Although TSBO achieves higher accuracy with less function

Table 17.2 Run time statistics of learning acquisition functions of TSBO for challenge functions

	PI		EI		UCB	
	Normalized gain	Number of calls	Normalized gain	Number of calls	Normalized gain	Number of calls
2D Branin	0.2669	17	0.3164	16	0.4167	67
3D Hartmann	0.3205	16	0.3475	68	0.3321	16
4D Shekel5	0.4074	22	0.4228	62	0.1698	16
6D Hartmann	0.3352	16	0.3461	68	0.3207	16

evaluations, it requires additional run time and memory to store and process the aforementioned $t2^d$ candidate points at each iteration. As IMGPO uses a partitioning scheme that has lighter weight on memory compared to TSBO, it requires lesser processing at each iteration, which results in faster run times. GP-UCB uses the least amount of memory among the algorithms considered as it does not store additional variables at each iteration; however, its run time is the highest due to the auxiliary optimization process.

In addition, run time statistics summarizing the effect of learning acquisition functions block of TSBO are provided in Table 17.2. Here, number of calls refer to number of times that particular $u(x)$ has been used to select x_{t+1}. It can be seen that each $u(x)$ contributes to the optimization process to find x^*; however, the amount of information gained from each $u(x)$ is different for each problem. As a result, the acquisition function providing the highest gain is called more by TSBO. Although the gains for each $u(x)$ are being dynamically updated after M observations as explained in Section 17.4.3, the choice of $u^*(x)$ at Mourad: M-th observation is only updated for the case of 4D-Shekel5 function as more observations are added.

The choice of hyperparameters can significantly affect the performance of black-box optimization algorithms, including TSBO. Here, problem specific knowledge can be leveraged to adjust these parameters. For instance, if the problem is believed to have large number of local extrema, the algorithm can be adjusted to give more importance to exploration than exploitation. For the case of BO-based algorithms, the effect of hyperparameters is a well-studied subject, resulting in rule of thumb values [13, 20, 21]. Counter-intuitively, these studies suggest dynamic modulation of these parameters have little to no-effect on empirical performance of the algorithms. Following these works, for these experiments, the hyperparameters of TSBO are chosen as: $\alpha = 0.1$ in Eq. (17.18); $\eta = 0.1$ in Eq. (17.11); $\zeta = 0$ and $\zeta = 0.01$ in Eqs. (17.9) and (17.10), respectively; $M = 50$ for learning acquisition functions and $N = 10$ for switching criteria. TSBO specific parameters such as N, M, and α can be modified accordingly to transfer domain expertise to the algorithm. For instance, if the design under consideration is to be fabricated using a process that cannot provide accuracy beyond micrometers, fine-tuning beyond this level using second stage of TSBO may result in invalid designs. Hence, N can be increased to force TSBO to spend more time in the first stage.

17.5 Co-optimization of Embedded Inductor and Integrated Voltage Regulator

The application chosen in this work is the multi-objective optimization of IVR with the goal of maximizing power conversion efficiency while minimizing area of embedded inductor. The architecture used is given in Fig. 17.8 and is a system in package (SiP) solution consisting of two chips (buck converter and LDO/load) with integrated inductor on an organic package [3]. The inductor structure is chosen as a solenoid with magnetic core as in Fig. 17.9a, b. Due to high levels of integration, maintaining high conversion efficiency while minimizing the area of inductor in the package becomes a major challenge that requires the handling of multiple trade-offs simultaneously.

In this section, we use TSBO to automate the design process. We briefly provide previously developed buck converter efficiency model followed by the embedded inductor characterization using two commercial EM solvers, namely Ansys HFSS [22] and Ansys Maxwell [23] for full-wave simulations at high frequency and at DC, respectively. Afterwards, we provide the optimization setup used and make a comparison of the optimized IVR design with hand-tuned design [3] as well as results generated using other non-ML and ML techniques such as non-linear solver, GP-UCB, and IMGPO.

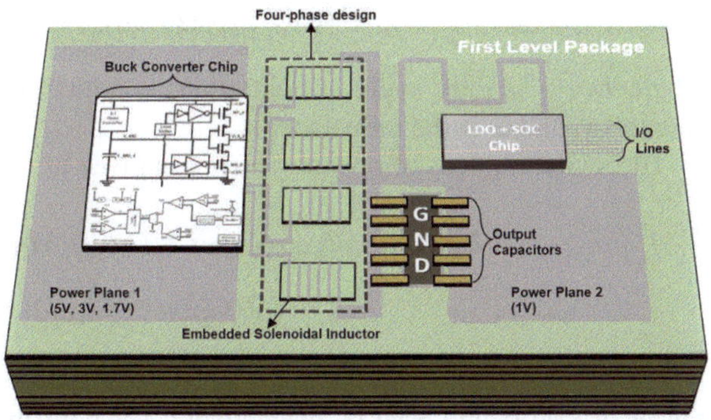

Fig. 17.8 Two-chip SiP IVR architecture

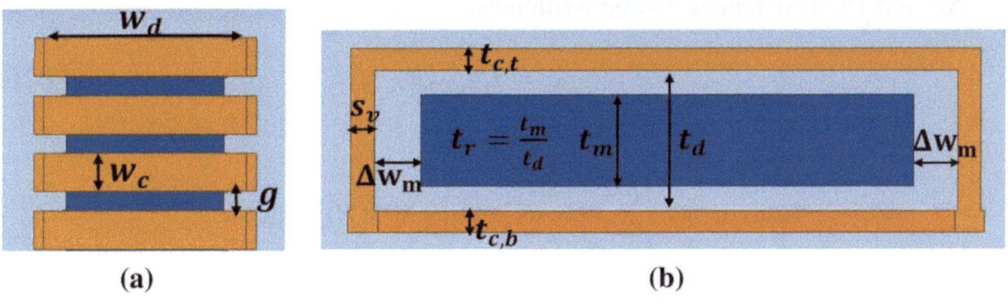

Fig. 17.9 Geometry of solenoidal inductor with magnetic core along with its control parameters. (**a**) Top view of inductor. (**b**) Side view of inductor

17.5.1 Buck Converter Efficiency Model

The buck converter shown in Fig. 17.8 is designed with stacked topology, using 130 nm GF process and consists of four phases (one master and three slaves) [3]. Efficiency calculations are based on the extensive model that accounts for switching and conduction losses of power switches, DC and AC losses of inductor, power delivery network (PDN), and output capacitance. Power switches used in the buck converter introduce two types of losses, namely switching and conduction. Conduction loss originates from the finite DC resistance of switches and can be written as:

$$L_{\mathrm{PS(COND)}} = R_{\mathrm{SW}}(I_{\mathrm{LOAD}}^2 + \Delta I_{\mathrm{RMS}}^2) \tag{17.20}$$

$$R_{\mathrm{SW}} = D_{\mathrm{CCM}} R_{\mathrm{PMOS}} + (1 - D_{\mathrm{CCM}}) R_{\mathrm{NMOS}} \tag{17.21}$$

where R_{NMOS} and R_{PMOS} are ON resistances of NMOS and PMOS, respectively, and ΔI_{RMS} is the RMS value of ripple current calculated as:

$$D_{\mathrm{CCM}} = V_{\mathrm{OUT}} / V_{\mathrm{IN}} \tag{17.22}$$

$$\Delta I_{pk-pk} = \frac{(V_{\mathrm{IN}} - V_{\mathrm{OUT}}) D_{\mathrm{CCM}}}{f_{\mathrm{SW}} L} \tag{17.23}$$

$$\Delta I_{\mathrm{RMS}} = \Delta I_{pk-pk} / 2\sqrt{3} \tag{17.24}$$

where D_{CCM} is duty cycle in continuous conduction mode, f_{SW} is the switching frequency, and L is the inductance.

The switching loss of power switches is given as:

$$\begin{aligned} L_{\mathrm{PS(SW)}} = & f_{\mathrm{SW}}\big[(C_{\mathrm{GSN}} + C_{\mathrm{GSP}})(V_{\mathrm{IN}}/2)^2 \ldots \\ & \ldots + 2(C_{\mathrm{GDN}} + C_{\mathrm{GDP}})(V_{\mathrm{IN}}/2)^2\big] \end{aligned} \tag{17.25}$$

where C_{GSP}, C_{GSN}, C_{GDP}, and C_{GDN} are gate-to-source and gate-to-drain capacitances of PMOS and NMOS, respectively. The contribution of the inductor in terms of AC and DC resistances can be written as:

$$L_{\mathrm{IND,DC}} = I_{\mathrm{LOAD}}^2 R_{L,\mathrm{DC}} \tag{17.26}$$

$$\begin{aligned} L_{\mathrm{IND,AC}} = & \sum_{f=-\infty}^{\infty} 2\frac{\Delta I_{pk-pk}^2}{4}\big[0.405^2 R_L(f_{\mathrm{SW}}) \ldots \\ & \ldots + 0.045^2 R_L(3 f_{\mathrm{SW}}) + 0.016^2 R_L(5 f_{\mathrm{SW}})\big] \end{aligned} \tag{17.27}$$

where $R_{L,\mathrm{DC}}$ and R_L are DC and frequency dependent effective series resistance of inductor, respectively; 0.405, 0.045, and 0.016 are Fourier series coefficients of fundamental frequency and its third and fifth order harmonics of a triangular current waveform. Resistive losses due to output capacitor and PDN are given as:

$$L_{\mathrm{CAP}} = \Delta I_{pk-pk}^2 \mathrm{ESR}_C \tag{17.28}$$

$$L_{\mathrm{PDN}} = I_{\mathrm{LOAD}}^2 R_{\mathrm{PDN}} \tag{17.29}$$

where R_{PDN} is effective DC resistance of PDN and ESR_C is effective series resistance of output capacitance. A more detailed description of the buck converter topology along with the model verification can be found in [3].

17.5.2 Embedded Inductor Characterization

Although there are approximate closed form expressions for preliminary design of embedded solenoidal inductors with magnetic core [24], a full-wave simulation is required to accurately account for eddy currents, proximity, and skin effect as well as demagnetization effect due to using magnetic cores.

Electrical characteristics of inductor have both direct and implied effects to the efficiency of the buck converter. The inductance needs to be sufficiently high in order to reduce the ripple current in Eq. (17.23), thereby reducing the conduction loss in Eq. (17.21). The ripple current also affects the AC loss in inductor along with ESR in Eq. (17.27). Although DC resistances of such inductors are in the range of milliohms, operating at higher currents (~10 A) introduces a substantial DC loss as in Eq. (17.26). Therefore, the complexity of the problem includes handling inductance, AC and DC resistance trade-offs and area constraints, while considering the direct and implied effects on buck converter efficiency. Even if the desired characteristics of the inductor is determined to handle these trade-offs, determining the dimensions of the inductor that will satisfy these characteristics requires substantial human intervention and CPU time due to the use of full-wave EM solvers.

In order to characterize the inductor behavior, the full-wave EM simulation is conducted to gather 2-port impedance matrix as a function of frequency. Using a pi-equivalent circuit as in Fig. 17.10, the series inductance, L_s, and AC resistance, R_s, are calculated from the Z-matrix as:

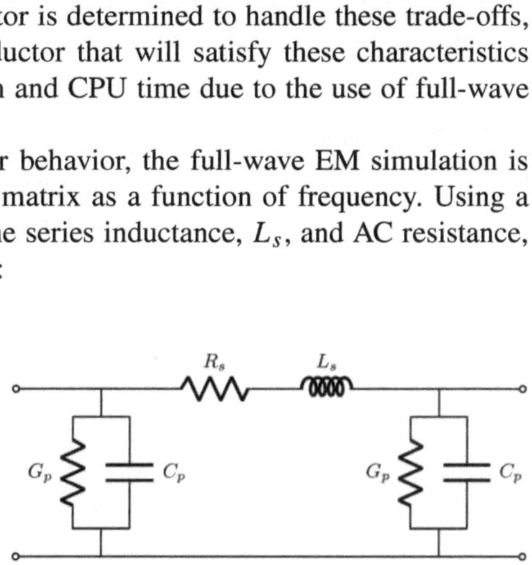

Fig. 17.10 π-equivalent inductor model

$$L_s = \frac{2Im\{Z_{11} - Z_{12}\}}{w}, \qquad R_s = 2Re\{Z_{11} - Z_{12}\} \tag{17.30}$$

where w is the angular frequency. In Fig. 17.10, both G_p and C_p are quite small and their effects are ignored. Finally, Ansys Maxwell [23] is used to accurately simulate the DC resistance of inductor to be used in Eq. (17.26) of the efficiency model.

In this work, we consider two different magnetic materials, namely carbonyl iron powder (CIP) and nickel–Zinc (NiZn). Between the two materials, NiZn has higher permeability ($\mu_r = 8.11 + j2.27$ at 100 MHz), which provides higher inductance values at reduced sizes, which decreases ripple current and DC resistance simultaneously and increases the efficiency. On the other hand, CIP has lower magnetic loss tangent ($\mu_r = 5.64 + j0.57$ at 100 MHz), which results in increased efficiency by reducing AC loss in Eq. (17.27). To accurately account for magnetic material effect on IVR efficiency, measured frequency dependent complex permittivity and permeability of both CIP and NiZn from [25] are imported into the EM solver for accuracy.

17.5.3 Optimization Setup

Among the components of the IVR, the focus in this work is on the optimization of inductor to maximize IVR efficiency. Ten control parameters are defined for the inductor as shown in Table 17.3 along with the range used for each parameter based on process capabilities. The objective function used for optimization is:

$$f(x) = \sum_{i=1}^{2} w_i y_i \tag{17.31}$$

where y_1 and y_2 are peak overall IVR efficiency calculated using the described model and area of inductor, respectively; $w_1 = 5$ and $w_2 = -2$ are corresponding

Table 17.3 Control parameters of solenoidal inductor

Parameter		Unit	Min	Max
Gap between windings	g	mil	2	20
Number of windings	N		3	13
Size of via	s_v	µm	50	103
Copper trace width	w_c	mil	2	20
Copper thickness bottom	$t_{c,b}$	µm	35	170
Copper thickness top	$t_{c,t}$	µm	35	170
Dielectric thickness	t_d	µm	50	650
Dielectric width	w_d	µm	50	350
Magnetic core thickness ratio	t_m		0.1	1
Magnetic core width offset	Δw_m	mil	0	100

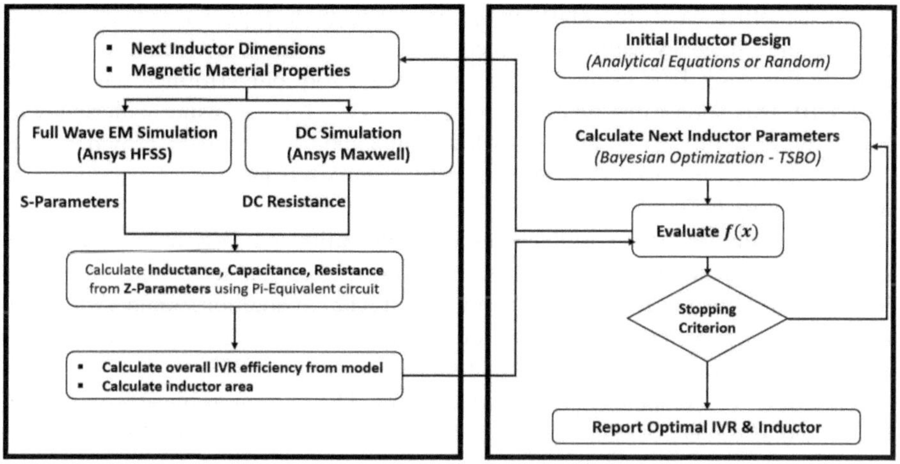

Fig. 17.11 Automated optimization setup used in IVR application

weights of multi-objective optimization. Since IVR efficiency is the main goal of the optimization process, it has higher weight compared to the inductor area. By choosing such weights, the trade-off between efficiency and area is adjusted to choose 2% of power efficiency over 5 mm^2 of the inductor area.

To investigate opportunities for reducing fabrication costs, two types of inductors are analyzed: one that uses only 1 oz thick copper (**Type I**) and the more costlier option that uses copper thickness up to 170 μm (**Type II**). In the case of Type II inductor, the optimization is done with 10 parameters and for Type I, only 8 parameters are used since $t_{c,b}$ and $t_{c,t}$ are fixed to 35 μm [26].

The automated optimization setup used is given in Fig. 17.11. Inductor dimensions are determined using TSBO and fed into the full-wave solver to extract the 2-port Z-matrix, which is then used by the inductor and buck converter models for calculating IVR efficiency. Calculated efficiency is combined with inductor area in Eq. (17.31) and fed back into TSBO to proceed to the next iteration.

17.5.4 Results

To make a direct comparison with hand-tuned design in [3] (which took several months to complete), power efficiencies are calculated assuming $R_{PDN} = 10\,m\Omega$ and $ESR_C = 10\,m\Omega$ in Eqs. (17.29) and 17.28. Among four optimized IVRs, the one using Type II inductor with CIP showed the best performance, providing efficiency of 85.1%, 93.1%, and 94.1% for 5 V:1 V, 3 V:1 V, and 1.7 V:1.05 V conversions, respectively, with an inductor area of 5.1 mm^2. Compared to hand-tuned design using NiZn in [3], the efficiency is increased by 5.7%, 4.5%, and 3% accompanied with 55.3% reduction in inductor area.

On the other hand, the cheaper option, IVRs with Type I inductors showed comparable efficiency with hand-tuned design but with 48.2% reduced inductor area. In this case, IVR using inductor with NiZn core outperformed the inductor with CIP. This shows that when copper thickness is limited, using materials with higher permeability as compared to lower magnetic loss tangent provides more opportunities to increase IVR efficiency by decreasing number of windings and hence, DC loss in Eq. (17.26).

Comparison for power efficiencies of each optimized IVR as a function of switching frequency at $I_{LOAD} = 10\,A$ (2.5 A per phase) and as a function of total load current at $f_{SW} = 10\,MHz$ is given in Fig. 17.12. In addition, Table 17.4 shows the inductor characteristics of each optimized IVR as well as peak conversion efficiencies and compares it to the hand-tuned design from [3].

Figure 17.13 compares the performance of TSBO to non-linear solver, GP-UCB, and IMGPO for the objective of maximizing Eq. (17.31), along with the corresponding value of hand-tuned design. Optimization using TSBO resulted in

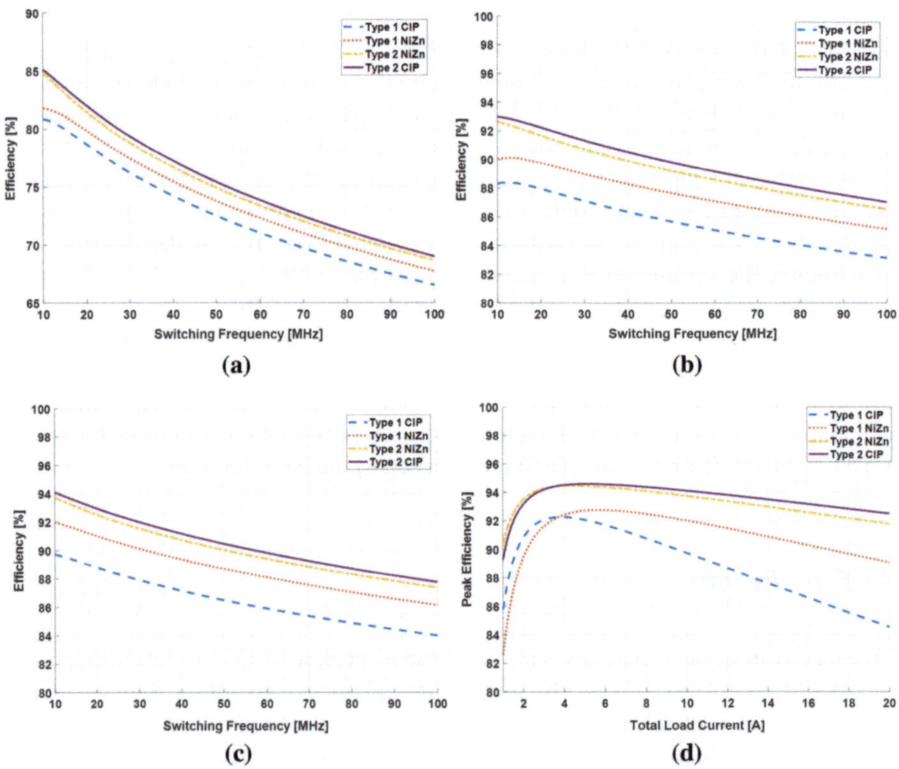

Fig. 17.12 Efficiency comparison of four optimized IVRs as a function of switching frequency and total load current. (**a**) 5 V:1 V conversion ($I_{LOAD} = 10\,A$). (**b**) 3 V:1 V conversion ($I_{LOAD} = 10\,A$). (**c**) 1.7 V:1.05 V conversion ($I_{LOAD} = 10\,A$). (**d**) Eff. vs load current ($f_{SW} = 10\,MHz$)

Table 17.4 Comparison of optimized IVRs and inductors to hand-tuned design

	Hand-tuned [3]	Type I NiZn	Type I CIP	Type II NiZn	Type II CIP
L [nH]	24.8	13.32	15.4	29.4	23.47
R_{AC} [Ω]	2.27	0.87	0.51	2.67	0.98
R_{DC} [mΩ]	14.7	17.0	30.1	10.5	8.7
Area [mm^2]	11.6	6.0	6.1	5.2	5.1
Peak eff. 5 V:1 V	79.4%	81.8%	80.9%	84.9%	85.1%
Peak eff. 3 V:1 V	88.5%	88.4%	90.1%	92.7%	93.0%
Peak eff. 1.7 V:1.05 V	91.1%	92.0%	89.7%	93.7%	94.1%

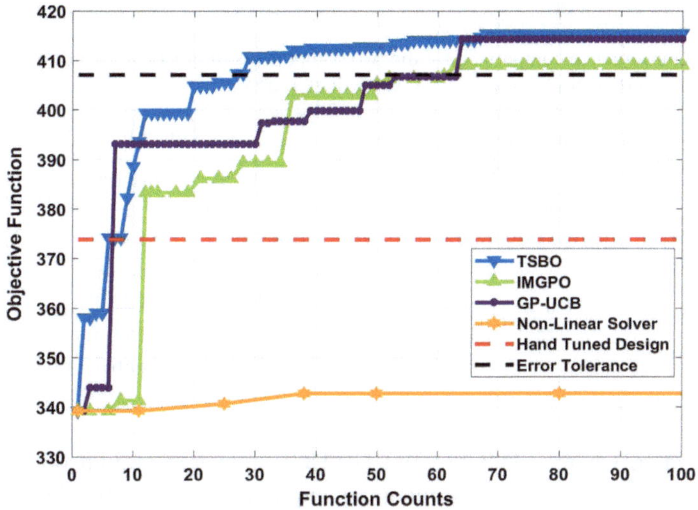

Fig. 17.13 Performance comparison of TSBO to non-linear solver, GP-UCB, and IMGPO on maximizing objective function at Eq. (17.31)

85.1% peak efficiency for 5V:1V conversion with the inductor area of 5.16 mm^2, compared to 78.6%, 84.4%, and 84.9% with 25.2 mm^2, 6.64 mm^2, and 5.18 mm^2 for non-linear solver, IMGPO, and GP-UCB, respectively. Though all algorithms started from the same initial point, TSBO reached the pre-determined error tolerance in Fig. 17.13 using 27 simulations (51.1 min), compared to 60 and 59 simulations (115.6 and 117.3 min) for IMGPO and GP-UCB, corresponding to a reduction of 56.7% and 57.4% in CPU time. Besides, optimization using non-linear solver resulted in 78.6% peak efficiency with the inductor area of 25.2 mm^2 and could not reach the error tolerance within 100 simulations (185 min) and lead to a significantly worse design compared to hand-tuned design.

Figure 17.14 shows the breakdown of objective function in Eq. (17.31) to its two components of peak efficiency and inductor area. This breakdown shows that the TSBO converges in around 30 simulations resulting in 84.5% efficiency with

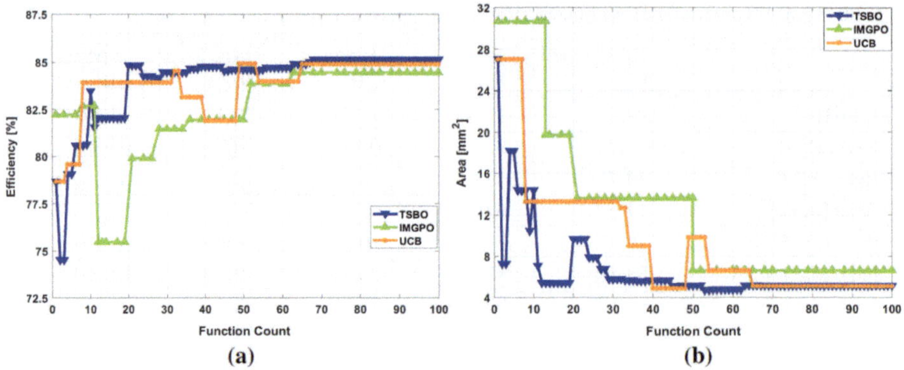

Fig. 17.14 Convergence comparison for the two components of objective function in Eq. (17.31). (**a**) Peak 5 V:1 V efficiency at I_{LOAD}=10 A. (**b**) Inductor area

Table 17.5 Optimization results for IVR using type II inductor with CIP

	Non-linear solver	GP-UCB	IMGPO	TSBO
Peak eff. 5V:1V (f_{SW}=10 MHz, I_{Load}=10 A)	78.6%	84.9%	84.4%	85.1%
Area	25.2 mm^2 (+79.6%)	5.18 mm^2 (+0.4%)	6.64 mm^2 (+22.3%)	5.16 mm^2
CPU time	>185 min (>+72.9%)	117.3 min (+57.4%)	115.6 min (+56.7%)	50.1 min

5.78 mm^2 area, whereas IMGPO and GP-UCB converge around 60 simulations. This shows that the number of simulations required for each algorithm to cross the tolerance level in Fig. 17.13 coincides with their convergence point, hence, validates the CPU time comparison made in Table 17.5.

17.6 Uncertainty Quantification

17.6.1 Introduction

Starting from initial information on the input parameters (i.e., with a very large range of values) of the system, the TSBO algorithm allows to identify the optimum design point of the system. However, input parameters of the system (e.g., temperature variations, geometrical and electrical tolerances, etc.), may present variability due to the manufacturing process. Therefore, the impact of these uncertain parameters has to be evaluated.

In this framework, the brute force Monte Carlo (MC) technique cannot be applied as it requires a large number of simulations to provide an accurate estimation of the output variability. To overcome this limitation, a surrogate model, which aims at replacing the numerical model by an analytic function, has to be developed. Among the solutions available, methods based on the polynomial chaos (PC) have been successfully applied in many engineering fields [27–30]. However, when the number of input random variables increases, the number of calls to the numerical model blows up. For this reason, advanced techniques have to be introduced to address the *curse of dimensionality* problem [9, 30–32].

In this work, we present a *sparse PC* approach based on the *least angle regression* (LARS) algorithm [9] in order to quantify uncertainties in high-dimensional problems. First, the method is introduced. Then, the approach is applied to the analysis of IVR efficiency taking into account uncertainties in the vicinity of the optimal design point identified by the TSBO technique. In addition, we provide a hierarchization of the uncertainties effect on the variability of the output.

17.6.2 Polynomial Chaos Expansion

17.6.2.1 Introduction

Let us consider a random vector $\boldsymbol{X}$ of joint probability density function (PDF) $f_{\boldsymbol{X}}(\boldsymbol{x})$, whose M random variables $(X_1, \ldots, X_M)$ are assumed to be independent and representing the input uncertainties of the problem. Suppose that $Y = \mathcal{M}(\boldsymbol{X})$ is the random scalar response of a numerical model $\mathcal{M}$ describing the physical system. Assuming that the random response Y has a finite variance, it may be written as [33]:

$$Y = \sum_{\boldsymbol{\lambda} \in \mathbb{N}^M} a_{\boldsymbol{\lambda}} \Phi_{\boldsymbol{\lambda}}(\boldsymbol{X}), \tag{17.32}$$

where $a_{\boldsymbol{\lambda}}$ are the unknown coefficients and $\Phi_{\boldsymbol{\lambda}}$ represent a basis of multivariate polynomials satisfying the orthonormality property regarding the joint PDF $f_{\boldsymbol{X}}(\boldsymbol{x})$, i.e., $\mathbb{E}\left[\Phi_{\boldsymbol{\lambda}}(\boldsymbol{X})\Phi_{\boldsymbol{\beta}}(\boldsymbol{X})\right] = \delta_{\boldsymbol{\lambda}\boldsymbol{\beta}}$, with $\delta_{\boldsymbol{\lambda}\boldsymbol{\beta}} = 1$ if $\boldsymbol{\lambda} = \boldsymbol{\beta}$ and 0 otherwise. In practice, families of orthonormal polynomials are associated in terms of probability distributions of input random variables (e.g., Legendre and Hermite polynomials are chosen for Gaussian and uniform input random variables, respectively).

Let $\mathcal{X} = \{\boldsymbol{x}^{(1)}, \ldots, \boldsymbol{x}^{(n)}\}$ be an *experimental design* (ED) of $\boldsymbol{X}$, and $\mathcal{Y} = \{\mathcal{M}(\boldsymbol{x}^{(1)}), \ldots, \mathcal{M}(\boldsymbol{x}^{(n)})\}$ be the associated set of model response quantities. Relying on this set of model evaluations, the PC coefficients $a_{\boldsymbol{\lambda}}$ may be estimated by using non-intrusive approaches such as projection and regression methods [30]. In many applications, the regression techniques showed a quicker convergence in terms of evaluation of the numerical model $\mathcal{M}$ [30]. Among them, the ordinary least square regression provides good results [34].

17.6.2.2 From Standard to Sparse PC Expansion

Since the coefficients of the PC expansion have to be computed, the infinite series of (17.32) has to be truncated. The standard truncation scheme adapted in the PC expansion retains the polynomial bases whose total degree is less than or equal to l, i.e., the truncation set $\mathscr{A}^{M,l} = \{\boldsymbol{\lambda} \in \mathbb{N}^M : \|\boldsymbol{\lambda}\|_1 = \sum_{i=1}^{M} \lambda_i \leq l\}$. With this truncation scheme, the cardinal of the set $\mathscr{A}^{M,l}$ is then $L = \dfrac{(M+l)!}{M!\,l!}$. We notice that number of coefficients in the set $\mathscr{A}^{M,l}$ becomes excessively high for large number of input random variables M and high degree l. For this reason, an improved truncation strategy is needed.

A hyperbolic truncation scheme based on a parameter k ($0 < k \leq 1$) is given by [9]:

$$\mathscr{A}^{M,l,k} = \{\boldsymbol{\lambda} \in \mathbb{N}^M : \|\boldsymbol{\lambda}\|_k = \left(\sum_{i=1}^{M} \lambda_i^k\right)^{1/k} \leq l\}. \tag{17.33}$$

This advanced truncation strategy favors the most relevant effects and low-order interactions, which are known to have the largest impact on the variability of the response according to the *sparsity-of-effects principle* [35]. It is important to note that the low the value of k, the more high-rank interactions will be neglected. Furthermore, when $k = 1$, this scheme is equivalent to the standard PC approximation defined by the truncation set $\mathscr{A}^{M,l}$. When $k < 1$, the retained terms of the polynomial basis can be significantly reduced compared to L [9]. This advanced truncation scheme for two input random variables ($M = 2$) is illustrated in Fig. 17.15a, b, with the squares representing all terms of the polynomial basis of degree less than or equal to $l = 5$, included in the set (17.33) for $k = 1$ (squares) and $k = 0.5$ (squares). Thus, Fig. 17.15b shows that when $k = 0.5$, this strategy consists of selecting a number of polynomials (squares) much smaller than those retained by the standard truncation set $\mathscr{A}^{M,l}$ (squares) in Fig. 17.15a.

The hyperbolic truncation scheme, given by a truncation set $\mathscr{A}^{M,l,k}$ of cardinal K, allows to reduce the number of coefficients to be estimated in the PC expansion. However, this may be still too costly in terms of calls to the numerical model when dealing with high-dimensional problems. In order to overcome this situation, the remaining number of terms of the polynomial basis may be further reduced by using a variable selection algorithm, such as the LARS algorithm [36].

The principle of LARS is briefly presented in the following paragraph while the reader may refer to [9] and [36] for additional details. From an iterative process, LARS builds up a sparse representation including from 1 to K elements of the polynomial basis (from one to all the squares in Fig. 17.15b), in accordance with their decreasing influence. The algorithm begins by choosing the polynomial basis $\Phi_{\boldsymbol{\lambda}_1}$, having the best correlation with the model response Y. In practice, the correlation is evaluated from a set of realizations (i.e., an ED) of the response $\mathscr{Y}$. After the identification of the first polynomial $\Phi_{\boldsymbol{\lambda}_1}$, the associated coefficient is

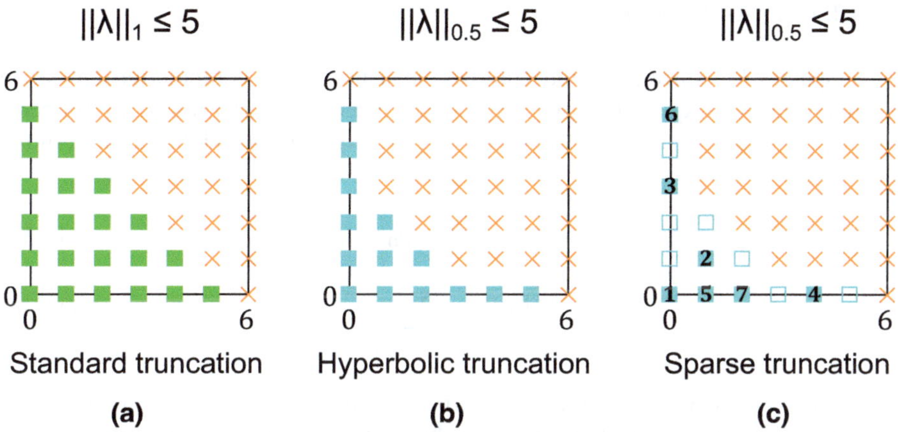

Fig. 17.15 Elements of the polynomial basis of degree up to $l = 5$ retained by the hyperbolic truncation strategy when (**a**) $k = 1$ (squares) and (**b**) $k = 0.5$ (squares). (**c**) Numbered squares are the polynomial basis elements identified by the LARS algorithm from the 1st to the 7th iteration [9]

estimated so that the residual $Y - a^{(1)}_{\boldsymbol{\lambda}_1}\Phi_{\boldsymbol{\lambda}_1}(\boldsymbol{X})$ becomes equi-correlated with another polynomial basis, considered as being $\Phi_{\boldsymbol{\lambda}_2}$. Then, the selection of a third polynomial basis will be carried out by moving along the direction $(\Phi_{\boldsymbol{\lambda}_1} + \Phi_{\boldsymbol{\lambda}_2})$, until the new residual becomes equi-correlated with a third polynomial basis $\Phi_{\boldsymbol{\lambda}_3}$, and so on. A representation of the selected polynomials by LARS after seven iterations is given by squares in Fig. 17.15c.

Estimating suitable terms of the polynomial basis using LARS is performed for each degree $l = 1, 2, \ldots, l_{\max}$. Finally, LARS produces a set of sparse expansions including an increasing number of polynomial terms. The quality of each expansion of degree l is then evaluated according to a leave-one-out cross-validation error ϵ_{LOO} as follows:

$$\epsilon_{\text{LOO}} = \frac{\sum_{i=1}^{N}\left(\mathcal{M}(\boldsymbol{x}^{(i)}) - \mathcal{M}^{\text{PC}}_{-i}(\boldsymbol{x}^{(i)})\right)^2}{\sum_{i=1}^{N}\left(\mathcal{M}(\boldsymbol{x}^{(i)}) - \frac{1}{N}\sum_{i=1}^{N}\mathcal{M}(\boldsymbol{x}^{(i)})\right)^2}, \tag{17.34}$$

where $\mathcal{M}^{\text{PC}}_{-i}(\boldsymbol{x}^{(i)})$ are N surrogate models built up on the ED $\mathcal{X} \setminus \boldsymbol{x}^{(i)} = \{\boldsymbol{x}^{(q)}, q = 1, \ldots, N,\ q \neq i\}$. The end result is that the retained degree l is the one minimizing the leave-one-out error ϵ_{LOO}.

In the following, the quality of the surrogate model is computed via the Q^2 coefficient defined by $Q^2 = 1 - \epsilon_{\text{LOO}}$, $0 \leq Q^2 \leq 1$. It is worth pointing out that the larger Q^2 is, the better is the prediction of the surrogate model.

17.6.2.3 Post-Processing

The (sparse) PC surrogate model allows, at a negligible computational cost, to estimate statistical quantities of the response. Indeed, the orthonormality property of the polynomial basis enables one to derive the expectation and the variance of the response Y as follows:

$$\mathbb{E}\left[Y\right] = a_0 \tag{17.35}$$

$$\mathbb{V}\left[Y\right] = \sum_{\boldsymbol{\lambda} \in \mathscr{A} \setminus \{\mathbf{0}\}} a_{\boldsymbol{\lambda}}^2 \tag{17.36}$$

Besides, it is also possible to obtain a sensitivity analysis from the *Sobol indices* [37], which aim at quantifying the impact of input uncertainties on the response variability. Thereby, the *PC-based Sobol indices* of *first-order* S_i allows the estimation of the variance of the model response Y with respect to the main effect of a single input random variable X_i as:

$$S_i = \frac{\sum_{\boldsymbol{\lambda} \in \mathscr{A}_i} a_{\boldsymbol{\lambda}}^2}{\mathbb{V}\left[Y\right]} \tag{17.37}$$

where $\mathscr{A}_i = \{\boldsymbol{\lambda} \in \mathscr{A} : \lambda_i > 0, \lambda_j = 0 \;\; \forall j \neq i\}$. The *total indices* $S_{T,i}$ of the response Y, which compute the influence of an input random variable X_i considering all interactions with other input random variables X_j $(j \neq i)$, are computed according to [38]:

$$S_{T,i} = \frac{\sum_{\boldsymbol{\lambda} \in \mathscr{A}_{T,i}} a_{\boldsymbol{\lambda}}^2}{\mathbb{V}\left[Y\right]} \tag{17.38}$$

where $\mathscr{A}_{T,i} = \{\boldsymbol{\lambda} \in \mathscr{A} : \lambda_i \neq 0\}$, respectively.

17.7 Uncertainty Quantification of the IVR Efficiency

17.7.1 Surrogate Model of the Model Response

In this section, we aim at evaluating, around the optimum design point identified by the TSBO algorithm (see Sect. 17.5.4), the effect of uncertainties on the IVR efficiency E_{ff}. For instance, we choose to analyze the inductor equipped with a Type 2 NiZn solenoid for 5 V:1 V conversion at the frequency of 100 MHz. For this structure, the TSBO algorithm provides an optimal efficiency of 68.7% as shown in Fig. 17.12. From this optimal design point, the randomness of the IVR is introduced through seven uniform random variables given in Table 17.6. Moreover, the number

Table 17.6 Uncertain parameters of solenoidal inductor

Uniform random variables		Unit	$\mathscr{U}$[Min; Max]
Gap between windings	g	mil	$\mathscr{U}$ [1.2; 3.5]
Size of via	s_v	µm	$\mathscr{U}$ [122.2; 183.4]
Copper trace width	w_v	mil	$\mathscr{U}$ [15.5; 18.0]
Copper thickness bottom	$t_{c,b}$	µm	$\mathscr{U}$ [122.9; 184.3]
Copper thickness top	$t_{c,t}$	µm	$\mathscr{U}$ [88.0; 132.0]
Dielectric thickness	t_d	µm	$\mathscr{U}$ [266.6; 325.8]
Dielectric width	w_d	mil	$\mathscr{U}$ [54.3; 56.5]

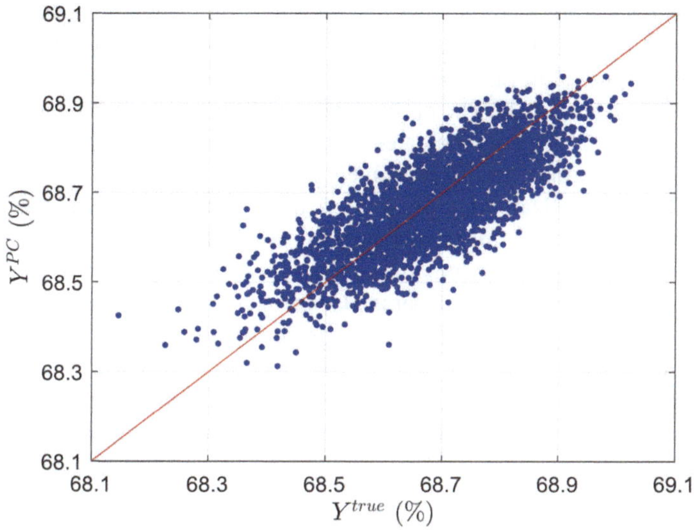

Fig. 17.16 Points represent the predictions obtained by the sparse PC surrogate model (Y-axis) against the real values of the numerical model (X-axis) given by MC sampling. The straight line highlights a perfect agreement between the two models

of windings, the magnetic core width offset, and the magnetic core thickness ratio are set to their optimal values, i.e., $N = 7$, $\Delta w_m = 0$ mil, and $t_r = 1$, respectively.

We thus build up a sparse PC with 200 realizations from Latin hypercube sampling (LHS) [39] and an adaptive degree h varying from 1 to 10 with [40]. The k-quasi-norms (17.33) parameter is fixed to 0.75, in order to significantly reduce the size of the polynomial basis and to properly approximate the model response [9].

The estimated quality of the sparse PC surrogate model obtained is $Q^2 = 0.70$. In order to illustrate it, we represent in Fig. 17.16, from 3000 MC realizations, the predictions Y^{PC} (Y-axis) of the sparse PC against the real values Y^{true} (X-axis) of the numerical model. We notice then that the general tendency of predictions obtained by the sparse PC surrogate model is relatively good as the samples (points) follow the straight line, which represents a perfect agreement between the surrogate and numerical models. We also observe that few samples are not well estimated in the area where Y^{true} varies between 68.1% and 68.3%, emphasizing a lower quality of the approximation of the surrogate model in that region. As the variability of the IVR efficiency is evaluated, we illustrate in Fig. 17.17, from 10,000 MC realizations, the PDFs of the efficiency E_{ff} obtained by the sparse PC (dashed line) and by MC simulation (solid line). Figure 17.17 shows that the main trend of the two curves is relatively close, in spite of differences at the level of the tails of PDFs. Indeed, we see that the sparse PC surrogate model overestimates the low values of the model response, which is illustrated by a shorter left tail of the PDF. In turn, the right tail of the PDF estimated by the sparse PC is more accurate, with a better estimation of the large values of the model response compared to the numerical model. Regarding the first two statistical moments of the IVR efficiency, the mean and the standard deviation of E_{ff} computed by sparse PC and by MC simulation are quite close, i.e., $\mu^{PC} = 68.68\%$, $\sigma^{PC} = 0.11\%$ and $\mu^{MC} = 68.68\%$, $\sigma^{MC} = 0.13\%$, respectively. The optimal degree of the sparse PC is 3 with a sparsity index $S = \frac{9}{43} \approx 21\%$. This indicates that few terms of the polynomial basis are sufficient to approximate the IVR efficiency E_{ff}. As regards to the computational time, 10,000 MC realizations took about 7 days and 4 h, whereas the sparse PC needed around 1 s, which highlights a speed-up of the surrogate model of about 619,200× compared to MC simulation. Note that this comparison does not include the training time required by the sparse PC surrogate model, which has taken about 3 h 27 min for the 200 LHS simulations (the surrogate model has been built up in less than 1 s).

Otherwise, the quantification of the output variability provided by the sparse PC surrogate model is very interesting from the design aspect. Indeed, the sparse PC

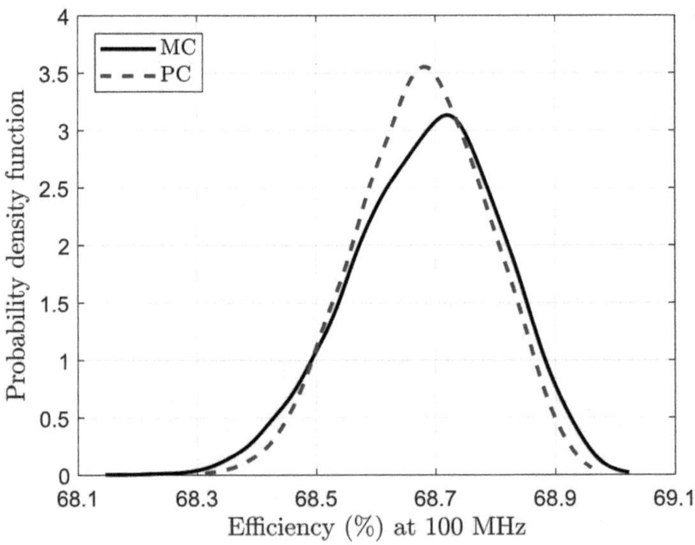

Fig. 17.17 PDFs of the efficiency E_{ff} obtained by sparse PC (dashed line) and by MC simulation (solid line) at the frequency of 100 MHz

allows to show that the design point identified by the TSBO algorithm, providing an IVR efficiency of 68.7%, is quite robust as the uncertainties of the system generate a relatively low variation of the IVR efficiency, i.e., from approximately 68.3% to 69% (see Fig. 17.17). Thus, the designer could be satisfied by its system as it will maintain a high level of performance.

17.7.2 Sensitivity Analysis of the Model Response

Apart from the estimation of the variability of the model response, the sparse PC surrogate model provides the sensitivity analysis of the output at a negligible computational cost. A histogram illustrating the total Sobol indices of E_{ff} at the frequency of 100 MHz is given in Fig. 17.18. This sensitivity analysis shows that the variability of E_{ff} is mainly related to the copper thickness top $t_{c,t}$ ($\approx$ 40%), the size of via s_v ($\approx$ 22%), the copper thickness bottom $t_{c,b}$ ($\approx$ 20%), and the copper trace width w_v ($\approx$ 18%) of the inductors. The rest of the variables, such as the dielectric thickness t_d, the dielectric width w_d, and the gap between windings g of the inductors, have a negligible effect on the variability of E_{ff}. Moreover, at 100 MHz, total losses of the IVR system are dominated by switching and conduction losses of power switches and the resistive loss due to the PDN [3]. Since the DC resistance of the inductor directly contributes to the finite resistance of the PDN, its AC losses become less significant. Therefore, the parameters of the inductor directly affecting its DC resistance become more important and have more effect on

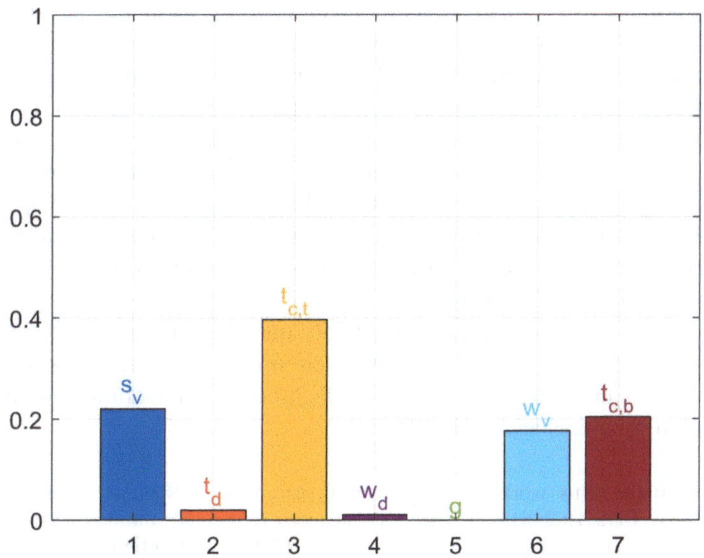

Fig. 17.18 Total Sobol indices of the efficiency E_{ff} at the frequency of 100 MHz

the IVR efficiency. For this reason, the sensitivity analysis illustrated in Fig. 17.18 is consistent as it identifies the more significant effect of the parameters that directly impact the DC resistance than the other parameters related to the magnetic material.

17.8 Conclusions

In this chapter, we have presented an EDA oriented Bayesian optimization (BO)-based global optimization algorithm, *two-stage Bayesian optimization*. Unlike conventional BO algorithms, the presented method is not prone to initial point selection and does not rely on other techniques for auxiliary optimization of the acquisition function. Furthermore, we have proposed and used a new hierarchical partitioning scheme that makes the algorithm EDA oriented in terms of substantially reducing number of simulations required to reach the global optima. Empirical analysis on a set of popular challenge functions with several local extrema and dimensions shows TSBO to have a faster convergence trend over other widely used methods.

For the IVR application considered in this chapter, we have shown that TSBO optimized IVR has 56.1% reduced inductor area compared to hand-tuned design and 5.7% increased peak efficiency for 5V:1V conversion at I_{LOAD} = 10 A and f_{SW}=10 MHz. Additionally, TSBO showed 72.4%, 57.4%, and 56.7% reduction in CPU time required to complete optimization compared to non-linear solver and the state-of-the-art ML-based GP-UCB and IMGPO techniques.

After the optimization phase, the sparse PC surrogate model was applied to estimate the IVR efficiency considering input uncertainties of the system. Based on a limited number of calls to the numerical model, the technique provided an estimation of the global variability of the output with a relative low computational cost in comparison with MC simulation. Moreover, it provides information associated with the model response such as the first two statistical moments and its sensitivity analysis, which can be useful for designers. Using this information, the engineer can quantify the variability of its output and identify the uncertain input variables having the largest impact. The variability of the parameters with the largest impact can then be tightened to ensure a robust design. For instance, our case study has shown that four uncertain input variables out of seven impact the variability of the inductor and IVR model response.

Although, the optimization and uncertainty quantification methods described in this chapter have been applied to IVR, their use can be quite generic. In fact, these methods can be applied to other black-box optimization and uncertainty quantification problems arising in EDA.

Acknowledgements This work was supported in part by the National Science Foundation under Grant No. CNS 16-24810—Center for Advanced Electronics through Machine Learning (CAEML) and Power Delivery for Electronic Systems (PDES) Consortium, Georgia Tech, USA. The authors would also like to acknowledge Dr. Mohamed L. F. Bellaredj, Dr. Anto K. Davis, and Dr. Colin Pardue for their assistance in simulation and analysis of the IVR.

References

1. N. Sturcken, E.J. O'Sullivan, N. Wang, P. Herget, B.C. Webb, L.T. Romankiw, M. Petracca, R. Davies, R.E. Fontana, G.M. Decad, I. Kymissis, A.V. Peterchev, L.P. Carloni, W.J. Gallagher, K.L. Shepard, A 2.5d integrated voltage regulator using coupled-magnetic-core inductors on silicon interposer. IEEE J. Solid-State Circuits **48**(1), 244–254 (2013)
2. N. Sturcken, M. Petracca, S. Warren, P. Mantovani, L.P. Carloni, A.V. Peterchev, K.L. Shepard, A switched-inductor integrated voltage regulator with nonlinear feedback and network-on-chip load in 45 nm SOI. IEEE J. Solid-State Circuits **47**(8), 1935–1945 (2012)
3. S. Mueller, K.Z. Ahmed, A. Singh, A.K. Davis, S. Mukhopadyay, M. Swaminathan, Y. Mano, Y. Wang, J. Wong, S. Bharathi, H.F. Moghadam, D. Draper, Design of high efficiency integrated voltage regulators with embedded magnetic core inductors, in *2016 IEEE 66th Electronic Components and Technology Conference (ECTC)* (2016), pp. 566–573
4. H.M. Torun, M. Swaminathan, Black box optimization of 3D integrated systems, in *Computational Modelling of Multi Uncertainty and Multi-Scale Problems (COMUS)* (2017)
5. H.M. Torun, M. Swaminathan, A.K. Davis, M.L.F. Bellaredj, A global Bayesian optimization algorithm and its application to integrated system design. IEEE Trans. Very Large Scale Integr. VLSI Syst. **26**(4), 792–802 (2018)
6. H.M. Torun, M. Swaminathan, Black box optimization of 3D integrated systems using machine learning, in *2017 IEEE 26th Conference on Electrical Performance of Electronic Packaging and Systems, EPEPS* (2017)
7. H.M. Torun, C.A. Pardue, M.L.F. Belladredj, A.K. Davis, M. Swaminathan, Machine learning driven advanced packaging and system miniaturization of IoT for wireless power transfer solutions, in *IEEE 68th Electronic Components Technology Conference, ECTC* (2018)
8. H.M. Torun, M. Larbi, M. Swaminathan, A bayesian framework for optimizing interconnects in High-Speed channels, in *2018 IEEE MTT-S International Conference on Numerical Electromagnetic and Multiphysics Modeling and Optimization (NEMO2018)*, Reykjavik, Iceland (2018)
9. G. Blatman, B. Sudret, Adaptive sparse polynomial chaos expansion based on least angle regression. J. Comput. Phys. **230**(6), 2345–2367 (2011)
10. M. Larbi, I.S. Stievano, F.G. Canavero, P. Besnier, Variability impact of many design parameters: the case of a realistic electronic link. IEEE Trans. Electromagn. Compat. **60**(1), 34–41 (2018). https://doi.org/10.1109/TEMC.2017.2727961
11. S.J. Park, B. Bae, J. Kim, M. Swaminathan, Application of machine learning for optimization of 3-D integrated circuits and systems. IEEE Trans. Very Large Scale Integr. VLSI Syst. **25**(6), 1856–1865 (2017)
12. K. Kawaguchi, L.P. Kaelbling, T. Lozano-Pérez, Bayesian optimization with exponential convergence. Adv. Neural Inf. Process. Syst. 2809–2817 (2015)
13. E. Brochu, V.M. Cora, N. de Freitas, A tutorial on Bayesian optimization of expensive cost functions, with application to active user modeling and hierarchical reinforcement learning. arXiv.org, arXiv:1012.2599 (2010)
14. N. Srinivas, A. Krause, S.M. Kakade, M. Seeger, Gaussian process optimization in the bandit setting: no regret and experimental design. arXiv:0912.3995 (2009)
15. J.R. Gardner, M.J. Kusner, Z.E. Xu, K.Q. Weinberger, J.P. Cunningham, Bayesian optimization with inequality constraints, in *Proceedings of the 31st International Conference on Machine Learning, ICML* (2014), pp. 937–945
16. Z. Wang, B. Shakibi, L. Jin, N. Freitas, Bayesian multi-scale optimistic optimization, in *Artificial Intelligence and Statistics* (2014), pp. 1005–1014
17. R. Munos, Optimistic optimization of a deterministic function without the knowledge of its smoothness, in *Advances in Neural Information Processing Systems* (2011), pp. 783–791
18. M.D. Hoffman, E. Brochu, N. de Freitas, Portfolio allocation for Bayesian optimization. in *Proceedings of the Twenty-Seventh Conference on Uncertainty in Artificial Intelligence, UAI* (2011), pp. 327–336

19. S. Surjanovic, Virtual library of simulation experiments [Online]. https://www.sfu.ca/~ssurjano/
20. D.J. Lizotte, Practical bayesian optimization. Ph.D Thesis, University of Alberta, 2008
21. D.R. Jones, A taxonomy of global optimization methods based on response surfaces. J. Glob. Optim. **21**(4), 345–383 (2001)
22. ANSYS, Ansys HFSS ver. 2015.2 [Online]. http://www.ansys.com
23. ANSYS, Ansys Maxwell ver. 2015.2 [Online]. http://www.ansys.com
24. D.S. Gardner, G. Schrom, F. Paillet, B. Jamieson, T. Karnik, S. Borkar, Review of on-chip inductor structures with magnetic films. IEEE Trans. Magn. **45**(10), 4760–4766 (2009)
25. M.L.F. Bellaredj, S. Mueller, A.K. Davis, P. Kohl, M. Swaminathan, Y. Mano, Fabrication, characterization and comparison of fr4-compatible composite magnetic materials for high efficiency integrated voltage regulators with embedded magnetic core micro-inductors, in *2017 IEEE 67th Electronic Components and Technology Conference (ECTC)* (2017)
26. M. Swaminathan, Power delivery for electronic system consortium (PDES). Georgia Institute of Technology (2017)
27. A. Rong, A.C. Cangellaris, Interconnect transient simulation in the presence of layout and routing uncertainty, in *2011 IEEE 20th Conference on Electrical Performance of Electronic Packaging and Systems* (2011), pp. 157–160
28. A.K. Prasad, M. Ahadi, B.S. Thakur, S. Roy, Accurate polynomial chaos expansion for variability analysis using optimal design of experiments, in *2015 IEEE MTT-S International Conference on Numerical Electromagnetic and Multiphysics Modeling and Optimization (NEMO)* (2015), pp. 1–4
29. D.V. Ginste, D.D. Zutter, D. Deschrijver, T. Dhaene, P. Manfredi, F. Canavero, Stochastic modeling-based variability analysis of on-chip interconnects. IEEE Trans. Compon. Packag. Manuf. Technol. **2**(7), 1182–1192 (2012)
30. B. Sudret, Uncertainty propagation and sensitivity analysis in mechanical models–contributions to structural reliability and stochastic spectral methods. Habilitation à diriger des recherches, Université Blaise Pascal, Clermont-Ferrand, 2007
31. Z. Zhang, T.W. Weng, L. Daniel, Big-data tensor recovery for high-dimensional uncertainty quantification of process variations. IEEE Trans. Compon. Packag. Manuf. Technol. **7**(5), 687–697 (2017)
32. A.K. Prasad, S. Roy, Accurate reduced dimensional polynomial chaos for efficient uncertainty quantification of microwave/RF networks. IEEE Trans. Microwave Theory Tech. (2017). https://doi.org/10.1109/TMTT.2017.2689742
33. C. Soize, R. Ghanem, Physical systems with random uncertainties: chaos representations with arbitrary probability measure. SIAM J. Sci. Comput. **26**(2), 395–410 (2004)
34. M. Berveiller, B. Sudret, M. Lemaire, Stochastic finite element: a non intrusive approach by regression. Eur. J. Comput. Mech. **15**(1–3), 81–92 (2006)
35. D.C. Montgomery, *Design and Analysis of Experiments* (Wiley, New York, 2004)
36. B. Efron, T. Hastie, I. Johnstone, R. Tibshirani, Least angle regression. Ann. Stat. **32**(2), 407–499 (2004)
37. I.M. Sobol, Sensitivity estimates for nonlinear mathematical models. Math. Model. Comput. Exp. **1**(4), 407–414 (1993)
38. B. Sudret, Global sensitivity analysis using polynomial chaos expansions. Reliab. Eng. Syst. Saf. **93**(7), 964–979 (2008)
39. M.D. McKay, R.J. Beckman, W.J. Conover, A comparison of three methods for selecting values of input variables in the analysis of output from a computer code. Technometrics **42**(1), 55–61 (2000)
40. S. Marelli, B. Sudret, UQLab: a framework for uncertainty quantification in matlab, in *Proceedings of 2nd International Conference on Vulnerability Risk Analysis and Management, Liverpool* (2014), pp. 2554–2563

Part V
Machine Learning for System Design and Optimization

The contrivance and the construction of tools must therefore ever stand at the head of the industrial arts.

Charles Babbage

Chapter 18
SynTunSys: A Synthesis Parameter Autotuning System for Optimizing High-Performance Processors

Matthew M. Ziegler, Hung-Yi Liu, George Gristede, Bruce Owens, Ricardo Nigaglioni, Jihye Kwon, and Luca P. Carloni

18.1 Introduction

Modern VLSI design is a quest to optimally tune and balance multiple objectives, such as power, performance, and reliability. The complexity of VLSI chips also continues to rise despite time-to-market pressures and the desire for cost savings via smaller design teams. These challenges have caused an industry shift from custom design techniques towards synthesis-centric methodologies for optimizing design quality and boosting design productivity. The complexity of modern designs coupled with multiple design objectives leads to the rise of difficult optimization problems beyond what even experienced designers can manage manually. Automated design space exploration (DSE) approaches, however, are well suited to handle many of these problems. In this chapter we present a new system called SynTunSys (synthesis tuning system) that automates the DSE process for the logic

M. M. Ziegler (✉) · G. Gristede
IBM T. J. Watson Research Center, Yorktown Heights, NY, USA
e-mail: zieglerm@us.ibm.com; gristede@us.ibm.com

H.-Y. Liu
Intel Technology and Manufacturing Group, Hillsboro, OR, USA
e-mail: hung-yi.liu@intel.com

B. Owens
IBM Systems and Technology Group, Rochester, MN, USA
e-mail: browens@us.ibm.com

R. Nigaglioni
IBM Systems and Technology Group, Austin, TX, USA
e-mail: nricardo@us.ibm.com

J. Kwon · L. P. Carloni
Department of Computer Science, Columbia University, New York, NY, USA
e-mail: jihyekwon@cs.columbia.edu; luca@cs.columbia.edu

I. M. Elfadel et al. (eds.), *Machine Learning in VLSI Computer-Aided Design*,
https://doi.org/10.1007/978-3-030-04666-8_18

and physical synthesis steps within an industrial VLSI design flow. SynTunSys was crucial for achieving timing closure as well as meeting power targets during the design of multiple high-performance server chips at IBM. In particular, we use the IBM z13 processor chip as a case study to illustrate quality of results (QoR) improvements that SynTunSys can achieve. This processor chip pushes the limits of high-performance server design and was designed in a 22 nm SOI technology, has a clock frequency of approximately 5 GHz, and is the heart of the IBM z13 mainframe system [1].

Synthesis tools continue to become more sophisticated and provide numerous parameters to improve design quality. As an example of the wide design space available from modifying synthesis parameters, Fig. 18.1 shows the scatterplot of achievable design points for a portion of a synthesized floating-point multiplier macro from the z13 processor.[1] Each point denotes the timing and power values achieved simply by tuning the input parameters of the synthesis program. The ultimate goal of this process is to reach timing closure at the lowest achievable power. Quite often the default values for the parameters are not ideal for a specific macro, which would benefit instead from parameter customization. Figure 18.1 also

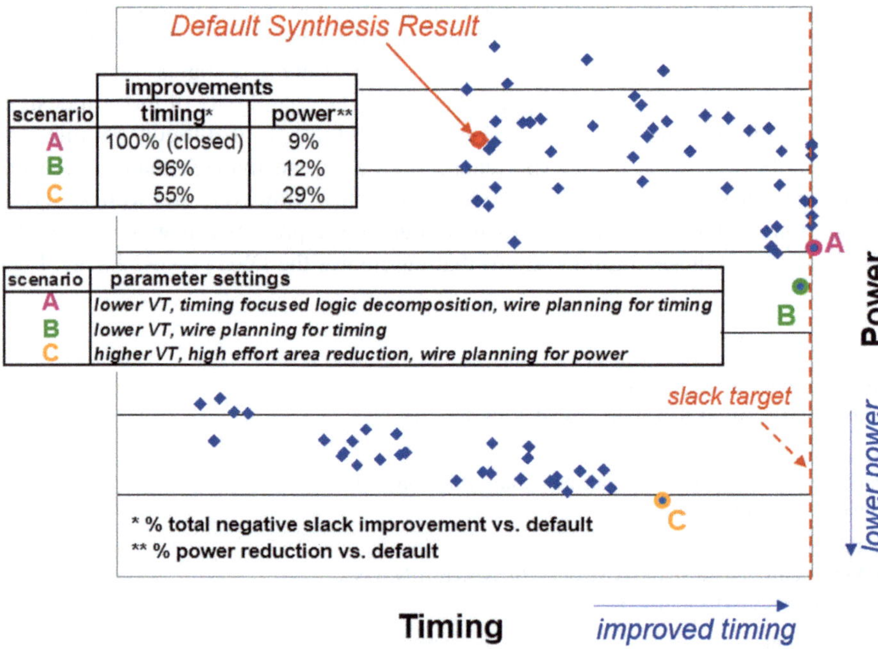

Fig. 18.1 An example of the available design space by modifying synthesis parameters

[1]A macro, also referred to as a partition, is a separately synthesized component that is integrated into a larger chip. Macros considered in this chapter may span in size from 1K to 1M gates.

highlights three scenarios (*A*, *B*, and *C*) along the Pareto frontier. These scenarios show the available tradeoffs between timing closure and power reduction, e.g., point *A* closes timing with a 9% power reduction, whereas point *C* improves timing by 55% with a 29% power reduction. These points along the Pareto set provide a number of potential steps towards the ultimate goal, depending on the additional techniques at the designer's disposal beyond parameter tuning. This example of a relatively simple macro underscores how significantly the parameters settings can affect a design.

The high flexibility and sophistication of advanced synthesis tools increases their complexity and makes navigating the design space difficult, and sometimes non-intuitive for their users. Since the number of parameters for synthesis tools can be in the thousands and synthesis runs may take several hours or even days, exhaustive parameter tuning is typically infeasible. Furthermore, while Fig. 18.1 portrays the design space of two metrics, designers often need to consider many more metrics, sometimes dozens, when evaluating the quality of a synthesis scenario. In summary, efficient DSE using advanced synthesis tools is increasingly challenging for even experienced professional designers and daunting for novice designers.

To efficiently and fully utilize the optimization potential of advanced synthesis tools, we present SynTunSys (STS), a new level of design automation abstraction that allows the human designer to offload the synthesis parameter tuning process to a learning-based system employing parallel computing. In summary, SynTunSys is a closed loop system that encapsulates the design optimization process by running multiple scenarios with unique CAD tool parameter settings through a design flow, analyzing results, and iteratively improving the results by modifying parameters. To navigate the vast parameter design space, SynTunSys employs parallel scenario submission and custom machine-learning algorithms to reach high-quality solutions in only a few iterations. This chapter provides an overview of the SynTunSys system architecture in Sect. 18.2 (Ref. [2] provides details), followed by a detailed account of custom decision algorithms in Sect. 18.3 (Ref. [3] provides details). Section 18.4 reports results from the usage of SynTunSys for optimizing industrial server chips. In particular, the IBM z13 22 nm server processor is provided as a case study. During the actual processor design cycle, a systematic study of ~200 macros shows that SynTunSys provides, on average, 36% improvement in total negative slack (TNS) and 7% power reduction. Further, macro internal negative slack improves by 60%, on average. Section 18.5 presents new SynTunSys enhancements and research directions. Section 18.6 profiles related work and Sect. 18.7 provides concluding remarks.

18.2 SynTunSys Architecture

In order to perform automatic tuning of synthesis parameters, SynTunSys constructs synthesis scenarios (synthesis parameter settings), runs the synthesis jobs, analyzes the results, and iteratively refines the solutions. This automated decision loop is

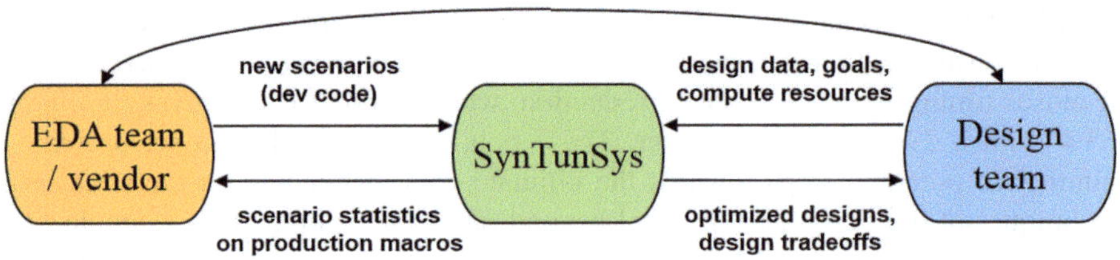

Fig. 18.2 Design team, EDA team, and SynTunSys interaction

a key component of the design process that is typically performed by human designers. The system employs parallel and iterative black-box search techniques to explore the design space in a manner that can efficiently scale to use the available resources in a compute cluster or cloud.

In addition to macro optimization, a secondary use model of SynTunSys provides a vehicle for in-house CAD teams or EDA vendors to deploy and test new synthesis optimizations on production macros, as Fig. 18.2 illustrates. SynTunSys is inherently tolerant of failures of individual synthesis jobs with little risk of degrading overall quality of results. This failure tolerance allows a small number of experimental parameters (or beta versions of the synthesis tool) to be run as part of DSE, enabling a DevOps deployment model. New experimental scenarios can be either provided to designers by the EDA team or run directly by the EDA team. SynTunSys collects synthesis run statistics and keeps a history of results for all designs that can be monitored to determine a scenario's effect on design quality as well as trends across multiple scenarios and designs.

Figure 18.3 shows the SynTunSys architecture. Although the system architecture is general and could be applied to earlier steps in a flow (e.g., HLS) or later steps (e.g., routing), we focus on the logical and physical synthesis steps because in our experience they offer a desired balance between (1) return on investment (ROI) of the compute resources and (2) modeling accuracy of the final design implementation. SynTunSys consists of a main tuning loop that, at each iteration, involves running multiple synthesis scenarios in parallel, monitoring the jobs in flight, analyzing the results of the jobs, and making a decision for the next set of scenarios. A secondary tuning loop archives the results of all runs from all macros, users, and projects. The *SynTunSys Archive* is a database that can be mined for historical trends across multiple macros and to provide feedback in terms of the performance of synthesis parameters.

The SynTunSys process begins at Step 1 in Fig. 18.3 where the initial synthesis scenarios are generated based on the following input data: (a) standard input data for circuit-level synthesis, which typically includes an RTL description, a physical abstract view providing macro boundaries, pin locations, and timing assertions; (b) a library of *primitives* that contains the detailed definitions of all potential exploration options; and (c) a SynTunSys *Rules* file describing the primitives for the design space exploration, a cost function describing the optimization goals, among other options. More details on (b) and (c) are provided later in this section. At Step

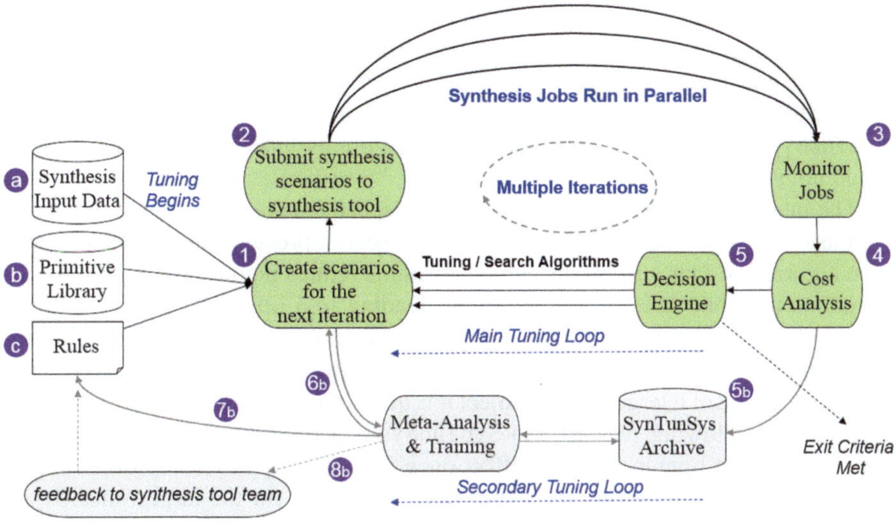

Fig. 18.3 Architecture of the SynTunSys process. The program employs a parallel and iterative tuning process to optimize macros

2 in the figure, multiple synthesis jobs are submitted in parallel to a compute cluster by issuing batch calls to the underlying synthesis tool. Following submission, SynTunSys starts a monitor process that tracks the synthesis jobs (Step 3). When either all jobs complete, or a large enough fraction of jobs complete, or a time limit is reached, the monitor process initiates the results collection and cost analysis of the parallel synthesis jobs (Step 4). Based on the collected results, a decision algorithm at Step 5 creates a new set of scenarios (synthesis parameter settings) to be run in the next iteration. These new jobs begin with the initial input data and are again run in parallel. The process iterates attempting to improve upon results until an exit criteria is met, e.g., the maximum number of user specific iterations is reached or the DSE algorithm experiences diminishing returns.

18.2.1 Human Design Time vs. Compute Time

In this work we make a key differentiation between compute time and human design time. SynTunSys significantly reduces human design time by offloading human designer effort to compute resources. Compute resources are scalable and more abundant over time, whereas there is a premium on training and retaining skilled designers.

The interaction between SynTunSys and a human designer can be viewed as a collaboration between human and computer, each with unique skill sets, rather than a conventional CAD point tool executing a fixed task. Freed from the parameter selection and tuning task, designers can pursue parallel and complementary design techniques, e.g., design structuring [4]. SynTunSys results can also provide hints about the macro design space for skilled human designers to pursue further. The SynTunSys DSE results can also help educate novice designers. Furthermore, SynTunSys often leads to non-intuitive design points that even a skilled designer could not achieve alone.

The inherent SynTunSys runtime and disk-space overhead compared to a single synthesis job motivates streamlining the DSE search process with intelligent decision algorithms. Our current decision algorithms reduce the number of SynTunSys iterations to about three to five iterations, leading to a little over a 3–5× runtime (latency) increase versus a single synthesis run. Over the three to five iterations, approximately 100–200 scenarios are run. Although this overhead may seem high, within the scope of a large design project it is quite tolerable and provides a desirable ROI for the following reasons: (1) SynTunSys is a fully autonomous system that does not require human designer effort once initiated. (2) A latency of around 3–5× still allows the total runtime to be less than a week for the largest macros that may require a day per iteration. A week may be the maximum allowable runtime for projects that follow a weekly update schedule for synthesis inputs. (3) Running SynTunSys is not necessary every time a macro is synthesized; SynTunSys needs only to be run at certain points in the design cycle to locate customized parameters for a specific macro; during subsequent synthesis runs, the SynTunSys scenarios can be reused such that subsequent synthesis runs have no runtime overhead. We should also note that research is underway to reduce the SynTunSys compute overhead, as Sect. 18.5 describes.

The following subsections provide an overview of SynTunSys specific terminology, beginning with the scenario composition that allows a reduction in the parameter tuning design space. The later subsections introduce the key components of the SynTunSys architecture.

18.2.2 Scenario Composition and Initial Design Space Reduction

Parameters Advanced synthesis tools can have a vast number of tunable parameters making an exhaustive design space search infeasible. In our specific case, the synthesis tool we employ has over 1000 parameters [5]. These parameters span the logic and physical synthesis space and the control settings for modifying the synthesis steps, such as logic decomposition, technology mapping, placement, estimated wire optimization, power recovery, area recovery, and/or higher-effort timing improvement. The parameters also vary in data type (Boolean, integer, floating point, and string). Considering that an exhaustive search of only 20 Boolean-type parameters leads to over one million combinations, it is clear that intelligent search strategies are required.

Primitives To reduce the ∼1000 multi-valued parameter space up front, we recast this DSE problem to have a space on the order of 100 Boolean parameters. This design space reduction involves an offline effort to create a library of primitives. A primitive contains one or more synthesis parameters set to specific values. Figure 18.4 shows an example of a small primitive library. The actual primitive library in our case consists of ∼300 primitives. In general, a primitive targets a singular action. Thus, a SynTunSys primitive is a binary decision, whereas setting many parameters individually may require many more decisions. The precise parameter settings within primitives are often products of expert knowledge from experienced designers or EDA developers.

Scenarios SynTunSys creates scenarios consisting of one or more primitives that contain parameters, as Fig. 18.5 shows. The construction of scenarios is not trivial, motivating the need for intelligent decision algorithms. In particular, some primitives are complementary, e.g., **restruct_t** and **wireopt_t** for timing optimization, which however may require additional **wireopt_c** to compensate for routability. Other primitives are non-complementary, such as **restruct_t** and **area_he**; the former may upsize standard cells for larger driving strength or duplicate downstream cells for smaller capacitance load, while the latter may do the opposite changes for maximal area reduction. In practice, the optimal scenarios are macro-specific with respect to a set of weighted design objectives, while default settings require balancing QoR across multiple macros (see Fig. 18.1).

Primitive Name	Primitive Description
restruct_a	Logic restructuring to reduce area
restruct_t	Logic restructuring to improve timing
rvt_lvt10	Native RVT, allow 10% LVT
rvt_lvt50	Native RVT, allow 50% LVT
vtr_he	High effort VT recovery
area_he	High effort area reduction
wireopt_t	Wire optimization for timing
wireopt_c	Wire optimization for congestion

Fig. 18.4 An example of primitive names and primitive descriptions

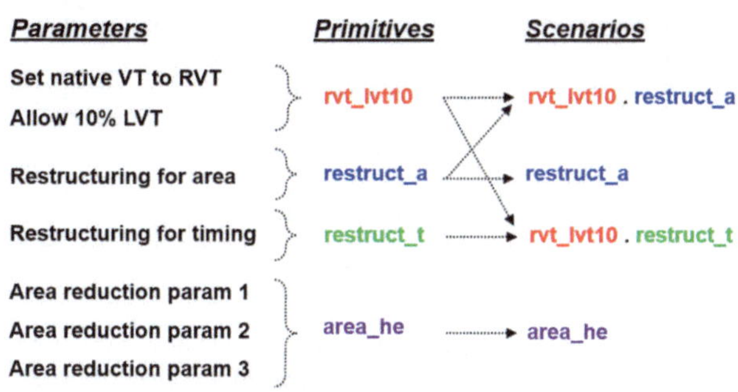

Fig. 18.5 Illustration of the interaction of parameters, primitives, and scenarios. Primitives consist of one or more parameters. Scenarios consist of one or more primitives

Remarks The primitive library was chosen to speed up and focus the design space search task, yet discretization of the design space via primitives may lead to unsearchable points. An alternative approach would be tuning individual parameters; however, this fine-grained approach would have high computation overhead and most likely would not yield strong results for quite some time. An attractive compromise may to first perform a coarse-grain search using SynTunSys primitives to quickly locate promising design points, followed by a finer-grain search of parameter settings to further refine the scenarios.

In practice, to further reduce the design space to a reasonable size, independent of the size of the primitive library, a subset of primitives are selected from the primitive library for a specific SynTunSys run based on the optimization goals for the specific tuning run. This process of selecting a subset of primitives per SynTunSys run allows scaling the specific tuning run for the available compute resources. Within a large design project with multiple designers sharing a compute cluster, the allowable compute quota per user and macro compute requirements often set a ceiling for the number of parallel compute jobs, which in turn provides guidance for the number of primitives to select.

18.2.3 SynTunSys Components

Rules File The SynTunSys *Rules* file contains configuration settings for a specific tuning run. Figure 18.6 shows examples of the key sections of the Rules file: (a) the initial primitives to be explored during the design space search; (b) the cost function guiding the DSE; (c) the overall search algorithm as well as the parameters to configure the search algorithm. These parameters can adjust the effort level of the search as well as the exit criteria, and are often modified depending on the available compute resources and/or the size of the macro. In practice, a set of default Rules files targeting various cost functions are provided to designers. Experienced designers may then further customize a Rules file for each macro.

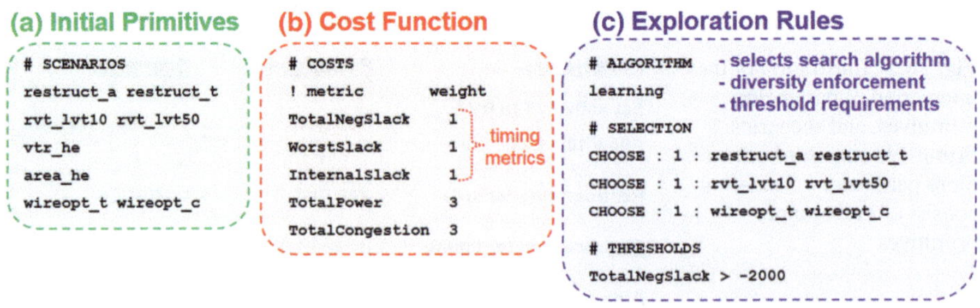

Fig. 18.6 Components of the Rules file: (**a**) the primitives to be explored during DSE; (**b**) the cost function guiding the DSE; (**c**) search algorithm, selection criteria, and configuration parameters

Cost Function The setting of the SynTunSys cost function conveys the optimization goals. It converts multiple design metrics into a single cost number, allowing cost ranking of scenarios. Examples of available metrics include: multiple timing metrics,[2] power consumption, congestion metrics, utilized area, electrical violations, runtime, etc. The selected metrics are assigned weights to signify their relative importance. The overall cost function is then a "normalized weighted sum" of the m selected metrics, expressed by Eq. (18.1) where $Norm(M_i)$ is the normalized M_i across all the scenario results in a SynTunSys run.

$$Cost = \sum_{i=1}^{m} W_i \cdot Norm(M_i) \tag{18.1}$$

Main Tuning Loop The main tuning loop of SynTunSys begins when the monitor triggers completion, i.e., Step 3 in Fig. 18.3. This process includes a step that eliminates scenarios with erroneous or missing results data and also scenarios that fail to meet minimum design criteria, as specified by the designer. The resulting scenarios are all valid for cost analysis. After cost ranking, an optional routine may be applied to ensure there is diversity among the parameters driving the DSE, e.g., primitives having higher cost than competitors across a dimension of the primitive space may be removed. The decision engine is the final step of the main tuning loop. Overall, the main tuning loop is an "online" process that uses information learned about a specific macro, within the tuning run in flight, for decisions.

Secondary Tuning Loop The collected results of all runs are stored in the *SynTunSys Archive* database, i.e., Step 5b in Fig. 18.3. The archived data provides a history of all previous runs that SynTunSys uses for multiple purposes. One use of the archived data is to generate DSE scenarios from historical primitive performance, represented by Arc 6b in Fig. 18.3. The default Rules files are also periodically updated based on the analysis of historical primitive performance (Arc 7b). The historical performance is also a useful feedback to the synthesis tool development team (Arc 8b). The secondary tuning loop is currently an "offline" process that allows learning across multiple or all macros, although integrating this capability into the online framework is currently being investigated. Section 18.5.1 also discusses SynTunSys enhancements that employ historical data.

18.2.4 Decision Engine Algorithms

The SynTunSys decision engine algorithms are key components of the system that determine which scenarios should be run at each iteration. The decision engine is also a component that can be upgraded independently and we constantly look

[2]The typical timing metrics of interest are: (1) internal slack or latch-to-latch slack (L2L), (2) worst negative slack (WNS), and (3) total negative slack (TNS).

to improve these algorithms in terms of QoR prediction accuracy and compute efficiency. The general problem the decision engine addresses can be described as black-box tuning or search, i.e., we treat the underlying synthesis tool as black-box software during the parameter tuning process. The black-box tuning problem appears in numerous applications and has also been approached using a number of techniques. Section 18.6 towards the end of this chapter will review related work. The following section details the evolution of the current production SynTunSys decision algorithm, which is a custom learning algorithm based on parallel and iterative exploration, and targets near-optimal results in only a few iterations.

18.3 The SynTunSys Decision Engine

The decision engine and tuning algorithms are key SynTunSys components that determine the total number of iterations and scenarios to be run during each iteration. It should be emphasized that for large-scale industrial processor chips it is crucial to reach near-optimal solutions in a few tuning iterations. This is particularly important for chips with new architectures and/or having frequent logic changes. Timing critical designs consisting of multiple macros also have frequent timing assertion feedback updates. Design schedules based on weekly or bi-weekly logic releases, timing assertion feedback, and macro rebuilds are common, thus a total tuning runtime of less than a week is a goal. This low-latency requirement drives the need to develop custom SynTunSys decision algorithms that can reach high-quality solutions in a few iterations.

One key property of any decision algorithm is the ability to adapt to a specific macro, i.e., "learn" the primitives that minimize the cost function. In practice, the optimal scenarios are macro-specific for these reasons: (1) the heuristic nature of the underlying synthesis algorithms requires different settings based on a macro's logic and (2) the objectives in a specific cost function will call for unique parameter settings. To date, the decision engine algorithms are "online" in nature, in that decisions are made based on feedback from the current run in flight. Although as discussed later in the chapter, there are opportunities to include knowledge from historical data from the SynTunSys Archive for decision making.

The decision engine is also a modular component that can be upgraded independently, allowing incremental refinement. The general objective of the decision engine is given a set of primitives, a decision algorithm returns a list of lowest-cost scenarios with respect to the designer-defined cost function (Eq. (18.1)) and generates a set of new scenarios for the next iteration. To avoid major SynTunSys architecture changes, we have designed the various decision engine algorithms to map to a common compute footprint as Fig. 18.7 illustrates. This common compute footprint allows reuse of job submission, job monitoring, results collection, and cost analysis components, requiring only recoding of the decision engine itself. The system was structured as such to meet the need dates of various industrial project schedules, i.e., it was a higher priority to first have a fully functional system that

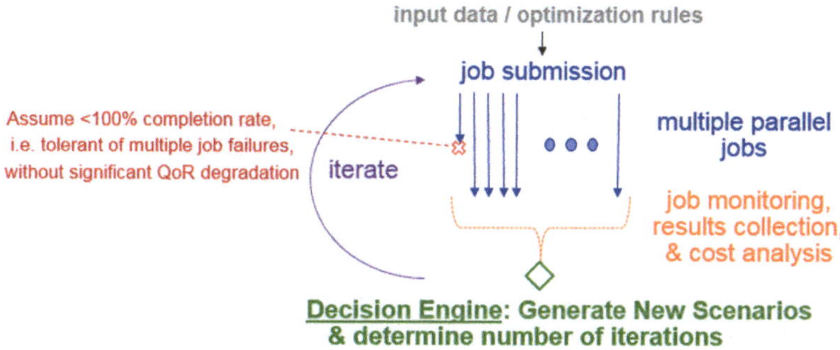

Fig. 18.7 A common compute footprint utilized by SynTunSys decision engines

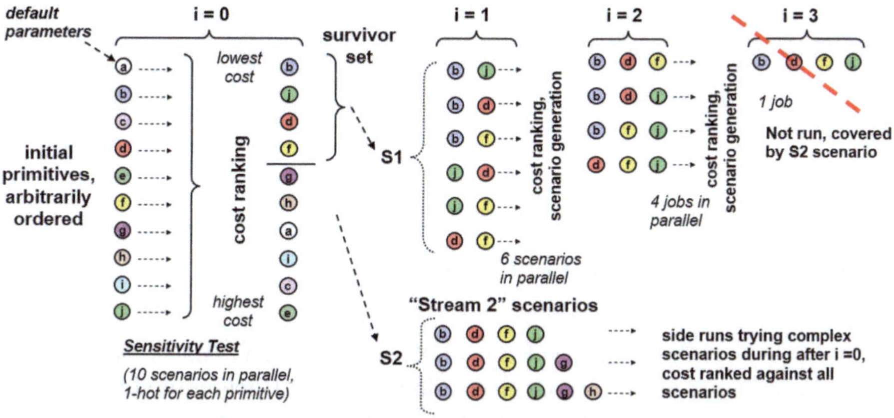

Fig. 18.8 The Base decision algorithm, where each colored bubble represents a primitive, a horizontal sequence of adjacent bubbles represents a scenario, and *i* denotes the iteration number

designers can use, even if results are less than optimal, rather than non-functional theoretically optimal system concept. The following subsections describe an initial version of the decision algorithm used for 22 nm projects, followed by an upgraded version for 14 nm projects and beyond.

18.3.1 The Base Decision Algorithm

The initial SynTunSys decision algorithm, which we call the *Base* algorithm, is illustrated in Fig. 18.8. This approach is a pseudo-genetic algorithm involving a survival of the fittest comparison (*sensitivity test*), followed by a dense search using the top primitives (pseudo-code is provided in Fig. 18.9, Algorithm 1).

```
Algorithm 1: Base
Input:
  P: set of primitives
  f(S): cost function (Equation (1)) on a set S of scenarios
  N: size of Survivor Set (1 ≤ N ≤ |P|)
  M: size of "Stream 2" scenarios (N < M < |P|)
Output:
  S*: sorted list of low-cost scenarios
// sensitivity test (i = 0)
 1: S ← {|P| scenarios, each including a unique primitive}
 2: run |S| synthesis jobs with S
 3: S* ← sort(S) with f(S)
 4: Survivor Set S' ← {combine top-N scenarios in S*}
// tuning iterations (i > 0)
 5: for (i ← 1; i < N-1; i++)
 6: |  if (i = 1) // prepare "Stream 2" scenarios
 7: |  |  S ← {the top-n scenarios in S' | n in {N, M, M+1}}
 8: |  else
 9: |  |  S ← Φ
10: |  S ← S U {C(N,i+1) scenarios: all combinations of i+1
    |            survivor primitives}
11: |  run |S| synthesis jobs with S
12: |  S* ← sort(S* U S) with f(S* U S)
13: return S*
```

```
Algorithm 2: Learning
Input:
  P: set of primitives
  f(S): cost function (Equation (1)) on a set S of scenarios
  k: size of Potential Set (1 ≤ k ≤ |P|)
  M: size of "Stream 2" scenarios (N < M < |P|)
  I: max # of tuning iterations
  O: combination order (O ≥ 2)
  β: parameter controlling cost estimation (0 < β < 1)
Output:
  S*: sorted list of low-cost scenarios
// sensitivity test (i = 0)
 1: same as in the Base algorithm (let N equal k)
// tuning iterations (i > 0)
 2: for (i ← 1; i ≤ I; i++)
 3: |  initialize set S as in Lines 6-9 of the Base algorithm
 4: |  for (j ← 2; j ≤ O; j++) // look-ahead combination
 5: |  |  S ← S U {C(k,j) scenarios: all combinations of j
    |  |            scenarios from S*}
 6: |  α ← β^(i-1) // α decreases as # iterations increases
 7: |  foreach scenario s in S // major learning iterations
    |  |  // to learn the inter-scenario interaction
 8: |  |  Sc ← set of the contributing scenarios of s
 9: |  |  Coarse-Cost ← average f(Sc)
    |  |  // to learn the inter-primitive interaction
10: |  |  Sr ← set of the reference scenarios of s
11: |  |  Fine-Cost ← average f(Sr)
    |  |  // dynamically weigh the two types of costs
12: |  |  estimated cost of s ← α×Coarse-Cost + (1-α)×Fine-Cost
13: |  Potential Set S' ← {top-k scenarios in S with the lowest
    |                       estimated costs}
14: |  run |S'| synthesis jobs with S'
15: |  S* ← sort (S* U S') with f(S* U S')
16: return S*
```

Fig. 18.9 Pseudo-code for the Base and Learning algorithms

The Base algorithm begins with SynTunSys launching an initial iteration ($i = 0$) consisting of one scenario for each given primitive, i.e., a 1-hot sensitivity test. Then, following cost analysis the algorithm selects the best N (lowest-cost N) scenarios as a "survivor set" and proceeds to iteration 1 ($i = 1$). Iteration 1 generates a stream **S1** of more complex scenarios, consisting of combinations of primitives from the survivor set (e.g., **b**, **j**, **d**, **f** in Fig. 18.8). The most common configuration is to generate all possible combinations of $i + 1$ primitives for each iteration i. The algorithm works on the premise that all primitives are *complementary*; therefore, combining the survivors would yield optimal or near-optimal scenarios. In the example of Fig. 18.8, the number of primitives is 10 and the size N of the survivor set is set to 4.

Although the Base algorithm fully searches the design space of the survivor set, its practical size is often constrained by compute resources and runtime latency, i.e., a large survivor set may lead to too many parallel scenarios in one iteration. To mitigate this problem and expand the survivor set, an **S2** stream of scenarios can also be added to the $i = 1$ iteration. The **S2** scenarios are rule-based guesses such as combining the M lowest-cost primitives, where $(M \geq N)$, e.g., $M = 4, 5$, and 6 in Fig. 18.8. These **S2** scenarios were a late addition to the Base algorithm to cover the known deficiency and provide a bridge until a more sophisticated algorithm could be implemented.

18.3.2 The Learning Algorithm

Within the compute footprint of the Base algorithm (i.e., a sensitivity test followed by iterative combination of scenarios), an enhanced decision algorithm was developed (see Fig. 18.9, Algorithm 2), which we call the *Learning* algorithm, to better address the deficiencies described in the previous subsection. The Learning algorithm selects a given number k of scenarios in each iteration as parallel synthesis jobs (i.e., maximized utilization of compute resource), and dynamically adapts to the k scenarios that are more likely to return lower costs (i.e., adaptive exploration).

Figure 18.10 illustrates the main idea of the Learning algorithm. Following the sensitivity test ($i = 0$) on the given primitives, the Learning algorithm *estimates* the cost of an unevaluated composite scenario by taking the average cost of its contributing scenarios as a cost predictor. For instance, to estimate the cost of a scenario that comprises three primitives (**b, j, d** in Fig. 18.10), the Learning algorithm calculates the average cost of the three contributing scenarios that respectively comprise the three primitives. Furthermore, the Learning algorithm can "look ahead" by a combination order $O > 1$, which allows combining up to O prior scenarios for cost estimation ($O = 3$ in Fig. 18.10). This look-ahead predictor allows complementary scenarios such as **restruct_t + wireopt_t + wireopt_c** (see description in Fig. 18.4) to be discovered for a cost function favoring timing and routability, earlier in tuning iteration $i = 1$ in the Leaning algorithm, as opposed to iteration $i = 2$ in the Base algorithm (compare Line 10 in Algorithm 1 and Lines 4–5 in Algorithm 2). The Learning algorithm uses the look-ahead predictor to learn the *inter-scenario* interaction (Algorithm 2, Lines 8–9).

After the cost estimation, the Learning algorithm selects the top-k composite scenarios with the lowest estimated costs to form a *potential* set and then submits k parallel synthesis jobs with the selected scenarios. Since the size k of the potential set is constrained, with a combination order greater than 1, the Learning algorithm can filter out non-promising scenarios early in the tuning loop and allocate instead

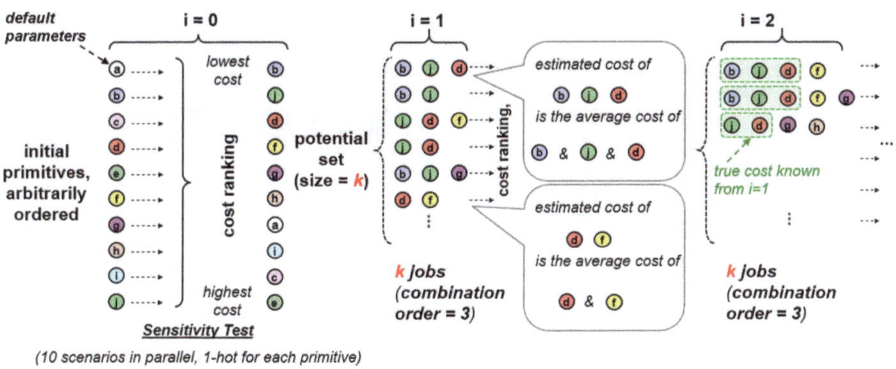

Fig. 18.10 Illustration of the Learning decision algorithm

the synthesis budget to the more promising scenarios. This estimation-selection-submission process repeats for every tuning iteration until an exit criterion is met (e.g., max iteration *i* is reached).

Furthermore, the Learning algorithm leverages the iterative process to continuously refine its cost-estimation accuracy on *non-complementary* combinations. Specifically, at any iteration *i*, whenever a composite scenario, say **restruct_t + area_he**, was predicted good (i.e., low timing and area costs) and selected for synthesis, but the synthesis result turns out to be not good (mediocre weighted cost because of conflicting underlying optimization mechanisms), then the algorithm can *learn* the actual effectiveness of combining scenarios **restruct_t** and **area_he**. Therefore, at any future iterations, it will demote any composite scenario that involves **restruct_t + area_he**. In summary, the Learning algorithm uses cost estimation to avoid non-promising scenarios and refines its estimation after learning actual synthesis results.

Moreover, to better estimate the cost based on *non-trivial* contributing scenarios (i.e., scenarios comprising more than one primitive), the Learning algorithm includes a fine-grained cost estimation (see Fig. 18.11). For instance, given two scenarios, **s1** = (**b** + **d**) and **s2** = (**b** + **j** + **f**), the algorithm not only regards the average cost of **s1** and **s2** as the *coarse-cost*, but also considers a *fine-cost*. The fine-cost aims to learn the *inter-primitive* interaction (such as **d** + **j**), in contrast to the coarse-cost that targets the inter-scenario interaction (such as **s1** + **s2**). To this end, we calculate the fine-cost of **s1** and **s2** using the average cost of their *reference scenarios*, which are the scenarios that have been run in the previous iterations and each include a pair of primitives, such that one primitive (e.g., **d**) comes from **s1** and the other (e.g., **j** or **f**) from **s2**. See pair-1 and pair-2 in Fig. 18.11 for illustration, where the scenarios listed to the left of pair-1 and pair-2 are the reference scenarios for **s1** and **s2**. Hence, the fine-cost of **s1** and **s2** is the average cost of these reference

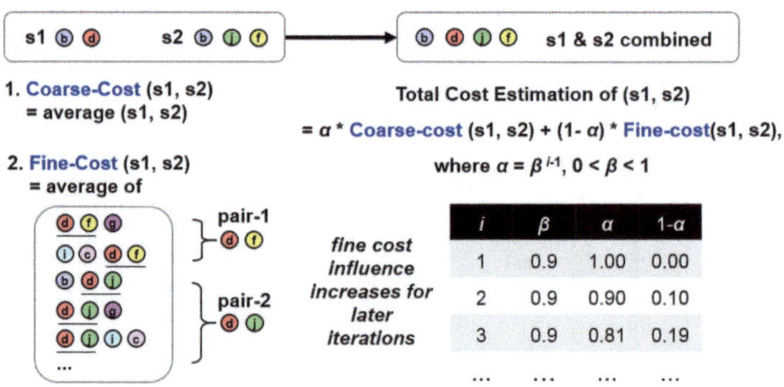

Fig. 18.11 Cost-estimation process for the Learning algorithm

scenarios. Thus, the previous example of non-complementary **restruct_t + area_he** combination can also be considered when estimating the cost of **s1 + s2**, if primitive **d** is **restruct_t** and primitive **j** (or **f**) is **area_he**. The learning of inter-primitive interaction is specified at Lines 10–11 in Fig. 18.9, Algorithm 2.

Overall, the Learning algorithm's cost-estimation function (Line 12 in Fig. 18.9, Algorithm 2) is a dynamic weighted sum of the coarse- and fine-cost with a changing weighting factor α for the coarse-cost and a factor $(1 - \alpha)$ for the fine-cost, where $0 \leq \alpha \leq 1$. For determining α we used the formula $\alpha = \beta^{(i-1)}$ where $0 < \beta < 1$ and i is the current iteration number. That is, we favor the fine-cost more in the later tuning iterations as there are more reference scenarios available.

18.4 SynTunSys Results

Synthesis parameter tuning was used during the design of the last three generations of IBM server chips, across multiple product lines, spanning 32, 22, and 14 nm technologies. An early, semi-automated version of SynTunSys was to design the IBM POWER7+ 32 nm processor [6]. This initial deployment primarily targeted power optimization during the second chip tapeout.[3] The first fully automated use of SynTunSys was for the IBM POWER8 22 nm processor [7] where a few timing critical macros were optimized during the first tapeout and multiple macros were optimized for timing and power during the second tapeout. The IBM z13 22 nm processor [1] was the first widespread usage of SynTunSys across nearly all macros for timing, power, and congestion during both tapeouts. In particular, a systematic study on SynTunSys QoR was performed on ~200 macros during the first z13 tapeout, the results of which are detailed in the next subsection. At the 22 nm node SynTunSys employed the Base decision algorithm, as described in Sect. 18.3.1.

More recently, SynTunSys was used to design the IBM POWER9 [8] and IBM z14 [9] processors, both designed in 14 nm finFET technology. For these chips SynTunSys was widely used on nearly all macros by both designers and a new automated scheduling system described in Sect. 18.5. In addition, SynTunSys was upgraded to the Learning algorithm for the 14 nm node. The following subsections present results from the systematic z13 study (Sect. 18.4.1), 14 nm results sampled from the SynTunSys Archive (Sect. 18.4.2), and a QoR comparison of the Base and Learning algorithms (Sect. 18.4.3).

[3]The server chips described in this chapter often go through two chip releases, a.k.a. tapeouts. The first tapeout is less optimized early version of the chip, whereas the second tapeout targets an optimized production quality chip.

18.4.1 *z13 Case Study Results*

SynTunSys was widely adopting early in the design flow for the design of the IBM z13 22 nm server processor. The processor underwent two chip releases (tapeouts) over a multi-year design cycle, during which SynTunSys was applied to macros over both releases. The chip consists of a few hundred macros that average around 30K gates in size, with larger macros in the 300K gate range. Although SynTunSys was used for earlier IBM chip designs, z13 was for the first to use the flow during the entire design cycle for timing closure, power reduction, and improving macro routability.

During the first chip release, a dedicated SynTunSys team performed tuning for hundreds of macros across the chip. In parallel, a number of designers further tuned the macros they owned, as needed. Based on the efforts of the dedicated tuning team we were able to track SynTunSys results on approximately 200 macros from the processor core. The "pass 1" rows of Fig. 18.12 show the average improvements achieved by SynTunSys over the best solution previously achieved by the macro owners for the first chip release. Note that these results are based on the routed macro timing and power analysis; in most cases the best known prior solutions included manual parameter tuning by the macro owner. The first pass of SynTunSys resulted in a 36% improvement in total negative slack, a 60% improvement in worst latch-to-latch slack (macro internal slack), and a 7% power reduction. The actual values of the metrics, summed across all the macros, underscore that the changes in the absolute numbers were significant, e.g., for pass 1, ~780,000 ps of total negative slack was saved across ~200 macros.

A second SynTunSys tuning run by the macro owners to further improve timing and power was performed during the second chip release. These second pass tuning runs build off the prior tuning run results to further search the design space. In

SynTunSys Pass (Chip Release)	Latch-to-Latch Slack	Total Negative Slack	Total Power
Improvement %	(%)	(%)	(%)
pass 1	60%	36%	7%
pass 2	24%	2%	3%
Sum of 200 macros	(ps)	(ps)	(arb. units)
pre-pass 1	-1,929	-21,50,385	17,770
post-pass 1	-765	-13,70,731	16,508
Sum of 25 macros	(ps)	(ps)	(arb. units)
pre-pass 2	-260	-1,85,087	3,472
post-pass 2	-198	-1,80,896	3,379

Fig. 18.12 Average SynTunSys improvement over best known prior solution based on post-route timing and power analysis

some cases the macro logic was also quite different in the second chip release, leading to a different design space. For the second tapeout we had a less controlled study, but wider usage by macro owners. Based on data from 25 macros (Fig. 18.12, "pass 2") we still see considerable improvements on macros after the second pass of SynTunSys.

18.4.2 14 nm Data Mining Results

In this subsection we describe SynTunSys improvements for macros from 14 nm server chips, i.e., POWER9 [8] and z14 [9] processors. The goal is to not only show the effectiveness of SynTunSys at a more advanced technology node, but also highlight the multiple design point options made available by SynTunSys.

SynTunSys ranks scenarios by a single cost value computed from multiple metrics and weights from a user's cost function. But the choice of scenario(s) to continue through the later stages of the design flow, e.g., routing steps, is ultimately in the hands of the designer. Thus, SynTunSys not only provides a suggested best scenario by cost, but also a number of other scenarios that may be more attractive with respect to specific metrics while having a higher overall cost. Figure 18.13 shows the average SynTunSys improvement percentages for over 150 macros from a 14 nm server chip. The results were mined from the SynTunSys Archive that stores data from all SynTunSys runs and are thus results from an actual processor design cycle. With SynTunSys more integrated into the design flow, we did not perform an explicit systematic study like was done for z13 in the previous section, but we can periodically mine the SynTunSys Archive for QoR statistics.

The first row of the table shows the average improvement percentages for the top scenario in terms of cost. The second through last rows of Fig. 18.13 provide average SynTunSys improvements for the top scenarios with respect to a specific metric. Although these scenarios often are not as well rounded as the scenario with the lowest cost, they provide a variety of design point options that may solve the challenges of a specific macro.

Scenario \ Metric	WNS	L2L	TNS	Cong	Power	Area
Overall Best (Lowest Cost)	19%	49%	40%	25%	2%	4%
WNS (worst negative slack) Best	25%	43%	38%	10%	1%	3%
L2L (latch-to-latch slack) Best	12%	60%	40%	0%	1%	3%
TNS (total negative slack) Best	19%	50%	48%	17%	2%	4%
Cong (congestion) Best	7%	20%	24%	30%	2%	3%
Power Best	0%	7%	9%	13%	11%	6%
Area Best	8%	27%	27%	16%	6%	9%

Fig. 18.13 Average SynTunSys improvements for a 14 nm processor (150 macros)

18.4.3 Decision Algorithm Comparisons

In this subsection we compare the Base and Learning algorithms. We begin by configuring settings for the Learning algorithm using a small test suite of macros. We then perform a thorough benchmarking of the Base and Learning algorithms using production macros from the IBM 22 nm z13 processor from [1].

The Learning algorithm provides a number of configuration settings that can be tuned to improve performance. To determine appropriate settings of the Learning algorithm we use a small test suite of four macros varying in logic functionality and create three Learning algorithm configurations shown in Fig. 18.14.[4] We use a balanced cost function considering timing, power, and congestion, and run the Base and three Learning algorithm variants for three iterations, in addition to the sensitivity test, on the four test macros. We compare these algorithms in terms of the average cost of their final top-*N* results, with *N* in {1, 5, 10}. Figure 18.15 shows the result of the comparison where the costs of all configurations of the Learning algorithms are relative to the Base algorithm costs. Overall, the winner is Learning3+, which consistently outperforms Learning3, which in turn outperforms

Fig. 18.14 Algorithm configurations

Algorithm	Survival / Potential Set Size	Comb. Order	Cost Estimation
Base	5	2	None
Learning2	20	2	Coarse
Learning3	20	3	Coarse
Learning3+	20	3	Coarse+Fine

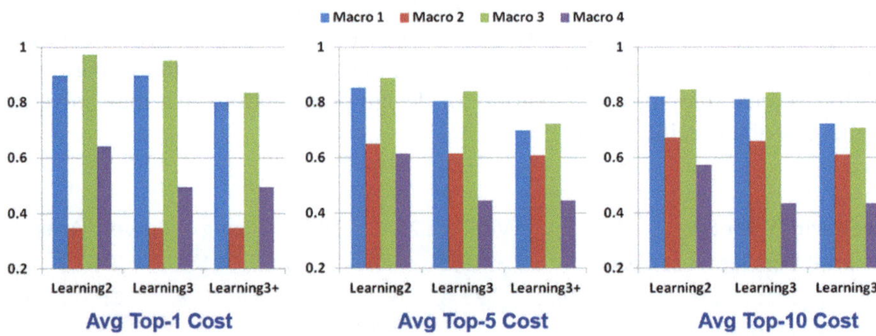

Fig. 18.15 Configuration comparison results for Base vs. Learning algorithms

[4]While many more configurations are possible, this experiment illustrates our initial attempt to configure the Learning algorithm parameters. Over time we have performed additional configuration experiments and designers running SynTunSys can also reconfigure the Learning algorithm as needed via the Rules file.

Learning2. We hypothesize that the Learning3+ performs the best for the following reasons: (1) the higher combination order allows exploring scenarios that combine more primitives during earlier iterations and (2) the addition of the fine-cost component in the cost estimation does indeed improve prediction accuracy. Thus, the winning algorithm, Learning3+, whose size of the potential set, max number of tuning iterations, combination order, and weight-changing parameter β are 20, 5, 3, and 0.9, respectively, was chosen for the production benchmark comparison.

For a more thorough Base and Learning algorithm comparison we selected a test suite of 12 macros from the z13 22 nm processor, ranging in size from 1K to 110K gates, with an average size of 31K gates (the same average macro size as the processor). The same cost function is used for both algorithms, which is balanced across timing, power, and congestion metrics. We use a large number of initial primitives (50) to emulate a realistic design scenario. For both algorithms, the maximum scenario count per iteration is 20 (after the sensitivity test, i.e., $i > 0$) and the total iteration count is 5, plus the sensitivity test.

The exploration results are summarized in Fig. 18.16. For this study we did not route the macros and, therefore, we report only post-physical synthesis statistics. In lieu of routing, we include a routability metric called *route-score*, in which a lower value denotes less congestion (i.e., a more routable macro). Overall, the results of Fig. 18.16 are in the same average improvement range for the Base algorithm as with the z13 processor tuning from [2]. However, the Learning3+ algorithm achieves an additional 20% total negative slack improvement over the Base results as well as significant improvements across all timing metrics. Figure 18.17 provides the macro-by-macro percentages of change for each metric category in the cost function. Only one macro sees degradation on one metric (routability of macro H) for the Learning3+ algorithm, while three macros see degradation for Base.

Figure 18.18 provides a visual representation of the explored design spaces for the two largest macros in the test suite. These plots show total power (top row) and a route-score (bottom row) vs. total negative slack (all values normalized). The

	Timing			**Power**	**Route-ability**
	Worst Slack	**Latch to Latch Slack**	**Total Negative Slack**	**Total Power**	**Route Score**
Improvement	(%)	(%)	(%)	(%)	(%)
Base	24%	43%	42%	11%	46%
Learning3+	35%	70%	62%	10%	43%
12 macro sum	(ps)	(ps)	(ps)	(a. u.)	(a. u.)
default	-256	-105	-77,505	1,125	129
Base	-195	-60	-45,104	1,003	69
Learning3+	-167	-32	-29,138	1,007	74

Fig. 18.16 Comparison of Learning and Base algorithms across a 12 macro test suite that is the representative of IBM 22 nm z13 processor

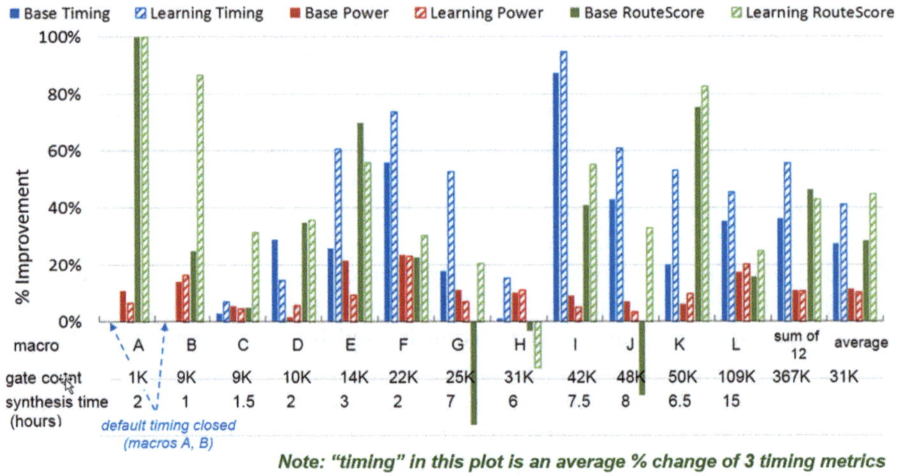

Fig. 18.17 Macro-by-macro breakdown of Learning vs. Base

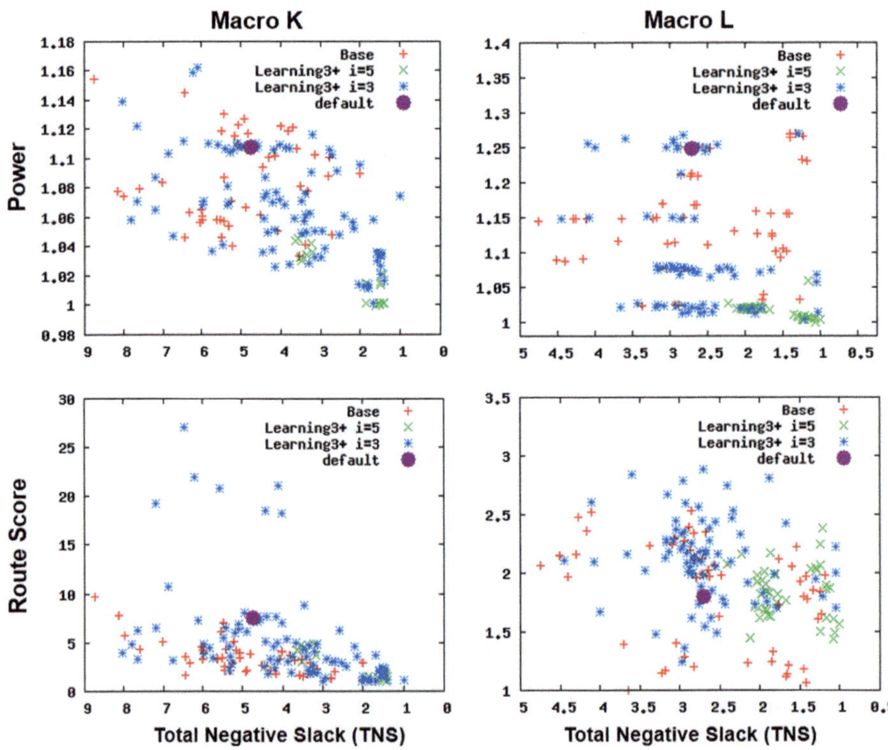

Fig. 18.18 Power and routability results for the largest test suite macro

less the power/route-score/negative slack, the better the quality of result (i.e., the optimization goal is the lower-right corner of the plots). We can see that Learning3+ effectively optimizes all the three metrics for macro K. The results also illustrate that Learning3+ indeed adapts to the more promising design space: the $i = 5$ dots are more focused on the lower-right region of the spaces than the $i = 3$ dots. On macro L, Learning3+ adapts to optimizing the timing and power, whereas the Base algorithm finds solutions with better route-score. If we compare the single best scenario (lowest-cost scenario) for macro L based on the balanced cost function (Fig. 18.18), Learning3+ actually outperforms Base in each metric, including route-score.

18.5 SynTunSys Enhancements and Future Work

There are many ongoing threads of research beyond the current production version SynTunSys. These efforts aim to improve QoR, compute efficiency, and allow the tuning process to efficiently scale to multiple macros. This section will preview a few of the promising SynTunSys enhancements that may soon be added to the production release.

18.5.1 Opportunities for History-Based Enhancements

The data within the SynTunSys Archive provides many promising opportunities for improving the SynTunSys flow. Saving SynTunSys run data was built into the initial production release of the system, although the precise usage of the data at that time was unclear. The first usage was an offline evaluation of relative primitive performance across multiple macros. This analysis was used to update the default Rules files and provide feedback to the EDA team, i.e., Arc 7b and Arc 8b in Fig. 18.3, respectively. These two usages, however, were based on average primitive performance across multiple macros and were not macro-specific, nor was the SynTunSys run in flight directly affected. More recently, work has begun to exploit the SynTunSys Archive data in a macro-specific manner. The following two subsections provide examples of how this data may be used.

18.5.1.1 The Jump-Start Algorithm

One enhancement to the SynTunSys decision algorithms targets the reduction of the overhead of the sensitivity test, i.e., iteration $i = 0$. The enhancement employs the archived data from prior SynTunSys runs on previously tuned macros.

In practice, our default sensitivity tests provide three levels of tuning effort: low, medium, and high, featuring 20, 35, and 50 initial primitives, respectively. In order to increase the efficiency of the sensitivity test (i.e., to approach the higher-effort exploration result based on a lower-effort sensitivity test), we present the *Jump-Start* test. The idea is to run a low-effort sensitivity test for the current macro as a pseudo test T_L, and to take T_L's ranking of the initial primitives as a signature. Then, the Jump-Start test retrieves the high-effort test T_H from the archive having the most similar primitive ranking to T_L's signature. The Jump-Start test adopts T_H's sensitivity test primitive ranking and proceeds to the search iterations ($i > 0$). The final exploration result should improve if the number of primitives in T_H is greater than that in T_L.

We measure the similarity between two signatures using "the Length of Longest Common Subsequence (LLCS)." Each signature is represented by a sequence in the order according to the primitive ranking. For example, given primitives in $[a \ldots z]$, the longest common subsequence of two signatures $\langle a, b, c, d \rangle$ and $\langle a, f, d \rangle$ is $\langle a, d \rangle$ with LLCS equal to two. Figure 18.19 illustrates a low-effort sensitivity test with six primitives, while in the archive, the high-effort sensitivity tests with more than six primitives are identified and sorted according to their LLCS with respect to T_L's signature. Among these high-effort tests, T_H #1 has the maximum LLCS (five). Thus, the complete sensitivity-test result of T_H #1 is taken to kick off the first search iteration ($i = 1$). As a tie-breaker for selecting among T_H's with equal LLCS, we favor the T_H that includes more high-ranking primitives, e.g., T_H #3 is considered more similar to T_L than T_H #2 in Fig. 18.19.

Next, to explore the effectiveness of the Jump-Start sensitivity test, we used the smaller test suite of macros that were employed for fine-tuning the Learning algorithm parameters and compare the Jump-Start results against the standard low-effort sensitivity test. The Jump-Start test is based on searching an archive of thousands of high-effort sensitivity tests. By default, the low-effort and high-effort sensitivity-test scenario counts are 20 and 50 initial primitives. Figure 18.20 shows the cost comparison between the default low-effort SynTunSys run (whose cost is normalized as 1.0) and a low-effort run employing the Jump-Start sensitivity test. The plot shows the average normalized cost of the top 1, 5, and 10 scenarios for the four macros. Although the same scenario count was used for both the standard low-effort run and low-effort run with Jump-Start, the Jump-Start algorithm achieved

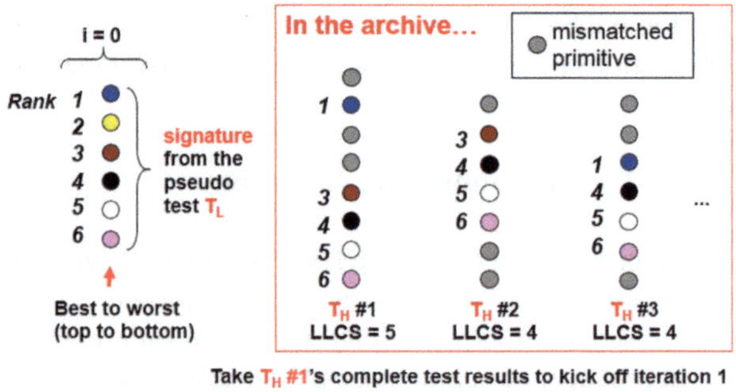

Fig. 18.19 Illustration of the Jump-Start sensitivity test

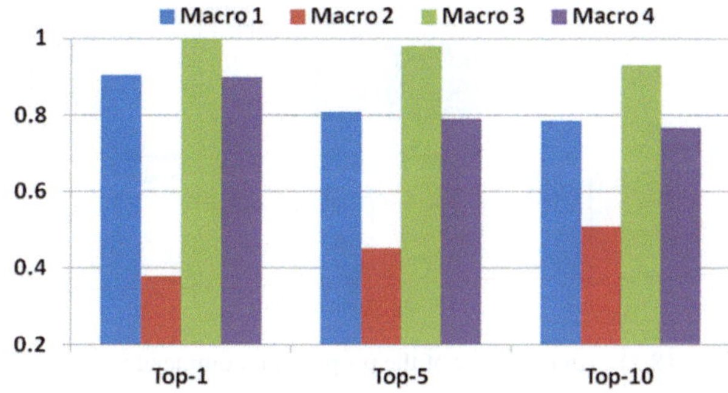

Fig. 18.20 Average cost comparison of the Jump-Start test vs. the standard low-effort SynTunSys run, which has a normalized cost of 1.0

lower cost (better QoR) for all four macros, thanks to the high-effort primitives retrieved from the archive.

18.5.1.2 The Recommender System

Another history-based enhancement to SynTunSys aims at automatically recommending near-optimal macro-specific scenarios at an earlier stage of the decision algorithm. In Sect. 18.4.3, 3 or 5 tuning iterations plus a sensitivity test were performed for each macro, submitting about 20–30 synthesis jobs per iteration. Additionally, the resulting search spaces were constrained by the list of initial primitives used in and survived from the sensitivity test. By exploiting the SynTunSys Archive and constructing a new macro-specific cost predictor, we can recommend scenarios at iteration $i = 0$ or $i = 1$, greatly reducing the number of synthesis jobs that are currently necessary to adapt to each macro. Furthermore, the recommended scenarios may include some primitives from the history that are not considered in the sensitivity test of a specific macro and thus omitted from the search space.

In many application areas outside synthesis parameter tuning, recommendation systems based on collaborative filtering have been actively suggested and implemented. For instance, in a movie recommender system, while each user reviewed only a small number of movies from the entire list, the system predicts the user's review scores for other movies by analyzing the collection of existing movie reviews by all users [10]. In this system, each user's preferences and each movie's characteristics are hidden, but revealed via collaborative filtering with techniques, such as matrix factorization. Similarly, in our SynTunSys scenario recommender system, we attempt to extract the hidden properties of macros and primitives from the history and use them in the cost prediction.

Figure 18.21 illustrates an overview of the proposed recommender system, composed of two engines: offline training and online recommendation. The training engine takes in normalized costs for (macro, scenario) combinations stored in the SynTunSys Archive, and outputs the trained cost predictor. The cost prediction model includes the hidden features of the macros and primitives from the Archive, and a combination generator. In [10], the movie review matrix is factorized into

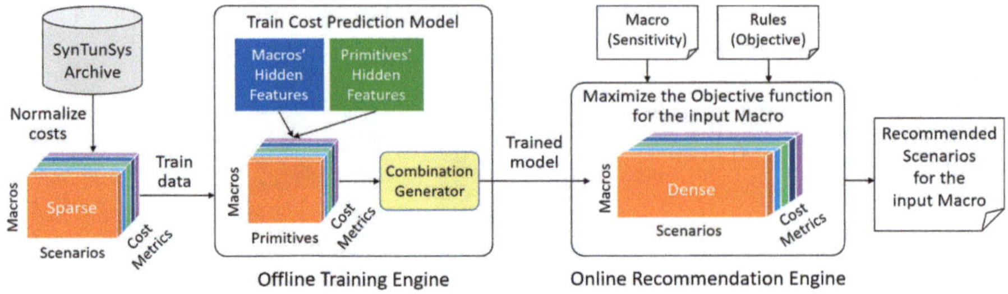

Fig. 18.21 An overview of the proposed recommender system

the user matrix and the movie matrix, containing hidden features of the users and movies, respectively. Similarly, the tensor of SynTunSys results can be decomposed into structures containing hidden features of macros and primitives. This part of the model can be used to predict the results for a (macro, primitive) combination. Then, the combination generator predicts the results for a (macro, scenario = (primitive 1, ..., primitive n)), given the predicted results for (macro, primitive 1), ..., (macro, primitive n). Given a trained cost predictor, the recommendation engine reads in the information about a target macro (the macro name or the sensitivity-test results) and the design objective function (from the Rules file), and outputs a list of recommended scenarios. These recommendations are made to maximize the target macro's design objective value estimated using the cost predictor.

18.5.2 Improving Compute Footprint Efficiency

Minimizing compute resources and disk space while maintaining quality of results is one area of continual research for SynTunSys. Streamlining the SynTunSys compute footprint (Fig. 18.7) is one thread of effort that would improve the compute resource usage, independent of the underlying decision algorithm.

One method to reduce SynTunSys resource requirements is to reuse the compute effort of scenarios that "overlap," i.e., share the same parameter settings over a subset of steps during execution of the synthesis program. Each synthesis scenario run by SynTunSys has been described thus far as a separate synthesis run. However, quite often many of the scenarios execute the same synthesis code up to a certain point and then diverge, e.g., when parameters are set to different values causing execution of different synthesis code branches. Figure 18.22 illustrates two overlapping operations that can be applied to realize this concept. The first technique is to "checkpoint" runs after each executing steps that are expected to run rerun in scenarios during future iterations. Figure 18.22-b1.1 illustrates that during the first iteration ($i = 0$) certain scenarios are checkpointed by writing the partially completed synthesis runs as well as information to resume to the run to disk. During the next iteration ($i = 1$), new scenarios can resume the partially completed

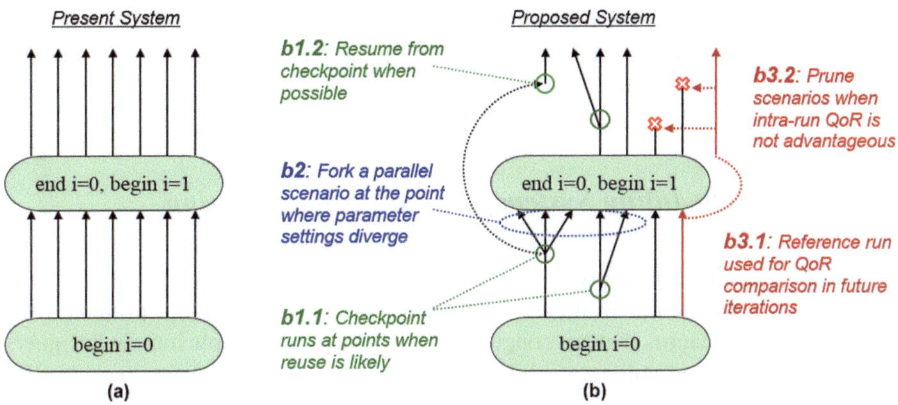

Fig. 18.22 Proposed SynTunSys compute efficiency enhancements: (**a**) the present system runs scenarios in parallel, through all steps of the flow. (**b**) The proposed enhancements would allow overlapping scenarios to reuse compute effort and non-promising scenarios to be pruned

checkpointed runs, i.e., Fig. 18.22-b1.2. Checkpointed runs can also be "forked" into multiple unique scenarios by resuming prior to the divergence of parameters as Fig. 18.22-b2 illustrates.

Pruning non-promising scenarios in flight is a third mechanism that can improve compute efficiency [11]. This technique relies on QoR statistics from partially completed synthesis runs to predict whether allowing a scenario to continue to the next step in the flow will lead to a competitive result. There are a number of potential policies for choosing whether to prune a scenario. Figure 18.22-b3 illustrates how a reference run can be used as a QoR comparison for scenarios in future iterations. Alternate policies such as only allowing a subset of the top scenarios to survive after certain steps or requiring a minimum QoR for survival are also possible. Another approach would be to train a model using historical data from the SynTunSys Archive to predict whether a scenario should be pruned. In terms of the expected improvements from pruning, an initial investigation into pruning techniques suggests that a 20% reduction in compute resources is achievable [11].

While these proposed mechanisms would allow improving compute efficiency, there are some potential drawbacks that we should discuss. First, for all the techniques, a more complex common compute footprint is needed, requiring additional bookkeeping, disk space, and inter-process communication, although these are technical challenges that would most likely be overcome with coding and testing effort. One other issue is the existence of run-to-run QoR variation in the synthesis tools. Although, it is generally expected that advanced synthesis tools will not provide exactly the same solution when running the same scenario multiple times, checkpointing may amplify issues with run-to-run variability if a run having a large deviation from the mean is checkpointed and reused for many future scenarios. In terms of scenario pruning, this approach would reduce the non-attractive scenarios logged in the SynTunSys Archive that may result in training data that has fewer occurrences of scenarios to avoid. Overall though, these compute efficiency improvement techniques should lead to more effective versions SynTunSys, if implemented wisely.

18.5.3 A Scheduling System for Multi-Macro Tuning

While the optimization approaches described in the previous sections are effective for achieving QoR improvements for a single macro, large design projects often consist of many macros that are concurrently designed by multiple human designers. Furthermore, limited compute resources, even in an industrial setting, require ROI considerations when investing effort into tuning the parameters of macros. For example, during the processor tuning described in [2] there were many times when the compute cluster was heavily loaded with parameter tuning jobs and other times when the cluster was relatively idle. This unpredictable pull on compute resources inherently arises when designers work independently without global project ROI considerations and/or without tightly coupled communication of compute needs. These challenges motivate a solution for enhancing the cumulative parameter tuning QoR for an entire design project that consists of multiple macros.

Our novel solution is the *SynTunSys Scheduler*, a.k.a., *STSS*, a system that manages multiple SynTunSys runs for multiple macros [3]. This system works at a higher level of abstraction that considers the ROI of SynTunSys runs at the project level. Figure 18.23 shows a diagram of the components and processes of STSS.

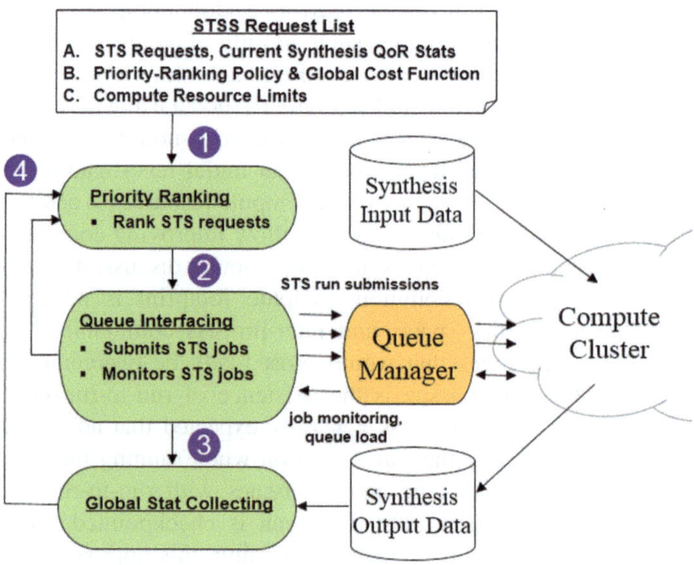

Fig. 18.23 Architecture of the SynTunSys Scheduler (STSS)

The general goal of STSS is to take a list of SynTunSys run requests and optimally determine the order in which to submit them to the queue manager, given resource limits. For our work we use the Platform LSF workload management system, which is the commercial software based on the Utopia project [12]. The more general problem STSS addresses is CAD-tool scheduling, which recently is receiving more attention, e.g., for scheduling architectural simulations for a single design [13].

The process begins at Step (1) in Fig. 18.23 where the following inputs are provided to STSS: (A) a list of SynTunSys run requests and a reference set of synthesis QoR stats for the multiple macros (note that if the reference QoR stats are not available, STSS will first schedule one synthesis run on all macros in the list to generate the reference stats), (B) a priority ranking policy and global cost function, which will be described below, and (C) compute resource limits. Given these input data, STSS creates a priority ordered list of the SynTunSys requests. There are multiple policies that can be employed for the priority ranking, as we describe later. Next, at Step (2), the queue interfacing component of STSS submits one or more SynTunSys runs to the queue manager (existing software, e.g., Platform LSF), which ultimately starts the synthesis jobs on the compute cluster. After submission, an STSS monitor process starts to interact with the queue manager to monitor progress of the SynTunSys jobs and to sample the compute cluster load. Step (3) in Fig. 18.23 is triggered whenever a SynTunSys run completes and involves a result collection process that adds the new SynTunSys tuning results to the list of the current best synthesis results for all macros. The feedback process is then initiated as Step (4) where the updated list of synthesis results for all macros, compute cluster load information, and compute resource limits are passed to the priority ranking algorithm. After updating the priority ranking, the queue interface determines if more SynTunSys runs should be submitted.

18.5.3.1 STSS Priority Ranking

The key component of STSS is the priority ranking that predicts which macros will provide the highest ROI from tuning. There are multiple policies for this prediction. Due to space limitations we describe only two possible ranking policies.

Policy 1 This policy ranks macros based on a cost analysis of existing synthesis run stats for each macro. The cost analysis can use the same "normalized weighted sum" cost function and metrics from Eq. (18.1). However, in this case the goal is to compare a single set of synthesis metrics for multiple macros, rather than compare multiple scenarios from a single macro. This policy effectively works on the worst macros based on the current QoR. One possible shortcoming is that this policy assumes that tuning macros with worse QoR will provide strong tuning improvements. In general, this policy is a depth-first approach to the multi-macro problem.

Policy 2 Alternatively, Policy 2 approaches the multiple macros in a breadth-first manner. The policy first performs a sensitivity test on all macros and then ranks macros based on QoR improvements from the sensitivity test. The sensitivity test could in fact be the first iteration of a SynTunSys run or a simplified test that is used to only rank macros. A low-effort *Jump-Start* test sensitivity as described in Sect. 18.5.1.1 would also work well in this context. The advantage of this policy is that it samples the actual tuning QoR potential before investing in a complete SynTunSys run. The downside is that the up-front effort invested in the sensitivity test could be used to directly tune macros.

18.5.3.2 Second Pass SynTunSys Runs

Another tradeoff to consider is that multiple SynTunSys runs can be applied to a macro to further explore the design space. These *second* pass runs build off the results of the first SynTunSys run by keeping top performing primitives and removing poor performing primitives from the search space. New unexplored primitives are then added to the second pass SynTunSys run. Thus, choosing whether to perform a second pass tuning run on a previously tuned macro or tune a new macro is a decision for the priority ranking algorithm. Although these multi-pass (second pass and beyond) SynTunSys runs can lead to diminishing returns, the next subsection will show the advantages.

18.5.3.3 STSS Results

To demonstrate the effectiveness of STSS we apply the system to the 12 macros from Sect. 18.4.3 and use the Learning3+ algorithm for all SynTunSys runs. For this example we use Policy 1 and the same balanced cost function is used for the tuning results from Sect. 18.4.3. Also for brevity, we assume that tuning runs are executed sequentially, i.e., one at a time. First, we consider tuning the 12 macros without allowing second pass tuning runs. Figure 18.24a shows results comparing cumulative improvement of three key metrics using the STSS Policy 1 ordering vs. a random macro ordering. We compare against random macro ordering because during an actual design project, without a centralized higher level system like STSS, designers effectively submit SynTunSys runs at will without considering macro priority, thus a random ordering reflects a realistic industrial setting. To simulate this, we generate 1000 random orderings of the 12 macros and then average the QoR improvements across them. The results show the distinct advantage of STSS in terms of total negative slack (TNS), power, and routability. We highlight one sample point in the plot where half the macros are tuned. At this point STSS has a 24% TNS advantage and a 6% power advantage over random tuning. Furthermore after tuning only half the macros, STSS has achieved 58% out of the total 62% TNS savings and 9% out of 10% power savings available from tuning all 12 macros. Thus, the STSS priority ordering allows us to evaluate ROI options and determine whether continuing to tune macros is beneficial or has diminishing returns.

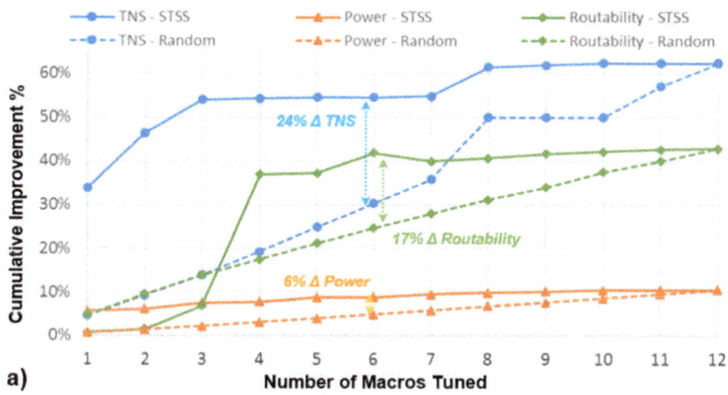

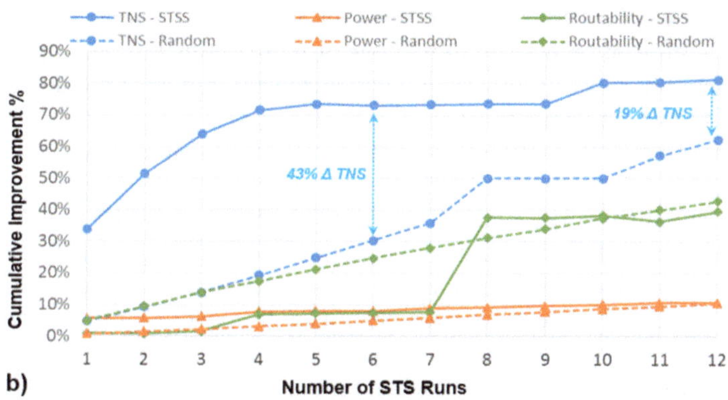

Fig. 18.24 Results of tuning 12 macros based on STSS priority ordering vs. random ordering. (**a**) Priority ranked tuning vs. random tuning. (**b**) Priority ranked tuning vs. random tuning, allowing second pass tuning runs

Next, we consider the same case study of the 12 macros but allow the option of second pass SynTunSys runs. Figure 18.24b shows the results of this experiment. We perform 12 total tuning runs, where the random ordering results are the same as in Fig. 18.24a. For the STSS case we perform 12 total tuning runs where with the second pass SynTunSys option not all macros are necessarily tuned. In this experiment, the priority ranking algorithm chooses to tune six macros twice, one macro once, and does not tune five macros. The ability to choose second pass SynTunSys runs leads to a 19% TNS improvement compared to random ordering after all 12 SynTunSys runs and a 43% TNS advantage after only 6 SynTunSys runs.

18.6 Related Work

The synthesis parameter tuning problem we address can be classified as a black-box optimization problem, i.e., we treat the synthesis program as black-box software by supplying input conditions (input data and parameter settings) and measuring the output response in terms of synthesis quality of results (QoR). Black-box problems are often approached using techniques from the field of simulation optimization [14], which is an umbrella term for optimization techniques that operate in the absence of an algebraic model of the system. Since each macro exhibits a unique input–output response to the synthesis parameter settings and digital logic can take on an intractable number of functionalities, the synthesis tool-flow of our focus is far too complex to be modeled algebraically. Black-box optimization techniques can also be employed for DSE purposes. However, unlike conventional DSE, the goal of black-box optimization is often to find one or more optimal or near-optimal design points without necessarily requiring a complete exploration of the design space to determine the whole Pareto frontier of tradeoff points.

Black-box parameter optimization is a common problem seen across a number of fields, e.g., compiler tuning [15] and software engineering [16, 17]. Hyperparameter tuning of neural networks is also a very active area of research [18]. With respect to VLSI design, DSE is becoming a more attractive solution for complex problems across various levels of abstraction. At the architectural level, many DSE studies based on models or simulators have been used to explore multi-objective design spaces, e.g., [19]. Architectural-level studies, however, typically do not result in implemented designs. DSE approaches have been used for high-level synthesis by leveraging machine-learning methods [20] and for FPGA synthesis by tuning parameters with genetic algorithms [21], Bayesian optimization [22], and ensemble approaches [23].

In comparison to the works mentioned above, SynTunSys is unique in that it operates on a general-purpose synthesis tool-flow targeting VLSI chip design. Also, SynTunSys has been proven in an industrial setting, being used for multiple high-performance processors, currently in production in advanced technology nodes.

Furthermore, when compared with recent works on FPGA parameter tuning, SynTunSys does not require the human designer hints that have been used for genetic algorithms in [21] and our in-house learning algorithm has advantages over the Bayes approaches in [22]. In particular, our Learning algorithm from Sect. 18.3.2 performs cost ranking after each iteration, as opposed to the classification that standard machine-learning algorithms, such as Bayes or SVM, would perform. The ranking allows predicting the top-k most promising scenarios for the next iteration, as opposed to random sampling followed by classification. Thus, we believe the Learning algorithm may converge faster than the classification approaches and is more appropriate for low-latency optimization. In fact, we often see high-quality scenarios emerge after only two iterations.

18.7 Conclusions

To the best of our knowledge, SynTunSys is the first self-evolving and autonomous system for tuning the input parameters of logic and physical synthesis tools. By taking over the process of tuning the input parameters and by learning automatically from the information of previous synthesis runs, SynTunSys realizes a new level of abstraction between designers and tool developers. SynTunSys has been applied to IBM server processors across three technology nodes (32, 22, and 14 nm). A systematic study of SynTunSys during the design of the IBM z13 22 nm server processor revealed SynTunSys provided, on average, a 36% improvement in total negative slack and a 7% power reduction. We have also presented enhancements to the SynTunSys system across various fronts, including a novel parameter tuning algorithm employing adaptive online learning that improves quality of results over our original algorithm. Results using this new algorithm were presented based on 22 and 14 nm high-performance industrial server components. Furthermore, we presented a number of additional threads of research in progress, covering topics such as history-based recommender systems, enhancements for computing efficiency, and a novel higher-level system that manages parameter tuning of multiple designs. Overall, we feel SynTunSys may be the first production example of a new breed of automation tools that targets not only design automation, but also automation of designer decisions that are conventionally performed by humans.

References

1. J.D. Warnock et al., 22nm next-generation IBM system z microprocessor, in *ISSCC*, 2015
2. M.M. Ziegler et al., A synthesis-parameter tuning system for autonomous design-space exploration, in *DATE*, 2016
3. M.M. Ziegler, H.-Y. Liu, L.P. Carloni, Scalable auto-tuning of synthesis parameters for optimizing high-performance processors, in *ISLPED*, 2016
4. M.M. Ziegler et al., POWER8 design methodology innovations for improving productivity and reducing power, in *CICC*, 2014
5. L. Trevillyan et al., An integrated environment for technology closure of deep-submicron IC designs. IEEE Des. Test Comput. **21**(1), 14–22 (2004)
6. M.M. Ziegler, G.D. Gristede, V.V. Zyuban, Power reduction by aggressive synthesis design space exploration, in *ISLPED*, 2013
7. E.J. Fluhr et al., POWER8: a 12-core server-class processor in 22nm SOI with 7.6Tb/s off-chip bandwidth, in *ISSCC*, 2014
8. C. Gonzalez et al., POWER9: a processor family optimized for cognitive computing with 25Gb/s accelerator links and 16Gb/s PCIe Gen4, in *ISSCC*, 2017
9. C. Berry et al., IBM z14: 14nm microprocessor for the next-generation mainframe, in *ISSCC*, 2018
10. Y. Koren, R. Bell, C. Volinsky, Matrix factorization techniques for recommender systems. Computer **42**, 30–37 (2009)
11. M. Anwar, S. Saha, M.M. Ziegler, L. Reddy, Early scenario pruning for efficient design space exploration in physical synthesis, in *International Conference on VLSI Design (VLSID)*, 2016

12. S. Zhou et al., *Utopia: A Load Sharing Facility for Large, Heterogeneous Distributed Computer Systems* (Wiley, New York, 1993)
13. G.P. Mariani et al., DeSpErate++: an enhanced design space exploration framework using predictive simulation scheduling. IEEE Trans. Comput. Aided Des. Integr. Circuits Syst. **34**(2), 293–306 (2015)
14. S. Amaran et al., Simulation optimization: a review of algorithms and applications. 4OR: Q. J. Oper. Res. **12**, 301–333 (2014)
15. G. Fursin et al., Milepost GCC: machine learning enabled self-tuning compiler. Int. J. Parallel Prog. **39**, 296–327 (2011)
16. A. Arcuri, G. Fraser, Parameter tuning or default values? An empirical investigation in search-based software engineering. Empir. Softw. Eng. **18**(3), 594–623 (2013)
17. H.H. Hoos, Programming by optimization. Commun. ACM **55**(2), 70–80 (2012)
18. G.I. Diaz, A. Fokoue-Nkoutche, G. Nannicini, H. Samulowitz, An effective algorithm for hyperparameter optimization of neural networks. IBM J. Res. Dev. **61**(4), 9:1–9:11 (2017)
19. O. Azizi et al., An integrated framework for joint design space exploration of microarchitecture and circuits, in *DATE*, 2010
20. H.-Y. Liu, L.P. Carloni, On learning-based methods for design-space exploration with high-level synthesis, in *DAC*, 2013
21. M.K. Papamichael, P. Milder, J.C. Hoe, Nautilus: fast automated IP design space search using guided genetic algorithms, in *DAC*, 2015
22. N. Kapre et al., Driving timing convergence of FPGA designs through machine learning and cloud computing, in *FCCM*, 2015
23. C. Xu et al., A parallel bandit-based approach for autotuning FPGA compilation, in *FPGA*, 2017

Chapter 19
Multicore Power and Thermal Proxies Using Least-Angle Regression

Rupesh Raj Karn and Ibrahim (Abe) M. Elfadel

19.1 Introduction

Power and thermal modeling of multicore processors using performance counters (PCs) has been widely researched in both academia and industry [1]. Most of these models are based on various forms of linear regression using an extensive set of architectural performance counters or micro-architectural activity monitors [2] as the regression variables. The industry has given the name of power proxies to these regression-based formulas because they are used to estimate or predict core power in the absence of per-core power measurements. The weights of the proxy's regression formula are typically computed off-line using workloads from industry-standard benchmarks such as SPEC CPU 2006, PARSEC, SPECPower, and SYSmark. The number of regression variables can be as high as 50 for multicore processors [2, 3]. In per-core proxies based on micro-architectural monitors, there could be as many as 30 activity monitors per core [3]. An activity-based power model is created in [4] for GPUs. Activity factors are identified to calculate power for micro-architectures on GPUs. The real-time power consumption is estimated using NVIDIA's Management Library. A 2-layer model of power consumption is generated in [5] for a hybrid supercomputer that combines CPU, GPU, and MIC technologies. The model takes the number and location of resources used by the programs and calculates the resulting system-level power consumption. Workload configurations are evaluated

R. R. Karn
Department of Electrical and Computer Engineering, Khalifa University of Science and Technology, Abu Dhabi, United Arab Emirates
e-mail: rupesh.karn@ku.ac.ae

I. M. Elfadel (✉)
Department of Electrical and Computer Engineering and Center for Cyber Physical Systems, Khalifa University, Abu Dhabi, UAE
e-mail: ibrahim.elfadel@ku.ac.ae

I. M. Elfadel et al. (eds.), *Machine Learning in VLSI Computer-Aided Design*,
https://doi.org/10.1007/978-3-030-04666-8_19

from a power perspective and more the efficient ones are selected. The model is also used for power-aware scheduling and anomalous behavior prediction. Often times, the same set of proxy variables is used across the full set of benchmarks and the same set of weights is used for power prediction and estimation.

In runtime power and thermal management, these proxies have shortcomings that can be summarized as follows:

1. The high number of regression variables, architectural or micro-architectural, severely limits the scalability of these proxies to a high number of cores.
2. Although they are optimized for a full database of workloads, there is no guarantee they will adequately predict power for new workloads.
3. They provide no guidance as to the regression variable most relevant or most correlated with the power consumed by the workload. Such guidance is crucial for selecting the control policy most appropriate to mitigate the most important sources of power dissipation as may be the case under power capping constraints.

Attempts have already been made to address the scalability issue by introducing ultracompact power proxies that are able to predict power with less than 3% error [6]. Unlike the traditional power proxies, these ultracompact ones are nonlinear and have the distinct benefit of predicting consumed power across idle and active states using regression variables modulated by active-state residency.

For workload adaptation, one possible approach is to use signal processing methods such as wavelet analysis [7] to detect workload phases and workload transitions and then select the regression weights that minimize the power estimation error for the newly detected workload. Another approach is to use a statistical metric model (SMM) that can pro-actively tune hardware or software to expected application characteristics [8]. For instance, an SMM can predict the memory access rate of different applications and thus guide the selection of power reduction techniques such as power gating and dynamic voltage and frequency scaling (DVFS). As for the selection of the control variable, one well-known ad-hoc method is to use principal component analysis (PCA) to find the linear combination of regression variables most correlated with the consumed power. This benefit of PCA is also its main disadvantage as it requires that several regression variables be linearly combined before taking any control action.

In this chapter, we propose the use of the least-angle regression (LAR) algorithm that first appeared in the statistical learning literature [9], to address the issues of workload adaptation and control variable selection. The main advantages of using LAR are correlation awareness, incremental model building, and the automatic selection of the regression subset.

We make use of the LAR advantages to develop an incremental proxy building algorithm for both power and temperature. The algorithm enables the progressive selection of a subset of regression variables from the full set of PC measurements. This subset can be interpreted as the workload signature in the PC space. Each workload will have a power signature and a thermal signature. Such signatures can be used to categorize, cluster, or differentiate workloads in real-time applications. In our work, we find that these correlations are essential to classify whether a set of given threads have similar workload signatures which, in turn, helps in mapping threads to cores to increase overall performance and reduce power consumption. They can also be used to refine any power or temperature management policies by making them adapt to the workload once its signature is ascertained.

The number of regression variables is determined by the desirable level of accuracy for the power or temperature proxy. The LAR algorithm constructs a nested set of regression models and orders the selected regression variables according to the correlation of the PC data vectors with the residual vector obtained from the predecessor regression model.

This chapter is a synthesis of our earlier work on multicore power and thermal proxies [10–12].

19.2 Preliminaries

In this section, we give the mathematical and algorithmic background needed to appreciate the role that least-angle regression plays as a regression subset selector in linear regression analysis. The generic situation is as follows. We are given a set of p input variables $X^T = (X_1, X_2, \ldots, X_p)$ and an output variable Y, then the linear regression model of Y given *all* the input variables is given by

$$Y = \beta_0 + \sum_{i=0}^{p} \beta_i X_i \tag{19.1}$$

The scalars $\beta_i, 0 \leq i \leq p$, are the regression coefficients that need to be estimated based on observations of the inputs and output. Very often, the input variables are centered around 0, and in this case, the bias regression coefficient $\beta_0 = 0$. Throughout this section, we will make the assumption that the input variables are centered around zero. To estimate the regression coefficients, we assume that N observations are taken of each input $\boldsymbol{x}_i = (x_{1i}, x_{2i}, \ldots, x_{Ni})^T, 0 \leq i \leq p$, and the output $\boldsymbol{y} = (y_1, y_2, \ldots, y_N)^T$. With these N observations, it is well known that the least-squares estimate of the regression coefficients is given by

$$\hat{\boldsymbol{\beta}} = (\boldsymbol{X}^T \boldsymbol{X})^{-1} \boldsymbol{X}^T \boldsymbol{y} \tag{19.2}$$

where the $N \times p$ matrix $\boldsymbol{X}$ is formed by the row stacking of all the N input observations. The estimated output will then be

$$\hat{\boldsymbol{y}} = \boldsymbol{X}\hat{\boldsymbol{\beta}} = \hat{\boldsymbol{\beta}}(\boldsymbol{X}^T \boldsymbol{X})^{-1} \boldsymbol{X}^T \boldsymbol{y} \tag{19.3}$$

It is well known that the estimated value $\hat{\boldsymbol{y}}$ is the orthogonal projection of the output measurements $\boldsymbol{y}$ on the subspace of $\Re^N$ that is spanned by the p columns of the input measurement matrix $\boldsymbol{X}$ and that $\hat{\boldsymbol{\beta}}$ is the optimal solution of the least-squares minimization problem

$$\hat{\boldsymbol{\beta}} = \underset{\boldsymbol{\beta}}{\operatorname{argmin}} \left[(\boldsymbol{y} - \boldsymbol{X}\boldsymbol{\beta})^T (\boldsymbol{y} - \boldsymbol{X}\boldsymbol{\beta}) \right] \tag{19.4}$$

The estimation error $\boldsymbol{y} - \hat{\boldsymbol{y}}$ is called the residual and is denoted $\boldsymbol{r}$. One important question in regression analysis is whether a subset of the input variables is sufficient for providing acceptable estimates of the output. Brute-force subset selection consists of picking up the subset size $k \in \{1, \ldots, p\}$, evaluating for each such subset, the residual $\boldsymbol{r}_k$ and selecting the subset that results in the smallest residual Euclidean norm. It is clear that such an approach has combinatorial complexity as the number of subsets that should be tried in the worst case is 2^p. The brute-force method of subset selection is therefore practical only for a small number of regression variables.

Another approach for controlling the number of regression variables entering the regression formula is indirectly by imposing constraints on the vector norm of the regression coefficients $\boldsymbol{\beta}$. This class of methods is known under the name of "shrinkage" methods because they are based on "shrinking" a vector norm of $\boldsymbol{\beta}$ below some given value. *Ridge regression* uses the Euclidean norm and results in the optimal solution of the constrained least-squares problem

$$\hat{\boldsymbol{\beta}}_{ridge} = \underset{\|\boldsymbol{\beta}\|_2^2 \leq t}{\operatorname{argmin}} \left[(\boldsymbol{y} - \boldsymbol{X}\boldsymbol{\beta})^T (\boldsymbol{y} - \boldsymbol{X}\boldsymbol{\beta}) \right] \tag{19.5}$$

where

$$\|\boldsymbol{\beta}\|_2^2 = \sum_{i=1}^{p} \beta_i^2$$

When the L_1 norm is used, we obtain the so-called *lasso*[1] regression estimate

$$\hat{\boldsymbol{\beta}}_{lasso} = \underset{\|\boldsymbol{\beta}\|_1 \leq t}{\operatorname{argmin}} \left[(\boldsymbol{y} - \boldsymbol{X}\boldsymbol{\beta})^T (\boldsymbol{y} - \boldsymbol{X}\boldsymbol{\beta}) \right] \tag{19.6}$$

where

$$\|\boldsymbol{\beta}\|_1 = \sum_{i=1}^{p} |\beta_i|$$

It is well known that for a given shrinkage factor t, the lasso regression results in a sparser set of regression coefficients than ridge regression, that is, a set where several of the regression coefficients are zero. The reader is referred to Chapter 3 in [13] for more details about the subset selection and shrinkage methods for regression modeling.

Least-angle regression (LAR) is closely related to *lasso*, and in one of its variations, provides a computationally efficient algorithm for computing $\hat{\boldsymbol{\beta}}_{lasso}$.

[1] LASSO stands for least absolute shrinkage and selection operator.

For the description of LAR, we will make the additional assumption that the measured input vectors $\boldsymbol{X}$ are not only centered but also normalized so that the $\|\boldsymbol{X}\|_2 = 1$. LAR starts by initializing $\beta_i = 0, 1 \le i \le p$ so that $\hat{\boldsymbol{y}}_0 = \boldsymbol{0}$, and setting the initial residual vector $\boldsymbol{r}_0 = \boldsymbol{y}$. LAR is an iterative algorithm in which at each step, a new regression variable enters the subset of active regression variables. At step k, we denote such subset by $\mathscr{A}_k$. Initially, $\mathscr{A}_0 = \phi$, the empty set. At Step 1, the regression variable that enters the active set is the one with the largest absolute correlation with the residual $\boldsymbol{r}_0$. The first-step correlations are computed as

$$\boldsymbol{c}_1 = \boldsymbol{X}^T \boldsymbol{r}_0 = \boldsymbol{X}^T \boldsymbol{y} \tag{19.7}$$

To simplify the algebra, we assume all the correlations non-negative, and let $\boldsymbol{x}_i$ be the input vector having the largest correlation with the output. Then based on this one regression vector, the output estimate is updated according to the rule

$$\hat{\boldsymbol{y}}_1 = \hat{\boldsymbol{y}}_0 + \hat{\gamma}_1 \boldsymbol{x}_i \tag{19.8}$$

In LAR, the choice of γ_1 is such that the current residual, namely

$$\boldsymbol{r}(\gamma_1) = \boldsymbol{y} - \gamma_1 \boldsymbol{x}_i \tag{19.9}$$

has equal correlation with $\boldsymbol{x}_i$ and its next competing variable $\boldsymbol{x}_j$. From the correlation equality

$$\boldsymbol{x}_i^T \boldsymbol{r}(\gamma_1) = \boldsymbol{x}_j^T \boldsymbol{r}(\gamma_1) \tag{19.10}$$

we get

$$\gamma_1 = \frac{\boldsymbol{y}^T (\boldsymbol{x}_i - \boldsymbol{x}_j)}{1 - \boldsymbol{x}_i^T \boldsymbol{x}_j} \tag{19.11}$$

The feature of *least angle* is due to the fact that the equality of the correlations in Eq. (19.10) amounts to the equality of the angles made between the vector $\boldsymbol{r}(\gamma_1)$ and each of the unit vectors $\boldsymbol{x}_i$ and $\boldsymbol{x}_j$. In other words, γ_1 is chosen so that $\boldsymbol{r}(\gamma_1)$ is a bisector of the angle between $\boldsymbol{x}_i$ and $\boldsymbol{x}_j$

To present the general case, we follow the treatment in Section 2 of Efron's seminal work [9] but with the simplification that all the correlations between the current residuals and the active regressors are non-negative. We also follow [9] and assume that the input measurements are linearly independent. Now let $\mathscr{A}$ be the subset of indices within $1, \ldots, p$ of the current LAR iteration. The input measurements corresponding to this subset are

$$\boldsymbol{X}_{\mathscr{A}} = \{x_j\}_{j \in \mathscr{A}}$$

The LAR algorithm takes the following steps:

1. Solve the least-squares problem with respect to the inputs $\boldsymbol{X}_{\mathscr{A}}$ and output $\boldsymbol{Y}$, and form the current estimate $\hat{\boldsymbol{y}}_{\mathscr{A}}$.
2. Compute the unit *equiangular* vector $\boldsymbol{u}_{\mathscr{A}}$ of the subset $\mathscr{A}$. This vector makes the same angle with each vector $\boldsymbol{x}_j$, $j \in \mathscr{A}$.
3. Update the output estimate $\hat{\boldsymbol{y}}_{\mathscr{A}}$ along $\boldsymbol{u}_{\mathscr{A}}$ by selecting the scalar $\gamma_{\mathscr{A}}$ so that

$$\hat{\boldsymbol{y}}_{\mathscr{A}+} = \hat{\boldsymbol{y}}_{\mathscr{A}} + \gamma_{\mathscr{A}} \boldsymbol{u}_{\mathscr{A}} \tag{19.12}$$

and that the current residual $\boldsymbol{r}(\gamma_{\mathscr{A}})$ has equal correlations with $\boldsymbol{x}_j$, $j \in \mathscr{A}$ and an additional input vector of index $\hat{j} \in \mathscr{A}^c$, the complement of $\mathscr{A}$.

The least-squares regression estimate based on the subset $\mathscr{A}$ is given by

$$\hat{\boldsymbol{y}}_{\mathscr{A}} = \boldsymbol{X}_{\mathscr{A}} \boldsymbol{\beta}_{\mathscr{A}} = \boldsymbol{X}_{\mathscr{A}} \boldsymbol{G}_{\mathscr{A}}^{-1} \boldsymbol{X}_{\mathscr{A}}^T \boldsymbol{y} \tag{19.13}$$

where

$$\boldsymbol{G}_{\mathscr{A}} = \boldsymbol{X}_{\mathscr{A}}^T \boldsymbol{X}_{\mathscr{A}}$$

To derive the equiangular vector, we proceed as follows. We denote by $\boldsymbol{e}_{\mathscr{A}}$ a vector of 1's whose size is equal to the size of $\mathscr{A}$ and use it to define the normalization scalar

$$\alpha_{\mathscr{A}} = \left(\boldsymbol{e}_{\mathscr{A}}^T \boldsymbol{G}_{\mathscr{A}}^{-1} \boldsymbol{e}_{\mathscr{A}} \right)^{-1/2}$$

With the above definitions, the equiangular vector is given by

$$\boldsymbol{u}_{\mathscr{A}} = \alpha_{\mathscr{A}} \boldsymbol{X}_{\mathscr{A}} \boldsymbol{G}_{\mathscr{A}}^{-1} \boldsymbol{e}_{\mathscr{A}} \tag{19.14}$$

Clearly, $\|\boldsymbol{u}_{\mathscr{A}}\|^2 = 1$, and furthermore,

$$\boldsymbol{X}_{\mathscr{A}}^T \boldsymbol{u}_{\mathscr{A}} = \alpha_{\mathscr{A}} \boldsymbol{e}_{\mathscr{A}}$$

which shows that $\boldsymbol{u}_{\mathscr{A}}$ is indeed equiangular. The third step is to update the least-squares estimate as in Eq. (19.12). The update is based on the selection of $\gamma_{\mathscr{A}}$ so that the current residual has identical correlations with $\boldsymbol{X}_{\mathscr{A}}$ and an additional input vector of index $\hat{j} \in \mathscr{A}^c$, the complement of $\mathscr{A}$. The current residual is given by

$$\boldsymbol{r}(\gamma_{\mathscr{A}}) = \boldsymbol{y} - \boldsymbol{y}_{\mathscr{A}+} = \boldsymbol{y} - \boldsymbol{y}_{\mathscr{A}} - \gamma_{\mathscr{A}} \boldsymbol{u}_{\mathscr{A}} \tag{19.15}$$

The correlation of the current residual with the variables of the active set $\mathscr{A}$ is given by

$$\begin{aligned}\boldsymbol{X}_{\mathscr{A}}^{T}\boldsymbol{r}(\gamma_{\mathscr{A}}) &= \boldsymbol{X}_{\mathscr{A}}^{T}\left(\boldsymbol{y}-\boldsymbol{y}_{\mathscr{A}}\right)-\gamma_{\mathscr{A}}\boldsymbol{X}_{\mathscr{A}}^{T}\boldsymbol{u}_{\mathscr{A}} \\ &= \hat{C}_{\mathscr{A}}\boldsymbol{e}_{\mathscr{A}}-\gamma_{\mathscr{A}}\alpha_{\mathscr{A}}\boldsymbol{e}_{\mathscr{A}} = \left(\hat{C}_{\mathscr{A}}-\gamma_{\mathscr{A}}\alpha_{\mathscr{A}}\right)\boldsymbol{e}_{\mathscr{A}}\end{aligned} \tag{19.16}$$

where the first term in Eq. (19.16) results from the LAR assumption at its $\mathscr{A}$ subset iteration that all the variables of the active set have the same correlation with the residual $\boldsymbol{y}-\boldsymbol{y}_{\mathscr{A}}$. The second term results from the definition of the equiangular vector $\boldsymbol{u}_{\mathscr{A}}$. This expression of the correlation with the variables of the active set is at the core of the numerical behavior of the LAR algorithm, namely that once the regression variables are members of the active set, they remain so until the algorithm stops, with all their correlations with the current residual tied up, and decreasing with each update of the active set. This will be further clarified with the numerical examples that will be provided later in this chapter.

Let us now select an index $j \in \mathscr{A}^c$, the complement of the active set, and let us find the correlation of $\boldsymbol{x}_j$ with the current residual $\boldsymbol{r}(\gamma_{\mathscr{A}})$. It is given by

$$\begin{aligned}\boldsymbol{x}_j^{T}\boldsymbol{r}(\gamma_{\mathscr{A}}) &= \boldsymbol{x}_j^{T}\left(\boldsymbol{y}-\boldsymbol{y}_{\mathscr{A}}\right)-\gamma_{\mathscr{A}}\boldsymbol{x}_j^{T}\boldsymbol{u}_{\mathscr{A}} \\ &\triangleq \hat{c}_j-\gamma_{\mathscr{A}}a_j\end{aligned} \tag{19.17}$$

where $\hat{c}_j$ is the correlation with the active-set residual. Note that we have $\hat{c}_j < \hat{C}_{\mathscr{A}}, j \in \mathscr{A}^c$. Equating the right-hand side of Eq. (19.17) with the common value of Eq. (19.16), we see that for each $j \in \mathscr{A}^c$, there is a candidate value of $\gamma_{\mathscr{A}} > 0$ at which the j-th variable enters the active set. The value is given by

$$\gamma_{\mathscr{A}}^{(j)} = \frac{\hat{C}_{\mathscr{A}}-\hat{c}_j}{\alpha_{\mathscr{A}}-a_j} \tag{19.18}$$

Of course, the index $\hat{j}$ that needs to be selected for the update of the active set corresponds to the very first index $j \in \mathscr{A}^c$ whose regression variable satisfies the correlation equality condition. This index $\hat{j}$ corresponds to the minimum value of $\gamma_{\mathscr{A}}^{(j)}$. We therefore have

$$\gamma_{\mathscr{A}} = \min_{j\in\mathscr{A}^c}\gamma_{\mathscr{A}}^{(j)} = \min_{j\in\mathscr{A}^c}\left\{\frac{\hat{C}_{\mathscr{A}}-\hat{c}_j}{\alpha_{\mathscr{A}}-a_j}\right\} \tag{19.19}$$

$$\hat{j} = \underset{j\in\mathscr{A}^c}{\operatorname{argmin}}\gamma_{\mathscr{A}}^{(j)} \tag{19.20}$$

It is instructive to compare the expressions of $\gamma_{\mathscr{A}}^{(j)}$ and $\gamma_{\mathscr{A}}$ with the expression of γ_1 in Eq. (19.10). In particular, in the case of just two regression variables, $\gamma_{\mathscr{A}}$ reduces to γ_1 with the following identification:

$$\boldsymbol{u}_{\mathscr{A}} = \boldsymbol{x}_i,\ \hat{C}_{\mathscr{A}} = \boldsymbol{y}^T \boldsymbol{x}_i,\ \hat{c}_j = \boldsymbol{y}^T \boldsymbol{x}_j,\ \alpha_{\mathscr{A}} = 1,\ a_j = \boldsymbol{x}_i^T \boldsymbol{x}_j \tag{19.21}$$

The LAR algorithm stops after all the regression variables enter the active set. The theoretical properties of LAR are all verified with the numerical results of Fig. 19.1. These results are based on our own data sets for system-level power and thermal modeling of multicore processors.

Typically, each data has six input variables and one output variable. The left-hand side plot in Fig. 19.1 shows the correlation profile in which the regression variable having the highest correlation with the output is selected, and a new variable enters into the LAR active set whenever its correlation with the current residual error is the same as the correlation between the latter and the variables of the LAR active set. This is also shown numerically in Table 19.1. The right-hand side plot of Fig. 19.1 shows that every time a new variable enters the LAR active set, the regression coefficients are updated. The final regression coefficients are calculated when all the variables enter the active set. This is also shown numerically in Table 19.2.

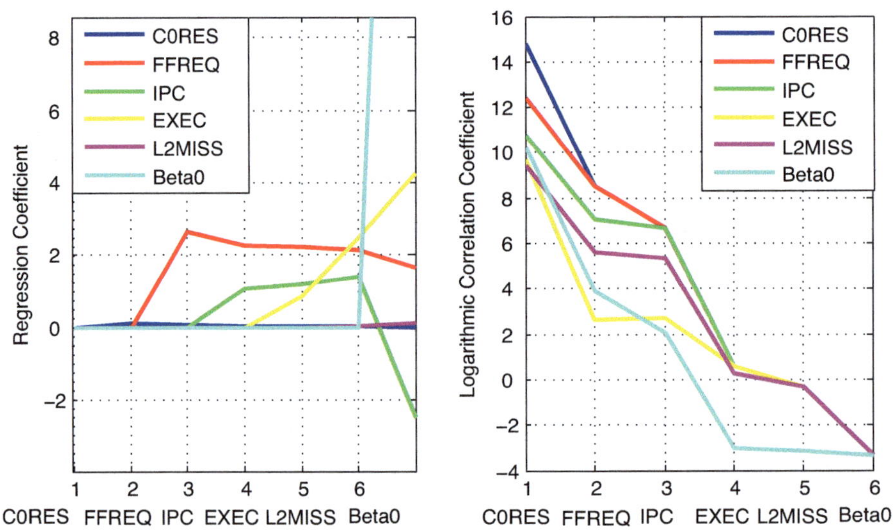

Fig. 19.1 Coefficient and correlation profile

Table 19.1 LAR logarithmic correlation coefficients

CORES	IPC	EXEC	L2MISS	FFREQ	β_0
14.8017	10.7795	9.6648	9.4089	12.4028	10.1969
8.5144	7.0718	2.6317	5.5878	8.5144	3.8989
6.6464	6.6464	2.6867	5.3351	6.6464	2.0224
0.5864	0.5864	0.5864	0.2486	0.5864	−3.0391
−0.3394	−0.3394	−0.3394	−0.3394	−0.3394	−3.1924
−3.3571	−3.3571	−3.3571	−3.3571	−3.3571	−3.3571

Table 19.2 LAR regression coefficients

CORES	FFREQ	IPC	EXEC	L2MISS	β_0
0	0	0	0	0	0
0.0930	0	0	0	0	0
0.0455	2.6252	0	0	0	0
0.0388	2.2626	1.067	0	0	0
0.0304	2.2236	1.1710	0.8448	0	0
0.0133	2.1390	1.3968	2.4581	0.0112	0
−0.0105	1.6538	−2.5136	4.2268	0.0933	38.5934

One advantage of LAR is that it constructs an incremental series of nested subsets of regression variables, each subset differing from its predecessor by one non-zero regression coefficient. The particular subset to choose is determined by model accuracy and complexity requirements.

19.3 Data Collection Platform

Before delving into the details of applying LAR to power and thermal modeling in multicore processors, we would like to briefly describe our data collection platform.

The hardware used is a Dell's Power Edge T620 server equipped with two sockets, each socket containing six cores. The 12 cores are arranged as follows: $Socket_0$ contains the even-numbered cores $k = 0, 2, 4, 6, 8, 10$ and $Socket_1$ contains the odd-numbered cores $k = 1, 3, 5, 7, 9, 11$. The server runs the Red Hat Linux Enterprise (RHLE) operating system. All the cores of a particular socket run on the same clock frequency. The frequency is selected by writing to mode specific registers (MSRs) or alternatively by using kernel modules such as CPUfreq. The voltage is selected internally by the voltage regulation module (VRM), which is internal to the processor. A summary of hardware specifications is shown in Table 19.3. Dell's Power Edge server uses Intel's Sandy Bridge processors

Table 19.3 System specification

Attribute	Specification
Processor	2X Intel Xenon E5-2630
Cores number	6 per socket. Total 2 socket
Available frequency (GHz)	1.2, 1.3, 1.4, 1.5, 1.6, 1.7, 1.8, 1.9, 2.0, 2.1, 2.2, 2.3
L1 I-cache	32 KB/core
L1 D-cache	32 KB/core
L2 cache	256 KB/core
L3 cache	15 MB
Main memory	16 GB DDR3-1333
Bus speed	1333 MHz

with digital energy metering and power limiting capabilities through the Running Average Power Limit (RAPL) interface, which is designed to access the MSRs used for socket energy measurement [14]. With a sampling time of 1 s, the difference between two consecutive MSR sample values gives power consumption. Intel's Performance Counter Monitor (PCM) is used to collect processor event data and other on-chip measurements, including temperature [15]. The PCM has an API and utilities to measure core and system-level internal resources.

The modeling work of this chapter uses performance counters that are common to all cores. They include EXEC (instruction per nominal CPU cycle), FREQ (clock rate normalized to maximum clock rate), AFREQ (clock rate normalized to maximum clock rate while the processor is in active state), IPC (instruction per CPU cycle), L2HIT and L3HIT (cache hit rates for the L2 and L3 caches, respectively), L2MISS and L3MISS (cache miss rates for the L2 and L3 caches, respectively), L2CLK (ratio of CPU cycles lost due to missing L2 cache but hitting L3 cache), L3CLK (ratio of CPU cycles lost due to L3 cache miss), FFREQ (processor clock frequency in GHz), and C0RES (residency in C0 state, i.e., the percentage of time during which the core is active). EXEC is not used in modeling because similar information can be obtained from IPC. Also the FREQ and AFREQ are not considered for modeling because they are simply related to CPU frequency and does not specify any core or system-level resource utilization. The remaining nine PCs are used for per-core power and thermal modeling in Sects. 19.4 and 19.5. As for PCs such as C3RES, C6RES, and C7RES that represent processor sleep states, they are always zero when the processor is running with sleep states disabled. These sleep-state PCs are used in Sect. 19.6 to take sleep-state residency into account in the power and thermal models.

The very first step in our modeling methodology is to consider the dependence of steady-state processor power and temperature on a variety of workloads. Various workloads from the PARSEC benchmarks have been run one at a time on a single core, and core frequency has been increased stepwise from 1.2 to 2.3 GHz, with every step lasting for 30 min. The power and temperature samples have been recorded, and samples of the last 15 min have been averaged for each frequency to get the steady-state values. Figure 19.2 shows the power and temperature profiles versus frequency for various workloads.

The first five workloads are from PARSEC and the remaining five are from the SPEC CPU benchmark. The plots show that temperature has stronger dependence on workloads than power at any given core frequency. Predicting core temperature therefore requires a model incorporating workload-dependent parameters that act as a thermal workload signature in the PC space. It is also intuitively clear that at a given level of accuracy, more PCs may be needed for a temperature proxy than for a power proxy. For both temperature and power, we use LAR to develop accurate, workload-adapted proxies based on power and thermal workload signatures in the PC space. For both cases, we first start with a single-core proxy and then extend it and validate it for a multicore processor.

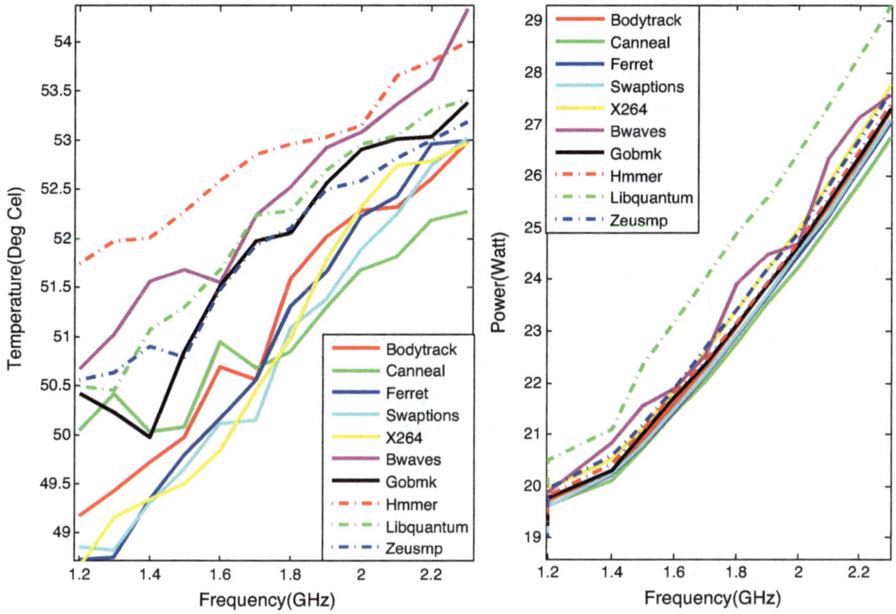

Fig. 19.2 Temperature and power profile vs. frequency and workload

19.4 Power Proxies

The total power consumed by a processor is the sum of dynamic and static power. The static power is due to leakage currents and is a function of temperature. In CMOS, the leakage current equation is

$$I_{lkg} = i_s \left(\exp\left(\frac{qV}{kT}\right) - 1 \right) \tag{19.22}$$

where i_s is reverse saturation current, V the supply voltage, T the temperature, k Boltzmann's constant ($1.38 \times 10^{-23}\,\mathrm{m^2\,kg\,s^{-2}\,K^{-1}}$), and q the electron charge (1.6×10^{-19} C). The static power, P_s, is given as the product of leakage current and supply voltage

$$P_s = \sum I_{lkg} V \tag{19.23}$$

Higher temperature leads to higher leakage current and therefore higher static power. While the temperature can be predicted through a mathematical model, the leakage current depends on the processor physical design and layout and therefore is more challenging to model.

The dynamic power is a function of clock frequency, the supply voltage, and the switching activity of the logic gates. The processor dynamic power is the sum of the power consumed by processor core *and* non-core components such as memory controllers, I/O circuits, and bus interfaces. Direct measurement of per-core power in a multicore processor is quite complex, especially that all the cores share the same power grid. The use of hardware power meters can ease per-core power measurement when the number of cores is small. When the number of cores increases, the data volume of the hardware power sensors will, in fact, become a bottleneck for efficient per-core power management and core-to-core coordination. In multicore platforms, it is often the case that power is measured on a per-socket basis rather than per-core basis. Dell's Power Edge $T620$ server is a case in point. A scalable approach to per-core power modeling is to use power proxies and performance counters that have been subselected according to LAR. We first describe the single-core modeling approach and then generalize it to multicore processors.

19.4.1 The Single-Core Case

For a processor with k cores, workloads from the PARSEC and SPEC CPU 2006 benchmarks have been run on core $k = 0$ while all other cores have been disabled. For collecting training data, workloads have been selected randomly from PARSEC = $\{bodytrack, blackscholes, fluidanimate, streamcluster\}$ and from SPEC CPU = $\{milc, gamess, zeusmp, perlbench\}$. The DVFS frequency has been stepped several times from 1.2 to 2.3 GHz. Data of all the experiments have been concatenated and fed into LAR. When only core $k = 0$ is active, the socket power given by the RAPL interface is the sum of the power consumed by core $k = 0$ and other non-core components in the processor package. Using the nine PCs described in Sect. 19.3, the power model is

$$P_0 = \beta_0 + \sum_{k=1}^{9} \beta_k X_k \tag{19.24}$$

where the X_k, $1 \le k \le 9$ are the PCs, β_0 is a bias term, and β_k, $1 \le k \le 9$ are the regression coefficients. The collected data is fed to LAR to obtain the correlation order. The results are given in Fig. 19.3.

The correlation profiles indicate that L3HIT, L2HIT, L3CLK, and L2CLK are less correlated with power than the offset term β_0. They can, therefore, be removed from the model. The reduced power proxy is then given by Eq. (19.25).

$$\begin{aligned} P &= \beta_0 + \beta_{\mathrm{CORES}} X_{\mathrm{CORES}} + \beta_{\mathrm{FFREQ}} X_{\mathrm{FFREQ}} \\ &+ \beta_{\mathrm{L2MISS}} X_{\mathrm{L2MISS}} + \beta_{\mathrm{IPC}} X_{\mathrm{IPC}} + \beta_{\mathrm{L3MISS}} X_{\mathrm{L3MISS}} \end{aligned} \tag{19.25}$$

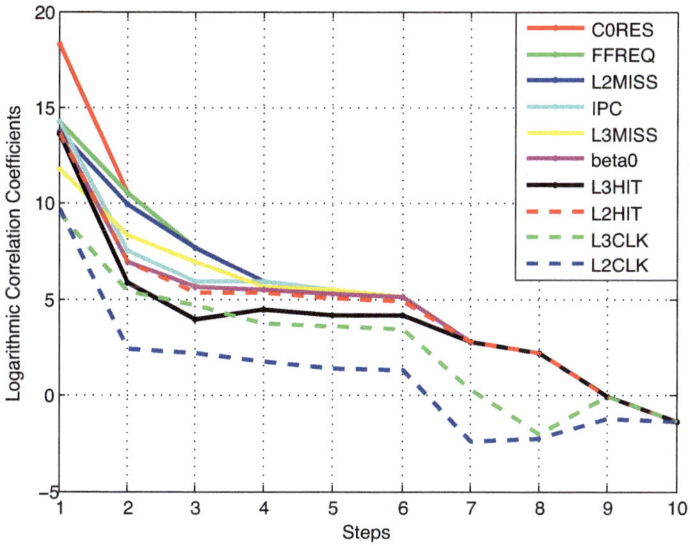

Fig. 19.3 Correlation coefficients of PCs with power

Table 19.4 Coefficients for single-core power model

IPC	L3MISS	L2MISS	CORES%	FFREQ	β_0
−0.0555	−0.0020	0.0333	0.0232	7.2896	7.8141

The final regression coefficients are shown in Table 19.4.

The accuracy of (19.25) is tested using workloads from the PARSEC and SPEC CPU benchmarks outside the training set. The Mean Absolute Percentage Error (MAPE) over N samples

$$\mathrm{MAPE} = \frac{100}{N} \sum_{i=1}^{N} \left| \frac{P_{\mathrm{Measured}} - P_{\mathrm{Calculated}}}{P_{\mathrm{Measured}}} \right| \tag{19.26}$$

ranges from 0.9% to 2.5%, as shown in Fig. 19.5. For the workload "gamess" of SPEC CPU, the comparison between measured and predicted power is shown in Fig. 19.4. The two plots in red and green show the difference in accuracy when a subset of PCs having high correlation with power is selected. The blue power curve is calculated using Eq. (19.24) while the green curve is calculated using Eq. (19.25).

19.4.2 The Multicore Case

For the multicore case, there is no measurement setup to measure per-core power consumption. One option is to estimate the total power consumption of one socket

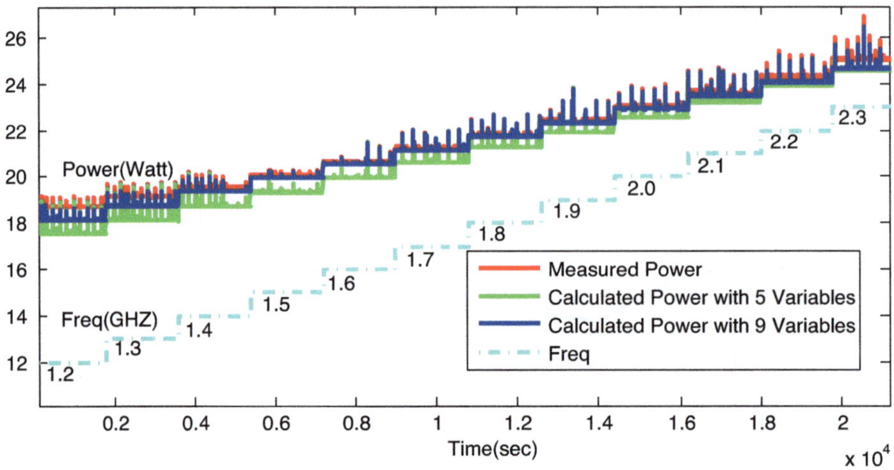

Fig. 19.4 Measured vs. predicted power comparison

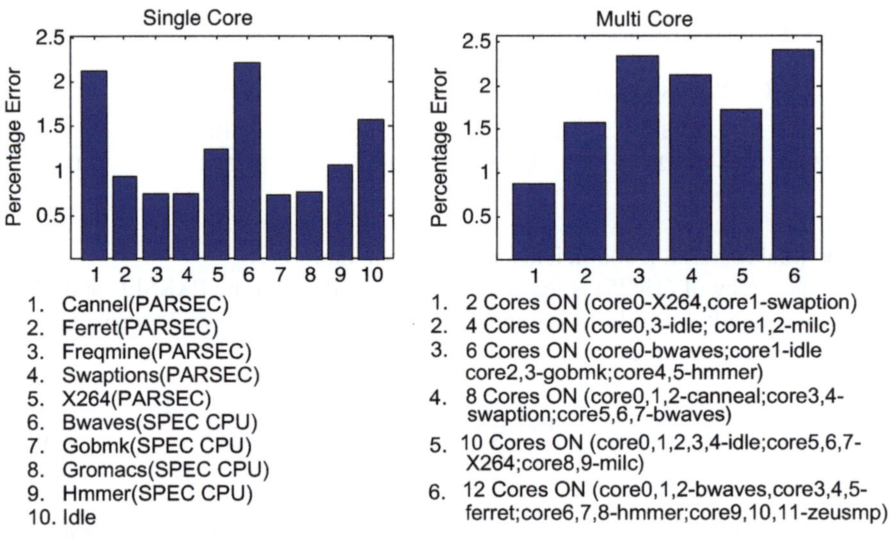

Fig. 19.5 Power model accuracy

using PCs and logically divide this power over the number of active cores based on the type of workload and the intervals of idleness (C0RES represents the percentage of idleness). Using PCs of the highest ranking in correlation order, the power of the k-th core is modeled as

$$P_k = \beta_{\text{C0RES}_k} X_{\text{C0RES}_k} + \beta_{\text{FFREQ}_k} X_{\text{FFREQ}_k} + \beta_{\text{L2MISS}_k} X_{\text{L2MISS}_k} + \beta_{\text{IPX}_k} X_{\text{IPX}_k} + \beta_{\text{L3MISS}_k} X_{\text{L3MISS}_k} \tag{19.27}$$

The sum of the power consumed by all the cores along with the power consumed by the non-core components gives the total power consumption of the processor. Let L be the subset of active cores, then the total power is given by

$$P = \beta_0 + \sum_{k \in L} P_k \tag{19.28}$$

To verify the multicore proxy, the randomly selected training workloads: *bodytrack, freqmine, streamcluster*, and *fluidanimate* from PARSEC and *gamess, gromacs, bzip2*, and *soplex* from SPEC CPU have been run and power consumption and PC data have been collected for different number of active cores ranging from 1 to 12. Data obtained from each experiment are concatenated and fed into LAR. To find the accuracy of the proxy, workloads from outside the training set have been randomly selected and run on one or several cores using linux command *taskset*. The results are shown in Fig. 19.5. As an example, one of our experiments has six active cores with the other cores turned off. Workloads from the PARSEC suite are run: canneal on core $k = 2$ and X264 on core $k = 5$. MAPE for the system power consumption is found to be 2.78%. The plot is shown in Fig. 19.6. It is quite logical that the active cores should consume more power than the idle ones. This is clear for cores $k = 2, 5$.

It is worth noting that the global proxy of Eq. (19.28) remains acceptable when the LAR regression coefficients of a *given* core are used for *all* active cores. The global model for total power consumption is given by Eq. (19.29).

$$P_{\text{total}} = \beta_0 + \beta_{\text{CORES}} \sum_{k \in L} X_{\text{CORES}_k} + \beta_{\text{FFREQ}} \sum_{k \in L} X_{\text{FFREQ}_k} + \beta_{\text{L2MISS}} \tag{19.29}$$

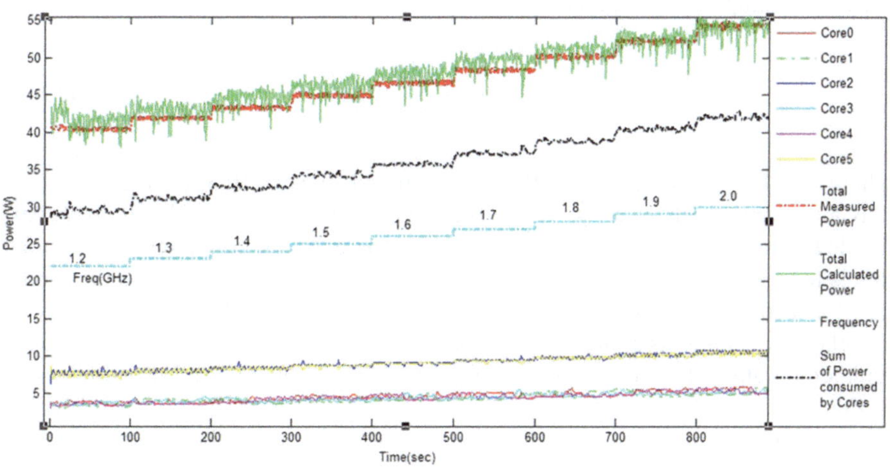

Fig. 19.6 Per-core power and total power comparison

Table 19.5 Coefficients for reduced power model

IPC	L3MISS	L2MISS	CORES%	FFREQ	β_0
−2.579	−0.612	0.32	0.032	2.208	18.113

$$\sum_{k\in L} X_{\mathrm{L2MISS}_k} + \beta_{\mathrm{IPC}} \sum_{k\in L} X_{\mathrm{IPX}_k} + \beta_{\mathrm{L3MISS}} \sum_{k\in L} X_{\mathrm{L3MISS}_k}$$

In Table 19.5, a set of core-based coefficients are shown for which the MAPE of the *global* model is in the range 5–7%.

19.5 Temperature Proxies

For thermal modeling, the core temperature is measured through Intel's PCM interface [15] which reads the maximum temperature across all thermal sensors located in each core. The LAR temperature proxy is compared with this measured temperature to evaluate its accuracy.

19.5.1 *The Single-Core Case*

For the single-core temperature proxy, $Core_0$ was active while all other cores were off. Note that $Core_0$ cannot be turned off because it runs the operating system. In analogy with the single-core power proxy, one possible thermal proxy for predicting the $Core_0$ temperature is given by

$$T_0 = \beta_0 + \sum_{k=1}^{9} \beta_k X_k \tag{19.30}$$

where $X_k, 1 \leq k \leq 9$, are the PCs as discussed in Sect. 19.3, page 579, β_0 is a bias term, and $\beta_k, 1 \leq k \leq 9$, are the LAR coefficients. Workloads have been divided randomly into two parts. The first part is used for collecting training data that is fed to LAR for computing the nested set of regression models and the second part for testing the accuracy of the model. Feeding training data to LAR gives the correlation coefficients along with the correlation ranking of the regression variables with temperature. The correlation coefficient profile is shown on a logarithmic scale in Fig. 19.7.

One experimental observation is that we can drop from the regression proxy those variables whose correlation ranking is below that of the offset term β_0 (labeled X_0 in Fig. 19.7, where $X_0 = 1$ for all samples). Examples of such parameters are L2HIT, L2CLK, L3HIT, and L3CLK. Another observation from Fig. 19.7 is the

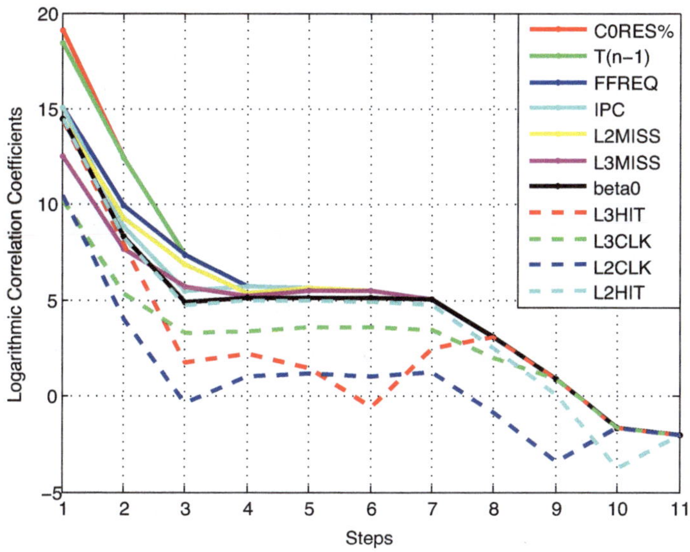

Fig. 19.7 Correlations between performance counters and temperature

correlation ranking of all the PCs with temperature throughout the nested regression process. The least-angle aspect is evident from the equality of certain correlations as the number of regression variables increases (e.g., the transition from three to four variables). A third observation is that the correlation between the temperature $T(n)$ at time instant n and the temperature $T(n-1)$ at time instant $n-1$ is systematically higher than that of the β_0 term. In the descending order of correlation ranking, the regression variables are COREs, $T(n-1)$, FFREQ, IPC, L2MISS, and L3MISS. This suggests that the initial thermal proxy of Eq. (19.30) should be modified from a static proxy to a dynamic one so as to capture the thermal transients

$$
\begin{aligned}
T(n) = {} & \beta_0 + \beta_{\text{COREs}} X_{\text{COREs}} + \beta_{\text{FFREQ}} X_{\text{FFREQ}} \\
& + \beta_{\text{IPC}} X_{\text{IPC}} + \beta_{T(n-1)} T(n-1) + \beta_{\text{L2MISS}} X_{\text{L2MISS}} \\
& + \beta_{\text{L3MISS}} X_{\text{L3MISS}}
\end{aligned}
\tag{19.31}
$$

The coefficients calculated from LAR are shown in Table 19.6. To validate Eq. (19.31), the LAR coefficients have been used in inference runs using testing workloads with the mean absolute percentage error (MAPE) over N samples as

$$
\text{MAPE} = \frac{100}{N} \sum_{i=1}^{N} \left| \frac{T_{\text{Measured}} - T_{\text{Calculated}}}{T_{\text{Measured}}} \right| \tag{19.32}
$$

Table 19.6 Coefficients for single-core temperature model

CORES%	$T(n-1)$	FFREQ	IPC	L2MISS	L3MISS	β_0
0.0190	0.5231	1.4268	0.6042	-0.0363	0.0962	18.4192

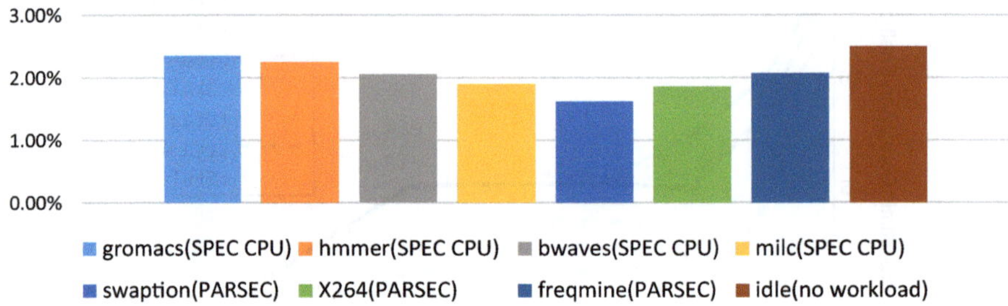

Fig. 19.8 Single-core temperature proxy accuracy

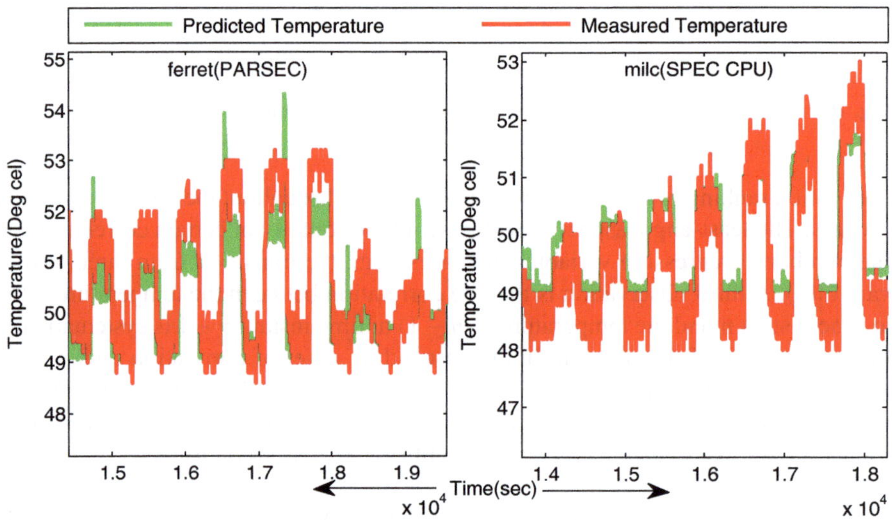

Fig. 19.9 Temperature transient response comparison

For the single-core case, MAPE is between 1.5% and 3%. Figure 19.8 shows the accuracy of the model on workloads from the PARSEC and SPEC CPU benchmark.

To verify the dynamic property of the thermal proxy, frequency has been changed several times between 1.2 and 2.3 GHz. The comparison of the transient response between the LAR predicted temperature and the measured temperature for two workloads is shown in Fig. 19.9.

19.5.2 The Multicore Case

The major objective of a multicore temperature proxy is to capture the spatial dependence of core temperature on core location in the chip floorplan, especially in the context of thermal exchanges with neighboring cores. For a processor with K cores, the temperature of the k-th core at instant, n, is given by

$$\begin{aligned} T_k(n) = {} & \beta_0 + \beta_{\text{CORES}} X_{\text{CORES}} + \beta_{\text{FFREQ}} X_{\text{FFREQ}} \\ & + \beta_{\text{IPC}} X_{\text{IPC}} + \beta_{\text{L2MISS}} X_{\text{L2MISS}} + \beta_{\text{L3MISS}} X_{\text{L3MISS}} \\ & + \beta_{T_k(n-1)} T_k(n-1) + \sum_{i \in L, i \neq k} \beta_{T_i(n-1)} T_i(n-1) \end{aligned} \tag{19.33}$$

where L is the subset of active cores. This set is strictly a function of time, and so it should be denoted $L(n-1)$ to signify that it is the subset of active cores at time instant $(n-1)$. In our model, we assume that the subset of active cores doesn't change throughout the workload run. This thermal proxy should work well for any number of cores. However, it is not easily scalable to a large number of cores that are all active. On the other hand, the temperature of a given core will mainly depend on temperatures of the active cores that are adjacent to it. To make Eq. (19.33) scalable, the subset L has to be intersected with the set A_k of cores adjacent to k. This assumes that the processor physical design is known at the system level, which is almost never the case. Interestingly, the core floorplan can be inferred using LAR by looking at the correlation ranking between T_k and each of the $T_i, i \neq k$. For Dell's 12-core Power Edge T620 platform, the correlation ranking is as shown in Fig. 19.10.

This figure shows that the temperature of Core $k = 0$ is correlated with almost all the other core temperatures. The temperatures of Cores $k = 4, 7, 9$ are next in the correlation ranking with the temperature of Core $k = 0$, while Core $k = 11$ is the least correlated. This correlation order can also be verified from the temperature profiles of the individual cores without running any workload. For Dell's 12-core Power Edge $T620$ platform, the temperature profile of each core at various DVFS frequencies is shown in Fig. 19.11. The temperature plot shows that Core $k = 0$ is at the highest temperature because even though no workload is running on it, the operating system is. Core $k = 11$ is at the lowest temperature, so it must be physically farthest from Core $k = 0$. Cores $k = 4, 7, 9$ have temperatures that are lower than that of $k = 0$ but higher than those of all the other cores. It follows that Cores $k = 4, 7, 9$ are closer to Core $k = 0$ than all others.

These temperature correlation rankings allow us to replace the set L in Eq. (19.33) with a much smaller subset that is core-specific and layout-dependent. For instance, for Cores $k = 4, 7, 9$, we have $L \cap A_k = \{0, 4, 7, 9\}$. Similarly, given that Core $k = 0$ is always running the OS and that the other cores almost never experience full utilization, its subset $L \cap A_0 = \{0\}$. Such considerations help make the model described in Eq. (19.33) scalable to a very large number of cores. For illustration, the temperature due to workload execution on Core $k = 4$ is given by

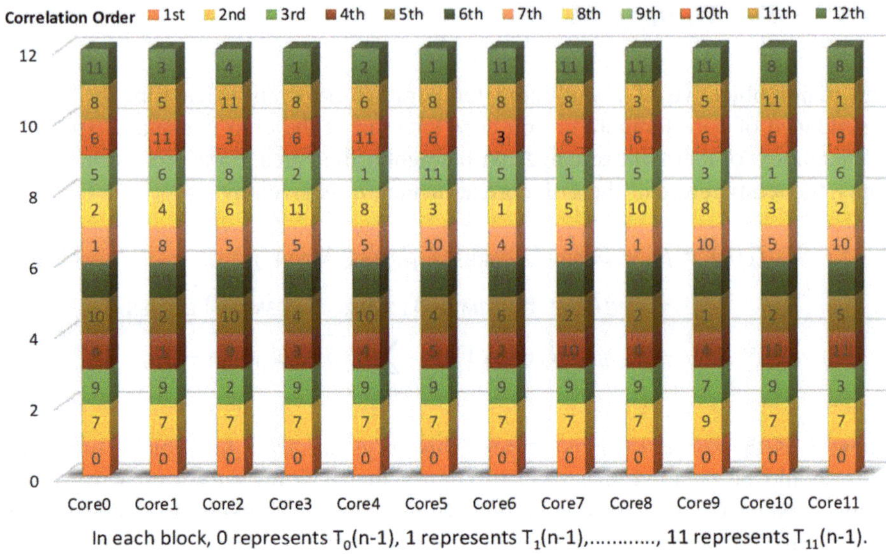

Fig. 19.10 Correlation rank matrix of 12 cores (Dell's Power Edge $T620$ server)

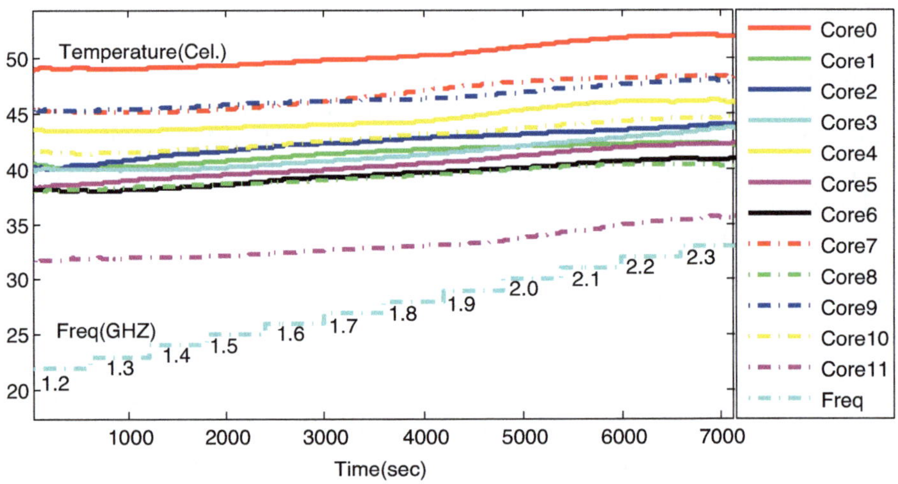

Fig. 19.11 Per-core temperature vs. frequency and time plot for 12 cores

$$
\begin{aligned}
T_4(n) = {} & \beta_0 + \beta_{\text{CORES}} X_{\text{CORES}} + \beta_{\text{FFREQ}} X_{\text{FFREQ}} \\
& + \beta_{\text{IPC}} X_{\text{IPC}} + \beta_{\text{L2MISS}} X_{\text{L2MISS}} + \beta_{\text{L3MISS}} X_{\text{L3MISS}} \\
& + \beta_{T_4(n-1)} T_4(n-1) + \beta_{T_0(n-1)} T_0(n-1) \\
& + \beta_{T_7(n-1)} T_7(n-1) + \beta_{T_9(n-1)} T_9(n-1)
\end{aligned} \tag{19.34}
$$

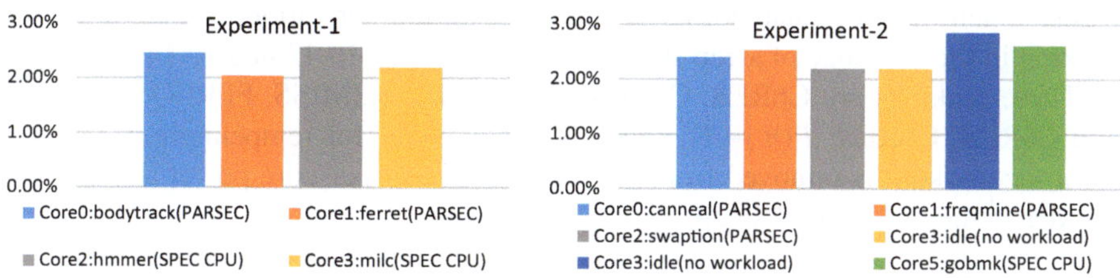

Fig. 19.12 Multicore temperature model accuracy

For the validation of this model, two experiments have been performed, one with four cores active and another with six cores active. The accuracy is shown in Fig. 19.12. This is the general proxy for predicting core temperatures. A workload-specific thermal proxy validation is presented in Sect. 19.6.

19.6 Proxies Incorporating Sleep States

In Sects. 19.4 and 19.5, the thermal and power proxies have been derived under the assumption that the cores never enter into sleep states. This assumption is quite reasonable for multi-threaded, high-performance computing applications but is not satisfied under productivity, media, and web workloads where power and thermal management benefits tremendously from the presence of sleep states. Sleep states are also important in the management of dark silicon [16–19]. There are several categories of sleep states that can be summarized as follows:

1. C_0: This is a fully operational state. The CPU is fully turned on.
2. C_1: This is the first idle state. The CPU stops its main internal clocks through software while the bus interface and the advanced programmable interrupt controller (APIC) remain functional at full frequency.
3. C_1^E: This state is similar to C_1 with the additional feature of the CPU lowering the frequency and supply voltage.
4. C_2: In this state, the CPU stops the main internal clock through hardware control.
5. C_3: In this state, the CPU stops all internal clocks. Data in the cache are not coherent.
6. C_6: This state starts as soon as the core saves the architectural state into an on-chip SRAM. The supply voltage to the core is reduced to zero.
7. C_7: This is the deep-sleep state. The content of the shared cache is written back to main memory. Once the CPU enters C_7, it takes higher latency to go back into C_0 state.

The Dell's Power Edge server has three levels of sleep states: C_3, C_6, and C_7. When they are disabled, the corresponding CnRES PCs for state residency are all zeros. When they are enabled, LAR again allows the derivation of the correlation ranking in the presence of sleep states as shown in Figs. 19.13 and 19.14.

These rankings are C6RES, C0RES, IPC, L2HIT, L3MISS, FFREQ for power; and C6RES, C0RES, $T(n-1)$, FFREQ, IPC, L3HIT for temperature. C7RES is always zero, so it is removed from the regression. The per-core power and thermal proxies in the presence of sleep states are given as

$$\begin{aligned} P = \beta_0 + \beta_{\text{C6RES}} X_{\text{C6RES}} + \beta_{\text{C0RES}} X_{\text{C0RES}} + \beta_{\text{IPC}} X_{\text{IPC}} \\ + \beta_{\text{L2HIT}} X_{\text{L2HIT}} + \beta_{\text{L3MISS}} X_{\text{L3MISS}} + \beta_{\text{FFREQ}} X_{\text{FFREQ}} \end{aligned} \tag{19.35}$$

$$\begin{aligned} T(n) = \beta_0 + \beta_{\text{C6RES}} X_{\text{C6RES}} + \beta_{T(n-1)} T(n-1) + \beta_{\text{C0RES}} X_{\text{C0RES}} \\ + \beta_{\text{IPC}} X_{\text{IPC}} + \beta_{\text{FFREQ}} X_{\text{FFREQ}} + \beta_{\text{L3HIT}} X_{\text{L3HIT}} \end{aligned} \tag{19.36}$$

At testing, these equations have resulted in a MAPE of 5%.

For a highly accurate proxy, all the variables shown in Figs. 19.13 and 19.14 need to be included in the regressions. Note that the correlation of C6RES is smaller than that of the offset term, β_0, and as can be experimentally verified, it can therefore be removed from both the power and thermal proxies. The final proxies are given by

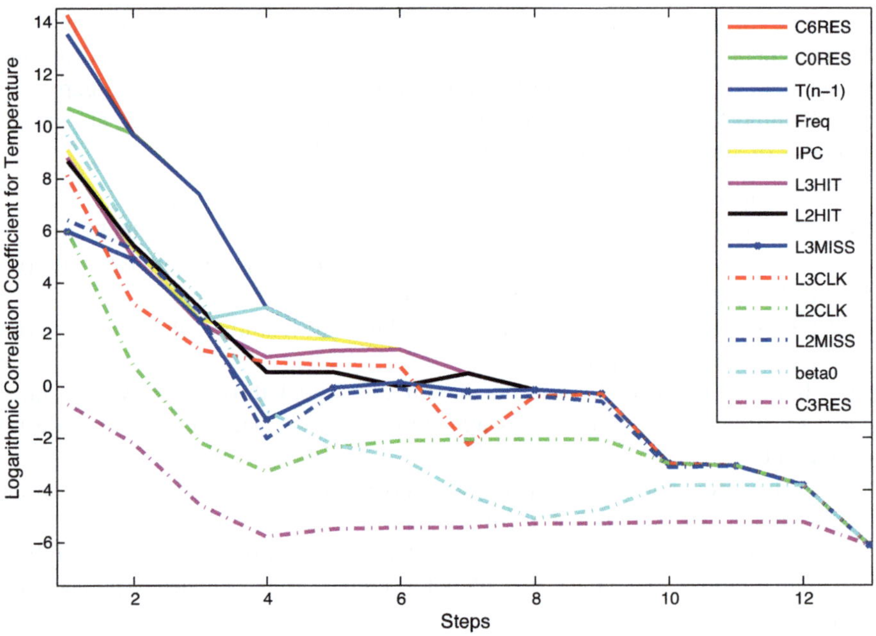

Fig. 19.13 Enabled sleep states: correlation coefficients between temperature and PCs

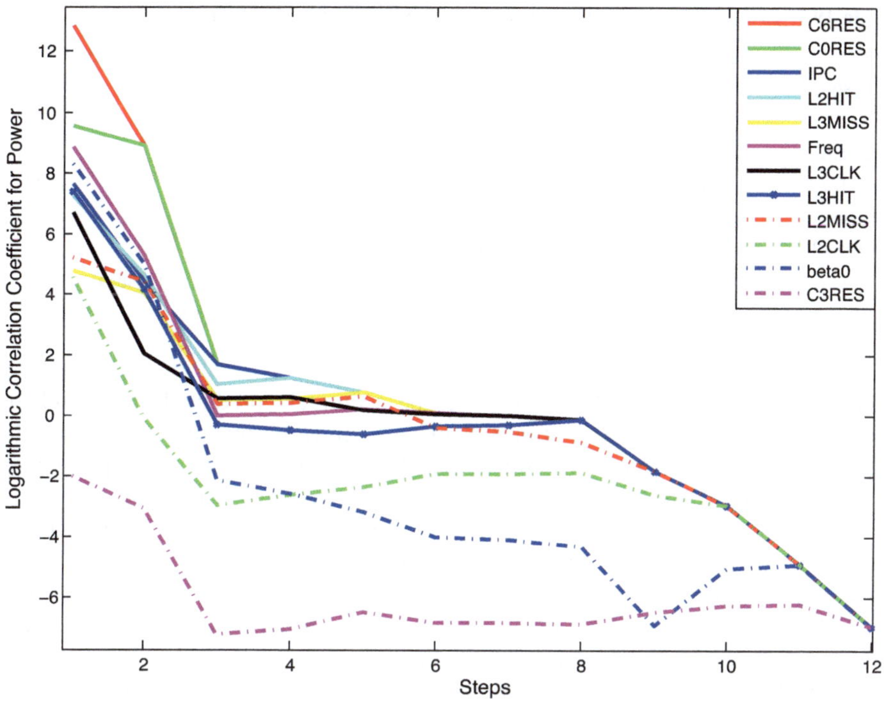

Fig. 19.14 Enabled sleep states: correlation coefficients between power and PCs

$$
\begin{aligned}
T(n) = \beta_0 &+ \beta_{\text{C6RES}} X_{\text{C6RES}} \\
&+ \beta_{\text{CORES}} X_{\text{CORES}} + \beta_{\text{IPC}} X_{\text{IPC}} + \beta_{\text{L2CLK}} X_{\text{L2CLK}} \\
&+ \beta_{\text{L3CLK}} X_{\text{L3CLK}} + \beta_{T(n-1)} T(n-1) \\
&+ \beta_{\text{L3HIT}} X_{\text{L3HIT}} + \beta_{\text{L2HIT}} X_{\text{L2HIT}} \\
&+ \beta_{\text{L3MISS}} X_{\text{L3MISS}} + \beta_{\text{FFREQ}} X_{\text{FFREQ}} + \beta_{\text{L2MISS}} X_{\text{L2MISS}}
\end{aligned}
\tag{19.37}
$$

$$
\begin{aligned}
P = \beta_0 &+ \beta_{\text{C6RES}} X_{\text{C6RES}} + \beta_{\text{CORES}} X_{\text{CORES}} + \beta_{\text{IPC}} X_{\text{IPC}} + \beta_{\text{L2CLK}} X_{\text{L2CLK}} \\
&+ \beta_{\text{L3CLK}} X_{\text{L3CLK}} + \beta_{\text{L3HIT}} X_{\text{L3HIT}} \\
&+ \beta_{\text{L2HIT}} X_{\text{L2HIT}} + \beta_{\text{L3MISS}} X_{\text{L3MISS}} \\
&+ \beta_{\text{FFREQ}} X_{\text{FFREQ}} + \beta_{\text{L2MISS}} X_{\text{L2MISS}}
\end{aligned}
\tag{19.38}
$$

The LAR coefficients are shown in Tables 19.7 and 19.8. Their proxies result in a modeling accuracy of more than 97%.

As expected, the proxies with enabled sleep states have more regression variables than the proxies with the sleep states disabled. This is because the sleep-state

Table 19.7 LAR coefficients of the temperature proxy with sleep states enabled

IPC	L3MISS	L2MISS	L3HIT	L2HIT	L3CLK
−0.8632	2.0046	−0.8575	−2.2401	1.3279	−1.0283

L2CLK	C0RES	C6RES	$T(n-1)$	FFREQ	β_0
−3.0772	0.5601	0.5586	0.2148	−0.7546	−16.8668

Table 19.8 LAR coefficients of the power proxy with the sleep states enabled

IPC	L3MISS	L2MISS	L3HIT	L2HIT	L3CLK	L2CLK	C0RES	C6RES	FFREQ	β_0
−1.1203	5.8039	−2.6608	−1.0318	1.8579	−1.0111	−3.6273	0.2435	0.1705	−0.1531	−5.4146

residency of the running thread needs to be accurately predicted along with its workload characteristics. The affinity between a given hardware platform and the sleep states is dependent on the micro-architecture, and in our own experimental platform, the CPU has higher affinity with the C_6 state than the C_3 state.

The LAR proxies are flexible enough to accommodate any type of workload and run environment, as well as the activity levels of the various cores. An alternative approach to accounting for sleep states is given in [6] where the sleep-state residency is used to modulate the other PCs of the workload in the power proxy. A similar idea can be applied in the LAR context although in this chapter, we have adopted the view of using the sleep-state residencies as regression variables in the power and thermal proxies. Another advantage of the LAR approach is that unlike [20], it uses one proxy, rather than several *separate* ones, to predict power for any number of active cores.

19.7 Workload Signature

One of the major benefits of using LAR is that it can be used to characterize the workload using the correlation ranking of the regression variables. Indeed, the top variables in the correlation ranking can be used as a workload signature in the PC space. Experiments have been conducted on four PARSEC workloads and two SPEC CPU workloads. The top five PCs in correlation ranking for each of these workloads are listed in Tables 19.9 and 19.10 for power and temperature, respectively.

The major conclusion is of course that these rankings vary from one workload to another. In Table 19.9, for memory-intensive workloads (*Canneal* and *X264*), the cache miss rate (L2MISS) is the one most correlated with power. On the other hand, for CPU-intensive workloads (e.g., *Ferret*), it is the IPC that is most correlated with power. These three-dimensional workload signatures are valid for active cores as is evident from the presence of C0RES in the signature. The latter PC is highly correlated with power for all workloads as it only depends on the proportion of time the core is active. Similar rankings can be observed for temperature in Table 19.10.

Table 19.9 Workload signature for power

	PCs correlation ranking				
Workload	1st	2nd	3rd	4th	5th
Bodytrack	CORES	FFREQ	L2MISS	IPC	L3MISS
Canneal	CORES	L2MISS	FFREQ	L3CLK	L2HIT
Ferret	CORES	IPC	L2MISS	FFREQ	L2HIT
X264	CORES	L2MISS	FFREQ	IPC	L3MISS
Gamess	CORES	FFREQ	IPC	L3HIT	L2MISS
Libquantum	CORES	L2MISS	IPC	FFREQ	L3CLK
Gobmk	CORES	FFREQ	IPC	L3HIT	L2MISS

Table 19.10 Workload signature for temperature

	PCs correlation ranking				
Workload	1st	2nd	3rd	4th	5th
Bodytrack	CORES	$T(n-1)$	FFREQ	L2MISS	IPC
Canneal	CORES	$T(n-1)$	L2MISS	FFREQ	IPC
Ferret	CORES	IPC	$T(n-1)$	FFREQ	L2HIT
X264	CORES	$T(n-1)$	L2MISS	FFREQ	IPC
Gamess	CORES	$T(n-1)$	FFREQ	IPC	L3HIT
Libquantum	CORES	$T(n-1)$	L2MISS	L3MISS	IPC
Gobmk	CORES	$T(n-1)$	FFREQ	IPC	L2MISS

Such workload signatures provide power and thermal management policies with a new capability, namely the ability to adapt the controlled variable in power or thermal management policies to the workload at hand.

Similar experiments can be performed across all workloads to get their PC signatures. As an application, one can use the workload signature as a pointer indexing into subsets of regression coefficients with one of these subsets being selected for each workload signature. Such indexing can help in reducing thermal management response to either changes in workloads or to transitions from one phase to another in a given workload.

One way of verifying the results that we have obtained for workload signatures with hardware event counters is to compare them with signatures obtained from software performance profiling tools such as the LINUX's *perf* command [21]. The command

$$perf\ \ stat\ \ -p\ \ < PID\ \ of\ \ workload >$$

lists all the events on the given process ID, PID. We have run this command on few workloads with the results shown in Table 19.11.

This table shows that for workload *canneal*, more time is spent on fetching cache data with a cache miss rate of 6.717 M/s and IPC of 0.626. On the other hand, for the workload *ferret*, the IPC is 3.636 with much less time spent fetching

Table 19.11 Performance profiling comparison of workloads

Ferret (PARSEC)	Canneal (PARSEC)
957,759,076 cache-misses # 1.604 M/sec	3,063,851,228 cache-misses # 6.717 M/sec
2,067,584,087 cache-references # 3.462 M/sec	4,156,217,347 cache-references # 9.112 M/sec
7,206,062,783 branch-misses # 4.461 %	490,077,756 branch-misses # 1.265 %
161,536,850,662 branches # 270.481 M/sec	38,734,624,196 branches # 84.919 M/sec
889,732,533,585 instructions # 3.636 IPC	342,447,868,569 instructions # 0.626 IPC
244,690,563,224 cycles # 409.715 M/sec	547,027,701,197 cycles # 1199.259 M/sec
169,405 page-faults # 0.000 M/sec	120,590 page-faults # 0.000 M/sec
1 CPU-migrations # 0.000 M/sec	1 CPU-migrations # 0.000 M/sec
127,846 context-switches # 0.000 M/sec	68,217 context-switches # 0.000 M/sec
597220.940384 task-clock-msecs # 542 CPUs	456137.911099 task-clock-msecs # 0.432 CPUs
1102.096362113 seconds time elapsed	1056.383648515 seconds time elapsed

Gamess (SPEC CPU)	Libquantum (SPEC CPU)
530,853 cache-misses # 0.002 M/sec	6,656,391,448 cache-misses # 19.424 M/sec
249,446,361 cache-references # 1.172 M/sec	20,301,608,325 cache-references # 59.242 M/sec
1,118,536,958 branch-misses # 1.087 %	436,068,369 branch-misses # 0.216 %
102,945,618,684 branches # 483.555 M/sec	201,614,775,384 branches # 588.328 M/sec
1,047,934,544,412 instructions # 2.140 IPC	1,153,471,633,587 instructions # 1.763 IPC
489,743,153,165 cycles # 2300.418 M/sec	654,397,954,984 cycles # 1909.585 M/sec
94 page-faults # 0.000 M/sec	0 page-faults # 0.000 M/sec
21 CPU-migrations # 0.000 M/sec	32 CPU-migrations # 0.000 M/sec
250 context-switches # 0.000 M/sec	427 context-switches # 0.000 M/sec
212893.086471 task-clock-msecs # 0.549 CPUs	342691.270723 task-clock-msecs # 0.999 CPUs
387.800184874 seconds time elapsed	1056.383648515 seconds time elapsed

cache information with a miss rate of 1.604 M/s. In other words, hardware PCs and software profiling tools are in agreement on workload signature characterization, with the PCs having the distinct advantage of being much more efficient to generate as they depend on direct access to hardware registers rather than on running operating system commands. This agreement confirms that *canneal* can be categorized as a memory-intensive workload and that *ferret* can be categorized as a CPU-intensive workload. Other workloads conform to this agreement as may be verified by comparing the workload signatures from Tables 19.9 and 19.10 with the *perf* profiling data from Table 19.11.

An important application of workload signature is in the area of control variable selection. In the context of thermal management and control, PCs that are most correlated with temperature are the most natural candidate to use as control knobs. As an example, for the *bodytrack, ferret,* and *X264* workloads, the PCs most correlated with temperature (after C0RES) are FFREQ, IPC, and L2MISS, respectively, as given in Table 19.10. So to control the temperature for *ferret* and *X264*, the controller should use IPC and memory access rate rather than core frequency. In other words, restricting thermal management to the use of DVFS may not be the most appropriate control action for these two workloads. As explained in Sect. 19.3 in page 579, Dell's Power Edge T620 server has two sockets, each containing six cores, and the clock frequency is global to the socket. The scaling of clock frequency in a particular core forces the same scaling on the other cores in the same socket. As a result having DVFS as a socket-based control variable greatly limits the per-core management policies. Consider now a multicore design that provides DVFS on a per-core basis. In such a design, assume we have two different threads: Th_1 for a low-performance, energy hungry process, and Th_2 for

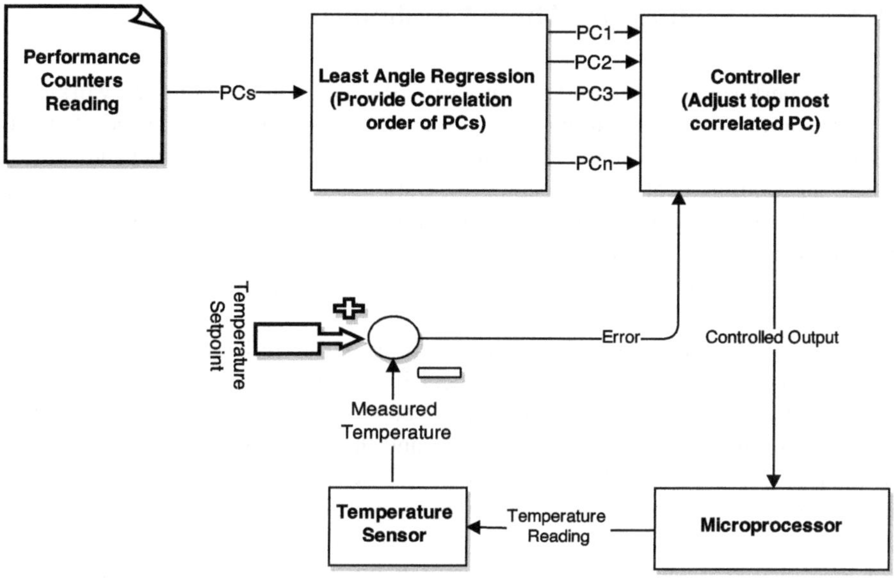

Fig. 19.15 Controller based on workload signature

a high-performance, low-energy process, executing on the same core. Core thermal management via DVFS will prioritize Th_1 over Th_2, thus resulting in an overall loss of performance. A finer-grain policy would look at the workload signatures of the threads and adopt the control variable of Th_1 that most impacts temperature while being absent in the workload signature in Th_2.

A controller based on memory access rate is described in [22] where a feedback controller is built using the L2 hit rate as a setpoint with the controller providing, dynamically, the sizes of L2 and L3 caches needed to satisfy a performance target across a variety of workloads. The implementation of such controllers typically requires modifications to the Linux kernel or, as in the case of the IPC, access to the micro-architecture hardware. The generic block diagram of such a controller is given in Fig. 19.15.

In conclusion, rather than always using DVFS settings as is the case in the most common thermal and power management policies, other control strategies may be considered such as miss-rate reduction in cache-intensive workloads and pipeline management in CPU-intensive workloads. In the next sections, we give further applications of workload signatures to fine-grained thermal and power management in multicore processors.

19.8 Core Scaling and Thread Assignment

For high-performance or parallel computing, the execution of complex workloads requires multiple cores. In multicore scenarios, the execution speed depends additionally on shared resources such as high-level cache memory, data and address buses, and communication between cores. Dark silicon management and switching cores between sleep and active states increase dynamic power consumption and raise multiprocessor temperature. To ease managing the trade-offs between power, temperature, and performance, fine-grained core scaling policies are required [16–19].

Traditionally, performance metrics such as CPU utilization and workload throughput are used to make scaling decisions. For instance, in [23], dynamic core scaling is implemented to minimize leakage power. A model linking power consumption to the number of active cores and the utilization of each core is used to manage core scaling. One disadvantage of such model is that it does not take into account memory throughput and memory power consumption for core scaling decisions. Another disadvantage is that the core scaling policy does not account for workload characteristics. These characteristics are important, especially when the workload has many threads and each thread demands large processor resources for high performance. Attempts have been made to address the scalability of large workloads such as the Enterprise Java application [24]. A "Thread Reinforce" framework is used in [25] for finding the number of threads at runtime and assigning individual thread to each core. However, these thread-allocation methods are limited to specific workload types and may not be appropriate for a workload with dependencies among its threads. Furthermore, these methods do not consider the inherent correlations among multiple threads of a given workload. Such correlations may open up opportunities to minimize the overall runtime and power consumption.

To address the shortcomings of the aforementioned core scaling techniques, we propose a real-time autoscaling methodology that can provide superior performance with comparatively lower power consumption and temperature. Our methodology accounts for metrics that have so far been neglected such as the cache access rate and sleep states. The proposed methodology is based on the following components:

1. Generation of workload signatures PCs that are highly correlated with power and temperature.
2. Optimization of the allocation of workloads and threads among cores to achieve high performance under given power and thermal budgets.
3. Implementation of *autoscaling* where the scaling decision is made autonomously or independently based on the type of workload, user-specified performance, and the temperature or power/energy setpoints.

Workloads of the *PARSEC* [26] and *SPEC CPU 2006* [27] benchmarks are used in the experiments. PARSEC provides various load levels labeled as *test, small, medium, large*, and *native*. These differ in the size of working set and thread

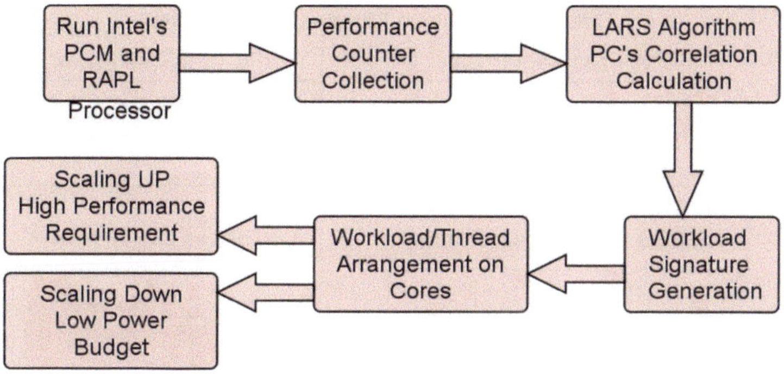

Fig. 19.16 Experiment flow diagram

parallelism. The block diagram shown in Fig. 19.16 represents the experimental methodology.

19.8.1 The Relation of PCs with Power Consumption

A first step towards improved autoscaling is to check the power state (P-state) of the processor and estimate the DRAM power. To calculate the correlation of different PCs with power, the workload *ferret* from the *PARSEC* suite [26] has been run on a single core (with other cores turned off). PC samples are collected every second.

The dependence of different PCs on processor power is shown in Fig. 19.17. The plots show that low P-states are prevalent for the processor while for the DRAM, high P-states are more prevalent. Also, the PC values remain for longer times at low P-states for the processor but at high P-states for the DRAM. It is clear that both memory power consumption *and* memory throughput must be considered when allocating workloads via autoscaling.

The correlation of PCs with core power consumption and their correlation rankings have already been shown in Tables 19.9 and 19.10, page 595. Recall that in Table 19.10, $T(n-1)$ represents the sampled temperature at instant $(n-1)$. The main observation there was that PC rankings and workload signatures vary notably from one workload to another. As already observed, for memory-intensive workloads, such as *canneal* and *x264*, the cache miss rate (L2MISS) is the one most correlated with power and temperature. On the other hand, for CPU-intensive workloads, such as *ferret*, it is the IPC that is most correlated. The difference in workload signatures for *ferret* and *canneal* is shown in Fig. 19.18. *Ferret* being CPU-intensive exhibits a greater IPC while *canneal* being memory-intensive results in higher DRAM power.

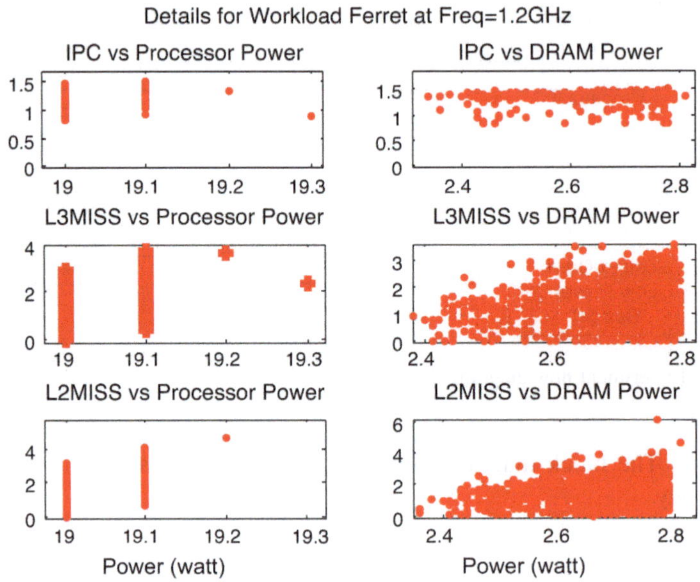

Fig. 19.17 Performance counter vs. power

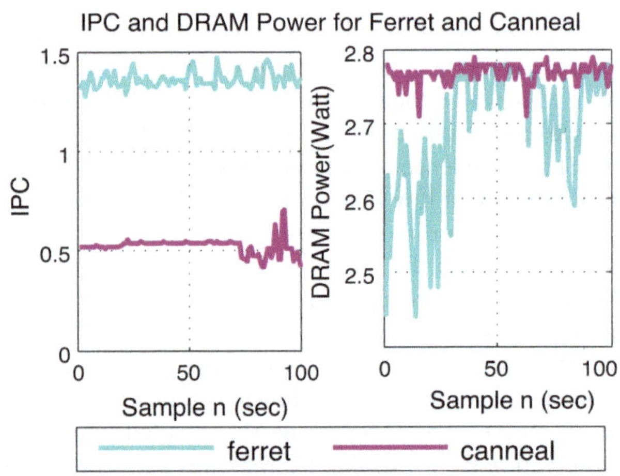

Fig. 19.18 Workload signatures of *ferret* and *canneal*

19.8.2 Energy-Aware Workload Execution

Multicore processors make heavy use of multi-threading and hyper-threading to execute multiple instructions concurrently. In case a workload requires shared resources such as the I/O bus or L3 cache, the scheduler manages the order of thread execution. As will be described next, we have conducted dedicated experiments on our RHLE platform to reveal the scheduler in action and, more importantly, to understand the impact of resource sharing on runtime performance.

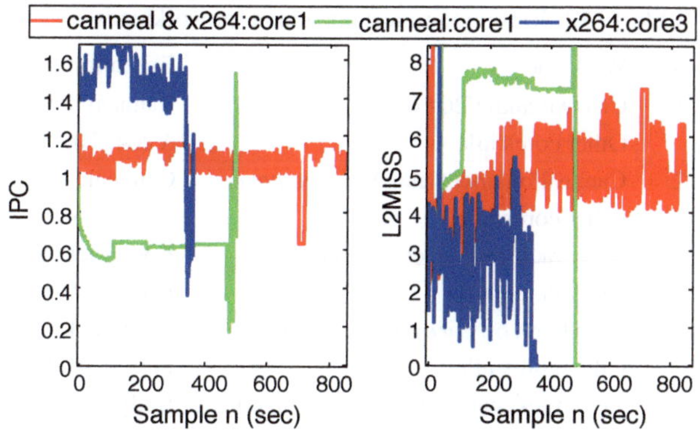

Fig. 19.19 IPC and L2MISS comparison for *ferret* and *canneal*

Workloads from the PARSEC benchmark suite have been run with load level set to *native*. In general, Cores $k = 0,\ 1,\ 2,\ 3$ are active while all the remaining cores are off. In one experiment, two workloads *canneal* and *x264* are run on Core $k = 1$ with all the other cores are set to idle. In another experiment, *canneal* is run on Core $k = 1$ while *x264* is run on Core $k = 3$.

The resulting plots for IPC and L2MISS are shown in Fig. 19.19. The red curve represents the intermediate value of IPC and L2MISS for the case when both workloads run on the same core. When the workloads are executed on different cores, one core exhibits a higher IPC while the other exhibits a higher L2MISS. It is clear that the availability of shared resources does impact the performance. The level of this impact can be extracted from the difference in workload runtimes. To quantify the impact, two cores, one from each socket, are run at the same frequency of 1.2 GHz and workloads are selected at random and executed on the two selected cores. The processor total power consumption and workload runtime are shown in Table 19.12.

Comparing ROWs 1 and 2, the completion time approximately doubles in case both workloads are run on the same core as compared with the case where both workloads run on two separate cores. An increment of power consumption is observed in case workloads are assigned to separate cores, but this increment is relatively small. Note that similar characteristics are observed for other experiments described in Table 19.12. Even though power consumption in ROW 2 (resp. ROW 4) is slightly larger than in ROW 1 (resp. ROW 3) the total energy consumption (Power $*$ Completion Time) is much smaller. For example, in ROW 1, Energy $= 36\,\text{W} * 843\,\text{s} = 30.348\,\text{kJ}$ while in ROW 2, Energy $= 36.8\,\text{W} * 497\,\text{s} = 18.289\,\text{kJ}$. It is therefore economical to run different workloads on separate cores as it allows us to trade-off a large gain in completion time with a small increase in power consumption.

Table 19.12 Execution time comparison for workloads

S.N	Workload	Execution time (s) Δt	Power consumption (W)
1.	Canneal and $X264$ on $core_1$. Other cores idle	Canneal: 843 $X264$: 708	36
2.	Canneal on $core_1$ and $X264$ on $core_3$. Other cores idle	Canneal: 497 $X264$: 362	36.8
3.	Bodytrack and ferret on $core_1$. Other cores idle	Bodytrack: 896 Ferret: 1419	36.4
4.	Bodytrack on $core_1$ and ferret on $core_3$. Other cores idle	Bodytrack: 407 Ferret: 1015	37.2
5.	Bodytrack, canneal, and ferret on $core_1$. Other cores idle	Bodytrack: 1291 Canneal: 1498 Ferret: 1907	37.5
6.	Bodytrack on $core_1$, canneal on $core_2$, and ferret on $core_3$. Other cores idle	Bodytrack: 408 Canneal: 494 Ferret:1013	39.9

Δt = time of workload final output – time of workload run request

19.8.3 Autoscaling for Separate Workloads

Another application of workload signature is to assess performance gains when separate workloads are run in multiple threads on a multicore processor. To investigate this scenario, workloads are first run on a single core with a clock frequency of 1.2 GHz and load level set to *large*. At runtime, the number of instantiated threads for a given workload can be obtained using the *Linux* commands

$$\%cat\ /proc/ < PIDofworkload > /stat$$

$$\%cat\ /proc/ < PIDofworkload > /status$$

All threads are continuously monitored and transferred to idle cores using the *Linux* command

$$taskset\ \ -c < corenumber >< threadID >$$

The overall workload completion time is measured. The results are shown in Fig. 19.20.

For the workload *ferret*, there is a notable reduction in completion time when increasing the number of cores from one to two. However, no significant reduction is observed when increasing from four to six cores. For the workloads *bodytrack* and *canneal*, the completion time varies only slightly for the different numbers of used cores. Recall from Tables 19.9 and 19.10 in page 595 that *bodytrack* and *canneal* are memory-intensive. This implies that the performance achieved for *bodytrack* and

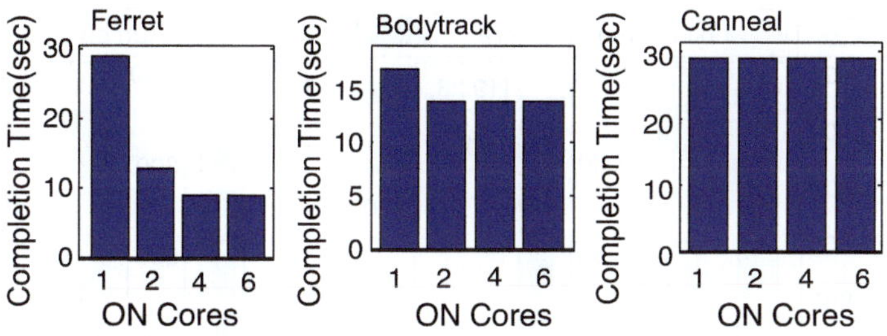

Fig. 19.20 Runtime comparison of workloads

canneal depends mainly on shared memory throughput, even in cases when multiple threads are executed on separate cores. On the contrary, *ferret* is a CPU-intensive workload, and its performance depends mainly on instruction throughput. Since the instruction pipeline is private to each core, *ferret* can fully utilize the core resources and runs its threads in parallel with high performance. In summary, when scaling up the number of cores, achievable performance is higher for CPU-intensive than for memory-intensive workloads. This should also be the case for all symmetric multiprocessors with shared memory.

Another important observation is that achieved performance typically saturates even for CPU-intensive workloads. For instance, the completion time is almost the same whether *ferret* is run on four or six cores. We note that it is important to determine the saturation point for each workload of interest. In many cases, scaling up the number of cores only increases power and temperature without significantly improving performance.

19.8.4 Autoscaling for Multiple Workloads

19.8.4.1 Scaling Down

For multiple workloads running on the same core, the overall performance can be improved by merging the workloads based on their signatures to run on particular cores. To illustrate this, we have conducted experiments with two different groups of workloads (whose signatures are stated in Tables 19.9 and 19.10 in page 595, core frequency set to 1.2 GHz, and load level set to *native* for *PARSEC*). Two CPU-intensive workloads, namely *ferret* and *gamess*, and two memory-intensive workloads, *canneal* and *X264*, have been selected for the experiments. The number of active cores has been scaled down, and two workloads have been executed in parallel on the same core. With four workloads in total, there are four possible combinations of workload assignments as shown in Fig. 19.21. The green bars show the combined completion time while the blue bars show the completion time for separate workloads run on different cores. The shortest completion time is observed

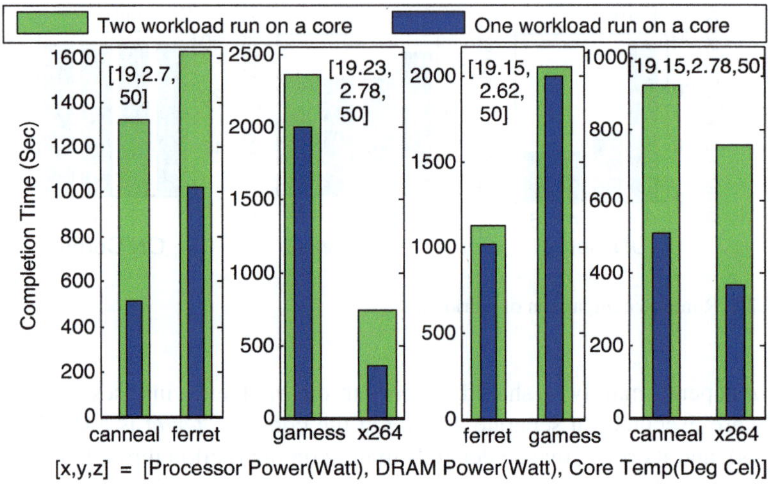

Fig. 19.21 Runtime comparison for multi-workload execution on single-core processor

in the third and fourth bar graphs of the top row. In the third bar graph, both workloads, *ferret* and *gamess*, are CPU-intensive and run on the same core. A similar case is the fourth bar graph, where the two memory-intensive workloads, *canneal* and *X264*, run on one core. These results demonstrate that two workloads running on the same core can provide relatively high performance as long as they share the same signature.

Similar experiments have been conducted for a multicore configuration. Two cores, $k = 0,\ 1$ have been used for executing the same four workloads. The workload completion times are shown in Fig. 19.22. The core temperature and power consumption are approximately the same for each run, but there are significant differences in completion time. The shortest completion time is observed in the third bar graph of the bottom row where the CPU-intensive workloads, *ferret* and *gamess*, are run on Core $k = 0$ and the memory-intensive workloads, *canneal* and *X264*, are run on Core $k = 1$. Under similar power budgets and thermal constraints, performance may be significantly increased by assigning workloads with the similar signatures to the same cores.

19.8.4.2 Scaling Up

In a multi-thread context, performance may be improved more effectively by arranging the threads of the various workloads based on *thread* rather than *workload* signature. This is because complex, large-scale workloads are typically comprised of many different computational phases. For example, the workload *ferret* mimics applications tailored for content-based similarity search in feature-rich data such as audio, video, and 3D visualization [26]. During the initial execution phase, data

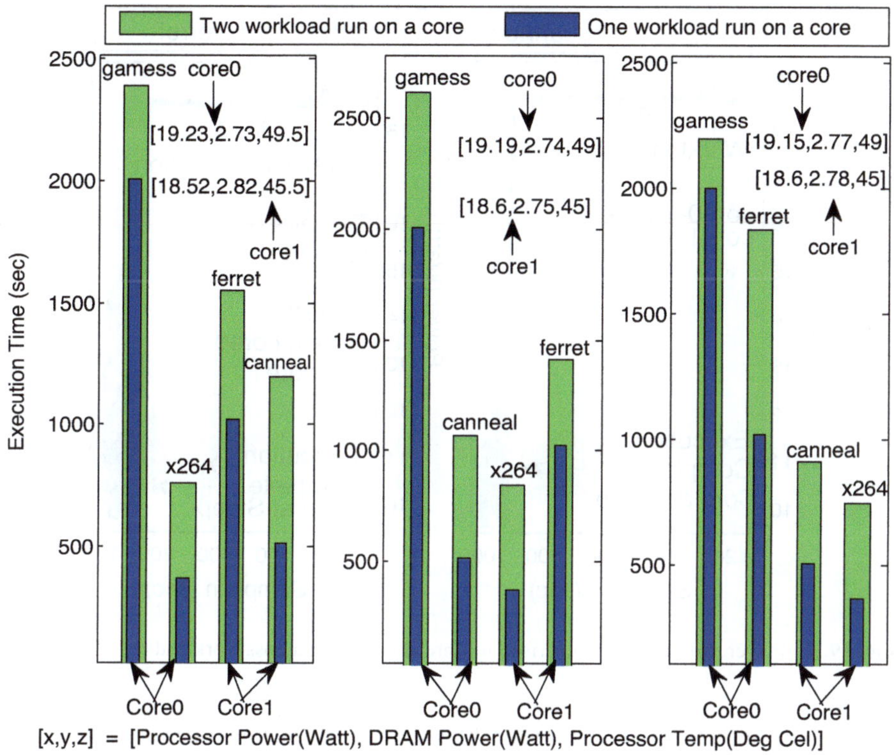

Fig. 19.22 Runtime comparison for multi-workload execution on a multicore processor

is requested from the user, at which time the threads are CPU intensive. During subsequent phases, *ferret* searches for pattern matches in its database and provides outputs. During these search phases, the threads become memory intensive. Similar behavior over time applies for most workloads.

Given the dynamic behavior of workloads, it may be advantageous for overall performance to run *same-signature* threads of *different* workloads on the same cores. To illustrate this advantage, the workloads, *ferret*, *gamess*, *canneal* and *X264*, have been run on Core $k = 0$, and the process IDs of all threads have been collected. The signature of each thread is determined in real-time based on the *Linux* profiling command

$$perf\ stat\ -p\ < threadID >$$

Sample signatures via *perf* have been shown in Table 19.11 in page 596. We assign threads with similar cache miss rates to Core $k = 1$ and threads with similar IPCs to Core $k = 0$. Threads are transferred using the *Linux* command

$$taskset\ -pc\ < corenumber > < threadID >$$

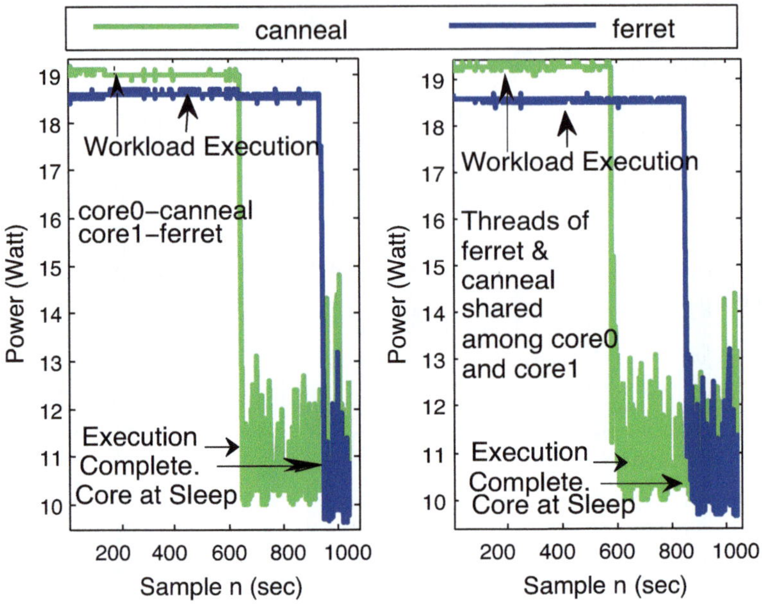

Fig. 19.23 Power and runtime comparison for thread swapping across workload

This simple but effective technique enables shorter completion times under the same power budget. This is illustrated graphically in Fig. 19.23 for workloads *canneal* and *X264*.

19.9 Conclusions

In this chapter, power and thermal proxies for multicore processors have been generated and validated using least-angle regression (LAR), a well-known modeling technique in statistical machine learning. In the thermal proxy, we have shown how LAR can be exploited to improve the scalability of the model while incorporating layout-dependent effects. The power and thermal proxies are validated for both active and idle cores with the idle cores residing at various sleep levels. Workload signatures defined using LAR performance counters have also been proposed to predict workload characteristics. More specifically, these signatures have been used to dynamically scale the number of cores and then arrange the workloads and their threads to run on specific cores to improve overall performance under the same power budgets and thermal constraints. An important outcome of our work is that aggregating workloads and their threads according to their signatures often results in significant savings in runtime and energy. We are currently investigating the use of similar techniques to develop *workload-driven autoscaling* policies of virtual machines on cloud servers.

Acknowledgements The authors would like to acknowledge very helpful discussions with Andrew Henroid from Intel, and with Pradip Bose, Alper Buyuktosunoglu, Canturk Isci, Prabhakar Kudva, and Charles Lefurgy from IBM. This work was supported by SRC under Contract 2011-TJ-2192 with customized funding from Mubadala, Abu Dhabi, UAE.

References

1. J.S. Lee, K. Skadron, S.W. Chung, Predictive temperature-aware DVFs. IEEE Trans. Comput. **59**(1), 127–133 (2010)
2. R. Kalla, B. Sinharoy, W.J. Starke, M. Floyd, Power7: IBM's next-generation server processor. IEEE Micro **30**(2), 7–15 (2010)
3. M. Floyd, M. Allen-Ware, K. Rajamani, B. Brock, C. Lefurgy, A.J. Drake, L. Pesantez, T. Gloekler, J.A. Tierno, P. Bose et al., Introducing the adaptive energy management features of the power7 chip. IEEE Micro **31**(2), 60–75 (2011)
4. K. Kasichayanula, D. Terpstra, P. Luszczek, S. Tomov, S. Moore, G.D. Peterson, Power aware computing on GPUs, in *2012 Symposium on Application Accelerators in High Performance Computing (SAAHPC)* (IEEE, Piscataway, 2012), pp. 64–73
5. A. Sîrbu, O. Babaoglu, Predicting system-level power for a hybrid supercomputer (2016). Preprint. arXiv:1605.09530
6. M. Yasin, A. Shahrour, I.M. Elfadel, Unified, ultra compact, quadratic power proxies for multi-core processors, in *Design, Automation and Test in Europe Conference and Exhibition (DATE), 2014* (IEEE, Piscataway, 2014), pp. 1–4
7. C.-B. Cho, T. Li, Using wavelet domain workload execution characteristics to improve accuracy, scalability and robustness in program phase analysis, in *IEEE International Symposium on Performance Analysis of Systems & Software, 2007. ISPASS 2007* (IEEE, Piscataway, 2007), pp. 136–145
8. R. Sarikaya, C. Isci, A. Buyuktosunoglu, Runtime application behavior prediction using a statistical metric model. IEEE Trans. Comput. **62**(3), 575–588 (2013)
9. B. Efron, T. Hastie, I. Johnstone, R. Tibshirani et al., Least angle regression. Ann. Stat. **32**(2), 407–499 (2004)
10. R.R. Karn, I.M. Elfadel, Extraction of thermal workload signatures in multicore processors using least angle regression, in *2015 International Conference on Communications, Signal Processing, and Their Applications (ICCSPA'15)*, Feb 2015, pp. 1–5
11. R.R. Karn, I.M. Elfadel, Multicore power proxies using least-angle regression, in *2015 IEEE International Symposium on Circuits and Systems (ISCAS)*, May 2015, pp. 2872–2875
12. R.R. Karn, I.M. Elfadel, Autoscaling of cores in multicore processors using power and thermal workload signatures, in *IEEE 59th International Midwest Symposium on Circuits and Systems (MWSCAS 2016)*, Oct 2016, pp. 1–4
13. T. Hastie, R. Tibshirani, J. Friedman, T. Hastie, J. Friedman, R. Tibshirani, *The Elements of Statistical Learning*, vol. 2, no. 1 (Springer, Berlin, 2009)
14. J. Demmel, A. Gearhart, Instrumenting linear algebra energy consumption via on-chip energy counters. UC at Berkeley, Tech. Rep. UCB/EECS-2012-168 (2012)
15. Intel PCM performance counter monitor description, https://software.intel.com/en-us/articles/intel-performance-counter-monitor. Accessed 30 March 2015
16. M. Shafique, S. Garg, J. Henkel, D. Marculescu, The EDA challenges in the dark silicon era: temperature, reliability, and variability perspectives, in *Proceedings of the 51st Annual Design Automation Conference* (ACM, New York, 2014), pp. 1–6
17. J. Henkel, H. Khdr, S. Pagani, M. Shafique, New trends in dark silicon, in *Design Automation Conference (DAC), 2015 52nd ACM/EDAC/IEEE* (IEEE, Piscataway, 2015), pp. 1–6

18. M. Shafique, D. Gnad, S. Garg, J. Henkel, Variability-aware dark silicon management in on-chip many-core systems, in *Proceedings of the 2015 Design, Automation & Test in Europe Conference & Exhibition*. EDA Consortium (2015), pp. 387–392
19. H. Khdr, S. Pagani, M. Shafique, J. Henkel, Thermal constrained resource management for mixed ILP-TLP workloads in dark silicon chips, in *Proceedings of the 52nd Annual Design Automation Conference* (ACM, New York, 2015), p. 179
20. I. Takouna, W. Dawoud, C. Meinel, Accurate multicore processor power models for power-aware resource management, in *2011 IEEE Ninth International Conference on Dependable, Autonomic and Secure Computing (DASC)* (IEEE, Piscataway, 2011), pp. 419–426
21. V.M. Weaver, Linux perf event features and overhead, in *The 2nd International Workshop on Performance Analysis of Workload Optimized Systems, FastPath*, 2013, p. 80
22. H. Zhao, A. Sharifi, S. Srikantaiah, M. Kandemir, Feedback control based cache reliability enhancement for emerging multicores, in *2011 IEEE/ACM International Conference on Computer-Aided Design (ICCAD)* (IEEE, Piscataway, 2011), pp. 56–62
23. E. Seo, J. Jeong, S. Park, J. Lee, Energy efficient scheduling of real-time tasks on multicore processors. IEEE Trans. Parallel Distrib. Syst. **19**(11), 1540–1552 (2008)
24. X. Guerin, W. Tan, Y. Liu, S. Seelam, P. Dube, Evaluation of multi-core scalability bottlenecks in enterprise java workloads, in *2012 IEEE 20th International Symposium on Modeling, Analysis & Simulation of Computer and Telecommunication Systems (MASCOTS)* (IEEE, Piscataway, 2012), pp. 308–317
25. K.K. Pusukuri, R. Gupta, L.N. Bhuyan, Thread reinforcer: dynamically determining number of threads via OS level monitoring, in *IEEE International Symposium on Workload Characterization (IISWC)*, November 2011, pp. 116–125
26. C. Bienia, S. Kumar, J.P. Singh, K. Li, The parsec benchmark suite: characterization and architectural implications, in *Proceedings of the 17th International Conference on Parallel Architectures and Compilation Techniques* (ACM, New York, 2008), pp. 72–81
27. J.L. Henning, Spec cpu2006 benchmark descriptions. ACM SIGARCH Comput. Archit. News **34**(4), 1–17 (2006)

Chapter 20
A Comparative Study of Assertion Mining Algorithms in GoldMine

Shobha Vasudevan, Lingyi Liu, and Samuel Hertz

20.1 Introduction

Design verification constitutes the major bottlenecks in the hardware design cycle, often requiring majority of the time and resources needed to design the chip. With the increase in complexity and heterogeneity of general purpose hardware and integration of hardware components in system-on-a-chip, the verification task remains a daunting challenge for the foreseeable future. Simulation-based verification, although ad-hoc and incomplete, is the most prevalent verification approach. Formal verification, despite its scaling limitations, is valued for its complete and exhaustive exploration of the design state space. In both cases, the purpose of design verification is to check if the design implementation satisfies the specification.

Assertions are logical statements expressed in a propositional or temporal logic that describe design intent. Assertions describe invariant properties of the design. These properties can be used for expressing combinational as well as sequential behavior of the design. Assertions can be used during simulation for monitoring behavior. They can also be used as properties that are checked against the design during formal verification. Due to their versatility and capacity to express more behavior than tests, assertions have become critical to most state-of-the-art verification environments.

Assertions have transitioned from complicated high level model specifications written by few experts to widely used artifacts in hardware designs and verification IP in system-on-a-chip (SoC) designs [23]. This is due to the burgeoning of assertion specification languages that presented a familiar interface to designers and verification engineers, as well as the support provided by commercial tools

S. Vasudevan (✉) · L. Liu · S. Hertz
Electrical and Computer Engineering Department, University of Illinois at Urbana-Champaign, Champaign, IL, USA
e-mail: shobhav@illinois.edu

I. M. Elfadel et al. (eds.), *Machine Learning in VLSI Computer-Aided Design*,
https://doi.org/10.1007/978-3-030-04666-8_20

for assertion-based verification. A survey [59] reported that assertion usage in the hardware design industry has increased from 37% to 69% between 2007 and 2010.

The singular aspect limiting assertion-based verification is in writing these assertions. Assertion generation is a typically manual task, with designers attempting to write assertions. Writing effective assertions that express relevant temporal behavior, inter-modular interactions, and have high behavioral coverage is a challenging task [23]. This ends up being a time and resource sink in most verification environments.

Recently, multiple solutions have emerged for automatically generating assertions [5, 12, 21, 29, 54]. Many of these techniques use dynamic simulation data to infer the design's behavior [29, 54, 66]. These mechanically generated assertions suffer from some issues as compared to human generated assertions. They typically are of non-uniform quality, with some assertions being more trivial than others. They are frequently redundant and not as succinct as human generated assertions. They are also numerous, making it a daunting task for a designer to parse through a long list of assertions to select from. They may also be incomprehensible from a human readability perspective.

GoldMine [32, 66] is an automatic assertion generation tool that uses data mining along with static source code analysis and formal verification. The principle in GoldMine is that the static analysis can be used to focus on the statistical, dynamic inferencing, i.e., data mining. It is observed in GoldMine that iterative, structured guidance from static source code analysis [32, 37, 39] and formal verification [40] can significantly enhance the results of data mining.

GoldMine works are shown in Fig. 20.1. A register transfer level (RTL) design is simulated in the data generator using random or constrained stimuli. The data mining phase (A-miner) mines the simulated data for candidate assertions. Static analysis information from RTL is provided to A-miner. The candidate assertions are then checked by a formal verification engine against the design. If the candidate assertion guessed by the data miner is incorrect, the formal verification will fail the candidate. The passed assertions are invariant in the design. A-val represents the evaluation and ranking phase of the assertions, which is based on subjective and objective metrics.

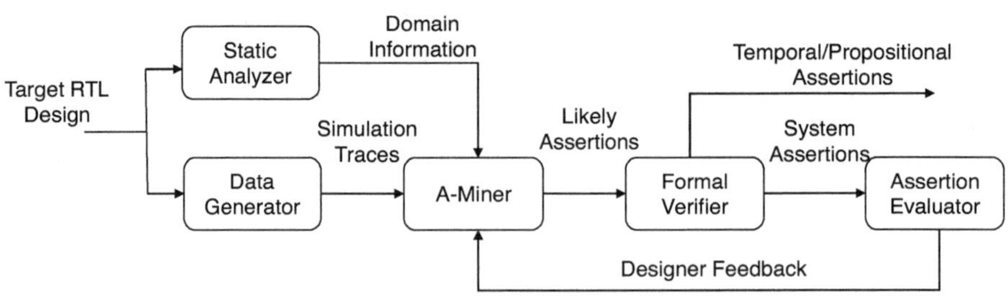

Fig. 20.1 The GoldMine methodology

In GoldMine, we are seeking to resolve the issues with automatic assertion generation outlined above, to make this a widely used technology that can evolve to increase the efficiency and productivity of the designer. To this end, we are investigating the extent to which each phase of GoldMine affects assertion quality.

We have researched the impact of static analysis such as logic cone-of-influence [32], word level feature and target generation [39], design domain knowledge [41], and counterexample guided assertion refinement [38]. We have also investigated evaluation algorithms [6] and [31] in the A-val phase.

In this work, we investigate the effect of applying different types of data mining algorithms on assertion generation, i.e., the A-miner. In [32, 40, 66], we have used a decision tree-based supervised learning algorithm in the data mining phase (A-miner). Broadly speaking, there are two kinds of mining algorithms—tree based and association rule based [26]. Decision tree algorithms [52, 53] are ideal for assertion generation because they are simple, scalable, and represent the data in a compact and intuitive way [26].

Decision tree algorithms perform a greedy search to quickly identify local regions in the data space. These algorithms recursively partition the dataset by assigning values to feature variables until the value of the target variable is consistent. Consequently, decision tree algorithms induce a rule by walking from the root of the tree to a leaf. Each edge in the walked path defines a proposition in the rule as antecedent and the leaf node defines its consequent.

Decision tree algorithms can introduce irrelevant propositions into a set of rules since they partition the dataset hierarchically. Tree-based algorithms are greedy in their search for the choice of splitting variable. Once they decide on a splitting variable, they continue to split further and add more levels to the tree based on the initial choice.

However, due to their greedy approach to partitioning the dataset and choosing splitting variables, they can generate overconstrained rules. In the context of assertions, redundancies in the antecedent of the assertion[1] will result in low coverage of the input space by the assertion.

In this work, we introduce the *best-gain decision forest*, as an alternative to the decision tree algorithm. Decision tree algorithms select one feature variable with maximum information gain to partition the dataset. In contrast, the best-gain decision forest (BGDF) algorithm partitions the dataset using all such variables. Consequently, the BGDF algorithm builds *all decision trees* optimized for maximum information gain. To maintain efficiency, the algorithm shares nodes between decision trees. The BGDF algorithm extracts all assertions from a decision forest and uses a set containment algorithm to discard those that are redundant. The BGDF algorithm is most useful in cases where there are many candidate variables with equal information gain, any of which could be chosen as the next feature variable

[1]GoldMine generates assertions in linear temporal logic (LTL) of the form $A => C$ where the antecedent A and consequent C can be propositional or temporal.

in the tree. It removes the artificial constraints imposed by the hierarchical nature of the decision tree by maintaining all valid trees that could be built.

Association rule learning induces rules exhaustively between all variables [3]. Association rules that fall beneath a minimum level of support are then discarded as insufficiently represented in the data. In [56, 57], we presented a novel algorithm for association mining guided by input space coverage of the assertions. Such a *coverage guided mining approach* iteratively eliminates assertions (rules) that do not satisfy a progressively shrinking minimum coverage threshold. Consequently, this method is optimized for high input space coverage, concise assertions. A drawback of association rule learning algorithms in machine learning is their inefficiency in performance and the large number of association rules they generate. The reliance on coverage filters out many of the irrelevant association rules, and emphasizes succinctness of generated assertions. Some preliminary results on small benchmarks were presented in [56]. We present an extensive set of results on this algorithm in our comparison experiments in this work.

We have also implemented the PRISM [11] algorithm in GoldMine for comparing our results against the other algorithms. PRISM is known to alleviate some of the problems of the decision tree. Unlike the decision tree and decision forest (BGDF), the PRISM algorithm does not select feature variables based on their ability to partition the dataset. Instead, the algorithm selects the feature variable, value pair that reduces the dataset entropy most. The PRISM algorithm recognizes that not all values of a feature variable are relevant to the target variable. Hence, it does not introduce irrelevant propositions into a set of assertions.

Our experiments compare the four data mining algorithms discussed using open source benchmark designs such as USB, PCI, and OR1200 [47]. We compare the algorithms with respect to objective and subjective metrics that provide a (partial) method to compare assertion quality. We provide an analysis of the comparisons and empirical results.

Our contributions to this work are as follows. We introduce the best-gain decision forest algorithm for generating more concise and succinct assertions than the decision tree. We provide a comparative study of four different static analysis guided data mining algorithms for assertion generation—decision tree, coverage guided association mining, best-gain decision forest, and PRISM. We provide an in-depth analysis of each algorithm. We present experimental results on open source benchmark designs such as the PCI, USB, and OR1200. We also hereby provide access to all these implementations in GoldMine that can be downloaded from [64] for research purposes.

20.1.1 Related Work

Work in deductive program verification [7, 10, 44] has studied assertion generation through static analysis of source code since the 1970s. Static analysis [49, 65] and dynamic analysis [4, 14, 46] techniques to generate assertions for program verification exist. Due to the size and complexity of software programs and constructs, these are not as wide in scope and applicability as hardware assertions.

In hardware, previous work [30, 50, 67] has used static analysis to generate hardware assertions. IODINE [29] generates low-level dynamic assertions for hardware designs, but it does not use data mining. Instead, IODINE analyzes dynamic program behavior with respect to standard property templates such as onehot encoding or mutex. The methodology in [54] uses dynamic simulation trace data to generate assertions, but it does not use data mining. Instead, the methodology tries to generalize design behavior based on simulation data. It has two phases. In the first phase, a predefined set of basic properties are inferred using template matching. In the second phase, the temporal dependency between the basic properties is checked to formulate more complex properties. There is no learning process and the derived properties are also limited by the predefined set of basic properties. Inferno [20] uses simulation data to extract the semantic protocol from a communication interface RTL design. Inferno infers a set of transaction diagrams which it uses to generate a set of assertions. For each vertex in the transaction diagram, it generates a disjunction of all the expressions on outgoing edges. Therefore, it discards the complex temporal relationship between different signals. Consequently, Inferno assertions might complicate the search for bugs in such designs. In addition, Inferno is limited to designs that implement a communication protocol, which constitute a small subset of hardware.

In [12], the authors use sequential pattern mining to infer causal relationships between frequently occurring sequences of input and output events. Since their methodology only seeks relationships between inputs and outputs, it might not generate meaningful assertions for designs with high temporal depth. In addition, since their methodology enumerates each unique input and output event, it might not scale for large designs.

In [15], the authors use symbolic simulation to extract assertions from test bench constraints. The authors use the generated assertions to optimize the design under test. Since their methodology extracts assertions from a constrained test bench, the assertions might reflect a bug in the test bench or overlook untested behavior. In addition, the generated assertions might have limited value since their methodology extracts them using simple templates and verifies them using bounded formal verification.

Previous work [9, 33, 68] has proposed methods to overcome the limitation of decision trees by constructing random subspace forests. These methodologies construct decision forests to avoid overfitting the dataset but do not address high quality rule induction.

Since GoldMine's inception, several commercial methodologies have emerged for assertion generation, including Atrenta's BugScope [5] and Jasper's Active-Prop [1]. The quality and number of assertions generated tend to be inconsistent with commercial tools, making them good point solutions. GoldMine is intended to be a comprehensive, integrated, generic solution to this problem.

In our own previous work, we have enhanced GoldMine to improve the quality of assertions in various ways [32, 38] from generating word level assertions [39] to mining system level assertions [37].

Many researchers have used GoldMine for various applications [18, 19, 22, 24, 27, 28, 60–62]. Some work [13, 19, 42] improves over the early version of GoldMine [66] and is comparable to more recent work in GoldMine [32, 39].

20.2 Summary of Comparison of Assertion Generation Algorithms in GoldMine

We provide an overview of how the algorithms discussed in this chapter compare. The four mining algorithms discussed here are at different points in the spectrum of inferencing. We quantify some common metrics to judge the quality of an assertion—(1) conciseness of the assertion, (2) predictive accuracy of the algorithm in generating likely assertions, (3) number of assertions generated, (4) complexity of behavior and importance of the assertions generated, and (5) how unexpected the generated assertions are. With respect to these metrics, the four algorithms presented here have different capabilities and trade-offs. The definitions of these metrics are provided in Sect. 20.4.6 and the quantitative comparisons are presented in Sect. 20.9.

The decision tree algorithm attempts to find an explanation for the given data as quickly as possible. This means that it optimizes for minimal depth in the tree structure. It also generates non-overlapping assertions, so that each one corresponds to a different behavior of the dataset. It also maintains a single tree data structure for each output. This is very restrictive in terms of the type of rules generated. The splitting variable choices made once in the past cannot be revised. This inclusion of all previous splitting variables, as well as frequent overfitting to explain the data results in redundant propositions in the antecedent. This makes decision tree assertions less concise and lower quality. The quickest explanation usually means that it is not subtle or capturing complex behaviors. The predictive accuracy of the decision tree is high, since it makes "safe" guesses. It does not attempt to make leaps in inferencing, since it is constrained by the tree structure and simply fits the data irrespective of the size or relevance of the assertion. Due to the overconstraining problem, the assertions capture design paths that are less expected from a uniform distribution. Resource utilization is minimal for the decision tree.

The decision forest, introduced in this chapter, is a decision tree with more freedom and flexibility. This forest structure circumvents the issue of being confined to a particular splitting variable choice. Multiple trees are built for exploring the equal information gain choices that would not be explored by the tree. In situations where there is a tie between splitting variable choices, the decision forest ensures that the redundant variables are eliminated. Hence it is better than the decision tree in conciseness. Since the decision forest is composed of many trees, the basic tree structure is maintained for inferencing every assertion. So within a tree structure, the attempt is to generate minimal depth explanations. The complexity of behaviors captured by the forest (quality) is higher than the tree due to more than one tree capturing alternate behaviors for the same output. However, it is less than the inferencing by free associations between variables. The predictive accuracy is high as in the decision tree. The resource utilization would normally be high, but due to key optimizations like node sharing, our implementation consumes very reasonable resources.

The PRISM algorithm also maintains a tree structure in inferencing, but it generates rules (assertions) for a single concrete value of a variable each time. In the Boolean space, this implies that assertion generation for variable f when $f = 0$ is distinct from when $f = 1$. The focused inferencing around the values of a variable results in very high quality assertions. PRISM assertions capture the most complex, subtle behaviors, especially in temporal assertions. However, due to the duplication in generating assertions for the same variable, the total number of assertions generated by PRISM is very high. Since every inference is made within the tree structure, the minimal depth principle is followed. Due to the external regulation of the target variables in each tree, there is less redundancy in the resulting assertions. PRISM is optimized for relevance and completeness of predictions with respect to the dataset. It can produce assertions that are concise, but it also generates verbose assertions infrequently. It means there are a lot of redundant propositions in generated assertions. PRISM has a lower predictive accuracy than the decision tree and forest, since it does not adopt the low risk, greedy approach of these algorithms. It makes more risky inferences, in favor of more quality in the assertions instead of finding the best existing explanation. Also, the high number of generated candidate assertions results in a smaller percentage being correct as attested by formal verification. However, the subtle assertions mean that they are unexpected as they cover rare behaviors. The resource utilization by PRISM is not high.

Coverage guided association mining is a domain-aware variation of association rule mining. Input space coverage is used as a guiding parameter to accept or discard an association rule. This approach does not infer using a tree structure, but with free associations between variables. The filtering and constraining due to input space coverage being above a certain threshold for this algorithm results in the free associations being regulated significantly. This alleviates the problem with association rule-based approaches generating large volumes of results. The coverage guided miner produces a reasonable number of assertions, which tend to be less than PRISM and more than decision tree. Since the input space coverage is used to filter the candidate assertions, conciseness is highest with this algorithm. If an assertion has higher input space coverage, it translates to fewer propositions in the antecedent and more concise (simpler) explanations. The resource utilization of the coverage guided miner is the highest, since it exhaustively searches the rule space to generate simpler explanations until it cannot find them. It has a higher percentage predictive accuracy than PRISM due to the fewer number of assertions generated. However, the unexpectedness factor is low in these assertions due to the focus on high input space coverage.

In summary, there is no single winner with respect to all metrics. The best algorithms for a task depend on the design and user-defined relevance for each metric.

20.3 The GoldMine Principle: Statistics Meet Static

In this section we present the intuition behind the scalability and effectiveness of our solution. This section can be skipped without loss in continuity to the reader.

Data mining is the process of deciphering knowledge from data [48, 63]. Data mining uses dynamic behavior in the form of simulation data or training sets to find statistical correlations and make inferences about the system. In typical data mining applications such as web-mining, online recommendation systems, health-care, and bioinformatics, there are three striking characteristics.

- The underlying systems, namely human psychology, human interest patterns, medical records, protein sequences, or web data are highly unstructured and prone to wild fluctuations.
- It is not possible to claim that the knowledge generated by data mining is *true*, since it relies purely on statistical evidence and there is no oracle to attest it. For highly unstructured systems (or those whose structure has not yet been deciphered), behavioral traces are the only information that can be gleaned from the system.
- The data mining suffers from the problem of not being able to simulate judgment, i.e., it will not be able to decide how "interesting" a piece of information (say, a rule) is to a domain expert [43, 58, 63]. It relies on iterative learning from a human domain expert and tries to predict interestingness, or relevance of a rule to the domain where it is applied.

In contrast to typical applications of data mining, hardware designs are extremely structured and their design is a very regulated process. This means that data mining algorithms would have higher accuracy and predictability than usual.

- Since these systems are designed by humans, there is a notion of "absolute truth" in these systems that can be attested by an oracle—either the specification document, a reference model, or the system architect. The data mining algorithms can get reinforcement for their correct statistical inferences as well as quick recovery from their wrong ones. Formal verification techniques are built to answer the questions of truth about a system automatically [25] as well as expose falsehood through a witness. A formal verification algorithm coupled with a data mining algorithm makes a statistical approach a rigorous method; it would also make an inherently intractable problem of state space exploration directed and restrained.
- Another reason is that measures of interestingness [43, 58] for hardware systems can, to a large extent, be identified and quantified. For example, in a digital

circuit, a signal with a high fan-in may be considered important, and any assertion about its behavior interesting. This type of steady and accurate high level guidance is atypical for data mining algorithms.

The existence of techniques that can extract domain knowledge from systems and guide the data mining is unique to computer systems. We have given examples of fan-in-based priority and cone-of-influence that can provide this knowledge. There are multiple such known techniques in hardware that can be used for analyzing the target system. We group all these techniques under the term *static analysis techniques*. In hardware, we use the term static analysis to mean methods that analyze design structure/function (analogous to program syntax and semantics in software). Examples of structural methods include cone-of-influence [8, 16, 17] and localization reduction [36]. Formal verification can be considered a static analysis of the semantics of a model. Formal verification, therefore, is also a type of static analysis in our terminology. We distinguish between the two by calling one "lightweight" and the other "formal."

20.4 Background on GoldMine

Figure 20.1 depicts each component of the methodology. GoldMine has different use cases in industry: (1) ranked list of assertions to be used directly by designers for legacy/stable designs, (2) providing hints to designers about assertions they can write, (3) enhanced design understanding from a different perspective, (4) corner case or rare event analysis by examining low support assertions, (5) code coverage and behavioral coverage analysis for each assertion, (6) coverage hole analysis of simulation data by examining failing assertions, (7) regression test suite over optimization phases of the design, (8) generating "specification" assertions from System C for functionality and performance, and (9) coverage analysis of tests on emulation/prototyping FPGA platforms.

20.4.1 GoldMine Assertions

In this chapter, *a target* is a Boolean RTL variable for which we want to generate assertions. A *feature* variable with respect to a target variable is a Boolean variable that can potentially affect the value of the target [26, 45]. In our implementation, the variables considered for mining assertions are primary inputs, registers, and the primary outputs. Output serves as target variable in mining algorithm and primary inputs and registers serve as feature variables. GoldMine can generate *propositional* as well as *temporal* assertions.

The generated propositional assertions are of the form $A \Rightarrow B$, where the *consequent* B is a single proposition with respect to a target variable. The consequent asserts that the target variable value is true or false. The *antecedent* A is a conjunction of propositions in terms of feature variables of the corresponding target. For example, in assertion $p \wedge \neg q \Rightarrow \neg r$, r is the target and p and q are feature variables with respect to r.

GoldMine generates bounded time assertions in LTL [51] using the "next" operator X. We provide the data miner a *mining window length*, or the duration of time cycles over which we capture temporal behavior. For example, if we want to consider the following behavior: once req is valid, the ack should be valid two cycles later, the mining window length should be set to 2. The generated assertion will span up to a maximum of two cycles, as in $req \Rightarrow XXack$. We do not generate unbounded assertions using the "eventually" operator F. Although "until" operator U has not been shown in the examples of this chapter, some of our assertions can be interpreted as a "bounded" until operation. For example, $p \wedge Xp \wedge XXp \Rightarrow XXXq$, we can potentially extend this to a contrived "until" proposition like pUq. It should be noted that it is not faithful to the true semantics of the "until" operation, since the observations are only within a bounded cycle.

For temporal assertions, we annotate a variable within a mining window with its corresponding cycle. In the example, variable req is annotated as $req(t)$, $req(t+1)$, and $req(t+2)$ within a mining window. Each time annotated variable can now be treated like a propositional variable, converting temporal assertion generation into a propositional assertion generation problem. When we finally report the generated assertions, the generated assertion $req(t) \Rightarrow ack(t+2)$ is transformed to $req \Rightarrow XXack$ in LTL. We will only explain propositional assertion generation for all algorithms for ease of understanding.

For ease of the explanation of the algorithms, we will use a variable value pair $\langle v, \alpha \rangle$, where $\alpha \in 0, 1$, to represent the propositions of the assertion, where v can be a target or a feature variable. $\langle v, 0 \rangle$ represents $\neg v$ and $\langle v, 1 \rangle$ represents v. Therefore, the antecedent can then be simply represented as a set of such pairs, and the consequent by a single such pair.

20.4.2 Static Analyzer

The static analyzer parses and extracts a variety of static information from a Verilog [35] design. It determines the top module in the design hierarchy, clock, reset signals, and the target variable. It also selects a set of feature variables for the target variable.

It uses the bounded cone-of-influence to select feature variables [16, 17]. The bounded cone-of-influence extends the classic cone-of-influence [36]. Instead of computing all of a variable's dependencies, it computes only those within a bounded number of temporal frames.

In [39], the static analyzer uses a modification of the weakest liberal precondition computation [34] to discover word level features from RTL design for mining word level assertions. The word level feature extraction is independent of and applicable to all the mining algorithms. Finally, the assertion evaluator A-val uses the static analyzer to compute evaluation metrics for an assertion, including the importance, complexity, and coverage of an assertion [31].

20.4.3 Data Generator

The data generator generates and parses simulation data. The data generator creates an unconstrained as well as a directed/constrained random test bench for the design. The unconstrained random test bench assigns random values to each input variable for user-specified number of cycles of simulation. We have found empirically that the optimal simulation length for generating interesting data mining results is between 10,000 and 20,000 cycles. The data generator uses the created (supplied) test bench to simulate the design. It unrolls the data into temporal frames of length t and compresses it by discarding duplicate frames.

20.4.4 A-Miner

A-miner is the data mining engine of GoldMine. A-miner searches for causal relationships between feature variables and the target variable in the simulation data. If the data miner finds a relationship with 100% confidence, then it will generate an assertion. Any suitable rule induction algorithm can mine a set of assertions from the simulation data. In this chapter, we discuss four algorithms GoldMine uses to generate assertions.

20.4.5 Formal Verifier

The data miner generates assertions from the simulated data. This means that the assertions are being generated on the subset of design behavior that was simulated. Even if we consider only the 100% confidence results, there is no guarantee that the generated assertions are true system invariants. These are still likely invariants. A formal verification phase is used to verify if the likely invariants are true system invariants. The likely invariants that are not system invariants are poor guesses made by the mining algorithms, mostly due to insufficient data. In a counterexample guided version of Goldmine [38, 40], we use the counterexamples from formal verification as feedback to generate better assertions. In this chapter, we will not be describing this algorithm. We constrain the design's reset signal to prevent the tool from generating trivial counterexamples. The tool reports assertions that pass formal verification as true system invariants.

20.4.6 Assertion Evaluator

GoldMine attempts to automate a traditionally manual process. Consequently, the assertion evaluator evaluates GoldMine's performance and the quality of the generated assertions. The original metrics for assertion evaluation in GoldMine included antecedent size, temporal length, input space coverage, and hit rate. *Antecedent size* refers to the number of propositions in an assertion's antecedent. *Temporal length* refers to the number of bounded temporal frames spanned by an assertion's antecedents. The assertion evaluator reports the average antecedent size and temporal length for a set of generated assertions to estimate their readability.

- *Input space coverage* estimates the functional coverage of an assertion. Suppose we create a truth table that computes the target variable's value as a function of the feature variables. The input space coverage of assertion $p : A \Rightarrow C$ is equal to the fraction of entries in the table that satisfy A. We can use the formula $\frac{1}{2^{|A|}}$ to compute the input space coverage p. Intuitively, an assertion with fewer propositions in its antecedent will cover a greater fraction of the input space. Though computing an assertion's input space coverage is trivial, doing so for an entire set of assertions is not since the coverage of two assertions might not be mutually exclusive.
- *Hit rate* refers to the fraction of generated assertions that passed formal verification. Hit rate evaluates the data miner's performance and completeness of the simulation data. In other words, we prefer a data mining algorithm that can find true assertions over one that can find high quality false assertions. Regardless, a data mining algorithm's performance will be limited by the completeness of the simulation data.
- *Statement coverage of assertion* estimates the fraction of RTL statement covered by an assertion. In [2], the authors propose a correctness-based coverage algorithm to compute RTL code coverage of an assertion. The correctness-based coverage algorithm computes the set of statements executed after the satisfaction of an assertion's antecedent. If an executed statement transitively satisfies the assertion's consequent, then the assertion covers that statement. The algorithm proceeds as follows. Let p denote an assertion with temporal length k. First, the algorithm initializes the dataset S so that it satisfies the antecedent of p. Next, the algorithm performs an event-based simulation of the design for k temporal frames. During simulation, the algorithm records all assignment statements triggered/executed in each temporal frame. Starting from the variable used in the consequent of p, we could identify the statement assigning to that variable. Then for each variable used in the statement, we can transitively identify the corresponding assignment statements. This final transitive set of statements is covered by assertion p.
- *Assertion expectedness* estimates the probability that an assertion's antecedent will be satisfied, assuming that the design inputs vary uniform-randomly. Assertions with low expectedness convey design behavior that occurs rarely. Con-

sequently, we rank assertions with low expectedness higher than those with high expectedness. We use simulation techniques to estimate an assertion's expectedness.

- *Assertion importance* estimates the importance of an assertion to a design using fan-in, fan-out cones and other design specific measures computed by static analysis as in [31]. A more important assertion is desirable.
  Intuitively, the *importance score* of an assertion estimates the *importance* of the design paths that are covered by the assertion. Assertion importance also considers temporal depth and spatial distance between the target variable and the variables in the antecedent. It ranks an assertion higher if the temporal depth and the spatial distance are higher. Consequently, an assertion with a high importance score covers behavior that is critical with respect to its design's functionality.
- *Assertion complexity* estimates the human understandability of an assertion. It quantifies the amount of effort in understanding an assertion in terms of levels of logic, temporality, etc.[31]. Lower complexity means ease of understanding, but higher complexity may also be desirable to identify subtle interactions in the design.
  Intuitively, the *complexity score* of an assertion estimates the "coverage of complex behaviors" of a design. It measures the spatial distance, temporal distance, and understandability of execution paths between uses of a variable in the antecedent and definitions of the target variable. A complex assertion will have high spatial distance, high temporal distance, will cover complex design paths between the satisfaction of its antecedent and consequent.

20.5 Decision Tree-Based Learning

The classical algorithm for building decision trees called ID3 employs a top-down, greedy search through the space of possible branches with no backtracking [52, 53]. ID3 uses entropy and information gain [45] to construct a decision tree. In each partition, the algorithm computes information gain for each feature variable and then selects one feature variable with maximum information gain to partition the dataset. If the entropy is zero, the algorithm stops splitting and exits. In the context of hardware, each variable is Boolean variable and the decision tree is thus a binary tree.

In [32], GoldMine uses decision tree-based learning algorithm to automatically generate assertions for hardware verification. Here, we present an example of the process. Consider the AND gate design and its decision tree in Fig. 20.2. The root node assigns multiple values to f. Therefore, the algorithm cannot generate an assertion and partitions the simulation data. The root's left child node always assigns $f = 0$. Therefore, the algorithm generates the assertion $\neg a \Rightarrow \neg f$. Again, the root's right child node assigns multiple values to f. Therefore the algorithm will partition the simulation data again. Each new child node assigns a single value to f.

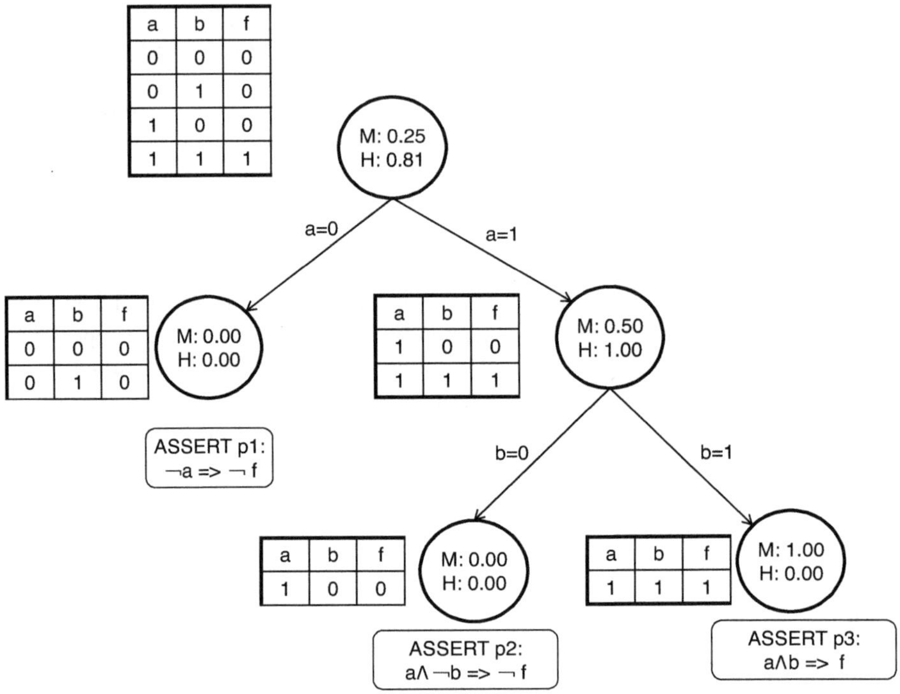

Fig. 20.2 A decision tree for an AND gate design. Each node is pictured with the data it represents and is labeled with its *mean* and *entropy*. Each branch is labeled with the feature variable and value used to partition the data represented by its parent

Therefore, the algorithm generates assertions $a \wedge \neg b \Rightarrow \neg f$ and $a \wedge b \Rightarrow f$. In each node, we calculate the mean value and the entropy value of target in the corresponding examples. We select the best splitting variable according to the highest information gain [26].

20.6 Best-Gain Decision Forest Algorithm

We introduce the *best-gain decision forest algorithm* (BGDF) to generate concise assertions.[2] Decision tree algorithms only select one feature variable with maximum information gain to partition the dataset, even if there are multiple feature variables with the same maximum information gain. As a result, the decision tree algorithm tends to introduce irrelevant propositions in the antecedent of generated assertions. Irrelevant proposition in assertion overconstrains the antecedent condition for

[2]We use the terms decision forest and acronym BGDF interchangeably through this chapter.

asserting the target variable, and thus degrades the readability and coverage of the assertion. Consider the assertion $a \wedge \neg b \Rightarrow \neg f$ in Fig. 20.2. Regardless of the value of a, $\neg b \Rightarrow \neg f$. Therefore, the proposition a in this assertion is irrelevant. As a result, the assertion is verbose and constrained—limiting its use and value.

Let us analyze why decision tree-based learning algorithm introduces irrelevant propositions in the generated assertion. At the root node in Fig. 20.2, the information gain for feature variables a and b is calculated. a and b have the same information gain. However, the algorithm is allowed to select only one feature variable for partitioning the data. Variable a is selected in this example. Variable b is selected to partitioning the dataset in the next level of decision tree. Inevitably, the proposition a is included in the assertion $a \wedge \neg b \Rightarrow \neg f$.

Our BGDF algorithm solves this problem by allowing multiple feature variables to partition the dataset. It breaks down the dataset using all such feature variables with the same maximum information gain simultaneously. If there are two feature variables with the same maximum information gain, both of them will be selected for split the dataset. The algorithm will build a decision forest, within which one decision tree is constructed for each selected feature variable at every internal decision node.

In order to maintain efficiency, the algorithm implements two optimizations during the construction of decision forest. First, when we split for multiple variables with the same maximum information gain on current node, we do not generate multiple copies of current node for every subtree. We retain only one copy of the current node. As shown in Fig. 20.3, the root node is shared when we split on variable a and variable b. From this perspective, the decision forest we finally construct is actually a tree, within which internal decision nodes might have more than two branches.

Another optimization tries to improve the running efficiency of the algorithm in a practical way. This occurs when multiple variables have the same maximum information gain. We are required to split on all these variables according to BGDF algorithm and construct a subtree for each. In order to improve the algorithm efficiency, we no longer retain all these variables as features for each subtree. If one such variable is previously used to grow a subtree of current node, then we do not retain it as feature variable for subsequent subtree of current node. Let's use the example in Fig. 20.3 to explain this optimization. In the root node, we split on both a and b. Variable a is first used to grow a subtree, which includes feature variables a and b. We subsequently construct subtree for variable b. Now variable a is not used as features in this subtree, since the first subtree has already included a and b as features. That's why we stop splitting on the rightmost node in Fig. 20.3, though the entropy is not 0. In Fig. 20.3, suppose we continue to split on the rightmost node using variable b. We will generate two assertions. One is p_2. Another one is $\neg a \wedge b \Rightarrow \neg f$, which is redundant since $\neg a \wedge b$ implies $\neg a$ in p_0.

When the entire decision forest is finally constructed, we can generate assertions from it. Due to multiple decision trees within the decision forest, we might generate the same assertion only with different order of propositions in the antecedent. Also the antecedent of one assertion may imply the antecedent of another assertion. For

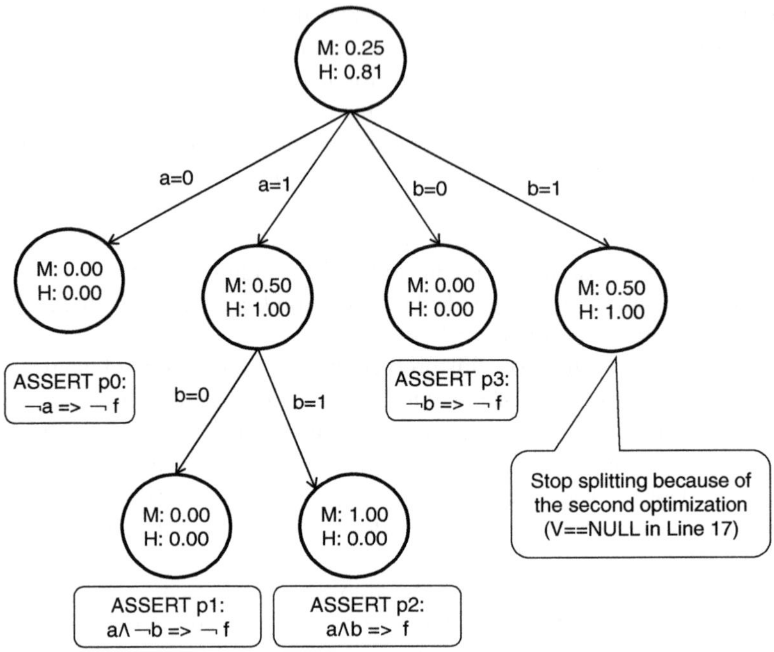

Fig. 20.3 A decision forest for the AND gate design. Each node is labeled with its mean and entropy. Each branch is labeled with the feature variable and value used to partition the data represented by its parent. The BGDF algorithm selects both a and b to partition the root node's data since they do so equally well

example, assertion p_1 and p_3 in Fig. 20.3 are derived from two different decision trees within the decision forest. Assertion p_1 is redundant since $a \wedge \neg b$ implies $\neg b$ in p_3. In such situation, the first assertion is redundant and should not be reported out. As a post-processing step of our implementation, we use set containment algorithm to detect the redundant assertions.

20.6.1 Algorithm

Algorithm 7 shows the best-gain decision forest algorithm. To simplify the algorithm explanation, the proposition represented by $\langle v, \alpha \rangle$ pair is also referred to as *literal*. The input parameters of the algorithm are V, S, and A. P is the global variable recording all generated assertions. A records all literals from the root node to current node. $G(S, v)$ is the information gain if feature variable v is selected to partition S. We record the set of feature variables having maximum information gain into V_{maxG}.

Algorithm 7 Best-gain decision forest algorithm

1: *Global* P : *Set of generated assertions*
 ▷ S : *Dataset for mining*
 ▷ v_t : *Target variable*
 ▷ V : *Set of feature variables of dataset*
 ▷ A : *Set of literals appearing in antecedent*
2: **procedure** $decision\_forest(V, S, A)$
3: **if** $H(S) = 0$ **then**
4: **if** $Mean(S) = 1$ **then**
5: $P \leftarrow P \cup ((\bigwedge_{lit_i \in A} lit_i) \Rightarrow v_t);$
6: **else**
7: $P \leftarrow P \cup ((\bigwedge_{lit_i \in A} lit_i) \Rightarrow \neg v_t);$
8: **return** P;

9: **for all** $v \in V$ **do**
10: $Calculate\ G(v, S);$
11: $V_{maxG} \leftarrow all\ v \in V\ having\ maximum\ G(v, S);$

12: **for all** $v \in V_{maxG}$ **do**
13: $V \leftarrow V \setminus v$
14: **if** $V! = NULL$ **then**
15: $decision\_forest(V, S_{v=0}, A \cup \langle v, 0 \rangle);$
16: $decision\_forest(V, S_{v=1}, A \cup \langle v, 1 \rangle);$

Lines 3–10 of the algorithm report the generated assertion if $H(S) = 0$. It means each row in dataset S has the same value: $Mean(S)$, on target variable v_t. The algorithm then stops splitting. If all rows in S have value 0 on target variable, the consequent of the generated assertion is $\neg v_t$. If all rows in S have value 1 on target variable, the consequent of the generated assertion is v_t. Since the set A records all literals from the root node to the current node, the conjunction of all these literals constitutes the antecedent of the generated assertion.

Lines 11–13 first calculate the information gain for each feature variable. Line 14 records the feature variables with maximum information gain. In the classical decision tree-based algorithm, only one feature variable with maximum information gain is selected to partition dataset. If two feature variables have the same maximum information gain, the decision tree-based algorithm randomly selects one feature variable. However, in our BGDF algorithm, we retain all feature variables with the same maximum information gain. Each such feature variable is used to partition the dataset simultaneously and leads to one decision tree. That's the reason why we name the structure as decision forest. Our BGDF algorithm is most useful when there is a tie between candidate variables with the same maximum information gain.

Lines 15–20 recursively construct the decision forest. For each feature variable with maximum information gain, one left subtree and one right subtree are recursively built. For left subtree, we also record into the literal set A the literal $\langle v, 0 \rangle$, which means $\neg v$. For right subtree, we record the literal $\langle v, 1 \rangle$ into the literal set A.

The two optimizations are also implemented between Line 15 and Line 20. In Line 18 and 19, we share the same variable v as the root node to recursively construct two different subtrees. Line 16 implements the second optimization. The variables with the same maximum information gain are removed from set V for further subtree construction.

Our BGDF algorithm simultaneously partitions the dataset using all the feature variables with maximum information gain. The assertions generated from different decision trees within the decision forest may contain redundancy. For example, in first decision tree, one feature variable v_1 is first selected to partition the dataset and another feature variable v_2 is then selected, while in the second decision tree, v_2 is first selected and then v_1 is then chosen to partition the dataset. Our BGDF algorithm will derive assertions from both decision trees. As a result, we may generate two completely same assertions. Moreover, the v_1 is probably not chosen at all in the second decision tree. As a result, the antecedent of the assertion from the first decision tree implies the antecedent of the assertion from the second decision tree. For example, we may derive the following two assertions: (1) $v_1 \wedge v_2 \Rightarrow v_t$ and (2) $v_2 \Rightarrow v_t$, from two decision trees in the decision forest. Since v_2 is implied by $v_1 \wedge v_2$, assertion $v_1 \wedge v_2 \Rightarrow v_t$ is redundant.

The BGDF algorithm minimizes redundancy in P using set containment. Assertion $p_x : (\bigwedge_{lit_i \in A_x} lit_i) \Rightarrow C_x$ ***contains*** assertion $p_y : (\bigwedge_{lit_i \in A_y} lit_i) \Rightarrow C_y$ if $A_x \subset A_y$ and $C_x \equiv C_y$. Here, A_x is the set of literals appearing in antecedent of assertion p_x and A_y is the set of literals appearing in antecedent of assertion p_y. That is, if assertion p_x contains assertion p_y, then p_x conveys the same behavior as p_y more concisely. Therefore, assertion p_y is redundant. Moreover, it is impossible for assertion p_x to contain assertion p_y if $|A_x| > |A_y|$. The BGDF algorithm uses this fact to reduce the number of set containment checks between assertions.

20.6.2 Example

We revisit the AND gate design to illustrate the BGDF algorithm. Let $V = \{a, b\}$ and let $v_t = f$. Let S be the dataset on the left of the root node in Fig. 20.2.

The best-gain decision forest algorithm begins by computing $H(S) = -\frac{3}{4}\log_2(\frac{3}{4}) - \frac{1}{4}\log_2(\frac{1}{4}) = 0.81$. Since the $H(S) \neq 0$, the algorithm partitions S. To do so, the algorithm computes each variable's information gain and records maximum information gain. $G(a, S) = 0.81 - 0.50 \cdot 0.00 - 0.50 \cdot 1.00 = 0.31$; $G(b, S) = 0.81 - 0.50 \cdot 0.00 - 0.50 \cdot 1.00 = 0.31$. Now, we have to feature variables having the maximum information gain. $V_{maxG} = a, b$.

Next, the algorithm recurses. Since both a and b have the same maximum information gain, the algorithm uses both of them to partition S. In the recursive call where $a = 0$, the mean of S is also equal to 0. Therefore, the algorithm adds assertion $p_0 : \neg a \Rightarrow \neg f$ to P and terminates.

In the recursive call where $a = 1$, the $H(S) \neq 0$. Therefore, the algorithm selects b to partition S since it is the only remaining variable in V. In both subsequent recursive calls, $H(S) = 0$. Therefore, the algorithm adds the assertions $p_1 : a \wedge \neg b \Rightarrow \neg f$ and $p_2 : a \wedge b \Rightarrow f$ to P.

Figure 20.3 depicts the decision forest constructed by the algorithm. Set P contains the following assertions:

$$p_0 : \neg a \Rightarrow \neg f$$

$$p_1 : a \wedge \neg b \Rightarrow \neg f$$

$$p_2 : a \wedge b \Rightarrow f$$

$$p_3 : \neg b \Rightarrow \neg f$$

Consider the p_1 and p_3. Both assertions have the same consequent, but p_3 is more concise. Therefore, the proposition $(a, 1)$ in the antecedent of p_1 is unnecessary. After the algorithm uses set containment to remove such assertions, P contains p_0, p_2, and p_3.

20.6.3 Analysis

We analyze the best-gain decision forest algorithm. First, we analyze the algorithm's complexity. Consider the construction of a worst case decision forest with variables $V = \{v_0, v_1, \ldots, v_n\}$. Since the BGDF will select all variables to partition S, it cannot select v_0 to partition S in the subforest of any other $v_i \in V$. Consequently, if we disregard v_0, then the algorithm will construct a worst case decision forest with $n-1$ feature variables. Similarly, each child of v_0 will be a worst case decision forest with $n-1$ feature variables. Let's denote the algorithm complexity for n variables as $T(n)$. Then we have the following recursive equation:

$$T(n) = 2 * T(n-1) + 2 * T(n-2) + \cdots + 2 * T(1) \qquad (20.1)$$

Solving the above equation, the worst case size of a decision forest is $O(3^{|V|})$. In practice, the complexity of the algorithm is much smaller than the theoretical bound since we generally do not have a tie of all feature variables in every node.

Finally, we show that the BGDF algorithm generates assertions that are more concise than those generated by the decision tree algorithm. Let P and P' denote the sets of assertions generated by the BGDF and decision tree algorithms, respectively. Since the BGDF algorithm builds all optimal decision trees, $P \supseteq P'$. Now, the BGDF algorithm discards only functionally redundant assertions and $P \supseteq P'$. Therefore, the assertions in P are either equivalent to or more concise than those in P'.

20.7 Coverage Guided Mining Algorithm

Gain is a data mining concept that refers to the value of adding some rule to the solution set of rules. In data mining, we only want to add a rule to our solution set if its gain is higher than any other potential rules. This concept fits well with our concept of input space coverage since we can define a notion of coverage gain. The coverage gain of a rule (assertion) refers to the change in total coverage of a set given that the rule is added to that set.

Typically, an *association mining algorithm* will try to exhaustively produce all possible rules relating all input variables to all output variables. To restrict the number of rules, we apply several constraints. Our first constraint, as in [66], is that only rules with 100% confidence can be considered as candidate assertions for association rule mining. We now include coverage feedback as a constraint. We impose a minimum coverage gain to drastically limit the number of candidate assertions. We then gradually relax this constraint until we have reached a desired coverage value. The *greedy set covering algorithm* will always choose the highest coverage assertions in each iteration.

Following the notation of previous algorithm, P is defined as the solution set of assertions. The total input space coverage of P is defined as $c(P)$. We define $g(P, P')$ as the input space coverage gain between two sets of assertions where $P' = P \cup p$ and p is an assertion. We also define g_{min} as the minimum coverage gain. The minimum coverage gain ensures that any assertion that is mined must raise the total coverage of P by g_{min}. We set a minimum coverage gain threshold $g_{\text{threshold}}$ and a maximum total coverage threshold $c_{\text{threshold}}$ which result in algorithm termination when reached. Our goal is to maximize the total input space coverage $c(P)$ by maximizing the $g(P, P')$ in each iteration while minimizing the total number of assertions and propositions in the antecedent of each assertion.

20.7.1 Overview of the Algorithm

The basic flow of the coverage guided mining algorithm [57] is shown in Fig. 20.4. We will apply the algorithm as it is explained to the simulation trace in Fig. 20.5. We set the maximum total coverage threshold to 99% and the minimum coverage gain threshold to 1%.

The algorithm starts by initializing the $g_{min} = 50\%$, $P = \{\}$, and $c(P) = 0\%$. We know that at least one proposition must be in the antecedent of the assertion which means that the maximum coverage gain must be 50%. We do not consider assertions without any propositions in the antecedent since those assertions are trivial.

In the next step, *gen_candidate*, the algorithm described in Algorithm 8, is invoked. In the *gen_candidates*, A refers to a set of $\langle var, val \rangle$ pairs representing the antecedent of a potential assertion p. R refers to the set of {var, val} pairs not in A, since we do not want to add the same $\langle var, val \rangle$ pair to an antecedent twice.

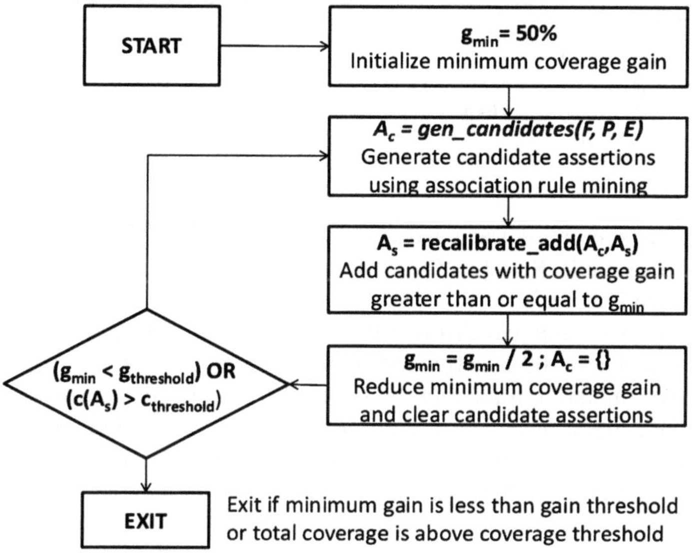

Fig. 20.4 Coverage guided association mining algorithm

Fig. 20.5 Dataset and function for example and candidate assertions

Simulation trace
f=(a|~c)&b

a	b	c	f
0	0	0	0
0	1	1	0
1	0	1	0
1	1	0	1

Candidate assertions
p1: ¬a => ¬f
p2: ¬b => ¬f
p3: c => ¬f
p4: a ∧ ¬b => ¬f
p5: a ∧ b => f
p6: a ∧ ¬c => f
p7: a ∧ c => ¬f

Algorithm 8 Association miner

$gen\_candidates(R, A, S)$

for each {var, val} pair r_i in R **do**
  if $g(P, P \cup assertion(A \cup r_i \Rightarrow \langle f, X\rangle)) \geq g_{min}$ **then**
    if $\forall s_j \in S, A \cup r_i \Rightarrow \langle f, 0\rangle$ **then**
      $P_c = P_c \cup assertion(A \cup r_i \Rightarrow \langle f, 0\rangle)$
    else if $\forall s_j \in S, P \cup r_i \Rightarrow \langle f, 1\rangle$ **then**
      $P_c = P_c \cup assertion(A \cup r_i \Rightarrow \langle f, 1\rangle)$
    else
      $gen\_candidates(R - r_i, A \cup r_i, S)$

Algorithm 9 recalibrate_add

$recalibrate\_add(P_c, P)$

for all $p \in P_c$ **do**
  if $g(P, P \cup p) \geq g_{min}$ **then**
    $P = P \cup p$

S refers to the simulation trace and is represented as a set of signal values at each cycle. In our example $R = \{\langle a, 0\rangle, \langle a, 1\rangle, \langle b, 0\rangle, \langle b, 1\rangle, \langle c, 0\rangle, \langle c, 1\rangle\}$, $A = \{\}$, and S is the data in Fig. 20.5.

Essentially what $gen\_candidates$ does is recursively add $\langle var, val\rangle$ pairs to A. If all pairs in A are 100% correlated with the target pair $\langle f, 0\rangle$, or $\langle f, 1\rangle$, in all cycles of the simulation trace represented by S, a candidate assertion is generated based on that correlation and the algorithm returns. The algorithm also returns when the coverage gain falls below the minimum coverage gain because adding more propositions to the antecedent can only decrease the coverage gain.

In line 1, $r_i = \langle a, 0\rangle$. The coverage gain of the assertion $(\neg a) \Rightarrow (f = X)^3$ is calculated to 50% in line 2, which is equal to g_{min}. At line 3, we can see that for the data in every cycle, s_j, $(\neg a) \Rightarrow (\neg f)$, which means that there is a correlation between $a = 0$ and $f = 0$ which indicates a candidate assertion. The candidate assertion $p1$ is added to P_c, the set of candidate assertions, in line 4.

Now, back at line 1, $r_i = \langle a, 1\rangle$. Even though the coverage gain of assertion $a \Rightarrow (f = X)$ is also 50%, neither the rule $a \Rightarrow (\neg f)$ nor $a \Rightarrow f$ is true for each cycle of data, e_j. This means that the conditions in lines 3 and 5 are not satisfied. The algorithm recurses at line 8 with $A = \{\langle a, 1\rangle\}$ and $R = \{\langle b, 0\rangle, \langle b, 1\rangle, \langle c, 0\rangle, \langle c, 1\rangle\}$.

Next the coverage gains of assertions $(a \wedge \neg b) \Rightarrow (f = X)$, $(a \wedge b) \Rightarrow (f = X)$, $(a \wedge \neg c \Rightarrow (f = X))$, and $(a \wedge c) \Rightarrow (f = X)$ are each 25% since each has two propositions in the antecedent. The minimum coverage gain is never satisfied in lines 2, and the algorithm returns.

The algorithm is continued from line 1 for the remaining $\langle var, val\rangle$ pairs resulting in the candidates $p2$ and $p3$ being added to P. The assertions in P are sorted by the number of propositions to keep the number of propositions per assertion to a minimum. In the example, the list remains unchanged since each candidate has the same number of propositions.

In the next step, $recalibrate\_add$ adds candidate assertions with coverage gain greater than or equal to g_{min} to the solution set as shown in Algorithm 9. Because coverage gain, $g(P, P')$, is relative to the solution set P, as soon as the solution set changes, the coverage gain of all assertions must be recalculated based on the new solution set. For this reason, even though all assertions in P_c must have coverage gain greater than or equal to g_{min} with respect to the P before this function is called, the coverage gain of any assertion may decrease below g_{min} as other assertions are

[3] X refers to a "don't care" value since the output does not affect the input space coverage.

added to P. Because of this, P_c must be recalibrated with regard to coverage gain of each assertion before an assertion may be added to P.

In our example $p1$ is added to the solution set, P, since P remains the same as before the function was called. After adding that candidate to the solution set, the coverage gain of next candidate, $p2$, is recalculated based on the new P. Since P contains assertion $p1$ with the antecedent $\langle a, 0\rangle$, it should be noted that the truth table entries where $\langle a, 0\rangle$ and $\langle b, 0\rangle$ are already covered. Therefore, the assertion $p2$ with antecedent $\langle b, 0\rangle$ can only cover the truth table entries where $\langle a, 1\rangle$ and $\langle b, 0\rangle$, resulting in decreased coverage gain of only 25%. By the same logic, the coverage gain of assertion $a3$ with antecedent $\langle c, 1\rangle$ is also reduced to 25%. Since both candidates have coverage gain less than g_{min}, they are both discarded.

In the final step of the first iteration, P_c is cleared and the minimum coverage gain, g_{min}, is reduced by half. In the example, g_{min} is reduced from 50% to 25%, which is still greater than the minimum gain threshold. The total coverage of P is 50%, which is less than the maximum total coverage threshold, $c_{\text{threshold}}$. Since neither threshold is passed, the algorithm continues to the second iteration.

In the second iteration, $gen\_candidates$ is performed again with the reduced g_{min}. This generates the following candidate assertions which are added to P_c: $p4$, $p5$, $p6$, $p7$, $p2$, and $p3$. These candidate assertions are added to P_c and then sorted by number of propositions per assertion with resulting order is $p2$, $p3$, $p4$, $p5$, $p6$, $p7$.

Assertion $p2$ is added to P. The coverage gain of the remaining candidate assertions is recalculated, causing $p3$, $p6$, and $p7$ each drop to 12.5% and $p4$ drop to 0. This leaves only the assertion $p5$ that remains at 25% which is also added to P. It should now be noted that the total input space coverage of P has reached 100%, which is above the total coverage gain threshold. This means that the algorithm can exit, producing the following assertions: $p1$, $p2$, and $p5$.

It should be noted that this algorithm can be applied to temporal assertions much like in the decision tree algorithm.

20.8 PRISM Algorithm

The PRISM algorithm [11] was initially proposed to resolve several disadvantages of the decision tree-based algorithm. The tree structure is difficult to manipulate. The entire tree needs to be traversed to extract information for any single classification [11]. Moreover, to sustain a tree structure, irrelevant feature variables are added to the derived rules. This is because a variable, once committed to a path, needs to be present for every rule that will be generated along that path. When a feature variable v is selected to partition the dataset, the decision tree algorithm recursively constructs both the left subtree and right subtree. Any assertion derived from the right subtree must include the proposition $\neg v$ and any assertion derived from the left subtree must include the proposition v. In the example of Fig. 20.2, the decision tree derives the rules: $a \Rightarrow f$ and $\neg a \wedge b \Rightarrow f$ since variable a is selected in the root node. Obviously, proposition $\neg a$ is irrelevant in assertion $\neg a \wedge b \Rightarrow f$. As a result, the rules generated from decision tree are too specific and there can be many irrelevant feature variables in some rules.

To solve the problem, PRISM algorithm does not maintain the tree structure. Also, it selects a variable, value pair $\langle v, \alpha \rangle$ to split the dataset each time. The PRISM algorithm generates rules for each value of the target independently. For current $\langle v_t, \beta \rangle$ pair, where $\beta \in \{0, 1\}$, it goes through the following steps:

1. It selects a $\langle v, \alpha \rangle$ pair, where $\alpha \in \{0, 1\}$, with maximum information gain. We will elaborate how to calculate the information gain for the $\langle v, 0 \rangle$ and $\langle v, 1 \rangle$ pair with respect to $\langle v_t, \beta \rangle$ pair.
2. The current dataset is checked and only the rows satisfying the proposition represented by the selected $\langle v, \alpha \rangle$ pair are retained for next step.
3. Steps 1–2 are repeated until each row of the current dataset has the consistent value β on target variable.
4. At this point, a rule can be derived from the structure. All rows covered by the induced rule are **eliminated** from the dataset. Steps 1–4 are repeated until all rows satisfying $v_t = \beta$ are removed.

When all datasets satisfying $v_t = \beta$ have been covered by the induced rules, the algorithm switches to another value of the target variable and the training set is restored to its initial state for the next iteration.

The PRISM algorithm does not branch on any node and thus resolves the disadvantages of the decision tree. Irrelevant feature variables are automatically excluded from the induced rules, since a rule (Steps 1–4) is induced for exactly one value of a target variable. No side effects are allowed to propagate.

The PRISM algorithm does not select one feature variable to partition the dataset. Instead, it selects $\langle v, 0 \rangle$ or $\langle v, 1 \rangle$ to partition the dataset. Therefore, we define the information gain for $\langle v, \alpha \rangle$ instead of variable v with respect to $\langle v_t, \beta \rangle$. We denote it as $G_\beta(S, v, \alpha)$, and compute it according to the following formula:

$$G_\beta(S, v, \alpha) = \frac{p_\beta(S_{v=\alpha}, v_t)}{p_\beta(S, v_t)}, \alpha \in 0, 1 \text{ and } \beta \in 0, 1. \tag{20.2}$$

We adapt the PRISM algorithm in [11] to generate assertions. The algorithm induces one assertion at a time for assignment $v_t = \beta$. In each iteration, the algorithm adds a proposition $(v \in V, \alpha)$ for which $G_\beta(S, v, \alpha)$ is maximum to the candidate rule until $H(S) = 0$. After the algorithm generates a rule, it discards vectors in S that satisfy the rule. The algorithm repeats this process until $S_{v_t=\beta}$ is empty.

Figure 20.6 shows an example of the PRISM algorithm for the AND gate design. First, the algorithm generates an assertion for $f = 0$. The algorithm computes the

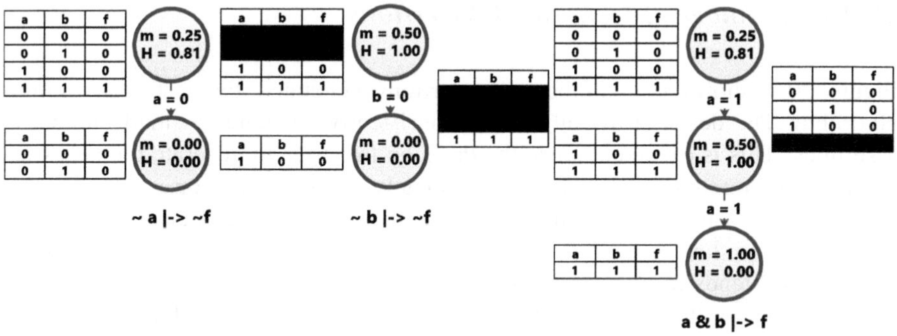

Fig. 20.6 An example of the PRISM algorithm for the AND gate design. Each node is labeled with its mean and entropy. Each branch is labeled with the feature variable and value used to partition the data represented by its parent

information gain of propositions $(a, 0)$, $(a, 1)$, $(b, 0)$, and $(b, 1)$ as follows: The algorithm adds $(a, 0)$ to the assertion's antecedent and generates the assertion p_0 : $\neg a \Rightarrow \neg f$ since $H(S) = 0$.

Next, the algorithm removes the first and second rows from S since they satisfy p_0. Since S contains another row where $f = 0$, the algorithm generates another assertion for $f = 0$. Again, the algorithm computes the information gain of propositions $(a, 0)$, $(a, 1)$, $(b, 0)$, and $(b, 1)$ and adds $(b, 0)$ to the assertion's antecedent. The algorithm generates the assertion $p_1 : \neg b \Rightarrow \neg f$ and removes the third row from S. Now, since S does not contain any rows where $f = 0$, the algorithm generates the assertion $p_2 : a \wedge b \Rightarrow f$ for $f = 1$ and terminates.

20.9 Experimental Results

We present experimental results to compare the assertion mining algorithms. We used each algorithm to generate assertions for three Verilog designs—the Universal Serial Bus (USB) protocol, master state machine from the Peripheral Component Interconnect (PCI) protocol, and the OpenRisc 1200 (OR1200) CPU. USB has 142 inputs, 100 outputs, and 868 lines of RTL code, PCI has 99 inputs, 133 outputs, and 669 lines of RTL code, and OR1200 has 41 inputs, 239 outputs, and 1533 lines of RTL code. For all experiments, the data generator simulated each module for 10,000 cycles using an unconstrained random test bench. We explicitly limited the antecedent size and temporal length of all assertions to 5 and 2, respectively. We conducted all experiments using a 2.67 GHz quad core Intel Core i5 with 16 gigabytes of memory. The results shown in this section are averaged across *all assertions for all outputs* in the three designs above.

20.9.1 Number of Generated Assertions

Figure 20.7 shows the number of assertions generated by each data mining algorithm. The decision tree algorithm consistently generates a small number of assertions. This is expected, since the decision tree tries to generate the necessary assertions required to minimally and completely describe the dataset. The decision tree generates a minimal depth tree to explain the data. The decision forest tries to optimize for low antecedent size within the tree structure. Hence the number of assertions tends to be on the lower side for the forest as well. For all designs, the PRISM algorithm generates more assertions than any other algorithm. This is because PRISM finds assertions for every (variable, value) pair. Hence, the number of assertions it generates tends to be high.

The average antecedent size per algorithm is shown in Fig. 20.8. This is a metric that captures the readability and comprehensibility of the assertions generated. The coverage guided mining algorithm consistently generates smaller, succinct antecedents, since it is optimized for that parameter.

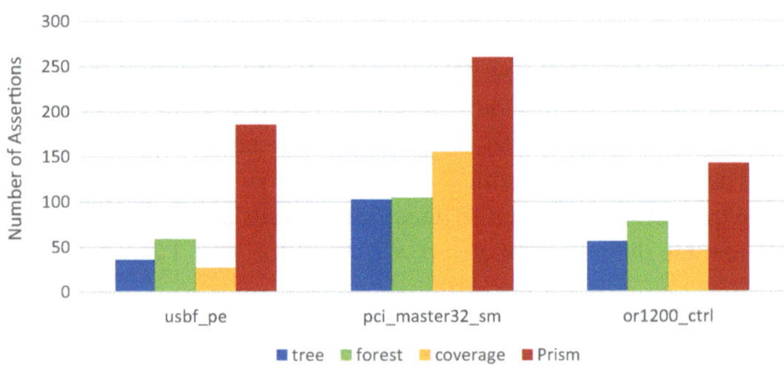

Fig. 20.7 The number of assertions generated by each data mining algorithm

Fig. 20.8 The average antecedent size of generated assertions for each data mining algorithm

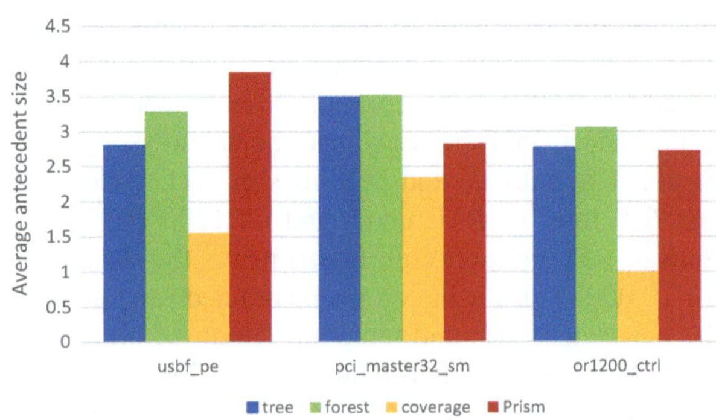

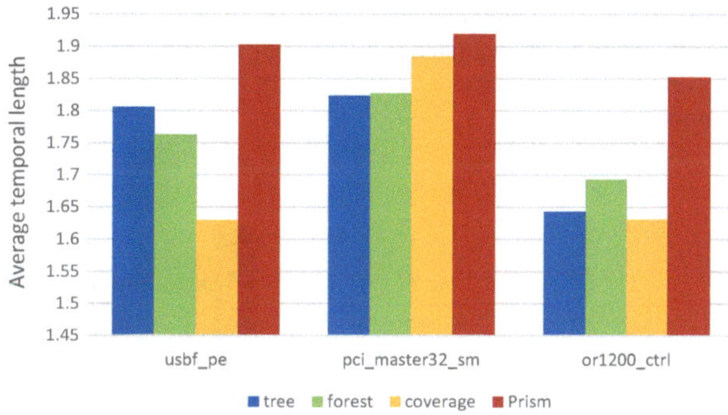

Fig. 20.9 The average temporal length of generated assertions for each data mining algorithm

20.9.2 Sequential Depth of Generated Assertions

In general, temporal behavior is difficult for the human mind to analyze, and temporal properties tend to be the most difficult to express correctly. We use average temporal length as a metric for comparison of algorithms, since it gives us an idea which algorithm produces more complex assertions.

Figure 20.9 shows the average temporal length of generated assertions for each data mining algorithm. For all designs, the PRISM algorithm generates assertions that have higher temporal length than those generated by any other algorithm.

For PCI and OR1200, the decision forest algorithm generates assertions that have higher temporal length than those generated by the decision tree algorithm.

For USB and OR1200, the coverage mining algorithm generates assertions that have least temporal length.

20.9.3 Input Space Coverage and Hit Rate of Generated Assertions

Figure 20.10 shows the average input space coverage of generated assertions for each data mining algorithm. Since input space coverage is inversely proportional to antecedent size, these results provide an inverse graph of the antecedent size graph. Here again, the coverage mining algorithm wins among the other algorithms, since it optimizes for high input space coverage.

Hit rate is a measure of the predictive accuracy of the data mining algorithm. If the data mining algorithm makes accurate predictions, the formal verification will pass the predicted assertion. If not, the assertion will fail formal verification. Figure 20.11 shows the hit rate of generated assertions for each data mining algorithm. The decision tree algorithm consistently generates a high percentage of true assertions.

The decision forest algorithm's hit rate is very close to the decision tree's hit rate, since the decision forest can be viewed as a collection of decision trees.

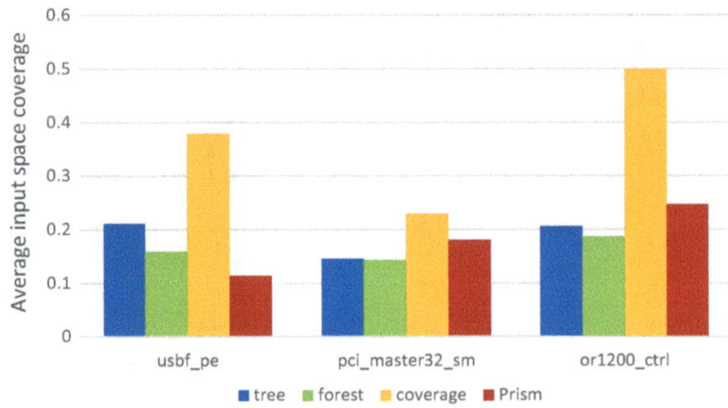

Fig. 20.10 The average input space coverage of generated assertions for each data mining algorithm

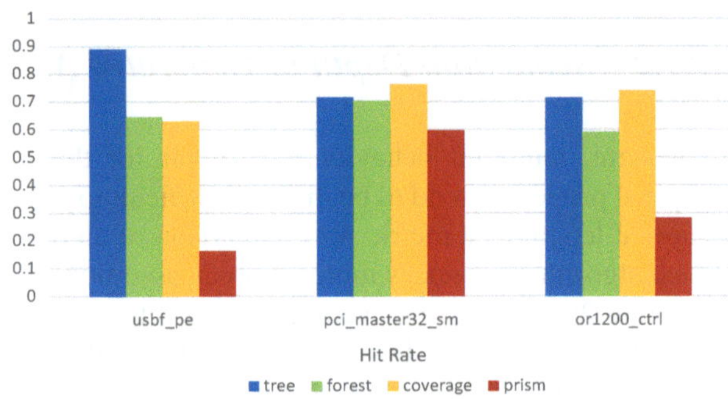

Fig. 20.11 Hit rate of generated assertions for each data mining algorithm

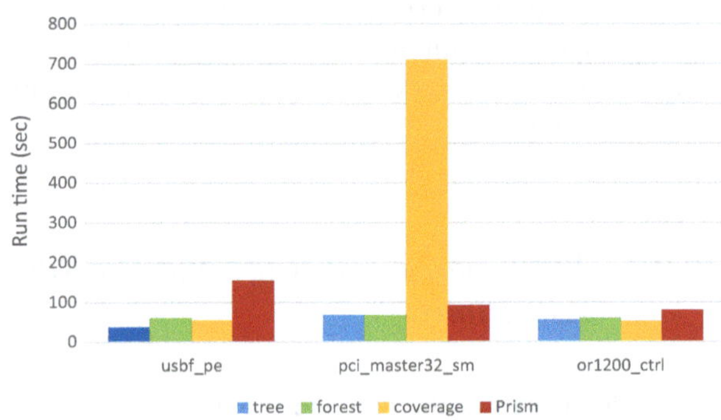

Fig. 20.12 Runtimes of each data mining algorithm

The coverage guided miner's hit rate seems to be quite high in general, since it exhaustively attempts to generate all possible associations and rules and cover a large fraction of the dataset.

20.9.4 Runtime and Memory Usage

Figures 20.12 and 20.13 show the runtime and memory use of each data mining algorithm. While the decision tree, decision forest, and PRISM algorithms utilize

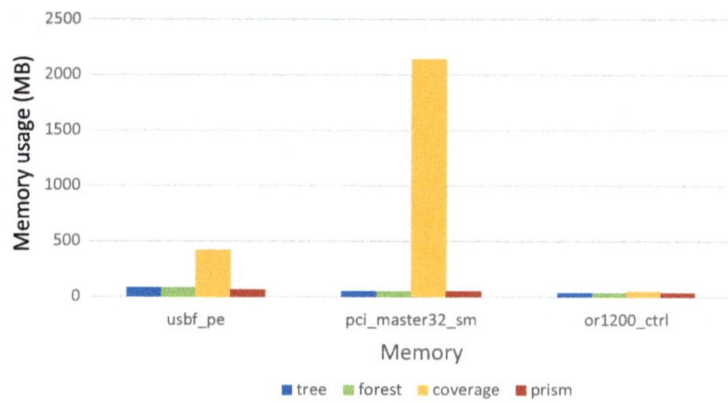

Fig. 20.13 The memory use of data mining algorithm

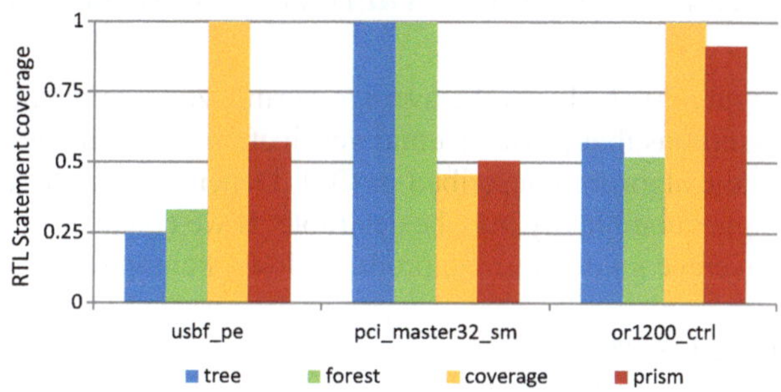

Fig. 20.14 The average statement coverage of the assertions that passed formal verification for each data mining algorithm

less runtime (less than 3 min) and memory (less than 100 Mb), the coverage guided miner is resource intensive.

20.9.5 RTL Statement Coverage of Generated Assertions

In this subsection, we present experimental results that compare GoldMine's data mining algorithms with respect to statement coverage and expectedness. RTL statement coverage is defined in Sect. 20.4.6. We show the results for the assertions that passed formal verification since they will be retained for further verification usage.

Figure 20.14 shows the average normalized code (statement) coverage of the 25% highest coverage assertions that passed formal verification for each data mining algorithm. The definition of code coverage is from [2]. We observe that there is no algorithm that always generates assertions with high statement coverage. This implies that the extent of statement coverage is not related to the mining approach. This might be due to code coverage being a different, unrelated metric as compared to input space or state space-based coverage. Hence, the code coverage itself is not strongly correlated to the type of mining algorithm used.

20.9.6 *Expectedness Analysis of Generated Assertions*

Figure 20.15 shows the average normalized expectedness of the 25% highest ranked assertions that passed formal verification for each data mining algorithm according to the ranking algorithm described in [31]. Lower expectedness is desirable, as unexpected scenarios are more interesting in many contexts. The coverage mining algorithm performs the worst with respect to unexpectedness, while the decision tree-based algorithm performs the best.

20.9.7 *Complexity Analysis of Generated Assertions*

Figure 20.16 shows the average normalized complexity of the 25% highest ranked assertions that passed formal verification for each data mining algorithm according to the algorithm described in [31]. Depending on the context, lower complexity or higher complexity may be desirable. If we consider lower complexity desirable, the coverage guided mining produces lowest complexity assertions.

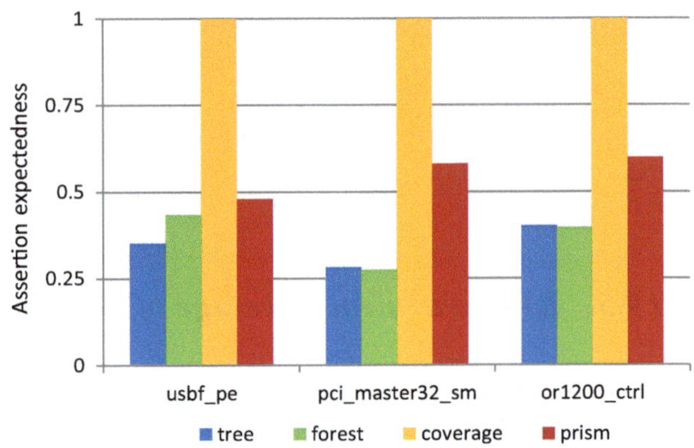

Fig. 20.15 The average normalized expectedness of the assertions that passed formal verification for each data mining algorithm

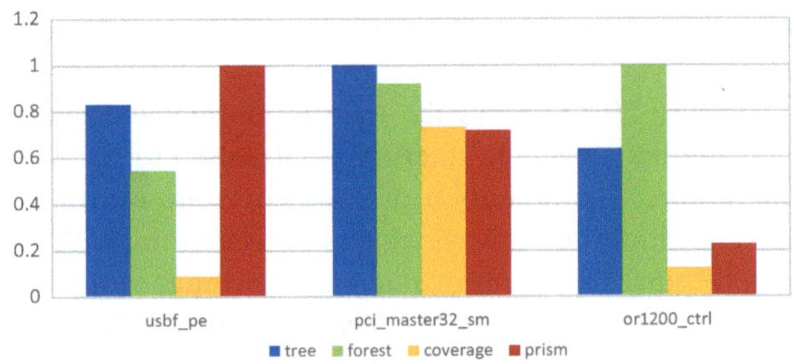

Fig. 20.16 The average normalized complexity of assertions that passed formal verification for each data mining algorithm

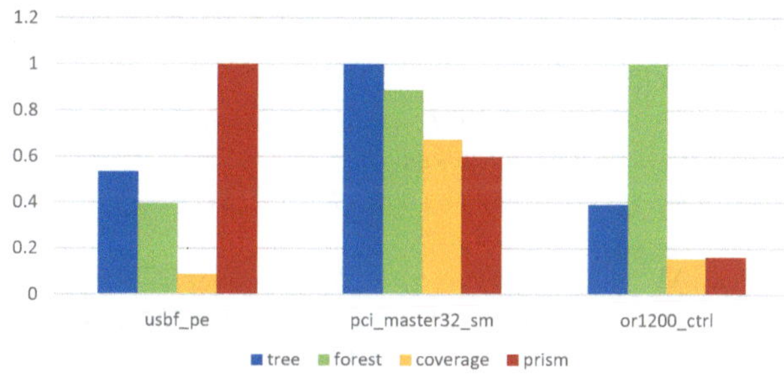

Fig. 20.17 The average normalized importance of assertions that passed formal verification for each data mining algorithm

20.9.8 Importance Analysis of Generated Assertions

Figure 20.17 shows the average normalized importance of the 25% highest ranked assertions that passed formal verification for each data mining algorithm according to the algorithm described in [31]. Higher importance is desirable. Since no algorithm produces important assertions consistently, the algorithms should be selected in a design specific manner. The coverage guided mining tends to generate less important assertions than the others.

20.9.9 Qualitative Analysis of Sample Assertions

20.9.9.1 Qualitative Comparison of Assertions Generated by Different Algorithms

In this experiment, we select two assertions generated by each mining algorithm and make a qualitative analysis of the assertions. The design used for analysis is $usbf\_pe$ module.

From the assertions shown in Table 20.1, it can be observed that the decision tree always contains too many propositions in each assertion due to its tree structure. For example, the proposition $\neg csr[0]$ in $tree\_A1$ is redundant if we compare the assertion $tree\_A1$ with the assertion $PRISM\_A1$. The assertion $PRISM\_A1$ does not include the redundant proposition $\neg csr[0]$ and is still a true assertion. Other algorithms avoid using the strict tree structure and generate assertion with fewer propositions.

In Table 20.1, coverage guided mining always generates the most succinct assertions, since it tries to maximize the input space coverage of each assertion during the generation process. In the assertions $coverage\_A1$ and $coverage\_A2$, there is only one proposition in each assertion. A false assertion: $pid\_SOF \Rightarrow X\neg send\_token$ is also generated by coverage guided mining method. However, the true assertion $tree\_A1$ is not generated by coverage guided mining method. The reason is that the input space coverage of $tree\_A1$ is too low and does not satisfy the input space coverage requirement.

Table 20.1 Sample assertions generated by each mining algorithm for qualitative analysis

Assertion ID	Assertions
tree_A1	$\neg pid\_SOF \wedge pid\_NG \wedge model\_hs \wedge \neg csr[22] \wedge \neg csr[0] \wedge match\_r \Rightarrow X send\_token$
tree_A2	$match \wedge buf0[8] \wedge X\neg pid\_SOF \wedge X pid\_PING \wedge X mode\_hs \wedge X\neg csr[3] \Rightarrow XX\neg rx\_dma\_en$
forest_A1	$match \wedge X\neg pid\_SOF \wedge X csr[27] \Rightarrow XX\neg tx\_dma\_en$
forest_A2	$\neg pid\_SOF \wedge \neg csr[23] \wedge csr[22] \wedge match\_r \Rightarrow X\neg send\_token$
coverage_A1	$\neg csr[15] \Rightarrow X\neg dma\_en$
coverage_A2	$\neg rx\_data\_done \Rightarrow X\neg int\_crc16\_set$
PRISM_A1	$\neg pid\_SOF \wedge pid\_PING \wedge model\_hs \wedge \neg csr[22] \wedge match\_r \Rightarrow X send\_token$
PRISM_A2	$\neg csr[15] \Rightarrow \neg dma\_en$

We pick up two assertions generated by each mining algorithm for comparison

Table 20.2 Manually written assertions for outputs send_token and rx_dma_en

Assertion ID	Assertions
A1	$(pid\_SETUP \,\&\, idma\_done \,\&\, \neg abort) \wedge X(state == IDLE) \Rightarrow (send\_token == 1)$
A2	$(csr[27:26] == 2'b01 \,\&\, no\_buf0\_dma) \wedge X(state == IDLE) \Rightarrow (send\_token == 1)$
A3	$(\neg csr[17]) \wedge X(to\_large == 1 \,\&\, match == 1) \wedge X(csr[27:26] == 2'b10 \,\&\, state == IDLE) \rightarrow (rx\_dma\_end == 1)$

From Table 20.1, it is hard to compare the quality of generated assertions by PRISM algorithm with the decision forest algorithm. They both resolve the over-constraining problem in decision tree based learning.

20.9.9.2 Qualitative Comparison of Automatically Generated Assertions with Manually Written Assertions

Since the effort of automatic assertion generation is to reduce the human effort of writing good assertions, our "gold standard" is manually generated assertions. We compare automatically and manually generated assertions with respect to quality.

We compare automatic assertions from Table 20.1 and manual assertions of Table 20.2. We analyze assertions for two outputs, *send_token* and *rx_dma_en* outputs of the USB protocol engine.

- **Succinctness**: We compare assertions from PRISM (PRISM_A1 from Table 20.1) and manual assertions (A1 from Table 20.2). PRISM generates assertions that incorporate internal registers (like $match\_r$, which is the input *match* delayed by one cycle). On the other hand, manual assertions tend to use primary inputs and outputs ($abort$ and $idma\_done$). Depending on the temporal distance between the target variable and the feature variables, the presence of internal design registers in the *automatically generated assertions capture behavior more succinctly* compared to manually created assertions.

- **Understandability**: A comparison between PRISM_A1 of Table 20.1 and A2 of Table 20.2 shows that *manually written assertions are more understandable.* For example, A2 incorporates $csr[27 : 26]$ in the assertion which can deterministically decide if the control status register is operating on a IN, OUT, or CONTROL token for the buffer. A single isolated bit such as $csr[22]$ does not provide sufficient contextual information regarding control status registers, making the assertion PRISM_A1 less expressive.
- **Temporal depth**: It is very challenging, even for an experienced verification engineer to craft an assertion that captures multi-cycle design behavior effectively. In Table 20.2, none of the assertions span more than 2–3 clock cycles whereas GoldMine can reason about large number of cycles (e.g., up to 7 clock cycles for tree_A2) as shown in Table 20.1. Hence, *automatic assertions capture complex temporal behavior* with ease.
- **Effort**: Each assertion in Table 20.2 took approximately 1 person-hour to construct (not counting time taken to understand module behavior). In contrast, GoldMine could generate assertions such as tree_A2 of Table 20.1 spanning over 7 clock cycles in less than a minute. Manual crafting of assertions involves understanding the cone-of-influence of the target variable, temporal relationship between target variable and feature variables, identifying the correct values of each of the feature variables—each of which is an involved task. This is over and above the effort spent in understanding the design intent and the design modules. *Automatic assertion generation is much more time and resource efficient.*

Overall, manually written assertions tend to be more comprehensible than automatically generated assertions. On the other hand, GoldMine generates succinct and multi-cycle assertions easily and quickly where the human mind struggles. We believe that automatically generated succinct assertions spanning over multiple cycles can compensate for understandability, by allowing the verification engineer to focus on debugging.

20.9.10 Scalability of GoldMine

In practical designs, data mining algorithms do not suffer from scalability issues. Our methodology's scalability is restricted by formal verification. Although formal verification technology is sensitive to state space, we find that in practice, we are able to effectively verify many modules of large designs, like PCI design. So far, we find that we cannot verify a large design with many memory elements. In these infrequent cases, there are several options. One option is to individually verify the submodules of the limiting module. Another option is to disable formal verification of candidate assertions. The candidate assertions can then be simulated to determine if they are valid.

20.10 Conclusions

In this chapter, we presented an in-depth analysis of the principal components of our automatic assertion generation tool, GoldMine. GoldMine was one of the first known applications of machine learning to hardware, and to verification. GoldMine has been applied to and adopted in several industrial, practical contexts as well as in research environments. The applications of GoldMine assertions have been in regression testing, equivalence checking between system level and RTL, debugging, root causing, and post-silicon validation. Since this software and its source has been made publicly available, it has been used considerably more for research.

While this research has a practical objective, the fundamental objective is to be able to mimic human intuition through computational analysis. A consistent observation in all our experiments is that statistical dynamic analysis, when constrained by the parameters of static, deterministic analysis, gives rise to insights that are not obvious otherwise. Machine learning, in the absence of static analysis guidance, does not generate sequentially deep, complex, and subtle behavior. Algorithms that combine deterministic guidance with statistical analysis are the keystone of GoldMine. We believe that this insight can be applied to generate causal knowledge in other domains where machine learning is being used.

We present four different mining algorithms used in GoldMine and compare these algorithm for generating hardware assertions: decision tree, best-gain decision forest, coverage guided association mining, and PRISM. Our comparisons do not produce a single winner across all metrics, since we intentionally use very diverse metrics. In practice, not all metrics may simultaneously be desirable. In our experience, only one or two metrics are of interest in realistic environments. The algorithm(s) that perform best with respect to the desired metrics can be used on a case by case basis. We believe that this versatility is a merit of the assertion generation technology of GoldMine.

In the future work, we will generate an ensemble method that combines all the approaches in GoldMine, for better predictive performance than what is available through any of the algorithms alone.

Acknowledgement We thank Debjit Pal for his assistance with editing this chapter.

References

1. ActiveProp assertion-based verification system, http://www.jasper-da.com/products/activeprop-assertion-based-verification-system
2. V. Athavale, S. Ma, S. Hertz, S. Vasudevan, Code coverage of assertions using RTL source code analysis, in *The 51st Annual Design Automation Conference 2014, DAC '14*, San Francisco, CA, June 1–5, 2014 (2014), pp. 61:1–61:6
3. R. Agrawal, T. Imielinski, A. Swami, Mining association rules between sets of items in large databases, in *Proceedings of SIGMOD* (1993), pp. 207–216
4. G. Ammons, R. Bodk, J.R. Larus, Mining specifications, in *POPL* (2002), pp. 4–16

5. Assertion synthesis, http://www.atrenta.com/about-assertion_based-verification.htm5
6. V. Athavale, S. Hertz, S. Vasudevan, Evaluating code coverage of assertions by static analysis of RTL. Technical report uilu-eng-11-2209 (2011)
7. S. Bensalem, Y. Lakhnech, H. Sadi, Powerful techniques for the automatic generation of invariants, in *Proceedings of CAV* (1996), pp. 323–335
8. A. Biere, E.M. Clarke, R. Raimi, Y. Zhu, Verifying safety properties of a power PC microprocessor using symbolic model checking without BDDs, in *Computer Aided Verification, 11th International Conference, CAV '99*, Trento, July 6–10, 1999, Proceedings (1999), pp. 60–71
9. L. Breiman, Random forests. Mach. Learn. **45**(1), 5–32 (2001)
10. M. Caplain, Finding invariant assertions for proving programs, in *Proceedings of the International Conference on Reliable Software* (ACM, New York, 1975), pp. 165–171
11. J. Cendrowska, Prism: an algorithm for inducing modular rules. Int. J. Man-Mach. Stud. **27**(4), 349–370 (1987)
12. P.-H. Chang, L.-C. Wang, Automatic assertion extraction via sequential data mining of simulation traces, in *Proceedings of ASPDAC* (2010), pp. 607–612
13. H. Chao, H. Li, X. Song, T. Wang, X. Li, On evaluating and constraining assertions using conflicts in absent scenarios, in *2017 IEEE 26th Asian Test Symposium (ATS)*, Nov 2017, pp. 195–200
14. X. Cheng, M.S. Hsiao, Simulation-directed invariant mining for software verification, in *Proceedings of DATE* (2008), pp. 682–687
15. C.-N. Chung, C.-W. Chang, K.-H. Chang, S.-Y. Kuo, Applying verification intention for design customization via property mining under constrained testbenches, in *Proceedings of ICCD* (2011), pp. 84–89
16. E.M. Clarke, M. Fujita, S.P. Rajan, T.W. Reps, S. Shankar, T. Teitelbaum, Program slicing of hardware description languages, in *Correct Hardware Design and Verification Methods, 10th IFIP WG 10.5 Advanced Research Working Conference, CHARME '99*, Bad Herrenalb, September 27–29, 1999, Proceedings (1999), pp. 298–312
17. E.M. Clarke, A. Biere, R. Raimi, Y. Zhu, Bounded model checking using satisfiability solving. Formal Methods Syst. Des. **19**(1), 7–34 (2001)
18. A.M. Cruz, R.B. Fernández, H.M. Lozano, M.A. Ramírez Salinas, L.A. Villa Vargas, Automated functional test generation for digital systems through a compact binary differential evolution algorithm. J. Electron. Test. **31**(4), 361–380 (2015)
19. A. Danese, F. Filini, G. Pravadelli, A time-window based approach for dynamic assertions mining on control signals, in *2015 IFIP/IEEE International Conference on Very Large Scale Integration (VLSI-SoC)*, Oct 2015, pp. 246–251
20. A. DeOrio, A. Bauserman, V. Bertacco, B. Isaksen, Inferno: streamlining verification with inferred semantics. IEEE Trans. Comput. Aided Des. Integr. Circuits Syst. **28**(5), 728–741 (2009)
21. M.D. Ernst, J.H. Perkins, P.J. Guo, S. McCamant, C. Pacheco, M.S. Tschantz, C. Xiao, The daikon system for dynamic detection of likely invariants. J. Sci. Comput. Program. **69**, 35–45 (2007)
22. F. Farahmandi, R. Morad, A. Ziv, Z. Nevo, P. Mishra, Cost-effective analysis of post-silicon functional coverage events, in *Design, Automation & Test in Europe Conference & Exhibition, DATE 2017*, Lausanne, March 27–31, 2017 (2017), pp. 392–397
23. H.D. Foster, A.C. Krolnik, D.J. Lacey, *Assertion-Based Design*, 2nd edn. (Springer Publishing Company, New York, 2010)
24. T. Ghasempouri, G. Pravadelli, On the estimation of assertion interestingness, in *2015 IFIP/IEEE International Conference on Very Large Scale Integration (VLSI-SoC)*, Oct 2015, pp. 325–330
25. O. Grumberg, H. Veith (eds.), *25 Years of Model Checking: History, Achievements, Perspectives* (Springer, Berlin, 2008)
26. J. Han, M. Kamber, *Data Mining: Concepts and Techniques* (Morgan Kaufmann Publishers Inc., San Francisco, 2000)

27. M. Hanafy, H. Said, A.M. Wahba, Complete properties extraction from simulation traces for assertions auto-generation, in *2015 IEEE 24th North Atlantic Test Workshop*, May 2015, pp. 1–6
28. M. Hanafy, H. Said, A.M. Wahba, New methodology for complete properties extraction from simulation traces guided with static analysis. J. Electron. Test. **32**(6), 705–719 (2016)
29. S. Hangal, N. Chandra, S. Narayanan, S. Chakravorty, Iodine: a tool to automatically infer dynamic invariants for hardware designs, in *Proceedings of DAC* (2005), pp. 775–778
30. A. Hekmatpour, A. Salehi, Block-based schema-driven assertion generation for functional verification, in *Proceedings of ATS* (2005), pp. 34–39
31. S. Hertz, Enhancing quality of assertion generation: methods for automatic assertion generation and evaluation. PhD thesis, The University of Illinois at Urbana-Champaign, 2013
32. S. Hertz, D. Sheridan, S. Vasudevan, Mining hardware assertions with guidance from static analysis. IEEE Trans. Comput. Aided Des. Integr. Circuits Syst. **32**, 952–965 (2013)
33. T.K. Ho, The random subspace method for constructing decision forests. IEEE Trans. Pattern Anal. Mach. Intell. **20**(8), 832–844 (1998)
34. C.A.R. Hoare, An axiomatic basis for computer programming. Commun. ACM **12**(10), 576–580 (1969)
35. IEEE standard for verilog hardware description language. IEEE Std 1364-2005 (Revision of IEEE Std 1364-2001) (2006), pp. 1–560
36. R.P. Kurshan, *Computer-Aided Verification of Coordinating Processes: The Automata-Theoretic Approach* (Princeton University Press, Princeton, 1994)
37. L. Liu, D. Sheridan, V. Athavale, S. Vasudevan, Automatic generation of assertions from system level design using data mining, in *Proceedings of MEMOCODE* (2011), pp. 191–200
38. L. Liu, D. Sheridan, W. Tuohy, S. Vasudevan, Towards coverage closure: using GoldMine assertions for generating design validation stimulus, in *Proceedings of DATE* (2011), pp. 173–178
39. L. Liu, C.-H. Lin, S. Vasudevan, Word level feature discovery to enhance quality of assertion mining, in *Proceedings of ICCAD* (2012), pp. 210–217
40. L. Liu, D. Sheridan, W. Tuohy, S. Vasudevan, A technique for test coverage closure using goldmine. IEEE Trans. CAD Integr. Circuits Syst. **31**(5), 790–803 (2012)
41. L. Liu, X. Zhong, X. Chen, S. Vasudevan, Diagnosing root causes of system level performance violations, in *Proceedings of ICCAD*, 2013
42. J. Malburg, T. Flenker, G. Fey, Property mining using dynamic dependency graphs, in *22nd Asia and South Pacific Design Automation Conference, ASP-DAC 2017*, Chiba, January 16–19, 2017 (2017), pp. 244–250
43. K. McGarry, A survey of interestingness measures for knowledge discovery. Knowl. Eng. Rev. **20**(1), 39–61 (2005)
44. J. Misra, Prospects and limitations of automatic assertion generation for loop programs. SIAM J. Comput. **6**(4), 718–729 (1977)
45. T.M. Mitchell, *Machine Learning*. McGraw Hill Series in Computer Science (McGraw-Hill, New York, 1997)
46. J.W. Nimmer, M.D. Ernst, Automatic generation of program specifications, in *ISSTA* (2002), pp. 229–239
47. Opencore web page, http://www.opencores.org
48. M.S. Pang-Ning Tan, V. Kumar, *Introduction to Data Mining* (Pearson, London, 2005)
49. C.S. Pasareanu, W. Visser, Verification of java programs using symbolic execution and invariant generation, in *Proceedings of SPIN* (2004), pp. 164–181
50. G. Pinter, I. Majzik, Automatic generation of executable assertions for runtime checking temporal requirements, in *Proceedings of HASE 2005* (2005), pp. 111–120
51. A. Pnueli, The temporal logic of programs, in *Proceedings of the 18th Annual Symposium on Foundations of Computer Science* (1977), pp. 46–57
52. J.R. Quinlan, Induction of decision trees. Mach. Learn. **1**(1), 81–106 (1986)
53. J.R. Quinlan, *C4.5: Programs for Machine Learning* (Morgan Kaufmann Publishers, San Francisco, 1993)

54. F. Rogin, T. Klotz, G. Fey, R. Drechsler, S. Rülke, Automatic generation of complex properties for hardware designs, in *Proceedings of DATE* (2008), pp. 545–548
55. C.E. Shannon, A mathematical theory of communication. SIGMOBILE Mob. Comput. Commun. Rev. **5**(1), 3–55 (2001)
56. D. Sheridan, H.-S. Kim, L. Liu, J. Han, S. Vasudevan, Guided association rule learning for assertion generation. Technical report (2012)
57. D. Sheridan, L. Liu, H. Kim, S. Vasudevan, A coverage guided mining approach for automatic generation of succinct assertions, in *2014 27th International Conference on VLSI Design and 2014 13th International Conference on Embedded Systems*, Mumbai, January 5–9, 2014 (2014), pp. 68–73
58. A. Silberschatz, A. Tuzhilin, What makes patterns interesting in knowledge discovery systems. IEEE Trans. Knowl. Data Eng. **8**(6), 970–974 (1996)
59. Systemverilog, 2012
60. P. Taatizadeh, N. Nicolici, Emulation-based selection and assessment of assertion checkers for post-silicon validation, in *2015 33rd IEEE International Conference on Computer Design (ICCD)*, Oct 2015, pp. 46–53
61. P. Taatizadeh, N. Nicolici, Automated selection of assertions for bit-flip detection during post-silicon validation. IEEE Trans. CAD Integr. Circuits Syst. **35**(12), 2118–2130 (2016)
62. P. Taatizadeh, N. Nicolici, Emulation infrastructure for the evaluation of hardware assertions for post-silicon validation. IEEE Trans. VLSI Syst. **25**(6), 1866–1880 (2017)
63. P. Tan, V. Kumar, J. Srivastava, Selecting the right interestingness measure for association patterns, in *Proceedings of the Eighth ACM SIGKDD International Conference on Knowledge Discovery and Data Mining*, July 23–26, 2002, Edmonton, AB (2002), pp. 32–41
64. The GoldMine Website, http://goldmine.csl.illinois.edu/
65. A. Tiwari, H. Rue, H. Sadi, N. Shankar, A technique for invariant generation, in *Proceedings of TACAS* (2001), pp. 113–127
66. S. Vasudevan, D. Sheridan, S.J. Patel, D. Tcheng, W. Tuohy, D.R. Johnson, Goldmine: automatic assertion generation using data mining and static analysis, in *Design, Automation and Test in Europe, DATE 2010*, Dresden, March 8–12, 2010 (2010), pp. 626–629
67. L.-C. Wang, M.S. Abadir, N. Krishnamurthy, Automatic generation of assertions for formal verification of powerpc microprocessor arrays using symbolic trajectory evaluation, in *Proceedings of DAC* (1998), pp. 534–537
68. H. Zhao, A. Sinha, An efficient algorithm for generating generalized decision forests. IEEE Trans. Syst. Man Cybern. A: Syst. Hum. **35**(5), 754–762 (2005)

Chapter 21
Energy-Efficient Design of Advanced Machine Learning Hardware

Muhammad Abdullah Hanif, Rehan Hafiz, Muhammad Usama Javed, Semeen Rehman, and Muhammad Shafique

21.1 Artificial Intelligence and Machine Learning

Over the past several years, artificial intelligence (AI) has evolved significantly because of the revolutionary advancements in its subdomains of machine learning (ML) and deep neural networks (DNNs). The AI gives machines the ability to learn from the provided set of examples called training data and use the experience gained from it for solving real-time problems. Today's AI systems are capable of outperforming professional human players at sophisticated strategic games, like chess and Go [8], and have also been reported to perform significantly well in many other real-world applications, like cancer detection [25], human activity recognition [30], autonomous driving [63], etc. Such algorithms are nowadays commonly being employed for assisting and training humans [26].

Out of all the machine learning algorithms, neural networks (NNs) offer the state-of-the-art accuracy on many AI applications, like computer vision, data analytics, speech and language processing, etc. The superior performance of NNs is usually attributed to their ability to learn features directly from raw sensory data using backpropagation algorithm, which is in contrast to the conventional ML techniques (illustrated in Fig. 21.1) that require explicitly designed features and pipeline for performing the intended tasks. However, the state-of-the-art accuracy offered by neural networks comes with high-computational and memory overheads which give rise to numerous other problems like enormous memory traffic and data

M. A. Hanif (✉) · S. Rehman · M. Shafique
Vienna University of Technology (TU Wien), Vienna, Austria
e-mail: muhammad.hanif@tuwien.ac.at; semeen.rehman@tuwien.ac.at; muhammad.shafique@tuwien.ac.at

R. Hafiz · M. U. Javed
Information Technology University (ITU), Lahore, Pakistan
e-mail: rehan.hafiz@itu.edu.pk; usama.javed@itu.edu.pk

I. M. Elfadel et al. (eds.), *Machine Learning in VLSI Computer-Aided Design*,
https://doi.org/10.1007/978-3-030-04666-8_21

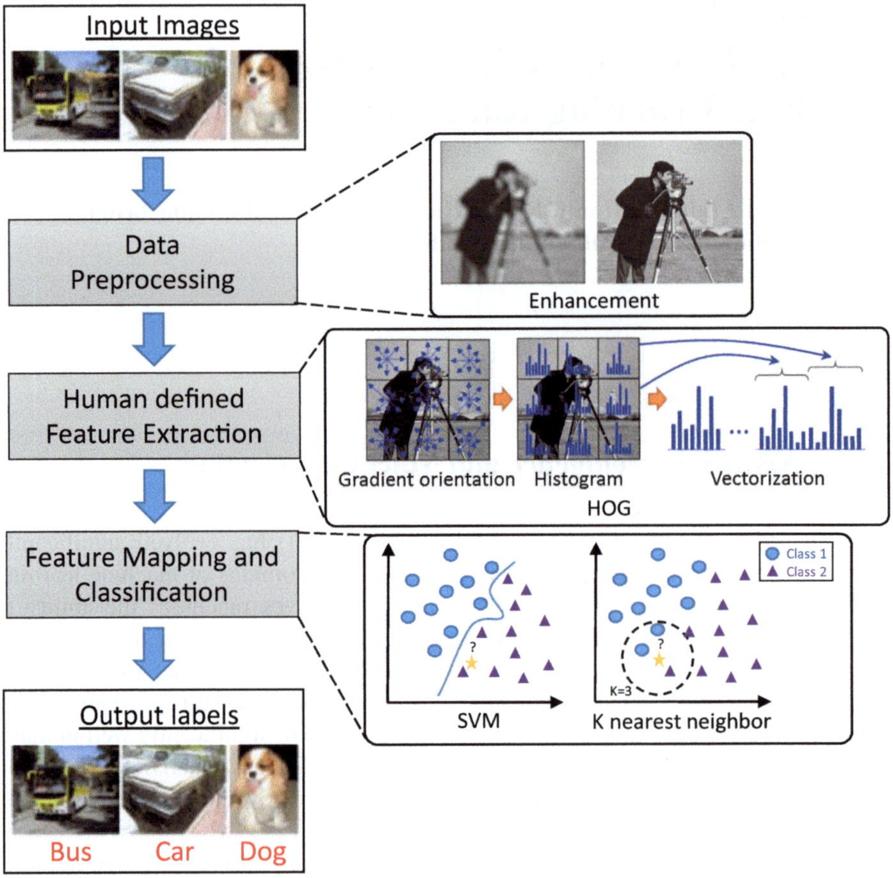

Fig. 21.1 Pipeline of conventional image/object classification algorithm

management for real-time processing applications. Since the design of efficient architectures demands insightful knowledge of the algorithm to be implemented, in the upcoming subsections, we provide an in-depth premier of neural networks, specifically in the context of deep convolutional neural networks.

21.1.1 Neural Networks

A neural network comprises a large number of interconnected units called neurons, where a neuron is the basic computational unit of a neural network. Each neuron in a network receives information in the form of signals from other neurons/inputs through connections called synapse and generates an output which is then transmitted to other subsequent neurons/output. The operation performed by these neurons is typically the weighted sum of inputs followed by a transformation function (commonly known as activation function), for example, rectified linear function. An illustration of a neuron along with a few example activation functions is shown in Fig. 21.2.

Typically, in NNs the neurons are organized in the form of layers where each layer receives input from the preceding layer(s), performs some operation, and transfers the processed information to the succeeding layer(s)/output. Multiple types of layers are used in state-of-the-art neural networks, namely, convolutional layer, fully connected layer, pooling layer, activation layer, and normalization layer. A brief summary of the layers is presented in Table 21.1. These layers are usually connected in series to perform a sequence of operations on the inputs to generate a meaningful output. Out of all the aforementioned types of layers, the convolution (CONV) and fully connected (FC) layers have the most computational significance. They can optionally be succeeded by pooling, activation, and/or normalization layer(s). A sample architecture of a convolutional neural network (CNN) is shown in Fig. 21.3. Here, we consider a CNN trained for handwritten digit recognition (MNIST Dataset [20]) for numbers from 0 to 9. Since there are 10 possible outcomes, we have 10 different classes to which a particular input can be mapped. As illustrated in Fig. 21.3, the first CONV layer receives the input and generates a series of feature maps, which are then fed to the subsequent layer for processing. One of the basic interpretations of a convolutional neural network is that the initial convolutional layers extract low-level features using small- and medium-sized filters which are then fed to the subsequent layers to extract higher-level features. In the end, the fully connected layers operate on the higher-level features to compute the final classification scores and, based on the maximum score, the input is associated with one of the available classes.

Thus far, several neural network architectures have been proposed in the literature to cover a wide variety of applications ranging from simple handwritten digit recognition [20] all the way to human activity recognition in videos [30]. These

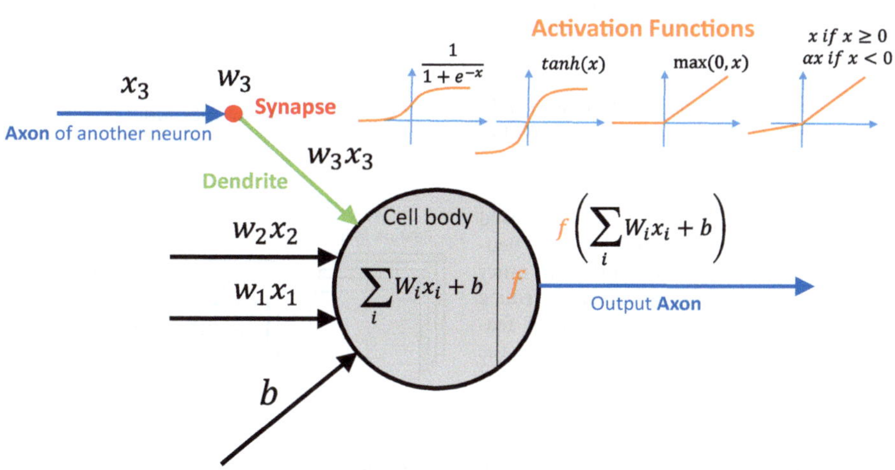

Fig. 21.2 A naive mathematical model of a biological neuron

Table 21.1 Building blocks of CNNs

Type of layer	Brief description	Illustration
Fully connected (FC)	FC layers are usually used at the end of convolutional neural networks as classification layers. Neurons in FC layers have full connections to all activations in the previous layer	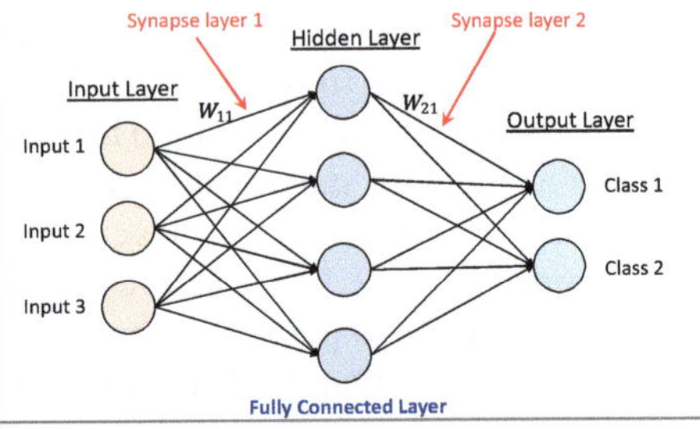
Convolution (CONV)	A convolution layer is composed of multiple filters where each filter defines a feature. The convolved outputs (a.k.a feature maps) of all the filters are concatenated together to form the output of the layer which us then used as input to the subsequent layer in a feed forward network	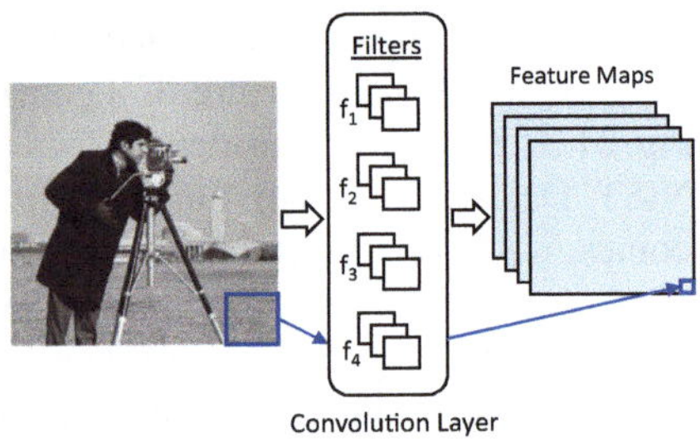
Activation	Activation layers are employed to introduce nonlinearity in neural networks. Usually this is considered a part of each convolutional and fully connected layer	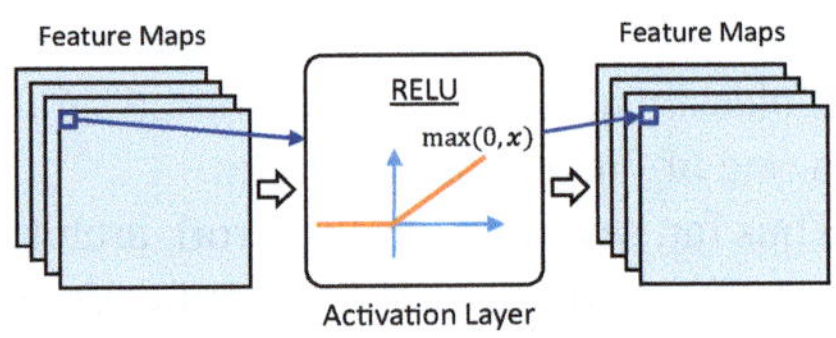
Normalization	Normalization is used in CNNs to realize 'lateral inhibition', i.e., to reduce the response of neighboring neurons of an excited neuron. The normalization is usually of two types 'Within Channel' and 'Across Channel'	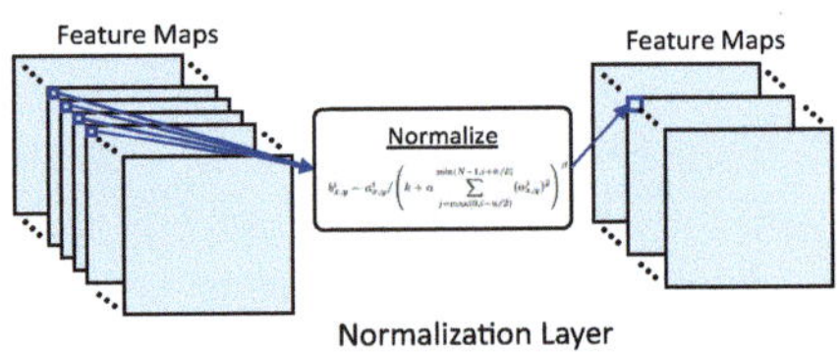
Pooling	Pooling layers are used in convolutional neural networks in order to periodically reduce the spatial size of feature maps to reduce the amount of parameters in order to avoid overfitting	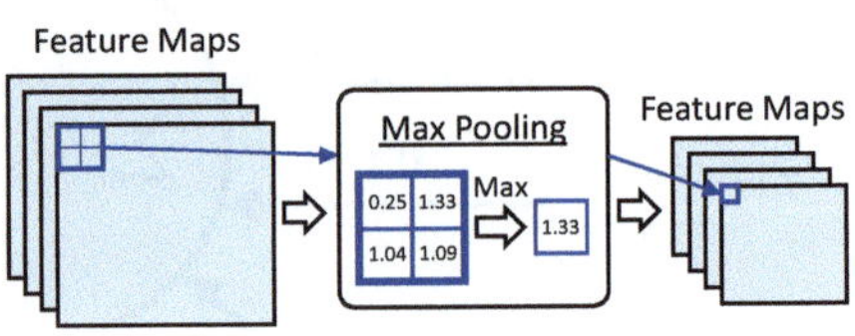

Deep CNNs typically comprise of multiple instances of such blocks

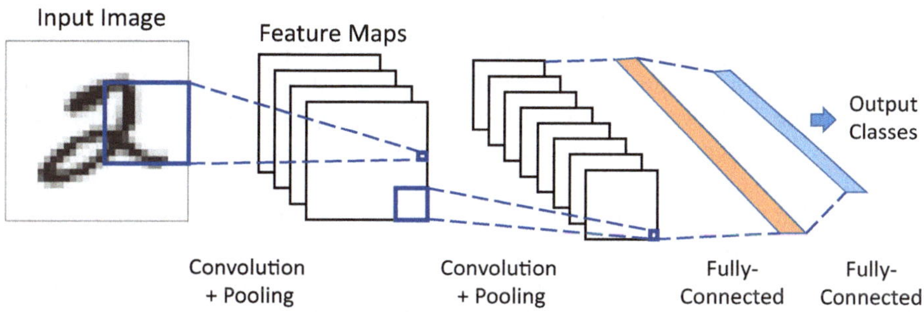

Fig. 21.3 An example convolutional neural network (CNN) architecture consisting of multiple convolution and fully connected layers

network architectures typically vary in terms of connectivity of neurons/layers, types of layers, and/or additional components used in each layer, e.g., cells for storing intermediate states in long short-term memory (LSTMs) networks. Among all the networks, the feed-forward neural networks (which includes the conventional fully connected networks and convolutional neural networks) along with recurrent neural networks (which covers fully recurrent networks and LSTMs) are the most widely used networks because of their wide applicability in the areas of computer vision and language processing.

21.1.2 Resource Requirements of State-of-the-Art Neural Networks

Over the past few years, neural networks have evolved greatly and a significant portion of this advancement can be related to the developments in hardware technology of the underlying computing devices. Because of these hardware advancements, specifically in SIMD and GPU processors, neural networks were able to scale significantly. Recently, it has also been reported that deeper networks demonstrate more accuracy as compared to shallower networks [60]. The main reason behind the high accuracy offered by deeper neural networks is that it extracts low-level features from the input samples and later in each subsequent layer extracts higher-level features considering the combination of the features extracted in the previous layer(s). This allows the deeper networks to classify samples using less number of parameters as compared to wider networks, which also lead to faster and robust training and lower memory footprint. However, to solve complex problems, a significant number of the layers (i.e., depth) with significant width per layer (i.e., number of neurons/filters) are required which eventually leads to significant computational and memory (and thereby also energy/power) requirements.

To highlight the resource requirements of state-of-the-art neural networks, we present a summary of few of the D-CNN architectures that, over the course of years, showed groundbreaking results on ImageNet Dataset [21] in Table 21.2. The table illustrates the evolution of the neural networks for image classification applications. The top-5 classification accuracy of the networks increased over the years and has a strong correlation with the depth of the network. It can also be observed from the table that even the simplest DNN architecture, i.e., AlexNet, which has a top-5 accuracy of 80.2% requires 61 million floating point weights and 724 million MAC operations to classify a single image of size $227 \times 227 \times 3$. Therefore, to reduce the computational complexity and the memory requirements associated with a neural network, various optimization/approximation techniques have been proposed in the literature. For ease of illustration, we partition these techniques into two broad categories, i.e., software-level and co-design optimization techniques, and hardware-level optimization techniques.

21.2 Software and Co-design Optimizations

In earlier days of deep neural networks, most of the research was concentrated on improving the accuracy without paying much attention to the computational and memory requirements. However, the recent surge of use of AI in battery-powered embedded computing devices demands design of dedicated hardware and hardware/software co-design for reducing the implementation complexity of the networks. For example, the architectural improvement in ResNet50 [34] resulted in a significant reduction in the number of weights and computations as compared to VGG [58] while providing significant improvement in classification accuracy, as illustrated in Table 21.2. This reduction in the total number of effectual parameters and the number of operations can directly be associated with the energy efficiency.

In this section, we discuss few of the most important software and hardware–software co-design optimization methods, which have shown remarkable results in state-of-the-art NN architectures.

Table 21.2 Summary of a subset of popular DNN architectures

Metrics	AlexNet	Overfeat fast	VGG	GoogLeNet	ResNet 50	ResNet 152
Top-5 error	19.8	17	8.8	10.7	7	6.16
Year released	2012	2013	2014	2014	2015	2015
Total MACs (billion)	0.724	2.8	15.5	1.43	3.9	–
Depth (# of layers)	8	8	16	22	50	152
# of weights (million)	61	146	138	7	25.5	–
Execution time (ms)	4.31	–	41.2	11.5	34.14	73.52

The reported accuracy is based upon top-5 error on ImageNet [21]. The forward pass execution time are computed by executing the model architecture on a machine with an Intel Core i7-7700 CPU, 64 GB RAM and GTX 1080Ti [1]

- The software-level optimizations discussed in this section include pruning, weight sharing, compact architectural designs, and knowledge distillation, which mainly focus on reducing the number of effectual parameters and the model size of a network.
- The co-design techniques work mainly on the principle of reducing the precision of operands and operations for achieving the energy and power efficiency.

21.2.1 Pruning

At present, neither there is a formal way of designing neural network architectures nor the knowledge of the optimal number of parameters required for solving a specific AI problem. Therefore, in order to facilitate the learning process, the networks are usually over-parameterized. As a result, a significant amount of parameters in trained networks are redundant and ineffectual and hence can be removed. This process of removing the ineffectual weight from a network is usually termed as pruning and is often followed by retraining (a.k.a fine-tuning) to regain a significant amount of the lost accuracy.

Pruning was first introduced by Cun et al. [18] in 1989, which showed a significant improvement in training as well as testing/classification speed. It was also observed that by removing redundant/unimportant weights followed by retraining (for regaining a significant portion of lost accuracy), the networks were able to generalize better, especially for a smaller dataset. A similar method was recently introduced by Han et al. [31]. However, instead of using a saliency-based approach, as used by Cun et al. in [18], a magnitude-based approach was adapted for pruning. Weights having magnitude below a certain threshold (close to zero) were considered unimportant and were pruned. This approach showed remarkable results for reducing the overall model size of the network, for example, it can reduce the total number of weights in AlexNet by 9× and in VGG-16 by 13× without affecting the overall classification accuracy. A simple example of pruning is illustrated in Fig. 21.4, where all the branches corresponding to weights that have magnitude below 0.05 are pruned.

One significant drawback of this type of pruning is that the removal of weights only on the basis of their absolute values leads to irregular/unstructured distribution of zeros in the network which in turn requires a special compressed format, e.g., compressed sparse row (CSR) or compressed sparse column (CSC), for storing the weights in memories. The compression formats usually have some overheads associated with them for keeping track of the locations of the effectual weights, which is specifically significant in case the sparsity is low.

Another disadvantage of irregular/unstructured pruning is that in most of the scenarios the processing of neural networks is carried out using matrix–matrix/matrix–vector multiplication, as illustrated in Fig. 21.5a, and in case the weights are stored in CSR format, complete input data is required to be fetched from the memory, despite the fact that a significant portion of it is not used. Similar is the case when

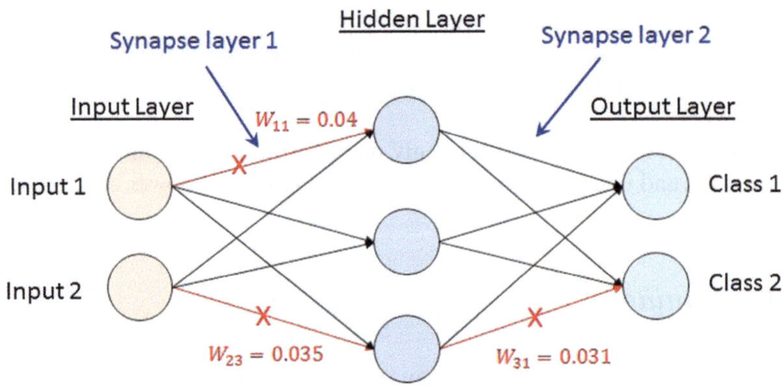

Fig. 21.4 An illustration of unstructured pruning where the highlighted branches are removed solely based upon their weight values regardless of the location

the weights are stored in CSC format except that in this scenario the output has to be accessed and updated multiple times [32]. In both the cases, a significant amount of benefit is lost and cannot be fully exploited by data-parallel hardware architecture (e.g., SIMD), which are commonly found in existing GPGPUs. Note that here it is assumed that the weights are stored in a matrix such that each row corresponds to a single neuron.

As specified above, unstructured pruning does not provide any significant advantage for regular hardware architectures like GPGPUs. One way to leverage it is by designing custom hardware that process data without decompressing, which is discussed in Sect. 21.3. Recently, researchers have also explored a more effective way of pruning to avoid the need for custom hardware, i.e., structured pruning. Instead of removing weights individually from a network, structured pruning involves pruning weights in a regular fashion (in the form of groups) from a network. Therefore, rather than computing the saliency of individual weights a combined saliency of a group of weights is computed and based upon that significance of the group it is eradicated. The major benefits of this technique over unstructured pruning are: (1) it requires lesser overhead for storing the weights; (2) as illustrated in Fig. 21.5b, it results in a significant reduction in the size of input and weight matrices that significantly reduces the number of operations and the overall size of data that have to be fetched from the memory for processing.

21.2.2 Weight Sharing

Weight sharing is another optimization technique used for alleviating the memory requirements of a neural network. The main concept behind weight sharing is to reduce the number of bits required to represent each weight. This is usually achieved by using a common set of a limited number of values (defined in the form of a dictionary) for representing the weights of a network/layer. For storing the model,

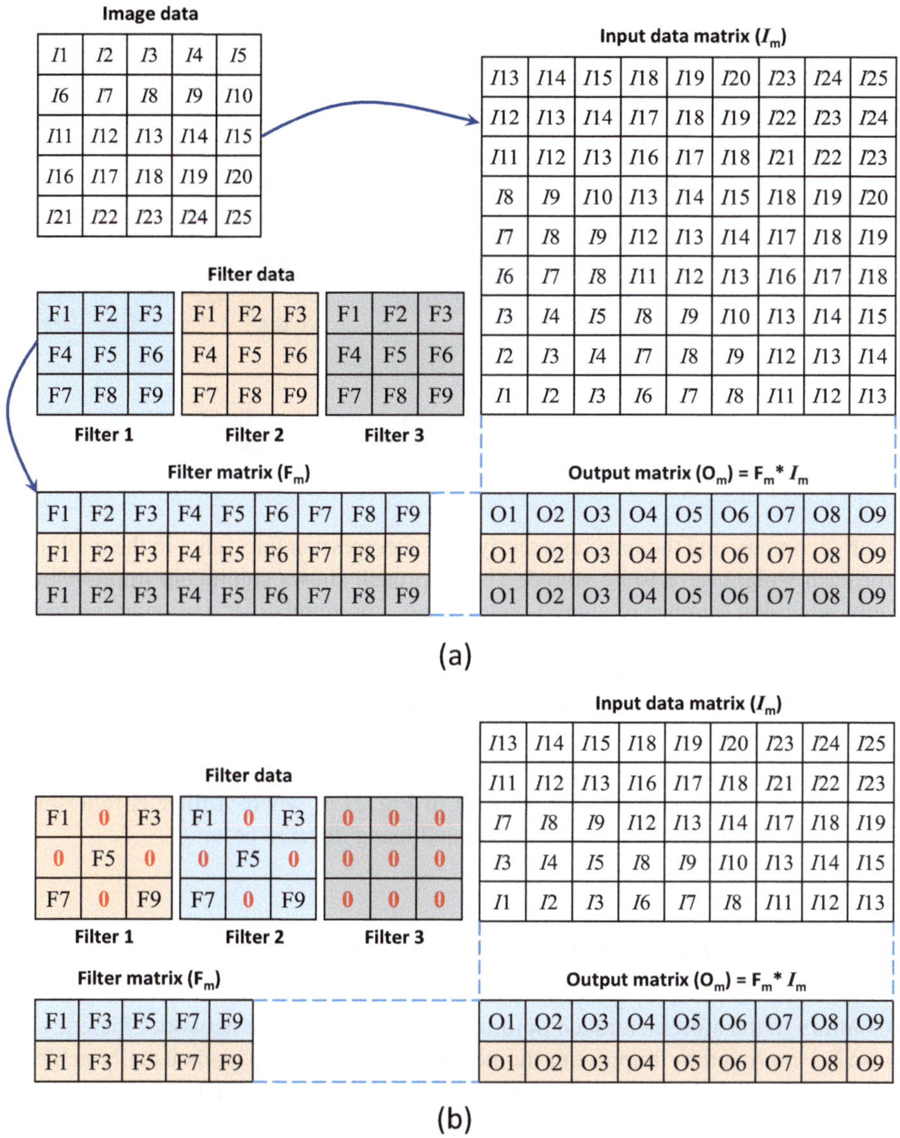

Fig. 21.5 Lowering convolution to matrix multiplication for pruned and unpruned networks. (**a**) illustrates the conversion of input data and filters into matrices that can be multiplied to generate the resultant output. (**b**) shows the reduction in the dimensions of matrices in case of structured pruning

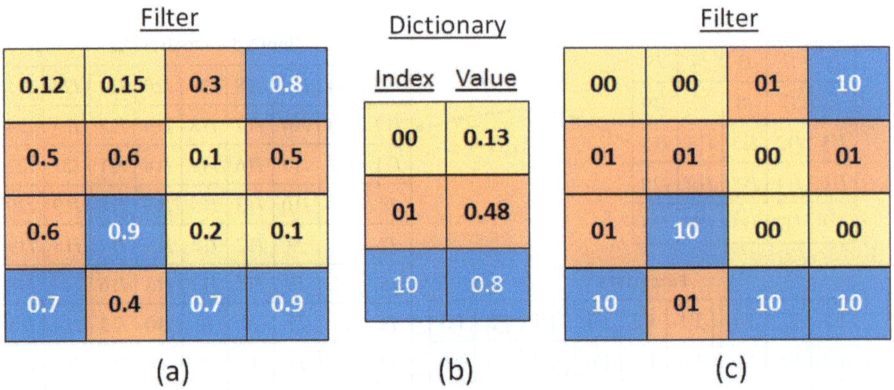

Fig. 21.6 Illustration of weight sharing in a filter. (**a**) Values of the filter before weight sharing. (**b**) Dictionary that stores the common weight values. (**c**) The corresponding binary indexes that are actually stored in memory along with the dictionary

both dictionary and the indexes specifying the corresponding value in the dictionary for each weight have to be stored. This approach shows significant improvements only when the number of shared weights, i.e., the number of dictionary elements, is significantly less. For example, in [31], through weight-sharing approach the number of unique weights per layer in AlexNet was reduced to 256 for convolutional layers and 16 for fully connected layers, thus required only 8-bit and 4-bit weight indexes, respectively. Note that in weight-sharing approach the precision of MAC units is not changed and they perform the same floating point operation. A naive example of weight sharing is illustrated in Fig. 21.6, where a filter is represented using just 3 floating point values (illustrated using differed colors).

21.2.3 Compact Network Architectures

Contrary to the aforementioned optimization techniques, the network efficiency can also be improved by improving the network architecture. This is usually achieved by replacing large filters with a series of smaller filters, which are implemented sequentially without the use of nonlinearity/activation layers in between. The core concept behind this is that the smaller filters, when applied sequentially, can traverse the same amount of input region (in images) as a larger filter and therefore can give the effects similar to that of the larger filter. For example, an 11×11 2D filter can be replaced with two 6×6 2D filters in series to produce a similar effect. Similarly, for 3D filters, the convolution can be broken down into multiple 2D channel-wise convolutions (depthwise convolution) followed by multiple of 1×1 3D convolutions (pointwise convolution). This type of convolution is usually termed as depthwise separable convolution. MobileNets [36] and Xception [14] are based upon these types of layers.

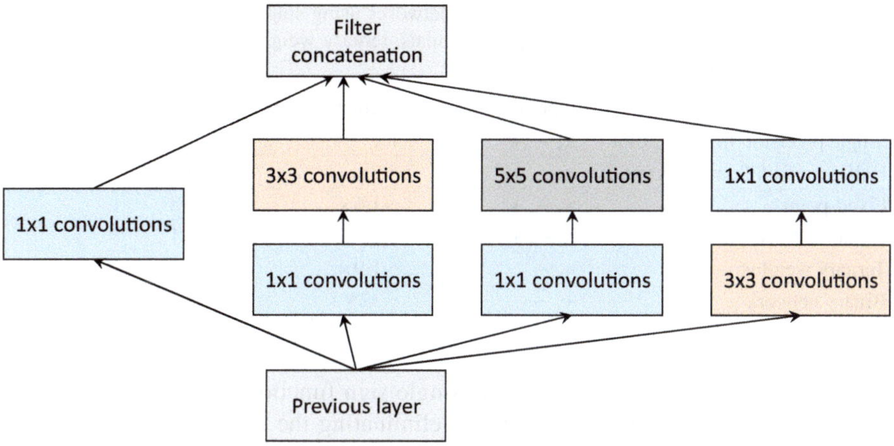

Fig. 21.7 Inception module with dimension reductions [60]

One drawback of this type of network is that although it reduces the number of weights significantly, the reduction in the computations is not proportional to the reduction in the number of parameters. For example, when compared to AlexNet, GoogLeNet utilizes only 11.5% of the weights while requiring almost twice the number of computations, as can be observed from Table 21.2. To address this issue, pointwise convolutional layers have been employed in "Inception module with dimension reductions," as shown in Fig. 21.7, for reducing the number of channels before expansion to reduce the total number of computations in the subsequent convolutional layers [60].

SqueezeNet [37], another state-of-the-art network architecture, is also based upon the aforementioned principles and reports to reduce the number of weights by 50×, while still consuming more energy as compared to AlexNet.

21.2.4 Hardware–Software Co-design

To address the energy efficiency and storage issues in neural networks, several techniques have been proposed that seek to reduce the precision of weights and/or activations. The most common technique, in this context, is the conversion of floating point numbers to fixed-point numbers, which achieves a significant reduction in the complexity of computational elements like adders and multipliers and thereby reduces the energy consumption of a network significantly. To further improve the energy efficiency, the precision of fixed-point numbers is also reduced and, usually, a modified version of fixed-point format, i.e., dynamic fixed-point format, is employed [29]. A summary of a few state-of-the-art techniques is presented in Table 21.3.

Table 21.3 Relative validation set accuracy of networks using single and half-precision floating point formats, fixed and dynamic fixed-point formats, ternary weights, and binary format on the CIFAR-10 dataset [15, 42]

	Operators used	Memory saving	Accuracy (CIFAR-10)
Single precision floating point	x, − ,+	1×	1
Half-precision floating-point	x, −, +	~2×	0.999
Fixed-point	x, −, +	~1.6×	0.978
Dynamic fixed-point	x, −, +	~2.9×	0.991
Ternary weights	+, −	~1.9×	0.997
Binary network	+, −	~32×	0.967

An extreme in this context is to use single sign function to binarize the weight values [16] to reduce the complexity by eliminating the need for multiplications. Extensions of the aforementioned [17, 48] have also been proposed that uses binary values for both activations and weights. A counterpart of binarized networks, i.e., ternary networks [42], have also been introduced that showed significant improvement in the learning capability as compared to a binarized network without the need for multiplications. Note that the aforementioned techniques require specialized training process for learning the quantized weights.

21.3 Hardware-Level Techniques

Multiple hardware-level architectures have been proposed recently to efficiently process neural networks in hardware. The primary motivation for building FPGA/ASIC-based systems is the large power overhead associated with GPU-based processing. For example, although a NVIDIA DGX-1 with Tesla V100 can train ResNet50 96× faster than a CPU server with two Intel Xeon E5-2699v4 CPUs, it consumes 32× more power as compared to a commodity GPU that typically consumes 100 W [61]. In this section, we discuss few of the most prominent architectures along with the underlying techniques that are designed to maximize the energy efficiency and minimize the execution time of a neural network.

21.3.1 Dataflows and Architectures for Accelerators

The key insight in DNN acceleration is to process in parallel to the maximum. Accordingly, numerous dataflows have been proposed with varied parallelism goals. One popular choice is systolic arrays. Systolic arrays are a mesh-like assembly of processing elements (PEs), where each processing element is connected to adjacent processing elements, as shown in Fig. 21.8. The data enters the array from the PEs

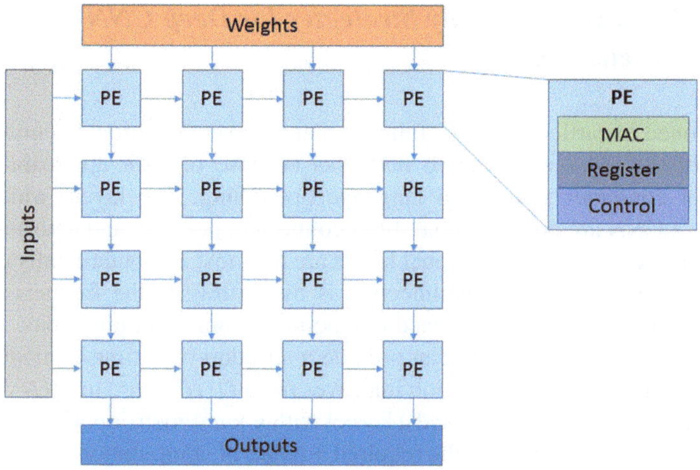

Fig. 21.8 An illustration of a typical systolic array where data and weights are fed from one side and the resultant partial sums are produced on the other side, after several iterations/clock cycles. Each processing element in the array is composed of a MAC unit, a control logic, and an internal register for storing the intermediate or stationary values

that are at the boundaries of the systolic array and is moved across the array for processing. The outputs are then generated on the other side of the array which is then transferred to the memory. Systolic arrays are inherently pipelined and can be implemented with ease. Recently, high-level mapping systems are being proposed that make use of systolic arrays [65]. Even though systolic arrays are easy to implement, they may not provide optimal performance across networks with varying parallelism opportunities. In practice, DNN architectures show varying degrees of parallelism, and using one dataflow for all types of workloads is not optimal. For example, systolic arrays only exploit parallelism in synapses, excluding the parallelisms in input and output feature maps [44], which makes them inflexible. As a solution, [44] FlexFlow dataflow offers flexible dataflow and enables a mix of synapse, input, and output feature maps parallelisms.

As DNNs show great diversity among their architectures, a scalable accelerator should be able to accommodate newer architectures with multiple filter sizes, strides, and channels. Eyeriss [11] is one such architecture. It has 168 reconfigurable processing elements which are placed spatially in a 2D mesh-like fashion. Each PE has its own MAC unit, control logic, and register/scratch pad memory (for caching intermediate results). To minimize the energy consumption of data processing and movement, Eyeriss uses row stationary (RS) dataflow in which the movement of weights, pixels, and partial sums is minimized. One limitation of using such spatial architecture is the communication cost among PEs that can lead to complex control circuitry and extra power consumption.

21.3.2 Hardware Friendly Strategies for Deep CNN Accelerators

As mentioned in earlier sections of the chapter, deep convolutional neural networks are one of the most widely used networks because of their applicability in the domain of computer vision and language processing. The layers which mainly constitute a CNN are convolutional, fully connected, pooling, and activation layers. Out of all these layers, convolutional layers are the most computationally expensive.

One way to optimize the architecture is by shifting to the frequency domain, where the convolution is transformed into point-to-point multiplications. Therefore, frequency-domain acceleration can provide a speedup over time-domain acceleration. The complexity of convolution improves from $O(N^2 \times n^2)$ to $O(N^2 \times \log_2 N)$ for an input data of size $N \times N$, and a kernel with $n \times n$ dimension. Use of Hermitian symmetry can further decrease the required number of computations, as components at negative frequencies are conjugate of those at positive frequencies. Overlap and Add method (OaA) [35] is another FFT-based algorithm that further improves the computational complexity of frequency-domain convolution to $O(N^2 \log_2 n)$. In OaA method, input arrays are first divided into parts. These subparts are convolved with each other (subparts of one input with subparts of second input) to generate partial products. Finally, these partial products are added. Zhang and Prasanna in [68] designed an accelerator while making use of OaA method. Earlier attempts only performed the convolutional layer in the frequency domain. Thus, FFT and IFFT steps were required for each convolutional layer, thus adding an overhead to move back and forth from the Fourier domain. Recently, Ko et al. [40] presented an end-to-end frequency-domain DNN training accelerator. This work requires FFT operation only once at the start, and then an IFFT towards the end of the DNN. All DNN modules including convolution, pooling, and nonlinear functions are performed in frequency domain. They reported $4.5\times$ and $6.0\times$ improvement for latency and energy as compared to spatial domain design for AlexNet CNN.

Note that the benefits of frequency-domain acceleration diminish for small filter sizes. As the trend in newer convolutional neural network architectures is to use 3×3 filters, frequency-domain acceleration is unlikely to improve the performance metrics. This motivates the use of another class of algorithms: minimal convolution algorithms. Minimal convolution algorithms are transforms that lead to a reduction in the number of required multiplications. Multipliers are the most power consuming blocks, and reduction in their use directly enhance the performance. One such technique is the Winograd convolution, which is inspired by Chinese Remainder theorem. It is useful as it requires less number of multiplications. Given a 3×3 filter and 2×2 output matrix, Winograd convolution require $2.25\times$ less multiplication [41]. Similarly, Wang et al. [64] proposed a fast convolution unit (FCU) for kernel sizes of five and three. It makes use of fast finite impulse response algorithm (FFA). FFA can save 36% energy for kernel size of five: A typical convolution requires

25 multiplications, and 24 additions, while a convolution using FFA requires 15 multiplications, and 34 additions, thus providing 1.6× reduction in multiplication operations.

Use of minimal convolution algorithms can lead to a decrease in the number of resources but can put a strain on memory, as more intermediate results are to be stored. Also, the benefits of these algorithms diminish with the increase in filter size. A practical use, as demonstrated in [66], includes the use of both conventional and Winograd convolution to avoid high memory traffic, and exploit available hardware resources, respectively. This is because conventional convolution is resource exhaustive, and minimal convolution is memory exhaustive.

21.3.3 Memory-Efficient Architectures

Memory transactions account for a significant portion of energy consumption and worsen the system latency and throughput. Data reuse is a solution to alleviate this problem. This approach includes maximizing the data reuse to minimize memory traffic and thereby improves latency, energy efficiency as well as the throughput of a system. DNNs offer several opportunities for data reuse. Some of them are discussed below:

1. Allocating on-chip memory for storing kernel values: In this approach, kernel values are stored using on-chip memory. Thus, they only need to be loaded once for several or all the associated computations.
2. Reusing overlapped data between filters: As there are multiple filters in a layer, a segment of an input image is required to be loaded only once for several/all filters.
3. Avoiding the data transfer of intermediate results: The intermediate results from previous computations can be kept on-chip and then summed with the new results (of associated values) to generate partial versions of the output feature maps. This provides with an opportunity to avoid intermediate data storage.
4. Reusing overlapped data between adjacent windows of an input: A window of input feature maps can be cached using on-chip memory. Afterwards, only those rows/columns of input need to be loaded which are not currently in the cache buffer. This is significantly useful for convolutional layers where the overlap between adjacent windows is very significant.
5. Sharing feature maps between adjacent layers: As multiple convolution layers are cascaded, the result of one layer can serve as the input of the next layer. This eliminates the need to move and then fetch intermediate results to/from off-chip memory. Using the approach, authors in [33] achieved 55% less communication-to-computation ratio.

A similar approach was adopted by Alwani et al. [5] for enabling data reuse. They proposed to process multiple convolution layers in a fused way at the same time, in contrast with existing methods which process each layer separately, to avoid memory transfer. Two or more convolution layers are fused, and the results of intermediate layers are not communicated back to memory and are reused for further processing. Using fused processing, they reported 95% decrease in memory transfers. Escher [57] made use of a batching strategy to effectively retrieve data for multiple inputs. In batching, accelerator avoids data transfers by storing weights on-chip. In [57], batching is applied to both fully connected and convolutional layers. Groups of weights are reused for given batches of images. Their design was optimized for throughput, making use of analytical models, incorporating computation resources, on-chip storage, input and output buffers, and memory bandwidth.

21.3.4 Hardware Architectural Techniques for Leveraging Sparsity in Neural Networks

A significant portion of weights in a DNN is zero-valued numbers or insignificant values. Alongside this, since ReLU (rectified linear unit) is one of the most popular activation functions which is used in almost all the state-of-the-art neural networks, a significant amount of activations are also zero valued. For example, according to [47], 50–70% of activations are zero valued and 20–80% weights are zero valued. Computation and memory transfer of such numbers can be skipped, as they do not contribute to classification accuracy. Overall, computation skipping shows a promising way to reduce computation and memory overhead as DNN architecture remains unchanged. This insight can lead to substantial savings in energy consumption and reduction in memory transfers. A few techniques exploit the particular response of the activation function. These techniques are usually effective for the networks that employ ReLU activation function for introducing the nonlinearity in the network, as mentioned above. ReLU function sets all the negative inputs to zero and thereby introduces a significant amount of sparsity in feature maps; for example, the feature maps in AlexNet have a sparsity between 19% and 63%. This sparsity can be exploited for improving the overall efficiency of the underlying hardware by skipping corresponding DRAM accesses and computations. Furthermore, the hardware can also be modified such that it skips the memory access to weights and also skips computation of MAC for zero-valued activations to improve the energy efficiency [12]. Rather than just gating the read and MAC computation, the hardware could also skip the cycle to increase the throughput by 1.37× [4]. Also, as described earlier applying pruning to a network results in a sparse network. However, due to the asymmetrical nature of sparsity, synchronization is required to keep track of valid data. Even with 89% weights removed, sparse AlexNet on CPU takes 25% more time than baseline dense model [67]. Therefore, application-specific hardware architectures are proposed that can deliver the improved performance for a sparse network.

EIE is an ASIC accelerator [32] that operates on the compressed DNN model acquired after applying Deep Compression [31] methodology. It keeps track of sparsity and skips zero-valued activations by storing only nonzeros values in compressed sparse column (CSC) format. Apart from a vector containing the nonzero values of the weights, a vector that contains the locations of nonzeros values is also required. The EIE delivers 103 GOPs/s computation power for a sparse network, which (due to high sparsity ratios offered by Deep Compression on state-of-the-art neural networks) is equivalent of 3 TOPs/s for a dense network. Furthermore, the CNV accelerator proposed in [4] is capable of skipping the computation when an activation value is zero. It is built on top of DaDianNao [10] and targets convolution layers. CNV skips the computation when the activation value is zero across all layers of a network except first layer. The zero skipping is achieved by using Zero-Free Neuron Array Format that removes the zero values of activations. An offset variable is used to determine the original positions of remaining nonzero values. CNV achieves 1.37× performance on average over DaDianNao accelerator, least being 1.24× on GoogleNet at 4.49% area overhead. Parashar et al. proposed another accelerator SCNN [47] that can skip computations corresponding to zero-valued weights and activations, in contrast to Cnvlutin [4] that only optimizes for zero-valued activations. It tracks location of nonzero values by encoding the nonzero-valued inputs, and weights. So, the data delivered to processing elements are only nonzeros values. Unlike EIE, SCNN works on both convolution and FC layer.

As mentioned earlier, a conventional DNN contains a large number of insignificant parameters which lead a large number of ineffectual computations and thereby can be removed. Apart from this, even the remaining parameters contain a majority of zero-valued bits, for example, only 18.1% of nonzero neurons are nonzero bits for AlexNet [3]. Such zero-valued bits result in ineffectual partial products during multiplication operations and eliminating the generation of such partial products can provide further energy savings. Pragmatic accelerator or PRA [3] aims to eliminate such ineffectual multiplications. PRA is also built to function as an extension of DaDianNao. It tracks the locations of 1's in a neuron value and uses shifters to calculate the desired partial product. PRA provides 1.48× energy efficiency on average as compared to DaDianNao with 1.35× area overhead.

21.4 Error Resilience Analysis: DNN-Specific Approximations for Low-Power Accelerators

As mentioned in the previous sections, various optimization techniques can be employed for reducing the power and energy requirements of an NN accelerator. To further improve the energy/power efficiency, implementations of DNNs can benefit from approximate computing-based low-power/low-energy computing modules, as most of the applications of DNNs are inherently error resilient. This error resilience can be associated with two main reasons:

1. In DNN architecture, all input samples are processed in the same manner regardless of their inherent difficulty for a particular classification application. There is a potential in identifying ineffectual computations/neurons and hence simplifying their implementation to reduce power/energy consumption without affecting the output quality of the system [55].
2. Training in NNs is an iterative process, which can be stopped when the desired accuracy is achieved. Moreover, retraining can be employed to mitigate the effects of approximations and can lead to even better results than with the accurate implementation [55].

In the following, we present a case study of approximate resilience analysis that selects an appropriate datapath for approximation within a particular DNN [56]. We consider a scenario where a reduced power budget demands the application of low-power approximations. The problem then reduces to identifying datapaths where approximations have least affect while offering a significant amount of reduction in the power consumption. Since filters in a trained DNN architecture are responsible for extracting specific features, we anticipate that their outputs shall have defined data distributions. Furthermore, the errors produced by the approximate modules are highly data dependent and, therefore, different approximate modules respond differently to different data distributions. An analysis of data distribution across various datapaths of NN and the sensitivity of approximate modules under consideration can help in choosing appropriate approximate circuits and the paths that result in the minimum loss in the output quality.

For this analysis, we consider a LeNet-5 [20]-based neural network, as shown in Fig. 21.9, for classifying handwritten digits in the MNIST dataset. The network is composed of two convolution layers (C_1 and C_2), two pooling layers (P_1 and P_2), and a fully connected layer (FC). The convolutional layers are used for extracting features and the fully connected layers, which are connected at the end, are used as the classifiers. C_1 layer is composed of six different filters and produces six different output Feature Maps (FM_1, FM_2, ..., FM_6) for a single input sample. The input size for this network is 28×28. Afterwards, P_1 layer applies average pooling on the output from C_1, assuming stride of one and a window of 2×2.

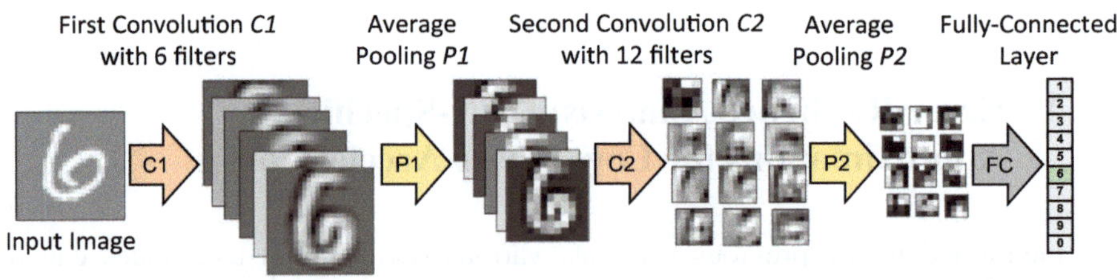

Fig. 21.9 Modified LeNet for Handwritten Digit Classification (MNIST dataset) used in our experiments. Figure illustrates the outputs of various layers. Adapted from [56]

Then, C_2 layer, composed of 12 filters of six kernels each, operates on the output from P_1 layer. Note that the output of P_1 is composed of six kernels. The filters are convolved with the output of P_1 to generate an output composed of 12 feature maps, i.e., one FM per filter. Finally, these 12 output feature maps are passed through P_2 layer and afterwards are concatenated together in a vector, which is then passed onto the final FC layer for classification. Output vector at the end of the FC layer holds the probability of the input belonging to each possible digit. In order to show the importance of the case study on the optimized network, we used fixed-point simulations with Q(6.7) bit format (6 integer and 7 fractional bits), which provided classification accuracy of 94%, when tested on 200 images.

In this, we analyze the data distribution of the six feature maps generated after convolution at the output of C_1. The analysis is then employed to select appropriate low-power approximate adder(s) for average pooling in P_1. Figure 21.10 shows the Feature Maps at the output of C_1, for four randomly selected input images from the MNIST dataset.

Figure 21.11 shows the corresponding data distributions of the feature maps illustrated in Fig. 21.10. It can be observed from the figure that histograms corresponding to the same kernels have strong correlation and are nearly identical, regardless of the input image.

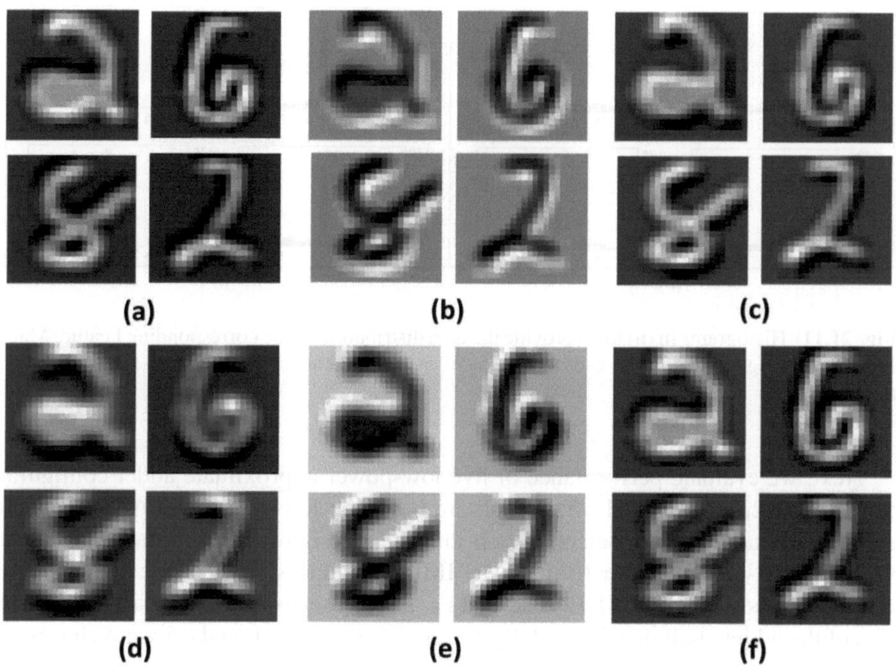

Fig. 21.10 (**a**) to (**f**) show the Feature Maps (FMs) at the output of six filters of C_1, obtained for four randomly selected input images. Adapted from [56]

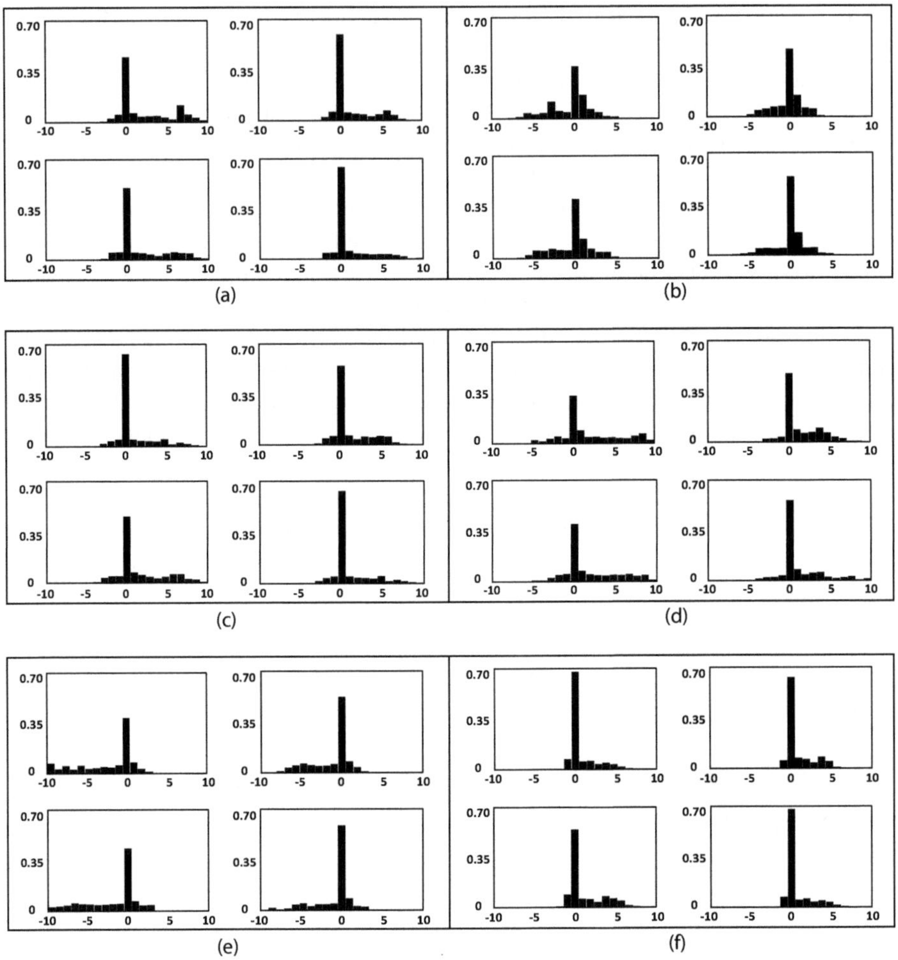

Fig. 21.11 Histograms in (**a**) to (**f**) provide the data distribution of the corresponding Feature Maps illustrated in Fig. 4 [56]. The x-axis and y-axis of each represent activation values and probability, respectively

Next, we evaluate performance of five low-power approximate adder configurations ($\alpha 1$, $\alpha 2$, $\alpha 3$, $\alpha 4$, and $\alpha 5$), which are based on low-power IMPACT [28] full adders. The adders were selected such that their design is in accordance with our Q (6.7) fixed-point number format. The IMPACT adders provide five approximate full-adder designs (LP_1 to LP_5). Here, $\alpha 1 = LP_1$:6 means that the 6 least significant bits of the adder shall be of type LP_1. Similarly, $\alpha 2$, $\alpha 3$, $\alpha 4$, and $\alpha 5$ were chosen to be LP_2:6, LP_3:6, LP_4:7, and LP_5:7, respectively. We evaluate the performance of each of these five adders for six Feature Maps using data generated with the help of

Table 21.4 Classification accuracy obtained by employing approximate adders ($\alpha 1$ to $\alpha 5$) in the P_1 layer for various datapaths [56]

Adder configuration	Approximation applied to individual feature maps						
	FM_1	FM_2	FM_3	FM_4	FM_5	FM_6	Average
α_1	92%	92%	92%	92%	90%	92%	~92%
α_2	56%	92%	60%	84%	92%	94%	~80%
α_3	56%	92%	60%	88%	93%	94%	~81%
α_4	92%	92%	92%	92%	91%	92%	~92%
α_5	94%	92%	92%	94%	92%	94%	~93%
Average	~78%	~92%	~79%	~90%	~92%	~93%	

50 randomly selected images from the training set. It can be seen from Table 21.4, different approximate adders have different tolerance for different data distributions. It can also be observed that the datapaths that belong to FM_2, FM_5, and FM_6 show better error resilience to approximations (when employed for average pooling in P_1 layer) as compared to rest of the datapaths. This is because of the fact that the input data distribution resulted in adder operands that were less sensitive to approximation and resulted in an accuracy of 92% at least. Based upon this analysis, we define a function $R(\alpha i)$ to rank the approximation resilience of a particular feature map for each of the provided approximate adder configuration. Thus, the FM that illustrates the lowest error for a particular adder configuration ranks first in terms of error resilience. Thus, for example, for the simulated data, $R(\alpha 3) = \{FM_6, FM_5, FM_2, FM_4, FM_3, FM_1\}$. Accordingly, if there is a power constraint requirement to apply $\alpha 2$ approximation to the P_2 layer of any one of the FMs, FM_6 shall be the most suitable while FM_1 be the least. Table 21.4 also highlights that $\alpha 1$, $\alpha 4$, and $\alpha 5$ provide results with higher accuracy, in general, as compared to other adder variants for all the FMs. This is because, as per their truth tables in [28], these adders produce accurate results in case both of their inputs are zero, which is the most probable case in the current scenario as evident from the data distributions illustrated in Fig. 21.11.

To evaluate our hypothesis, we consider a scenario where it is required to apply approximation to at least two datapaths between C_1 and P_1. Table 21.4, in general, predicts that the highest accuracy is achieved when datapaths corresponding to FM_5 and FM_6 are approximated. Similarly, for the scenario where the same approximation has to be used for all the datapaths, adder configuration $\alpha 1$, $\alpha 4$, and $\alpha 5$ provides better results as compared to $\alpha 2$ and $\alpha 3$. Thus, we demonstrate that due to the fixed architectural design of a neural network, each datapath has a fixed data distribution. Since the error of approximate modules is also a function of data distribution, prior analysis on training data can thus provide useful insight for selecting appropriate approximation modules, given that there is a strong correlation between the testing and training data.

21.5 Energy-Efficient Hardware Accelerator Design Methodology for Neural Networks

In this section, we provide a holistic design methodology, from training to hardware realization, for the on-chip realization of neural networks. The suggested methodology is expected to result in hardware designs with significant energy reductions. As illustrated in Fig. 21.12, the methodology comprises four major stages. Below, we provide a description of each of these.

1. **Training:** Libraries such as TensorFlow and Caffe are typically employed to train networks for a particular application [12]. The decision to select a particular network architecture (AlexNet, Googlenet, SqueezeNet, etc.) is a fundamental one. The selection is guided by the application requirements (required accuracy level, number of classes, and computational resources at hand). It has been recently demonstrated that thinner and deeper CNNs with fewer weights may not necessarily consume less energy than shallower CNNs with more weights. Thus, while SqueezeNet provides a model with low memory footprint (33× lower than AlexNet), its energy consumption was reported to be 33% greater than that of AlexNet. Thus, an estimate of number and type of memory accesses along with an estimate of computational load has to be considered when selecting a particular DNN. The result of this stage is a trained neural network described by its filter weights and layer/connectivity information.
2. **Model Compression:** Once a particular model/DNN architecture is selected (such as AlexNet and ResNet), the next step is to prepare the model for its implementation on an embedded platform such as ASICs and FPGAs. Since embedded platforms are typically low on resources (battery and storage), the selected models are typically compressed to make them hardware friendly. Two pre-dominantly employed schemes are Pruning and Quantization [31]. As described in Sect. 21.2.1, pruning involves removing the unwanted portion of the network. This may involve removing a few of the filter coefficients [31], complete filters [6], or even a few of input activations [50]. When applied to filters, pruning is typically performed by removing the filters weights that have lower magnitude. Note, however, that in case the filter weights have correlated values, large error shall be introduced. It has been thus recently suggested to perform pruning while minimizing the error in the output feature maps. At algorithmic/software level, MAC operations are carried out using floating point multiplications. Quantization of floating point is thus suggested to reduce the number of possible values in which the filter weights can be represented [31]. Such a quantization leads to a representation of filters using a fewer subset of floating point values (quantized subset). This effectively results in multiple filter weights sharing the same value from the quantized subset. When considering implementation on embedded processors, this step is proceeded by a codebook generation followed by Huffman coding to reduce the memory requirements. Here, we are focusing on custom dedicated designs and hence codebook generation is carried out at a later stage.

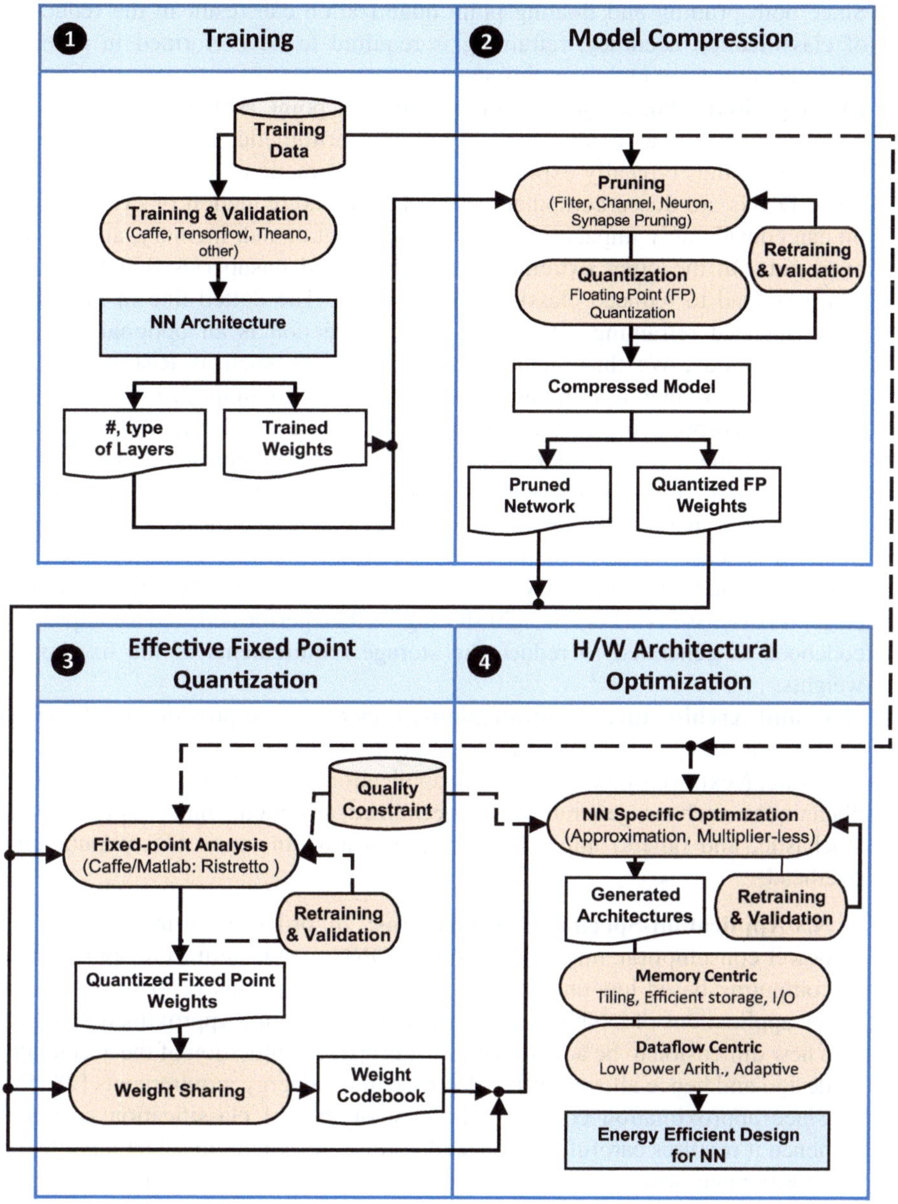

Fig. 21.12 A systematic and comprehensive design methodology for designing of energy-efficient hardware accelerators for deep neural networks (DNNs)

Since both pruning and floating point quantization can result in the reduction of classification accuracy, retraining is required to be performed in order to minimize the accuracy loss at this stage.

3. **Effective Fixed-Point Quantization:** Floating point arithmetic is computationally expensive as compared to fixed-point arithmetic. Dedicated hardware accelerators thus typically exploit fixed-point arithmetic to leverage its benefits. Since DNNs are dense computational architectures, reduction of even a single bit can considerably impact the associated computational, memory, and storage overhead. On the other extreme, uncontrolled and unsupervised reduction in bits may lead to reduced classification accuracy. The dotted line in Fig. 21.12 illustrates that retraining after fixed-point conversion is an optional but recommended step. We thus suggest the use of tools, such as Ristretto, Caffe, and MATLAB, that retrain the network in order to minimize the effect of reduced fixed-point representation. A quality constraint is typically provided that assists the selection of best possible fixed-point representation within a quality bound. Since, fixed-point representation affects all the three aspects (computation, memory bandwidth, and storage). This shall provide a convenient trade-off in case an adaptive design, supporting multiple accuracy levels, is required. Note that multiple fixed-point representations may be obtained at this stage. As discussed earlier, "weight sharing" is then performed, and a compressed codebook is generated to reduce the storage requirements of the fixed-point weights.
4. **H/W and Architectural Optimizations:** Previous stage provided the finalized NN architecture along with the fixed-point weights to be implemented on hardware. Next, we suggest to perform application-specific design optimizations. Techniques such as generation of specialized arithmetic units (approximate, stochastic, and others) allow further optimization in terms of computational overhead.

   - **NN/Application-Specific Optimizations:** In order to further reduce the power consumption, implementations of DNNs can benefit from approximate computing-based low-energy computing units [56]. This is possible because the applications of DNNs are inherently error resilient. Approximation allows a new dimension to be added to the design space exploration of the accelerator design and hence allows further decrease in the energy requirements [28, 55]. Since approximation comes at the cost of loss of classification accuracy, hence it requires careful analysis and at times retraining to select a particular approximate arithmetic configuration. In [22], authors evaluated the use of low-power imprecise multipliers using the UCI Machine Learning repository.

They reported that inexactness could be overcome by retraining the network. They reported an energy savings of up to 62.49% over accurate face recognition. A similar approximate multiplier design was also explored [51], where a multiplier-less neuron structure was trained to alleviate the accuracy loss incurred due to approximation. Another strategy is to incorporate a selective rule to apply approximation only to multiplication where both operands are nonzero. The power consumption can be significantly reduced by such strategies for a negligible decrease in classification accuracy [45].

- **Memory Centric Optimizations:** Memory operations are responsible for a considerable portion of energy consumption. In order to reduce this energy, one can optimize memory cells, memory hierarchy, memory access patterns (e.g., tiling), data storage, refresh rates of DRAMs, and other properties of the memory subsystem [54]. For example, a hybrid 8T-6T SRAM cell was proposed to store the synaptic weights, wherein the sensitive MSBs are stored in 8T bit-cells and resilient LSBs are stored in 6T bit-cells [54]. If voltage scaling is applied, errors can appear in 6T cells, while 8T cells remain operating correctly. Memory compression schemes are further suggested to reduce the overall memory traffic and memory energy.
- **Dataflow Centric:** FPGA- and ASIC-based designs allow dedicated architectures to be built. This allows the use of custom arithmetic units (for low-precision arithmetic), pipelining, and even allows design of adaptive architectures [56]. Dataflow design is the most important part of accelerator design as it determines the parallelism, data reuse, and defines memory requirement. To cope up with varying degrees of parallelism in modern DNN models, multiple data flows are proposed [10, 12, 23, 44, 47]. For instance, Eyeriss employs Row-major type dataflow which processes each row using a processing element. On the other hand, DaDianNao and SCNN operate on a tile of the input data to reduce active memory footprint. Since 50–70% of activations and 20–80% are zero-valued [47], computation and memory transfer of such numbers can be skipped, as they generally do not contribute significantly to classification accuracy. This can lead to substantial savings in energy consumption and reduction in memory transfer.

21.6 Efficient Machine Learning Architectures: Challenges and the Way Forward

In this section, we highlight few of the key research challenges and opportunities that will help progress towards designing energy-efficient systems for machine learning-based applications.

21.6.1 Optimizing Memory vs. Computations

DNNs are highly compute intensive and at the same time have a significant memory footprint. Studies have shown that the average energy consumption associated with memory access, especially off-chip memory, in state-of-the-art DNNs can grow up to 50–80% of the total energy consumption of the system. Therefore, optimizing the size of the models and reducing the memory footprint across the complete memory hierarchy should be one of the key objectives for developing modern systems for machine learning applications. To achieve this, all possible optimization techniques, like pruning, weight sharing, quantization, efficient data storage, refresh rate control, optimizing memory cells and access patterns, and devising fine-grained memory power management policies, can be considered. In future, emerging technologies like nano-wires-based 3D-stacked architectures could also provide a very effective solution for reducing the energy/power consumption associated with memory accesses. This can be achieved by connecting a sea of accelerators directly to dense memory layers through these high-bandwidth nano-wires. One significant effort in this direction is N3XT1000x [62]. Spintronic devices [9], Memristors, and Graphene-based design are few of the other parallel emerging technologies that possess the capability to reshape the course of energy-efficient ML architectures.

21.6.2 Neuromorphic Computing

It is a well-known fact that the neural networks are mainly inspired from the brain. However, the neuronal model which is widely being used at the current point in time is very naive and barely mimics the functionality of the actual neurons. Apart from this, the general computing systems process instructions at regular intervals, i.e., based on a clock, which is significantly different from the communication mechanism used by real neurons, i.e., asynchronous spiking [24]. This spiking phenomenon is one of the key concepts which makes the neuromorphic systems highly energy and power efficient, as it encodes information in impulses rather than regular-interval signals. Another main functionality in neuromorphic systems which is significantly different from the general computing systems is that they process data in an analog, rather than a digital fashion. A number of systems/chips have been proposed over the past few years, for example, TrueNorth [2], Neurogrid [7], and Loihi [19]. However, the biggest challenge for such systems is learning algorithms, as the conventional algorithms are not optimal for training networks for such devices. Another alternative which has received a significant amount of attention because of limitations of spiking-based neuromorphic systems is in-memory computing using nonvolatile memory devices. Few of the most significant works in this direction are PRIME [13], PipeLayer [59], and ISAAC [52]. However, the use of this technology is limited because of the lack of development tools and the sophistication in underlying technology so far.

21.6.3 Accuracy vs. Energy Trade-off

Approximate computing is one of the emerging paradigms that adds an additional dimension to the design space by offering design and run-time trade-offs between accuracy and performance metrics. Particularly, DNN architectures may require different precision for different set of ML-based applications. Thus, an efficient hardware for such applications should be able to optimize the performance and/or energy consumption by adaptively regulating the accuracy of the accelerator while meeting the user-defined quality constraints. However, to achieve this, there is a substantial need to analyze the effects of approximations on the accuracy of the DNNs. This requires development of efficient simulation frameworks and evaluation methodologies. Furthermore, due to the layered nature of the NNs, it may be advantageous to distribute the available budget across multiple layers while considering the sensitivity of each layer/neuron.

21.6.4 Adaptability, (Re-)configurability, and Scalability

DNN architectures are composed of repetitive arrangement of limited types of layers, for example, pooling, fully connected, convolutional, activation, and normalization. However, the arrangement and the number of layers are different for different applications and also vary depending upon the complexity of the applications. Developing a generic hardware/accelerator for DNN-based applications requires a highly configurable and scalable design. This can lead to different design choices, like time-shared vs. dedicated designs and choice of scratchpad vs. caches, etc. In this direction, selection of reconfigurable fabric, such as FPGA and CGRA, is also an interesting research question, as both offer particular advantages related to reconfiguration granularity and associated performance and energy overheads. Orthogonal to this, memory bandwidth requirements also scale with model sizes and input data dimensions. Therefore, data compression techniques and efficient storage patterns can also be employed for the energy-efficient implementation of these networks.

21.6.5 Run-Time Evolutionary Algorithms for Designing and Optimizing DNN Architectures

General-purpose optimization algorithms can also be used to optimally trade off accuracy in particular layers/whole DNN architectures for improving performance, power, and/or energy efficiency of the system. This can be achieved by optimizing the DNN structure, memory access pattern, or the overall footprint of the network. Evolutionary algorithms have shown remarkable success in the selection of suitable NN models for particular applications, and a number of hyper-parameter tuning works are based on such algorithms. Apart from the design time optimizations, such algorithms can also be used for run-time adaptation for achieving the optimal trade-off between accuracy and performance characteristics of a network.

21.6.6 Correct Benchmarking with Fairness and High Fidelity

Due to rapid developments of fields, a fair evaluation methodology has not yet been established across different communities that are working on these challenging problems. For validation, various different metrics are required to estimate the standard test conditions, the fidelity of results, etc. to compare approximation methods. Each benchmark method/circuit is requested to provide a Pareto-front containing the points that offer the best trade-off for key circuit parameters, like throughput, latency, area, accuracy, and power/energy consumption, assuming that a fixed time budget is available for the procedure. Several works provided a detailed comparison between different approximations, for example, for adders and multipliers [38, 39]. However, if the models/codes of appropriate modules, details of experimental setup, simulation platforms, and test conditions are not available online, it is significantly hard to reproduce/validate their results and thereby achieve fair comparisons between the state-of-the-art and the newly proposed techniques.

21.6.7 Open-Source Contributions

Alongside aforementioned challenges and future directions, there is a strong need for researchers and developers to open-source their contributions, to facilitate others. The open-source libraries not only offer valid candidates for comparison but also facilitate the students and rookie researchers and developers to start directly from state-of-the-art models/implementations without reinventing the wheel. Towards this, we contribute through our open-source libraries of approximate arithmetic modules which are available at: GeAr-DAC15 [27, 53] and lpACLib-ICCAD16 [43, 49] and [46].

References

1. https://github.com/jcjohnson/cnn-benchmarksAccessedo13thNov2017. Accessed 13th Nov 2017
2. F. Akopyan, J. Sawada, A. Cassidy, R. Alvarez-Icaza, J. Arthur, P. Merolla, N. Imam, Y. Nakamura, P. Datta, G.J. Nam, B. Taba, Truenorth: design and tool flow of a 65 mw 1 million neuron programmable neurosynaptic chip. IEEE Trans. Comput. Aided Des. Integr. Circuits Syst. **34**(10), 1537–1557 (2015)
3. J. Albericio, P. Judd, A. Delmás, S. Sharify, A. Moshovos, Bit-pragmatic deep neural network computing (2016). Preprint. arXiv:1610.06920
4. J. Albericio, P. Judd, T. Hetherington, T. Aamodt, N.E. Jerger, A. Moshovos, Cnvlutin: ineffectual-neuron-free deep neural network computing, in *2016 ACM/IEEE 43rd Annual International Symposium on Computer Architecture (ISCA)* (IEEE, Piscataway, 2016), pp. 1–13
5. M. Alwani, H. Chen, M. Ferdman, P. Milder, Fused-layer CNN accelerators, in *2016 49th Annual IEEE/ACM International Symposium on Microarchitecture (MICRO)* (IEEE, Piscataway, 2016), pp. 1–12
6. S. Anwar, K. Hwang, W. Sung, Structured pruning of deep convolutional neural networks. ACM J. Emerg. Technol. Comput. Syst. **13**(3), 32 (2017)
7. B.V. Benjamin, P. Gao, E. McQuinn, S. Choudhary, A.R. Chandrasekaran, J.M. Bussat, R. Alvarez-Icaza, J.V. Arthur, P.A., Merolla, K. Boahen, Neurogrid: a mixed-analog-digital multichip system for large-scale neural simulations. Proc. IEEE **102**(5), 699–716 (2014)
8. S. Borowiec, T. Lien, Alphago beats human go champ in milestone for artificial intelligence. Los Angeles Times **12** (2016)
9. C-spin, http://cspin.umn.edu/
10. Y. Chen, T. Luo, S. Liu, S. Zhang, L. He, J. Wang, L. Li, T. Chen, Z. Xu, N. Sun et al., DaDianNao: a machine-learning supercomputer, in *Proceedings of the 47th Annual IEEE/ACM International Symposium on Microarchitecture* (IEEE Computer Society, Washington, 2014), pp. 609–622
11. Y.H. Chen, J. Emer, V. Sze, Eyeriss: a spatial architecture for energy-efficient dataflow for convolutional neural networks, in *2016 ACM/IEEE 43rd Annual International Symposium on Computer Architecture (ISCA)* (IEEE, Piscataway, 2016), pp. 367–379
12. Y.-H. Chen, J. Emer, V. Sze, Eyeriss: a spatial architecture for energy-efficient dataflow for convolutional neural networks, *in ACM SIGARCH Computer Architecture News*, vol. 44, no. 3 (IEEE Press, Piscataway, 2016)
13. P. Chi, S. Li, C. Xu, T. Zhang, J. Zhao, Y. Liu, Y. Wang, Y. Xie, PRIME: a novel processing-in-memory architecture for neural network computation in ReRAM-based main memory. SIGARCH Comput. Archit. News **44**(3), 27–39 (2016)
14. F. Chollet, Xception: deep learning with depthwise separable convolutions (2016). Preprint. arXiv:1610.02357
15. M. Courbariaux, Y. Bengio, J.P. David, Training deep neural networks with low precision multiplications (2014). Preprint. arXiv:1412.7024
16. M. Courbariaux, Y. Bengio, J.P. David, Binaryconnect: training deep neural networks with binary weights during propagations, in *Advances in Neural Information Processing Systems* (2015), pp. 3123–3131
17. M. Courbariaux, I. Hubara, D. Soudry, R. El-Yaniv, Y. Bengio, Binarized neural networks: training deep neural networks with weights and activations constrained to +1 or −1 (2016). Preprint. arXiv:1602.02830
18. Y.L. Cun, J.S. Denker, S.A. Solla, Optimal brain damage, in *Advances in Neural Information Processing Systems 2* (Morgan Kaufmann Publishers, San Francisco, 1990), pp. 598–605. http://dl.acm.org/citation.cfm?id=109230.109298

19. M. Davies, N. Srinivasa, T. Lin, G. Chinya, Y. Cao, S.H. Choday, G. Dimou, P. Joshi, N. Imam, S. Jain et al., Loihi: a neuromorphic manycore processor with on-chip learning. IEEE Micro **38**(1), 82–99 (2018)
20. L. Deng, The MNIST database of handwritten digit images for machine learning research [best of the web]. IEEE Signal Process. Mag. **29**(6), 141–142 (2012)
21. J. Deng, W. Dong, R. Socher, L.J. Li, K. Li, L. Fei-Fei, Imagenet: a large-scale hierarchical image database, in *IEEE Conference on Computer Vision and Pattern Recognition, 2009. CVPR 2009* (IEEE, Piscataway, 2009), pp. 248–255
22. Z. Du, A. Lingamneni, Y. Chen, K. Palem, O. Temam, C. Wu, Leveraging the error resilience of machine-learning applications for designing highly energy efficient accelerators, in *2014 19th Asia and South Pacific Design Automation Conference (ASP-DAC)* (IEEE, Piscataway, 2014), pp. 201–206
23. Z. Du, R. Fasthuber, T. Chen, P. Ienne, L. Li, T. Luo, X. Feng, Y. Chen, O. Temam, Shidiannao: shifting vision processing closer to the sensor, in *ACM SIGARCH Computer Architecture News*, vol. 43 (ACM, New York, 2015), pp. 92–104
24. S.K. Esser, R. Appuswamy, P. Merolla, J.V. Arthur, D.S. Modha, Backpropagation for energy-efficient neuromorphic computing, in *Advances in Neural Information Processing Systems* (2015), pp. 1117–1125
25. R. Fakoor, F. Ladhak, A. Nazi, M. Huber, Using deep learning to enhance cancer diagnosis and classification, in *Proceedings of the International Conference on Machine Learning* (2013)
26. K. Finley, Ai fighter pilot beats a human, but no need to panic (really) (2016). https://www.wired.com/2016/06/ai-fighter-pilot-beats-human-no-need-panic-really/
27. Gear adder library, https://sourceforge.net/projects/approxadderlib
28. V. Gupta, D. Mohapatra, S.P. Park, A. Raghunathan, K. Roy, Impact: imprecise adders for low-power approximate computing, in *Proceedings of the 17th IEEE/ACM International Symposium on Low-Power Electronics and Design* (IEEE Press, Piscataway, 2011), pp. 409–414
29. P. Gysel, M. Motamedi, S. Ghiasi, Hardware-oriented approximation of convolutional neural networks (2016). Preprint. arXiv:1604.03168
30. S. Ha, J.M. Yun, S. Choi, Multi-modal convolutional neural networks for activity recognition, in *2015 IEEE International Conference on Systems, Man, and Cybernetics (SMC)* (IEEE, Piscataway, 2015), pp. 3017–3022
31. S. Han, H. Mao, W.J. Dally, Deep compression: compressing deep neural networks with pruning, trained quantization and Huffman coding (2015). Preprint. arXiv:1510.00149
32. S. Han, X. Liu, H. Mao, J. Pu, A. Pedram, M.A. Horowitz, W.J. Dally, EIE: efficient inference engine on compressed deep neural network, in *Proceedings of the 43rd International Symposium on Computer Architecture* (IEEE Press, Piscataway, 2016), pp. 243–254
33. X. Han, D. Zhou, S. Wang, S. Kimura, CNN-MERP: an FPGA-based memory-efficient reconfigurable processor for forward and backward propagation of convolutional neural networks, in *2016 IEEE 34th International Conference on Computer Design (ICCD)* (IEEE, Piscataway, 2016), pp. 320–327
34. K. He, X. Zhang, S. Ren, J. Sun, Deep residual learning for image recognition, in *Proceedings of the IEEE Conference on Computer Vision and Pattern Recognition* (2016), pp. 770–778
35. T. Highlander, A. Rodriguez, Very efficient training of convolutional neural networks using fast Fourier transform and overlap-and-add (2016). Preprint. arXiv:1601.06815
36. A.G. Howard, M. Zhu, B. Chen, D. Kalenichenko, W. Wang, T. Weyand, M. Andreetto, H. Adam, Mobilenets: efficient convolutional neural networks for mobile vision applications (2017). https://arxiv.org/abs/1704.04861
37. F.N. Iandola, S. Han, M.W. Moskewicz, K. Ashraf, W.J. Dally, K. Keutzer, Squeezenet: Alexnet-level accuracy with 50x fewer parameters and< 0.5 mb model size (2016). Preprint. arXiv:1602.07360
38. H. Jiang, J. Han, F. Lombardi, A comparative review and evaluation of approximate adders, in *Proceedings of the 25th edition on Great Lakes Symposium on VLSI* (ACM, New York, 2015), pp. 343–348

39. H. Jiang, C. Liu, N. Maheshwari, F. Lombardi, J. Han, A comparative evaluation of approximate multipliers, in *2016 IEEE/ACM International Symposium on Nanoscale Architectures (NANOARCH)* (IEEE, Piscataway, 2016), pp. 191–196
40. J.H. Ko, B. Mudassar, T. Na, S. Mukhopadhyay, Design of an energy-efficient accelerator for training of convolutional neural networks using frequency-domain computation, in *Proceedings of the 54th Annual Design Automation Conference 2017* (ACM, New York, 2017), p. 59
41. A. Lavin, S. Gray, Fast algorithms for convolutional neural networks, in *Proceedings of the IEEE Conference on Computer Vision and Pattern Recognition* (2016), pp. 4013–4021
42. F. Li, B. Zhang, B. Liu, Ternary weight networks (2016). Preprint. arXiv:1605.04711
43. lpaclib library, https://sourceforge.net/projects/lpaclib/
44. W. Lu, G. Yan, J. Li, S. Gong, Y. Han, X. Li, Flexflow: a flexible dataflow accelerator architecture for convolutional neural networks, in *2017 IEEE International Symposium on High Performance Computer Architecture (HPCA)* (IEEE, Piscataway, 2017), pp. 553–564
45. V. Mrazek, S.S. Sarwar, L. Sekanina, Z. Vasicek, K. Roy, Design of power-efficient approximate multipliers for approximate artificial neural networks, in *2016 IEEE/ACM International Conference on Computer-Aided Design (ICCAD)* (IEEE, Piscataway, 2016), pp. 1–7
46. V. Mrazek, R. Hrbacek, Z. Vasicek, L. Sekanina, Evoapproxsb: library of approximate adders and multipliers for circuit design and benchmarking of approximation methods, in *2017 Design, Automation & Test in Europe Conference & Exhibition (DATE)* (IEEE, Piscataway, 2017), pp. 258–261
47. A. Parashar, M. Rhu, A. Mukkara, A. Puglielli, R. Venkatesan, B. Khailany, J. Emer, S.W. Keckler, W.J. Dally, SCNN: an accelerator for compressed-sparse convolutional neural networks, in *Proceedings of the 44th Annual International Symposium on Computer Architecture* (ACM, New York, 2017), pp. 27–40
48. M. Rastegari, V. Ordonez, J. Redmon, A. Farhadi, XNOR-Net: Imagenet classification using binary convolutional neural networks, in *European Conference on Computer Vision* (Springer, Berlin, 2016), pp. 525–542
49. S. Rehman, W. El-Harouni, M. Shafique, A. Kumar, J. Henkel, J. Henkel, Architectural-space exploration of approximate multipliers, in *2016 IEEE/ACM International Conference on Computer-Aided Design (ICCAD)* (IEEE, Piscataway, 2016), pp. 1–8
50. M. Rhu, M. O'Connor, N. Chatterjee, J. Pool, S.W. Keckler, Compressing DMA engine: leveraging activation sparsity for training deep neural networks (2017). Preprint. arXiv:1705.01626
51. S.S. Sarwar, S. Venkataramani, A. Raghunathan, K. Roy, Multiplier-less artificial neurons exploiting error resiliency for energy-efficient neural computing, in *Design, Automation & Test in Europe Conference & Exhibition (DATE), 2016* (IEEE, Piscataway, 2016), pp. 145–150
52. A. Shafiee, A. Nag, N. Muralimanohar, R. Balasubramonian, J.P. Strachan, M. Hu, R.S. Williams, V. Srikumar, ISAAC: a convolutional neural network accelerator with in-situ analog arithmetic in crossbars. SIGARCH Comput. Archit. News **44**(3), 14–26 (2016)
53. M. Shafique, W. Ahmad, R. Hafiz, J. Henkel, A low latency generic accuracy configurable adder, in *2015 52nd ACM/EDAC/IEEE Design Automation Conference (DAC)* (IEEE, Piscataway, 2015), pp. 1–6
54. M. Shafique, F. Sampaio, B. Zatt, S. Bampi, J. Henkel, Resilience-driven STT-RAM cache architecture for approximate computing, in *Workshop on Approximate Computing (AC)* (2015)
55. M. Shafique, R. Hafiz, S. Rehman, W. El-Harouni, J. Henkel, Cross-layer approximate computing: from logic to architectures, in *2016 53rd ACM/EDAC/IEEE Design Automation Conference (DAC)* (IEEE, Piscataway, 2016), pp. 1–6
56. M. Shafique, R. Hafiz, M.U. Javed, S. Abbas, L. Sekanina, Z. Vasicek, V. Mrazek, Adaptive and energy-efficient architectures for machine learning: challenges, opportunities, and research roadmap, in *2017 IEEE Computer Society Annual Symposium on VLSI (ISVLSI)* (IEEE, Piscataway, 2017), pp. 627–632
57. Y. Shen, M. Ferdman, P. Milder, Escher: a CNN accelerator with flexible buffering to minimize off-chip transfer, in *2017 IEEE 25th Annual International Symposium on Field-Programmable Custom Computing Machines (FCCM)* (IEEE, Piscataway, 2017)

58. K. Simonyan, A. Zisserman, Very deep convolutional networks for large-scale image recognition (2014). Preprint. arXiv:1409.1556
59. L. Song, X. Qian, H. Li, Y. Chen, Pipelayer: a pipelined ReRAM-based accelerator for deep learning, in *IEEE International Symposium on High Performance Computer Architecture (HPCA)* (IEEE, Piscataway, 2017), pp. 541–552
60. C. Szegedy, W. Liu, Y. Jia, P. Sermanet, S. Reed, D. Anguelov, D. Erhan, V. Vanhoucke, A. Rabinovich, Going deeper with convolutions, in *Proceedings of the IEEE Conference on Computer Vision and Pattern Recognition* (2015), pp. 1–9
61. The NVIDIA DGX-1 deep learning system, https://www.nvidia.com/content/dam/en-zz/Solutions/Data-Center/dgx-1/dgx-1-ai-supercomputer-datasheet-v4.pdf
62. A. Todri-Sanial, A. Magnani, M. De Magistris, A. Maffucci, Present and future prospects of carbon nanotube interconnects for energy efficient integrated circuits, in *2016 17th International Conference on Thermal, Mechanical and Multi-Physics Simulation and Experiments in Microelectronics and Microsystems (EuroSimE)* (IEEE, Piscataway, 2016), pp. 1–5
63. C. Urmson, J. Anhalt, D. Bagnell, C. Baker, R. Bittner, M. Clark, J. Dolan, D. Duggins, T. Galatali, C. Geyer et al.: Autonomous driving in urban environments: boss and the urban challenge. J. Field Rob. **25**(8), 425–466 (2008)
64. J. Wang, J. Lin, Z. Wang, Efficient convolution architectures for convolutional neural network, in *2016 8th International Conference on Wireless Communications & Signal Processing (WCSP)* (IEEE, Piscataway, 2016) , pp. 1–5
65. X. Wei, C.H. Yu, P. Zhang, Y. Chen, Y. Wang, H. Hu, Y. Liang, J. Cong, Automated systolic array architecture synthesis for high throughput CNN inference on FPGAs, in *2017 54th ACM/EDAC/IEEE Design Automation Conference (DAC)* (IEEE, Piscataway, 2017), pp. 1–6
66. Q. Xiao, Y. Liang, L. Lu, S. Yan, Y.W. Tai, Exploring heterogeneous algorithms for accelerating deep convolutional neural networks on FPGAs, in *Proceedings of the 54th Annual Design Automation Conference 2017* (ACM, New York, 2017), p. 62
67. J. Yu, A. Lukefahr, D. Palframan, G. Dasika, R. Das, S. Mahlke, Scalpel: customizing DNN pruning to the underlying hardware parallelism, in *Proceedings of the 44th Annual International Symposium on Computer Architecture* (ACM, New York, 2017), pp. 548–560
68. C. Zhang, V.K. Prasanna, Frequency domain acceleration of convolutional neural networks on CPU-FPGA shared memory system, in *FPGA* (2017), pp. 35–44

Index

I. M. Elfadel et al. (eds.), *Machine Learning in VLSI Computer-Aided Design*,
https://doi.org/10.1007/978-3-030-04666-8

MIX
Papier aus verantwortungsvollen Quellen
Paper from responsible sources
FSC® C105338

If you have any concerns about our products,
you can contact us on
ProductSafety@springernature.com

In case Publisher is established outside the EU,
the EU authorized representative is:
Springer Nature Customer Service Center GmbH
Europaplatz 3, 69115 Heidelberg, Germany

Printed by Libri Plureos GmbH
in Hamburg, Germany